The Manual of
Bridge Engineering

The Manual of Bridge Engineering

Edited by

M.J. RYALL, G.A.R. PARKE and J.E. HARDING

Thomas Telford

Published by Thomas Telford Publishing, Thomas Telford Ltd, 1 Heron Quay, London E14 4JD.
URL: http://www.thomastelford.com

Distributors for Thomas Telford books are
USA: ASCE Press, 1801 Alexander Bell Drive, Reston, VA 20191-4400, USA
Japan: Maruzen Co. Ltd, Book Department, 3–10 Nihonbashi 2-chome, Chuo-ku, Tokyo 103
Australia: DA Books and Journals, 648 Whitehorse Road, Mitcham 3132, Victoria

First published 2000

Also available from Thomas Telford Books
Bridge management four. M. J. Ryall, G. A. R. Parke & J. E. Harding (eds). ISBN: 0 7277 2854 7
Cable stayed bridges (2nd edn). R. Walther, B. Houriet, W. Isler, P. Moïa & J. Klein. ISBN: 0 7277 2773 7
Current and future trends in bridge design, construction and maintenance. Institution of Civil Engineers & Highways Agency. ISBN: 0 7277 2841 5
Integral bridges: a fundamental approach to the time temperature loading problem. G. L. England, N. C. M. Tsang & D. I. Bush. ISBN: 0 7277 2845 8
Management of highway structures. P. C. Das. ISBN: 0 7277 2775 3
Post-tensioned concrete bridges. Highways Agency, Service d'Etudes Techniques des Routes et Autoroutes, Transport Research Laboratory and Laboratoire Central des Ponts et Chaussées. ISBN: 0 7277 2760 5
Prototype bridge structures: analysis and design. M. Y. H. Bangash. ISBN: 0 7277 2778 8
Safety of bridges. P. C. Das. ISBN 0 7277 2591 2

A catalogue record for this book is available from the British Library

ISBN: 0 7277 2774 5

© Thomas Telford Limited 2000

Typeset by Gray Publishing, Tunbridge Wells, Kent
Printed and bound in Great Britain by MPG Books, Bodmin, Cornwall

Contents

3   Structural analysis

N.E. Shanmugam and R. Narayanan

5 Design of prestressed concrete bridges
N.R. Hewson

6 Design of steel bridges
G.A.R. Parke and J.E. Harding

7 Composite construction
D. Collings

8 Design of arch bridges
C. Melbourne

9   Seismic response and design

A.S. Elnashai

10   Cable stayed bridges

D.J. Farquhar

11 Suspension bridges

V. Jones and J. Howells

14   Substructures

P. Lindsell

15   Bridge accessories
P. Thayre, D.E. Jenkins, R.A. Broome and D.J. Grout

16   Protection
M. Mulheron

17   Bridge management
P. Vassie

18   Inspection, monitoring, and assessment
C. Abdunur

19   Repair, strengthening and replacement
J. Darby with contributions from G. Cole, S. Collins and P. Brown

Foreword

Very occasionally books are written which can be classed as landmark publications and I believe that this manual is one of them. Such a publication is long overdue. As a British publication, albeit from an international base, it is unique in drawing together such a wealth of information into one comprehensive work. As bridge design becomes ever more innovative and the demands placed on bridges increase, the bridge engineering industry faces a constantly growing variety of challenges.

With contributions from some of the most respected experts in the field, this publication addresses these challenges. Not only does it provide a broad overview of the whole subject of bridge engineering from concept to completion, but it focuses on detailed aspects of analysis, design, construction and maintenance.

I am sure that it will quickly establish itself as an invaluable reference work for practising bridge engineers, researchers and students alike. I would like to pay tribute to the editors who have had the foresight and tenacity to make this manual a reality.

Professor Patrick J. Dowling DL, FREng, FRS
Vice-Chancellor and Chief Executive
University of Surrey

Preface

This book was conceived in 1988 and now, after a gestation period of more than a decade, it has been born. It was clear to us in 1988 that there was an obvious gap in the market for a compendium of good practice and a single source of information pertinent to bridge engineers – both expert and novice alike. It was a daunting challenge and progress was very slow at the beginning, mainly due to the fact that prospective well-known and expert authors are inevitably under heavy workloads. Gradually, however, the manual has come together and while we are very pleased with the result, it remains with those in the bridge engineering fraternity to finally judge its worth.

This manual, as with any manual or handbook, is for use as a reference book and not as a textbook, although reading it as such will give an invaluable initial oversight of the subject. It presents a comprehensive overview of the fundamental principles which govern the concept, analysis, design, construction and maintenance of many types of bridge. Throughout the book emphasis has been placed on simplicity (or elegance) in that the author of each chapter has attempted to present a text which is simple and straightforward and not obscured with unnecessary detail, though where apposite (for example in design procedures) some topics are considered in detail. Each chapter is written in sufficient depth to enable budding engineers to gain an awareness of the subject matter and more seasoned engineers to refresh their memories or discover something new. Numerous references are available which can be explored for more refined or extended information.

We would like to thank Thomas Telford for their perseverance, patience and foresight and, of course, all of the contributing authors for their excellent and selfless work.

M.J. Ryall, G.A.R. Parke, J.E. Harding

Contributors

Charles Abdunur

After receiving his civil engineering degree, Charles worked on the design and supervision of several special building construction projects. After obtaining his doctorate from the Paris Academy of Science, he turned to research and devoted his main activities to bridges. Since then, he has directed studies on the local and general behaviour of both sound and damaged structures, at the Laboratoire Central des Ponts et Chaussées, Paris. Charles also leads the committee in charge of organizing research on the properties and mechanical behaviour of materials and structures.

David Bennett

Is a civil engineer and consultant who has written a number of technical publications including the highly successful books *Skyscraper* for Simon & Schuster; *The Architecture of Bridge Design* for Thomas Telford and *The Creation of Bridges* for Quintet. As a consultant with special interest in site-cast concrete, he advises on the use of visual concrete. His principle business interest is construction marketing, researching and writing press articles for professional trade journals.

Charles Birnstiel

Studied civil engineering at New York University, earning at the doctorate in 1962. His engineering experience includes 20 years of teaching and structural mechanics research at New York and Polytechnic University and 23 years leading a multidisciplinary engineering firm specializing in the design, inspection and testing of moveable bridge machinery. He is now with Hardestry & Hanover, LLP, a prominent moveable bridge engineering firm headquartered in New York City. Birnstiel is a licensed Professional Engineer and active on technical committees of engineering societies.

Robert Broome

Is a structural engineer employed by WS Atkins Consultants Ltd, as a senior engineer specializing in the management of highway projects constructed under conventional, design and build and design, build, finance and operate forms of contract. He has in excess of forty years experience in the design, construction, assessment and refurbishment of highway structures and has been responsible for the designs of numerous bridges on the UK motorway and trunk road system.

Peter Brown

Is Group Manager of the Bridges Section at Oxfordshire County Council taking up the position in 1996 after a number of years working on bridges in both the private and public sectors. He is been responsible for the Assessment and Strengthening Programme within the county and has been involved in bridge strengthening on a variety of roads, often using innovative techniques and methods.

Graham Cole

Graduated from The City University in 1979. He has been involved in the structural design of a variety of major highway schemes as well as a large number of road and rail bridge inspections for both local authorities and consultants. Graham joined Surrey County Council in 1990 where he is now the Principal Bridge Engineer responsible for the management of the stock of 2100 highway structures which has included the strengthening and maintenance of a large number of masonry arch bridges.

David Collings

Was born in London in 1957, studied at Thames Polytechnic and is currently a Technical Director at Robert Benaim & Associates London office. David has been an active bridge designer since 1980 and has been responsible for a number of major bridges world-wide. He has published and presented a number of papers and is an acknowledged expert in the design and construction of bridges in many materials. He lectures regularly on composite structures at the University of Surrey and is a member of the working party drafting advice on prestressed concrete bridges in the UK.

Simon Collins

Is a Divisional Director of Mouchel Consulting Limited, responsible for the management and technical direction of bridge engineering work carried out at Mouchel's Head Office. He has over 20 years' experience in bridge and structures engineering. Recent bridge modification projects for which he has been responsible have included the strengthening of a 400 m suspension bridge in the Caribbean, and the independent checking of the widening and strengthening of the Tamar Suspension Bridge.

John Darby

Was Chief Bridge Engineer of Oxfordshire County Council until 1996, with responsibility for design of new structures and maintenance of Trunk and County Bridges. He is now a consultant, specializing in bridge management, advanced composite materials, post-tensioned grouting, testing and monitoring, and life cycle costing.

Amr Elnashai

Is professor of earthquake engineering and head of the Engineering Seismology and Earthquake Engineering Section at Imperial College, London. He is a fellow of the Royal Academy of Engineering, the Institution of Structural Engineers and the American Society of Civil Engineers. Amr has written more than 150 research publications, and is currently Vice President of the European Association of Earthquake Engineering and a member of the drafting panel of Eurocode 8. He is also Principle Consultant to EQE International Ltd and has been visiting professor at the University of Tokyo, University of Southern California, Polytechnic of Milan and University of Pavia. He is editor of the *Journal of Earthquake Engineering* (IC Press) and Director of the Japan–UK Seismic Risk Forum.

Daniel Farquhar

Is an Associate with Mott MacDonald with over 30 years' experience in the design and construction of bridges and other major civil engineering projects. He has specialized in the design and construction of cabled stayed and suspension bridges. Notable projects include cable stayed bridges for the Rama VIII and the Industrial Ring Road crossings over the Chao Phraya River in Bangkok, Thailand and the Tsing Ma and Tsing Lung suspension bridges in Hong Kong.

David Grout

Is a civil engineer employed by WS Atkins Consultants Ltd as a senior engineer specializing in bridge design. He has over 25 years' experience in the design, construction, assessment, strengthening and refurbishment of civil engineering structures, in particular highway and railway structures. He has been responsible for the designs of several bridges and cut-and-cover tunnels in the UK and overseas.

John Harding

Is Professor of Structural Engineering and Head of the Civil Engineering Department at the University of Surrey, Guildford. He is also a Pro-Vice-Chancellor at the University, and has been at Surrey for some 15 years having previously been at Imperial College in London. Professor Harding is editor of the *International Journal of Constructional Steel Research*, author of numerous journal and conference publications and editor of numerous books. His specialist interests are mainly related to the behaviour and design of steel plated structures, particularly in the context of steel bridges and offshore oil rigs. He has contributed to the work of international code and research groups in these areas and his research underpinned the strength clauses relating to the behaviour of stiffened plating in the British Standard for the design of steel bridges. He is a Chartered Engineer and Fellow of the Institutions of Civil and Structural Engineers and has chaired and been a member of a number of Institution committees.

Nigel Hewson

Graduating from Loughborough University in 1978, Nigel worked on the construction of the first matchcast post-tensioned concrete bridge to be built in the UK. In 1980 he joined Freeman Fox & Partners and designed several major prestressed concrete bridges in the UK and Internationally. After spending 9 years' overseas on such projects as the Bangkok Second Stage Expressway and the Malaysia Singapore Second Crossing, Nigel returned to the UK and is currently the Group Director – Bridges and Civil Structures for Hyder Consulting Ltd.

Len Hollaway

Leads the Composite Structures Research Unit of the Department of Civil Engineering, University of Surrey. He has been engaged on research into fibre/matrix composites for 30 years and has been intimately involved in the numerical and experimental analysis of plates and skeletal systems made from fibre/polymer composites, in upgrading conventional structural materials with advanced polymer composites (APC) and in combining APCs with conventional materials to form optimized structural systems; he has published over 160 technical papers. Professor Holloway has, over the years, served on a number of committees associated with the composites and has served on numerous advisory boards of International Conferences.

John Howells

After graduation John Howells worked on aircraft structural analysis. In 1965 he joined Dorman Long, with whom he worked on the construction of the Humber Suspension bridge, following which he worked on the initial design of the Tsing Ma bridge. After joining High-Point Rendel in 1985, he has worked on many projects including the widening of the Rodenkirchen suspension bridge, provision of advice on cable construction technology for the Storebaelt suspension bridge, and has assisted with construction of the Jiangyin suspension bridge.

Paul Jackson

Worked for a consultant and contractor before joining the Cement and Concrete Association. There he ran courses and contributed to the advisory service as well as undertaking research, obtaining a PhD for work on bridge deck behaviour. He joined Gifford & Partners in 1988 and was appointed an Associate in 1994. He has been responsible for a wide variety of design, assessment, checking and research work on bridges. He serves on British, European and American Bridge code committees.

David Jenkins

Is a civil engineer employed by WS Atkins Consultants Ltd as a principle engineer with responsibility for managing the bridges department in their Epsom office. He has 15 years' experience in the design, construction, assessment and refurbishment of highway and railway structures. He has been responsible for the design of a number of bridges in the UK and overseas, many of which have been procured under design and build forms of contract.

Vardiman Jones

Joined consultants High-Point Rendel after graduating from University College London and finishing a Master's in Steel Structures at Imperial College. He has worked on the Rion–Antirion suspension bridge preliminary design, the East London River Crossing and Jamuna bridge detailed design, the Rodenkirchen suspension bridge widening, the Aswan cable stayed bridge and the Storebaelt suspension bridge and approach viaducts erection engineering amongst others. He is currently Director responsible for Infrastructure Engineering.

Peter Lindsell

Graduated from Oxford University in 1967 and spent the next 7 years in bridge design and construction. He then returned to academic life to teach prestressed concrete and bridge design at Salford and Surrey Universities. He was awarded a PhD degree in 1982 for research into the design of bridge abutments. Dr Lindsell was elected a Fellow of New College, Oxford in 1987 and formed his own consultancy practice in 1993.

Clive Melbourne

Is Professor of Structural Engineering at the University of Salford. Professor Melbourne obtained his BEng and PhD degrees from the University of Sheffield. He then practised as a bridge engineer for 12 years before returning to academia to study the behaviour of masonry arch bridges. He has published over 30 papers and organized the first Arch Bridge Conference in 1995. He is a Fellow of both the Institution of Civil Engineers and the Institution of Structural Engineers and serves on several committees.

Mike Mulheron

Is lecturer in Construction Materials within the Department of Civil Engineering at the University of Surrey. He has practical experience of the construction, inspection and renovation of steel and concrete structures, both in the UK and abroad and has a particular interest in the durability and protection of bridges. His research interests include non-invasive repair methods and the application of modern analytical techniques, such as magnetic resonance imaging, for assessing the mobility of water and surface treatments in porous construction materials.

Rangachari Narayanan

Graduated in Civil Engineering from Annamalai University (India) in 1951 and subsequently obtained his MSc and PhD degrees from Imperial College and Manchester University respectively. In a varied professional career spanning 50 years, he has served in UK and USA at Manchester, Cardiff and Duke Universities and also headed the Education and Publications Division at the Steel Construction Institute. Professor Narayanan is the winner of Benjamin Baker Gold Medal and George Stephenson Gold Medal, both from the Institution of Civil Engineers. Currently he directs an Indian Government Project on Structural Steel Design Education.

Gerard Parke

Is Head of the Structures Research Group at the University of Surrey. He is particularly interested in the analysis and design of steel structures, specializing in assessing the collapse behaviour of space structures. Dr Parke has edited seven books and written over 50 papers on the design of steel structures. He is Chairman of the Institution of Structural Engineers Study Group on Space Structures and a member of the Institution's working group on the Dynamic Performance of Stadia.

Michael Ryall

Graduated from Loughborough University in 1964 with a first-class honours degree in civil engineering. He has practised as a bridge engineer in the USA, and both taught and practised in Africa and the UK. He is currently lecturing at the University of Surrey where he directed the MSc course in bridge engineering for over 10 years. He gained his PhD at Surrey researching *in situ* stress measurement in concrete bridges and has an interest in bridge management and distribution factors for bridge deck analysis.

N.E. Shanmugam

Currently with the Department of Civil Engineering, The National University of Singapore as Professor in Civil Engineering. He has taught in the University of Madras, Delhi University, University of Wales and Polytechnic of Wales before moving to Singapore. He has published extensively on steel-plated structures in international journals and conference proceedings, with research interest including various aspects of steel-plated structures and steel–concrete composite construction. Professor Shanmugam is a Chartered Engineer, Fellow of the Institution of Structural Engineers, Royal Institution of Naval Architects, American Society of Civil Engineers, The Institution of Engineers, Singapore and The Institution of Engineers, India. He is a Member-at-Large of the Structural Stability Research Council (SSRC), USA.

Peter Thayre

Is a structural engineer employed by WS Atkins Consultants Ltd as a senior engineer specializing in bridge design. He has 30 years' experience in the design, construction, assessment and refurbishment of highway structures and has been responsible for the design of numerous bridges on the UK motorway and trunk road system. For many years he has been involved in the design, testing and specification of bridge parapets and acts as an adviser to the UK Highways Agency.

Perry Vassie

Has worked at the Transport Research Laboratory for the last 25 years. He has worked on the diagnosis, detection and control of deterioration in bridges. He has also worked on various aspects of bridge management such as inspection, maintenance strategies, deterioration algorithms, optimization and prioritization of maintenance programmes and whole-life cost models. Perry was appointed visiting professor in the School of Architecture and Civil Engineering at the South Bank University London in 1993.

1 The history and aesthetic development of bridges

D. BENNETT

1.1 The early history of bridges

1.1.1 The age of timber and stone

The bridge has been a feature of human progress and evolution, ever since man the hunter–gatherer became curious about the world beyond the horizon and the fertile land, the animals and fruit flourishing on trees on the other side of a river or gorge. Early man had to devise ways to cross a stream and a deep gorge to survive. A boulder or two dropped into a shallow stream works well as a stepping stone, as many of us have discovered – but for deeper flowing streams, a tree dropped between banks is a more successful solution. So the primitive idea of a simple beam bridge was born.

In the forests of Peru today and the foothills of the Himalayas, crude rope bridges span deep gorges and fast-flowing streams to maintain pathways from village to village for hill tribes. Such primitive rope bridges evolved from the vine and creeper that early man would have used to swing through the forest and to cross a stream. Here is the second basic idea of a bridge – the suspension bridge.

For thousands of years during the Palaeolithic period that lasted to about 8000BC, we know that man was living as a nomad and wanderer, hunting and gathering food. Slowly it dawned on early humans that following herds of deer or buffalo or just foraging for plant food haphazardly, could be better managed if the animals were kept in herds nearby and plants were grown and harvested in fields. If there was a shorter way to travel between two places, man's new enterprise spirit would find a way to bridge a river or to cut a clearing through a forest.

In this period the simple log bridge had to serve many purposes. It needed to be broad and strong enough to take cattle, it needed to be level and a solid platform to transport food and other materials, and it needed to be movable so that it could be withdrawn to prevent their enemies from using it. Narrow tree trunk bridges were inadequate and were replaced by double log beams spaced wider apart on which short lengths of logs were placed and tied down to create a pathway. The pathways were planed by sharp scraping tools or axes in the Bronze Age and any gaps between them plugged with branches and earth to create a level platform. For crossings over wide rivers, support piers were formed from piles of rocks in the stream. Sometimes stakes were driven into the riverbed to form a circle and then filled with stones creating a crude coffer dam. Around 4000BC, early Bronze Age 'lake dwellers' lived in timber houses built out over the lakes, in the area which is now Switzerland. To ensure their houses did not sink early humans evolved ways to drive timber piles into the lake bed. From the discovery of this came the timber pile and the trestle bridge.

So primitive bridges were essentially post and lintel structures, either made from timber or stone or a combination of both. Sometime later, the simple rope and bamboo suspension bridge was devised, which developed into the rope suspension bridges that are in regular use today in the mountain reaches of China, Peru, Columbia, India and Nepal.

But it took man until 4000BC to discover the secrets of arch construction. In the Tigris–Euphrates valley the Sumerians began building with adobe – a sun-dried mud brick – for their palaces, temples, ziggurats and city defences. Stone was not plentiful in this region and had to be imported from Persia, so was used sparingly. The brick module dictated the construction principles they employed, to scale any height and to bridge any span. And through trial and error it was the arch and the barrel vault that was devised to build their monuments and grand architecture at the peak of their civilization. The ruins of the magnificent barrel-vaulted brick roof at Ptsephon and the Ishtar Gate at Babylon, are a reminder of Mesopotamian skill and craftsmanship. By the end of Third Dynasty around 2475BC, the Egyptians had also mastered the arch and used it frequently in constructing relieving arches and passageways for their temples and pyramids.

Without doubt, the arch is one of the greatest discoveries of humankind. The arch principle was the essential element in all building and bridge technology over later centuries. Its dynamic and expressive form gave rise to some of the greatest bridge structures ever built.

1.1.2 Earliest records of bridges

The earliest written record of a bridge appears to be a bridge built across the Euphrates around 600BC as described by Herodotus, the fifth century Greek historian. The bridge linked the palaces of ancient Babylon on either side of the river. It had a hundred stone piers which supported wooden beams of cedar, cypress and palm to form a carriageway 35 ft wide and 600 ft long. Herodotus mentions that the floor of the bridge would be removed every at night as a precaution against invaders.

In China it would appear that bridge building evolved at a faster pace than the ancient civilizations of Sumeria and Egypt. Records exist from the time of Emperor Yoa in 2300BC on the traditions of bridge building. Early Chinese bridges included pontoons or floating bridges and probably looked like the primitive pontoon bridges built in China today. Boats called sampans about 30 ft long were anchored side by side in the direction of the current and then bridged by a walkway. The other bridge forms were the simple post and lintel beam, the cantilever beam and rope suspension cradles. Timber beam bridges, probably like those of Europe, were often supported on rows of timber piles of soft fir wood called 'foochow poles', so called because they were grown in Foochow. A team of builders would hammer the poles into the riverbed using a cylindrical stone fitted with bamboo handles. A short crosspiece was fixed between pairs of poles to form the supports that would carry timber boards which were then covered in clay to form the pathway over the river.

In later centuries Chinese bridge building was dominated by the arch, which they copied and adapted from the Middle East as they travelled the silk routes which opened during the Han Dynasty around 100AD.

Through Herodotus we learn about the Persian ruler Xerxes and the vast pontoon bridge he built, consisting of two parallel rows of 360 boats, tied to each other and to the bank and anchored to the bed of the Hellespont, which is the Dardanelles today. Xerxes wanted to get his army of two million men and horses to the other bank to meet the Greeks at Thermopohlae. It took seven days and seven nights to get the army across the river. Sadly for Xerxes, his massive army was defeated at the Battle of Thermophalae in 480BC, the remnants of which retreated back over the pontoon bridge to fight another day. The Persians were great bridge builders and built many arch, cantilever and beam bridges. There is a bridge still standing in Khuzistan at Dizful over the river Diz which could date anywhere from 350BC to 400AD. The bridge consists of 20 voussoir arches which are slightly pointed and has a total length of 1250 ft. Above the level of the arch springing are small spandrel arches, semicircular in length, which gives the entire bridge an Islamic look, hence the uncertainty of it Persian origins.

The Greeks did not do much bridge building over their illustrious history, being a seafaring nation that lived on self-contained islands and in feudal groups scattered across the Mediterranean. They exclusively used post and lintel construction in evolving a classical order in their architecture, and built some of the most breathtaking temples, monuments and cities the world has ever seen, such as the Parthenon, the Temple of Zeus, the city of Ephesus, Miletus, and Delphi, to name but a few. They were quite capable of building arches like their forbears the Etruscans when necessary. There are examples of Greek voussoir arch construction that compare with the Beehive Tomb at Mycenea, such as the ruins of an arch bridge with a 27 ft span at Pergamon in Turkey.

1.1.3 The Romans
The Romans on the other hand were the masters of practical building skills. They were a nation of builders who took arch construction to a science and high art form during their domination of Mediterranean Europe. Their influence on bridge building technology and architecture has been profound. They conquered the world as it was then know, built roadways, canals and cities that linked Europe to Asia and North Africa and produced the first true bridge engineers in the history of humankind. The Romans understood that the establishment and maintenance of their empire depended on efficient and permanent communications. Building roads and bridges was therefore a high priority.

The Romans also realized, as did the Chinese in later centuries, that timber structures, particularly those embedded in water had a short life, were prone to decay, insect infestation and fire hazards. Prestigious buildings and important bridge structures were therefore built in stone. But the Romans had also learnt to preserve their timber structures by soaking timber in oil and resin as a protection against dry rot, and coating them with alum for fireproofing. They learnt that hardwood was more durable than softwoods, and that oak was best for substructure work in the ground, alder for piles in water; while fir, cypress and cedar were best for the superstructure above ground.

They understood the different qualities of the stone that they quarried. Tufa, a yellow volcanic stone, was good in compression but had to be protected from weathering by stucco – a lime wash. Travertine was harder and more durable and could be left exposed, but was not very fire resistant. The most durable materials such as marble had to imported from distant regions of Greece and even as far as Egypt and Asia Minor (Turkey). The Romans big breakthrough in material science was the discovery of lime mortar and pozzolanic cement, which was based on the volcanic clay which was found in the village of Puzzoli. They used it as mortar for laying bricks or stones and often mixed it with burnt lime and stones to create a waterproof concrete.

The Romans realized that voussoir arches could span further than any unsupported stone beam, and would be more durable and robust than any other structure. They ought to have known because the early Roman settlers were Etruscans. Semi-circular arches were always built by the Romans, with the thrust from the arch going directly down on to the support pier. It meant that piers had to be large. If they were built

Figure 1.1 Pont du Gard, Nîmes.

wide enough at about one-third of the arch span, then any two piers could support an arch without shoring or propping from the sides. In this way it was possible to build a bridge from the shore to shore, a span at a time, without having to form the entire substructure across the river before starting the arches. They developed a method of constructing the foundation on the riverbed within a coffer dam or water-tight dry enclosure, formed by a double ring of timber piles with clay packed into the gap between them to act as a water seal. The water inside the coffer dam was then pumped out and the foundation substructure was then built within it. The massive piers often restricted the width of the river channel, increasing the speed of flow past the piers and increasing the scour action. To counter this the piers were built with cutwaters, which were pointed to cleave the water so it would not scour the foundations.

The stone arch was built on a wooden framework built out from the piers and known as centring. The top surface was shaped to the exact semi-circular profile of the arch. Parallel arches of stones were placed side by side to create the full width of the roadway. The semi-circular arch meant that all the stones were cut identically and that no mortar was needed to bind them together once the keystone was locked into position. The compression forces in the arch ensured complete stability of the span. Of course, the Romans did build many timber bridges, but they have not stood the test of time and today all that remains of their achievement after 2000 years are a handful of stone bridges in Rome, and a few scattered examples in France, Spain, North Africa, Turkey and other former Roman colonies. But what still stands today, whether it is a bridge or an aqueduct, rank among the most inspiring and noble of bridge structures ever built, considering the limitations of their technology.

1.1.4 The Dark Ages and the brothers of the bridge

When the Roman Empire collapsed it seemed that the light of progress around the world went out for a long while. The Huns, the Visigoths, Saxons, Mongols and Danes did not do much building in their raids across Europe and Asia to plunder and destroy. It was left to the spread of Christianity and the strength of the church to start the next boom in road building and bridge building around 1000AD.

It was the church who had preserved and developed both spiritual understanding and the practical knowledge of building during this period. And not surprisingly it was bridge building among the many skills and crafts that became associated with it.

A group of friars of the Altopascio order near Lucca north Italy, lived in a large dwelling called the Hospice of St James'. The friars were skilled at carpentry and masonry having built their own priory and no doubt helped with others. The surrounding countryside was wild and dangerous, and the refuge they built was a popular resting place for pilgrims and travellers using the ancient road from Tuscany to Rome. In 1244 Emperor Frederick II required that the hospice build a proper bridge across the White Arno for pilgrims and travellers. With their skills and practical knowledge the friars set up a co-operative to build the bridge. After completing the bridge over the White Arno their fame spread through Italy and France. It sparked off an interest in bridge building among other ecclesiastical orders. In France, a group of Benedictine monks established the religious order of the Frères Pontiffs (brothers of the bridge) to build a bridge over the Durance.

And so the 'brothers of the bridge' order became established among Benedictine monks and spread from France to England by the thirteenth century. The purpose of the order, apart from its spiritual duties, was to aid travellers and pilgrims, to build bridges along pilgrimage routes or to establish boats for their use and to receive them in hospices built for them on the bank. The brothers of the bridge were great teachers, who strove to emulate and continue the magnificent work of the Roman bridge builders.

The most famous and legendary bridge of this period was built by the Order of the Saint Jacques du Haut Pas, whose great hospice once stood on the banks of the Seine in Paris on the site of present church of that name. They built the Pont Esprit over the Rhône but their masterpiece was the neighbouring bridge at Avignon. It was truly a magnificent and record-breaking achievement for its time. Its beauty has inspired writers, poets, and musicians over the centuries. Sadly all that remains today at Avignon are just four out of the 20 spans of the bridge and the chapel where the supposed creator of the bridge was interred and later canonized as Saint Benezet.

While Pont d'Avignon was being built in France, another monk of the Benedictine order in England, Peter of Colechurch was planning the building of the first masonry bridge over the Thames. A campaign for funds was launched with enthusiasm, not only the rich town people, the merchants and money lenders made generous donations, but the common people of London all gave freely. Until the sixteenth century a list of donors could be seen hanging in the chapel on the bridge. The structure that was built in 1206 was Old London Bridge and ranks after Pont d'Avignon in fame. It was such a popular bridge that buildings and warehouses were soon erected on it. It became so fashionable a location that the young noblemen of Queen Elizabeth's household resided in a curious four-storey timber building imported piece by piece from the Netherlands, called the 'Nonesuch House'.

Towns continued to sponsor and promote the building of stronger and better bridges and roads. They did not always get the brothers of the bridge to build them, because they were often committed to other projects for many years. Instead guilds of master masons and carpenters were formed and spread across Europe offering their services. Even government officials were united in this community enterprise and began to grasp the initiative and drive for better road and bridge networks across the country. Soon the vestiges of the Dark Ages and feudalism were transformed to the age of enlightenment and the Renaissance. The Ponte de Vecchio in Florence, built towards the end of this period, marks the turning point of the Dark Ages. It was a covered bridge erected in 1345, lined with jewellery shops and galleries, with an upper passageway added later, that was a link between the royal and government palaces, the Uffizi and Pitti Palaces. The piers, which are 20 ft thick, support the overhanging building as well as the bridge spans. The most innovative feature of the bridge are the arch spans which are extremely shallow compared with any previous arches

Figure 1.2 Old London Bridge.

ever built or indeed many contemporary European bridges. It was built as a segmental arch, which is unusual for bridge builders of that period because they could not possibly determine the thrust from the arches mathematically with the knowledge they had. How they achieved this is not known; just like segmental arches of Pont d'Avignon. The architect of this radical design was Taddeo Gaddi, who had studied under the great painter Giotto, and was regarded as one of the great names of the Italian Renaissance that followed.

1.1.5 The Renaissance

Not since the days of Homer, Aristotle and Archimedes in Hellenistic times have such great feats of discovery in science and mathematics, and such works of art and architecture been achieved, as during the Renaissance. Modern science was born in this period through the enquiring genius of Copernicus, Da Vinci, Francis Bacon and Galileo and in art and architecture through Michelangelo, Brunellesci and Palladio. During the Renaissance there was a continual search for the truth, explanations of natural phenomena, greater self-awareness and rigorous analysis of Greek and Roman culture. As far as bridge building was concerned, particularly in Italy, it was regarded as a high art form. Much emphasis was placed on decorative order and pleasing proportions as well as the stability and permanence of its construction. Bridge design was architect-driven for the first time with Da Vinci, Palladio, Brunelleschi and even Michelangelo all experimenting with possibility of new bridge forms. The most significant contribution of the Renaissance was the invention of the truss system, developed by Palladio from the simple king post and queen post roof truss, and the founding of the science of structural analysis with the first book ever written on the subject by Galileo Galilei entitled 'Dialoghi delle Nuove Scienze' (*Dialogues on the New Science*) published in 1638.

Palladio did not build many bridges in his lifetime, many of his truss bridge ideas were considered too daring and radical and his work lay forgotten until the eighteenth century. His great treatise published in 1520 *Four Books of Architecture* in which he applied four different truss systems for building bridges, was destined to influence bridge builders in future years when the truss replaced the arch as the principal form

Figure 1.3 Monmow Bridge, Monmouth – an example of a medieval fortified bridge.

of construction. Bridge builders during the Renaissance were clever material tech-
nologists who were preoccupied with the art of bridge construction and how they could
build with less labour and materials. It was a time of inflation when the price of building materials and labour was escalating. The most famous bridge builders in this era
were Amannati, Da Ponte, and Du Cerceau.

Which bridge of the Renaissance is the most beautiful: Florence's Santa Trinita,
Venice's Rialto or Paris' Pont Neuf? Arguably the most famous and celebrated bridge
of the Renaissance was the Rialto bridge designed by Antonio Da Ponte in Venice.

John Ruskin siad of the the Rialto: 'The best building raised in the time of the
Grotesque Renaissance, very noble in its simplicity, in its proportions and its masonry'. Its designer was 75 years old when he won the contract to build the Rialto, and
was 79 when it was finished. It was a single segmental arch span of 87 ft 7 in, which
rises 25 ft 11 in at the crown. The bridge is 75 ft 3in wide, with a central roadway,
shops on both sides and two small paths on the outside, next to the parapets. Two
sets of arches, six each of the large central arch, support the roof and enclose the 24
shops within it. It took three and half years to build and kept all the stone masons
in the city fully occupied in work for two of those years.

Equally innovative and skilful bridge construction was progressing across Europe.
In the state of Bohemia across the Moldau at Prague was built the longest bridge
over water, the Karlsbrucke in 1503 and the most monumental and imperial bridge
of the Renaissance. It took a century and half to completely finish. It was adorned

with statues of saints and martyrs and terminates on each bank with an imposing tower gateway. In France at this time a fine example of the early French Renaissance, the Pont Neuf, was being designed. It was the second stone bridge to be built in Paris and although its design and construction did not represent a great leap forward in bridge building, it occupies a special place in Parisian hearts. Designed by Jacques Androuet Du Cerceau, the two arms of the Pont Neuf that join the Ille de la Cité to the left and right bank was a massive undertaking. Although all the arches are semi-circular and not segmental, no two spans are alike, as they vary from 31 to 61 ft in span and also differ on the downstream and upstream sides of each arch and were built on a skew of 10%. Du Cerceau wanted the bridge to be a true unencumbered thoroughfare bereft of any houses and shops. But the people of Paris demanded shops and houses which resulted in modification to the few short-span piers that had been constructed.

The Pont Neuf has stood now for 400 years and was the centre of trade, and the principal access to and from the crowded island when it was built. The booths and stalls on the bridge became so popular that all sorts of traders used it including book-sellers, pastry cooks, jugglers, and peddlers. They crowded the roadway until there were some 200 stalls and booths packed into every niche along the pavement. The longer left bank of the Pont Neuf was extensively reconstructed in 1850 to exactly the same details, after many years of repairs and attention to its poor foundations. The right bank with the shorter spans has been left intact. The entire bridge has been cleared of all stalls and booths and is used today as a road bridge.

Figure 1.4 Pont Neuf, Paris (courtesy of JL Michotey).

The finest examples of late French Renaissance bridge built during the seventeenth century are the Pont Royale and Pont Marie bridge both of which are still standing today. The Pont Royale was the first bridge in Paris to feature elliptical arches and the first to use an open caisson to provide a dry working area in the river bed. The foundations for the bridge piers were designed and constructed under the supervision of Francosi Romain, a preaching brothers from the Netherlands who was an expert in solving difficult foundation problems. The bridge architect François Mansart and the builder Jacques Gabriel called on Romain after they ran into foundation problems. Romain introduced dredging in the preparation of the riverbed for the caisson using a machine that he had developed. After excavations were finished the caisson was sunk to the bed, but the top was kept above the water level. The water was then pumped out and the masonry work of the pier was then built inside the dry chamber. The five arch spans of the Pont Royale increase in span towards the centre, and although it has practically no ornamentation, it blends beautifully into it river setting and the bank-side environment.

The Renaissance brought improvements in both the art and science of bridge building. For the first time bridges began to be regarded as civic works of art. The master bridge builder had to be an architect, structural theorist and practical builder, all rolled in one. The bridge which was without doubt the finest exhibition of engineering skills in this era, was slender elliptical arched bridge of Santa Trinita at Florence, designed by Bartholomae Ammannati in 1567. Many scholars are still mystified to this day as to how Ammannati arrived at the such pleasing, slender curves to the arches.

Figure 1.5 Pont Royale, Paris (courtesy of J Crossley).

1.2 Eighteenth-century bridge building

1.2.1 The Age of Reason

In this period, masonry arch construction reached perfection, due to a momentous discovery by Perronet and the innovative construction techniques of John Rennie. Just as the masonry arch reached its zenith 7000 years after the first crude corbelled arch in Mesopotamia, it was to be threatened by a new building material – iron – and the timber truss, as the principal construction for bridges in the future.

This was the era when civil engineering as a profession was born, when the first school of engineering was established in Paris at the Ecole de Paris during the reign of Louis XV. The director of the school was Gabriel who had designed the Pont Royal. He was given the responsibility of collecting and assimilating all the scientific information and knowledge there was on the science and history of bridges, buildings, roads and canals.

With such a vast bank of collective knowledge it was inevitable that building architecture and civil engineering should be separated into the two fields of expertise. It was suggested it was not possible for one man in his brief life to master the essentials of both subjects. Moreover, it also became clear that the broad education received in civil engineering at the Corps des Ponts et Chaussées at the Ecole de Paris was not sufficient for the engineering of bridge projects. More specialized training was needed in bridge engineering. In 1747 the first school of bridge engineering was founded in Paris at the historic Ecole des Pont De Chaussées. The founder of the school was Trudiane, and the first teacher and director was a brilliant young engineer named Jean Perronet.

Jean Perronet has been called the father of modern bridge engineering for his inventive genius and design of the greatest masonry arch bridges of the century. In his hands the masonry arch reached perfection. The arch he chose was the curve of a segment of a circle of larger radius, instead of the familiar three-centred arch. To express the slenderness of the arch he raised the haunch of the arch considerably above the piers. He was the first person to realize that the horizontal thrust of the arch was carried through the spans to the abutments and that the piers, in addition to the carrying the vertical load, also had to resist the difference between the thrusts of the adjacent spans. He deduced that if the arch spans were about equal and all the arches were in place before the centring was removed, the piers could be greatly reduced in size.

What remains of Perronet's great work? Only his last bridge, the glorious Pont de la Concorde in Paris, built when he was in his eighties. It is one of the most slender and daring stone arch bridges ever built in the world. 'Even with modern analysis', suggests Professor James Finch the author of *Engineering and Western Civilisation* 'we could not further refine Perronet's design'.

With France under the inspired leadership of Gabriel and then Perronet, the rest of Europe could only admire and copy these great advances in bridge building. In England, a young Scotsman, John Rennie, was making his mark following in the footsteps the great French engineers. He was regarded as the natural successor to Perronet, who was a very old man when Rennie started on his career. Rennie was a brilliant mathematician, a mechanical genius and pioneering civil engineer. In his early years he worked for James Watt to build the first steam-powered grinding mills at Abbey Mills in London, and later designed canals and drainage systems to drain the marshy fens of Lincolnshire. He built his first bridge in 1779 across the Tweed at Kelso. It was a modest affair with a pier width-to-span ratio of one to six with a conservative elliptical arch span. He picked up the theory of bridge design from textbooks and from studies and discussion about arches and voussoirs with his mentor Dr Robison of Edinburgh University. He designed bridges with a flat, level roadway and not the characteristic hump of most English bridges. It was radical departure from convention and was much admired by all the town's people, farmers and traders who transported material and cattle across them. This bridge was a modest forerunner to the many famous bridges that Rennie went on to build: Waterloo, Southwark and New London Bridge. What then was Rennie's contribution to bridge building? For Waterloo bridge, the centring for the arches was assembled on the shore then floated out on barges into position. So well and efficiently did this system work that the framework for each span could be put into position in a week. This was a fast erection speed and as a result Rennie was able to halve bridge construction time. So soundly were Rennie's bridges built that 40 years later Waterloo bridge had settled only 5 in. Rennie's semi-elliptical arches, sound engineering methods and rapid assembly technique, together with the Perronet segmental arch, divided pier and understanding of arch thrust, changed bridge design theory for all time.

Figure 1.6 John Rennie's New London Bridge – under construction.

1.2.2 The carpenter bridges

The USA with its vast expansion of roads and waterways, following in the wake of commercial growth in the eighteenth century, was to become the home of the timber bridge in the nineteenth century.

The USA had no tradition or history of building with stone, and so early bridge builders used the most plentiful and economical materials that was available: timber. The Americans produced some of the most remarkable timber bridge structures ever seen, but they were not the first to pioneer such structures. The Grubenmann brothers of Switzerland were the first to design quasi-timber truss bridges in the eighteenth century. The Wettingen bridge over the Limmat just west of Zürich was considered their finest work. The bridge combines the arch and truss principle with seven oak beams bound close together to form a catenary arch to which a timber truss was fixed. The span of the Wettingen was 309 ft and far exceeded any other timber bridge span.

Of course, there were numerous timber beam and trestle bridges built in Europe and the USA. But in order to bridge deep gorges, broad rivers and boggy estuaries such as those that ran through North America and support the heavy loads of chuck wagons and cattle, something more robust was needed. The answer according to the Grubenmanns was a timber truss arch bridge, but it was not a true truss.

Figure 1.7 Example of the Bollman truss, Central Park Bridge (courtesy of E Deloney).

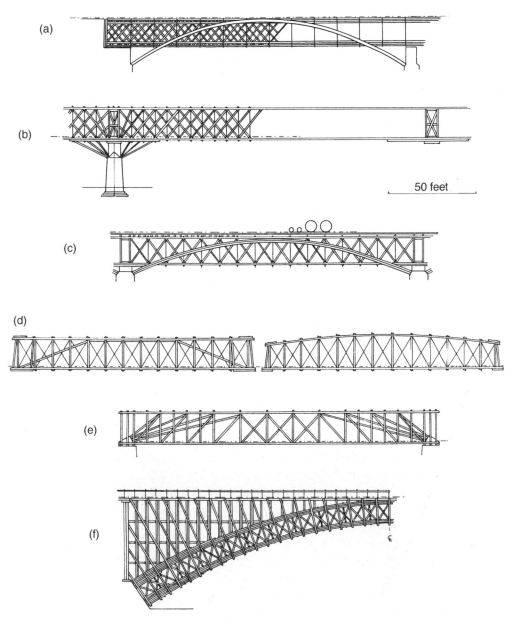

Figure 1.8 US patent truss types: (a) 1839 Wilton, (b) 1844 Howe, (c) 1849 Stone-Howe, (d) 1844 Pratt, (e) 1846 Hassard, (f) 1848 Adams.

Palmer, Wernwag and Burr – the so-called American carpenter bridge builders of North America, who designed more by intuition than by calculation – developed the truss arch to span further than any other wooden construction. This was the third and last of the three basic bridge forms to be discovered. The first person who made the truss arch bridge a success in the USA and who patented his truss design was Timothy Palmer. In 1792 Palmer built a bridge consisting of two trussed arches over the Merrimac – it looked very like one of Palladio truss designs, except the arch was the dominant supporting structure. His 'Permanent Bridge' over the Schuylkill built in 1806 was his most celebrated. When the bridge was finished the president of the bridge company suggested that it would be a good idea to cover the bridge to preserve the timber from rot and decay in the future. Palmer went further than that and timbered the sides as well, completely enclosing the bridge. Thus, America's distinctive covered bridge was established. By enclosing the bridge it stopped snow getting in and piling up on the deck causing it a collapse from the extra load.

Wernwag was a German immigrant from Pennsylvania, who built 29 truss type bridges in his lifetime. His designs integrated the arch and truss into one composite structure rather more successfully than Palmer's. Werwnag's famous bridge was the Collossus over the Schuylkill just upstream from Palmer's 'Permanent Bridge' and composed of two pairs of parallel arches, linked by a framing truss, which carried the roadway. The truss itself was acting as bracing reinforcement and consisted of heavy verticals and light diagonals. The diagonal elements were remarkable because they were iron rods, and were the first iron rods to be used in a long span bridge. In its day the Collossus was the longest wooden bridge in the USA, having a clear span of 304 ft. Fire destroyed the bridge in 1838. It was later replaced by Charles Ellet's pioneering suspension bridge.

Theodore Burr was the most famous of the illustrious triumvirate. Burr developed a timber truss design based on the simple king and queen post truss of Palladio. He came closest to building the first true truss bridge, but in doing so, it proved unstable under moving loads. Burr then strengthened the truss with an arch. It was significant that here the arch was added to the truss rather than the other way round. Burr arch–trusses were quick to assemble and modest in cost to build, and for a time they were the most popular timber bridge form in the USA.

By 1820 the truss principle had been well explored and although the design theory was not understood, in practice it had been tested to the limit. It was left to Ithiel Town to develop and build the first true truss bridge, which he patented and called the Town Lattice. It was a true truss because it was free from arch action and any horizontal thrust. It was so simple to build that it could be nailed together in a few days and cost next to nothing compared to other alternatives. Town promoted his timber structure with the slogan ' built by the mile and cut by the yard'. He did build the Town Lattice truss bridges himself, but issued licences to local builders to use his patent design instead. He collected a dollar for every foot built and two if a bridge was built without his permission. By doubling the planking and wooden pins to fasten the structure together Town made his truss carry the early railroads.

1.2.3 The railroad and the truss bridge

With arrival of the railways in the USA, bridge building continued to develop along two separate ways. One school continued to evolve stronger and leaner timber truss structures, while the other experimented with cast iron and wrought iron, slowly replacing timber as the principal construction material.

The first patent truss to incorporate iron into a timber structure was the Howe Truss. It had top and bottom chords and diagonal bracing in timber and vertical members of iron rod in tension. This basic design with modifications, continued right into the next century. The first fully designed truss was the Pratt Truss which reversed the forces of the Howe Truss by putting the vertical timber members in compression and the iron diagonal members, in tension. The Whipple truss in 1847 was the first all-iron truss – a bowstring truss – with the top chord and vertical compression members made from cast iron and the bottom chord and diagonal bracing members made from wrought iron. Later Fink, Bollman, Bow, and Haupt in the USA, along with Cullman and Warren in Europe developed the truss to a fine art, incorporating wire-strand cable, timber and iron to form lightweight but strong bridges that could carry railways.

The stresses and fatigue loading from moving trains in the late nineteenth century caused catastrophic failure of many timber truss and many iron truss bridges. The world was horrified by the tragedy and death toll from collapsing bridges. At one stage as many as one bridge in every four used by the railway network in the USA, had a serious defect or had collapsed. By the turn of the century the iron truss railway bridges had been replaced by stronger and more durable structures. Design codes and safety regulations were drawn up and professional associations were incorporated to train, regulate and monitor the quality of bridge engineers

In the nineteenth century the truss, the last of the three principle bridge forms, had at long last been discovered. With the coming of the industrial revolution, and the rapid growth of the machine age – dominated by the railway and motor car – a huge burden was placed on civil engineering, material technology and bridge building. Many new and daring ideas were tried and tested, many innovative bridge forms were built. There were some spectacular failures. As many as seven major new bridge types were to emerge during this period: the box girder, the cantilever truss girder, the reinforced and prestressed concrete arch, the steel arch, glued segmental construction, cable stayed bridges and stressed ribbon bridges.

1.3 The past 200 years: bridge development in the nineteenth and twentieth centuries

The industrial revolution which began in Britain at the end of the eighteenth century, gradually spread and brought with it huge changes in all aspects of everyday life. New forms of bulk transportation, by canal and rail, were developed to keep pace with the increasing exploitation of coal and the manufacture of textiles and pottery. Coal fuelled the hot furnaces to provide the high temperatures to smelt iron. Henry Bessemer invented a method to produce crude steel alloy by blowing hot air over smelted iron. Seimens and Martins refined this process further to produce the low carbon steels of today. High temperature was also essential in the production of cement which Joseph Aspdin discovered by burning limestone and clay on his kitchen stove in Leeds in 1824. Wood and stone were gradually replaced by cast iron and wrought iron construction, which in turn was replaced by first steel and then concrete; the two primary materials of bridge building in the twentieth century.

Growing towns and expanding cities demanded continuous improvement and extension of the road, canal and railway infrastructure. The machine age introduced the steam engine, the internal combustion engine, factory production lines, domestic appliances, electricity, gas, processed food and the tractor. Faster assembly of bridges was essential, and this meant prefabricating lightweight but tough, bridge components. The heavy steam engines and longer goods train imposed larger stresses on bridge structures than ever before. Bridges had to be stronger and more rigid in construction and yet had to be faster to assemble to keep pace with progress. Connections had to be stronger and more efficient. The nut and bolt was replaced by the rivet, which was replaced by the high strength friction grip bolt and the welded connection.

When the automobile arrived it resulted in a road network that eventually crisscrossed the entire countryside from town to city, over mountain ranges, valleys, streams, rivers, estuaries and seas. Even bigger and better bridges were now needed to connect islands to the mainland and countries to continents in order to open up major trading routes. The continuous search and development of high strength materials of steel, concrete, carbon fibre and aramids today combined with sophisticated computer analysis and dynamic testing of bridge structures against earthquakes, hurricane wind, and tidal flows has enabled bridges to span even further. In the last two centuries bridge spans have leapt from 350 ft to over 6000 ft. This is the age of the mighty suspension bridges, the elegant cable stayed bridges the steel arch truss, the glued segmental and cantilever box girder bridges.

The key events and achievements of this large output of bridge building are briefly summarized to illustrate the rapid pace of change and many bridge ideas that were advanced. In the past two centuries more bridges were built than in the entire history of bridge building prior to that!

1.3.1 The age of iron (1775–1880)

Of all the materials used in bridge construction – stone, wood, brick, steel and concrete – iron was used for the shortest time. Cast iron was first smelted from iron ore successfully by Dud Dudley in 1619. It was another century before Abraham Derby devised a method to economically smelt iron in large quantities. However, the brittle quality of cast iron made it only safe to use in compression in the form of an arch. Wrought iron, which replaced cast iron many years later, was a ductile material that could carry tension. It was produced in large quantities after 1783 when Henry Cort developed a puddling furnace process to drive impurities out of pig iron.

But iron bridges suffered some of the worst failures and disasters in the history of bridge building. The vibration and dynamic loading from a heavy steam locomotive and from goods wagons, create cyclic stress patterns on the bridge structure as the wheels roll over the bridge, going from zero load to full load then back to zero. Over a period of time these stress patterns can lead to brittle failure and fatigue in cast iron and wrought iron. In one year alone in the USA, as many as one in every four iron and timber bridge had suffered a serious flaw or had collapsed. Rigorous design codes, independent checking and new bridge building procedures were drawn up, but it was not soon enough to avert the worst disaster in iron bridge history over the Tay estuary in 1878. It marked the end of the iron bridge for good.

Significant bridges
1779   Iron Bridge in Coalbrookdale, the first cast-iron bridge, designed as an arch structure by Pritchard for owner and builder Abraham Darby, the third.
1790   Buildwas Bridge, the second cast-iron bridge built in Coalbrookdale, designed by Thomas Telford used only half the weight of cast iron of the Iron Bridge.
1807   James Finlay builds first elemental suspension bridge – the Chain Bridge – in wrought iron in 1807 over the Potomac.
1821   Guinless Bridges, George Stephenson's wrought iron 'lentilcular' girder bridge for the Stockton to Darlington Railway.
1826   Menai Straits Bridge, famous eye bar, wrought iron chain suspension bridge over the Menai Straits, by Thomas Telford.
1834   The Fribourg Bridge, the world's longest iron suspension bridge.
1841   Whipple patents the cast iron 'bowstring' truss bridge.
1846   Wheeling Suspension Bridge, Charles Ellet's record-breaking 1000 ft span, iron wire suspension bridge
1850   Britannia Bridge, first box girder bridge concept, built in wrought iron by Robert Stephenson
1853   Murphy designs a wrought iron Whipple truss, with pin connections.
1858   Royal Albert Bridge, Saltash, Brunel's famous tubular iron bridge, over the Tamar.
1876   The Ashtabula Bridge disaster in USA, 65 people die when this iron modified Howe truss collapses plunging a train and its passengers into the deep river gorge below.
1878   The Tay Bridge disaster, Dundee, Scotland where a passenger train with 75 people plunges into the Tay estuary, as the supporting wrought iron girders collapse in high winds.

1.3.2 The arrival of steel
Steel is a refined iron where carbon and other impurities are driven off. Techniques for making steel are said to have been known in China in 200BC and in India in 500BC. But the process was very slow and laborious and after a great deal of time and energy only minute amounts were produced. It was very expensive, so it was only used for edging tools and weapons until the nineteenth century. In 1856, Henry Bessemer developed a process for bulk steel production by blowing air through molten

iron to burn off the impurities. It was followed by the open hearth method patented by Charles Siemens and Pierre Emile Martins in Birmingham, England in 1867, which is the basis for modern steel manufacture today. It took a while for steel to supersede iron, because it was expensive to manufacture. But when the world price of steel dropped by 75% in 1880, it suddenly was competitive with iron. It had vastly superior qualities, both in compression and tension – it was ductile and not brit-

Figure 1.9 Iron Bridge in Coalbrookdale (courtesy of J Gill).

Figure 1.10 Britannia Bridge, Anglesey.

tle like iron, and was much stronger. It could be rolled, cast or even drawn; to form rivets, wires, tubes, and girders. The age of steel opened the door to tremendous advances in long-span bridge building technology. The first bridges to exploit this new material were in the USA, where the steel arch, the steel truss and the wire rope suspension bridges were pioneered. Later, Britain led the world in the cantilever truss bridge and the steel box girder bridge deck.

The historical progress of the principle of building bridges in steel covering the period from 1880 to the present is described below.

Figure 1.11 Royal Albert Bridge, Saltash.

Figure 1.12 The Tyne Bridge, Newcastle upon Tyne.

The steel truss arch

When the steel prices dropped in the 1870s and 1880s the first important bridges to use steel were all in the USA. The arches of St Louis Bridge over the Mississippi and the five Whipple trusses of the Glasgow Bridge over the Missouri, were the first to incorporate steel in truss construction. St Louis, situated on the Mississippi and near the confluence of the Missouri and Mississippi was the most important town in mid-west USA, and the focal point of north–south river traffic and east–west overland routes.

1874   The St Louis Bridge, St Louis – James Eads builds the first triple arch steel bridge.

1884   The Garabit Viaduct, St Flour, France – Gustav Eiffel's truss arch in wrought iron, was the prototype for future steel truss construction. Eiffel would have preferred steel but chose wrought iron because it was more reliable in quality and cheaper.

1916   The Hell Gate Bridge, New York – the first 977 ft steel arch span in the world, was designed by Gustav Lindenthal.

1931   The Bayonne Bridge, New York – the first bridge to be built with a cheaper carbon manganese steel, rather than nickel steel, and which is the composition of most modern steel.

1932   Sydney Harbour Bridge, Sydney – this famous steel arch was built using 50 000 tons of nickel steel. Its design was based on the Hell Gate Bridge.

1978   New River Gorge Bridge, West Virginia – currently the world's longest steel arch span.

The cantilever truss

Arch bridges had been constructed for many centuries in stone, then iron, and steel when it became available. Steel made it possible to build longer span trusses than cast iron without any increase in the dead weight. Consequently it made cantilever long-span truss construction viable over wide estuaries. The first and most significant cantilever truss bridge to be built was the rail bridge over the Firth of Forth near Edinburgh, Scotland in 1890. The cantilever truss was rapidly adopted for the building of many US railroad bridges until the collapse of the Quebec Bridge in 1907.

1886   The Fraser River Bridge, Canada – believed to be the first balanced cantilever truss bridge to be built. All the truss piers, links, and lower chord members were fabricated from Siemens–Martin steel. It was dismantled in 1910.

1890   The Forth Rail Bridge, Edinburgh, Scotland – the world's longest spanning bridge at 1700 ft, when it was finished.

1891   The Cincinnatti Newport Bridge,Cincinnati – with it long through cantilever spans and short truss spans, was the prototype of many rail bridges in the USA.

1902   The Viaur Viaduct, France – this rail bridge between Toulouse and Lyons, was an elegant variation of the balanced cantilever, with no suspended section between the two cantilever arms.

1917   The Quebec Bridge – completion of the second Quebec bridge, the world's longest cantilever span.

1927   Carquinez Bridge – the last of the long cantilever truss bridges to be built in the US although second identical bridge was built alongside it in 1958 to increase traffic flow.

The suspension bridge

The early pioneers of chain suspension bridges were James Finlay, Thomas Telford, Samuel Brown and Marc Seguin, but they had only cast and wrought iron available in the building of their early suspension bridges. It was not until Charles Ellet's Wheeling Bridge had shown the potential of wire suspension using wrought iron that the concept was universally adopted. Undoubtedly the greatest exponent of early wire suspension construction and strand spinning technology was John Roebling. His Brooklyn Bridge was the first to use steel for the wires of suspension cables.

Suspension bridges are capable of huge spans, bridging wide river estuaries and deep valleys and have been essential in establishing road networks across a country. They have held the record for longest span almost unchallenged from 1826 to the present day and only interrupted between 1890 to 1928, when the cantilever truss held the record.

1883   Brooklyn Bridge – following the completion of the Wheeling suspension bridge pioneered by Charles Ellet; John Roebling went on to design the Brooklyn Bridge, the first steel wire suspension bridge in the world.

1931   George Washington – the heaviest suspension bridge to use parallel wire cables rather than rope strand cable, and the longest span in the world for nearly a decade.

Figure 1.13 The Quebec Bridge, Canada.

Figure 1.14 The George Washington Bridge, New York.

1950   Tacoma Narrows – the second Tacoma Narrows rebuilt after the collapse of
        the first bridge with a deep stiffening truss deck, set the trend for future sus-
        pension bridge design in the USA.
1957   Mackinac – Big Mac is the longest overall suspension bridge in the
        USA.
1965   Verazzano Bridge – the last big suspension bridge to be built in the USA,
        also held the record for the longest span until 1981.
1967   Severn Bridge – the first bridge to have a slim, aerodynamic bridge deck, elim-
        inating the need for deep stiffening trusses like those of US suspension
        bridges. It set the trend for future suspension bridge construction.
1981   Humber Bridge – the longest span in the world when it was completed, with
        supporting strands that were inclined in a 'zig-zag' fashion rather than the
        parallel arrangement preferred by the Americans.
1998   Great Belt, and the East Bridge – the Great Belt crossing is now complete,
        and is the longest bridge in Europe. For a short while the main span of the
        East Bridge held the record for the longest span in the world.
1998   Akashi Kaikyo – is one of a family of long-span bridges linking the islands
        of Honshu and Shikoku, now well under construction. Its main span of 6529
        ft makes it the longest span in the world.

Steel plate girder and box girder
Since the development of steel and the I-beam, many beam bridges were built using
a group of beams in parallel which were interconnected at the top to form a roadway. They were quick to assemble but they were only practical over relatively short
spans for rail and road viaducts. The riveted girder I-beam was later superseded by
the welded and friction grip bolted beam. However, relatively long spans were not
efficient as the depth of the beam could become excessive. To counter this, web plate
stiffeners were added at close intervals to prevent buckling of the beam. Another solution was to make the beam into a hollow box which was very rigid. In this way the
depth of the beam could be reduced and material could be saved. The steel box girder beams could be quickly fabricated and were easy to transport. Their relatively shallow depth meant that high approaches were not necessary. Most of this pioneering
work was carried out during and after the second world war when there was a huge
demand for fast and efficient bridge building for spans of up to 1000 ft. The major
rebuilding programme in Germany witnessed the construction of many steel box
girder and concrete box girder bridges in the 1950s and 1960s. For spans greater
than 1000 ft the suspension and cable stay bridge are generally more economical to
construct.

    In the 1970s the world's attention was focused on the collapse of four steel box
girder bridges under construction. The four bridges were in Vienna over the Danube,
in Milford Haven in Wales when four people were killed, a bridge over the Rhine in
Germany and the West Gate bridge in Melbourne over the Lower Yarra River. By
far the worst collapse was on the West Gate bridge, a single cable stay structure with
a continuous box girder deck. A deck span section 200 ft long and weighing 1200 tons,

buckled and crashed off the pier support on to some site huts below, where work-men and engineers were having their lunch. Thirty-five people were killed in the tragedy. After this accident, further construction of steel box girder deck bridges was halted until better design standards, new site checking procedures and a fabrication specification was agreed internationally.

1936 Elbe Bridge – one of the early plate girder bridges on the German autobahn.
1948 Bonn Beuel – a later development of the plate girder into a flat arch, to reduce material weight.
1952 Cologne Deutz Bridge – first slender steel box girder bridge in the world.
1970s Failure of box girders at Milford Haven in Wales and Westgate Bridge in Aus-tralia, halted further building of the steel box girder bridge decks for a time.

1.3.3 Concrete and the arch

Although engineers took longer to realize the true potential of concrete as a build-ing material, today it is used everywhere in a vast number of bridges and building applications. Concrete is a brittle material, like stone, good in compression, but not in tension so if it starts to bend or twist it will crack. Concrete has to be reinforced with steel to give it ductility, so naturally its emergence followed the development of steel. In 1824 Joseph Aspdin made a crude cement from burning a mixture of clay and limestone at high temperature. The clinker that was formed was ground into a powder, and when this was mixed with water it reacted chemically to harden back into a rock. Cement is combined with sand, stones and water to create concrete, which remains fluid and plastic for a period of time, before it begins to set and hardens. It can be poured and placed into moulds or forms while it is fluid, to create bridge beams, arch spans, support piers – in fact a variety of structural shapes. This gives concrete special qualities as a material, and scope for bold and imaginative bridge ideas.

François Hennebique was the first to understand the theory and practical use of steel reinforcement in concrete, but it was Robert Maillart (1872–1940) who was first to pioneer and build bridges with reinforced concrete. Eugene Freyssinet, Maillart's contemporary, was also keen to experiment with concrete structures and went on to discover the art of prestressing and gave the bridge industry one of the most effi-cient methods of bridge deck construction in the world. Both these men were great engineers and champions of concrete bridges. What they achieved, set the trend for future developments in concrete bridges – precast bridge beams, concrete arch, box girder and segmental cantilever construction. Concrete box girder bridge decks are incorporated on many modern cable stay and suspension bridges.

Jean Muller and contractors Campenon Bernard were responsible for building the first match cast, glued segmental, concrete box girder bridge in the world. It is a tech-nique that is used by many bridge builders across the world. The box girder span can be precast in segments or cast in place using a travelling formwork system. They can be built as balanced cantilevers each side of a pier or launched from one span to the next.

Concrete has been used in building most of the world's longest bridges. The relative cheapness of concrete compared to steel, the ability to rapidly precast or form prestressed beams of standard lengths, has made concrete economically attractive. Lake Ponchetrain Bridge, a precast concrete segmental box girder bridge, in Louisiana is the longest bridge in the US with an overall length at 23 miles.

The concrete arch
1898   Glenfinnian Viaduct – the first concrete arch bridge to be built in England.
1905   Tavanasa Bridge – a breakthrough in the stiffened arch slab.
1922   St Pierre de Vouvray – early concrete bowstring arch of Freysinnet.
1929   Plougastel Bridge – unique construction concept which used prestressing, for the first time.
1930   Salgina Gorge Bridge – one of the most aesthetic arch spans of Maillart.
1936   Alsea Bay Bridge – completion of one of Conde McCullough's fine 'art deco' bridges in Oregon (demolished).
1964   Gladesville – use of precast prestressed segmental construction for the arch span
1964   KRK (Croatia) – the longest concrete arch span in the world.

Concrete box girders
1950s–60s Many motorway bridges and viaducts were built in Europe and USA using concrete box girder construction. Some were precast segmental construction, some were cast in place.
1952   Shelton Road Bridge – first match cast, glue segmental, box girder construction in the world developed by Jean Muller.

Figure 1.15 Tavanasa Bridge: a stiffened concrete arch bridge by Robert Maillart (courtesy of EH).

Figure 1.16 Salgina Gorge Bridge – one of the most aesthetic arch spans of Maillart (courtesy of EH).

Figure 1.17 The Medway bridge, Kent.

1956   Lake Ponchartrain Bridge – the second longest bridge in the world, is a
              precast segmental box girder bridge with 2700 spans and runs for 23 miles
              across Lake Ponchetrain near New Orleans. The second identical bridge,
              built alongside the original one in 1969 and was 69 m longer.
1972   Medway Bridge – the first European river bridge to be built using con-
              crete box girder construction.

1.3.4 Cable stay bridges

Cable stays are an adaptation of the early rope bridges, and guy ropes for securing
tent structures and the masts of sailing ships. When very rigid, trapezoidal box girder bridge decks were developed for suspension bridges, it allowed a single plane of
stays to support the bridge deck directly. This meant that fewer cables were needed

Figure 1.18 The Brotonne Bridge, Sotteville, France (courtesy of J Crossley).

than a conventional suspension system, there was no need for anchorages and therefore it was cheaper to construct. Cost and time have always been the principal motivators for change and innovation in the bridge engineering.

The first modern cable stay bridges were pioneered by German engineers just after the second world war, led by Fritz Leonhardt, Rene Walter and Jörge Schlaich. The cable stay bridge is probably the most visually pleasing of all modern long span bridge forms. In recent times the development of the cable stay and box girder bridge deck has continued with the work of Swedish engineers COWI consult; bridge engineers Carlos Fernandez Casado of Spain; R Greisch of Belgium; Jean Muller International, Sogelerg, and Michel Virloguex of France.

Cable stay history

1955 Störmstrund, Norway – the first cable stay bridge.

1956 North Bridge, Dusseldorf – early harp arrangement for a family of cable stay bridges over the Rhine – it was the prototype for many cable stay bridges

1959 Severins Bridge – the first to adopt an A-frame tower and the first bridge to use a fan configuration for the stays – a very efficient bridge form

1962 Lake Maracaibo Bridge – an unusual composite cable stay and concrete frame support structures for a bridge built in Venezuela, using local labour.

1962 Nordelbe Bridge, Hamburg – the first bridge to use a single plane of cables; the deck was a stiffened rectangular box girder

1966 Wye Bridge, England – a single cable stay from the mast supports the continuous steel box girder bridge deck. Erskine Bridge built in 1971 was a better example of this construction.

1974 Brotonne Bridge – the first cable stay bridge to use a precast concrete box girder deck and a single plane of cable stays.

1984 Coatzacoalcos II Bridge, Mexico – elegant pier and mast tower combining the rigidity of the A-frame with the economy of a single foundation.

1995 Pont de Normandie – breakthrough in the design of very long cable stay spans.

1.4 Aesthetic design in bridges

1.4.1 Introduction

Is it possible there is a universal law or truth about beauty on which we can all agree? We can probably argue that no matter what out aesthetic taste in art, literature or music, certain works have been universally acclaimed as masterpieces because they please the senses, evoke admiration and a feeling of well-being. Music, literature and painting can appeal to an audience directly, unlike a building or bridge whose beauty has to be 'read' through its structural form, which has been designed to serve another more fundamental purpose. Judging what is great from many competent

Figure 1.19 An example of Fritz Leonhardt's work: Maintelbrücke Gamunden Bridge (courtesy of F Leonhardt).

examples must come from an individual's own experience and understanding of past and contemporary styles of expression. The desire to please or to shock is not fundamental in the design of bridges whose primary purpose is to provide a safe passage over an obstacle, be it a river or gorge or another road way. A bridge taken in it purest sense is no more than an extension of a pathway, a roadway or a canal. We do not regard roads, paths and canals as 'art forms' that evoke aesthetic pleasure as we do with buildings. Hence, it is reasonable to ask why should a bridge be an art form? In the very early years of civilization, bridges were built to breach a chasm or stream to satisfy just that purpose. They had no aesthetic function. Later on when great civilizations placed a temporal value on the quality of their buildings and heightened their religious and cultural beliefs through their architecture, these values transferred to bridges. And like all the important buildings of a period, when stone and timber were the principal sources of construction material, work was done by skilled craftsmen. Masons would cut, chisel and hew stones: carpenters would saw, plane and connect pieces of timber falsework or centring to support the masonry structure. It took many years to 'fashion' a bridge. Each stone was carefully cut to fit precisely into position. Hundreds of stone masons would be employed to work on the important bridges. Voussoirs and key stones were sculptured and tooled in the architectural style of the period. Architecture was regarded as an integral part of bridge construction and this tradition continued into the age of iron, where highly decorative wrought iron and cast iron sections were expressed on the external faces of the bridge. Well into the middle of the twentieth century arch bridges in concrete and steel were cloaked in masonry panels to imitate the Renaissance, Classical and Baroque periods.

But gradually as the pace of industrial change intensified, by the expansion of the railways, and by the building of road networks, a radical step change in the design and construction of bridges occurred. Bridges had to be functional, they had to be quick to build, low in cost, and structurally efficient. They had to span further and use fewer materials in construction. Less excavation for deep piers and foundations underwater meant faster construction, whereas short continuous trestle supports across a wide valley were simple to construct and required shallow foundations. Under these pressures, standardization and prefabrication of bridges displaced aesthetic consideration in bridge design. Of course, there were exceptions when prestigious bridges were commissioned in major commercial centres to retain the quality and character of the built environment. And sometimes even these considerations were sidelined in the name of progress and regeneration, as was the case in the aftermath of the two world wars. When economic stability returns to a nation after the ravages of war, and living standards start to rise, so does interest in the arts and quality of the built environment.

After the second world war, for example, rebuilding activity had to be fast and efficient, with great emphasis placed on prefabrication, system-built housing and the tower block to re-house as many people as possible. In Germany rebuilding the many bridges that were demolished, led to the development of the plate girder and box girder structure. Box girder bridge structures with standardized sections, proliferated the road network and motorways of Europe, over viaducts, interchanges, flyovers, and river crossings. In this period the shape and form of the bridge was dictated by the contractor's preference for repetition and simplicity of construction.

Given this history it is hardly surprising to find that many of our towns and urban areas and motorway network are blighted by ugly, functional bridging structures whose presence now causes a public outcry.

1.4.2 Bridge aesthetics in the twentieth century

Over the centuries as the various forms of bridges evolved in the major towns and cities, the architectural style of the period was superimposed on them, to create order and homogeneity. Classical, Romanesque, Byzantine, Islamic, Renaissance, Gothic, Baroque, Georgian and Victorian architectural styles adorn many historic bridges today, such as the Renaissance Rialto Bridge in Venice, the Romanesque Pont Saint Angelo in Rome, the French Gothic of the Pont de la Concorde in Paris. They are recognizable symbols of an era, of imperialist ambition and nationhood, where the dominant form of construction was the arch. But with the arrival of steel and concrete in the early part of the twentieth century, new structural forms emerged in building and bridge design that radically changed both the architecture and visual expression of bridges. The segmental arch was replaced by the flat arch, the flat plate girder and box girder beam; the cantilever truss was replaced by the cable stay and the suspension bridge. The decorative stone-clad bridges of the past were slowly replaced by the minimalism of highly engineered structures.

Undoubtedly, during the period from the 1920s to the 1940s the greatest concentration of bridge building was in the USA. It was in step with the massive industrial and commercial expansion throughout the country, and emergence of the high-rise building – the skyscraper. And in building bridges – the great suspension, steel arch and cantilever truss bridges – those that were important were the subject of much debate about appearance, and harmony with its surrounding environment. Champions of aesthetic bridge design emerged – David Steinman, Condo McCullough, Gustav Lindenthal and Othmar Ammann. All of them were engineers. Steinman was the most flamboyant and outspoken individual among this group and wrote books and articles on the subject. Condo McCullough's 'art deco' bridges – inspired by the bridges of Robert Maillart – were aesthetic masterpieces of the concrete arch and steel cantilever truss bridge.

In the 1950s and 1960s the bridge building boom moved to Europe following the war years, with a plethora of utilitarian structures built in the name of economy. Architectural and aesthetic considerations were reduced to a minor role. Bland, insensitive and crude bridge structures and viaducts appeared across the open countryside, and through towns and across cities. Concern about the impact these bridges would have on the built environment brought Fritz Leonhardt, one of Germany's leading bridge engineers, to Berlin in the 1950s. He was part of a small team who the government highways department made responsible for incorporating aesthetics into bridge design. He worked with a number of leading German architects, particularly Paul Bonatz and through this association and from extensive field studies of bridges,

he evolved a set of criteria on the design of good looking bridges. He set this out in
his book on bridge aesthetics *Brucken – Bridges*.

Although bridge design was dominated by civil engineers in the twentieth century,
somehow the aesthetic vision of the early pioneer's such as Roebling, Eiffel and Maillart and later by Steinman, McCullough *et al.*, was never seriously addressed in contemporary bridge design in the UK during the middle to later half of the twentieth
century. Education and training of British civil engineers it appears, generally did
not include one iota of understanding on the architecture of the built environment.

Was this also true in other parts of Europe after the war? It is possible that in France
with the emergence of bridges such as Plougastel, Orly Airport Viaduct, Tan Carville
and Brotonne and more recently examples such as Isère, the second Garabit Viaduct
and Pont de Normandie, a conscious effort has been made to build beautiful bridges.
In conversation with Jean Muller and Michel Virloguex, comparing their education
background and training with that of the great Eugene Freyssinet, it would seem that
all of them had some education and teaching on bridge aesthetics at university.
It might explain why their bridges look elegant and thoroughly well engineered.
It also appears that senior personnel in government bridge departments in France who
appoint consultants and commission the building of the major bridges, have the same
commitment to build visually pleasing bridge structures. Many of them have been
schooled in bridge engineering at the University of Paris. Awareness of bridge aesthetics
at engineering school is a critical factor. And having developed a design which fully
reconciles aesthetics, it is then sent out for tendering. Contractors in France are not
given the opportunity to propose cheaper alternative designs, only the opportunity to

Figure 1.20 Plougastel, France (courtesy of JMI).

Figure 1.21 Pont Isère, Romans, France (courtesy of NCE/Grant Smith).

propose construction innovations in building the chosen design economically. Not surprisingly, aesthetically designed bridges are competitive on price, as the major constructors in France over the years have invested in new technology and sophisticated erection techniques to build efficiently.

In England in the 1990s two unconnected, yet controversial, events marked a watershed in bridge aesthetics and gave recognition to the role of architects in bridge design. The 'Bloomers Hole Bridge' competition run by the Royal Fine Arts Commission (RFAC) on behalf of the District Council of Thamesdown. The competition which was run on RIBA rules, was open to anyone – bridge engineers, architects, civil engineers and so on. The entrants had to submit an artistic impression of the bridge and accompany it with notes explaining its construction, how it would be built and describing its special qualities for the location. The bridge was to be a new pedestrian crossing over the upper reaches of the Thames in a very unspoilt setting in Lechlade. Each entrant was given a reference number, so that the judges had no knowledge of

the name of the entrant. The winning design out of 300 entries was done by an architect. The president of the RFAC speaking on behalf of the judging panel, described
the winning design as a 'beautiful solution of great simplicity and elegance entirely
appropriate to its rural setting' – but it was not built. The residents of Lechlade
labelled the design a 'yuppic tennis racket from hell' and planning permission was
withheld. Nevertheless, the imaginative design ideas that resulted from this
competition prompted many local authorities and development corporations,
particularly the London Docklands Development Corporation (LDDC) to follow
suit. Coincident with the competition was a design study for the proposed East
London River Crossing by Santiago Calatrava, that took the bridge world by storm.
Calatrava's dramatic, rapier slim bridge concept arching over the Thames, showed
how a well-engineered bridge design can produce a pleasing aesthetic – it seemed
that everyone wanted Calatrava to design a bridge for them.

In the past three decades in the UK, architectural style has been a confusing cock-
tail of past and present influences, high-tech and neo-classical, romantic modernism
and minimalism which has in some ways marginalized the influence and appreciation of architecture. As a result, highway authorities that commission bridges, have
paid more attention to structural efficiency, cost control and long-term durability.
Aesthetic consideration, if addressed at all, was treated as an appendage like
decoration, and the first item to be dropped if the tender price was high. The
reason for this was simple: both the client and design consultant were civil engineers
with little empathy towards modern architecture and the aesthetic judgement of
architects on bridge design. Unfortunately a recent exhibition on 'living bridges' at
the Royal Academy has confirmed this point of view. The architecture-inspired ideas
tended to make bridges look and function like buildings ... and failed. But despite
this setback the 'old school' attitudes of civil and bridge engineers are slowly being
replaced by a new generation of engineers and clients who have recognized the value
of working with architects.

1.4.3 The search for aesthetic understanding

Why have architecture and bridge engineering not found a common language over
the centuries as has happened in building structures? There have been periods of bridge
building when both ideals were combined in bridges. Engineers like Lindenthal,
Ammann, Steinman and McCullough in the USA were advocates of visually pleasing
bridges. In Europe individuals like Freyssinet, Maillart, Leonhardt, Menn, Muller and
Caltrava and consultant groups like Arup, Cowi and Cassado were recognized for their
aesthetic design of bridges. All of them will own up to the fact that they employed or
worked alongside architects. Ammann worked closely with Cass Gilbert, the architect of the gothic Woolworth Tower, arguably the most beautiful skyscraper ever built.

Steinnam built many great bridges, and tried hard to add flair and style to his
designs, but he had to teach himself aesthetics at university. 'In my student days when
we were taught bridge design, I never heard the word "beauty" mentioned once. We
concentrated on stress analysis, design formulae and graphic methods, strength of
materials, locomotive loading and influence lines, pin connections, gusset plates and

Figure 1.22 David Steinman (courtesy of Steinman Consulting).

lattice bars, estimating, fabrication and erection and so on ... But not a word was said about artistic design, about the aesthetic considerations in the design of engineering structures. And there was no whisper of thought that bridges could be beautiful' writes Steinman in an article on the beauty of bridges that appeared in the *Hudson Engineering Journal*. So how did Steinman learn to develop his skill in aesthetic design? 'For my graduation thesis in 1908 at Columbia University I chose to design the Henry Hudson Memorial Bridge as a steel arch. I worked on the idea for a year and a half before my graduation. I was determined to make this design a model of technical and analytical excellence. But this was not all ... I was further determined to make my design a model of artistic excellence.' Steinman read everything he could on the subject of beauty, and aesthetics in design. He discussed the subject with friends who were studying architecture, but they could not help him much. 'They were trained in masonry architecture, in classic orders, ornamentation and mouldings ... steel was an unfamiliar material.' Instead, he visited existing bridges, to observe and reason why some were ugly and others were thrilling to look at. He spent a lot of time climbing and walking over the Washington Arch Bridge, to study its design from every artistic angle because it inspired him. 'In my thesis I included a thorough discussion and analysis of the artistic merits of my design. When I finished the thesis

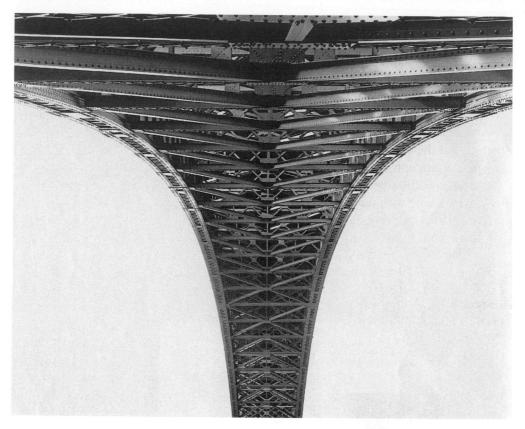

Figure 1.23 Henry Hudson Memorial Bridge, New York (courtesy of Steinman Consulting).

and turned it over to Professor William Burr ... he gave me the unusual mark of 100 percent.'

Perronet in the eighteenth century took exception to any design that was not pleasing to the eye. In discussing the Nogent bridge on the upper Seine, he remarked 'some engineers, finding that the arches ... do not rise enough near the springing, have given a large number of degrees and a larger radius to this part of the curve ... but such curves have a fault disagreeable to the eye.' The proportions, the visual line and aesthetic of the arch were important factors in Perronet's mind. He was trained as an architect. Pont Neuilly built over the Seine in Paris in 1776 was one of the most admired bridges in architecture. James Finch author of *Engineering and Western Civilisation* called Nueilly 'the most graceful and beautiful stone bridge ever built'. Sadly it was demolished in the 1956 to make way for a new steel arch bridge. It would have been useful to have studied Nueilly today, but one has to recognize that the stone arch is an obsolete technology and has been replaced by the cable stay, steel arch and concrete box girder bridge.

Leonhardt in the 1960s suggests the use of the Greek 'golden section' to solve the problem of 'good order' and harmony of proportions. He blames the lack of education for the poor understanding of the importance of aesthetics. There is too much emphasis placed on material economy and that is why there are so many ugly structures. 'The whole of society, especially the public authority, the owner builders, the cost consultants and clients are just as much to blame as the engineers and architects' argues Leonhardt. In his search for an explanation on good aesthetics he referred to the work of Vitruvius and Palladio and believed that in architecture the idea of good proportion, order and harmony are very appropriate in bridge design. Many engineers have regarded his book *Brucken* as the definitive guide to bridge aesthetics, but the majority may not have fully appreciated the moral, philosophical and esoteric arguments that he explored. The section on the origins of the golden mean and golden section will generally appeal to the more numerate engineers, who are used to working with mathematical formulae to find solutions.

The Greek philosophers tried to define aesthetic beauty through geometric proportion after years of study and observation. The suggestion was that a line should be divided so that the longer part is to the short part as the longer part is to the whole. The resulting section was know as the golden section and was roughly divided into irrational ratios of 5:8, 8:13 13:21 and so on. The ratios must never be exact multiples.

It is a dangerous precedent to set, as the golden section can be applied to a bridge just as deflection or stress calculations are done. What Leonhardt concluded in his book, after considering how aesthetics in design were assimilated in both buildings and bridges, was that aesthetics could only be learnt by practice and by the study of

Figure 1.24 St John's Bridge, Portland, Oregon (courtesy of Steinman Consulting).

good-looking bridges. He warned that designers must not assume that the simple application of rules on good design will in itself lead to beautiful bridges. He recommends that models are made of the bridge to visualize the whole design in order to appraise it aesthetic values. Ethics and morality play a part in good design according to Leonhardt. Perhaps the words that he was searching for were integrity and purity of form. There has been a tendency to design gigantic and egotistic statements for bridge structures out of the vanity and ambition of the client. The recent competition for Poole Harbour Bridge was a case in point. It may never be built because of its high cost and because of its lack of integration into the local community it must also serve. One solution that was modest in ambition, but was high on community value with small shops, houses and light industrial building built along the length of a new causeway, was entirely appropriate, but alas it was not designed as a 'gateway' structure and did not win.

Jon Wallsgrove of the Highways Agency in the UK suggests that the proportions of a bridge – the relationship of the parts to each other and to the whole – could be distilled down to the number seven. He made this observation after researching many books written on aesthetics and beauty over the centuries. The reason for this is that the brain apparently can recognize ratios and objects up to a maximum of seven without counting. He suggests that the ratio of say the span to the height of a bridge, or the span to overall length for example, should not exceed seven – for example 1:7; 2:3; 1:2:4 and so on. When the proportions are less than seven they are instantly recognized and appear right and beautiful. The use of shadow line, edge cantilevers and modelling of the surface of the bridge can improve the aesthetic proportion by reducing the visual line of the depth or width of a section, since the eye will measure the strongest visual line of the section, not the actual structural edge.

Fred Gottemoeller – a bridge engineer and architect – concurs with the view that in the USA today aesthetics in bridge design has largely been ignored by the bridge profession and client body. In his book *Bridgescape*, Gottemoeller sums up the dilemma facing many bridge engineers on the question of aesthetics. 'Aesthetics is a mysterious subject to most engineers, not lending itself to the engineers usual tools of analysis, and rarely taught in engineering schools. Being both an architect and engineer, I know that it is possible to demystify the subject in the mind of the engineer. The work of Maillart, Muller, Menn and others prove that engineers can understand aesthetics. Unfortunately such examples are too rare. The principle of bridge aesthetics should be made accessible to all engineers.' Gottemoeller has written a clearsighted, practical book on good bridge design, in a style and language that should appeal to any literate bridge engineer. It is not a book full of pretty reference pictures – the ideas have to work on the intellect through personal research.

It may take time before the new generation of bridge engineers with greater awareness and sensitivity of bridge aesthetics will soften attitudes towards working with architects out of choice. It is doubtful that the basic training and education of civil engineers will change very much in the coming decade. Many academics will feel there is no need for aesthetics to be included in a degree course and that it should be something an individual should learn in practice. Like it or not, those that are attracted to bridge design and civil engineering do so because they have good analytical and numerate skills. It is pointless putting a paintbrush in the hands of someone who hates painting and then expect them to awaken to aesthetic appreciation. In general, the undergraduate engineer has taken the civil engineering option because calculus is preferred to essay writing, technical drawing to abstract art, and scientific experiment to an appraisal of a Thomas Hardy novel. Encouragement in the visual arts and aesthetics will come with practice, and from working alongside architects who are more able to sketch ideas on paper, model the outline of bridge shapes and look for the visible clues to see if a scheme fits well with the surrounding landscape. Architects can help with aesthetic proportion – of structural depth-to-span length, pier shape and spacing, the detailing of the abutment structure, the colour and texture of finished surface of a bridge, and the preparation of scale model. After all they have been trained to do this.

The growing trend today is to appoint a team of designers from partnerships between engineers and architects to ensure that aesthetics in design is fully considered. This is a healthy sign. The LDDC successfully forged partnerships between architects and engineers in the design of a series of innovative and creative footbridges that are sited in London's Docklands. Architects like the Percy Thomas Partnership, Sir Norman Foster & Partners, Leifschutz Davidson and Chris Wilkinson in particular, have made the transfer from building architecture to bridge architecture effortlessly. In France, the architect Alain Speilman has specialized in bridge architecture for nearly 30 years, and has worked with many of France's leading bridge consultants and been involved in the design of over 40 bridge schemes. He is following a

Figure 1.25 The A75 Clermont–Ferrand highway bridge (courtesy of Grant Smith).

tradition in France, where architects like Arsac and Lavigne have worked closely with bridge engineers. Without doubt the most significant bridge project of the decade, the Milau Viaduct in central France, which was won in competition by architect Sir Norman Foster & Partners and a team of leading French bridge designers, will re-define the role of the architect and bridge engineer for the future, when it is completed.

Each period in history will no doubt uncover monster and marvels of bridge engineering, as they have done with buildings. Succeeding generations can learn to distinguish between good and bad design. What is an example of bad design? We may look on Tower Bridge today as a wonderful, monumental structure, the gateway into the Pool of London, but as a bridge it is ostentatious, with grossly exaggerated towers for such a short span. Some might regard it as a building with a drawbridge, but as a building it serves no real function other than to glorify the might of the British Empire. It would have made more sense to have built two great towers rising out of the water some way upstream of an elegant bridge, located where the bridge is now sited. And if individuals care about the quality of architecture of the built environment, they should voice their opinion and express their views on good and bad design. Silent disapproval is no better that bored indifference. It's worth reflecting that when Tower Bridge was being designed the Garabit Viaduct and the Brooklyn Bridge had been built. Both bridges and their famous designers were to inspire the engineering world for many decades, but alas not the Victorians.

Civic pride has over the centuries compelled governments and local highway authorities to attempt to build pleasing bridges in our cities and important towns in order to maintain the quality of the built environment. We all agree that the linking of places via bridges symbolizes co-operation, communication and continuity and that the bridge is one of the most important structures to be built. It is the modest span bridges over motorways, across canals and waterways in built-up urban areas that are most devoid of any sensitivity with their surroundings – the built environment and the urban fabric of our community. These featureless structures are in such profusion – plate girder bridge decks carrying trains over a busy high street and dirt-stained urban motorway overbridges – that they are the only bridges most of see as we journey through a town or a city. The cause of this blight stems largely from legislative doctrine on bridge design imposed by highway authorities, whose remit is to ensure that the design conforms to a set of rules on how it should perform and how little it will cost. It encourages the mediocre, the mundane and unimaginative design to be passed as 'fit for purpose'. What can be done to improve things? The way forward has already been shown by the footbridges commissioned by LDDC in the UK, by the bridges built by Caltrans along the west coast of the USA in the 1960s, by the bridges built by the Oregon Highways Department in the 1930s and 1940s and the bridges commissioned by SETRA in France in the 1980s, for example along the A75 Clermont–Ferrand highway. So it can be done.

Vitruvius identified three basic components of good architecture as firmness, commodity and delight. Many subsequent theorists have proposed different systems or arguments by which the quality of architecture can be analysed and their meaning understood. The tenets Vitruvius identified provides a simple and valid basis for judging the quality of buildings and bridge structures today. 'Firmness' is the most basic quality a bridge must posses and relates to the structural integrity of the design, the choice of material, and the durability of the construction. 'Commodity' refers to the function of the bridge, and how it serves the purpose for which it was designed. This quality is rarely lacking in any bridge design, whether it is ugly or good to look at. 'Delight' is the term for the effect of the bridge on the aesthetic sensibilities for those who come in contact with it. It may arise from the chosen shape and form of the bridge, the proportion of the span to the pier supports, the rhythm of the span spacing and how well the whole structure fits in with the surrounding environment. It is the component that is most lacking in bridges built in the middle half of this century.

The argument that good design costs more is facile ... good design requires a good design team. Look at the bridges of Roebling, Steinman, Maillart and Freyssinet – they were won in competition because they were economic to build and because the designer had considerable knowledge about construction and a gift for visual delight. They also worked closely with talented architects.

The fact that bridges have been designed by bridge engineers and civil engineers for only 300 out of past 4000 years in the history of bridge building has not been lost on those who lobby for better-looking bridges. Before that is was the domain of the architect and master builder. It is reassuring to know that as we enter the new millennium we seem to be learning from the lessons of the past.

Figure 1.26 Bedford – The Butterfly Bridge (courtesy of Wilkson Eyre).

2 Loads and load distribution

M.J. RYALL

2.1 Introduction

The predominant loads on bridges are gravity loads due to self-weight and that of moving traffic using the bridge and its dynamic effects. Other loads include those due to wind, earthquakes, snow, temperature and construction and so on as shown in Figure 2.1.

Most of the research and development has, understandably, been concentrated on the specification of the live traffic loading model for use in the design of highway bridges. This has been a difficult process, and the aim has been to produce a simplified static load model which has to account for the wide range and distribution of vehicle types, and the effects of bunching and vibration.

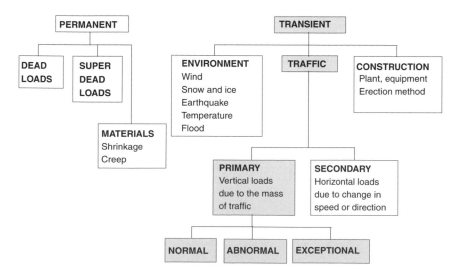

Figure 2.1 Loads on bridges.

2.2 Brief history of loading specifications

2.2.1 Early loads

Prior to the industrial revolution in the UK most bridges were single- or multiple-span masonry arch bridges. The live traffic loads consisted of no more than pedestrians; herds of animals, and horses and cart and were insignificant compared with the self-weight of the bridge.

The widespread construction of roads introduced by JL McAdam in the latter half of the eighteenth century and the development of the traction engine brought with them the necessity to build bridges able to carry significant loads (Rose, 1952–1953). In 1875, for the first time in the history of bridge design, a live loading was specified for the design of *new* road bridges.

This was proposed by Professor Fleming Jenkins (Henderson, 1954) and consisted of '1 cwt per sq. foot [approximately 5 kN/m^2] plus a wheel loading of perhaps ten tons on each wheel on one line across the bridge'. In the early part of the twentieth century, Professor Unwin suggested '120 lbs. per sq. foot [approximately 5.4 kN/m^2] or the weight of a heavily loaded wagon, say 10 to 20 tons on four wheels. In manufacturing districts this should be increased to 30 tons on four wheels'.

The development of the automobile and the heavy lorry introduced new requirements. The numbers of vehicles on the roads increased, as did their speed and their weight. In 1904 this prompted the British government to specify a rigid axle vehicle with a gross weight of 12 tons. This was the 'Heavy Motor Car Order' and was to be considered in all new bridge designs.

2.2.2 Standard loading train

The period between 1914 and 1918 marked a new era in the specification of highway loading. The military made demands for heavy mechanical transport. The Ministry of Transport (MOT) was created immediately after the first world war, and in June 1922 introduced the standard loading train (see Figure 2.2) which consisted of a 20-ton tractor plus pulling three 13-ton trailers (similar to loads actually on the roads at the time, as in Figure 2.3) and included a flat-rate allowance of 50% on each axle to account for the effects dynamic impact. This train was to occupy each lane

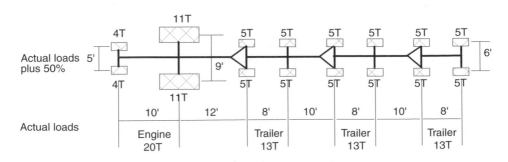

Figure 2.2 Standard load for highway bridges.

Figure 2.3 Traction engine plus three trailers.

width of 10 ft, and where the carriageway exceeded a multiple of 10 ft, the excess load was assumed to be the standard load multiplied by the excess width/10. The load was therefore uniform in both the longitudinal and transverse directions.

2.2.3 Standard loading curve
This loading prevailed until 1931 when the MOT adopted a new approach to design loading. This was the well-known *MOT loading curve*. It consisted of a uniformly distributed load (UDL) considered together with a single invariable knife-edge load (KEL). Although based on the standard loading train, it was easier to use than a series of point wheel loads. The KEL represented the excess loading on the rear axle of the engine, (that is 2×11 tons $- 2 \times 5$ tons $= 12$ tons).

In view of the improvement in the springing of vehicles at the time, and the advent of the pneumatic tyre, the total impact allowance was considered to diminish as the loaded length increased, while a reduction in intensity of loading with increasing span was recognized, hence the longitudinal attenuation of the curve. The loading was constant from 10 to 75 ft and thereafter reduced to a minimum at 2500 ft. For loaded lengths less than 10 ft a separate curve was produced to cater for the probability of high loads due to heavy lorries occupying the whole of the span where individual wheel loads exert a more onerous effect. (It also included a table of recommended amounts of distribution steel in reinforced concrete slabs.) A reproduction of the curve is shown in Figure 2.4.

The UDL was applied to each lane in conjunction with a single 12-ton KEL (per lane) to give the worst effect. The MOT also introduced Construction and Use (C&U) Regulations for lorries or trucks, which indicated the legally allowed loads and dimensions for various types of vehicle.

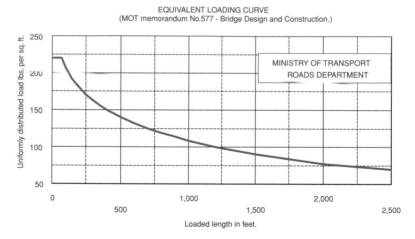

Figure 2.4 Original MOT loading curve.

After the second world war, Henderson (1954) observed that, in reality, the actual vehicles on the roads differed from the standard loading train or standard loading curve. There were those that could be described as 'legal' (that is those conforming to the C&U Regulations), and those carrying *abnormal* indivisible loads outside the regulations where special permission was required for transportation. The weight limits in effect at the time were 22 tons for the former and 150 tons for the latter, although it was possible for hauliers to obtain a special order to move greater loads.

Henderson observed that the *abnormal* load-carrying vehicles were generally well-deck trailers having one axle front and rear for the lighter loads and a two-axle bogie at each end for heavier loads – of which there were about three examples in existence – and each axle had four wheels and was about 10 ft long. A typical example is shown in Figure 2.5.

Figure 2.5 Example of an early abnormal load c. 1928 carrying a 60-ton cylinder of paper.

Henderson's (1954) conclusion was that 'both ordinary traffic and abnormal vehicles are *dissimilar in weight and arrangement of wheels* to those represented by the former loading trains'. He therefore proposed the idea of defining traffic loads as *normal* (everyday traffic consisting of a mix of cars, vans and trucks); and *abnormal*, consisting of heavy vehicles of 100 tons or more. The abnormal loading could consist of two types, namely those conforming to the current C&U regulations and those less-frequent loads in excess of 200 tons. The latter loads would be confined to a limited number of roads and would be treated as special cases. Bridges *en route* could be strengthened and precautions taken to prevent heavy normal traffic on the bridge at the same time.

In conjunction with the MOT and the British Standards Institute (BSI), he proposed the idea of considering two kinds of loading for design purposes, namely: *normal and abnormal,* and that 'designs should be made on the basis of normal loading and checked for abnormal traffic'.

2.2.4 Normal loading

The widely adopted MOT loading curve with a UDL plus a KEL would constitute *normal* loading defined as *HA loading.* Experience showed the extreme improbability of more than two carriageway lanes being filled with the heaviest type of loading, and although no qualitative basis was possible he proposed that two lanes should be loaded with full UDL and the reminder with one half UDL as shown in Figure 2.6.

Any attempt to state a sequence of vehicles representing the worst concentration of ordinary traffic which can be expected, must be a guess, but it seemed reasonable to propose the following:

- 20 ft (6 m) to 75 ft (22.5 m)   lines of 22-ton lorries in two adjacent lanes and 11-ton lorries in the remainder
- 75 ft (22.5 m) to 500 ft (150 m)   five 22-ton lorries over 40 ft (12 m) followed and preceded by four 11-ton lorries over 35 ft (10.5 m) and 5-ton vehicles over 35 ft (10.5 m) to fill the span.

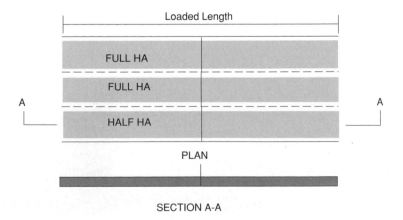

Figure 2.6 Normal loading.

These were found to correspond well to the MOT loading curve. For spans in excess of 75 ft (22.5 m), an equivalent UDL (in conjunction with a KEL) was derived by equating the moments and shear per lane of vehicles with the corresponding effects under a distributed load. Henderson emphasized that these loadings could be looked on only as a guide. A 25% increase was considered appropriate for the impact of suspension systems.

A more severe concentration of load was considered appropriate for short-span members and units supporting small areas of deck. A heavy steamroller had wheel loads of about 7.5 tons similar to the weight of the then 'legal' axle, and adding 25% for impact gave 9 tons. It seemed suitable to use two 9 tons loads at 3 ft (0.915 m) spacing on such members. Separate loading curves were proposed to give a UDL on the basis of this loading.

2.2.5 Abnormal loading

Henderson (1954) proposed that abnormal loading be referred to as *HB Loading* defined by the now familiar HB vehicle which, although, hypothetical, was based on existing well deck trailers such as the one shown in Figure 2.5 having two bogies each with two axles and four wheels per axles. Each vehicle was given a rating in *units* (one unit being 1 ton) and referred to the load per axle. Thus 30 units meant an axle load of 30 tons. Henderson proposed 30 units for main roads and at least 20 units on other roads. In 1955, because of the increasing weights of abnormal loads, the upper limit was increased to 45 units. Since abnormal vehicles travel slowly no impact allowance was made.

2.2.6 Variations

The standard loading curve has undergone several revisions over the years as more precise information about traffic volumes and weights has been gathered and processed. The basic philosophy of the normal and abnormal loads has been retained, indeed a colloquium convened at Cambridge in 1975 to examine the basic philosophy concluded that the *status quo* should be maintained (Cambridge, 1975). This is still the current view and the major changes which have taken place are reflected in BS153 (1954); BE1/77 (1961); BS5400 (1978) and Memorandum BD37/88 (1989) which each contain the HA loading model of a UDL in conjunction with a KEL.

One interesting phenomenon which has occurred over the years is that the maximum permitted lorry load to be included in the HA loading has increased significantly from the original 12 tons to the current value of 40 tonnes. The increase with time is illustrated in Figure 2.7. If this trend continues then the next likely load limit will be 45 tonnes in the year 2005. (In Denmark, 48-tonne vehicles are already in existence.)

2.3 Current load specifications

2.3.1 Introduction

The basic philosophy of the normal and abnormal loading is common throughout the world, but there are, of course, variations to account for the range and weights of vehicles in use in any given country.

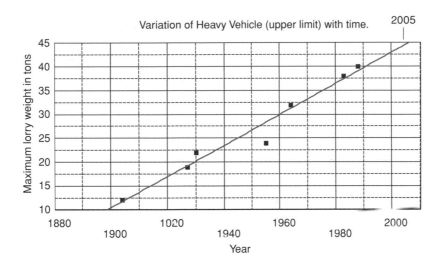

Figure 2.7 Variation of heavy vehicle load with time.

In this section normal and abnormal traffic loads specified in the UK, US, and Eurocodes will be referred to.

2.3.2 British specification

The current UK code is, by agreement with the British Standard Institution, Department of Transport Standard BD37/88 (1989) which is based on BS5400: Part 2 (1978).

Normal load application

The normal load consists of a lane UDL plus a lane KEL. The UDL (HAU) is based on the loaded length and is defined by a two-part curve as shown in Figure 2.8, each defined by a particular equation, one up to 50 m loaded length and the other for the reminder up to 1600 m. The KEL (HAK) has a value of 120 kN per lane.

The application and intensity of the traffic loads depends on:

• the carriageway width
• the loaded length
• the number of loaded lanes.

The carriageway width is essentially the distance between kerb lines and is described in Figure 1 of BD 37/88. It includes the hard strips, hard shoulders and the traffic lanes marked on the road surface.

The two most prominent load applications are defined as HA only, and HA + HB. HA is applied as described as previously to every (notional) lane across the carriageway attenuated as defined in Table 14 of BD 37/88.

The attenuation of the curve in Figure 2.8 takes account of vehicle bunching along the length of a bridge. Lateral bunching is taken account of by applying lane factors

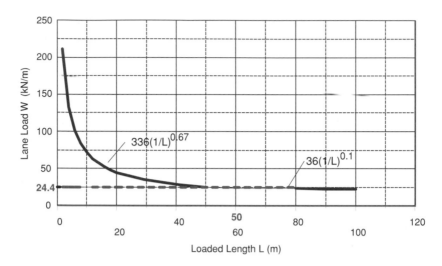

Figure 2.8 British Standard normal loading curve.

β to the load in each lane (both the UDL and the KEL). Generally this amounts to $\beta = 1.0$ for the first two lanes and $\beta = 0.6$ for the remainder. Thus nominal lane load $= \beta HAU + \beta HAK$.

The number of lanes (called *notional* lanes, and not necessarily the same as the actual traffic lanes defined by carriageway marking) is based on the total width (*b*) of the carriageway (the distance between kerbs in metres) and is given by Int[(*b*/3.65) + 1], where 3.65 is the standard lane width in metres. Notional lanes are numbered from a free edge.

Local effects

For parts of a bridge deck which are susceptible to the local effects of traffic loading, an alternative normal wheel load of 100 kN is applied. The wheel load is assumed to exert a pressure of 1.1 N/mm^2 to the surfacing and is generally considered as a square of 300-mm side. Allowance can also be made for dispersal of the load through the surfacing and the structural concrete if required.

Abnormal loading

The loading for the abnormal vehicle is concentrated on 16 wheels arranged on four axles as shown in Figure 2.9. Its weight is measured in units per axle, where 1 unit $= 10$ kN.

The maximum number of units applied to all motorways and trunk roads is 45 (equivalent to a total vehicle weight of 1800 kN), and the minimum number is 30 units applied to all other public roads. The inner axle spacing can vary to give the worst effect, but the most common value taken is 6 m. (It is worth noting that vehicles with this configuration are not considered in the Construction and Use regulations because it is a hypothetical vehicle and used only as a device for rating a bridge.) Each wheel area is based on a contact pressure of 1.1 N/mm^2.

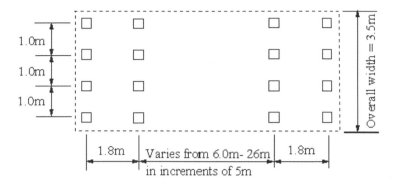

Figure 2.9 Abnormal HB vehicle.

Load application

All bridges are designed for HA loading and checked for a combination of HA + HB loading. HA and HB are applied according to Figure 13 of BD 37/88 with the HB vehicle placed in one lane or straddled over two lanes (depending on the width of the notional lane). Since such a load would normally be escorted by police, an unloaded length of 25 m in front and behind is specified, with HA loading occupying the remainder of the lane. The other lanes are loaded with an intensity of HA appropriate to the loaded length and the lane factor.

Exceptional loads

Road hauliers are often called on to transport very heavy items of equipment such as transformers or parts for power stations which can weigh as much as 750 tonnes (7500 kN) or more. Special flat-bed trailers are used with multiple axles and many wheels to spread the load so that the overall effect is generally no more than that of HA loading, and contact pressures are no more than 1.1 N/mm^2, but where this is not possible, then any bridges crossed *en route* have to be strengthened. The loads on the axles can be relieved by the use of a central air cushion which raises the axles slightly and redistributes some of the load to the cushion. Heavy diesel traction engines placed in front and to the rear are used to pull and push the trailer. Some typical dimensions are shown in Figure 2.10.

Figure 2.11 shows a catalytic cracker installation unit 41 m long and 15.3-m diameter weighing 825 tonnes being transported for Ellesmere Post to Stanlow Refinery via the M53 in 1984. The load was spread over 26 axles and 416 wheels.

2.3.3 US specification

The US highway loads are based on American Association of State Highway and Transportation Officials (AASHTO) Standard Specification for Highway Bridges (1996). These specify standard lane and truck loads.

Lane loading

The commonly applied lane loading consists of a UDL plus a KEL on 'design lanes' typically 3.6-m wide placed centrally on the 'traffic lanes' marked on the road

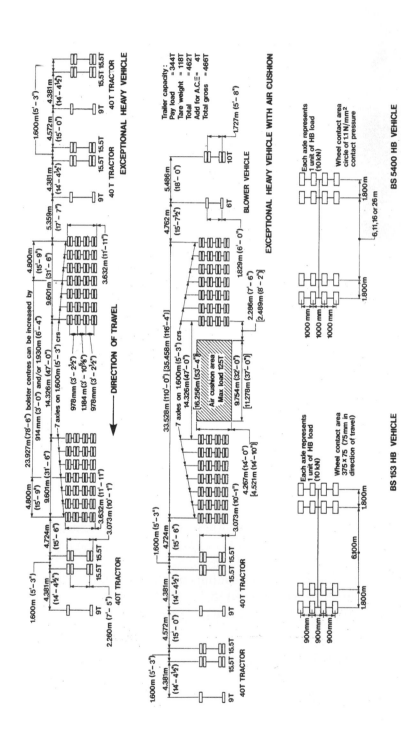

The abnormal loading stipulated in BS153 is applied to most public highway bridges in the UK: 45 units on motorway under-bridges, 37½ units on bridges for principal roads and 30 units on bridges for other roads.

Some bridges are checked for special heavy vehicles which can range up to 466 tonnes gross weight. Where this is needed the gross weight and trailer dimensions are stated by the authority requiring this special facility on a given route.

Figure 2.10 Typical vehicles used to transport exceptional loads (after Pennells).

Figure 2.11 Transportation of an exceptionally heavy (825 tonnes) load.

surface. The number of 'design lanes' is the integer component of the carriageway width/3.6. Traffic lanes less than 3.6 m wide are considered as design lanes with the same width as the traffic lanes. Carriageways of between 6 and 7.3 m are assumed to have two design lanes.

The *lane* load is constant regardless of the loaded length and is equal to 9.3 kN/m and occupies a region of 3 m transversely as indicated in Figure 2.12.

Acting with the lane loading there are three different design *truck* loadings namely:

- tandem
- truck
- lane.

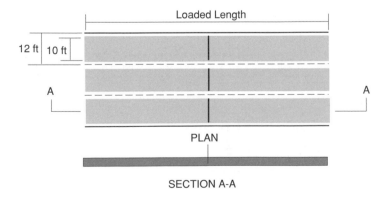

Figure 2.12 Simple lane loading.

Table 2.1 Class of loading

Class of loading – 1944	Class of loading – 1993
H 20-44	HL 20-93
H 15-44	HL 15-93
HS 20-44	HLS 20-93
HS 15-44	HLS 15-93

The prefix H refers to a standard two-axle truck followed by a number that indicates the gross weight of the truck in tons, and the affix refers to the year the loading was specified. The prefix HS refers to a three-axle tractor (or semitrailer) truck. The dimensions and wheel loadings of the two types of truck are shown in Figure 2.13, where W is the gross weight in tons.

The *new* (AASHTO, 1994) tandem and the truck loadings are shown in Figure 2.13 compared with the *old* (AASHTO, 1977) standard H and HS trucks.

To account for the fact that trucks will be present in more than one lane, the loading is further modified by a *multiple presence factor*, *m*, according to the number of design lanes and ranges from 1.2 for one lane to 0.65 for more than three lanes (AASHTO, 1994).

The actual intensity of loading is dependent on the class of loading as indicated in Table 2.1.

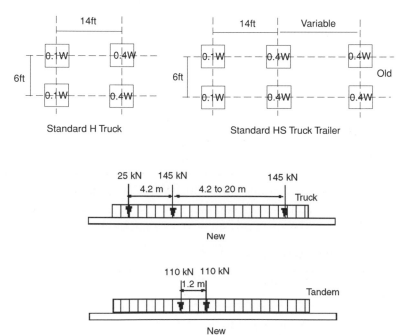

Figure 2.13 New and old AASHTO truck loadings.

Dynamic effects

Dynamic effects due to irregularities in the road surface and different suspension systems magnify the static effects from the live loads and this is accounted for by an *impact factor* called a dynamic load allowance (DLA) defined as:

$$\text{DLA} = D_{\text{dyn}}/D_{\text{sta}} \tag{2.1}$$

where D_{sta} is the static deflection under live loads, and D_{dyn} is the additional dynamic deflection under live loads. This is applied to the static live load effect using the following equation:

$$\text{dynamic live load effect} = (\text{static live load effect}) \times (1 + \text{DLA}) \tag{2.2}$$

Values of the DLA are given in AASHTO (1996) for individual *components* of the bridge such as deck joints, beams, bearings, etc. and the global effects are not considered at all. This is a departure from the *old* practice where the basic static live load was multiplied by an impact factor:

$$I = 50/(L + 125) \tag{2.3}$$

where L = loaded length in feet and the maximum value of I allowed was 0.3.

The variable spacing of the trailer axles in the HS truck trailer is to allow for the actual values of the more common tractor trailers now in use.

2.3.4 European specification

The European models for traffic loading are embodied in Eurocode 1 (1993) and are identified in Table 2.2.

General loading

The general loading comprises a UDL in kN/m^2 plus a double axle tandem per lane. (The tandem is dispensed with on the fourth lane and above, on carriageways of four lanes or more.)

The notional lane width is generally taken as 3 m, and the number of notional lanes as Int($w/3$), where w is the carriageway width. Areas other than those covered by notional lanes are referred to as *remaining areas*. The first lane is the most heavily loaded with a UDL of 9 kN/m^2 (equivalent to a lane loading of 27 kN/m for a 3 m

Table 2.2 European load definitions

Load model	Definition
LM1	General (normal) loading due to lorries or lorries plus cars
LM2	A single axle for local effects
LM3	Special vehicles for the transportation of exceptional loads
LM4	Crowd loading

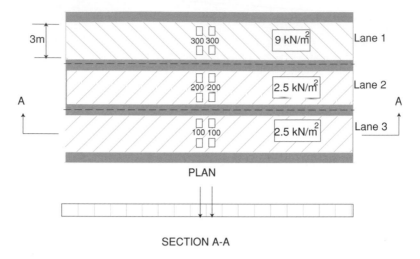

Figure 2.14 General loading model (LM1) to European Code EC1.

notional lane) plus a single tandem with axle loads of 300 kN each. The loads on remaining lanes reduce as indicated in Figure 2.14.

Local loads

To study local effects, the use of a 400 kN tandem axle is recommended as shown in Figure 2.15. In certain circumstances this can be replaced by a single wheel load of 200 kN.

Abnormal loads

Abnormal loads are considered in a similar manner to the British Code, with a special abnormal load (model LM3) placed in one lane (or straddling two lanes) with a 25 m clear space front and back and normal LM1 loading placed in the other lanes. The vehicle may be specified by the particular load authority involved, or alter-

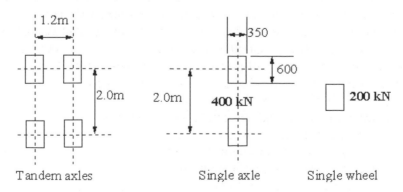

Figure 2.15 Tandem axle used in LM1 and single-axle or wheel-load used in LM2.

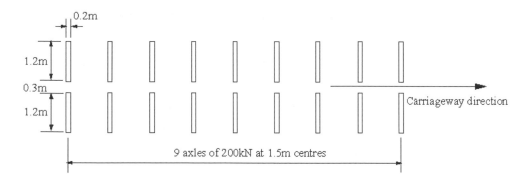

Figure 2.16 Typical LM3 vehicle (in this case 1800 kN).

natively it may be as defined in EC1 which specifies eight load configurations with varying numbers of axles, and loads from 600 to 3600 kN. Wheel areas are assumed to merge to form long areas of 1.2×0.15 m. Axle lines are spaced at 1.5 m and may consist of two or three merged areas. A typical configuration for an 1800 kN vehicle is shown in Figure 2.16.

Crowd loading
Most countries specify a nominal crowd loading of about 5 kN/m^2 (EC1 model LM4) to be placed on the footways of highway bridges or across pedestrian and cycle bridges. In some instances reduction of loading is allowed for loaded lengths greater than 10 m.

2.3.5 Modern trends
The modern trend towards traffic loading is to try and model the movement, distribution and intensity of loading in a probability-based manner (Bez *et al.*, 1991). Stopped traffic is considered which represents a traffic jam situation consisting of semi-trailers, tractor trailers and trucks, and which are then related to the response of the bridge structure in a random manner. From this it is possible to determine the mean value and standard deviation of the maximum bending moment in the bridge. Different models are considered at both the ULS and SLS conditions. Vrouwenvelder *et al.* (1993) have carried out similar research in order to construct a probabilistic traffic flow model for the design of bridges at the ultimate limit state, both long term and short term. The loading that they arrived at is able to be transformed into a uniform load in combination with one or more movable truck loads. Bailey *et al.* (1996) studied the effect of traffic actions on existing load bridges with the idea of developing the concept of *site-specific* traffic loads. Their study considered the random nature of the traffic and the simulation of maximum traffic action effects and developed correction factors for application to the Swiss design traffic loads. Studies have also been carried out in the UK (Cooper, 1997; Page, 1997) by the collection of traffic data and the application of reliability methods for both assessment and design, but for the foreseeable future the simple lane loading of a UDL plus a KEL is set to continue to be the model adopted in practice.

2.4 Secondary loads (UK)

2.4.1 Braking

This is considered as *group* effect as far as HA loads are concerned, and assumes that the traffic in one lane brakes simultaneously over the entire loaded length. The effect is considered as longitudinal force applied at the road surface.

There is evidence to suggest that the force is dissipated to a considerable extent in plan, and for most concrete and composite shallow deck structures it is reasonable to consider the loads spread over the entire width of the deck.

The braking of an HB vehicle is an *isolated* effect distributed evenly between eight wheels of two axles only of the vehicle and is dissipated as for the HA load.

The significance of the braking load on the structure is twofold, namely:

- the design of the bridge abutments and piers where it is applied as an horizontal load at bearing level, thus increasing the bending moments in the stem and footings
- the design of the bridge bearings if composed of an elastomer resisting loads in shear.

The code specifies these loads as:

- 8 kN/m of loaded length + 250 kN for HA but not greater than 750 kN
- nominal HB load × 0.25 for HB.

2.4.2 Secondary skidding load

This is an *accidental* load consisting of a single point load of 300 kN acting horizontally in any direction at the road surface in a single notional lane. It is considered to act with the primary HA loading in Combination 4 only.

2.4.3 Secondary collision load

A vehicle out of control may collide with either the bridge parapets or the bridge supports, and guidance is given in the code for the intensity of loads expected in Clauses 6.7 and 6.8.

2.4.4 Secondary centrifugal loads

These loads are important only on elevated curved superstructures with a radius of less than 1000 m, supported on slender piers.

The forces are based on the centrifugal acceleration ($a = \text{velocity}^2/\text{radius of curve}$) which, when substituted in Newton's second law gives:

$$F = mv^2/r \tag{2.4}$$

which acts at the centre of mass of the vehicle in an outward horizontal direction. If the weight of the vehicle is W, then:

$$F = Wv^2/gr \tag{2.5}$$

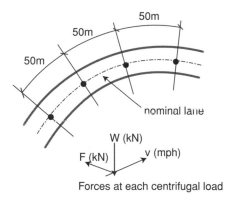

Forces at each centrifugal load

Figure 2.17 Centrifugal forces.

The code suggests a nominal load of:

$$F_c = 40\ 000/(r + 150) \qquad (2.6)$$

which approximates to a 40 tonne (400 kN) vehicle travelling at 70 mph.

Each centrifugal force acts as a point load in a radial direction at the surface of the carriageway and parallel to it and should be applied at 50 m centres in each of two nominal lanes, each in conjunction with a vertical live load component of 400 kN.

2.5 Other loads

2.5.1 Introduction

All of the loads that can be expected on a bridge at one time or another are shown in Figure 2.1. Different authorities deal with these loads in slightly different ways but the broad specifications and principles are the same all over the world. Actual values will not be given as they vary with each highway authority.

2.5.2 Permanent loads

Permanent loads are defined as *dead* loads from the self-weight of the structural elements (which remains essentially unchanged for the life of the bridge) and *superimposed dead loads* from all other materials such as road surfacing; waterproofing; parapets; services; kerbs; footways; lighting standards, etc. Also included are loads due to permanent imposed *deformations* such as differential settlement and loads imposed due to shrinkage and creep.

2.5.3 Differential settlement

Differential settlement can cause problems in continuous structures or wide decks which are stiff in the lateral direction. It can occur due to differing soil conditions in the vicinity of the bridge; varying pressures under the foundations or due to subsidence of old mine workings. Whenever possible, expert advice should be sought from geotechnical engineers in order to assess their likelihood and magnitude.

2.5.4 Material behaviour loads

The *shrinkage and creep* characteristics of concrete induce internal stresses and deformations in bridge superstructures. Both effects also considerably alter external reactions in continuous bridges. The implications are critical at the serviceability limit state and affect not only the main structural members but the design of expansion joints and bearings. The drying out of concrete due to the evaporation of absorbed water causes *shrinkage*. The concrete cracks and where it is restrained due to re-inforcing steel, or a steel or precast concrete beam tension stresses are induced whilst compression stresses are induced in the restraining element. A completely symmetrical concrete section will shorten only, resulting in horizontal deformation and a uniform distribution of stresses; but a singly reinforced; unsymmetrically doubly re-inforced or composite section will be subjected to varying stress distribution and also curvatures which could exceed the rotation capacity of the bearings. *Creep* is a long term effect and acts in the same sense as shrinkage. The effect is allowed for by modifying the short-term Young's modulus of the concrete E_c by a reduction (creep) factor ϕ_c. As for shrinkage, both stresses and deformations are induced.

Shrinkage

Shrinkage stresses are induced in all concrete bridges whether they consist of pre-cast elements or constructed *in situ*. Generally the stresses are low and are consid-ered insignificant in most cases.

However, where a concrete deck is cast in-situ onto a prefabricated member (be it steel of concrete) it becomes composite and shrinkage stresses can be significant. Figure 2.18 illustrates how shrinkage of the *in situ* concrete deck affects the com-posite section.

Shrinkage produces *compression* in the top region of the precast concrete beam. When the concrete deck slab is poured it flows more or less freely over the top of the precast beam and additional stresses are induced in the beam due to the wet con-crete. As it begins to set, however, it begins to bond to the top of the precast beam and because it is partially restrained by the precast beam below, shrinkage stresses are induced in both the slab and the beam. *Tensile stresses are induced in the slab and compressive stresses in the top region of the beam.* For the purposes of analysis a fully composite section is assumed, and the same principles applied as when calculating temperature stresses.

The total restrained shrinkage force is assumed to act at the centroid of the slab and results in a uniform restrained stress throughout the depth of the slab only. Since the composite section is able to deflect and rotate, balancing stresses are induced due to a direct force and a moment acting at the centroid of the composite section (see Figure 2.18):

$$\text{restrained shrinkage force } F = -EA\varepsilon_{cs} \qquad (2.7)$$

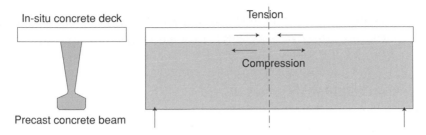

Figure 2.18 Effect and development of shrinkage stresses on composite section.

Table 2.3 Shrinkage strains and creep reduction factors

Environment	ε_{cs}	ϕ_c
Very humid, e.g. directly over water	-100×10^{-6}	0.5
Generally in the open air	-200×10^{-6}	0.4
Very dry, e.g. dry interior enclosures	-300×10^{-6}	0.3

where E is the Young's modulus of the *in situ* concrete; A, the area of the slab, and ε_{cs} is the shrinkage strain and depends on the humidity of the air at the bridge site. In the UK, guidance is given in Table 2.3.

Shrinkage modified by creep
Creep is a long-term effect and modifies the effects of shrinkage in that the apparent modulus of the concrete is reduced, which in turn reduces the modular ratio, which in turn affects the final stresses in the section. The effect of creep is defined by the creep coefficient ϕ:

$$\phi = \text{long-term creep strain/initial elastic strain} \\ \text{due to constant compressive stress} \tag{2.8}$$

$$\text{total long-term strain } \varepsilon_c^{\,0} = (1 + \phi)\,\varepsilon_c = (1 + \phi)f_{cc}/E_c \tag{2.9}$$

where E_c is the *short-term* modulus for concrete.

$$\text{Long-term modulus of concrete } E_c' = f_{cc}/\varepsilon_c^{\,0} = E_c/(1 + \phi) \tag{2.10}$$

$$E_c' = \phi_c E_c \tag{2.11}$$

where $\phi_c = 1/(1 + \phi)$ is defined as the reduction factor for creep. Therefore:

$$\alpha_{eL} = E_{cb}/E_c' = (E_{cb}/E_c)\,(1/\phi_c) \tag{2.12}$$

where E_{cb} is the modulus of concrete in the beam. (Note that this assumes that all of the shrinkage has taken place in the beam.) Normally ϕ_c is taken as 0.5, but guidance is given in Table 2.3. For a steel beam the long-term modular ratio is αeL, where $\alpha eL = E_s/(\phi_c E_c)$ and E_s and E_c are the Young's modulii of the steel and concrete, respectively. (See Appendix A1 for an example.)

2.5.5 Transient loads

Transient loads are all loads other than permanent loads and are of a varying duration such as traffic; temperature; wind, and loads due to construction. Transient traffic loads have been referred to in Section 2.4.

Wind

Wind causes bridges – particularly long, relatively light bridges – to *oscillate*. It can also produce large *wind forces* in the transverse, longitudinal and vertical directions of all bridges.

The estimation of wind loads on bridges is a complex problem because of the many variables involved, such as the size and shape of the bridge; the type of bridge construction; the angle of attack of the wind; the local topography of the land and the velocity–time relationship of the wind.

Although wind exerts a dynamic force, it may be considered as a static load if the time to reach peak pressure is equal to or greater than the natural frequency of the structure. This is the usual condition for a majority of bridges. Wind is not usually critical on most small to medium span bridges but some long-span beam-type bridges on high piers are sensitive to wind forces.

The greatest effects occur when the wind is blowing at right angles to the line of the bridge deck, and the nominal wind load can be defined as:

$$P = qAC_D \tag{2.13}$$

where q is the the dynamic pressure head; A, the solid projected area, and C_D, the drag coefficient. Guidance is given in the various bridge codes on the calculation of these three quantities for different bridge types.

The velocity of the wind varies parabolically with height, similar to that shown in Figure 2.19. Then:

$$q = \rho v_c^2/2 \tag{2.14}$$

where ρ is the density of air normally taken as 1.226 N/m^3 and v_c is the maximum gust speed based on the mean hourly wind speed v and modified by a *gust factor* K_g (which increases with height above ground level, but decreases with increased loaded length) and an *hourly speed factor* K_s (*which increases with height above ground level*) for particular loaded lengths, and thus:

$$q = 0.613v_c^2/10^3 \text{ kN/m}^2 \tag{2.15}$$

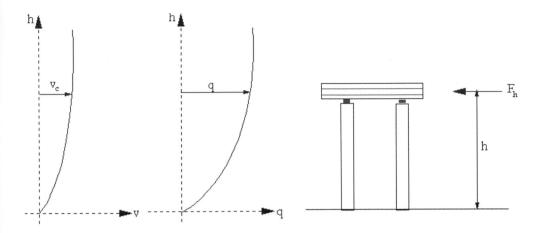

Figure 2.19 Variation of wind speed and pressure with height.

The value of v is normally obtained from local data in the form of isotachs in m/s, and values of the gust factor (K_g) and the hourly speed factor (K_s) are quoted in the codes of practice, thus:

$$v_c = v\,K_g\,K_s \qquad (2.16)$$

The value of the force acting at deck level (and at various heights up the piers) can thus be determined for design purposes.

In the UK, both isotachs and drag coefficients for various cross-sectional shapes are given in BD37/88 (see Appendix A2.1). In the USA these are found in AASHTO (1994).

Cable-supported bridges such as cable-stayed and suspension bridges are subject to vibrations induced by varying wind loads on the bridge deck. The total wind load on the deck is given by Dyrbe *et al.* (1996) as:

$$F_{tot} = F_q + F_t + F_m \qquad (2.17)$$

where F_q is the time-averaged mean wind load; F_t is the fluctuating wind load due to air turbulence (buffeting) and F_m is the motion-induced wind load.

2.6 Long bridges

The main effects on *long, light* bridges (such as cable-stayed or suspension) are:

• vortex excitation
• galloping and stall hysteresis
• classical flutter.

Random changes of speed and direction of incidence can cause dynamic excitation.

2.6.1 Vortex excitation

Due to vortex shedding – alternately from upper and lower surfaces – causes periodic fluctuations of the aerodynamic forces on the structure. These are proportional to the wind speed, thus a resonant response will occur at a specific speed. In extreme cases (witness the Tacoma Narrows bridge in 1940) this can result in vertical and torsional deformations leading to the failure of the bridge.

Structural damping can decrease the maximum amplitude and extent of wind speed range, but it will not affect the critical speed.

Truss girder stiffened suspension bridges are generally free of vortex excited oscillations, but *plate* girder and *box* girder stiffened bridges are prone to such oscillations. Appropriate modification of the size and shape of box girders can considerably reduce these effects and that is why wind tunnel tests are essential.

Wherever there is a surface of velocity discontinuity in flow, the presence of viscosity causes the particles of the fluid (wind) in the zone to spin. A vortex sheet is then produced which is inherently unstable and cannot remain in place and so they roll up to form vortices that increase in size until they are eventually 'washed' off and flow away. To replace the vortex, another vortex is generated and under steady-state conditions it is reasonable to expect a *periodic generation* of vortices.

The most likely places for them to appear are at discontinuities such as sharp edges, and they form above and below the body concerned – in the case of a bridge it is the deck which is subjected to this phenomenon.

The frequency of the shedding of the vortices can be related to the wind speed by means of a Stroudal number S which is dimensionless.

$$S = n_v D/v \tag{2.18}$$

where n_v is the vortex shedding frequency, v is the wind speed and D is a reference dimension of the cross-section.

The worst situation occurs where the frequency of oscillation (f) is equal to the natural frequency (f_n) when the wind is at its critical speed V_{crit}. Thus:

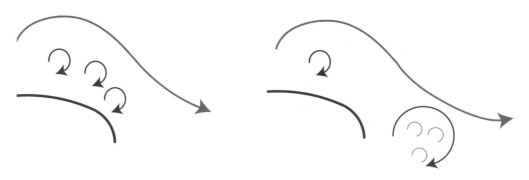

Figure 2.20 Vortex shedding.

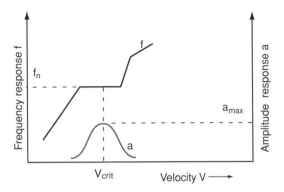

Figure 2.21 Relationship response frequency and response amplitude.

$$V_{crit} = f_n D/S \qquad (2.19)$$

At this point the response amplitude is a maximum as shown in Figure 2.21.

If there is more than one mode of vibration (generally the case), then there will be several critical wind speeds each with a different corresponding amplitude.

Design must ensure that V_{crit} is kept outside of the normally expected range at the bridge site. Design features that will minimize the depth of the *wake* (the turbulent air leeward of the deck) are found to reduce the power of vortex excitation (see Figure 2.22).

Some practical details which have been found to work are:

- shallow sections (compared with width)
- perforation of beams to vent air into wake
- soffit plate to close off spaces between main girders
- tapered fairings or inclined web panels
- avoidance of high solidity fixings and details such as fascia beams near the edges of the deck
- the use of deflector flaps or vanes on deck edges to obviate vortices or promote re-attachment of surfaces.

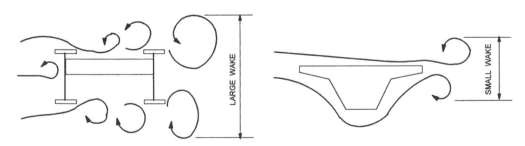

Figure 2.22 Effect of shape on depth of the wake.

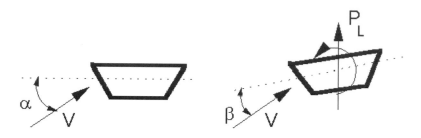

Figure 2.23 Apparent change of wind direction due to movement of deck.

2.6.2 Galloping

This is large-amplitude, low frequency oscillation of a long linear structure in transverse wind at the natural frequency f_n of the structure. It is a phenomenon that does not require high wind speeds when the cross-section has certain aerodynamic characteristics. Such was the case with the Tacoma Narrows bridge which began to gallop in wind speeds of only 40 mph and is why it got its nickname of 'Galloping Gertie'.

Once the critical wind speed has been reached, an oscillating motion begins at the natural frequency f_n, but the amplitude of vibration increases with increasing wind speed, apparently due to changes in the direction of the wind due to the motion of the structure – essentially a constant changing of the angle of incidence (see Figure 2.23).

The lifting force;

$$P_L = [0.5\rho V^2]\,[d^2]\,[C_L] \tag{2.20}$$

where $[0.5\rho V^2]$ is the wind pressure; $[d^2]$ is a unit of deck area and $[C_L]$ is the coefficient of alternating wind force which depends on α.

2.6.3 Flutter

This is caused by a stalling air flow, and causes an aero-elastic condition in which a two degree of freedom – *rotation and vertical translation* – couple together in a flow driven oscillation. This was first observed on aerofoil sections used in the aeroplane industry, and is usually confined to suspension bridges.

2.6.4 Buffetting

Unsteady loading by velocity fluctuations in the oncoming (windward) flow. In can also occur in the wake and cause problems with an adjacent bridge.

2.6.5 Transverse wind loads

In this section the real dynamic wind forces are converted to *equivalent statical forces* acting transversely, longitudinally and vertically on short to medium span bridges

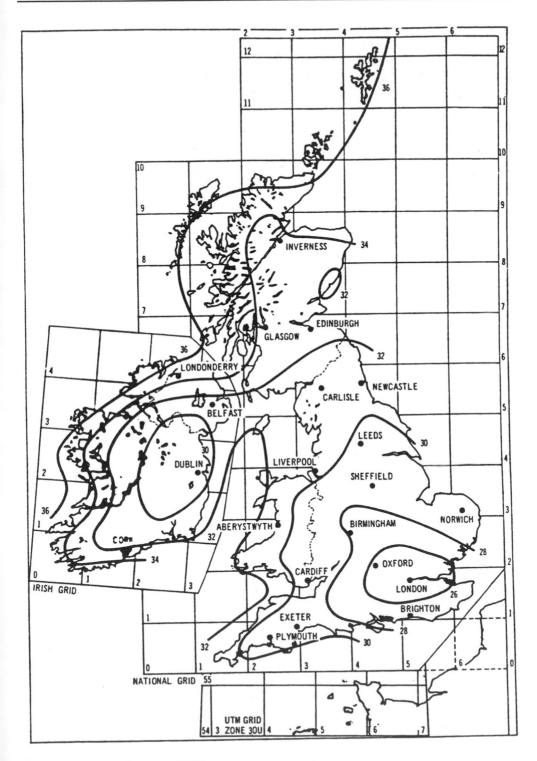

Figure 2.24 Isotach map of UK.

which comprise the majority of the nations bridge stock. Cable-stayed and suspension bridges subject to dynamic forces and movements are *not* considered.

Detailed analysis requires first of all that an isotach map is available for the country or region where the bridge is to be constructed. For the UK this is Figure 2 of BD 37/88 reproduced in Figure 2.24.

This enables the determination of the mean hourly wind speed, v, from the equation:

$$v_c = vS_1S_2K_1 \tag{2.21}$$

Values of S_2 and K_2 are derived from Table 2 of BD 37/88, and are related to the height (H) of the bridge above sea level. H is defined in Figure 2.25.

The forces acting on the bridge are then calculated from the equation:

$$P_t = qA_1C_D \tag{2.22}$$

where the dynamic pressure head:

$$q = 0.613v_c^2/10^3 \text{ kN/m}^2 \tag{2.23}$$

$$A_1 = \text{projected unshielded area (m}^2) \tag{2.24}$$

$$C_D = \text{drag coefficient} \tag{2.25}$$

2.6.6 Solid bridges

For bridges presenting a *solid* elevation to the wind, A_1 is derived by determining the *solid projected depth* (d) from Table 4 of BD 37/88 (reproduced opposite as Figure 2.26) thus:

$$A_1 = d \times 1 \text{ per unit metre along the bridge} \tag{2.26}$$

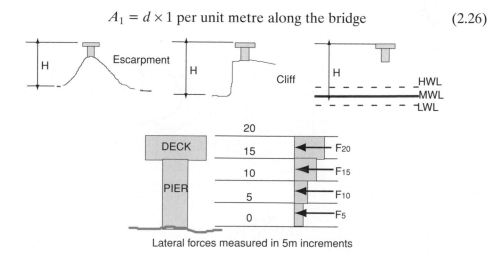

Figure 2.25 Definition of H.

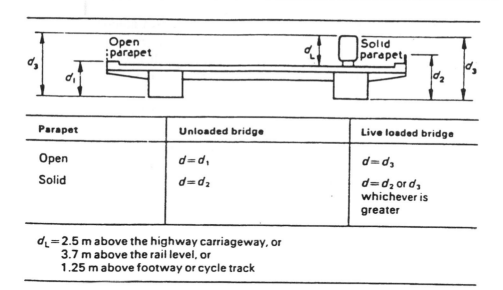

Figure 2.26 Depth d to be used for calculating A_1.

2.6.7 Truss girder bridges

For truss girder bridges A_1 is the solid (net) area presented to the wind by the girder members, that is the sum of the projected areas of the truss, thus:

$$A_1 = \Sigma \text{member areas} \qquad (2.27)$$

For *windward* girders, the value of C_D is taken from Table 6 of BD 37/88 according to the *solidity* of the truss defined by a *solidity ratio*:

$$\sigma = \text{net area of truss/overall area of truss} \qquad (2.28)$$

For *leeward* girders some *shielding* is inevitable from the windward girder, and this is taken into account by a *shielding factor* η derived from Table 7 of BD 37/88 based on the spacing ratio SR:

$$SR = \text{spacing of trusses/depth of windward truss} \qquad (2.29)$$

and the drag coefficient is given by ηC_D.

Note that for both solid and truss bridges two cases have to be considered:

- wind acting on the superstructure alone
- wind acting on the superstructure plus live loading from the traffic (maximum wind speed allowed is 35 m/s).

The worst case is taken for design purposes.

2.6.8 Parapets and safety fences

The drag coefficient for parapets and safety fences is taken from Table 8 depending on the *shape* of the structural sections used. For a bridge with two parapets, the force calculated for the windward and leeward parapets is normally assumed to be equal.

2.6.9 Piers

The drag coefficients for piers are taken from Table 9 of BD 37/88 depending on the cross-sectional shape. Normally no shielding is allowed for.

2.6.10 Longitudinal wind loads

Superstructure

As with transverse wind loads, the worst of wind load on the superstructure alone (P_{LS}) and wind on the superstructure plus live loading (P_{LL}) is taken for design purposes.

All structures with a solid elevation:

$$P_{LS} = 0.25qA_1C_D \tag{2.30}$$

All truss girder structures:

$$P_{LS} = 0.5qA_1C_D \tag{2.31}$$

Live load on all structures:

$$P_{LL} = 0.5qA_1C_D \tag{2.32}$$

where $C_D = 1.45$.

Parapets and safety fences

(i)	With vertical infill	$P_L = 0.8P_t$	(2.33)
(ii)	With two or three rails	$P_L = 0.4P_t$	(2.34)
(iii)	With mesh	$P_L = 0.6P_t$	(2.35)

Piers

The longitudinal wind load is given by:

$$P_L = qA_2C_D \tag{2.36}$$

where A_2 is the transverse solid area and C_D is taken from Table 9 with b and d interchanged.

Uplift

The vertical uplift wind force on the deck (P_v) is given by:

$$P_v = qA_3C_L \tag{2.37}$$

where A_3 is the plan area of the deck, and C_L (the lift coefficient) is taken from Figure 6 if the angle of elevation is less than 1°. For angles between 1° and 5°, C_L is taken as ±0.75. For angles >5° tests must be carried out.

2.6.11 Load combinations
The are four wind load cases to consider in load combination 2.

(1) P_t
(2) $P_t \pm P_v$
(3) P_L
(4) $0.5P_t + P_L \pm P_v$.

2.6.12 Overturning effects
For narrow piers it is necessary to check the stability of the structure when subject to heavy vehicles on the outer extremities of the deck. This is illustrated in Figure 2.27.

2.6.13 Concluding remarks
Wind loads affect bridges in one of two ways: globally or locally.

Global wind forces induce overall bending, shear and twisting forces, and these loads are transferred to the tops of piers and abutments via the bearings and expansion joints; and they are also transferred to the foundations. Stability is also a factor to be considered, especially in the case of continuous bridges with long spans and having a degree of curvature.

Local wind forces are resisted by parapets and safety fences; and may have fatigue consequences in steel bridges in the cables and hangers of tension bridge structures.

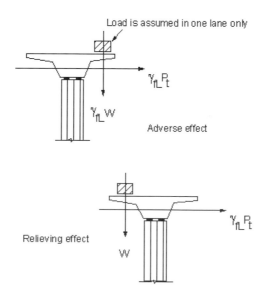

Figure 2.27 Stability requirements.

For small- to medium-span bridges wind loads are not normally critical, but in the case of long-span suspended structures, wind forces are dominant and can cause collapse (see the example in Appendix A.2).

2.7 Temperature

The temperature of both the bridge structure and its environment changes on a daily and seasonal basis and influences both the *overall movement* of the bridge deck and the *stresses* within it. The former has implications for the design of the bridge bearings and expansion joints, and the latter on the amount and disposition of the structural materials.

The *daily* effects give rise to temperature variations within the depth of the superstructure which vary depending on whether it is *heating or cooling*, and guidance is normally given in the form of idealized linear temperature gradients to be expected when the bridge is *heating or cooling* for various forms of construction (concrete slab; composite deck, etc.) and blacktop surface thickness. The temperature gradients result in *self-equilibrating internal stresses.* Two types of stress are induced – *primary and secondary* – the former due to the temperature differences throughout the superstructure (whether simply supported or continuous), and the latter due continuity. Both must be assessed and catered for in the design.

The temperature of a bridge deck varies throughout its mass. The variation is caused by:

- the position of the sun
- the intensity of the suns rays
- thermal conductivity of the concrete and surfacing
- wind
- the cross-sectional make-up of the structure.

The effects are complex and have been investigated in the UK by the Transport Research Laboratory (TRL).

Changes occur on a *daily* (short-term) and *annual* (long-term) basis. Daily there is *heat gain* by day and *heat loss* by night. Annually there is a variation of the ambient (surrounding) temperature.

On a daily basis, *temperatures near the top are controlled by incident solar radiation, and temperatures near the bottom are controlled by shade temperature*. The general distribution is indicated in Figure 2.28. Positive represents a rapid rise in temperature of the deck slab due to direct sunlight (solar radiation). Negative represents a falling

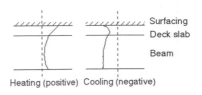

Figure 2.28 Typical temperature distributions.

ambient temperature due to heat loss (re-radiation) from the structure.

Research has indicated that for the purposes of analysis the distributions (or thermal gradients) can be idealized for different 'groups' of structure as defined in Figure 9 of BD37/88 Cl. 5.4. The critical parameters are the thickness of the surfacing; the thickness of the deck slab and the nature of the beam. Concrete construction falls within group 4. Temperature differences cause curvature of the deck and result in internal *primary* and *secondary* stresses within the structure.

2.7.1 Primary stresses

These occur in both simply supported and continuous bridges and are manifested as a variation of stress with depth. They develop due to the redistribution of restrained temperature stresses which is a *self-equilibrating* process. They are determined by balancing the restrained stresses with an equivalent system of a couple and a direct force acting at the neutral axis position. The section is divided into slices, and the restraint force in each slice determined. The sum of the moments of each force about the neutral axis and the sum of the forces gives the couple and the direct force respectively. These are shown in Figure 2.29.

2.7.2 Secondary stresses

These occur in continuous bridges only and result due to a change in the global reactions and bending moments. They are determined by applying the couple and the force at each end of the continuous bridge and determining the resulting reactions and moments. These are then added to the self-weight and live load reactions and moments.

Primary stresses are not necessarily larger than secondary stresses. Both can be significant and depend on a whole range of variables. Once calculated they are included in Combination 3 defined in BD37/88.

2.7.3 Annual variations

Annual (or seasonal) changes result in a *change in length* of the bridge and therefore affect the design of both bearings and expansion joints. Movement is related to the *minimum* and *maximum* expected ambient temperatures. This information is normally available in the form of *isotherms* for a particular geographic region. The

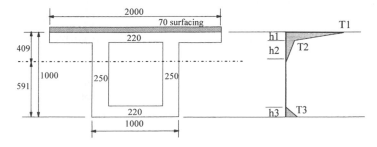

Figure 2.29 Primary stresses due to temperature gradient through bridge (see the example in Appendix A.3).

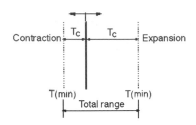

Figure 2.30 Setting of an expansion joint.

total expected movement (Δ) takes place from a fixed point called the *thermal centre or stagnant point* and is given by:

$$\Delta = \text{thermal strain} \times \text{span} = \beta_L TL \qquad (2.38)$$

where β_L is the coefficient of thermal expansion and T the temperature change and is based on a total possible range of movement given by the difference of the *maximum* and *minimum* shade temperatures and specified in the code for a given bridge location as Isotherms in Figures 7 and 8 of BD37/88. These are further modified to take account of the bridge construction in Tables 10 and 11 of BD37/88 to give the *effective bridge temperature.*

The bearings and expansion joints are set in position to account for actual movements which will depend on the time of year in which they are installed. This is shown graphically in Figure 2.30 in relation to the 'setting' of an expansion joint.

2.8 Earthquakes

Until recently the effect of earthquakes on buildings has received more attention than bridges probably because the social and economic consequences of earthquake damage in buildings has proved to be greater than that resulting from damage to bridges. In a recent study of seismic shock, Albon (1998) observes that 'Bridges should be designed to absorb seismic forces without collapse to ensure that main arterial routes remain open after major seismic events. This helps the movement of aid and rescue services in the first instance and underpins the ability of the local community to recover in the long term'.

Observations over many years indicate that bridge failures due to earthquake forces on bridges are not caused by collapse of any single element of the superstructure but rather by two effects:

• the superstructure being shaken off the bearings and falling to the ground, and
• structural failure due to the loss of strength of the soil under the superstructure as a result of the vibrations induced in the ground.

The effect of an earthquake depends on the elastic characteristics and distribution of the self-weight of the bridge, and the usual procedure is to consider that the earthquake produces *lateral* forces acting in any direction at the centre of gravity of the structure and having a magnitude equal to a percentage of the weight of the structure or any part of the structure under consideration. These loads are then treated as static.

The design lateral force applied at deck level is given by:

$$F = C_D \gamma_i W_i \tag{2.39}$$

where C_D is the seismic design coefficient which depends on the soil conditions; the risk against collapse; the ductility of the structure and an amplification factor; γ_i is a distribution factor depending on the height of the deck from foundation level and W_i is the permanent load plus a given percentage of the live load.

Modern codes such as the current European Code EC8 (1998) allow three different methods of analysis, namely:

- fundamental mode method (static analysis)
- response spectrum method
- time–history representation.

The first two are linearly elastic analyses and the last is non-linear.

The recent high profile earthquakes in Northridge, Los Angeles in January 1994 and Kobe, Japan in January 1995 have proved invaluable to the understanding of the behaviour of bridge structures under earthquake loading, and no doubt more refined and reliable design procedures will result.

2.9 Snow and ice

In certain parts of the world snow and ice are in evidence for considerable periods and in the case of cable-stayed and suspension bridges can contribute significantly to the dead weight by forming around the cables, parapets and on the supporting towers. Complete icing of the parapets also means that lateral wind forces are increased due to the solid area exposed to the wind. Expansion joints and bearings can also become locked resulting in large restraining forces to the deck and substructures.

2.10 Water

Rivers in flood represent a serious threat to bridges both from the point of view of lateral forces on the abutments, piers and superstructures and the possible undermining of the foundations due to the scouring effect of the water.

The lateral hydrodynamic forces are calculated in a similar manner to those due to wind. Thus from:

$$q = \rho v_c^2 / 2 \tag{2.40}$$

where v_c is the velocity of flow in m/s. If the density of water is taken as 1000 N/m^3 then the water pressure:

$$q = 500v_c^2/10^3 \text{ kN/m}^2 \tag{2.41}$$

and

$$P = qAC_D \text{ kN} \tag{2.42}$$

(as for wind). Values of C_D for various shaped piers in the USA arc given in AASHTO (1994) and in the UK are found in BA59 (1994).

The degree of scour depends on many factors such as the geometry of the pier; the speed of flow and the type of soil (Hamill, 1998).

The total depth of bridge scour is due to a combination of *general* scour due to the constriction of the waterway area leading to an increase in the flow velocity, and *local* scour adjacent to a pier or abutment from turbulence in the water. Many models are available for dealing with these phenomena (Melville *et al.*, 1988; Federal Highways Administration, 1991; Faraday *et al.*, 1995; Hamill, 1998) all with particular points of merit. Scour is one of the major causes of bridge failure (Smith, 1976), and proper design and protection is essential to guard against such catastrophic events.

There is also the danger from *fast moving debris* hitting the piers or the deck, and also the possibility of *accumulated debris* blocking the bridge opening – both need to be considered in design.

2.10.1 Normal flow
Bridge piers are designed to resist lateral forces from water in *normal flow* conditions. The forces induced are calculated using the same formulae as for moving air, namely (2.44) and (2.45), but with the density of air replaced by that for water:

$$\rho_w = 1000 \text{ kg/m}^3 = 10 \text{ kN/m}^3 \tag{2.43}$$

the dynamic pressure head:

$$q = 500v_c^2/10^3 \text{ kN/m}^2 \tag{2.44}$$

$$P_{tw} = qA_1C_D \tag{2.45}$$

Values of C_D can be determined from Table 9 of BD37/88.

In the UK some guidance is given in Departmental Memorandum BA 59/94 for the design of *Highway Bridges for Hydraulic Action*, which also considers forces due to *ice, debris and ship collision*.

2.10.2 Flood conditions
Flood waters exert forces many times those under normal conditions. Very often the waters top the bridge (*negative freeboard*) and both the *deck and piers* are subject to the full force of water and debris. Areas of turbulance cause high local forces and scour of the river bed around the piers. Estimation of the forces involved is complex

Figure 2.31 Effects of scour.

and unreliable (for example, estimating the speed and height of the flood waters), and most countries have their own procedures in place which take into account local topography and experience from previous floods.

2.10.3 Scour

Scour is not classed as a load, but it is caused by erosion of the river bed around the piers and foundations, and can cause undermining of the foundations and eventual collapse. Just how bad it can be is shown in Figure 2.31 – fortunately the piles prevented collapse!

BA59/94 considers three types of scour: *general, local and combined* and leans heavily on US reports FHWA-IP-90-017 and FHWA-IP-90-014. An example is provided to illustrate the use of the several equations and is reproduced in Figure 2.32.

2.11 Construction loads

Temporary forces occur in the construction at each stage of construction due to the self-weight of plant, equipment and the method of construction. Generally these forces are more significant in bridges built by the method of serial construction such as post-tensioned concrete box girders where long unsupported cantilever sections induce forces which are substantially different than those in the completed bridge both in magnitude and distribution. Cable-stayed and suspension bridges are also susceptible to the method of erection where the deck sections are built up piecemeal

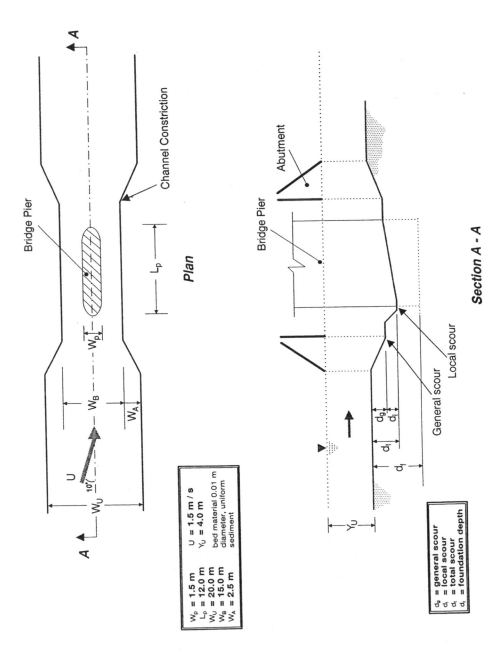

Figure 2.32 Example bridge, showing definition of symbols used in the equations (courtesy of Department of Transport).

Table 2.4 UK load combinations

Combination	UK (see Cl.3.2 for definitions)
1	Permanent + primary live
2	Permanent + primary live + wind + (temporary erection loads)
3	Permanent + primary live + temperature restraint + (temporary erection loads)
4	Permanent + secondary live + associated primary live
5	Permanent + bearing restraint

from the towers or supporting pylons. In all cases construction loads and method of erection should be closely examined to ensure that accidents do not happen and that the serviceability condition of the final structure is not impeded.

2.12 Load combinations

In the UK *five* combinations of loading are considered for the purposes of design: *three principal* and *two secondary*. These are defined in Cl.4.4 and Table 1 of BD37/88. It is usual in practice to design for Combination 1 and to check other combination if necessary.

2.13 Use of influence lines

2.13.1 Introduction

Influence lines are a useful visual aid at the analysis stage to enable the determination of the distribution of the primary traffic loads on the decks of continuous structures and trusses to give the *worst possible effect*. Although they can be used to calculate actual values of stress resultants, bridge engineers generally use them in a qualitative manner so that they can see where the critical regions are at a glance.

2.13.2 Continuous structures

Continuous concrete or composite bridges and the decks of cable-stayed and suspension bridges fall within this category as shown in Figure 2.33.

The influence lines (IL) for a typical five span arrangement for the bending moment at a mid-span region and a support region, are shown on Figure 2.34. The shapes (rather than the actual values) are the dominant feature of each line. The IL for the bending moment at an internal support always consists of two adjacent concave sections followed by alternate convex and concave sections, and the IL for the bending moment in the mid-span region always consists of a cusped section followed by alternate convex and concave sections. These patterns enable the influence line for any number of equal (or unequal) spans to be sketched out.

Modern bridge software programs are able to plot the IL for stress resultants and displacements for any member in a given bridge.

It is clear that the placement of the load in each case to maximize the moment is given by the shaded areas which are called *adverse* areas. The other (unshaded) areas

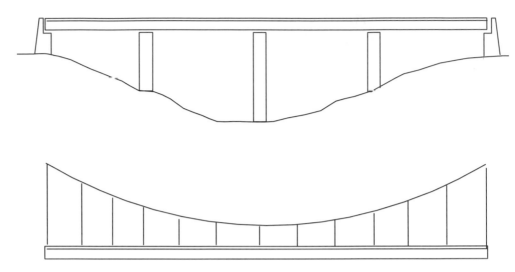

Figure 2.33 Beam and suspension bridges.

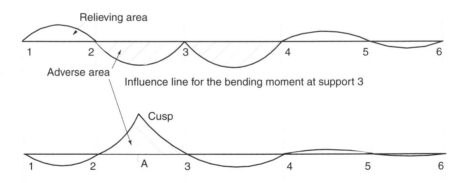

Influence line for the bending moment at a mid-span section of span 2 - 3

Figure 2.34 Typical influence lines for bending moment in continuous decks.

are called *relieving* areas, since loads placed on these spans will minimize the moment. *In codes which specify a decreasing load intensity as the loaded length increases, then the maximum moment is generally given by loading adjacent spans only for internal supports, and the single span only for midspan regions.* For codes which specify a constant load intensity regardless of span, then *all* adverse areas should be loaded.

2.13.3 Trusses

The axial forces in members of bridge trusses vary as moving loads cross the bridge, and influence lines are useful in determining the loaded length to give the worst effect. The Warren truss shown in Figure 2.35 illustrates the principle. For member A, the force remains positive for all positions of load, while for member B the sign changes

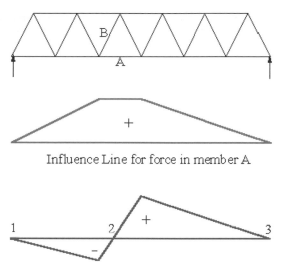

Figure 2.35 Typical influence lines for truss girder bridges.

as loads cross the panel containing B. Both the positive and the negative forces can be found by applying the load to the relevant adverse area of the IL, that is L_{12} for the maximum negative force and L_{23} for the maximum positive force.

This is true for codes which specify a constant UDL regardless of loaded length, but for codes with a varying intensity of UDL with loaded length the worst effect may be when only part of the adverse area is considered. Point loads from abnormal vehicles are considered in the same way. Shapes and ordinate values of ILs for different types of trusses can be found in any standard text books on the subject.

ILs can also be used to calculate the moments and forces in bridge members directly. For point loads these are given by:

$$M = Px_i \tag{2.46}$$

where P is the load and x_i is the value of the ordinate under the load. For UDLs these are given by:

$$M = w \times a_i \tag{2.47}$$

where w is the intensity of the UDL and a_i is the area of the IL under the loaded length.

2.14 Load distribution

2.14.1 Introduction

Traffic loads on bridge decks are distributed according to the stiffness, geometry and boundary conditions of the deck. The deflection of a typical beam-and-slab deck under an axle load is shown in Figure 2.36.

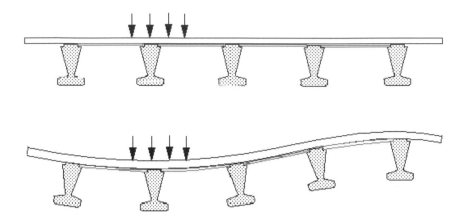

Figure 2.36 Typical transverse bending due to eccentric traffic load.

For a single-span right deck on simple supports with different stiffnesses in two orthogonal directions, it is possible, using classical plate theory to determine the load distributed to each member. If the amount of load carried by the most heavily loaded member can be found then the bending moment can be calculated easily.

The very first attempts at analysing bridge decks pioneered by Guyon (1946) and Massonet (1950) were aimed at simplifying the process for practising engineers by the method of *distribution coefficients*, that is, the calculation of the distribution of live loads to a particular beam (or portion of slab) as a fraction of the total. The method was developed in the UK by Morice *et al.* (1956), Rowe (1962), and Cusens *et al.* (1975). It was later refined by Bakht *et al.* (1985) of Canada, and has actually codified in the USA, AASHTO (1977) and Canada OHBDC (1983).

The basic assumption of the Distribution Coefficient (or D-Type) method is that the distribution pattern of longitudinal moments, shears and deflections across a transverse section is independent of the longitudinal position of the load and the transverse section considered, Bakht *et al.* (1985) and Ryall (1992). This is illustrated in Figure 2.37.

The implication is that:

$$M_{x1}/M_1 = M_{x2}/M_2 \tag{2.48}$$

where M_1 and M_2 are the gross moments at sections 1 and 2, respectively. For convenience, the maximum longitudinal bending moment M_{sw} from a single line of wheels of a standard vehicle (acting on a line beam) is determined and this is multiplied by a load fraction S/D to give the design moment for the beam; thus S/D is the proportion of the bending moment(or wheel lines) from a single line of wheels carried by a particular beam. This can be seen by reference to Figure 2.38 where each co-ordinate of the distribution diagram represents the longitudinal moment per unit width of the deck.

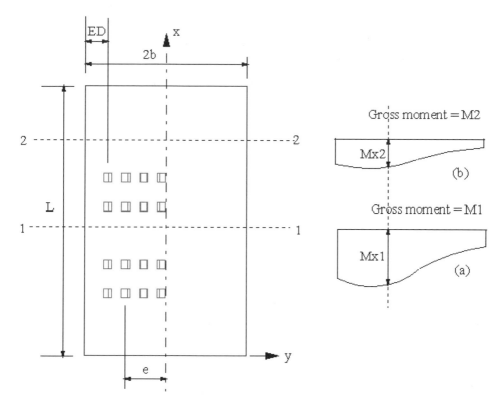

Figure 2.37 Transverse distribution of longitudinal moments at two sections due to traffic.

If the total area under the curve represents the gross bending moment at the section due to the design vehicle, then the total moment sustained by girder 2, for example, is represented by the hatched area, such that:

$$M_{g2} = M_x dy \qquad (2.49)$$

and:

$$M_{av} = M_{g2}/S \qquad (2.50)$$

If it assumed that for a particular bridge and design vehicle, a factor D (in terms of width) is known, such that:

$$D = M_{sw}/M_{av} \qquad (2.51)$$

then substituting for M_{av} gives:

$$M_{g2} = SM_{sw}/D \qquad (2.52)$$

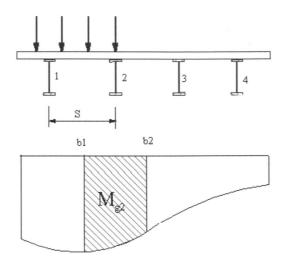

Figure 2.38 Proportion of total moment to be resisted by a given beam.

The calculation of M_{sw} is trivial, so that if the value of D is known, then it is a simple matter to determine M_{g2}.

The distribution coefficients (D) are calculated by solving the well-known partial differential plate equation:

$$D_x + 2H + D_y = p(x, y) \qquad (2.53)$$

where:

$$2H = D_{xy} + D_{yx} + D_1 + D_2 \qquad (2.54)$$

Solution is achieved numerically by satisfying the boundary conditions with the use of harmonic functions to represent the load and by assuming a sinusoidal deflection profile (Bakht *et al.*, 1985). This is tantamount to idealizing the deck as a continuum.

2.14.2 Controlling parameters

Past research (Bakht *et al.*, 1985) has shown that, apart from the pattern of live loads, the main factors affecting the transverse distribution of longitudinal bending moments are the flexural and torsional rigidities, the width of the deck ($2b$), and the edge distance (ED) of the standard vehicle. Furthermore, bridge decks in general can be defined by two non-dimensional *characterizing parameters* namely:

$$\alpha = H/(D_x D_y)^{0.5} \qquad (2.55)$$

and:

$$\theta = b(D_x/D_y)^{0.25}/L \qquad (2.56)$$

where D_x, D_y, D_1, D_2, D_{xy} and D_{yx} are the flexural and torsional rigidities of the deck per metre length. In defining a particular bridge, all that is required are the values of α and θ. The distribution coefficient D can be easily obtained from either pre-prepared tables or from a computer programme (see the example in Appendix A.4).

2.15 US practice

In the USA, the distribution method has been in use for many years and a widely adopted tool for the global analysis of simply supported right bridges. The latest distribution factors (DFs) – referred to as *mg* – are presented in AASHTO (1994) and provide equations for calculating the DFs for slab-on-girder bridges for the *maximum bending moment and shear force on an interior girder and an external girder.*

The distribution factors contained in AASHTO (1994) are for specifically defined bridge types and geometry. For example, the distribution factor for the interior girder of a spaced beam and slab deck with two or more loaded lanes is:

$$mg = 0.075 + (S/2900 \text{ mm})^{0.6}(S/L)^{0.2}(K_g/Lt_s^3)^{0.1} \tag{2.57}$$

where S is the girder spacing; L is the span length; K_g is the longitudinal stiffness parameter (nI_{comp}). This is then used as a basis for the moment in an exterior girder. (Note: the multiple presence factor is implicitly included in *mg*.) Typical examples are given in Barker *et al.* (1997).

Bibliography

Albon, JM *A study of the damage caused by seismic shock on highway bridges and ways of minimising it.* MSc dissertation, University of Surrey, 1998.

American Association of State Highway and Transportation Officials. *Standard specification for highway bridges,* AASHTO, Washington, DC, 1977.

American Association of State Highway and Transportation Officials. *LFRD, Bridge Design Specification.* AASHTO, Washington, DC, 1994.

American Association of State Highway and Transportation Officials. *Standard specification for highway bridges*, 16th edn. AASHTO, Washington, DC, 1996.

BA59 (1994) The design of highway bridges for hydraulic action. *Design Manual for Roads and Bridges*, Vol. 1, Section 3, Part 6. Stationery Office, London.

Bakht, B and Jaeger, LG. *Bridge Analysis Simplified.* McGraw-Hill, New York, 1985.

Bailey, S and Bez, R. Considering actual traffic during bridge evaluation. *Proceedings of the Third International Conference on Bridge Management*, pp. 795–802, Thomas Telford, London, 1996.

Barker, RM and Puckett, JA. *Design of Highway Bridges – Based on AASHTO LRFD Bridge Design Specifications.* John Wiley, New York, 1996.

Bez, R and Hirt, MA. Probability-based load models of highway bridges. *Structural Engineering International, IABSE* 2, 37–42, 1991.

BS153. *Girder Bridges. Part 3A:Loads.* British Standards Institution, London, 1954.

BE1/77. *Technical Memorandum (Bridges): Standard Highway Loadings*, 1961.

BS5400. *Steel, Concrete and Composite Bridges. Part 2: Specification for Loads.* British Standards Institution, London, 1978.

BD37. Loads for highway bridges. *Design Manual for Roads and Bridges*. HMSO, London, 1988.

Cambridge. Highway bridge loading. *Report on the Proceedings of a Colloquium*, 7–10 April, 1975.

Cooper, DI. Development of short span bridge-specific assessment live loading. In Das PC (Ed.) *Safety of Bridges*, pp. 64–89. Highways Agency, London, 1997.

Cusens, AR and Pama, RP. *Bridge Deck Analysis*. John Wiley, London, 1975.

Dyrbye, C and Hansen, SO. *Wind Loads on Structures*. John Wiley, 1996.

Eurocode 1. *Basis of Design and Actions on Structures*, Vol. 3, *Traffic Loads on Bridges*. CEN Brussels, 1993.

Eurocode 8. *Design Provisions for Earthquake Resistance of Structures – Part 2: Bridges*, CEN Brussels, 1998.

Faraday, RV and Charlton, FG. *Hydraulic Factors in Bridge Design*. Hydraulics Research Ltd, Wallingford, 1983.

Federal Highways Administration. *Evaluating Scour at Bridges*. *HEC-18*. US Department of Transportation, 1991.

Guyon, Y. Calcul des ponts larges a poutres multiples solidarisées par des entretoises. *Annals des Ponts et Chausees* 24, 553–612, 1946.

Hamill L. *Bridge Hydraulics*. E & FN Spon, London, 1998.

Henderson, W. British highway bridge loading. *Proc. Inst. Civil Engineers, Part II* 3, 325–373, 1954.

Massonet C. Method de calcul des ponts a poutres multiples tenant compte de leur resistance a la torsion. *Proc. Int. Association for Bridge and Structural Engineering* 10, 147–182, 1950.

Melville, BW and Sutherland, AJ. Design method for local scour at bridge piers. *Journal of Hydraulic Engineering, ASCE* 114, 1210–1226, 1988.

Ministry of Transportation and Communications. *Ontario Highway Bridge Design Code (OHBDC)*, Ministry of Transportation and Communications, Downsview, Ontario, 1983.

Morice, PB and Little, G. *The Analysis of Right Bridge Decks Subjected to Abnormal Loading*. Report Db 11, Cement and Concrete Association, London, 1956.

Ministry of Transport. MOT Memo No. 577 (1931) *Bridge Design and Construction – Loading*, 1931.

Page J. Traffic data for highway bridge loading rules. In Das PC (Ed.), *Safety of Bridges*, pp 90–98. Highways Agency, London, 1997.

Rose AC. *Public Roads of the Past*, two volumes, 1952–3.

Rowe RE. *Concrete Bridge Design*. Applied Science, London, 1962.

Ryall MJ. Application of the D-Type method of analysis for determining the longitudinal moments in bridge decks. *Proc. Inst. Civil Engrs. Structs and Bldgs* 94, 157–169, 1992.

Smith DW. Bridge failures. *Proc. Instn. Civil Engrs, Part I* August, 367–382, 1976.

Vrouwenvelder, ACWM and Waarts PH. Traffic loads on bridges. *Structural Engineering International, IABSE* 3, 169–177, 1993.

Appendix A.1: shrinkage stresses

A contiguous composite bridge is located over a waterway, and consists of a series of Y8 precast pre-stressed concrete beams at 2 m centres and with a 220 mm deep *in situ* concrete slab. Young's modulus for the Y-beam concrete is 50 N/mm^2 and for the *in situ* slab it is 35 N/mm^2. Determine the stresses induced in the section due to shrinkage of the top slab.

1. Calculate properties of section

Modular ratio = 50/35 = 1.429. Therefore effective width of slab = 2000/1.429 = 1400 mm.

$$I_x \text{ (slab)} = 140 \times 22^3/12 = 124\,227 \text{ cm}^4.$$

Distance of neutral axis from top = 607 471/8927 = 68 cm.

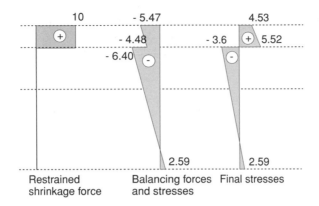

Restrained shrinkage force	Balancing forces and stresses	Final stresses

Figure A1.1 Final stress distribution

Table A1 Section properties

Section	A (cm²)	y (cm)	Ay
Slab	3080	11	33 880
Y8 beam	5847	98.1	573 591
	8927		607 471

$$I_x \text{ (comp)} = 124\ 227 + 3080 \times (68 - 11)^2$$
$$+ 118.86 \times 10^5 + 5847 \times (76.1 + 22 - 68)^2$$
$$= 273 \times 10^5 \text{ cm}^4$$

2. Calculate restrained shrinkage stresses

$$F = -50 \times 1400 \times 220 \times (-200 \times 10^{-6}) = 3080 \text{ kN}$$

$$M = 3080 \times (0.68 - 0.11) = 1756 \text{ kN m.}$$

Restrained shrinkage stress $f_0 = 3080 \times 10^3/308\ 000 = 10$ N/mm².

3. Calculate balancing stresses

Direct stress $f_{10} = -3080 \times 10^3/892\ 700 = -3.45$ N/mm²

Bending stresses $= My/I$, Balancing stresses;

$$f_{21} = -3.45/1.429 - [(1756 \times 10^6 \times 680)/(273 \times 10^9)/1.429$$
$$= -2.41 - 3.06 = -5.47 \text{ N/mm}^2$$

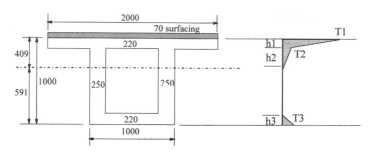

Figure A2.1 Box girder dimensions and temperature distribution.

$$f_{22} = -2.41 - (1756 \times 10^6 \times [680 - 220])/(273 \times 10^9 \times 1.429)$$
$$= -2.41 - 2.07 = -4.48 \text{ N/mm}^2$$

$$f_{23} = -4.48 \times 1.429 = -6.40 \text{ N/mm}^2$$

$$f_{24} = [-3080 \times 10^3/892\ 700 + (1756 \times 10^6 \times 940)/(273 \times 10^9)$$
$$= -3.45 + 6.04 = 2.59 \text{ N/mm}^2$$

It is clear that there is a substantial level of tension in the top slab which cannot only cause cracking, but also results in a considerable shear force at the slab/beam interface which has to be resisted by shear links projecting from the beam.

Appendix A.2: primary temperature stresses (BD37/88)

Determine the stresses induced by both the positive and reverse temperature differences for the concrete box girder bridge shown in Figure A2.1 ($A = 940\,000$ mm^2, $I = 102\,534 \times 10^6$ mm^4, depth to NA = 409 mm, $\beta_T = 12 \times 10^{-6}$, $E = 34$ kN/mm^2).

1. Calculate critical depths of temperature distribution

From BD37/88 Figure 9 this is a Group 4 section, therefore:

$$h1 = 0.3h = 0.3 \times 1000 = 300 > 150, \text{ thus } h1 = 150 \text{ mm}$$
$$h2 = 0.3h = 0.3 \times 1000 = 300 > 250, \text{ thus } h2 = 250 \text{ mm}$$
$$h3 = 0.3h = 0.3 \times 1000 = 300 > 170, \text{ thus } h3 = 170 \text{ mm}.$$

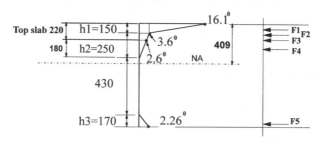

Figure A2.2 Element forces.

2. Calculate temperature distribution

Basic values are in Figure 9 which are modified for depth of section and surface thickness by interpolating from Table 24.

$$T1 = 17.8 + (17.8 - 13.5)20/50 = 16.1°C$$
$$T1 = 4.0 + (4.0 - 3.0)20/50 = 3.60°C$$
$$T1 = 2.1 + (2.5 - 2.1)20/50 = 2.26°C.$$

3. Calculate restraint forces at critical points

This is accomplished by dividing the depth into convenient elements corresponding to changes in the distribution diagram and/or changes in the section (see Figure 3.2):

$$F = E_c\beta_T T_i A_i$$

$F1 = 34\,000 \times 12 \times 10^{-6} \times (16.1 - 3.6) \times 2000 \times 150 / 1000$
$ = 765$ kN

$F2 = 34\,000 \times 12 \times 10^{-6} \times (3.6) \times 2000 \times 150 / 1000$
$ = 441$ kN

$F3 = 34\,000 \times 12 \times 10^{-6} \times [(3.6+2.6)/2] \times 2000 \times (220 - 150)/1000$
$ = 177$ kN

$F4 = 34\,000 \times 12 \times 10^{-6} \times (2.6/2) \times 2 \times (250 - 70) \times 250 / 1000$
$ = 48$ kN

$F5 = 34\,000 \times 12 \times 10^{-6} \times (2.26/2) \times 1000 \times 170 / 1000$
$ = 78$ kN

Total $F = 1509$ kN (tensile)

4. Calculate restraint moment about the neutral axis

$$M = [765(409 - 50) + 441(409 - 75) + 177(409 - 185)$$
$$+ 48(409 - 270) - 78(591 - 170 \times 2/3)]/1000$$

$$M = 431 \text{ kNm (hogging)}$$

5. Calculate restraint stresses

$$f = E_c\beta_T T_i$$

$f_{01} = -34\,000 \times 12 \times 10^{-6} \times 16.1 = -6.56$ N/mm^2
$f_{02} = -34\,000 \times 12 \times 10^{-6} \times 3.6 = -1.47$ N/mm^2

$$f_{03} = -34\,000 \times 12 \times 10^{-6} \times 2.6 = -1.06 \text{ N/mm}^2$$

$$f_{04} = -34\,000 \times 12 \times 10^{-6} \times 0 = 0.00 \text{ N/mm}^2$$

$$f_{05} = -34\,000 \times 12 \times 10^{-6} \times 0 = 0.00 \text{ N/mm}^2$$

$$f_{06} = -34\,000 \times 12 \times 10^{-6} \times 2.26 = -0.92 \text{ N/mm}^2$$

6. Calculate balancing stresses

Direct stress $f_{10} = 1509 \times 10^3/940\,000 = 1.61 \text{ N/mm}^2$.

Bending stresses $f_{2i} = My/I$:

$$f_{21} = \frac{431 \times 10^6}{102\,534 \times 10^6} \times 409 = 1.71 \text{ N/mm}^2$$

$$f_{22} = \frac{431 \times 10^6}{102\,534 \times 10^6} \times 259 = 1.08 \text{ N/mm}^2$$

$$f_{23} = \frac{431 \times 10^6}{102\,534 \times 10^6} \times 180 = 0.75 \text{ N/mm}^2$$

$$f_{24} = \frac{431 \times 10^6}{102\,534 \times 10^6} \times 9 = 0.06 \text{ N/mm}^2$$

$$f_{25} = \frac{431 \times 10^6}{102\,534 \times 10^6} \times 421 = -1.76 \text{ N/mm}^2$$

$$f_{26} = \frac{431 \times 10^6}{102\,534 \times 10^6} \times 591 = -2.47 \text{ N/mm}^2$$

7. Calculate final stresses

Table A2.1 Summary of stresses

	Restraint stresses	Balancing direct stress	Balancing bending stresses	Final stresses
1	−6.56	1.61	1.71	−3.24 (C)
2	−1.47	1.61	1.08	1.14 (T)
3	−1.06	1.61	0.75	1.3 (T)
4	0	1.61	0.06	1.67 (T)
5	0	1.61	−1.76	−0.15 (C)
6	−0.92	1.61	−2.47	−1.78 (C)

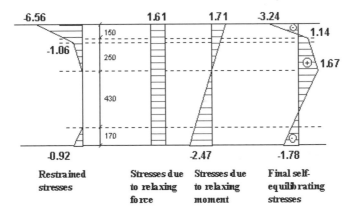

Figure A2.3 Final stress distribution (positive).

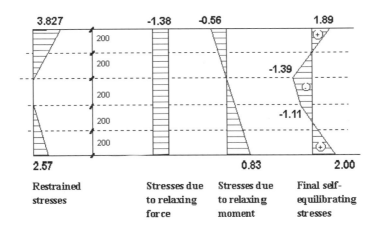

Figure A2.4 Final stress distribution (Negative).

The final stress distribution is shown in Figure A2.3. Similar calculations for the cooling (reverse) situation are shown in Figure A2.4.

Appendix A.3: wind loads (BD37/88)

Calculate the worst transverse wind loads on the structure shown in Figure A3.1. Assume that $v = 28$ m/s; span $= 33$ m; $H = 10$ m.

$$S_1 = K_1 = 1.0. \text{ From Table 2, } S_2 = 1.54.$$

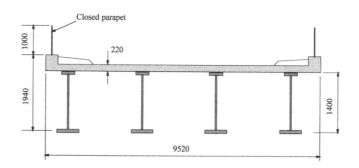

Figure A3.1

(i) Unloaded deck:

$$v_t = 28 \times 1 \times 1 \times 1.54 = 43.13 \text{ m/s}$$

$$q = 43.13^2 \times 0.613/10^3 = 1.14 \text{ kN/m}^2$$

From Table 4, $d = d_2 = 1 + 1.94 = 2.94$ m
From Table 5, $d_2 = 1.94$ m, thus $b/d_2 = 9.52/2.94 = 3.24$, and Figure 5, $C_D = 1.4$.

$$A_1 = 2.94 \times 33 = 97.02 \text{ m}^2.$$

Thus $P_t = 1.14 \times 97.02 \times 1.4 = 154.84$ kN

(ii) Loaded deck:

$$v_t = 35 \text{ m/s (maximum allowed in the code)}$$

$$q = 35^2 \times 0.613 \times 10^3 = 0.75 \text{ kN/m}^2 .$$

$$d_2 = 2.94 \text{ m} > d_L = 2.5 \text{ m}$$

From Table 5, $d = d_2$ thus $b/d_2 = 9.52/2.94 = 3.24$, and from Figure 5, $C_D = 1.4$.

From Table 4, $d = d_3 = d_L + $ slab thickness $+$ depth of steel beams
$$= 2.5 + 0.22 + 1.4$$
$$= 4.12 \text{ m}$$

$$P_t = 0.75 \times 1.4 \times (4.12 \times 33) = 142.76 \text{ kN}$$

Thus design force $=$ greater of (i) and (ii) $= 154.84$ kN.

Appendix A.4: D-type method

A reinforced concrete slab on prestressed concrete Y-beams will illustrate the method.
The span is 11 m; the carriageway is 9 m and it is subject to 45 units of an HB

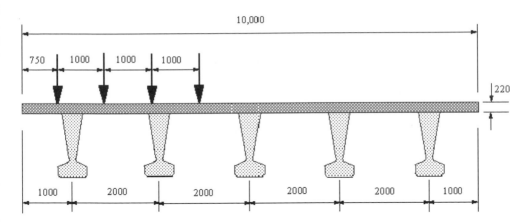

Figure A4.1 Prestressed concrete beam and reinforced concrete deck.

Table A4.1 Deck properties ($\times 10^{12}$ kN mm²/m) for example

Deck	D_x	D_y	D_1	D_2	D_{xy}	D_{yx}	H	α	θ
PSC/concrete	4.69	0.03	0.01	0.01	0.13	0.02	0.08	0.23	0.82

Table A4.2 Table of distribution coefficients D: 9 m carriageway; 0.75 m edge distance; 45 units of HB only

α	θ							
	0.2	0.3	0.4	0.5	0.6	0.7	0.8	0.9
0.05	1.22	1.11	1.07	1.05	1.05	1.05	1.06	1.09
0.1	1.41	1.22	1.14	1.1	1.08	1.07	1.08	1.1
0.15	1.49	1.28	1.18	1.13	1.1	1.08	1.09	1.1
0.2	1.6	1.37	1.25	1.18	1.14	1.11	1.11	1.12
0.25	1.69	1.45	1.31	1.23	1.17	1.14	1.13	1.14
0.3	1.76	1.52	1.37	1.27	1.21	1.17	1.16	1.16

vehicle. The deck is shown in Figure A4.1. The values of D_x, D_y, D_{xy} and D_{yx} are based on the beam and slab longitudinally, but on the slab only transversely, from which α and θ are calculated and are shown in Table A4.1.

Tables of D can be generated very simply to account for the number of lanes; the type and intensity of loading and the edge distance of the of the loading such as Table 2.5 which is the appropriate one for this example.

The load in each case was that from 45 units of an HB vehicle taken from BD37/88, and placed in an outside lane so as to induce the worst possible longitudinal bending moment. From Table 2.6, the distribution coefficients for the deck can be interpolated as 1.12. Then if the moment at the critical mid span section due to a single line of wheels from the reference vehicle (ie the HB vehicle) is calculated as 1644 kN m, then the moments in the most heavily loaded girder (the edge girder) $M_{g1} = 2 \times 1644/1.12 = 2936$ kN m. Using the grillage method the maximum moment was calculated as 3065 kN m.

The maximum difference between each of the methods is 4.2% which by any standards is quite acceptable. It could be argued that the distribution method is more accurate as it more closely models the deck as a continuum.

The main advantage of the D-Type method over the traditional grillage and FEA methods is its speed and simplicity. The initial data required is minimal, and if the computer option is utilized, the output data will not require more than a single sheet of paper.

The method can be used for design or assessment purposes for determining the value of critical moments under any load specification such as the UK, AASHTO and EC1 loadings.

3 Structural analysis

N.E. SHANMUGAM AND R. NARAYANAN

3.1 Fundamental concepts

3.1.1 General introduction

Structural analysis consists essentially of mathematical modelling of the response of a structure to the applied loading. Such models are based on idealizations of the structural behaviour of the material and are, therefore, imperfect to some extent, depending upon the simplifying assumptions in modelling. Consequently the assessment of structural responses is the best estimate that can be obtained in view of the assumptions implicit in the modelling of the system. Some of these assumptions are necessary in the light of inadequate data: others are introduced to simplify the calculation procedure to economic levels. Examples of idealizations introduced in the modelling process are as follows:

- The physical dimensions of the structural components are idealized. For example, skeletal structures are represented by a series of line elements and the joints are assumed to be of negligible size. The imperfections in the member straightness are ignored or at best idealized.
- Material behaviour is simplified. For example, the stress–strain characteristic of steel is assumed to be linearly elastic, and then perfectly plastic. No account is taken of the variation of yield stress along or across the member.
- The implications of actions which are included in the analytical process itself are frequently ignored. For example, the possible effects of change of geometry causing local instability are rarely, if ever, accounted for in the analysis.

However, it must be recognized that the design loads employed in assessing structural response are themselves approximate. The analysis chosen should be adequate for the purpose and capable of providing the solutions at an economical cost.

Structural design involves the arrangement and proportioning of structures and their components in such a way that the assembled structure is capable of supporting the designed loads over a designed life span within the allowable limit states.

In practice, all structures are three-dimensional. Structures that have regular layout and are rectangular in shape, can be idealized into two-dimensional frames. Joints in a structure are locations where two or more members are connected. A structural system in which joints are capable of transferring end moments is called a frame. Members in this system are assumed to be capable of resisting bending moments, axial forces and shear forces. A structure is said to be a two-dimensional or planar if all the members lie in the same plane.

Beams are members that are subjected to bending and shear. Ties are members that are subjected to axial tension only whilst struts are members subjected to axial compression only.

A hinge represents a pin connection to a structural assembly and it does not allow translational movements (Figure 3.1a). It is assumed to be frictionless and to allow rotation of a member with respect to the others. A roller represents a support that permits the attached structural part to rotate freely with respect to the foundation or base and to translate freely in the direction parallel to the base (Figure 3.1b). No translational movement in any other direction is allowed. A fixed support (Figure 3.1c) does not allow rotation or translation in any direction. A rotational spring represents a support that provides some rotational restraint but does not provide any translational restraint (Figure 3.1d). A translational spring can provide partial restraints along the direction of deformation (Figure 3.1e).

3.1.2 Loads and reactions
Loads may be broadly classified as permanent loads that are constant in magnitude and remain in one position and variable loads that may change in position and magnitude. Permanent loads are also referred to as dead loads which include the self

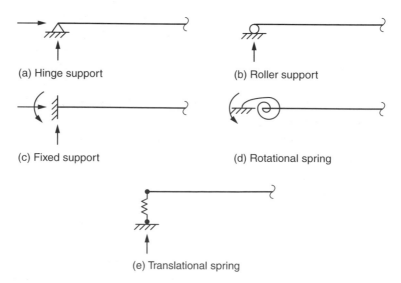

Figure 3.1 Various boundary conditions.

weight of the structure and other loads permanently attached to the structure. Variable loads are also referred to as live (or imposed) loads, and include those caused by construction operations, wind, rain, earthquakes, snow, blasts, and temperature changes in addition to those that move (i.e. vehicles, trains, etc.).

Wind loads act as pressures on windward surfaces and pressures or suctions on leeward surfaces. Impact loads are caused by suddenly applied loads or by the vibration of moving or movable loads. Earthquake loads are forces caused by the acceleration of the ground surface during an earthquake.

A structure that is initially at rest and remains at rest when acted upon by applied loads is said to be in a state of equilibrium. The resultant of the external loads on the body and the supporting forces or reactions is zero. If a structure or part thereof is to be in equilibrium under the action of a system of loads, it must satisfy the six equations of static equilibrium described below.

Equations of static equilibrium

When a body (or structure) remains in a state of static equilibrium the following two conditions must be satisfied:

- The sum of the components of all forces acting on the body, resolved along any arbitrary direction is equal to zero. This condition is completely satisfied if the components of all forces resolved along the x, y, z directions individually add up to zero. (This can be represented by $\Sigma P_x = 0$, $\Sigma P_y = 0$, $\Sigma P_z = 0$, where P_x, P_y and P_z represent forces resolved in the x, y, z directions.) These three equations represent the condition of zero translation.
- The sum of the moments of all forces resolved in any arbitrarily chosen plane about any point in that plane is zero. This condition is completely satisfied when all the moments resolved into xy, yz and zx planes all individually add up to zero ($\Sigma M_{xy} = 0$, $\Sigma M_{yz} = 0$ and $\Sigma M_{zx} = 0$). These three equations provide for zero rotation about the three axes. In general, therefore, there are a total of *six* equations of static equilibrium. If a structure is planar and is subjected to a system of coplanar forces, the conditions of equilibrium can be simplified to *three* equations as detailed below:
- The components of all forces resolved along the x and y directions will individually add up to zero ($\Sigma P_x = 0$ and $\Sigma P_y = 0$).
- The sum of the moments of all the forces about any arbitrarily chosen point in the plane is zero (i.e. $\Sigma M = 0$).

The principle of superposition

This principle is only applicable when the displacements are linear functions of applied loads. For structures subjected to multiple loading, the total effect of several loads can be computed as the sum of the individual effects calculated by applying the loads separately. This principle is a very useful tool in computing the combined effects of many load effects (e.g. moment, deflection, etc.). These can be calculated separately for each load and then summed.

3.1.3 Element analysis

Any complex structure can be looked upon as being built up of simpler units or components termed 'members' or 'elements'. Broadly speaking, these can be classified into three categories:

- Skeletal structures consisting of members whose one dimension (say, length) is much larger than the other two (namely breadth and height). Such a line element is variously termed as a bar, beam, column or tie. A variety of structures are obtained by connecting such members together using rigid or hinged joints. Should all the axes of the members be situated in one plane, the structures so produced are termed plane structures. Where all members are not in one plane, the structures are termed space structures.
- Structures consisting of members whose two dimensions (namely length and breadth) are of the same order but much greater than the thickness, fall into the second category. Such structural elements are called plated structures. Such structural elements are further classified as plates and shells depending upon whether they are plane or curved. In practice these units are used in combination with beams or bars.
- The third category consists of structures composed of members having all the three dimensions (namely length, breadth and depth) or the same order. The analysis of such structures is extremely complex, even when several simplifying assumptions are made.

For the most part the structural engineer is concerned with skeletal structures. Increasing sophistication in available techniques of analysis has enabled the economic design of plated structures in recent years. Recent advances in finite element method and the availability of fast and powerful computers have enabled the economic analysis of complex structures. Three-dimensional analysis of structures is only rarely carried out. Under incremental loading, the initial deformation or displacement response of a member is largely elastic. Continued application of load results in large displacements with yield spreading through the cross section before ultimate collapse occurs.

Line elements

The deformation response of a line element is dependent on a number of cross-sectional properties such as area, A, second moment of area ($I_{xx} = \int y^2 dA; I_{yy} = \int x^2 dA$) and the product moment of area ($I_{xy} = \int xy dA$). The two axes xx and yy are orthogonal. For doubly symmetric sections, the axes of symmetry are those for which $\int xy dA = 0$. These are known as principal axes. For a plane area, the principal axes may be defined as a pair of rectangular axes in its plane and passing through its centroid, such that the product moment of area $\int xy dA = 0$, the co-ordinates referring to the principal axes. If the plane area has an axis of symmetry, it is obviously a principal axis (by symmetry $\int xy dA = 0$). The other axis is at right angles to it, through the centroid of the area.

Table of properties of the section (including the centroid and shear centre of the section) are available as published data.

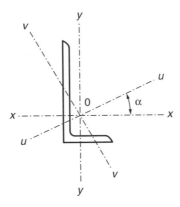

Figure 3.2 Angle section (no axis of symmetry).

If the section has no axis of symmetry (e.g. an angle section) the principal axes will have to be determined. Referring to Figure 3.2 if $u0u$ and $v0v$ are the principal axes, the angle a between the uu and xx axes is given by:

$$\tan 2\alpha = \frac{-2I_{xy}}{I_{xy} - I_{yy}} \tag{3.1}$$

$$I_{uu} = \frac{I_{xx} + I_{yy}}{2} + \frac{I_{xx} - I_{yy}}{2} \cos 2\alpha - I_{xy} \sin 2\alpha$$

$$I_{vv} = \frac{I_{xx} + I_{yy}}{2} - \frac{I_{xx} - I_{yy}}{2} \cos 2\alpha + I_{xy} \sin 2\alpha \tag{3.2}$$

The values of α, I_{uu} and I_{vv} are available in published design guides (e.g. Steel Construction Institute, 1997).

Elastic analysis of line elements under axial loading

When a cross-section is subjected to a compressive or tensile axial load, P, the resulting stress is given by the load/area of the section, i.e. P/A. Axial load is defined as one acting at the centroid of the section. When loads are introduced into a section in a uniform manner (e.g. through a heavy end-plate), this represents the state of stress throughout the section. On the other hand, when a tensile load is introduced via a bolted connection, there will be regions of the member where stress concentrations occur and plastic behaviour may be evident locally, even though the mean stress across the section is well below yield.

If the force P is not applied at the centroid, the longitudinal direct stress distribution will no longer be uniform. If the force is offset by eccentricities of e_x and e_y measured from the centroidal axes in the y and x directions, the equivalent set of actions are: (1) an axial force P, (2) a bending moment $M_x = Pe_x$ in the yz plane and

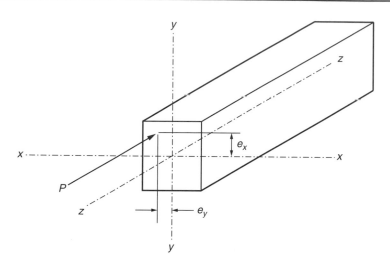

Figure 3.3 Compressive force applied eccentrically with reference to the centroidal axis.

(3) a bending moment $M_y = Pe_y$ in the zx plane (see Figure 3.3). The method of evaluating the stress distribution due to an applied moment is given in a later section. The total stress at any section can be obtained as the algebraic sum of the stresses due to P, M_x and M_y.

Elastic analysis of line elements in pure bending

For a section having at least one axis of symmetry and acted upon by a bending moment in the plane of symmetry, the Bernoulli equation of bending may be used as the basis to determine both stresses and deflections within the elastic range. The assumptions which form the basis of the theory are:

- the beam is subjected to a pure moment (i.e. shear is absent). (Generally the deflections due to shear are small compared with those due to flexure; this is not true of deep beams.)
- plane sections before bending remain plane after bending
- the material has a constant value of modulus of elasticity (E) and is linearly elastic.

The following equation results (see Figure 3.4).

$$\frac{M}{I} = \frac{f}{y} = \frac{E}{R} \tag{3.3}$$

where M is the applied moment; I is the second moment of area about the neutral axis; f is the longitudinal direct stress at any point within the cross section; y is the distance of the point from the neutral axis; E is the modulus of elasticity; R is the radius of curvature of the beam at the neutral axis.

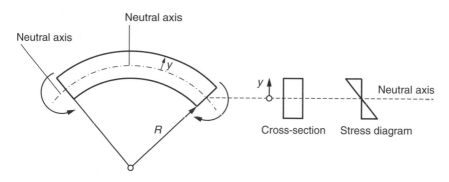

Figure 3.4 Pure bending.

From the above, the stress at any section can be obtained as:

$$f = \frac{M\,y}{I}$$

For a given section (having a known value of I) the stress varies linearly from zero at the neutral axis to a maximum at extreme fibres on either side of the neutral axis.

$$f_{max} = \frac{M\,y_{max}}{I} = \frac{M}{Z} \qquad (3.4)$$

where

$$Z = \frac{I}{y_{max}}$$

The term Z is known as the elastic section modulus and is tabulated in section tables for steel members (see Steel Construction Institute, 1997). The elastic moment capacity of a given section may be found directly as the product of the elastic section modulus, Z, and the maximum allowable stress.

If the section is doubly symmetric, then the neutral axis is mid-way between the two extreme fibres. Hence, the maximum tensile and compressive stresses will be equal. For an unsymmetric section, this will not be the case as the value of y for the two extreme fibres will be different.

For a monosymmetric section, such as the T-section shown in Figure 3.5, subjected to a moment acting in the plane of symmetry, the elastic neutral axis will be the centroidal axis. The above equations are still valid. The values of y_{max} for the two extreme fibres (one in compression and the other in tension) are different. For an applied sagging (positive) moment shown in Figure 3.5, the extreme fibre stress in the flange will be compressive and that in the stalk will be tensile. The numerical values of the maximum tensile and compressive stresses will differ. In the case sketched in Figure 3.5, the magnitude of the tensile stress will be greater, as y_{max} in tension is greater than that in compression.

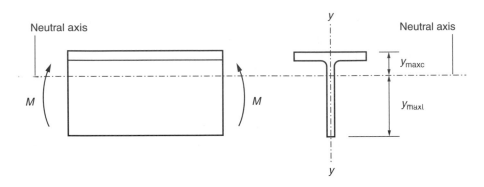

Figure 3.5 Monosymmetric section subjected to bending.

Caution has to be exercised in extending the pure bending theory to asymmetric sections. There are two special cases where no twisting occurs:

• bending about a principal axis in which no displacement perpendicular to the plane of the applied moment results
• the plane of the applied moment passes through the shear centre of the cross-section.

When a cross section is subjected to an axial load and a moment such that no twisting occurs, the stresses may be determined by resolving the moment into components M_{uu} and M_{vv} about the principal axes uu and vv and combining the resulting longitudinal stresses with those resulting from axial loading:

$$f_{u,v} = \pm \frac{P}{A} \pm \frac{M_{uu}v}{I_{uu}} + \frac{M_{vv}u}{I_{vv}} \tag{3.5a}$$

For a section having two axes of symmetry (see Figure 3.3) this simplifies to:

$$f_{y,x} = \pm \frac{P}{A} \pm \frac{M_{xx}y}{I_{rr}} + \frac{M_{yy}x}{I_{vv}} \tag{3.5b}$$

Pure bending does not cause the section to twist. When the shear force is applied eccentrically in relation to the shear centre of the cross section, the section twists and initially plane sections no longer remain plane. The response is complex and consists of a twist and a deflection with components in and perpendicular to the plane of the applied moment. This is not discussed in this chapter. A simplified method of calculating the elastic response of cross sections subjected to twisting moments is given in an SCI publication by Nethercot *et al.* (1998).

Elastic analysis of line elements subject to shear
Pure bending discussed in the preceding section implies that the shear force applied on the section is zero. Application of transverse loads on a line element will, in gen-

eral, cause a bending moment which varies along its length, and hence a shear force which also varies along the length is generated.

If the member remains elastic and is subjected to bending in a plane of symmetry (such as the vertical plane in a doubly symmetric or monosymmetric beam), then the shear stresses caused vary with the distance from the neutral axis.

For a narrow rectangular cross section of breadth b and depth d, subjected to a shear force V and bent in its strong direction (see Figure 3.6(a)), the shear stress varies parabolically from zero at the lower and upper surfaces to a maximum value, q_{max}, at the neutral axis given by:

$$q_{max} = \frac{3V}{2bd}$$

i.e. 50% higher than the average value.

For an I-section (Figure 3.6(b)), the shear distribution can be evaluated from:

$$q = \frac{V}{IB} \int_{y=h}^{y=h_{max}} by \, dy \qquad (3.6)$$

where B is the breadth of the section at which shear stress is evaluated. The integration is performed over that part of the section remote from the neutral axis, i.e. from $y = h$ to $y = h_{max}$ with a general variable width of b.

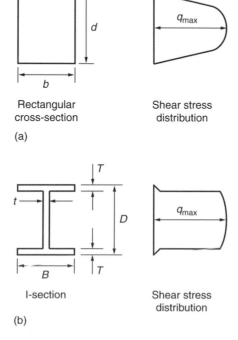

Rectangular
cross-section

Shear stress
distribution

(a)

I-section

Shear stress
distribution

(b)

Figure 3.6 Shear stress distribution (a) in a rectangular cross-section, (b) in an I-section.

Clearly, for the I- (or T-) section, at the web–flange interface, the value of the integral will remain constant. As the section just inside the web becomes the section just inside the flange, the value of the vertical shear abruptly changes as B changes from web thickness to flange width.

3.2 Flexural members

One of the most common structural elements is a *beam*; it bends when subjected to loads acting transversely to its centroidal axis or sometimes by loads acting both transversely and parallel to this axis. The discussions given in the following subsections are limited to idealized straight beams in which the centroidal axis is a straight line with shear centre coinciding with the centroid of the cross-section. The loads and reactions are assumed to lie in a plane that also contains the centroidal axis of the flexural member and the principal axis of every cross section. If these conditions are satisfied, the beam will only bend in the plane of loading without twisting.

3.2.1 Axial force, shear force and bending moment

Axial force at any transverse cross section of a straight beam is the algebraic sum of the components acting parallel to the axis of the beam, of all loads and reactions applied to the portion of the beam on either side of that cross-section. Shear force at any transverse cross-section of a straight beam is the algebraic sum of the components acting transverse to the axis of the beam, of all the loads and reactions applied to the portion of the beam on either side of the cross-section. Bending moment at any transverse cross-section of a straight beam is the algebraic sum of the moments on either side of cross-section, taken about an axis passing through the centroid of the cross section. The axis about which the moments are taken is, of course, normal to the plane of loading.

3.2.2 Relation between load, shear, bending moment, and deflection

When a beam is subjected to transverse loads, there exist certain relationships between load, shear and bending moment. Let us consider, for example, the beam shown in Figure 3.7 subjected to some arbitrary loading, p.

Let S and M be the shear and bending moment, respectively for any point 'm' at a distance x, which is measured from A, being positive when measured to the right. The corresponding values of shear and bending moment at point 'n' at a small distance dx to the right of m are $S + dS$ and $M + dM$, respectively. It can be shown, neglecting the second order quantities, that:

$$p = \frac{dS}{dx} \tag{3.7}$$

and:

$$S = \frac{dM}{dx} \tag{3.8}$$

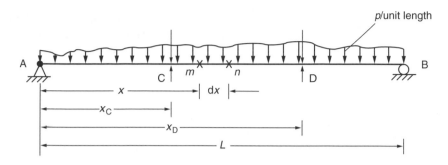

Figure 3.7 A beam under arbitrary loading.

$$m = EI \frac{d^2 y}{dx^2} \qquad (3.9a)$$

and:

$$p = EI \frac{d^4 y}{dx^4} \qquad (3.9b)$$

3.2.3 Beam deflection

There are several methods for determining beam deflections: (1) moment-area method, (2) conjugate-beam method, (3) virtual work and (4) Castigliano's second theorem, among others. The first two methods are described below.

The elastic curve of a member is the shape taken by the neutral axis when the member deflects under load. The inverse of the radius of curvature at any point of this curve is obtained as:

$$\frac{1}{R} = \frac{M}{EI} \qquad (3.10)$$

in which M is the bending moment at the point and EI is the flexural rigidity of the beam section. Since the deflection is small, $1/R$ is approximately taken as $d^2 y/d^2 x$, and Equation (3.10) may be rewritten as:

$$M = EI \frac{d^2 y}{dx^2} \qquad (3.11)$$

In Equation (3.11), y is the deflection of the beam at distance x measured from the origin of coordinate. The change in slope in a distance dx can be expressed as $M\,dx/EI$ and hence the slope in a beam is obtained as:

$$\theta_B - \theta_A = \int_A^B \frac{M}{EI} dx \qquad (3.12)$$

where θ_A and θ_B are the slopes at A and B of the deflected beam.

Equation (3.12) may be stated as follows: the change in slope between the tangents to the elastic curve at two points is equal to the area of the M/EI diagram between the two points.

Once the change in slope between tangents to the elastic curve is determined, the deflection can be obtained by integrating further the slope equation. In a distance dx the neutral axis changes in direction by an amount $d\theta$. The deflection of one point on the beam with respect to the tangent at another point due to this angle change is equal to $d\delta = x d\theta$, where x is the distance from the point at which deflection is desired to the particular differential distance.

To determine the total deflection from the tangent at one point A to the tangent at another point B on the beam, it is necessary to obtain a summation of the products of each $d\theta$ angle (from A to B) times the distance to the point where deflection is desired or:

$$\delta_B - \delta_A = \int_A^B \frac{Mx\, dx}{EI} \qquad (3.13)$$

The above equation can be stated as follows: 'The deflection of a tangent to the elastic curve of a beam with respect to a tangent at another point is equal to the moment of M/EI diagram between the two points, taken about the point at which deflection is desired'.

Moment area method

Moment area method is most conveniently used for determining slopes and deflections for beams in which the direction of the tangent to the elastic curve at one or more points is known, such as cantilever beams, where the tangent at the fixed end does not change in slope. The method is applied easily to beams loaded with concentrated loads, because the moment diagrams consist of straight lines. These diagrams can be broken down into single triangles and rectangles. Beams supporting uniform loads or uniformly varying loads may be handled by integration.

It should be noted that the slopes and deflections that are obtained using the moment area theorems are with respect to tangents to the elastic curve at the points being considered. The theorems do not directly give the slope or deflection at a point in the beam as compared to the horizontal axis (except in one or two special cases); they give the change in slope of the elastic curve from one point to another or the deflection of the tangent at one point with respect to the tangent at another point. There are some cases in which beams are subjected to several concentrated loads or the combined action of concentrated and uniformly distributed loads. In such cases it is advisable to separate the concentrated loads and uniformly distributed loads and the moment-area method can be applied separately to each of these loads. The final responses are obtained by the principle of superposition.

The use of the moment-area method is illustrated by considering a simply supported beam subjected to uniformly distributed load q as shown in Figure 3.8. The tangent to the elastic curve at each end of the beam is inclined. The deflection δ_1 of the tan-

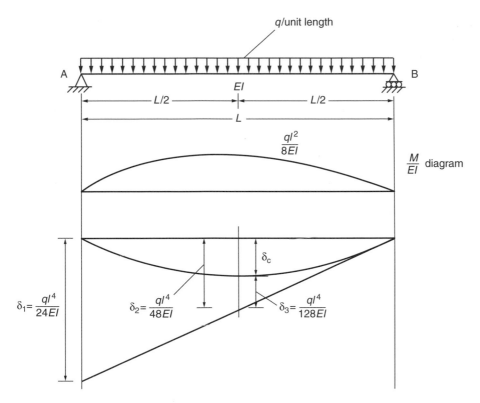

Figure 3.8 Deflection – simply supported beam under UDL.

gent at the left end from the tangent at the right end is found as $ql^4/24EI$. The distance from the original chord between the supports and the tangent at right end, δ_2, can be computed as $ql^4/48EI$. The deflection of a tangent at the centre from a tangent at right end, δ_3 is determined in this step as $ql^4/128EI$. The difference between δ_2 and δ_3 gives the centreline deflection as $5/384 \times ql^4/EI$.

In general, the bending moment, shear force and deflected shape diagrams can be drawn for a given loading by employing Equations (3.7)–(3.13). Examples of application to cantilevers and simply supported beams for some of the commonly encountered loading cases are given in Figure 3.9.

Conjugate beam method

The conjugate beam method (first developed by Otto Mohr in 1860) requires the same amount of computation as the moment–arca theorems to determine the slope or deflection of a beam. It relies on the similarity between several computations:

- the equation for the shear (V) compares with that for the slope (θ)
- the equation for the moment (M) compares with that for the displacement
- the application of the external load (w) to compute the moment (M) compares with that for the application of the moment (M) to compute displacement.

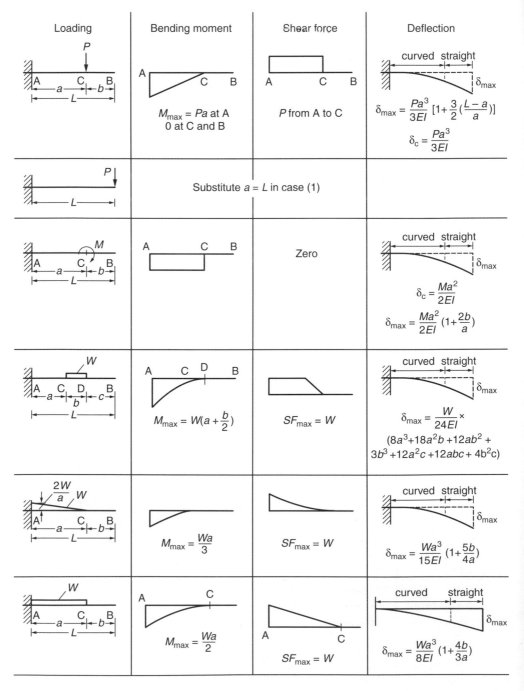

Loading	Bending moment	Shear force	Deflection

$M_{max} = Pa$ at A
0 at C and B

P from A to C

$\delta_{max} = \dfrac{Pa^3}{3EI}\left[1+\dfrac{3}{2}\left(\dfrac{L-a}{a}\right)\right]$

$\delta_c = \dfrac{Pa^3}{3EI}$

Substitute $a = L$ in case (1)

Zero

$\delta_c = \dfrac{Ma^2}{2EI}$

$\delta_{max} = \dfrac{Ma^2}{2EI}\left(1+\dfrac{2b}{a}\right)$

$M_{max} = W\left(a+\dfrac{b}{2}\right)$

$SF_{max} = W$

$\delta_{max} = \dfrac{W}{24EI}\times$
$(8a^3+18a^2b+12ab^2+3b^3+12a^2c+12abc+4b^2c)$

$M_{max} = \dfrac{Wa}{3}$

$SF_{max} = W$

$\delta_{max} = \dfrac{Wa^3}{15EI}\left(1+\dfrac{5b}{4a}\right)$

$M_{max} = \dfrac{Wa}{2}$

$SF_{max} = W$

$\delta_{max} = \dfrac{Wa^3}{8EI}\left(1+\dfrac{4b}{3a}\right)$

Figure 3.9 Bending moment shear force and deflected shape diagrams.

Loading	Bending moment	Shear force	Deflection

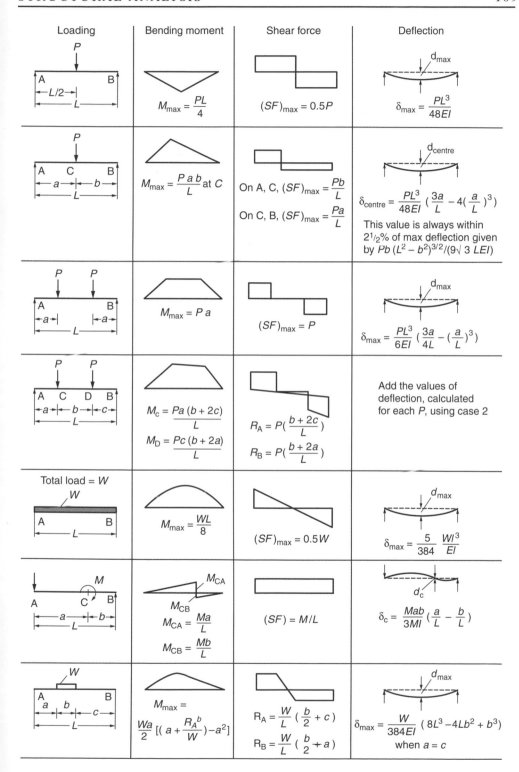

Row 1:

$M_{max} = \dfrac{PL}{4}$

$(SF)_{max} = 0.5P$

$\delta_{max} = \dfrac{PL^3}{48EI}$

Row 2:

$M_{max} = \dfrac{Pab}{L}$ at C

On A, C, $(SF)_{max} = \dfrac{Pb}{L}$

On C, B, $(SF)_{max} = \dfrac{Pa}{L}$

$\delta_{centre} = \dfrac{PL^3}{48EI} \left(\dfrac{3a}{L} - 4\left(\dfrac{a}{L}\right)^3 \right)$

This value is always within $2^{1}/_{2}\%$ of max deflection given by $Pb\,(L^2 - b^2)^{3/2}/(9\sqrt{3}\,LEI)$

Row 3:

$M_{max} = Pa$

$(SF)_{max} = P$

$\delta_{max} = \dfrac{PL^3}{6EI} \left(\dfrac{3a}{4L} - \left(\dfrac{a}{L}\right)^3 \right)$

Row 4:

$M_C = \dfrac{Pa\,(b + 2c)}{L}$

$M_D = \dfrac{Pc\,(b + 2a)}{L}$

$R_A = P\left(\dfrac{b + 2c}{L}\right)$

$R_B = P\left(\dfrac{b + 2a}{L}\right)$

Add the values of deflection, calculated for each P, using case 2

Row 5:

Total load = W

$M_{max} = \dfrac{WL}{8}$

$(SF)_{max} = 0.5W$

$\delta_{max} = \dfrac{5}{384} \dfrac{Wl^3}{EI}$

Row 6:

$M_{CA} = \dfrac{Ma}{L}$

$M_{CB} = \dfrac{Mb}{L}$

$(SF) = M/L$

$\delta_c = \dfrac{Mab}{3MI} \left(\dfrac{a}{L} - \dfrac{b}{L} \right)$

Row 7:

$M_{max} = \dfrac{Wa}{2}\left[\left(a + \dfrac{R_A b}{W} \right) - a^2 \right]$

$R_A = \dfrac{W}{L}\left(\dfrac{b}{2} + c \right)$

$R_B = \dfrac{W}{L}\left(\dfrac{b}{2} + a \right)$

$\delta_{max} = \dfrac{W}{384EI}\,(8L^3 - 4Lb^2 + b^3)$ when $a = c$

Figure 3.9(Continued) Bending moment shear force and deflected shape diagrams.

To make use of this comparison we employ a 'conjugate beam' having the same length as the 'real' beam and the computation follows the application of these two theorems:

1. the slope at a point in the 'real' beam is equal to the shear at the corresponding point in the 'conjugate' beam
2. the displacement of a point in the 'real' beam is equal to the moment at the corresponding point in the 'conjugate' beam.

This method is valuable for evaluating deflections of more complex cases like fixed beams. When employing the conjugate beam method it is important that proper boundary conditions be used. These are given in Figure 3.10.

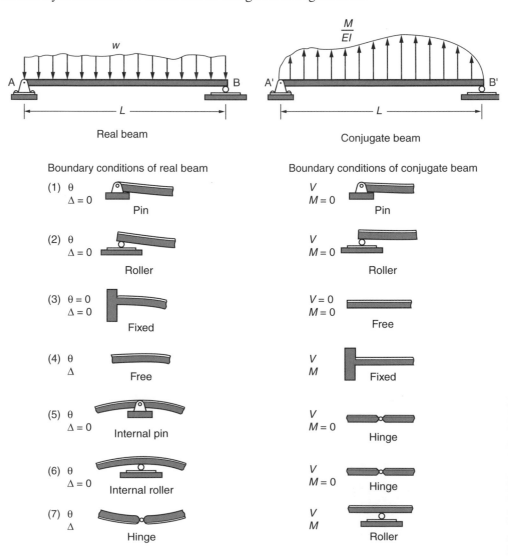

Figure 3.10 Real and conjugate beams.

3.2.4 Fixed-ended beams

When the ends of a beam are held firmly and are not free to rotate under the action of applied loads, the beam is known as a built-in or encastré beam and is statically indeterminate. The bending moment diagram for such a beam can be considered to consist of two parts namely the free (bending) moment diagram obtained by treating the beam as if the ends are simply supported and the fixing (moment) diagram resulting from the restraints imposed at the ends of the beam. It can be shown that:

1. the area of the fixing (bending) moment diagram is equal to that of the free (bending) moment diagram
2. the centres of gravity of the two diagrams lie in the same vertical line, i.e. are equidistant from either end of the beam.

The bending moment diagram for a fixed beam can be obtained as illustrated in the example shown in Figure 3.11. **P Q U T** is the free moment diagram, M_s and **P Q R S** is the fixing moment diagram M_i. The net bending moment diagram, M is shown shaded. If A_s is the area of the free bending moment diagram and A_i the area of the fixing moment diagram then, from (1) above, $A_s = A_i$ and:

$$\frac{1}{2} \times \frac{Wab}{L} \times L = \frac{1}{2}(M_A + M_B)L$$

$$M_A + M_B = \frac{Wab}{L} \tag{3.14}$$

Equating the moment about A of A_s and A_i, we have:

$$M_A + 2M_B = \frac{Wab}{L^3}(2a^2 + 3ab + b^2) \tag{3.15}$$

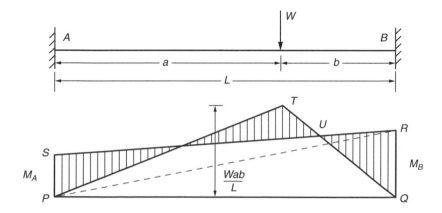

Figure 3.11 Fixed beam.

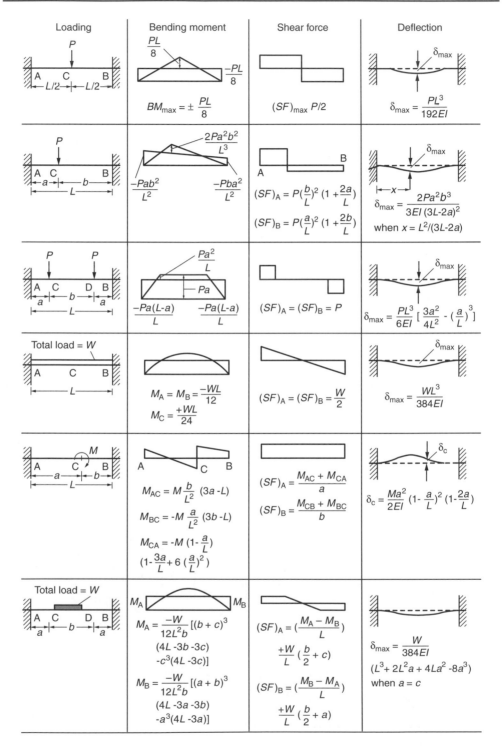

Figure 3.12 Bending moment, shear force, deflection diagrams of typical loading in fixed beams.

Solving these two equations for M_A and M_B, we obtain:

$$\left.\begin{array}{l} M_A = \dfrac{Wab^2}{L^2} \\[4mm] M_A = \dfrac{Wa^2b}{L^2} \end{array}\right\} \qquad (3.16)$$

Shear force can be determined once the bending moment is known. The shear force at the ends of the beam, A and B are:

$$S_A = \frac{M_A - M_B}{L} + \frac{Wb}{L}$$

$$S_B = \frac{M_B - M_A}{L} + \frac{Wa}{L}$$

Bending moment and shear force diagrams for some typical loading cases are shown in Figure 3.13.

3.2.5 Continuous beams

Three moment theorem

Continuous beams are statically indeterminate. Bending moments in these beams are functions of the geometry, moments of inertia and modulus of elasticity of individual members besides the load and span. They may be determined by Clapeyron's theorem of three moments, moment distribution method or slope deflection method.

The theorem of three moments is applied to two adjacent spans at a time and the resulting equations in terms of unknown support moments are solved. The theorem states that:

$$M_A L_1 + 2M_B(L_1 + L_2) + M_C L_2 = 6\left(\frac{A_1 x_1}{L_1} + \frac{A_2 x_2}{L_2}\right) \qquad (3.17)$$

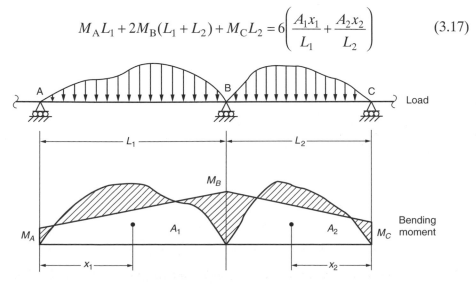

Figure 3.13 Continuous beams.

in which M_A, M_B and M_C, are the hogging moment at the supports A, B and C, respectively of two adjacent spans of uniform cross section and lengths L_1 and L_2 (Figure 3.13); A_1 and A_2 are the area of bending moment diagrams produced by the vertical loads on the simple spans AB and BC, respectively; x_1 is the centroid of A_1 from A, and x_2 is the distance of the centroid of A_2 from C. If the beam section is constant within a span but remains different for each of the spans, Equation (3.17) is written as:

$$M_A \frac{L_1}{I_1} + 2M_B\left(\frac{L_1}{I_1} + \frac{L_2}{I_2}\right) + M_C \frac{L_2}{I_2} = 6\left(\frac{A_1 x_1}{L_1 I_1} + \frac{A_2 x_2}{L_2 I_2}\right) \tag{3.18}$$

in which I_1 and I_2 are the moments of inertia of beam section in span L_1 and L_2, respectively.

Similar equations for two adjacent spans are set up at time, until all the spans have been covered, two at a time. These provide sufficient number of additional equations, which together with the equations of statics, enable the calculation of bending moments at all sections.

Values of bending moments and reactions at supports for two, three and four span continuous beams are presented for several simple cases of loading in Figure 3.14. These will be of value to designers for carrying out calculations and checks rapidly.

(a)   Two span continuous beams

Loading	Max +ve BM	Max −ve BM	Reactions at supports		
			1	2	3
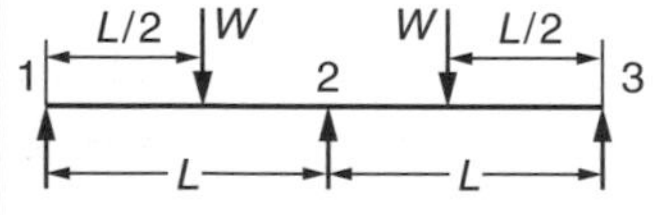 $L/2$, W, W, $L/2$ (spans L, L)	+0.156 WL	−0.188 WL	0.313 W	1.375 W	0.313 W
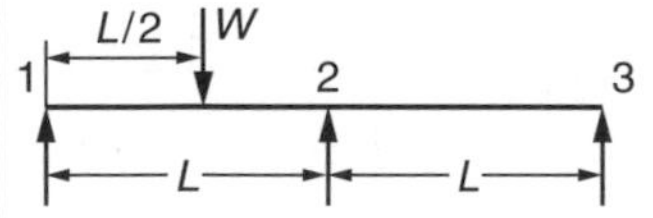 $L/2$, W (spans L, L)	+0.203 WL	−0.094 WL	0.406 W	0.688 W	−0.094 W (downwards)
Total load on each span = W 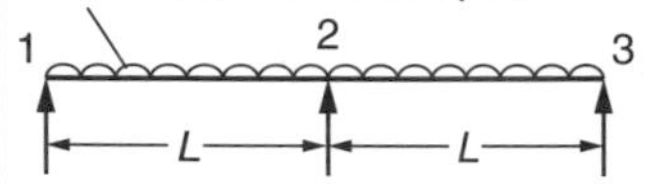 (spans L, L)	+0.070 WL	−0.125 WL	0.375 W	1.25 W	0.375 W
Total load on = W 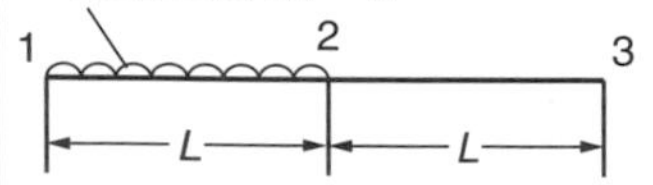 (spans L, L)	+0.063 WL	−0.063 WL	0.438 W	0.625 W	−0.063 W (downwards)

Figure 3.14 Bending moments and reactions of continuous beams.

(b)

Three span continuous beams

Loading	Max +ve BM	Max −ve BM	Reactions at supports			
			1	2	3	4
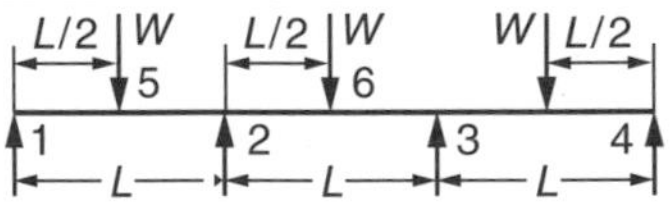 $L/2$ W_5 $L/2$ W_6 W $L/2$; supports 1,2,3,4; spans L,L,L	$0.175WL$ at ⑤	$-0.150WL$ at ②	$0.35W$	$1.15W$	$1.15W$	$0.35W$
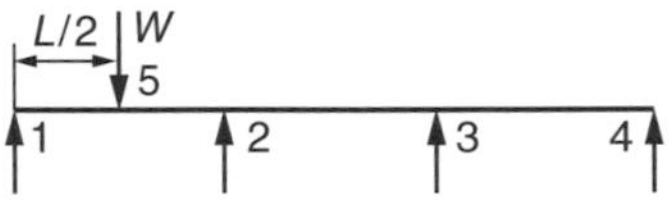 $L/2$ W_5 ; supports 1,2,3,4	$0.200WL$ at ⑤	$-0.100WL$ at ②	$0.4W$	$0.725W$	$-0.15W$ (downward)	$-0.025W$ (downward)
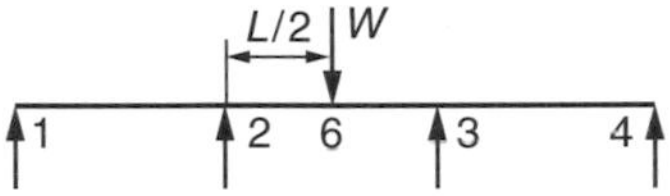 $L/2$ W ; supports 1, 2 6, 3, 4	$0.175WL$ at ⑥	$-0.075WL$ at ② & ③	$-0.075W$	$0.575W$	$0.575W$	$-0.075W$ (downward)
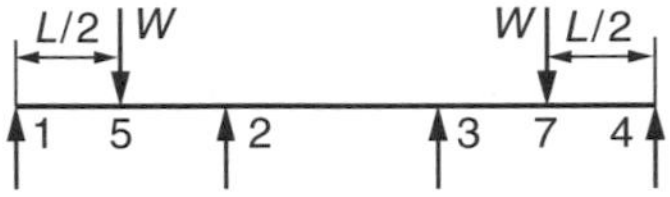 $L/2$ W W $L/2$; supports 1 5, 2, 3 7, 4	$0.213WL$ at ⑤ & ⑦	$-0.075WL$ at ② & ③	$0.425W$	$0.575W$	$0.575W$	$0.425W$
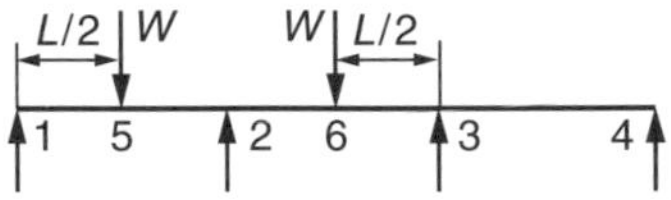 $L/2$ W W $L/2$; supports 1 5, 2 6, 3, 4	$0.163WL$ at ⑤	$-0.175WL$ at ②	$0.325W$	$1.3W$	$0.425W$	$-0.5W$ (downward)

Loading	Max +ve BM	Max −ve BM	Reactions at supports			
			1	2	3	4
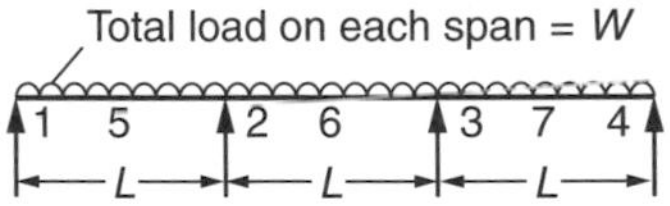 Total load on each span $= W$; supports 1 5, 2 6, 3 7, 4; spans L,L,L	$0.080WL$ at ⑤	$-0.100WL$ at ② & ③	$0.40W$	$1.10W$	$1.10W$	$0.40W$
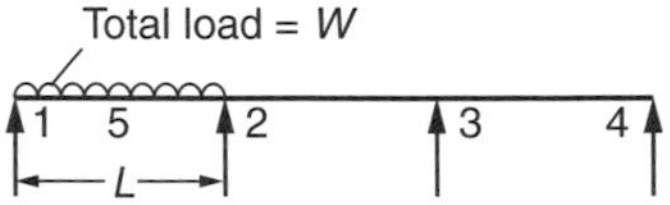 Total load $= W$; supports 1 5, 2, 3, 4; span L	$0.094WL$ at ⑤	$-0.067WL$ at ②	$0.433W$	$0.65W$	$-0.10W$ (downward)	$0.017W$
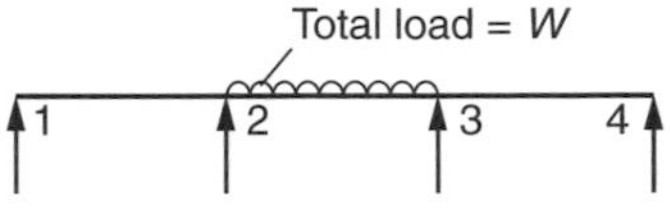 Total load $= W$; supports 1, 2, 3, 4	$0.075WL$ at ⑥	$-0.05WL$ at ② & ③	$-0.05W$	$0.55W$	$0.55W$	$-0.05W$ (downward)
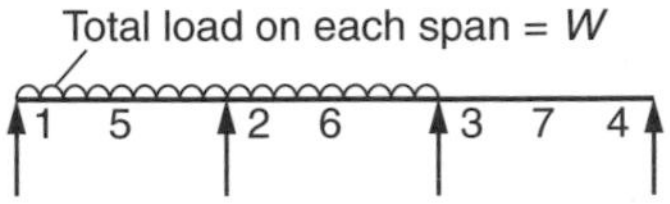 Total load on each span $= W$; supports 1 5, 2 6, 3 7, 4	$0.073WL$ at ⑤	$-0.117WL$ at ②	$0.383W$	$1.2W$	$0.45W$	$-0.033W$ (downward)
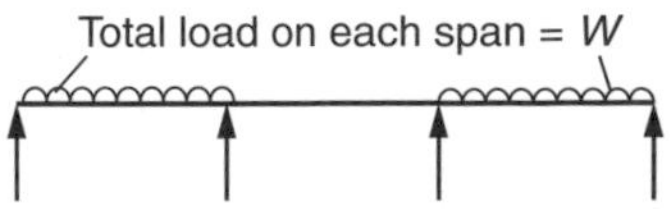 Total load on each span $= W$; supports 1,2,3,4	$0.101WL$	$-0.05WL$	$0.45W$	$0.55W$	$0.55W$	$0.45W$

Figure 3.14 (Continued) Bending moments and reactions of continuous beams.

Four span continuous beams

Loading	Max +ve BM	Max −ve BM	Reactions at support				
			1	2	3	4	5
	0.17WL at ⑥	−0.161WL at ②	0.339W	1.214W	0.893W	1.214W	0.339W
	0.20WL at ⑥	−0.10WL at ②	0.4W	0.728W	−0.161W	0.04W	−0.07W
	0.173WL at ⑦	−0.08WL at ③	−0.074W	0.567W	0.607W	−0.121W	0.02W
	0.21WL at ⑥	−0.08WL at ②	0.42W	0.607W	0.446W	0.607W	−0.08W
	0.16WL at ⑥	−0.181WL at ②	0.319W	0.335W	0.286W	0.647W	0.413W
	0.143WL at ⑦	−0.161WL at ③	−0.054W (downward)	0.446W	1.214W	0.446W	−0.054W (downward)

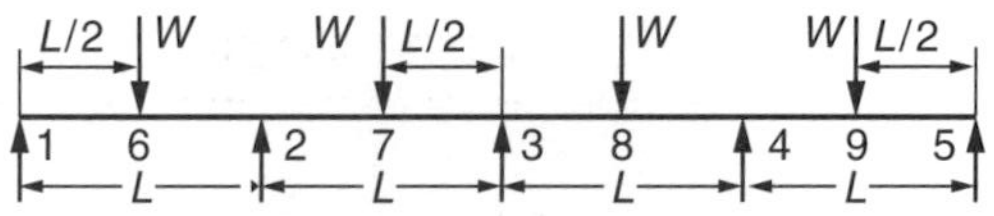

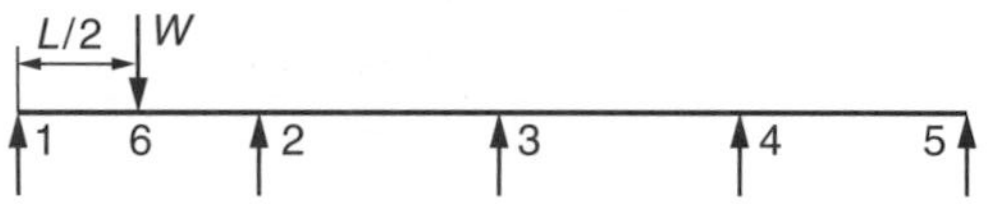

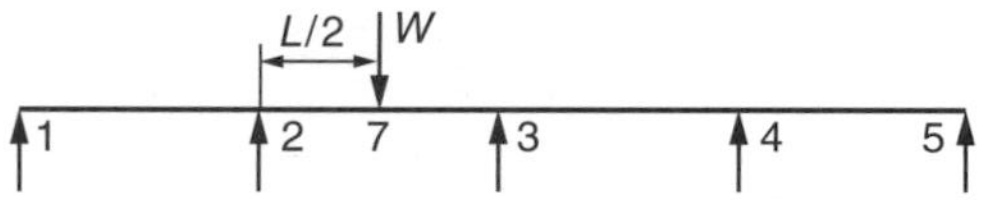

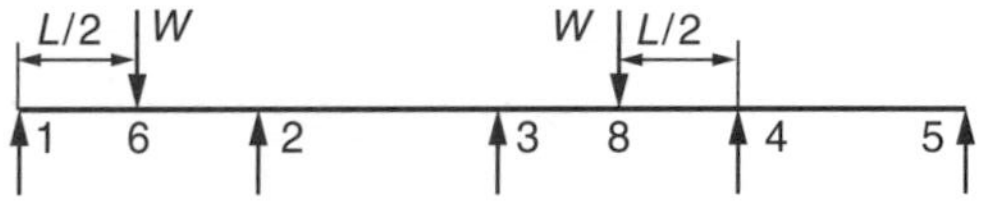

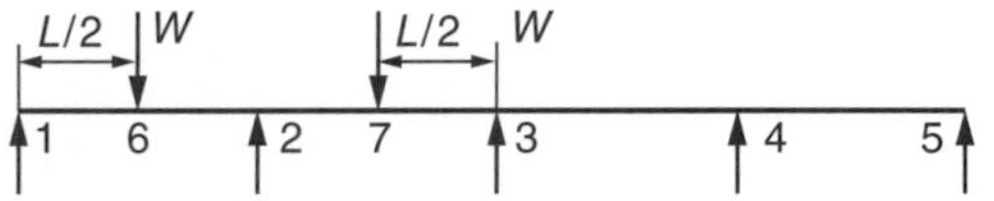

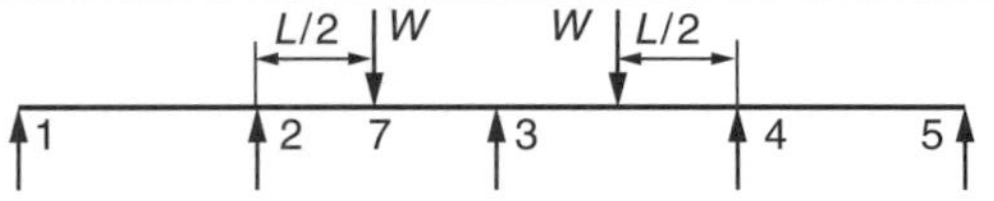

(c)

Figure 3.14 (Continued) Bending moments and reactions of continuous beams.

(d)

Four span continuous beams

Loading	Max +ve BM	Max −ve BM	Reactions at supports				
			1	2	3	4	5
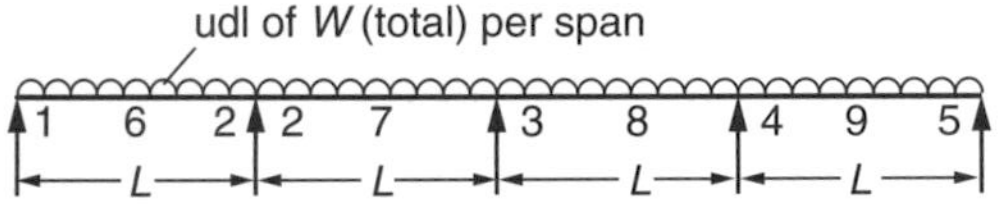udl of W (total) per span 1 6 2 2 7 3 8 4 9 5 L L L L	0.077 *WL*	−0.107 *WL*	0.393 *W*	1.143 *W*	0.929 *W*	1.143 *W*	0.393 *W*
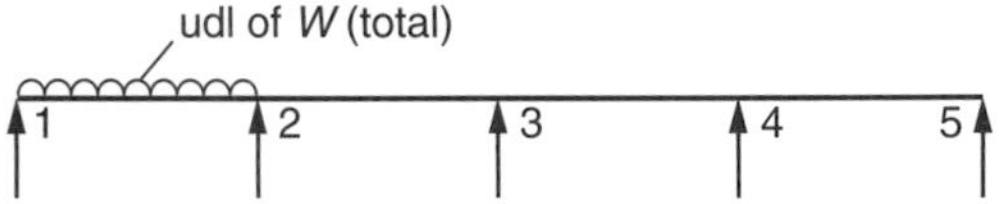udl of W (total) 1 2 3 4 5	0.094 *WL*		0.433 *W*	0.652 *W*	−0.107 *W*	0.027 *W*	−0.05 *W*
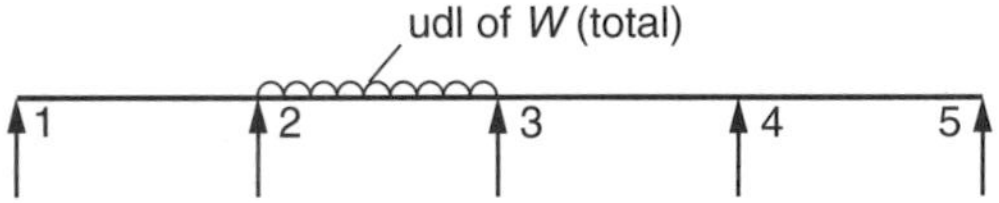udl of W (total) 1 2 3 4 5	0.074 *WL* at (7)	−0.054 *WL* at (3)	−0.049 *W*	0.545 *W*	0.571 *W*	−0.08 *W*	0.013 *W*
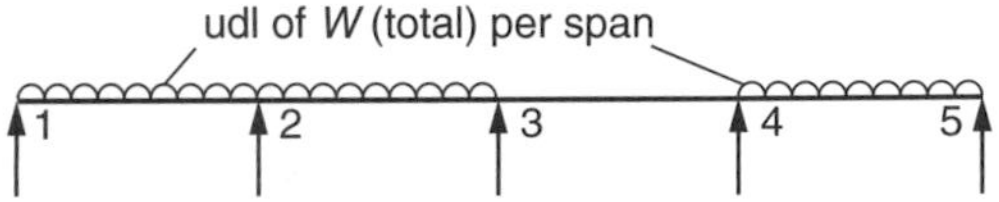udl of W (total) per span 1 2 3 4 5	0.098 *WL* at (9)	−0.058 *WL* at (4)	0.38 *W*	1.223 *W*	0.357 *W*	0.598 *W*	0.442 *W*
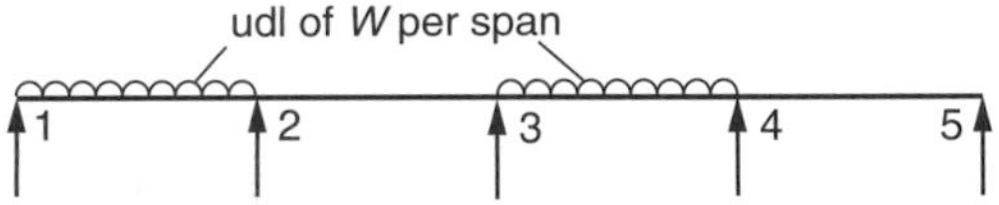udl of W per span 1 2 3 4 5	0.10 *WL* at (6)	−0.054 *WL* at (2) & (4)	0.446 *W*	0.572 *W*	0.464 *W*	0.572 *W*	−0.054 *W*
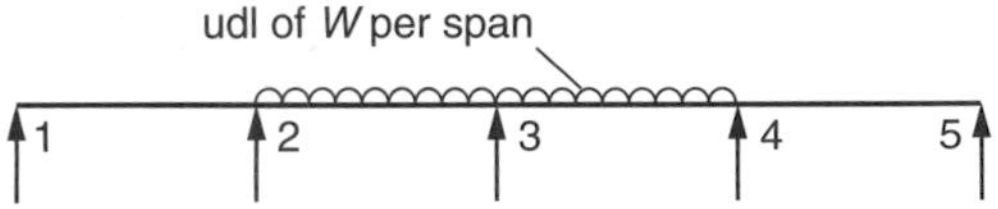udl of W per span 1 2 3 4 5	0.056 *WL* at (7)	−0.107 *WL* at (3)	−0.036 *W*	0.464 *W*	1.143 *W*	0.464 *W*	−0.036 *W*

Figure 3.14 (Continued) Bending moments and reactions of continuous beams.

Slope deflection method

This method is a special case of the stiffness method of analysis, and it is convenient for hand analysis of small structures. Members are assumed to be of constant section between each pair of supports. It is further assumed that the joints in a structure may rotate or deflect, but the angles between the members meeting at a joint remain unchanged. Moments at the ends of frame members are expressed in terms of the rotations and deflections of the joints.

The member force–displacement equations that are needed for the slope deflection method are written for a member AB in a frame. This member, which has its undeformed position along the x axis is deformed into the configuration shown in Figure 3.15. The positive axes, along with the positive member-end force components and displacement components, are shown in the figure (M_{AB} is the end moment at A on AB and M_{BA} is the end moment at B on BA).

The equations for end moments are written as:

$$M_{AB} = \frac{2EI}{l}(2\theta_A + \theta_B - 3\psi_{AB}) + M_{FAB}$$

$$M_{BA} = \frac{2EI}{l}(2\theta_B + \theta_A - 3\psi_{AB}) + M_{FBA}$$

(3.19)

in which M_{FAB} and M_{FBA} are fixed-end moments at supports A and B, respectively due to the applied load (see Figure 3.13). ψ_{AB} is the rotation as a result of the relative displacement between the member ends A and B given as:

$$\psi_{AB} = \frac{\Delta_{AB}}{l} = \frac{y_A + y_B}{l}$$

(3.20)

where Δ_{AB} is the relative deflection of the beam ends. y_A and y_B are the vertical displacements at ends A and B. The slope-deflection equations (3.19) show that the moment at the end of a member is dependent on member properties EI, dimension l, and displacement quantity. The fixed end moments reflect the transverse loading on the member.

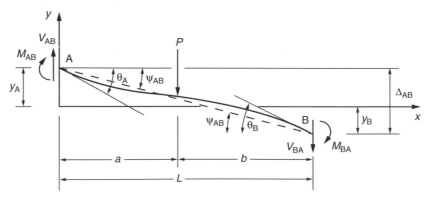

Figure 3.15 Deformed configuration of a beam.

Application of slope deflection method to frames

The slope-deflection equations may be applied to statically indeterminate frames with or without sidesway. A frame may be subjected to sidesway if the loads, member properties and dimensions of the frame are not symmetrical about the centreline. Application of slope deflection method can be illustrated by the following example

Example

Consider the frame shown in Figure 3.16. The support at D is assumed to sink downward by 12 mm and rotate in the clockwise direction by 0.002 radians. Equation (3.19) can be applied to each of the members of the frame and the member end moments may be obtained as follows:

Member AB:

$$M_{AB} = \frac{2EI}{8}\left(2\theta_A + \theta_B - \frac{3\Delta}{8}\right) + M_{FAB}$$

$$M_{BA} = \frac{2EI}{8}\left(2\theta_B + \theta_A - \frac{3\Delta}{8}\right) + M_{FBA}$$

$$\theta_A = 0,\ M_{FAB} = M_{FBA} = 0$$

as there are no externally applied loads.

Hence:

$$M_{AB} = \frac{2EI}{8}(\theta_B - 3\psi)$$

$$M_{BA} = \frac{2EI}{8}(2\theta_B - 3\psi)$$

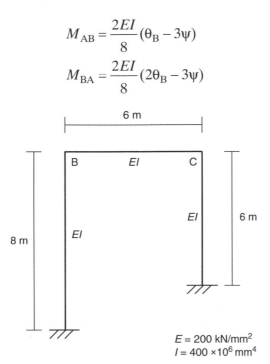

$$E = 200 \text{ kN/mm}^2$$
$$I = 400 \times 10^6 \text{ mm}^4$$

Figure 3.16 Portal frame example.

in which;

$$\psi = \frac{\Delta}{8}$$

Member BC:

$$M_{BC} = \frac{2EI}{6}\left(2\theta_B + \theta_C - 3 \times \frac{\Delta'}{6}\right) + M_{FBC}$$

$$M_{CB} = \frac{2EI}{6}\left(2\theta_C + \theta_B - 3 \times \frac{\Delta'}{6}\right) + M_{FCB}$$

$$\Delta' = 0.012 \text{ m}$$

$$M_{FBC} = M_{FCB} = 0, \text{ as before.}$$

Hence:

$$M_{BC} = \frac{2EI}{6}(2\theta_B + \theta_C - 0.006)$$

$$M_{CB} = \frac{2EI}{6}(2\theta_C + \theta_B - 0.006)$$

Member CD:

$$M_{CD} = \frac{2EI}{6}\left(2\theta_C + \theta_D - \frac{3\Delta}{6}\right) + M_{FCD}$$

$$M_{DC} = \frac{2EI}{6}\left(2\theta_D + \theta_C - \frac{3\Delta}{6}\right) + M_{FDC}$$

$$\theta_D = 0.002 \text{ radians}$$
$$M_{FCD} = M_{FDC} = 0, \text{ as before.}$$

Hence:

$$M_{CD} = \frac{2EI}{6}\left(2\theta_C + 0.002 - \frac{\Delta}{2}\right)$$

$$M_{DC} = \frac{2EI}{6}\left(\theta_C + 0.004 - \frac{\Delta}{2}\right)$$

The joint moment conditions are:

$$M_{BA} + M_{BC} = 0$$
$$M_{CB} + M_{CD} = 0$$

Shear conditions is:

$$\frac{M_{AB} + M_{BA}}{8} + \frac{M_{CD} + M_{DC}}{6} = 0$$

Substituting for moments M_{BA}, M_{BC}, M_{CB}, M_{CD}, M_{BA}, M_{AB}, M_{BA}, M_{CD} and M_{DC} and solving for θ_B, θ_C and Δ:

$$\theta_B = 2.0441 \times 10^{-3} \text{ rad}$$
$$\theta_C = 1.8676 \times 10^{-3} \text{ rad}$$
$$\Delta = 10.587 \times 10^{-3} \text{ m}$$

Substituting the above values of joint rotations and translation back into the slope-deflection equations:

$$M_{AB} = -38.52 \text{ kNm}$$
$$M_{BA} = +2.36 \text{ kNm}$$
$$M_{BC} = -2.36 \text{ kNm}$$
$$M_{CB} = -11.76 \text{ kNm}$$
$$M_{CD} = +11.76 \text{ kNm}$$
$$M_{DC} = +15.36 \text{ kNm}$$

This example illustrates that even without applying a load, the frame can develop substantial moments due to the settlement of supports.

3.2.6 Torsion

When a structural element is loaded such that the loading axis passes through the shear centre of the cross section, no torsion will result. When the loading axis and the centroidal axis of a beam do not coincide, torsion will inevitably result. Unsymmetrically loaded slabs supported on beams, interconnected girders, off-axis loading on beams and loading away from the shear centre are all instances where torsion is a significant factor in design.

When a circular cross section is subjected to torsion (Figure 3.17), plane sections remain plane and the torsional behaviour is described by an equation similar to the bending equation and takes a form similar to Equation (3.3).

$$\frac{T}{J} = \frac{\tau}{r} = \frac{N\theta}{l} \tag{3.21}$$

where T is the applied torque, J = polar moment of inertia, τ = shear stress at a radius of r, N = modulus of rigidity and θ/l is the angle of rotation per unit length of the bar.

In this case, the shear strain is directly proportional to the distance from the neutral axis.

On the other hand, when the cross section is rectangular the originally plane sections warp during twisting. The simple torsion equation (Equation 3.21) no longer

Figure 3.17. Torsional of a circular section.

applies as both axial and circumferential shear stresses are produced. The maximum shear stresses occur on the periphery at the middle of the longer sides of the rectangle. The torsional shear stress is given by:

$$\tau = \gamma\, a\, N\, \theta \qquad\qquad (3.22)$$

where γ is a function of the (longer side b/shorter side, a) of the rectangle, a = length of the shorter side, and θ is the angle of twist in radians per unit length.

The twisting moment resistance is obtained by:

$$T = \beta a^3 \gamma N \theta$$

where β = another function of b/a.

The torsional resistance of a section composed of a number of rectangles (e.g. Tee, L or I sections) is obtained approximately as the sum of torsional resistances of the component rectangles, i.e.

$$T = N\,\theta\,\Sigma\frac{1}{3}a^3 b.$$

(a)

Cross-section	Shear stress	Torsional constant J
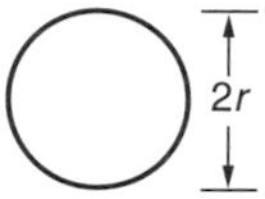 circle, diameter $2r$	$\dfrac{2T}{\pi r^3}$	$\dfrac{1}{2}\pi r^4$
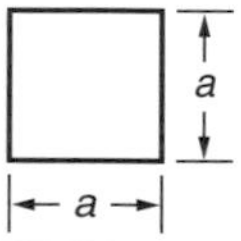 square, side a	$\dfrac{T}{0.208a^2}$ at midpoint each side	$0.141a^4$
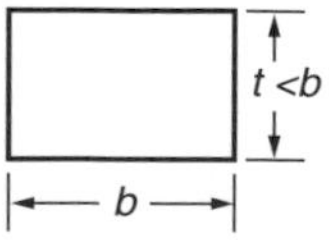 rectangle, $t<b$, width b	$\dfrac{3T}{bt^2}\left(1+0.6\,\dfrac{t}{b}\right)$ at midpoint each long side	$\dfrac{bt^3}{3}\left[1-0.63\,\dfrac{t}{b}+0.052\left(\dfrac{t}{b}\right)^2\right]$
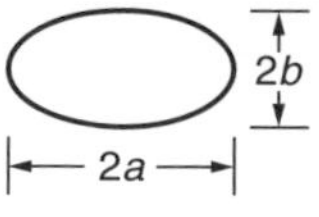 ellipse, $2b$, $2a$	$\dfrac{2T}{\pi ab^2}$ at ends of minor axis	$\dfrac{\pi a^3 b^3}{a^2+b^2}$
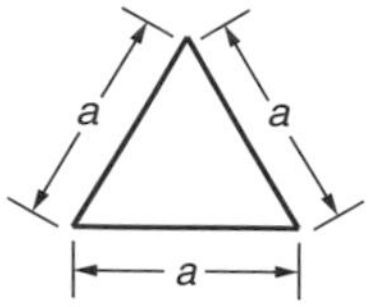 triangle, sides a, base a	$\dfrac{20T}{a^3}$ at midpoint each side	$\dfrac{a^4\sqrt{3}}{80}$

*$T = GJ\theta$, where T = torque, G = shearing modulus of elasticity, J = torsional stiffness,
θ = angle of twist, radians per unit length

Figure 3.18 Torsional properties of solid cross-sections.

The summation will cover all the separate rectangles which, when combined will result in the composite cross section. Torsional properties of solid cross sections and of open cross sections are tabulated in Figure 3.18. Detailed treatment of torsion in non-circular sections may be found in Timoshenko and Goodier (1970).

(b)

Cross-section	Warping constant C	Location e of shear centre S
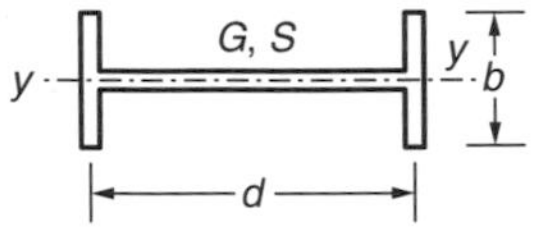	$\dfrac{d^2 I_y}{4}$	
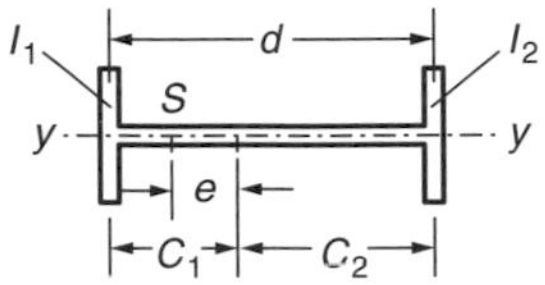	$\dfrac{d^2 I_1 I_2}{I_y}$	$\dfrac{C_1 I_1 - C_2 I_2}{I_y}$
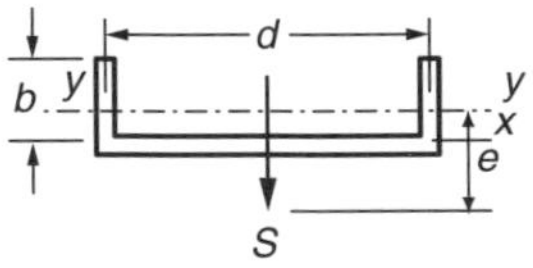	$\dfrac{d^2 I_y}{4}\left[1 - \dfrac{\bar{x}(e - \bar{x})}{r_y^2}\right]$	$\dfrac{\bar{x}}{4}\left(\dfrac{d}{r_x}\right)^2$
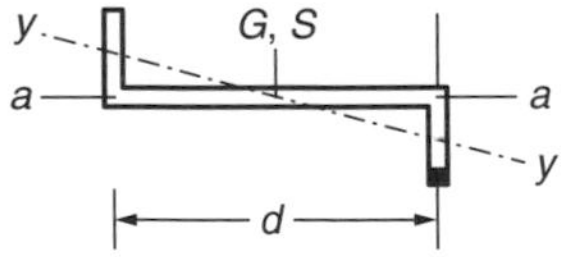	$\dfrac{d^2}{4} I_a$	
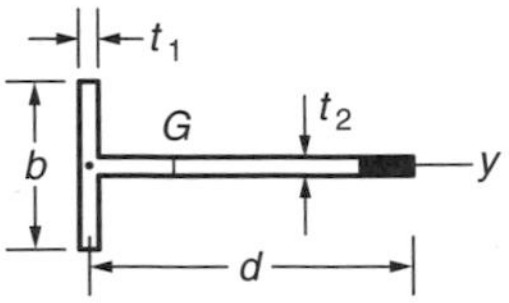	$\dfrac{t_1^3 b^3}{144} + \dfrac{t_2^3 d^3}{36}$	
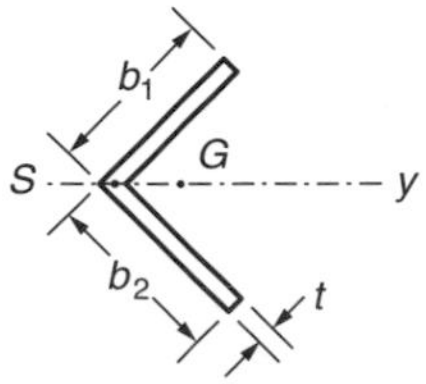	$(b_1^3 + b_2^3)\dfrac{t^3}{36}$	

Note: The torsional stiffness J for cross sections in this table can be determined closely enough for most applications by $J = \Sigma b t^3/3$. The warping constant C is usually negligible for the angle and the T.

Figure 3.18 (Continued) Torsional properties of solid cross-sections.

In general a structural component subjected to uniform pure torsion (or St Venant's torsion) develops only shear stresses. This implies that the section has no warping restraint. When there is restraint to longitudinal warping (for example by providing a fixed end), there will be both shear and longitudinal stresses. Generally non-uniform torsion can be computed as a combination of uniform torsion and warping restraint stresses.

More detailed treatment of torsion is provided later in this chapter in the section on under Box girders.

3.3 Trusses

A structure that is composed of a number of bars pin connected at their ends to form a stable framework is called a truss. If all the bars lie in a plane, the structure is a planar truss. It is generally assumed that loads and reactions are applied to the truss only at the joints. The centroidal axis of each member is straight, coincides with the line connecting the joint centres at each end of the member, and lies in a plane that also contains the lines of action of all the loads and reactions. Many truss structures are three-dimensional and a complete analysis would require consideration of the full spatial interconnection of the members. However, in many bridge structures the three-dimensional framework can be subdivided into planar components for analysis as planar trusses without seriously compromising the accuracy of the results. Figure 3.19 shows some typical idealized planar truss structures.

There exists a relation between the number of members, m, number of joints, j and reaction components, r. The expression is:

$$m = 2j - r \tag{3.23}$$

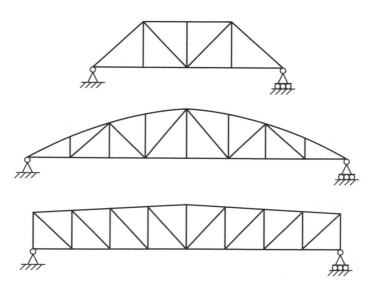

Figure 3.19 Typical planar trusses.

which must be satisfied if it is to be statically determinate internally. The least number of reaction components required for external stability is r. If m exceeds $(2j - r)$, then the excess members are called redundant members and the truss is said to be statically indeterminate.

Truss analysis gives the bar forces in a truss; for a statically determinate truss, these bar forces can be found by employing the laws of statics to assure internal equilibrium of the structure. The process requires repeated use of free-body diagrams from which individual bar forces are determined. The method of joints is a technique of truss analysis in which the bar forces are determined by the sequential isolation of joints – the unknown bar forces at one joint are solved and become known bar forces at subsequent joints. The other method is known as method of sections in which equilibrium of a part of the truss is considered.

3.3.1 Method of joints

An imaginary section may be completely passed around a joint in a truss. The joint has become a free body in equilibrium under the forces applied to it. The equations $\Sigma H = 0$ and $\Sigma V = 0$ may be applied to the joint to determine the unknown forces in members meeting there. It is evident that no more than two unknowns can be determined at a joint with these two equations.

3.3.2 Method of sections

If only a few member forces of a truss are needed, the quickest way to find these forces is by the method of sections. In this method, an imaginary cut (section) is drawn through a stable and determinate truss. Thus, a section subdivides the truss into two separate parts. Since the entire truss is in equilibrium, any part of it must also be in equilibrium. Either of the two parts of the truss can be considered and the three equations of equilibrium $\Sigma F_x = 0$, $\Sigma F_y = 0$ and $\Sigma M = 0$ can be applied to solve for member forces.

The example of a truss sketched in Figure 3.20(a) is considered. To calculate the force in the member 3–5, F_{35}, a section AA should be run to cut the member 3–5 as shown in the figure. It is only required to consider the equilibrium of one of the two parts of the truss. In this case, the portion of the truss on the left of the section is

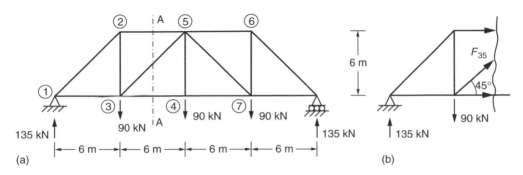

Figure 3.20 Example – method of sections, planar truss.

considered. The left portion of the truss as shown in Figure 3.20(b) is in equilibrium under the action of the external and internal forces. Considering the equilibrium of forces in the vertical direction one obtains:

$$135 - 90 + F_{35} \sin 45° = 0$$

Therefore, F_{35} is obtained as:

$$F_{35} = -45\sqrt{2} \text{ kN}$$

The negative sign indicates that the member force is compressive. The other member forces cut by the section can be obtained using the other equilibrium equations, namely $\Sigma M = 0$. More sections can be taken in the same way so as to solve for other member forces in the truss. The most important advantage of this method is that one can obtain the required member force without solving for the other member forces.

3.3.3 Compound trusses

A compound truss is formed by interconnecting two or more simple trusses. Examples of compound trusses are shown in Figure 3.21. A typical compound roof truss is shown in Figure 3.21(a) in which two simple trusses are interconnected by means of a single member and a common joint. The compound truss shown in Figure 3.21(b) is commonly used in bridge construction and in this case, three members are used

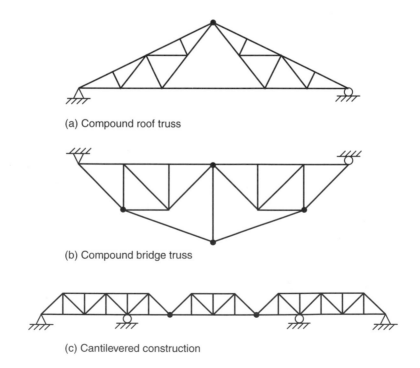

(a) Compound roof truss

(b) Compound bridge truss

(c) Cantilevered construction

Figure 3.21 Compound trusses.

to interconnect two simple trusses at a common joint. There are three simple trusses interconnected at their common joints as shown in Figure 3.21(c).

The method of sections may be used to determine the member forces in the interconnecting members of compound trusses similar to those shown in Figure 3.21(a) and (b). However, in the case of cantilevered truss the middle simple truss is isolated as a free body diagram to find its reactions. These reactions are reversed and applied to the interconnecting joints of the other two simple trusses. After the interconnecting forces between the simple trusses are found, the simple trusses are analysed by the method of joints or the method of sections.

3.3.4 Stability and determinacy

A stable and statically determinate plane truss should have at least three members, three joints and three reaction components. To form a stable and determinate plane truss of n joints, the three members of the original triangle plus two additional members for each of the remaining $(n - 3)$ joints are required. Thus, the minimum total number of members, m, required to form an internally stable plane truss is $m = (2n - 3)$. If a stable, simple, plane truss of n joints and $(2n - 3)$ members is supported by three independent reaction components, the structure is stable and determinate when subjected to a general loading. If the stable, simple, plane truss has more than three reaction components, the structure is externally indeterminate. That means not all of the reaction components can be determined from the three available equations of statics. If the stable, simple, plane truss has more than $(2n - 3)$ members, the structure is internally indeterminate and hence all of the member forces cannot be determined from the $2n$ available equations of statics in the method of joints. The analyst must examine the arrangement of the truss members and the reaction components to determine if the simple plane truss is stable. Simple plane trusses having $(2n - 3)$ members are not necessarily stable.

3.4 Influence lines

Bridge decks are required to support both static and moving loads. Each element of a bridge must be designed for the most severe conditions that can possibly be developed in that member. Live loads should be placed at the positions where they will produce severe conditions. The critical positions for placing live loads will not be the same for every member. A useful method of determining the most severe condition of loading is by using 'influence lines'.

An influence line for a particular response such as reaction, shear force, bending moment and axial force is defined as a diagram in which the ordinate at any point equals the value of that response attributable to a unit load acting at that point on the structure. Influence lines provide a systematic procedure for determining how the force (or a moment or shear) in a given part of a structure varies as the applied load moves about on the structure. Influence lines of responses of statically determinate structures consist only of straight lines whereas this is not true of indeterminate structures. They are primarily used to determine where to place live loads to cause maximum effect.

3.4.1 Influence lines for shear in simple beams

Figure 3.22 shows influence lines for shear at two sections of a simply supported beam. It is assumed that positive shear occurs when the sum of the transverse forces to the left of a section is in the upward direction or when the sum of the forces to the right of the section is downward. A unit force is placed at various locations and the shear force at sections 1–1 and 2–2 are obtained for each position of the unit load. These values give the ordinate of influence line with which the influence line diagrams for shear force at sections 1–1 and 2–2 can be constructed. Note that the slope of the influence line for shear on the left of the section is equal to the slope of the influence line on the right of the section. This information is useful in drawing shear force influence line in other cases.

3.4.2 Influence lines for bending moment in simple beams

Influence lines for bending moment at the same sections, 1–1 and 2–2 of the simple beam considered in Figure 3.22, are plotted as shown in Figure 3.23. For a section, when the sum of the moments of all the forces to the left is clockwise or when the sum to the right is counter-clockwise, the moment is taken as positive. The values of bending moment at sections 1–1 and 2–2 are obtained for various positions of unit load and plotted as shown in the figure.

It may be noted that a shear or bending moment diagram shows the variation of shear or moment across an entire structure for loads fixed in one position. On the

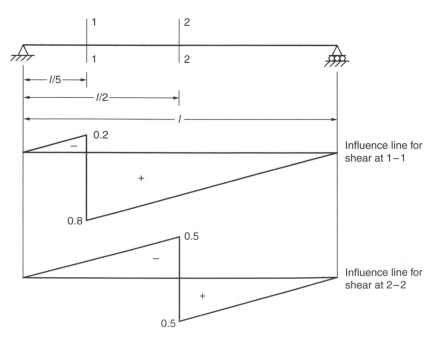

Figure 3.22 Influence line for shear force.

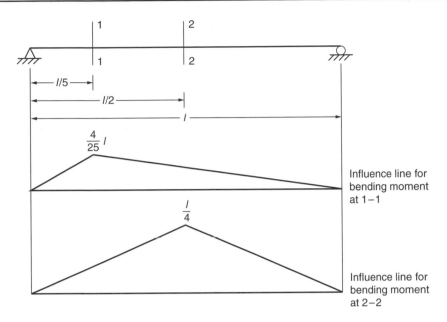

Figure 3.23 Influence line for bending moment.

other hand an influence line for shear or moment shows the variation of that response at one particular section in the structure caused by the movement of a unit load from one end of the structure to the other.

Influence lines can be used to obtain the value of a particular response for which it is drawn when the beam is subjected to any particular type of loading. If, for example, a uniform load of intensity q_o per unit length acts over the entire length of the simple beam shown in Figure 3.22, then the shear force at section 1–1 is given by the product of the load intensity, q_o and the net area under the influence line diagram. The net area is equal to $0.3l$ and the shear force at section 1–1 is, therefore, equal to $0.3q_ol$. In the same way, the bending moment at the section can be found as the area of the corresponding influence line diagram times the intensity of loading, q_o. The bending moment at the section is, therefore, $(0.08l^2 \times q_o =) \, 0.08q_ol^2$.

3.4.3 Influence lines for trusses

Influence lines for support reactions and member forces may be constructed in the same manner as those for various beam functions. They are useful to determine the maximum load that can be applied to the truss. As the unit load moves across the truss, the ordinates for the responses under consideration may be computed for the load at each panel point. Member force, in most cases, need not be calculated for every panel point, because certain portions of influence lines can readily be seen to consist of straight lines for several panels. One method used for calculating the forces in a chord member of a truss is by the method of sections discussed earlier.

The truss shown in Figure 3.24 is considered for illustrating the construction of influence lines for trusses.

The member forces in U_1U_2, L_1L_2 and U_1L_2 are determined by passing a section 1–1 and considering the equilibrium of the free body diagram of one of the truss segments. Unit load is placed at L_1 first and the force in U_1U_2 is obtained by taking moment about L_2 of all the forces acting on the right hand segment of the truss and dividing the resulting moment by the lever arm (the perpendicular distance of the force in U_1U_2 from L_2). The value thus obtained gives the ordinate of the influence diagram at L_1 in the truss. The ordinate at L_2 obtained similarly represents the force in U_1U_2 for unit load placed at L_2. The influence line can be completed with two other points, one at each of the supports. The force in the member L_1L_2 due to unit load placed at L_1 and L_2 can be obtained in the same manner and the corresponding influence line diagram can be completed. By considering the horizontal component of force in the diagonal of the panel the influence line for force in U_1L_2 can be constructed.

Figure 3.24 shows the respective influence diagram for member forces in U_1U_2, L_1L_2 and U_1L_2. Influence line ordinates for the force in a chord member of a 'curved-chord'

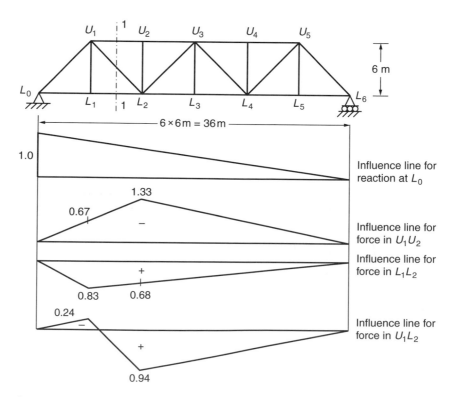

Figure 3.24 Influence line for member force in a truss.

truss may be determined by passing a vertical section through the panel and taking moments at the intersection of the diagonal and the other chord.

3.4.4 Qualitative influence lines: Muller–Breslau principle

One of the most effective methods of obtaining influence lines is by the use of Müller–Breslau's principle, which states that 'the ordinates of the influence line for any response in a structure are equal to those of the deflection curve obtained by releasing the restraint corresponding to this response and introducing a corresponding unit displacement in the remaining structure'. In this way, the shape of the influence lines for both statically determinate and indeterminate structures can be easily obtained, especially for beams.

To draw the influence lines of:

- support reaction: remove the support and introduce a unit displacement in the direction of the corresponding reaction to the remaining structure as shown in Figure 3.25 for a symmetrical overhang beam
- shear: make a cut at the section and introduce a unit relative translation (in the direction of positive shear) without relative rotation of the two ends at the section as shown in Figure 3.26
- bending moment: introduce a hinge at the section (releasing the bending moment) and apply bending (in the direction corresponding to positive moment) to produce a unit relative rotation of the two beam ends at the hinged section as shown in Figure 3.27.

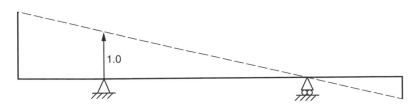

Figure 3.25 Influence line for support reaction.

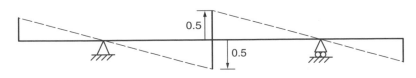

Figure 3.26 Influence line for midspan shear force.

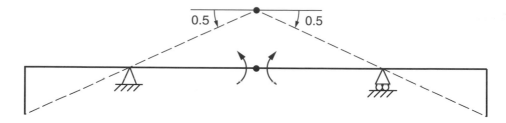

Figure 3.27 Influence line for midspan bending moment.

3.4.5 Influence lines for continuous beams

Using Müller–Breslau's principle, the shape of the influence line of any response of a continuous beam can be sketched easily. One of the methods for beam deflection can then be used for determining the ordinates of the influence line at critical points. Figures 3.28–3.30 show the influence lines for bending moment at various points of two, three and four span continuous beams, respectively.

(a)

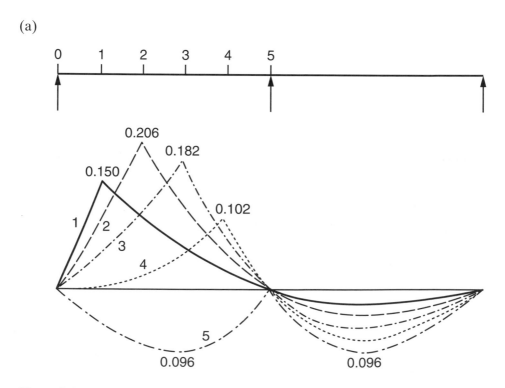

Figure 3.28 Influence lines for bending moment – two span continuous beams.

(b)

Influence lines for bending moment (two span continuous beams)

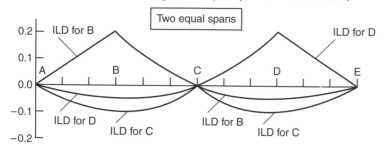

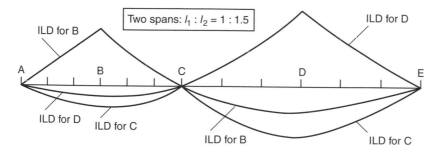

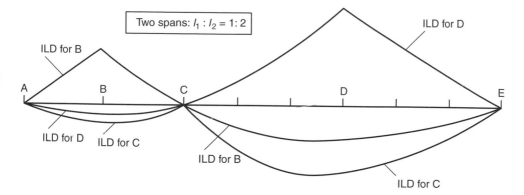

To obtain the max b.m. at any of the critical section due to a train of loads
(1) Plot on the diagram, the train of loads placed in the most adverse position
(2) Bending Moment due to a given load = Influence line ordinate × Load
(3) Algebraically add all the values of the b.m. calculated as above

ILD for	Max ordinate in	$l_1 : l_2 = 1 : 1$		$l_1 : l_2 = 1 : 1.5$		$l_1 : l_2 = 1 : 2$	
		Span AC	Span CE	Span AC	Span CE	Span AC	Span CE
B		0.203	−0.047	0.213	−0.084	0.219	0.125
C		−0.094	−0.094	−0.075	−0.169	−0.063	−0.250
D		−0.047	0.203	−0.038	0.291	−0.031	−0.375

Figure 3.28 (Continued) Influence lines for bending moment – two span continuous beams.

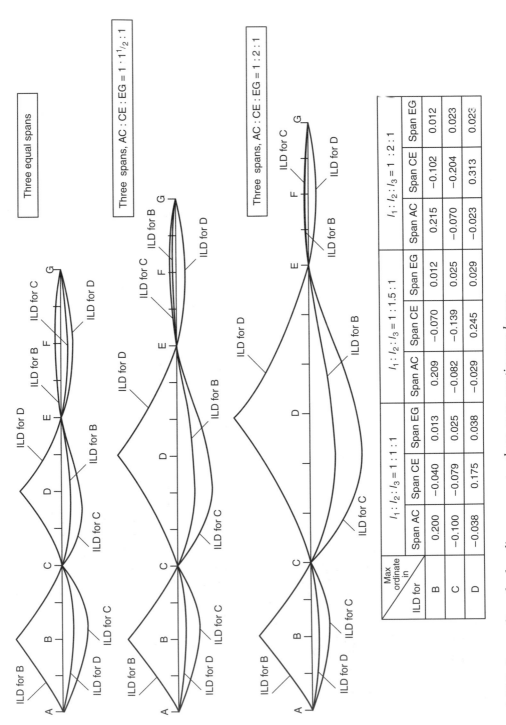

Max ordinate in	$l_1 : l_2 : l_3 = 1 : 1 : 1$			$l_1 : l_2 : l_3 = 1 : 1.5 : 1$			$l_1 : l_2 : l_3 = 1 : 2 : 1$		
ILD for	Span AC	Span CE	Span EG	Span AC	Span CE	Span EG	Span AC	Span CE	Span EG
B	0.200	−0.040	0.013	0.209	−0.070	0.012	0.215	−0.102	0.012
C	−0.100	−0.079	0.025	−0.082	−0.139	0.025	−0.070	−0.204	0.023
D	−0.038	0.175	0.038	−0.029	0.245	0.029	−0.023	0.313	0.023

Figure 3.29 Influence lines for bending moment – three span continuous beams.

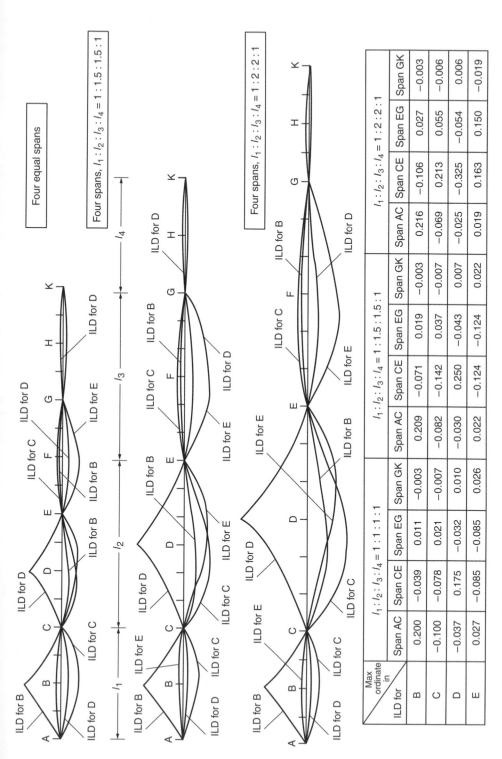

Figure 3.30 *Influence lines for bending moment – four span continuous beams.*

| Max ordinate in | $l_1:l_2:l_3:l_4 = 1:1:1:1$ | | | | | | | $l_1:l_2:l_3:l_4 = 1:1.5:1.5:1$ | | | | | | | $l_1:l_2:l_3:l_4 = 1:2:2:1$ | | | | | | |
ILD for	Span AC	Span CE	Span EG	Span GK	Span AC	Span CE	Span EG	Span GK	Span AC	Span CE	Span EG	Span GK	Span AC	Span CE	Span EG	Span GK
B	0.200	−0.039	0.011	−0.003	0.209	−0.071	0.019	−0.003	0.216	−0.106	0.027	−0.003
C	−0.100	−0.078	0.021	−0.007	−0.082	−0.142	0.037	−0.007	−0.069	0.213	0.055	−0.006
D	−0.037	0.175	−0.032	0.010	−0.030	0.250	−0.043	0.007	−0.025	−0.325	−0.054	0.006
E	0.027	−0.085	−0.085	0.026	0.022	−0.124	−0.124	0.022	0.019	0.163	0.150	−0.019

3.5 Plates and plated bridge structures

3.5.1 Bending of thin plates

Steel and steel – concrete composite plates are extensively employed in steel bridges. A 'plate' is defined as a structural component having two of its dimensions significantly larger than the third (namely its thickness). The structural analysis of a plate is carried out by considering the state of stress at the middle plane of a plate. All the stress component are expressed in terms of the deflection w of the plate (where w is a function of the two coordinates (x, y) in the plane of the plate). This deflection function has to satisfy a linear partial differential equation which together with its boundary condition completely defines w.

Figure 3.31 shows a plate element cut from a plate whose middle plane coincides with xy plane. The middle plane of the plate is subjected to a lateral load of intensity q. It can be shown, by considering the equilibrium of the plate element, that the stress resultants are given as:

$$M_x = -D\left(\frac{\partial^2 w}{\partial x^2} + v\frac{\partial^2 w}{\partial y^2}\right)$$

$$M_y = -D\left(\frac{\partial^2 w}{\partial y^2} + v\frac{\partial^2 w}{\partial x^2}\right)$$

$$M_{xy} = -M_{yx} = D(1-v)\frac{\partial^2 w}{\partial x \partial y}$$

$$V_x = \frac{\partial^3 w}{\partial x^3} + (2-v)\frac{\partial^3 w}{\partial x \partial y^2}$$

$$V_y = \frac{\partial^3 w}{\partial y^3} + (2-v)\frac{\partial^3 w}{\partial y \partial x^2}$$

$$Q_x = -D\frac{\partial}{\partial x}\left(\frac{\partial^2 w}{\partial x^2} + \frac{\partial^2 w}{\partial y^2}\right)$$

$$Q_y = -D\frac{\partial}{\partial y}\left(\frac{\partial^2 w}{\partial x^2} + \frac{\partial^2 w}{\partial y^2}\right)$$

$$R = 2D(1-v)\frac{\partial^2 w}{\partial x \partial y}$$

(3.24)

where M_x and M_y are the bending moments per unit length in the x and y directions, respectively, M_{xy} and M_{yx} are the twisting moments per unit length, Q_x and Q_y are the shearing forces per unit length in the x and y directions, respectively, V_x and V_y are the supplementary shear forces in the x and y directions, respectively, R is the corner force, $D = Eh^3/12(1 - v^2)$ which is flexural rigidity of the plate per unit length, E is the modulus of elasticity, h is the thickness of plate and v is Poisson's ratio.

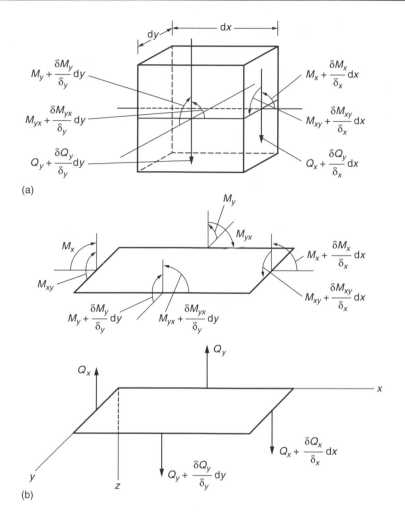

Figure 3.31 (a) Plate element, (b) stress resultants.

The governing equation for the plate is obtained as:

$$\frac{\partial^4 w}{\partial x^4} + 2\frac{\partial^4 w}{\partial x^2 \partial y^2} + \frac{\partial^4 w}{\partial y^4} = \frac{q}{D} \tag{3.25}$$

Any plate problem should satisfy the governing Equation (3.25) and boundary conditions of the plate.

3.5.2 Boundary conditions

There are three basic boundary conditions for plate problems. These are the clamped edge, simply supported edge and the free edge.

Clamped edge

For this boundary condition, the edge is restrained such that the deflection and slope are zero along the edge. If we consider the edge $x = a$ to be clamped, we have:

$$(w)_{x=a} = 0 \qquad \left(\frac{\partial w}{\partial x}\right)_{x=a} = 0 \qquad (3.26a)$$

Simply supported edge
If the edge $x = a$ of the plate is simply supported, the deflection w along this edge must be zero. At the same time this edge can rotate freely with respect to the edge line. This means that:

$$(w)_{x=a} = 0; \qquad \left(\frac{\partial^2 w}{\partial x^2}\right)_{x=a} = 0 \qquad (3.26b)$$

Free edge
If the edge $x = a$ of the plate is entirely free, there are no bending and twisting moments or vertical shearing forces. This can be written in terms of w, the deflection, as:

$$\left(\frac{\partial^2 w}{\partial x^2} + v\frac{\partial^2 w}{\partial y^2}\right)_{x=a} = 0 \qquad (3.26c)$$

$$\left(\frac{\partial^3 w}{\partial x^3} + (2-v)\frac{\partial^3 w}{\partial x\partial y^2}\right)_{x=a} = 0$$

3.5.3 Bending of simply supported rectangular plates
A number of the plate bending problems may be solved directly by solving the differential Equation (3.25). The solution, however, depends upon the loading and boundary condition. Consider a simply supported plate subjected to sinusoidal loading (Figure 3.32) given by:

$$q(x,y) = q_0 \sin\frac{\pi x}{a}\sin\frac{\pi y}{b} \qquad (3.27)$$

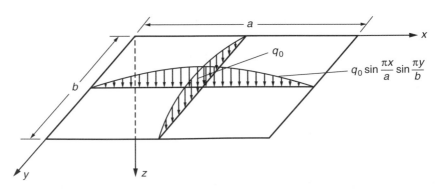

Figure 3.32 Rectangular plate under sinusoidal loading.

The governing differential equation which satisfies the boundary conditions is:

$$\frac{\partial^4 w}{\partial x^4} + 2\frac{\partial^4 w}{\partial x^2 \partial y^2} + \frac{\partial^4 w}{\partial y^4} = \frac{q_0}{D}\sin\frac{\pi x}{a}\sin\frac{\pi y}{b} \tag{3.28}$$

The boundary conditions for the simply supported edges are:

$$w = 0, \partial^2 w/\partial x^2 = 0 \text{ for } x = 0 \text{ and } x = a$$
$$w = 0, \partial^2 w/\partial y^2 = 0 \text{ for } y = 0 \text{ and } y = b \tag{3.29}$$

If the simply supported rectangular plate is subjected to any kind of loading given by:

$$q = q(x, y) \tag{3.30}$$

the function $q(x, y)$ is represented in the form of a double trigonometric series in order to satisfy the boundary conditions:

$$q(x, y) = \sum_{m=1}^{\infty}\sum_{n=1}^{\infty} q_{mn}\sin\frac{m\pi x}{a}\sin\frac{n\pi y}{b} \tag{3.31}$$

in which q_{mn} is given by:

$$q_{mn} = \frac{4}{ab}\int_0^a\int_0^b q(x, y)\sin\frac{m\pi x}{a}\sin\frac{n\pi y}{b}dxdy \tag{3.32}$$

This results in the expression for deflection as follows:

$$w = \frac{1}{\pi^4 D}\sum_{m=1}^{\infty}\sum_{n=1}^{\infty}\frac{q_{mn}}{\left(\dfrac{m^2}{a^2} + \dfrac{n^2}{b^2}\right)^2}\sin\frac{m\pi x}{a}\sin\frac{n\pi y}{b} \tag{3.33}$$

If the applied load is uniformly distributed of intensity q_0, we have:

$$q(x, y) = q_0$$

we obtain:

$$q_{mn} = \frac{4q_0}{ab}\int_0^a\int_0^b\sin\frac{m\pi x}{a}\sin\frac{n\pi y}{b}dxdy = \frac{16q_0}{\pi^2 mn} \tag{3.34}$$

in which m and n are odd integers. $q_{mn} = 0$ if m or n or both of them are even numbers. We can, therefore, write the expression for deflection of a simply supported plate subjected to uniformly distributed load as:

$$w = \frac{16q_0}{\pi^6 D} \sum_{m=1}^{\infty} \sum_{n=1}^{\infty} \frac{\sin\dfrac{m\pi x}{a} \sin\dfrac{n\pi y}{b}}{mn\left(\dfrac{m^2}{a^2} + \dfrac{n^2}{b^2}\right)^2} \tag{3.35}$$

where $m = 1, 3, 5, \ldots$ and $n = 1, 3, 5, \ldots$

The maximum deflection occurs at the centre and it can be written by substituting $x = a/b$ and $y = b/2$ in Equation (3.35) as:

$$w_{max} = \frac{16q_0}{\pi^6 D} \sum_{m=1}^{\infty} \sum_{n=1}^{\infty} \frac{(-1)^{\frac{m+n}{2}-1}}{mn\left(\dfrac{m^2}{a^2} + \dfrac{n^2}{b^2}\right)^2} \tag{3.36}$$

This equation is a rapid converging series and a satisfactory approximation can be obtained by taking only the first term of the series; for example, in the case of a square plate:

$$w_{max} = \frac{4q_0 a^4}{\pi^6 D} = 0.00416 \frac{q_0 a^4}{D}$$

Assuming $v = 0.3$ we get for the maximum deflection as:

$$w_{max} = 0.0454 \frac{q_0 a^4}{Eh^3} \tag{3.37}$$

The expressions for bending and twisting moments can be obtained by substituting Equation (3.35) into Equation (3.24). Figure 3.33 shows some loading cases and the corresponding loading functions.

Solution of plates with arbitrary boundary conditions are complicated. It is possible to make some simplifying assumptions for plates with same boundary conditions along two parallel edges in order to obtain the desired solution. Alternately, energy method can be applied efficiently to solve plates with complex boundary conditions. However, it should be noted that the accuracy of results depends upon the deflection function chosen. These functions must be so chosen that they satisfy at least the kinematic boundary conditions.

Figure 3.34 gives formulas for deflection and bending moments of rectangular plates with typical boundary and loading conditions.

3.5.4 Strain energy of simple plates

The strain energy expression for a simple rectangular plate is given by:

$$U = \frac{D}{2} \iint_{area} \left\{ \left(\frac{\partial^2 w}{\partial x^2} + \frac{\partial^2 w}{\partial y^2}\right)^2 - 2(1-v)\left[\frac{\partial^2 w}{\partial x^2}\frac{\partial^2 w}{\partial y^2} - \left(\frac{\partial^2 w}{\partial x \partial y}\right)^2\right] \right\} dxdy \tag{3.38}$$

(a)

No.	Load $q(x, y) = \Sigma m \Sigma n\, q_{mn} \sin \dfrac{m \pi x}{a} \sin \dfrac{n \pi y}{b}$	Expansion coefficients q_{mn}
1	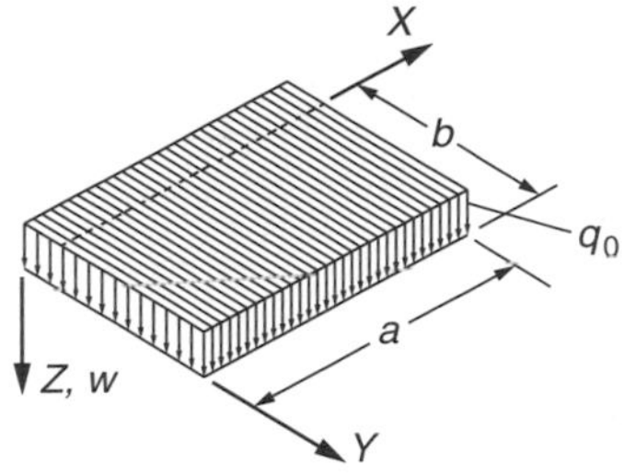	$q_{mn} = \dfrac{16 q_0}{\pi^2 mn}$ $(m, n = 1, 3, 5, \ldots)$
2	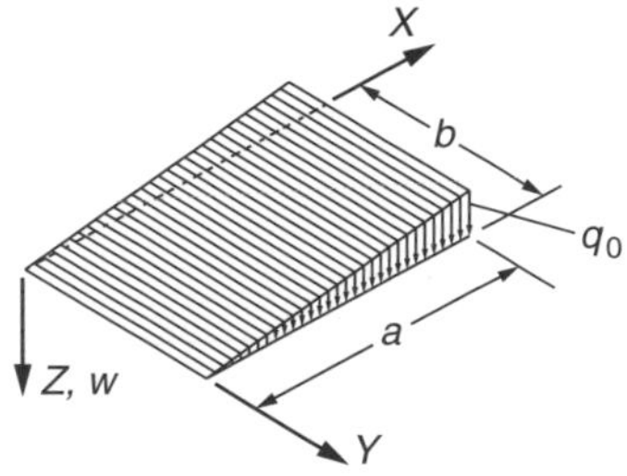	$q_{mn} = \dfrac{-8 q_0 \cos m \pi}{\pi^2 mn}$ $(m, n = 1, 3, 5, \ldots)$
3	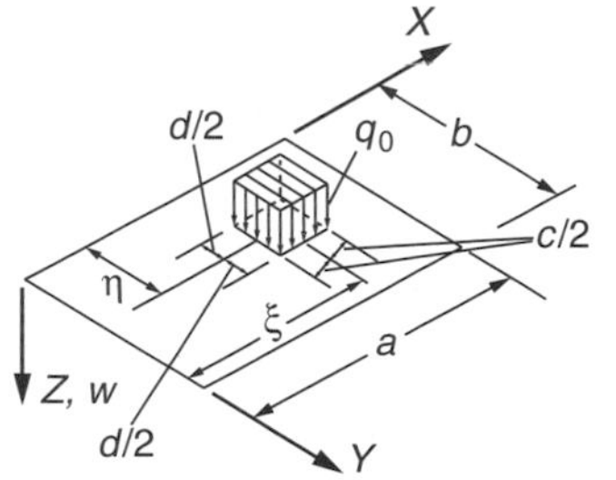	$p_{mn} = \dfrac{16 q_0}{\pi^2 mn} \sin \dfrac{m \pi \xi}{a} \sin \dfrac{n \pi \eta}{b}$ $\times \sin \dfrac{m \pi c}{2a} \sin \dfrac{n \pi d}{2b}$ $(m, n = 1, 3, 5, \ldots)$
4	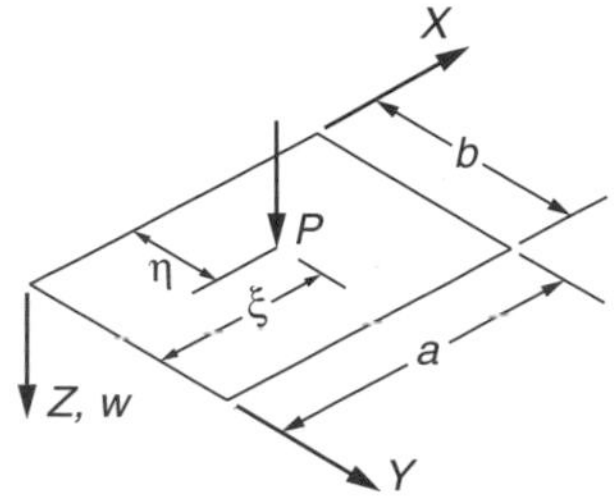	$q_{mn} = \dfrac{4 q_0}{ab} \sin \dfrac{m \pi \xi}{a} \sin \dfrac{n \pi \eta}{b}$ $(m, n = 1, 3, 5, \ldots)$

Figure 3.33 Typical loading on plates and loading functions for rectangular plates (Szilard, 1974). (Continued overleaf)

(b)

No	Load $q(x, y) - \Sigma m \Sigma n \, q_{mn} \sin \dfrac{m \pi x}{a} \sin \dfrac{n \pi y}{b}$	Expansion coefficients q_{mn}
5	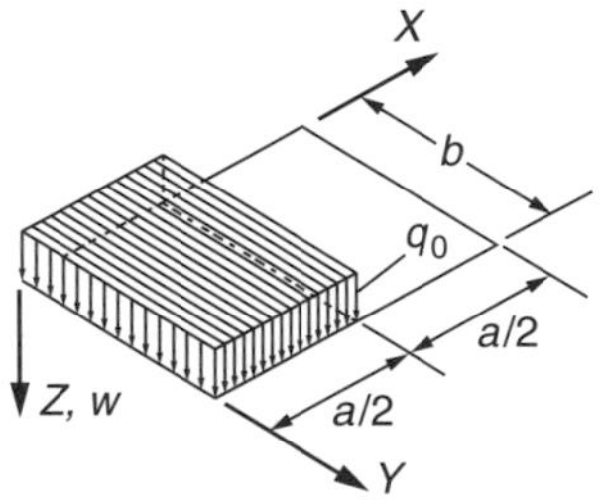	$q_{mn} = \dfrac{8q_0}{\pi^2 mn}$ for $m, n = 1, 3, 5, \dots$ $q_{mn} = \dfrac{16q_0}{\pi^2 mn}$ for $\begin{cases} m = 2, 6, 10, \dots \\ n = 1, 3, 5, \dots \end{cases}$
6	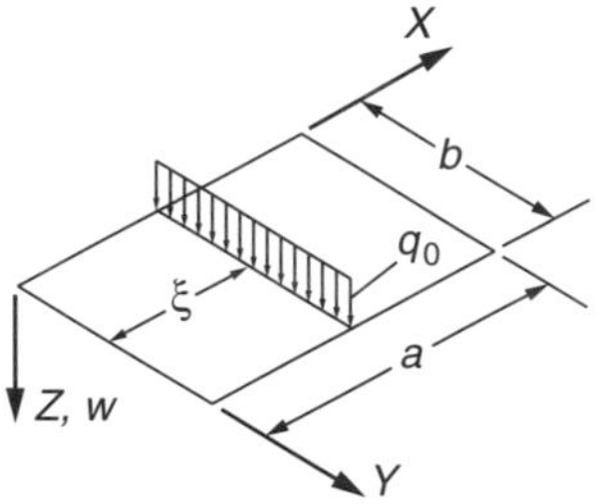	$q_{mn} = \dfrac{4q_0}{\pi an} \sin \dfrac{m \pi \xi}{a}$ $(m, n = 1, 3, 5, \dots)$

Figure 3.33 (Continued) Typical loading on plates and loading functions for rectangular plates (Szilard, 1974).

Suitable deflection function $w(x, y)$ satisfying the boundary conditions of the given plate may be chosen. The strain energy U, and the work done by the given load, $q(x, y)$:

$$W = -\int \int_{\text{area}} q(x, y)w(x, y)\mathrm{d}x\mathrm{d}y \qquad (3.39)$$

can be calculated. The total potential energy is, therefore, given as $V = U + W$. Minimizing the total potential energy the plate problem can be solved.

$$\left[\frac{\partial^2 w}{\partial x^2} \frac{\partial^2 w}{\partial y^2} - \left(\frac{\partial^2 w}{\partial x \partial y} \right)^2 \right]$$

is known as the Gaussian curvature.

If the function $w(x, y) = f(x)\, \phi(y)$ (product of a function of x only and a function of y only) and $w = 0$ at the boundary are assumed, then the integral of the Gaussian curvature over the entire plate equals zero. Under these conditions:

Case no.	Structural system and static loading	Deflection and internal forces
1	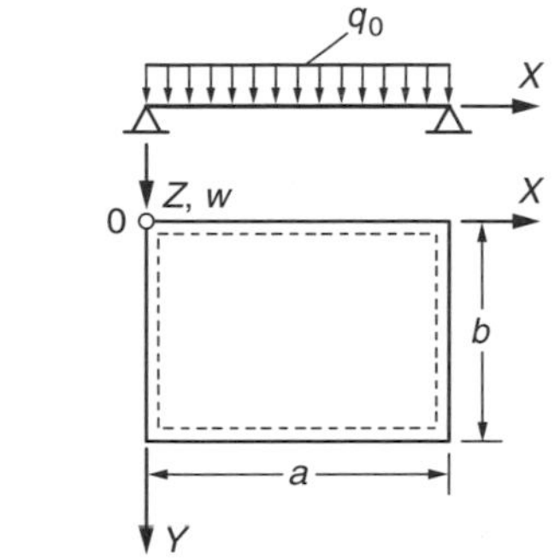	$$w = \frac{16 q_0}{\pi^6 D} \sum_m \sum_n \frac{\sin \frac{m\pi x}{a} \sin \frac{n\pi y}{b}}{mn \left(\frac{m^2}{a^2} + \frac{n^2}{b^2} \right)^2}$$ $$m_x = \frac{16 q_0 a^2}{\pi^4} \sum_m \sum_n \frac{\left(m^2 + \nu \frac{n^2}{\varepsilon^2} \right) \sin \frac{m\pi x}{a} \sin \frac{n\pi y}{b}}{mn \left(m^2 + \frac{n^2}{\varepsilon^2} \right)^2}$$ $$m_y = \frac{16 q_0 a^2}{\pi^4} \sum_m \sum_n \frac{\left(\frac{n^2}{\varepsilon^2} + \nu m^2 \right) \sin \frac{m\pi x}{a} \sin \frac{n\pi y}{b}}{mn \left(m^2 + \frac{n^2}{\varepsilon^2} \right)^2}$$ $\varepsilon = \frac{b}{a}$, $m = 1, 3, 5, \ldots, \infty$; $n = 1, 3, 5, \ldots, \infty$
2	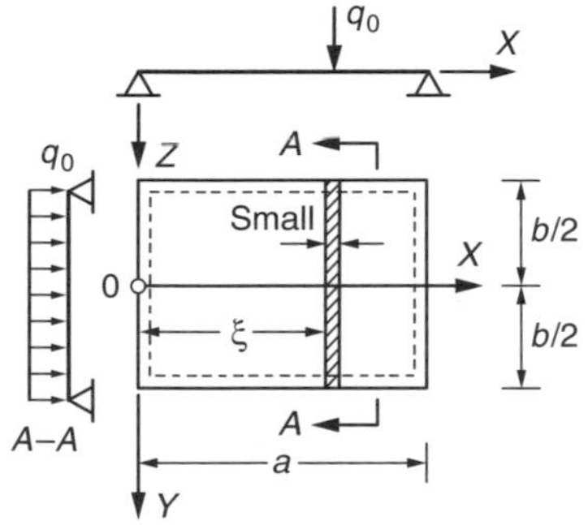	$$w = \frac{a^4}{D\pi^4} \sum_{m-1}^{\infty} \frac{P_m}{m^4} \left(1 - \frac{2 + \alpha_m \tanh \alpha_m}{2 \cosh \alpha_m} \cosh \lambda_m y \right.$$ $$\left. + \frac{\lambda_m y \sinh \lambda_m y}{2 \cosh \alpha_m} \right) \sin \lambda_m x$$ where $$P_m = \frac{2 q_0}{a} \sin \frac{m\pi\xi}{a} \qquad \lambda_m = \frac{m\pi}{a}$$ $$m = 1, 2, 3, \ldots. \qquad \alpha_m = \frac{m\pi b}{2a}$$
3	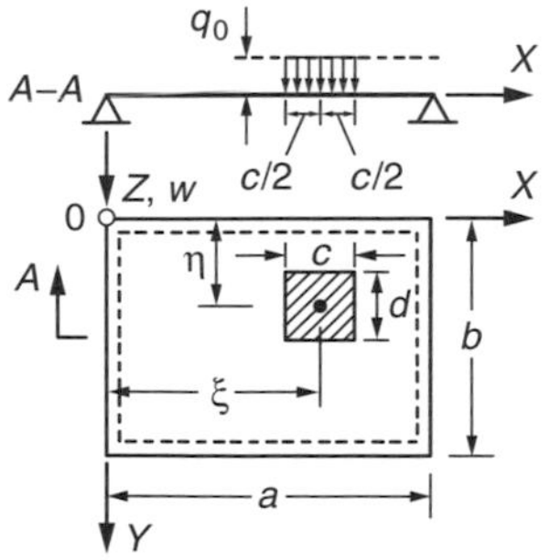	$$w = \frac{16 q_0}{D\pi^6} \sum_m \sum_n \frac{\sin \frac{m\pi\xi}{a} \sin \frac{n\pi\eta}{b} \sin \frac{m\pi c}{a} \sin \frac{n\pi d}{2b}}{mn \left(\frac{m^2}{a^2} + \frac{n^2}{b^2} \right)^2}$$ $$\times \sin \frac{m\pi x}{a} \sin \frac{n\pi y}{b}$$ $m = 1, 2, 3, \ldots.$ $n = 1, 2, 3, \ldots.$
4	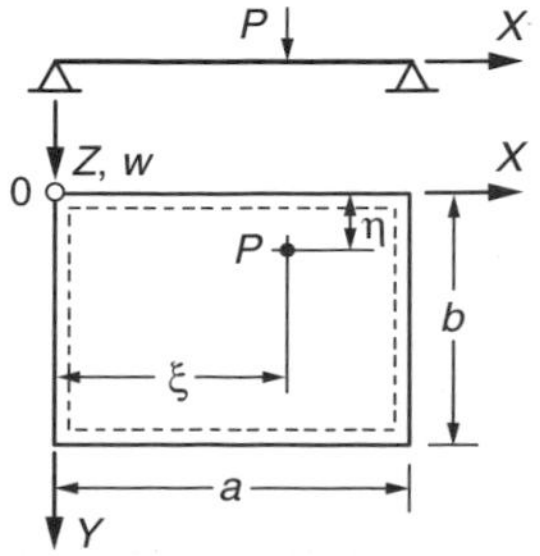	$$w = \frac{4P}{D\pi^4 ab} \sum_m \sum_n \frac{\sin \frac{m\pi\xi}{a} \sin \frac{n\pi\eta}{b} \sin \frac{m\pi x}{a} \sin \frac{n\pi y}{b}}{\left(\frac{m^2}{a^2} + \frac{n^2}{b^2} \right)^2}$$ $m = 1, 2, 3, \ldots.$ $n = 1, 2, 3, \ldots.$

Figure 3.34 Typical loading and boundary conditions for rectangular plates (Szilard, 1974).

$$U = \frac{D}{2} \iint_{arca} \left(\frac{\partial^2 w}{\partial x^2} + \frac{\partial^2 w}{\partial y^2} \right)^2 dxdy$$

Detailed treatment of plate bending problems may be found in Timoshenko and Kreiger (1959).

3.5.5 Orthotropic plates

Plates of anisotropic materials have important applications owing to their exceptionally high bending stiffness. A non-isotropic or anisotropic material displays direction-dependent properties. Simplest among them are those in which the material properties differ in two mutually perpendicular directions. A material so described is orthotropic, for example wood. A number of manufactured materials are approximated as orthotropic. Examples include corrugated and rolled metal sheets, fillers in sandwich plate construction, plywood, fibre reinforced composites, reinforced concrete and gridwork. The latter consists of two systems of equally spaced parallel ribs (beams), mutually perpendicular, and attached rigidly at the points of intersection.

The governing equation for orthotropic plate is similar to that of isotropic plate and takes the form:

$$D_x \frac{\delta^4 w}{\delta x^4} + 2H \frac{\delta^4 w}{\delta x^2 \delta y^2} + D_y \frac{\delta^4 w}{\delta y^4} = q \tag{3.40}$$

In which:

$$D_x = \frac{h^3 E_x}{12}, D_y = \frac{h^3 E_y}{12}, H = D_{xy} + 2G_{xy}, D_{xy} = \frac{h^3 E_{xy}}{12}, G_{xy} = \frac{h^3 G}{12}$$

The expressions for D_x, D_y, D_{xy} and G_{xy} represent the flexural rigidities and the torsional rigidity of an orthotropic plate, respectively. E_x, E_y and G are the orthotropic plate moduli. Practical considerations often lead to assumptions, with regard to material properties, resulting in approximate expressions for elastic constants. The accuracy of these approximations is generally the most significant factor in the orthotropic plate problem. Approximate rigidities for some cases that are commonly encountered in practice are given in Figure 3.35.

General solution procedures applicable to the case of isotropic plates are equally applicable to the orthotropic plates as well. Deflections and stress-resultants can thus be obtained for orthotropic plates of different shapes with different support and loading conditions. These problems have been researched extensively and solutions concerning plates of various shapes under different boundary and loading conditions may be found in Smith (1953), Timoshenko and Kreiger (1959), Tsai and Cheron (1968), Timoshenko and Goodier (1970), Lee *et al.* (1971), Shanmugan *et al.* (1988, 1989).

Geometry	Rigidities
(a) Reinforced concrete slab with x and y directed reinforcement steel bars 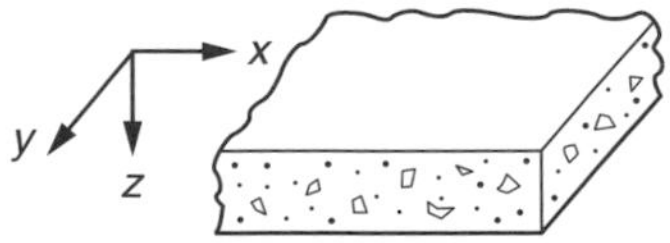	$D_s = \dfrac{E_c}{1-v_c^2}\left[I_{cx} + \left(\dfrac{E_s}{E_c} - 1\right)I_{sx}\right]$ $D_r = \dfrac{E_c}{1-v_c^2}\left[I_{cy} + \left(\dfrac{E_t}{E_c} - 1\right)I_{sy}\right]$ $G_{sy} = \dfrac{1-v_c}{2}\sqrt{D_xD_y}$ $H = \sqrt{D_xD_y}$ $D_{xy} = v_c\sqrt{D_xD_y}$ v_c: Poisson's ratio for concrete E_c, E_s: Elastic modulus of concrete and steel, respectively I_{cx}, (I_{sx}), I_{cy} (I_{sy}): Moment of inertia of the slab (steel bars) about neutral axis in the section x = constant and y = constant, respectively
(b) Plate reinforced by equidistant stiffeners 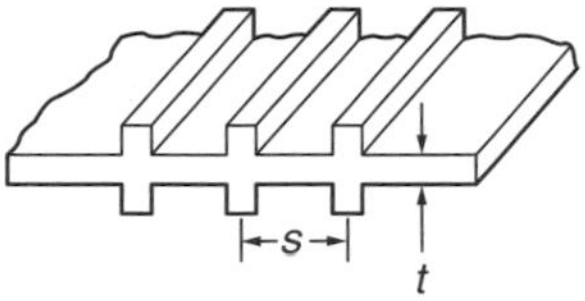	$D_s = H\ \dfrac{Et^3}{12(1-v^2)}$ $D_y = \dfrac{Et^3}{12(1-v^2)} + \dfrac{E'I}{s}$ E, E' : Elastic modulus of plating and stiffeners, respectively v : Poisson's ratio for plating S : Spacing between centrelines of stiffeners I : Moment of inertia of the stiffener cross-section with respect to midplane of plating
(b) Plate reinforced by a set of equidistant ribs 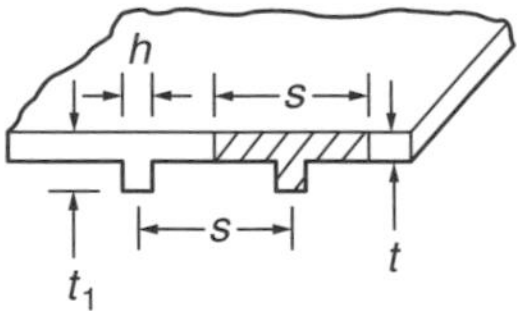	$D_s = \dfrac{Est^3}{12[s - h + h\,(t\,t_1)^2]}$ $D_y = \dfrac{EI}{s}$ $H = 2G'_{sy} + \dfrac{C}{s}$ $D_{sy} = 0$ C : Torsional rigidity of one rib I : Moment of inertia about neutral axis of a T- section of width s (shown above) G'_{sy} : Torsional rigidity of the plating E : Elastic modulus of the plating

Figure 3.35 Various orthotropic plates.

3.5.6 Buckling of thin plates
Rectangular plates
Buckling of a plate involves out-of-plane movement of the plate and results in bending in two planes. A significant difference between axially compressed columns and plates is apparent if one compares their buckling characteristics. For a column, buckling terminates the ability of the member to resist axial load, and the critical load is thus the failure load of the member. However, the same is not true for plates due to the membrane action of the plate. Subsequent to reaching the critical load, plates under compression will continue to resist increasing axial force, and will not fail until a load considerably in excess of the critical load is reached. The critical load of a plate is, therefore, not its failure load. Instead, one must determine the load-carrying capacity of a plate by considering its postbuckling behaviour.

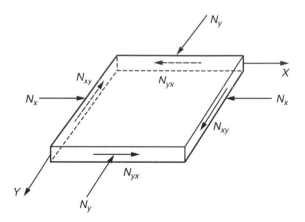

Figure 3.36 Plate subjected to in-plane forces.

To determine the critical in-plane loading of a plate by the concept of neutral equilibrium a governing equation in terms of biaxial compressive forces N_x and N_y and constant shear force N_{xy} as shown in Figure 3.36 can be derived as:

$$D\left(\frac{\delta^4 w}{\delta x^4} + 2\frac{\delta^4 w}{\delta x^2 \delta y^2} + \frac{\delta^4 w}{\delta y^4}\right) + N_x\frac{\delta^2 w}{\delta x^2} + N_y\frac{\delta^2 w}{\delta y^2} + 2N_{xy}\frac{\delta^2 w}{\delta x \delta y} = 0 \qquad (3.41)$$

The critical load for uniaxial compression can be determined from the differential equation:

$$D\left(\frac{\delta^4 w}{\delta x^4} + 2\frac{\delta^4 w}{\delta x^2 \delta y^2} + \frac{\delta^4 w}{\delta y^4}\right) + N_x\frac{\delta^2 w}{\delta x^2} = 0 \qquad (3.42)$$

which is obtained by setting $N_y = N_{xy} = 0$ in Equation (3.41).

For example, in the case of a simply supported plate, Equation (3.41) can be solved to give:

$$N_x = \frac{\pi^2 a^2 D}{m^2}\left(\frac{m^2}{a^2} + \frac{n^2}{b^2}\right)^2 \qquad (3.43)$$

The critical value of N_x, i.e. the smallest value can be obtained by taking n equal to one. The physical meaning of this is that a plate buckles in such a way that there can be several half-waves in the direction of compression but only one half-wave in the perpendicular direction. Thus the expression for the critical value of the compressive force becomes:

$$(N_x)_{cr} = \frac{\pi^2 D}{a^2}\left(m + \frac{1}{m}\frac{a^2}{b^2}\right)^2 \qquad (3.44)$$

The first factor in this expression represents the Euler load for a strip of unit width and of length a. The second factor indicates in what proportion the stability of the continuous plate is greater than the stability of an isolated strip. The magnitude of this factor depends on the magnitude of the ratio a/b and also on the number m, which gives the number of half-waves into which the plate buckles. If a is smaller than b, the second term in the parentheses of Equation (3.44) is always smaller than the first and the minimum value of the expression is obtained by taking $m = 1$, i.e. by assuming that the plate buckles in one half-wave. The critical value of N_x can be expressed as:

$$N_{cr} = \frac{k\pi^2 D}{b^2} \tag{3.45}$$

The factor k depends on the aspect ratio a/b of the plate and m, the number of half-waves into which the plate buckles in the x direction. The variation of k with a/b for different values of m can be plotted as shown in Figure 3.37. The critical value of N_x is the smallest value that is obtained for $m = 1$ and the corresponding value of k is equal to 4.0. This formula is analogous to Euler's formula for buckling of a column.

In the more general case in which normal forces N_x and N_y and the shearing forces N_{xy} on the boundary of the plate, the same general method can be used. The critical stress for the case of a uniaxially compressed simply supported plate can be written as:

$$\sigma_{cr} = 4\frac{\pi^2 E}{12(1-v^2)}\left(\frac{h}{b}\right)^2 \tag{3.46}$$

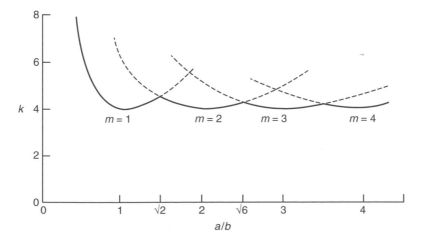

Figure 3.37 Buckling stress coefficients for uniaxially compressed plate.

The critical stress values for different loading and support conditions can be expressed in the form:

$$f_{cr} = k \frac{\pi^2 E}{12(1-v^2)} \left(\frac{h}{b} \right)^2 \tag{3.47}$$

in which f_{cr} is the critical value of different loading cases. Values of k for plates with different boundary and loading conditions are given in Figure 3.38.

Case	Boundary condition	Type of stress	Value of k for long plate
(a)	s.s. / s.s. s.s. / s.s.	Compression	4.0
(b)	Fixed / s.s. s.s. / Fixed	Compression	6.97
(c)	s.s. / s.s. s.s. / Free.	Compression	0.425
(d)	Fixed / s.s. s.s. / Free	Compression	1.277
(e)	Fixed / s.s. s.s. / s.s.	Compression	5.42
(f)	s.s. / s.s. s.s. / s.s.	Shear	5.34
(g)	Fixed / Fixed Fixed / Fixed	Shear	8.98
(h)	s.s. / s.s. s.s. / s.s.	Bending	23.9
(i)	Fixed / Fixed Fixed / Fixed	Bending	41.8

Figure 3.38 Values of K for plate with different boundary and loading conditions.

Detailed treatment of elastic buckling in plates may be found in Timoshenko and Gere (1961).

Buckling of plates in thin plated structures

A major advantage of thin-plated steel sections is that their use leads to highly efficient and weight-effective members and structures. Figure 3.39 shows two cases of local buckling in typical plated structures. Case (b) is a plate girder in which the top flange under compression is supported by the web along one edge and the flange is free to wave at its other longitudinal edge. In the case of a box girder top flange, the plate is supported by webs at both longitudinal edges as shown in case (a).

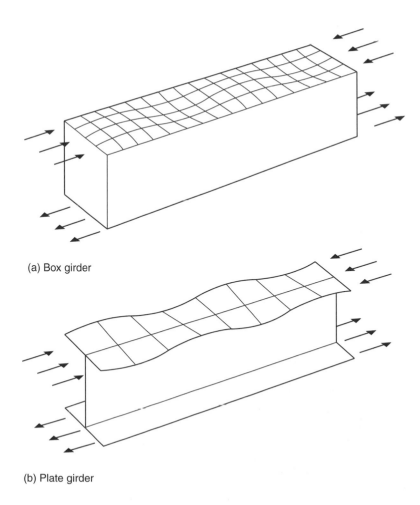

(a) Box girder

(b) Plate girder

Figure 3.39 Buckling of flanges in box and plate girders.

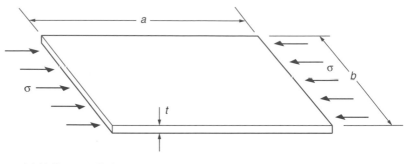

(a) Uniform applied stress

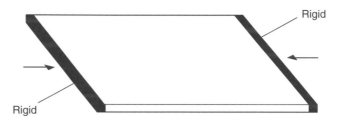

(b) Uniform applied displacement

Figure 3.40 Axially loaded plate.

In these two plates under uniform compression the elastic critical stress can be expressed (Figure 3.40), as given in Equation (3.47):

$$\sigma_{cr} = \frac{K\pi^2 E}{12(1-v^2)}\left(\frac{t}{b}\right)^2$$

in which t = thickness of the plate. The case when all four edges are simply supported approximates to the situation in the top flange of the box girder in Figure 3.39(a) and the value of K in the above equation is equal to four; the mode of buckling will be as shown in Figure 3.41 depending upon the aspect ratio (a/b) of the plate. The value of K for a plate supported on three sides as in the case of flange plate in a plate girder is 0.425 and the elastic critical stress corresponding to other types of in-plane loading can be computed by a formula similar to the one given in Equation (3.46) with different values for K. The postbuckling reserve strength of a plate can be envisaged in terms of its axial stiffness; for an ideally flat plate, the axial stiffness drops quickly to a much smaller value after buckling and remains relatively constant thereafter. Practical plates, on the other hand, exhibit a gradual loss of stiffness under in-plane loading as shown in Figure 3.42.

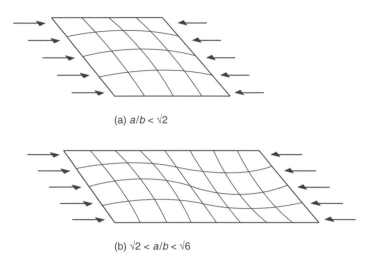

(a) $a/b < \sqrt{2}$

(b) $\sqrt{2} < a/b < \sqrt{6}$

Figure 3.41 Buckled shapes in axially loaded plates.

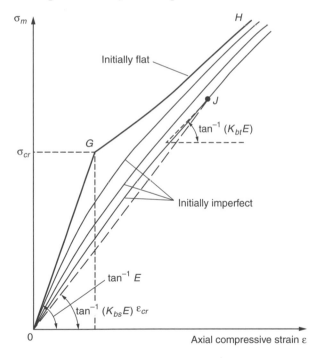

Figure 3.42 Axial stiffness of plates in compression.

Numerical methods of analysis of plates, which include both geometric and material non-linearities are available now and these analyses are capable of assessing the ultimate strength and postcollapse stiffness of plates with fabrication imperfections. However, most of the popular design methods are based on 'effective width' approach, a concept originally proposed by Von Karman *et al.* (1932) and subsequently improved by Winter (1947) to account for the effect of boundary conditions and plate imperfections. In accordance with this concept, the non-uniform stress distribution across the width of the buckled plate could be replaced by two uniform stress blocks of σ_{max} acting over two equal end strips of width $0.5b_{eff}$, where b_{eff} is the effective width as shown in Figure 3.43.

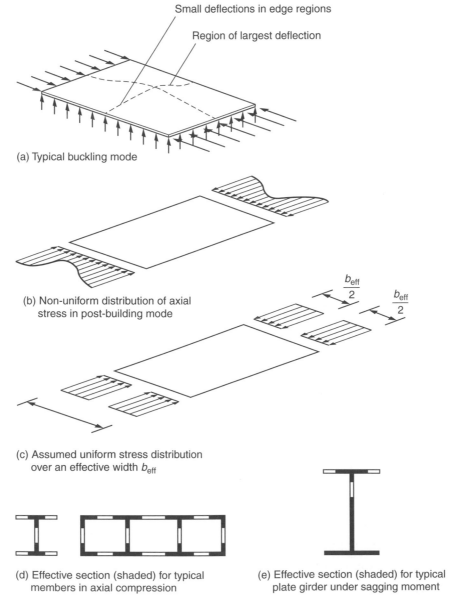

(a) Typical buckling mode

(b) Non-uniform distribution of axial stress in post-building mode

(c) Assumed uniform stress distribution over an effective width b_{eff}

(d) Effective section (shaded) for typical members in axial compression

(e) Effective section (shaded) for typical plate girder under sagging moment

Figure 3.43 Effective width concept in axially compressed plates.

The effective width may be obtained as:

$$\frac{b_{eff}}{b} = \sqrt{\frac{\sigma_{cr}}{\sigma_{ys}}}\left[1 - 0.22\sqrt{\frac{\sigma_{cr}}{\sigma_{ys}}}\right] \qquad (3.48)$$

A number of effective width formulae to account for variations in boundary conditions, stress distribution and plate imperfection have been proposed by various researchers during the last several years (Faulkner, 1975; Horne and Narayanan, 1976a; Narayanan and Shanmugam, 1979a; Mulligan and Pekoz, 1984; Shanmugam *et al.*, 1989).

In an analogous manner, the critical buckling stress of a rectangular panel loaded in shear (Figure 3.44) is given by:

$$\tau_{cr} = \frac{K\pi^2 E}{12(1-v^2)}\left(\frac{t}{d}\right)^2 \qquad (3.49)$$

As shown in Figure 3.44 a square element whose edges are oriented at 45° to the plate edges, experiences tensile stresses on two opposing edges and compressive ones on the other two. These compressive stresses induce a form of local buckling; the mode involves elongated bulges oriented at about 45° to the plate edges. As with the compressive loading, a thin plate loaded in shear can support an applied stress well in excess of the elastic critical one. This is due again to the resistance to in-plane deformation. As the applied shear stress is increased beyond τ_{cr} the plate buckles elastically and retains little stiffness in the direction in which the compressive component acts. However, the inclined tensile component is still resisted fully by the plate. The inclined buckles become progressively narrower and the plate acts like a series of bars in the tension direction, developing a so-called tension field. Further increase of applied stress causes plastic deformation in part of the tension field, which rotates to line up more closely with the plate diagonal. Tension field action is particularly important in plate and box girders, in which the function of the web plates is primarily to resist shear.

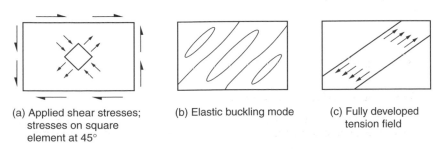

(a) Applied shear stresses; stresses on square element at 45°

(b) Elastic buckling mode

(c) Fully developed tension field

Figure 3.44 Web panels in shear; buckling and tension field action.

Stiffened compression flanges

Stiffened compression flanges of steel plate or box girders usually consist of a steel flange plate stiffened longitudinally by either open or closed stiffeners spanning between transverse stiffeners. Such compression flanges are subjected to the longitudinal stresses due to overall bending moment and axial force on the main girder; these stresses may vary across the width of the girder due to shear lag, and along the length due to the variation of bending moment along the span. When loaded in-plane, the plate elements are extremely stiff. Even the relatively simple case of a stiffened panel consisting of a bed plate with equally spaced stiffeners has to be analysed for a number of conditions (Chatterjee and Dowling, 1976; Horne, 1976b, 1977; Horne and Narayanan, 1976). The ultimate capacity of such a stiffened plate is the smallest load causing either: (i) overall buckling of the panel between cross frames, or (ii) the local buckling of the base plate between stiffeners or (iii) local buckling of the stiffeners. A simple and approximate method proposed by Horne and Narayanan can be used to predict the ultimate capacity of a stiffened plate. The method is based on the effective width concept to account for local buckling in the base plate and Perry–Robertson formula to account for the overall column buckling of stiffened plate. The Perry–Robertson formula is given as:

$$\frac{\sigma}{\sigma_{ys}} = \frac{1}{2}\left[1+(1+\eta)\frac{\sigma_e}{\sigma_{ys}}\right] - \sqrt{\frac{1}{4}\left[1+(1+\eta)\frac{\sigma_e}{\sigma_{ys}}\right]^2 - \frac{\sigma_e}{\sigma_{ys}}} \qquad (3.50)$$

in which, σ is the limiting applied axial stress on the effective strut section, σ_{ys} is the yield stress on the extreme compressive fibre, σ_e is the Euler stress of the effective stress, $\eta = \Delta y/r^2$, Δ is the maximum initial imperfection and eccentricity, y is the distance of the extreme compressive fibre, and r is the radius of gyration of the effective section. The ultimate capacity can, therefore, be estimated by aggregating the strength of the component stiffeners, along with the associated 'effective' plating as shown in Figure 3.45. Loads causing local yield in base plates between stiffeners or in the stiffener outstands can be separately evaluated (Narayanan and Shanmugam, 1979b). A further consideration is the local torsional buckling of thin walled open stiffeners; as the stiffener-induced collapse is catastrophic, it is usual to specify relatively stocky stiffeners than to allow for such torsional buckling. The bridge design code BS 5400 (British Standards Institution, 1982) allows for initial imperfection and compressive welding residual stresses. Detailed methods to design compression flanges and the longitudinal stiffeners (flat stiffeners, angle stiffeners, tee stiffeners) are included in the code.

Stiffened flange panel between the webs and cross frames of main girder can be assumed to behave as an orthotropic plate. This assumption is particularly beneficial in the case of stiffened flanges with stocky stiffeners and the buckling at the ultimate load is of stable nature. In this case non-uniform distribution of flange stresses due to shear lag or restrained warping may be neglected in the check for the ultimate state. Box girder compression flanges are normally subjected to longitudinal stress due to

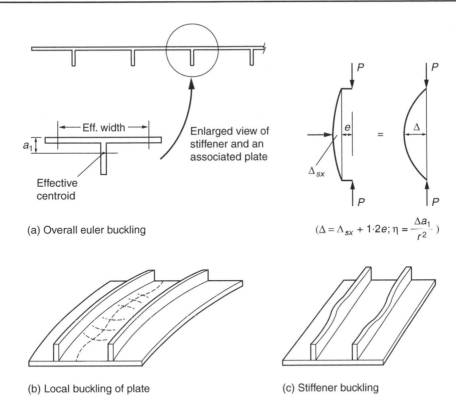

Figure 3.45 Ultimate load of a stiffened plate.

overall bending moment and axial force caused by vertical forces acting on the flange plate. The design procedures have to provide for all the combinations of in-plane compressive stresses, in-plane shear stresses and in-plane transverse stresses in the flange plate.

Web plates

Webs in plate and box girders are normally reinforced by intermediate transverse stiffeners and, longitudinal stiffeners when the depth is large. Such reinforcement will increase the ultimate load capacity of a girder significantly, but the introduction of stiffeners will obviously increase fabrication cost and the self-weight of the girder. The ultimate strength of a stocky web is, of course, limited to its yield strength. An unstiffened girder with a slender web may be designed simply by neglecting any post-buckling reserve strength and assuming shear capacity of the web to be equal to the shear buckling resistance (V_{cr}), given by:

$$V_{cr} = \tau_{cr}\, dt \qquad\qquad (3.51)$$

where τ_{cr} is taken as the elastic critical shear buckling stress. Webs of intermediate slenderness may be designed with transverse stiffeners. The modern codes recommend

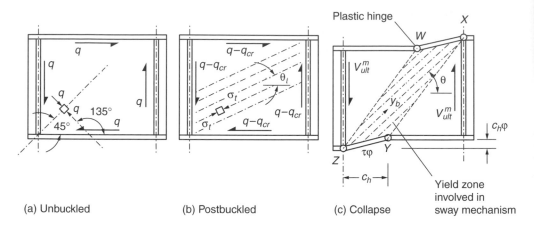

(a) Unbuckled   (b) Postbuckled   (c) Collapse

Figure 3.46 Buckling and postbuckling behaviour of web in shear.

the use of tension field action (Rockey *et al.*, 1977, 1978; Chatterjee, 1980; Evans, 1983) to determine the shear capacity in the case of slender webs. The behaviour of a plate/box girder subjected to an increasing shear load may be divided into three phases as shown in Figure 3.46. Prior to buckling, equal tensile and compressive principal stresses are developed in the plate as shown in Figure 3.46(a). In the post-buckling range no more increases in compressive stress are possible and only an inclined tensile membrane stress field is developed, as shown in Figure 3.46(b). The magnitude of the tensile membrane stress is indicated by σ_t in Figure 3.46(b) and its inclination to the horizontal is shown as θ. Since the flanges of the girder are flexible they will begin to bend inwards under the pull exerted by the tension field. Further increase in the load will result in yield occurring in the web under the combined effect of the membrane stress field and the shear stress at buckling. The value of σ_t at which yield occurs is identified as the basic tension field strength. Failure of the girder occurs when four plastic hinges form in the flanges of the gird-er, as shown in Figure 3.46(c). The resulting collapse mechanism then allows a shear displacement to occur as indicated. The ultimate shear capacity (V_b) can be obtained from the expression:

$$V_b = \tau_{cr} + y_b \sin^2 \theta \left(\cot \theta - \frac{b}{d} \right) dt + 4dt \sin \theta \sqrt{K_f p_{yw} y_b} \qquad (3.52)$$

in which, p_{yw} is the yield stress of web material; b is the width and d the depth of web panel. y_b is given by:

$$y_b = [p_{yw}^2 - 3q_{cr}^2 + 2.25q_{cr}^2 \sin^2 2\theta]^{1/2} - 1.5q_{cr} \sin 2\theta \qquad (3.53)$$

$K_f = M_{pf}/(4M_{pw})$, M_{pf} and M_{pw} being the plastic moment capacity of flange plate and web plate.

The expression for the ultimate shear capacity is also written as:

$$V_b = (q_b + q_f \sqrt{K_f})dt \qquad (3.54)$$

where:

$$q_b = q_{cr} + \frac{y_b}{2\{a/d + \sqrt{[1+(a/d)^2]}\}} \qquad (3.54a)$$

and:

$$q_f = \left[4\sqrt{3} \sin\left(\frac{\theta}{2}\right)\sqrt{\frac{y_b}{P_{yw}}}\right]0.6p_{yw} \qquad (3.54b)$$

In the above equation q_b represents the basic shear strength of the web panel and it combines the critical shear strength of the panel and the postbuckling strength derived from that part of the web tension field supported by the transverse stiffeners. The term $q_f\sqrt{K_f}$ represents the contribution made by the flanges to the postbuckling strength. The flanges support part of the web tension field that pulls against them and finally develop plastic hinges at collapse. The relative importance of the various terms in the above equation for V_b is illustrated for a typical case in Figure 3.47.

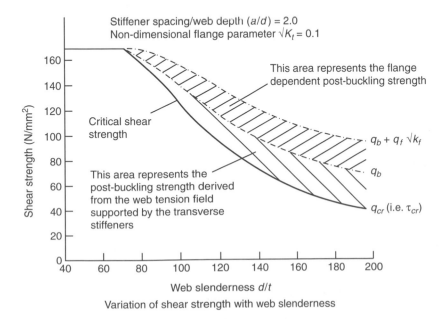

Variation of shear strength with web slenderness

Figure 3.47 Variation of shear strength with web slenderness.

The method must be used with caution in box girders in which the flange/web boundaries may have insufficient strength to develop adequate tension field capacity; moreover, due to the bending of the box in the prebuckling stage, the web panels will carry both longitudinal and shear stresses. The interaction between shear and in-plane compression (or tension) and bending stresses becomes important while examining the likely load combinations in the webs of box girder bridges, which are generally multiply stiffened. Based on an extensive parametric study Harding (1983) has suggested interaction diagrams for the design of such plate panels:

$$\frac{\sigma_c}{S_c \sigma_{ys}} + \left[\frac{\sigma_b}{S_b \sigma_{ys}}\right]^2 + \left[\frac{\tau}{S_s \tau_{ys}}\right]^2 \leq 1 \tag{3.55}$$

where σ_c, σ_b and τ are the direct, bending and shear stresses and S_c, S_b and S_c are the knock-down factors which allow for the influence of b/t on the stability of plates.

Openings in webs

Openings in the webs of plate and box girders are necessary to facilitate inspection and for providing services. The introduction of openings in a web alters the stress distribution within the member and causes a reduction in its strength and stiffness. Approximate design methods are now available as a result of extensive research work on openings in plated structures. The first paper of relevance to thin perforated webs was by Hoglund (1971) who reported on statically loaded plate girders containing circular and rectangular holes and subjected to transverse loading. The web plates of these girders were slender having d/t values ranging from 200 to 300; they would, therefore, buckle before failure. In these experiments, the girders with holes which were in the high shear zone failed at significantly lower loads than those in the zone of high bending and low shear. These tests indicated the relative importance of shear failure criteria in plate girders with perforated webs. The tension field concept has been extended by Narayanan and Der Avenessian (Narayanan, 1983; Narayanan and Der Avenessian, 1983a, b) to the cases of web plates containing centrally located circular and rectangular openings. Approximate equilibrium solutions have been proposed by them based on further assumption that the applied loading in the postcritical stage is carried by the membrane stresses developed along a diagonal band, the effective width of which is governed by the largest dimension of the hole as shown in Figure 3.48. For central circular openings the ultimate load can be obtained as:

$$V_{ult} = (\tau_{cr})_{red}dt + \sigma_r^y t \sin^2\theta[2c + d(\cot\theta - b/d)] - \sigma_r^y tD \sin\theta \tag{3.56}$$

where τ_{cr} is the reduced value of elastic critical shear stress of the web containing hole. When the central opening is a rectangle of size $b_o \times d_o$, the corresponding equilibrium equation is:

$$V_{ult} = (\tau_{cr})_{red}dt + \sigma_t^y t \sin^2\theta[2c + d(\cot\theta - b/d)] - \sigma_t^y t \sqrt{(b_o^2 + d_o^2)} \sin\theta \sin(\alpha + \theta) \tag{3.57}$$

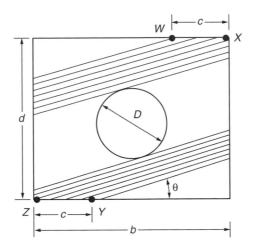

Figure 3.48 Opening in a plate girder web.

where $\alpha = \text{arc } \tan(d_o/b_o)$. The basis of the solution is the hypothetical tension band developed in the postcritical stage. If the loss of strength implicit in cutting a web hole is unacceptable, the web opening will need to be reinforced around its periphery, so that the opening can still be introduced. It would then be necessary to assess the strength of the reinforced web, with a view to examining its adequacy. Equilibrium solutions for the prediction of the ultimate capacity of girders containing a reinforced circular or rectangular hole has been proposed by Narayanan and Der Avenessian (1984). Methods to account for the presence of web openings in multi-cell box girders have also been proposed (Evans and Shanmugam, 1979a, 1981; Shanmugam and Evans, 1979).

Patch loading

The collapse or crippling of plate/box girders subjected to patch loading and distributed edge loading is a complex problem involving both material and geometric non-linearity. Solutions are now available for the elastic critical loads of idealized web panels subjected to a variety of combined loading but such solutions show little or no correlation with experimental collapse loads. The localized stress distribution due to patch loading will be, in general, combined with global bending and shear stresses. Whether it is the local or global stress distribution which is predominant will depend on the overall structural form and loading. Although the problem of patch loading may seem to be of only minor significance in the overall design, it is however, important to ensure that girders are not overstressed locally (e.g. during launching operations) since localized failure may precipitate overall structural failure. A large number of model tests in the past have shown that failure of slender plate girder subjected to patch loading occurs due to the formation of plastic hinges in the flange accompanied by yield lines in the web. This type of failure, which is very localized, has become known as web crippling. Simple approximate solutions, based on, mechanism solutions have been proposed by Roberts (1983) for predicting the collapse or crippling load of plate girders subjected to patch loading, and these have been adopted in Codes of Practice.

3.5.7 Plate and box girder analysis

Introduction

The high bending moments and shearing forces associated with the carrying of large loads over long spans as in the case of bridges frequently necessitates the use of fabricated plate and box girders. The high torsional strength of box girders makes them ideal for girders curved in plan. In their simplest form plate and box girders can be considered as an assemblage of webs and flanges as shown in Figure 3.49. In order to reduce the self-weight of these girders and thus achieve economy, slender plate

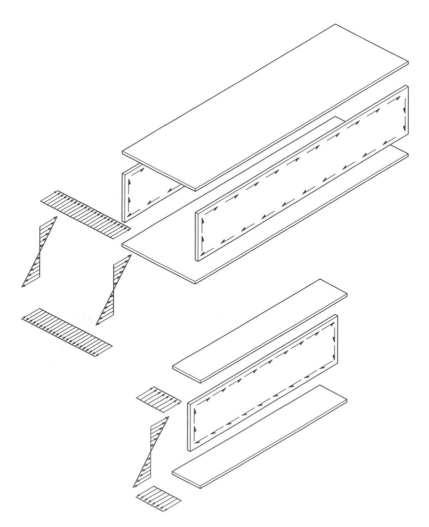

Figure 3.49 Box girder and plate girder as assemblages of plate elements.

sections (having large lateral dimensions compared with their thicknesses) are employed. Hence local buckling and post buckling reserve strengths of plates are important design criteria. Flanges in a box girder and webs in plate and box girders are often reinforced with stiffeners to allow for efficient use of thin plates. The designer has to find a combination of plate thickness and stiffener spacing that will result in the most optimal section with reduced weight and fabrication cost.

Limit states design codes have placed greater emphasis on development of new approaches based on the ultimate strength of plate and box girders and their components. Significant advances have been made on the understanding of the behaviour of steel plated structures as a result of continuing research during the last 25 years. The initial stimulus for this interest in steel plated structures was provided by the collapse of four steel box girder bridges during erection in the early 1970s. CIRIA research guide for Structural actions in box girders (Horne, 1975) lists the following difficulties that are usually encountered by the designers of plated structures.

- The engineer's simple 'plane sections remain plane' theory of bending is no longer adequate, even for linear elastic analysis.
- Non-linear elastic behaviour caused by the buckling of plates can be of great importance and must be allowed for.
- Because of this complex non-linear elastic behaviour, and also because of stress concentration problems, some yielding may occur at loads which are quite low in relation to ultimate collapse loads. While such yielding may not be of great significance as regards either rigidity or strength, it means that simple maximum stress criteria are not sufficient.
- Because of the buckling problem in plates and stiffened panels, complete 'plastification' is far from being realized at collapse. Hence simple plastic criteria are not sufficient either.
- Complex interactions occur between flanges, webs and diaphragms and the pattern of this interaction can change as the level of load increases.

Linear elastic analysis
In the case of plated beams such as the one shown in Figure 3.50 the vertical (figure a) and horizontal (figure d) components of applied loads produce elastic bending stresses (figure b and e) and shear stresses (figure c and f) when they act through the shear centre. These forces produce torque on the cross-section if they act eccentrically with respect to the shear centre. Box sections are very strong compared to I sections in resisting torsion. The torque T in box-sections could be resisted entirely by a 'shear flow' q acting round the box and the shear stress in any part of the wall of the box would be equal to q/t, where t is the thickness of the plate forming the wall. The shear flow q is obtained as:

$$q = \frac{T}{2A_B}$$

A_B being the area enclosed by the box.

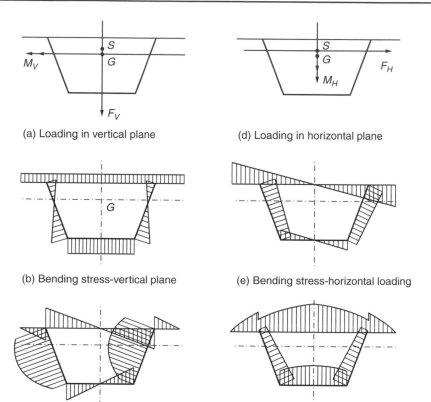

Figure 3.50 Bending of box girders.

The angle of twist of the box per unit length due to the shear flow is given by:

$$\theta = \frac{1}{2A_B G} \oint \tau \, ds$$

or:

$$\theta = \frac{1}{2A_B^2 G} \oint \frac{ds}{\tau}$$

where G is the elastic shear modulus.

In the case of open sections such as a plate girder, the angle of twist per unit length is expressed as:

$$\theta = \frac{T}{GJ}$$

where:

$$J = \frac{st^3}{3} \quad \text{or} \quad J = \frac{1}{3} \int_{section} t^3 ds$$

where J is the torsional constant of the section, s the length along the centre line of rectangular element, namely web or a flange, and t thickness of the element.

The assumption of simple bending theory is reasonably accurate if the span to depth ratio of the plate or box girders exceeds about four. However, because of differential bending in vertical planes, the cross-sections are subjected to warping. Another aspect of behaviour not allowed for in the simple treatment of torsion is the possible distortion of the cross-section. Distortion introduces additional stresses of various types, and these have to be allowed for. However, distortion effect can be controlled by the rigidity and spacing of cross-frames or diaphragms. In the case of sections in which the width between webs are very large and nearly equal to the span, the question arises about the effectiveness of the complete width of the flange. There obviously has to be some limitation on the 'effective width' of flanges of higher width/span ratios. These limitations are due to the intrusion of shear lag.

Shear lag effects
Box girder structures in bridges are subjected to shear and bending so that the normal stress distribution in a cross-section must take account of the shear lag phenomenon, which influences both tension and compression. Because of the action of the large in-plane shear strain in the flanges, the longitudinal strains in the central areas are less than in those areas adjacent to the web–flange junction, and the vertical deflection of the central areas are less than the deflection at the web–flange junction. The result is that the distribution of compressive stresses across the flange is not uniform during the early stages of loading, and the effect of this is to reduce the elastic stiffness of the girder in bending. This first order phenomenon induces a non-uniform normal stress distribution across the breadth of the flanges, the stresses being greatest along the web–flange junction because of compatibility requirements. The non-uniform stress distribution across the flange width is illustrated in Figure 3.51. The procedure commonly adopted in design is to replace the actual width of flange by an effective breadth which, when used in conjunction with simple beam theory, leads to a correct estimate of the maximum flange stresses or beam deflections as required. The effective breadth factor for a stiffened flange depends on the geometry of the box, the ratio of stiffener area to plate area (A_s/bt_f), support conditions and the loading distribution. Moffat and Dowling (1975) have provided, based on extensive studies using the finite element approach, a comprehensive picture of the shear lag effect on the behaviour of box girders. A detailed parametric study that has been carried out is very helpful to the designers for computing the effective flange breadths at all positions on the span of a box girder of given plan dimensions and cross-sectional proportions, subjected to point or distributed loading. The variation of effective breadth factor ψ with aspect ratio B/L for the effective breadth at mid-span due to a point load at mid-span and due to a uniformly distributed load is shown in Figure 3.52 for two values of A_s/bt_f (0 and 1.0).

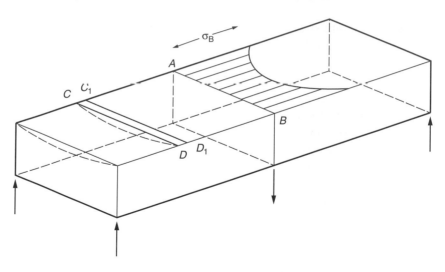

Figure 3.51 Shear lag in a box girder.

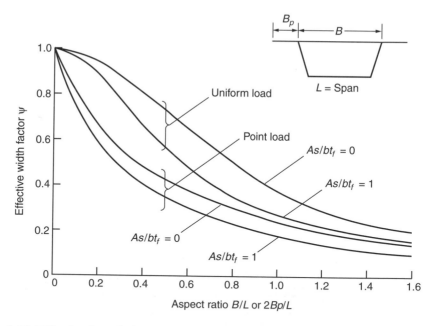

Figure 3.52 Effective breath factors for shear lag at midspan of simply supported box.

The most significant effect of shear lag on the behaviour of the point loaded girders is to reduce the overall stiffness. If the girder continues to be loaded after yielding has occurred near the flange–web interface, then the yielding will spread across the flange giving a more uniform stress distribution. Test results have shown that full redistribution of stress across the flange will have taken place as the girder

approaches its ultimate strength and the neglect of shear lag effect has no significant weakening effect on the ultimate strength of the stiffened compression flanges.

Torsion

Vertical loads that act eccentrically with respect to the centre line in a box girder
results in twisting of the box section. Twisting moment is resisted by pure shear stresses
in the walls of a box which is known to possess very high torsional resistance.
Longitudinal normal stresses arising from the relative warping of the section under
torsion are not considered in theory of pure torsion; however, these stresses can attain
very large values when the closed cross-sections are flexible. For example, consider
a general loading on a box section as shown in Figure 3.53 in which single vertical
eccentric load is replaced by sets of forces representing vertical, torsional and
distortional loading. The general loading in Figure 3.53 can be represented as two
different components of loading, one causing bending and the other causing torsion
as shown in Figures 3.53(b) and (c). The torsional loading component can be sub-
divided further into a pure torsional component and a distortional component as
shown in Figures 3.53(d) and (e). Although the pure torsional component will
normally result in negligible longitudinal stresses, the distortional component
will always tend to deform the cross-section, thus leading to distortional stresses in

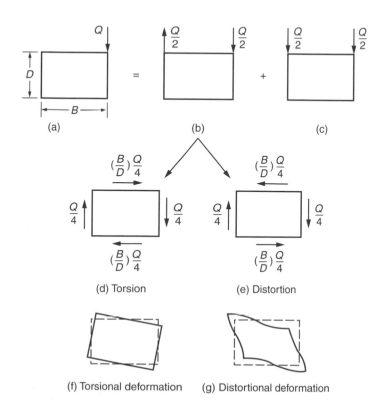

Figure 3.53 Idealization of eccentric loading in box girder.

the transverse direction and warping stresses in the longitudinal direction. The distortion of the cross-section will be resisted by cross frames and diaphragms and hence an accurate analysis involves evaluating the distortional warping and shear stresses and the associated distortional bending stresses in the transverse frames. The problem of torsion in box girders can be analysed more accurately by employing the well-known beam-on-elastic foundation analogy.

3.6 Grillage analysis

The approximate representation of bridge decks by a grillage of interconnected beams is a convenient way of determining the general behaviour of the bridge under load. A grillage is a structure of rigidly connected longitudinal and transverse beams each with a bending and torsional stiffness. At each junction of beams deflection and slope compatibility equations can be set up. The general availability of high-speed computers has enabled the development of efficient grillage analysis. The method of analysis involves idealization of the bridge deck through its representation as a plane grillage of discrete interconnected beams. Although the method is necessarily approximate, it has the great advantage of almost complete generality. At the joints of the grillage any normal form of restraint to movement may be applied so that any support condition may be represented. Simply-supported, built-in, continuous, discrete column support conditions, elastic foundation may all be represented without any difficulty. The plan form of the deck presents no real problems. West (1971) has made recommendations, based on extensive investigation, on the use of grillage analysis for bridge decks.

There are basically three steps in the analysis procedure when analysing a bridge deck using grillage approach: (i) idealization of the physical deck into an equivalent grillage, (ii) mathematical analysis of the grillage and (iii) interpretation of the results and their use in the design of the deck. The following sections deal with the methods of idealization and they are based on the recommendations by West (1971).

3.6.1 Grillage idealization of slab structure

Grillage method of analysis, pioneered by Lightfoot and Sawko (1959) represents the deck by an equivalent grillage of beams as shown in Figure 3.54. The dispersed bending and torsion stiffnesses in every region of the slab are assumed for purpose of analysis to be concentrated in the nearest equivalent beam. The longitudinal stiffnesses of the slabs are concentrated in the longitudinal beams while the transverse stiffnesses in the transverse beams. Ideally the beam stiffnesses should be such that when prototype slab and equivalent grillage are subjected to identical loads, the two structures should deflect identically and the moments, shear forces and torsions in any grillage beam should equal the resultants of the stresses on the cross-section of the part of the slab the beam represents. There are, however, some shortcomings one has to compromise since the grillage is an approximation of the real structure. Firstly, equilibrium of any element of the slab requires that torques are identical n orthogonal direction and also the twist is the same in orthogonal directions. In the equivalent grillage, it depends upon the type of mesh, fine or coarse, and it has been found that this condition is satisfied only when the grillage is sufficiently fine. Even so it is

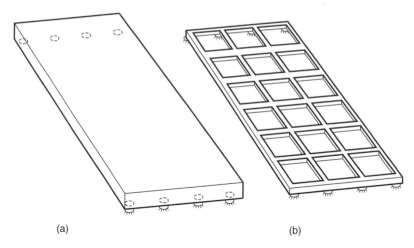

(a)   (b)

Figure 3.54 Grillage idealization of a slab deck.

often found that a coarse mesh is sufficient for design purposes. Secondly, the moment in any grillage beam is only proportional to the curvature in it, while in the proto-type slab the moment in any direction depends on the curvatures in that direction and the orthogonal direction. Bending stresses deduced from grillage results for dis-tributed moments have, however, been shown to be sufficiently accurate for most design purposes.

It is common to have the deck in this type of structures wider than the span. A choice of nine longitudinal beams and five transverse beams is recommended for bridge decks consisting of slabs. The beams will, therefore, be of rectangular cross-section having width equal to width or span of deck/number of beams chosen. The second moment of area and torsional inertia for the beam cross-section can be determined in the usual way. Torsional inertia of a rectangular section is calculated as $C = kbt^3$ in which b is the length of the long side of the cross-section, t the length of the short side and k a factor depending on the b/t ratio. The value of k can be obtained from any book on elasticity (Timoshenko and Goodier, 1970) and it is taken as 1/3 for b/t ratio of more than 10.

Example 3.6.1

Consider a reinforced concrete slab of 20 m long, 12 m wide and 1 m deep. The rein-forcement is assumed to be same order of magnitude in the longitudinal and trans-verse directions and hence the slab may be taken as isotropic. The proposed layout of the grillage consists of five longitudinal members and eight transverse members as shown in Figure 3.55. Widths of the longitudinal and transverse members are, respectively, 3 and 2.86 m. Second moment of area and torsion constant per unit width of the grillage member are calculated as:

$$i_x = i_y = 1.0^3/12 = 0.0834 \text{ per meter width of the grillage member}$$
$$c_x = c_y = 1.0^3/6 = 0.167 \text{ per meter width of the grillage member}$$

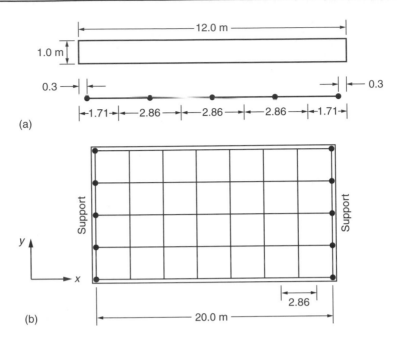

Figure 3.55 Example of an idealized slab deck (after Hambly, 1991).

For internal grillage members in the longitudinal direction, with 3 m width:

$$I_x = 3.0 \times 0.0834 = 0.25 \text{ m}^4, \quad C_x = 3.0 \times 0.167 = 0.501 \text{ m}^4$$

For edge members in the longitudinal direction:

$$I_x = 1.5 \times 0.0834 = 0.125 \text{ m}^4, \quad C_x = 1.5 \times 0.167 = 0.251 \text{ m}^4$$

Values of I and C for transverse members are calculated similarly and the grillage analysis can be carried out.

A shear-key deck can be represented by the grillage with longitudinal members coincident with the centre lines of the beams of the prototype. Each longitudinal grillage member has transverse outriggers which are stiff. The shear keys are represented by the pinned joints between the outriggers of adjacent beams. More detailed treatment of this subject may be found in the book *Bridge Deck Analysis* (Hambly, 1991).

3.6.2 Beam and slab decks
Selection of a suitable grillage mesh for a beam-and-slab deck depends upon the structural behaviour of a particular deck rather than on a set of rules. Figure 3.56 shows four examples of suitable grillage arrangement for four different types of decks (Cusens and Pama, 1975). The deck shown in Figure 3.56(a) is virtually a grid of longitudinal and transverse beams. The grillage simulates the prototype closely by having its members coincident with the centre lines of the prototypes beams. For the deck given in Figure 3.56(b), it is convenient and physically reasonable to place

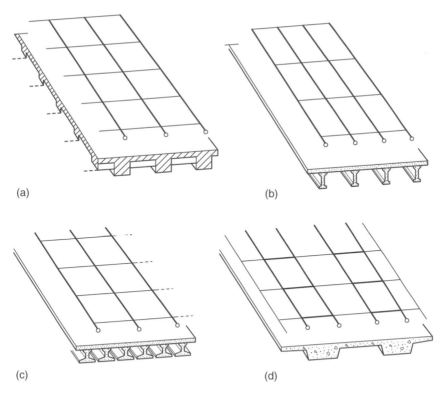

(a)   (b)

(c)   (d)

Figure 3.56 Different grillage idealization of beam-and-slab deck (after Hambly, 1991).

longitudinal grillage members coincident with the centre lines of the prototype's beams. With no midspan transverse diaphragms, the spacing of transverse grillage members is somewhat arbitrary. Where there is a diaphragm in the prototype such as over a support, then a grillage member should be coincident. In a deck with contiguous beams very closely as shown in Figure 3.56(c) it becomes unmanageable to assume grillage beam members coincident with all beams. Therefore, it can be expedient to represent more than one beam by each longitudinal grillage member. It should also be noted that it is generally unwise to use a single grillage member to represent two beams of markedly different stiffnesses. The deck in Figure 3.56(d) has large beams with width forming a significant fraction of the distance between the centre lines. The longitudinal beams are then best represented as slabs with two longitudinal members per beam.

The general form of construction for the beam and slab shown in Figure 3.57 is to have relatively few (10 or lower) longitudinal beams at 1.5–2.5 m spacing connected by a top slab. There will invariably be transverse diaphragms at the support as well as within the span. The exception in this structural form is the use of inverted T beams with top slab only; these will be placed closer together and there will therefore be no more longitudinal beams. The natural choice of longitudinal grillage beams for the I beam decks is to have them coincident with the physical beams. However, when considering inverted T beam decks, if there are many more than nine physical beams these should be replaced by about nine equally-spaced grillage beams positioned such that the centre-lines of the edge grillage beams are coincident with the centre-lines of the edge physical beams. It is recommended that the number of beams should be an odd number. Should the physical edge beams be different from the internal they must be replaced by grillage beams of equivalent stiffness and the internal uniform section treated as above but with a reduced number of beams. If the deck is extremely wide and is formed from many physical beams it may be necessary to increase the number of longitudinal grillage beams so that, as a general rule, one grillage beam does not replace more than two physical beams.

The next step is to determine the inertia for the beams. For the longitudinal beams the moment of inertia, I can be calculated for a section consisting of a physical beam and its associated width of top slab. It is recommended by West (1971) that it can be proportioned to all the grillage beams as ($I \times$ number of physical beams)/number of grillage beams. The torsional inertia of the beam cross-section is determined by idealizing it as a section consisting of a number of rectangles. The total inertia of the section can be considered to be the sum of the inertia of the individual rectangles. The torsional inertia for an individual rectangles can be obtained in the same way as that explained in the case of slab structure.

The transverse grillage beams within the span are considered to consist of diaphragm plus slab. Other alternative such as considering top slab of width corresponding to the width of a transverse grillage beam is also recommended by West (1971). The torsional inertia of such sections may then be obtained as an I section or a plain slab. It should, however, be ensured that torsional stiffness used in the analysis is truly representative of the physical beam.

Example 3.6.2

West (1971) suggests a simple grillage arrangement for a bridge structure consisting of I beams with *in situ* concrete top slab as shown in Figure 3.57. The actual structure of 25 m span and 17 m wide consists of nine precast I beams at 2 m spacing, *in situ* reinforced concrete slab and *in situ* post-tensioned abutment diaphragms. It is proposed to choose the longitudinal beams to coincide with the physical beams, i.e. nine beams of equal stiffness with the inertias calculated as for an internal beam with

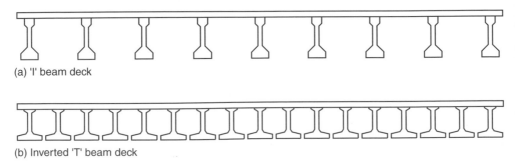

(a) 'I' beam deck

(b) Inverted 'T' beam deck

Figure 3.57 Grid-type structures.

2 m width of top slab. Using a ratio of 1.5:1 the spacing of transverse beams should be approximately 3 m. Nine beams at 3.125 m spacing are chosen in the transverse direction. The internal beams are of rectangles each of 3.125 m wide while the abutment beams are assumed to be of an L beam consisting full diaphragm concrete section with 1.56 m slab.

If the above structure consists, in addition to post-tensioned abutments, of pre-stressed diaphragms positioned at quarter points it is proposed that five transverse beams coincident with the physical beams can be used. The three internal beams are T-beams with 6.25 m of top slab, and the abutment diaphragms are L beams with 3.125 m of top slab.

A grillage beam of MoT/C&CA standard beam M7 with top slab only as shown in Figure 3.58(a) can be idealized as a section consisting of rectangles as shown in Figure 3.58(b) for the purpose of torsional inertia calculations. The total inertia can be considered to be the sum of the inertias of the individual rectangles as follows:

$$J_1 = 0.3 \times 160^3 \times 1000 \times 0.5 = 0.614 \times 10^9 \text{ mm}^4$$
$$J_2 = 0.294 \times 75^3 \times 400 = 0.050 \times 10^9 \text{ mm}^4$$
$$J_3 = 0.292 \times 160^3 \times 815 = 0.975 \times 10^9 \text{ mm}^4$$
$$J_4 = 0.292 \times 185^3 \times 950 = 1.756 \times 10^9 \text{ mm}^4$$

The total inertia, $J = (0.614 + 0.05 + 0.975 + 1.756) \times 10^9 = 3.395 \times 10^9 \text{ mm}^4$.

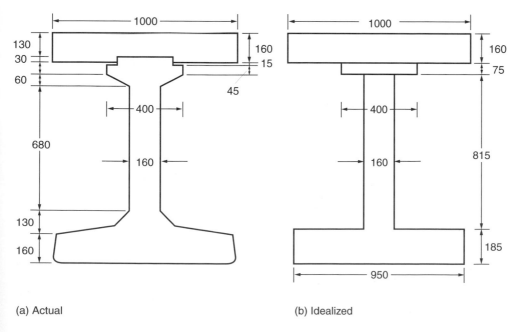

(a) Actual   (b) Idealized

Figure 3.58 T-beam and top slab (after West, 1971).

Example 3.6.3

Figure 3.59 shows part of a composite deck constructed of reinforced concrete slab on steel beams. Hambly (1991) suggests that longitudinal grillage members are placed coincident with the centre lines of steel beams, and each represents the part of the concrete slab as shown in the figure. With modular ratio $m = 7$ for steel (short-term loading) the second moment of area and the torsional constant for the grillage beam are, respectively, obtained as:

$$I_x = 0.21 \text{ m}^4 \text{ (steel beam and concrete slab) and}$$

$$C_x = (0.000031 \times 7) + (2.2 \times 0.2^3)/6 = 0.0032$$

Second moment area I_x and torsional constant C_x are calculated for the composite section consisting of the steel beam and the associated width of the concrete slab as shown in Figure 3.59(b).

Example 3.6.4

For a beam-and-slab deck constructed of spaced prestressed precast concrete box beams and reinforced concrete slab as shown in Figure 3.60, Hambly (1991) suggests that longitudinal grillage members are placed coincident with centre line of beams, with additional 'nominal' members running along centre lines of slab strips. The section properties of the nominal members are calculated for width of slab midway to neighbouring beams, hence:

$$I_x = 1.4 \times (0.25^3/12) = 0.0018 \text{ and } C_x = (1.4 \times 0.25^3/6) = 0.0036$$

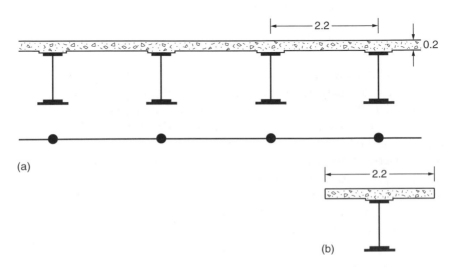

Figure 3.59 Steel-beam and concrete slab deck (after Hambly, 1991).

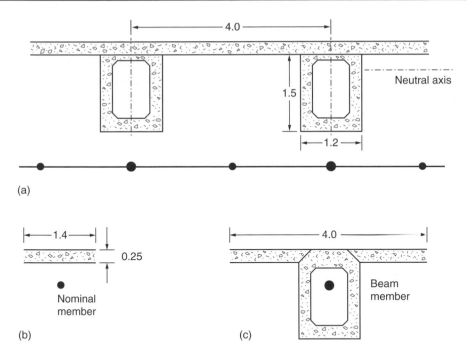

Figure 3.60 Cross-section of a deck with spaced box-beams (after Hambly, 1991).

The properties of the beam members are calculated for the sections with flanges including the area in nominal members (unless shear lag has reduced the effective width of flanges to less), but with previously calculated properties of 'nominal' members deducted. If the beams are much wider than those in Figure 3.60 in comparison with the beam spacing, account must be taken of the variation in transverse flexural stiffness between slab and beam. If the walls are thin, distortion of the cross-section must be considered.

3.63 Pseudo-slab structure

Pseudo slab structure may be formed either by inverted T-beams acting in conjunction with top slab (Figure 3.61a) or by box beams with in-situ voided slabs (Figure 3.61b). It is recommended that grillage beams for this type of bridge decks can be chosen such that the beams are coincident with the centre line of the physical beams as in the case of beam and slab decks. The cross-section of the beams may be assumed as box section and second moment of area for each of the grillage beam is calculated as in the other cases. Torsional inertia for this type of structures should be carefully determined in order to avoid underestimation of the torsional resistance of this type of cross-sections. West (1971) has recommended that in the case of decks with with-in span diaphragms the longitudinal and transverse inertia are calculated first as boxes and the values of C thus obtained be proportioned to all grillage beams as $0.5 \times C \times$ number of physical beams/number of grillage beams. For decks without

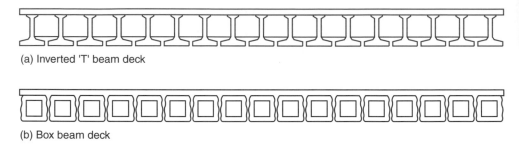

(a) Inverted 'T' beam deck

(b) Box beam deck

Figure 3.61 Pseudo-slab structures.

diaphragms, however, West views the requirement of the grillage simulation against the actual behaviour of the deck. Two twisting inertia, one longitudinal and one transverse, are provided in order to reflect the overall twisting inertia values of the actual deck. The twisting inertia for transverse and longitudinal beams are computed as $C_t = aC(1 + a)$ and $C_l = C(1 + a)$, respectively. In this, C represents the twisting inertia of the longitudinal cross-section and aC the twisting inertia of the transverse cross-section.

The torsional inertia, C for a box section is obtained as:

$$C = \frac{4A^2}{\oint \dfrac{ds}{t}}$$

in which A is the area inside the median line of the concrete walls and the denominator represents the sum of the lengths of the sides around the median line each divided by the corresponding wall thickness. This expression is for thin-walled boxes but will give sufficiently accurate results for box sections where both the void dimensions are greater than the total thickness of concrete in the same direction.

Example 3.6.5

The expression for torsional stiffness referred above is for thin-walled boxes but will give sufficiently accurate results for box sections where both the void dimensions are greater than the total thickness of concrete in the same direction. Let us consider the MoT/C&CA standard beam M7 used in pseudo-box construction shown in Figure 3.62 (West, 1971). The idealized section is shown in Figure 3.62(b). The thickness of the bottom *in situ* concrete is taken as the maximum thickness and the calculations are based on the median line.

$$A = 170 \times 800 + 775 \times 920 = 0.849 \times 10^6 \text{ mm}^2$$

$$\oint \frac{ds}{t} = \frac{170}{200} \times 2 + \frac{775}{80} \times 2 + \frac{800}{160} + \frac{920}{130} = 33.15$$

$$C = \frac{4 \times (0.849 \times 10^6)^2}{33.15} = 86.97 \times 10^9 \text{ mm}^4$$

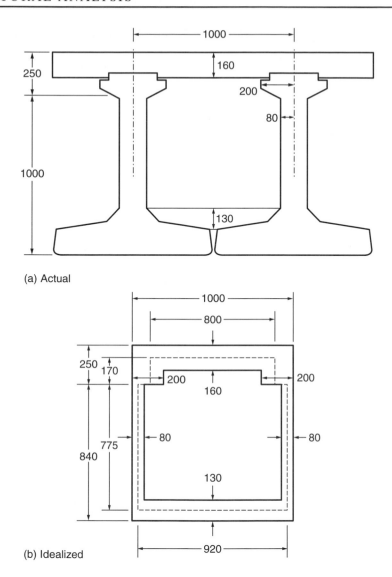

(a) Actual

(b) Idealized

Figure 3.62 Inverted T-section as pseudo-box.

3.6.4 Skew decks

The recommendations made in the preceding sections on the choice of equivalent grillage beams apply to skew decks also. There is clearly a choice between the selection of an orthogonal or skew system of grillage beams although the directions chosen should correspond with the reinforcement layout. There are three possibilities (Cusens and Pama, 1975) for a wide skew deck as shown in Figure 3.63. The skew system of beams in which transverse members are positioned parallel to the supports (Figure 3.63a) and the structural parameters are calculated using the orthogonal distance between the grillage beams. This system provides easier data preparation but, the

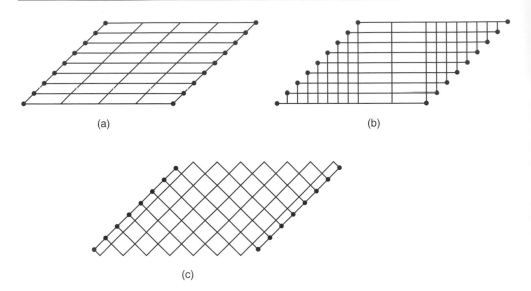

Figure 3.63 Different types of skew decks.

output will be in an inconvenient form as moments will relate to axes in the directions of the members. Alternatively, an orthogonal grid system (Figures 3.63b and c) in which transverse beams are positioned orthogonal to the longitudinal beams may be employed. West (1971) has suggested that diaphragm in the deck must be represented by an equivalent beam in the grillage simulation and that if they are within the span this will define the direction of the transverse members.

3.6.5 Grillage analysis for steel cellular structures

The application of a simplified grillage technique to the analysis of multi-cellular structures has been presented by Evans and Shanmugam (1979b) with reference to steel structures. It is proposed that a muti-cellular structure can be idealized as an equivalent grillage as shown in Figure 3.64. The centre lines of the grillage members are assumed to coincide with those of the physical webs and the beam cross-section is assumed as an I-section consisting of a web plate and an assumed effective width of flange. The problems of shear lag and torsional stiffness in idealizing the cellular structure into a grillage discrete beam elements have been solved by incorporating suitable effective width and torsional constant to the beam elements. It has been proposed that the flange effective width can be calculated in accordance with shear lag factors given in the Merrison Design Rules (Committee of Inquiry into the Basis for Design and Method of Erection of Steel Box Girder Bridges, 1973) for bridges. These rules were based on an extensive finite element studies carried out by Moffat and Dowling (1975). Additional information on shear lag effect in box girders were provided by Evans and Taherian (1977). Torsional constant for grillage beams was assumed to be equal to that of a closed cell in the multi-cell structure. Several examples of multi-cell structures were analysed by finite element method and the

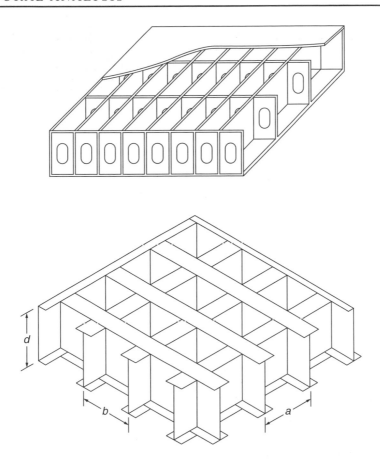

Figure 3.64 Grillage idealization of a multi-cell steel structures.

proposed grillage method in order to establish the accuracy of the method. Effect of web openings also has been considered in the Grillage analysis by Evans and Shanmugam (1979c, d). Further references to grillage analysis can be made to Hambly (1991) and, Cusens and Pama (1975).

3.7 Finite element method

The advent of high-speed electronic digital computers has given tremendous impetus to numerical methods for solving engineering problems. Finite element methods form one of the most versatile classes of such methods which rely strongly on the matrix formulation of structural analysis. The application of finite element dates back to the mid-1950s with the pioneering work by Argyris (1960), Clough and Penzien (1993) and others.

The finite element method was first applied to the solution of plane stress problems and subsequently extended to the analysis of axisymmetric solids, plate bending problems and shell problems. A useful listing of elements developed in the past is documented in text books on finite element analysis.

Stiffness matrices of finite elements are generally obtained from an assumed displacement pattern. Alternative formulations are equilibrium elements and hybrid elements. A more recent development is the so-called strain based elements. The formulation is based on the selection of simple independent functions for the linear strains or change of curvature; the strain–displacement equations are integrated to obtain expressions for the displacements.

3.7.1 Concept of the finite element method

The finite element method is based on the representation of a body or a structure by an assemblage of subdivisions called finite elements as shown in Figure 3.65. These elements are considered to be connected at nodes. Displacement functions are chosen to approximate the variation of displacements over each finite element. Polynomials are commonly employed to express these functions. Equilibrium equations for each element are obtained by means of the principle of minimum potential energy. These equations are formulated for the entire body by combining the equations for the individual elements so that the continuity of displacements is preserved at the nodes. The resulting equations are solved satisfying the boundary conditions in order to obtain the unknown displacements.

The entire procedure of the finite element method involves the following steps: (i) the given body is subdivided into an equivalent system of finite elements, (ii) suitable displacement function is chosen, (iii) element stiffness matrix is derived using variational principle of mechanics such as the principle of minimum potential energy, (iv) global stiffness matrix for the entire body is formulated, (v) the algebraic equations thus obtained are solved to determine unknown displacements and (vi) element strains and stresses are computed from the nodal displacements.

3.7.2 Basic equations from theory of elasticity

Figure 3.66 shows the state of stress in an elemental volume of a body under load. It is defined in terms of three normal stress components σ_x, σ_y and σ_z and three

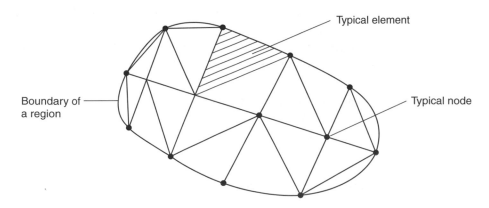

Typical element

Boundary of
a region

Typical node

Figure 3.65 Assemblage of subdivisions.

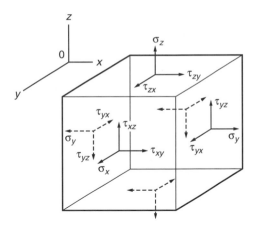

Figure 3.66 State of stress in an elemental volume.

shear stress components τ_{xy}, τ_{yz} and τ_{zx}. The corresponding strain components are three normal strain ε_x, ε_y and ε_z and three shear strain γ_{xy}, γ_{yz} and γ_{zx}. These strain components are related to the displacement components u, v and w at a point as follows:

$$\varepsilon_x = \frac{\partial u}{\partial x} \qquad \gamma_{xy} = \frac{\partial v}{\partial x} + \frac{\partial u}{\partial y}$$

$$\varepsilon_y = \frac{\partial v}{\partial y} \qquad \gamma_{yz} = \frac{\partial w}{\partial y} + \frac{\partial v}{\partial z} \qquad (3.58)$$

$$\varepsilon_z = \frac{\partial w}{\partial z} \qquad \gamma_{zx} = \frac{\partial u}{\partial z} + \frac{\partial w}{\partial x}$$

The relations given in Equation (3.58) are valid in the case of the body experiencing small deformations. If the body undergoes large or finite deformations, higher order terms must be retained.

The stress–strain equations for isotropic materials may be written in terms of the Young's modulus and Poisson's ratio as:

$$\sigma_x = \frac{E}{1-v^2}[\varepsilon_x + v(\varepsilon_y + \varepsilon_z)]$$

$$\sigma_y = \frac{E}{1-v^2}[\varepsilon_y + v(\varepsilon_z + \varepsilon_x)]$$

$$\sigma_z = \frac{E}{1-v^2}[\varepsilon_z + v(\varepsilon_x + \varepsilon_y)] \qquad (3.59)$$

$$\tau_{xy} = G\gamma_{xy}, \qquad \tau_{yz} = G\gamma_{yz}, \ \tau_{zx} = G\gamma_{zx}$$

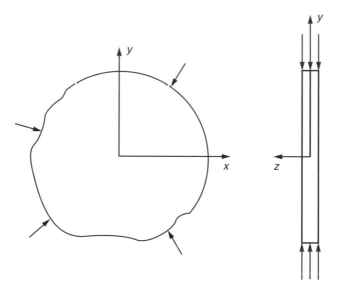

Figure 3.67 Plane-stress problem.

3.7.3 Plane stress

When the elastic body is very thin and there are no loads applied in the direction parallel to the thickness, the state of stress in the body is said to be plane stress. A thin plate subjected to in-plane loading as shown in Figure 3.67 is an example of a plane stress problem. In this case, $\sigma_z = \tau_{yz} = \tau_{zx} = 0$ and the constitutive relation for an isotropic continuum is expressed as:

$$\begin{bmatrix} \sigma_x \\ \sigma_y \\ \tau_{xy} \end{bmatrix} = \frac{E}{1-v^2} \begin{bmatrix} 1 & v & 0 \\ v & 1 & 0 \\ 0 & 0 & \dfrac{1-v}{2} \end{bmatrix} \begin{bmatrix} \varepsilon_x \\ \varepsilon_y \\ \gamma_{xy} \end{bmatrix} \qquad (3.60)$$

3.7.4 Plane strain

The state of plane strain occurs in members that are not free to expand in the direction perpendicular to the plane of the applied loads. Examples of some plane strain problems are retaining walls, dams, long cylinder, tunnels, etc. as shown in Figure 3.68. In these problems ε_z, γ_{yz} and γ_{zx} will vanish and hence:

$$\sigma_z = v(\sigma_x + \sigma_y)$$

The constitutive relations for an isotropic material is written as:

$$
\begin{bmatrix} \sigma_x \\ \sigma_y \\ \tau_{xy} \end{bmatrix} = \frac{E}{(1-v)(1-2v)} \begin{bmatrix} (1-v) & v & 0 \\ v & (1-v) & 0 \\ 0 & 0 & \dfrac{1-2v}{2} \end{bmatrix} \begin{bmatrix} \varepsilon_x \\ \varepsilon_y \\ \gamma_{xy} \end{bmatrix} \tag{3.61}
$$

3.7.5 Element shapes, discretization and mesh density

The process of subdividing a continuum is an exercise of engineering judgement. The choice by an analyst depends upon the geometry of the body. A finite element generally has a simple one, two or three-dimensional configuration. The boundaries of elements are often straight lines and the elements can be one-dimensional, two-dimensional or three-dimensional as shown in Figure 3.69. While subdividing the continuum one has to decide the number, shape, size and configuration of the elements in such a way that the original body is simulated as closely as possible. Nodes must be located in locations where abrupt changes in geometry, loading, and material properties occur. A node must be placed at the point of application of a concentrated load because all loads are converted into equivalent nodal-point loads.

It is easy to subdivide a continuum into a completely regular one having same shape and size. But problems encountered in practice do not involve regular shape; they may have regions of steep gradients of stresses. A finer subdivision may be necessary in regions where stress concentrations are expected in order to obtain a useful approximate solution. Typical examples of mesh selection are shown in Figure 3.70.

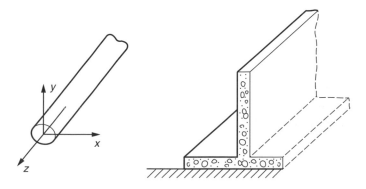

Figure 3.68 Practical examples of plane-strain problems.

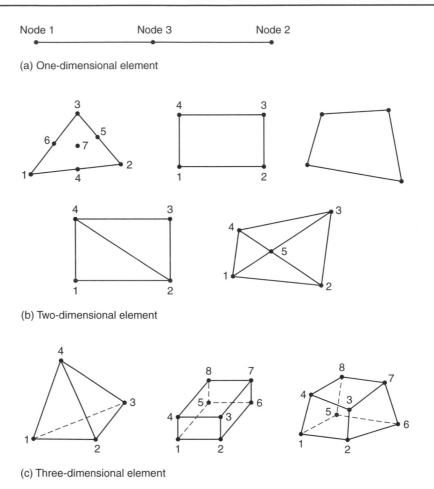

(a) One-dimensional element

(b) Two-dimensional element

(c) Three-dimensional element

Figure 3.69 (a) One-dimensional element. (b) Two-dimensional elements. (c) Three-dimensional elements.

3.7.6 Choice of displacement function

Selection of displacement function is the important step in the finite element analysis, since it determines the performance of the element in the analysis. Attention must be paid to select a displacement function which: (i) has the number of unknown constants as the total number of degrees of freedom of the element, (ii) does not have any preferred directions, (iii) allows the element to undergo rigid-body movement without any internal strain, (iv) is able to represent states of constant stress or strain and (v) satisfies the compatibility of displacements along the boundaries with adjacent elements. Elements which meet both the third and fourth requirements are known as *complete elements*.

A polynomial is the most common form of displacement function. Mathematics of polynomials are easy to handle in formulating the desired equations for various elements and convenient in digital computation. The degree of approximation is

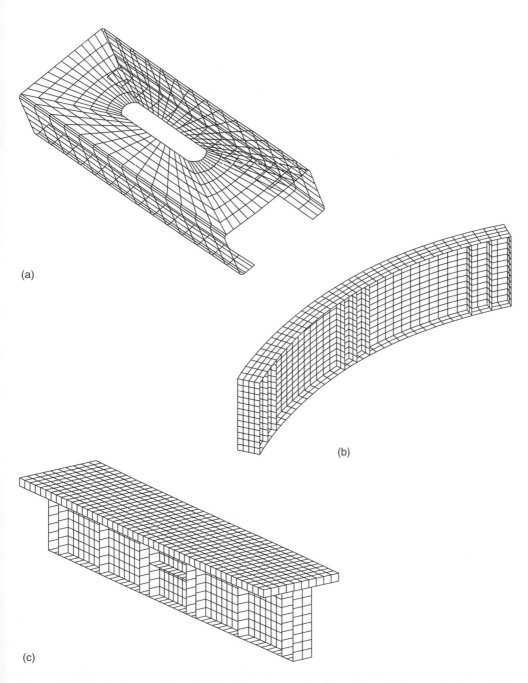

(a)

(b)

(c)

Figure 3.70 Typical examples of finite element mesh: slotted lipped channel, plate girder curved in plan, and steel–concrete composite plate girder.

governed by the stage at which the function is truncated. Solutions closer to exact solutions can be obtained by including more number of terms. The polynomials are of the general form:

$$w(x) = a_1 + a_2x + a_3x^2 + \ldots a_{n+1}x^n \qquad (3.62)$$

The coefficients a are known as generalized displacement amplitudes. The general polynomial form for a two-dimensional problem can be given as:

$$u(x, y) = a_1 + a_2x + a_3y + a_4x^2 + a_5xy + a_6y^2 + \ldots a_my^n$$

$$v(x, y) = a_{m+1} + a_{m+2}x + a_{m+3}y + a_{m+4}x^2 + a_{m+5}xy + a_{m+6}y^2 + \ldots + a_{2m}y^n$$

in which:

$$m = \sum_{i=1}^{n+1} i \qquad (3.63)$$

These polynomials can be truncated at any desired degree to give constant, linear, quadratic or higher order functions. For example, a linear model in the case of two dimensional problem can be given as:

$$u = a_1 + a_2x + a_3y$$

$$v = a_4 + a_5x + a_6y \qquad (3.64)$$

A quadratic function is given by:

$$u = a_1 + a_2x + a_3y + a_4x^2 + a_5xy + a_6y^2$$

$$v = a_7 + a_8x + a_9y + a_{10}x^2 + a_{11}xy + a_{12}y^2 \qquad (3.65)$$

3.7.7 Nodal degrees of freedom
The deformation of the finite element is specified completely by the nodal displacement, rotations and/or strains which are referred to as degrees of freedom. Convergence, geometric isotropy and potential energy function are the factors which determine the minimum number of degrees of freedom necessary for a given element. Additional degrees of freedom beyond the minimum number may be included for any element by adding secondary external nodes and such elements with additional degrees of freedom are called higher order elements. The elements with more additional degrees of freedom become more flexible.

3.7.8 Isoparametric elements
The scope of finite element analysis is also measured by the variety of element geometries that can be constructed. Formulation of element stiffness equations

requires the selection of displacement expressions with as many parameters as there are node-point displacements. In practice, for planar conditions, only the four-sided (quadrilateral) element finds as wider application as the triangular element. The simplest form of quadrilateral, the rectangle, has four node points and involves two displacement components at each point for a total of eight degrees of freedom. In this case one would choose four-term expressions for both u and v displacement fields. If the description of the element is expanded to include nodes at the mid-points of the sides an eight-term expression would be chosen for each displacement component.

The triangle and rectangle can approximate the curved boundaries only as a series of straight line segments. A closer approximation can be achieved by means of isoparametric coordinates. These are non-dimensionalized curvilinear coordinates whose description is given by the same coefficients as are employed in the displacement expressions. The displacement expressions are chosen to insure continuity across element interfaces and along supported boundaries, so that geometric continuity is assured when the same forms of expressions are used as the basis of description of the element boundaries. The elements in which the geometry and displacements are described in terms of the same parameters and are of the same order are called isoparametric elements. The isoparametric concept enables one to formulate elements of any order which satisfy the completeness and compatibility requirements and which have isotropic displacement functions.

3.7.9 Element shape functions

The finite element method is not restricted to the use of linear elements. Most finite element codes, commercially available, allow the user to select between elements with linear or quadratic interpolation functions. In the case of quadratic elements fewer elements are needed to obtain the same degree of accuracy in the nodal values. Also, the two-dimensional quadratic elements can be shaped to model a curved boundary. Shape functions can be developed based on the following properties: (i) each shape function has a value of one at its own node and is zero at each of the other nodes, (ii) the shape functions for two-dimensional elements are zero along each side that the node does not touch and (iii) each shape function is a polynomial of the same degree as the interpolation equation.

3.7.10 Formulation of stiffness matrix

It is possible to obtain all the strains and stresses within the element and to formulate the stiffness matrix and a consistent load matrix once the displacement function has been determined. This consistent load matrix represents the equivalent nodal forces which replace the action of external distributed loads.

As an example, let us consider a linearly elastic element of any of the types shown in Figure 3.71. The displacement function may be written in the form:

$$\{f\} = [P]\{A\} \tag{3.66}$$

in which $\{f\}$ may have two components $\{u, v\}$ or simply be equal to w, $[P]$ is a function of x and y only, and $\{A\}$ is the vector of undetermined constants. If Equation

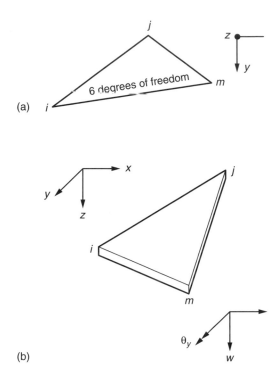

Figure 3.71 Degrees of freedom: (a) triangular plane-stress element, (b) triangular bending element.

(3.66) is applied repeatedly to the nodes of the element one after the other, we obtain a set of equations of the form:

$$\{D^*\} = [C]\{A\} \tag{3.67}$$

in which $\{D^*\}$ is the nodal parameters and $[C]$ is the relevant nodal coordinates. The undetermined constants $\{A\}$ can be expressed in terms of the nodal parameters $\{D^*\}$ as:

$$\{A\} = [C]^{-1}\{D^*\} \tag{3.68}$$

Substituting Equation (3.68) into (3.66):

$$\{f\} = [P][C]^{-1}\{D^*\} \tag{3.69}$$

Constructing the displacement function directly in terms of the nodal parameters one obtains:

$$\{f\} = [L]\{D^*\} \tag{3.70}$$

where $[L]$ is a function of both (x, y) and $(x, y)_{i,j,m}$ given by:

$$[L] = [P][C]^{-1} \tag{3.71}$$

The various components of strain can be obtained by appropriate differentiation of the displacement function. Thus:

$$\{\varepsilon\} = [B]\{D^*\} \tag{3.72}$$

$[B]$ is derived by differentiating appropriately the elements of $[L]$ with respect to x and y. The stresses $\{\sigma\}$ in a linearly elastic element are given by the product of the strain and a symmetrical elasticity matrix $[E]$. Thus:

$$\{\sigma\} = [E]\{\varepsilon\}$$

or:

$$\{\sigma\} = [E][B]\{D^*\} \tag{3.73}$$

The stiffness and the consistent load matrices of an element can be obtained using the principle of minimum total potential energy. The potential energy of the external load in the deformed configuration of the element is written as:

$$W = -\{D^*\}^T \{Q^*\} - \int_a \{f\}^T \{q\} \, da \tag{3.74}$$

In Equation (3.74) $\{Q^*\}$ represents concentrated loads at nodes and $\{q\}$ the distributed loads per unit area. Substituting for $\{f\}^T$ from Equation (7.13) one obtains:

$$W = -\{D^*\}^T \{Q^*\} - \{D^*\}^T \int_a [L]^T \{q\} \, da \tag{3.75}$$

Note that the integral is taken over the area a of the element. The strain energy of the element integrated over the entire volume v, is given as:

$$U = 1/2 \int_v \{\varepsilon\}^T \{\sigma\} \, dv$$

Substituting for $\{\varepsilon\}$ and $\{\sigma\}$ from Equations (3.72) and (3.73), respectively:

$$V = 1/2 \{D^*\}^T \left(\int_v [B]^T [E][B] \, dv \right) \{D^*\} \tag{3.76}$$

The total potential energy of the element is:

$$V = U + W$$

(or):

$$V = 1/2\{D^*\}^T (\int_v[B]^T[E][B]\mathrm{d}v)\{D^*\} - \{D^*\}^T\{Q^*\} - \{D^*\}^T\int_a[L]^T\{q\}\mathrm{d}a \tag{3.77}$$

Using the principle of minimum total potential energy, we obtain:

$$(\int_v[B]^T[E][B]\mathrm{d}v)\{D^*\} = \{Q^*\} + \int_a[L]^T\{q\}\mathrm{d}a$$

or:

$$[K]\{D^*\} = \{F^*\} \tag{3.78}$$

where:

$$[K] = \int_v[B]^T[E][B]\mathrm{d}v \tag{3.79}$$

and:

$$\{F^*\} = \{Q^*\} + \int_a[L]^T\{q\}\mathrm{d}a \tag{3.80}$$

Stiffness matrix can now be developed for problems such as plates subjected to in-plane forces or beam elements or plates in bending as the case may be. As mentioned before, the accuracy is dependent on choosing an appropriate mesh. Many refined elements giving greater accuracy are described in standard books on finite element methods (Zienkiewicz and Cheung, 1967; Desai and Abel, 1972; Nath, 1974). Finite element methods are nowadays adopted for solving a wide range of problems. It is possible these days to buy finite element software which can be employed to solve many complex analytical problems.

3.7.11 Non-linear elements

Various behaviours are called 'non-linear'. Stress–strain relations may be non-linear in either a time-dependent or time-independent way. Displacements may cause loads to alter their distribution or magnitude. Connected parts may slip or stick. Gaps in a structure may be open or closed. The problem may be static or dynamic. Non-linear analysis is harder to understand and more involved because of the mathematical complexity. Nevertheless, non-linear analyses are becoming more common because of stringent design requirements and because finite elements and the computer have made non-linear analysis a practical possibility. Most analysis programs allow the user to choose among various kinds of non-linearity and solution algorithms. A user must understand the problem and the analytical tools well enough to make an intelligent choice. Even then, several analyses may be needed to get a satisfactory result. An incremental analysis gives an answer, but another analysis with a different step length is needed to estimate the quality of the answer. An iterative analysis may fail to converge because of a bug in the program or in the data, numerical error, greater non-linearity than the algorithm can accommodate, or a prescribed load greater than the structure's collapse load.

Non-linear problems are usually solved by taking a series of linear steps. In structural terms, the process of non-linear analysis is explained by writing equilibrium equations in the incremental form:

$$[K]\{\Delta D\} = \{\Delta R\}$$

Here the stiffness matrix $[K]$ is a function of displacements $\{D\}$ because the problem is non-linear. In turn, the current $\{D\}$ is the sum of preceding $\{\Delta D\}$s. The current $[K]$, called the tangent stiffness, is used to compute the next step, $\{\Delta D\}$. Then we update $\{D\}$, update $[K]$, and are ready to take another step. In this way a load versus displacement curve is approximated by a series of straight-line segments and the accuracy of the results depends upon the number of segments. Similar solutions can apply to both geometric non-linearity and material non-linearity and computer programs available can allow both to be active at the same time.

The essential feature of geometric non-linearity is that equilibrium equations must be written with respect to the deformed geometry which is not known in advance. Only if the nature of the problem is substantially unchanged by deformation do we call the problem 'linear' and so presume that equilibrium equations can refer to the initial configuration. A large-displacement problem can be analysed in Lagrangian coordinates or in Eulerian coordinates. If stress–strain relations are linear, or non-linear but elastic, there is a unique relation between stress and strain. But if there are plastic strains, the stress–strain relation is path dependent, not unique: a given state of stress can be produced by many different straining procedures. In addition, different materials require different material theories. The essential computational problem of material non-linearity is that equilibrium equations must be written using material properties that depend on the strains, but the strains are not known in advance. For detailed treatment of geometric non-linearity, material non-linearity, formulation, solution algorithm and choice of solution method readers may refer to the reference books (Oden, 1972; Cook, 1981).

3.7.12 Finite strip method

The finite strip method combines some of the benefits of the series solution of orthotropic plates with the finite element concept. Bridge decks having the same cross-section from one end to the other can be analysed more economically and efficiently by this method. When it is not possible to find a displacement function applicable to all regions of a plate, it is suggested that the plate may be divided into discrete longitudinal strip spanning between supports. The strips are connected by nodes which run along the sides of the strips as shown in Figure 3.72. Simple displacement interpolation functions may be used to represent displacement fields within and between individual strips. For a longitudinal strip I (Figure 3.72) simply supported at its ends, the displacement function may be assumed as a third-degree polynomial in the form:

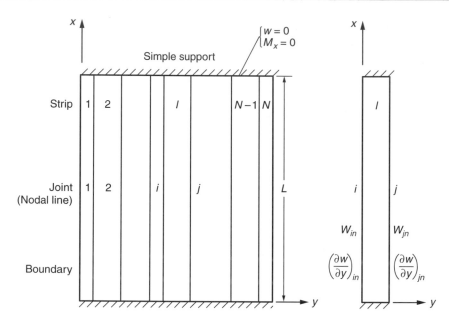

Figure 3.72 Finite strip simulation for plate bending.

$$w(x,y) = \sum_{n=1}^{\infty} (a_1 + a_2 y + a_3 y^2 + a_4 y^3) \sin \frac{n \pi x}{L}$$

satisfying the boundary conditions of the strip. The displacement function is assumed to be a simple polynomial in the transverse direction resulting in discontinuity of stresses at the strip interfaces. The above equation may be written in terms of the nodal line displacements in order to establish continuity at the boundaries with adjacent strips. The formulation and solution is similar to the finite element method and the detailed account of the method may be found in the references (Cheung and Cheung, 1971; Cheung et al., 1971; Cusens and Pama, 1975). Any form of loading may be conveniently handled by this method and the method is well-suited for computer use. The method has, however, some limitations in the sense that it is effectively applicable only to prismatic structures with simply supported ends.

3.8 Stiffness of supports: soil–structure interaction

3.8.1 General behaviour

Bridge decks are supported by piers and abutments, which are again supported by foundations resting on soil. The stiffness of piers, abutments, foundations and of soil are all significant in analysing the bridge for safety.

Piers and abutments together termed the sub-structure are subjected to many types of loading which include live and dead loads from the superstructure, dead load of substructure, soil pressure, wind load on the superstructure, substructure and on vehicles, pressure resulting from stream flow and ice and earthquake load. In addition, settlement of supports gives rise to secondary effects that result in significant changes to the distribution of forces within the bridge deck. It is, therefore, essential for a bridge designer to devote sufficient attention to assess the stiffness of sub-structures. The distribution of forces in super-structure and sub-structure of a bridge depends on relative stiffnesses of all its components. Continuous structures can be very effective in distributing forces, but they are also very sensitive to the effects of compressibility of supports and foundations. Any reasonable combination of these loads can be applied in the design of substructures.

The substructure is designed for stability, strength, and limit of soil pressure. One load combination may be most critical for stability while another loading may produce maximum stresses in the concrete or reinforcing steel, and a third combination may give the greatest soil pressure. If a portion of the pier is submerged, the buoyancy effect must be considered in determining the stability of the pier. Because of the many possible load combinations, substructure design is complicated. Careful soil investigation should be carried out in order to determine the stability, earth pressures, and permissible soil pressures. If the bridge is to be located in an earthquake region, locations of fault zones, slide regions, or any possible large unstable ground areas should be carefully studied before the bridge location and type are selected.

The distribution of forces throughout the foundation system of the structure is significantly influenced by the relative stiffnesses of all the components of the bridge, including the individual bearings. Continuous structures, though effective at distributing forces, are very sensitive to the effects of compressibility of supports and foundations. For example, a box girder deck, which is very stiff against torsion and distortion is very sensitive to differential settlement and compression of bearings.

Depending on the geometry of the substructure, loads from the superstructure will be transmitted to the foundation either by direct compression or by compression and bending. Piers shown in Figure 3.73(a)–(c) transmit the deck loads to the foundations by direct compression. On the other hand, structures which involve cross-head beam or cantilever as shown in Figure 3.73(d) and (e) may have significant flexibility (Hambly, 1991). If a bridge deck on such supports is analysed with a grillage approximation, it may be necessary to model the cross-head as well as the deck. It may be easier to model the structure with a space frame that reproduces the shape of the pier. It is then relatively simple to model compressible bearings with vertical members between deck and cross-head, and to model the foundation stiffnesses. For example, the deck shown in Figure 3.74(a) supported on pier (Figure 3.73d) can be idealized by a space frame model as shown in Figure 3.74(b).

3.8.2 Foundation stiffness and interaction with soil

A foundation subjected to dynamic loading will experience a motion at the same frequency as the applied force. It is essential to impose a limit on the dynamic force that the foundations may experience. There is a prescribed limit for the dynamic motions to be permitted, and also a prescribed limit upon the settlement that may develop. Methods based on the theory for an elastic half space have been proposed by researchers (Barkan, 1962; Gorbunov-Possadov; Serebrajanyi, 1961) to determine

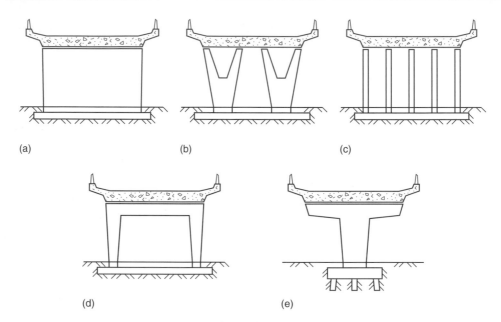

(a)   (b)   (c)

(d)   (e)

Figure 3.73 Bridge supports having different stiffnesses (after Hambly, 1991).

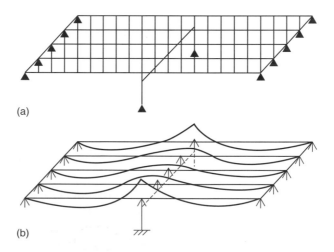

(a)

(b)

Figure 3.74 Space frame model of a bridge deck and the bending moment diagram (after Hambly, 1991).

the stiffnesses of shallow footing foundations (Figure 3.75). Equations from these methods have been simplified and approximate estimates of stiffnesses obtained by Hambly (1991). These simplified equations are reproduced below for shallow footings that may slide or tilt across the shorter direction.

Shear modulus: $$G = \frac{E}{2(1+v)} \tag{3.81}$$

Vertical stiffness:
$$K_z = \frac{2.5G\, A^{0.5}}{(1-v)} \qquad (3.82)$$

Horizontal stiffness:
$$K_x = 2G(1+v)A^{0.5} \qquad (3.83)$$

Rocking stiffness:
$$K_m = \frac{2.5G\, Z}{(1-v)} \qquad (3.84)$$

where G is the shear modulus of soil, E is the Young's modulus of soil, v = Poisson's ratio of soil, A is a foundation area $(b \times d)$ and Z is the foundation section modulus $(bd^2/6)$. Vertical and rotational stiffnesses can be more conveniently represented by two parallel springs as shown in Figure 3.75(e) for which the stiffness is taken as:

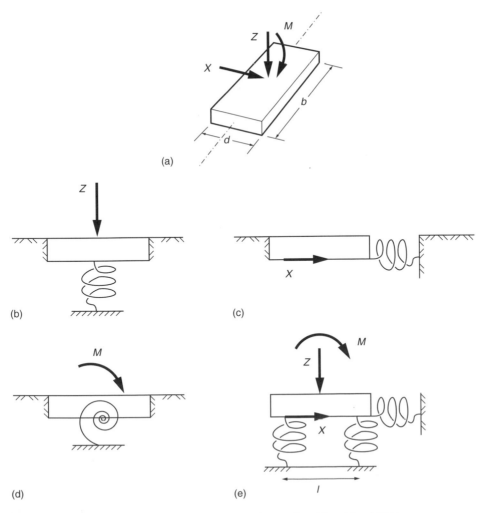

Figure 3.75 Spring models for stiffness of footing (after Hambly, 1991).

$K = 0.5K_z$. The parallel springs are spaced at a distance l given by $l = 2(K_m/K_z)^{0.5}$ $= 0.82b^{0.25}d^{0.75}$. These equations are approximate; however, they provide a quick means of determining the order of magnitude of foundation stiffnesses.

The stiffness of the ground under a foundation depends on the load paths and paths of reacting forces. As can be expected the vertical loads on the foundations will produce vertical soil reactions at great depths. On the other hand, under arching action, the horizontal force and moment on one foundation will be nullified by an equal and opposite force and moment from the other foundation and this may result in an increase in the horizontal stiffness of the soil, if the foundations are relatively close. When the horizontal breaking forces apply unbalanced reactions to the foundations, the horizontal stiffness of the foundation is reduced by concurrent loading on neighbouring foundation. Due to these and similar complexities, it is not always practicable to allow for these subtleties, when selecting foundation stiffness.

Detailed investigations of the stiffnesses of foundations for bridges over complicated ground conditions can be carried out with three-dimensional finite element analysis. Finite element programs which can represent the soils with very sophisticated non-linear stress–strain behaviour with coexisting porewater pressures are now available. Soil data of quality essential for the sophisticated computer programs is not frequently available. Where the soil support is applied some depth below the usable portion of the structure, pile foundation is the most commonly used. Wooden piles, concrete piles, steel piles, steel tubes filled with concrete are common; Figure 3.76 shows the

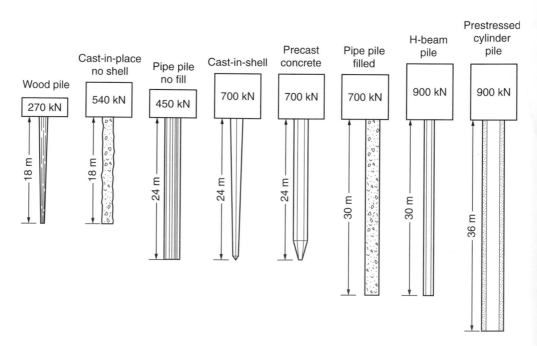

Figure 3.76 Usual maximum length and maximum design load for various types of piles.

values of usual maximum length and maximum design load for various types of piles (Lambe and Whitman, 1979). The computations for stiffness of pile foundations are more complicated, particularly if they obtain their stiffnesses from interaction of pile bending and lateral forces from the soil, rather than by axial compression of the piles. There are several computer programs available which can be used to calculate the stiffnesses of pile groups, either using finite elements, or the theory for an elastic half space (Poulos and Davis, 1980).

Let us consider for example the foundations for portal frame shown in Figure 3.77. The footings for the portal frame have $d = 3$ m (parallel to span) and $b = 12$ m wide. Hence:

$$\text{Footing area, } A = 12 \times 3 = 36 \text{ m}^2$$

$$\text{Section modulus of the footing, } Z = 12 \times 3^2/6 = 18 \text{ m}^3$$

The stiffnesses, with Poisson's ratio in the order of 0.3 to 0.5, can be calculated approximately as:

$$K_z = 1.5E(A)^{0.5} = 1.5E(36)^{0.5} = 9E$$

$$K_x = E(A)^{0.5} = E(36)^{0.5} = 6E$$

$$K_m = 1.5EZ = 1.5E(18) = 27E$$

For an assumed value of $E = 120$ MPa, we find that $K_z = 1080$ MN/m, $K_x = 720$ MN/m and $K_m = 3240$ MN/m. The total width of the foundation is 12 m and stiffnesses per unit width are calculated as:

$$K_z = 1080/12 = 90 \text{ MN/m/m}$$

$$K_x = 720/12 = 60 \text{ MN/m/m}$$

$$K_m = 3240/12 = 270 \text{ MN/m/rad/m}$$

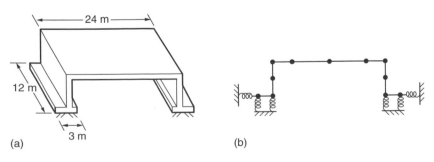

(a)   (b)

Figure 3.77 Arrangement and plane frame model of a portal-type bridge (after Hambly, 1991).

Stiffnesses K_z and K_m are represented by two parallel springs of:

$$K = 0.5K_z = 45 \text{ MN/m/m}$$

at a spacing of:

$$l = 2(K_m/K_z) = 2(270/90)^{0.5} = 3.464 \text{ m}$$

The influence of more complicated ground conditions can be investigated with a finite element analysis in which any degree of sophistication can be included.

3.8.3 Stiffness moduli of soils

Foundation stiffnesses are functions of many variables that include Young's modulus of soil beneath the foundation. The elastic modulus of granular soils based on effective stresses is a function of grain size, gradation, mineral composition of the soil grains, grain shape, soil type, relative density, soil particle arrangement, stress level and prestress. Typical values of the elastic modulus and Poisson's ratio for normally consolidated granular soils (Das, 1990) are given in Table 3.1. The undrained elastic modulus of cohesive soils is a function primarily of soil plasticity and overconsolidation. It can be determined from the slope of a stress–strain curve obtained from an undrained triaxial test (Holtz and Kovacs, 1981)

   In order to obtain ground stiffnesses one has to determine, by site investigation, the soil properties. This is not straightforward, because tests on small samples seldom provide reliable information of large soil mass. Stiffnesses are best estimated from large-diameter plate bearing tests and from back analyses of observations of comparable structures on similar ground conditions. Modulus and Poisson's ratio are not constants for a soil, but rather are quantities which approximately describe the behaviour of a soil for a particular set of stresses. Different values of modulus and Poisson's ratio will apply for any other set of stresses. Good judgment is needed when choosing values of these parameters. It is difficult to estimate values of modulus with much accuracy, and test data for the particular soil will be necessary whenever an accurate estimate is needed. The terms tangent modulus and secant modulus are used frequently. Specific guidance and information, available in text books, are helpful for preliminary assessments of soil–structure interaction. However, at final design stage calculations should be supported by investigations of the specific site and foundation conditions.

Table 3.1 Typical values of elastic modulus and Poisson's ratio for granular soils (Das, 1990)

Type of soil	Elastic modulus (MPa)	Poisson's ratio
Loose sand	10–24	0.20–0.40
Medium dense sand	17–28	0.25–0.40
Dense sand	35–55	0.30–0.45
Silty sand	10–17	0.20–0.40
Sand and gravel	69–170	0.15–0.35

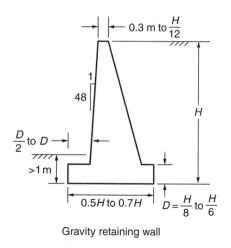

Gravity retaining wall

Figure 3.78 Typical dimensions of a gravity retaining wall.

3.8.4 Retaining walls

It is not uncommon to design bridge abutments as gravity type retaining walls typical dimensions of which are shown in Figure 3.78. Lateral active earth pressures for design of low retaining walls are usually estimated using conservative design charts. Some designers use also equivalent fluid pressures to compute the active earth pressure as $p_a = \gamma_{eq}H$ in which γ_{eq} equals the product of the minimum active earth pressure coefficient, K_a and the unit weight, γ of the backfill material. More rigorous methods using Rankine or Coulomb theories can also be used to determine the earth pressure. All earth retention structures should be designed to sustain potential surcharge loadings. It is assumed that retaining walls are constructed with free-draining backfill materials and with effective drainage systems so that ponding of water which may result in additional load on the retaining structure can be minimized or eliminated. If it is not possible to preclude ponding the wall should be designed to account for higher total pressures. Design of retaining walls involves several steps that include checks for overturning stability, bearing capacity, sliding, excessive settlement and overall stability of the earth mass that contains the retaining structure.

3.8.5 Stiffness from lateral earth pressure

Horizontal forces exerted on the bridge supports are influenced by the earth pressure, thermal expansion and vehicle braking forces. The distribution of these forces on abutments and piers depends upon their relative horizontal stiffnesses. Great care should be exercised while estimating these stiffnesses. Earth pressure–displacement diagrams can be employed to estimate the lateral stiffness of an abutment (Hambly and Burland, 1979). The following example (Hambly. 1991) demonstrates how the horizontal stiffness can be calculated from earth pressure resistance and rocking rotation of an abutment.

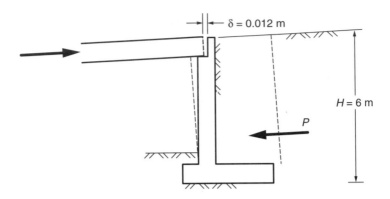

Figure 3.79 Bridge abutment resisting horizontal forces from deck (after Hambly, 1991).

Let us consider the abutment of 6 m high shown in Figure 3.79. Initially guess a displacement of 12 mm at deck level, i.e. 0.002 of the height. The earth pressure coefficient would correspondingly change from about 0.4 to 1.5. The earth resistance force per unit width of abutment, with retained fill of density γ, is:

$$P = \tfrac{1}{2}K\gamma H^2$$

inclined to the horizontal at the angle of the wall friction. Along the wall all components of stress increase linearly with depth, and so the resultant thrust acts at the third-point of the wall. With a uniform surcharge q_s (Figure 3.79), the total active thrust against the wall is given by:

$$P = \tfrac{1}{2}K\gamma H^2 + q_s HK$$

Note that the horizontal stress resulting from the surcharge is distributed uniformly with depth, and hence the resultant force corresponding to the surcharge is located at mid-height of the wall. Thus the resultant of the total thrust, reflecting the effects of surcharge and weight of soil, will lie between mid-height and the third point. The location of the resultant of the total thrust is found by vectorial addition of the thrusts for each of the two components.

The change in force, due to K changing from 0.4 to 1.5 with $\gamma = 0.020$ MN/m^3, is:

$$P = (1.5 - 0.4) \times 0.020 \times 6^2/2 = 0.40 \text{ MN/m}$$

The resultant earth pressure force acts at one-third height (2 m) up the abutment wall. Thus the resistance provided at deck level is (by taking moments about the foundation):

$$R = 0.40 \times 2/6 = 0.13 \text{ MN/m}$$

Since the assumed displacement at deck level is 0.012 m, the effective stiffness K_x per meter width is:

$$K_x = 0.13/0.012 = 11 \text{ MN/m/m}$$

An abutment 12 m wide would provide a stiffness of 132 MN/m, that is, a resistance force of 1.6 MN for the displacement of 12 mm.

3.8.6 Integral bridges

Bridges constructed without any movement joints between spans or between spans and abutments are called integral bridges. Integral bridges are becoming more widespread as designers endeavour to avoid the maintenance problems that develop at joints between simply supported decks (Hambly and Nicholson, 1990). The road surfaces are continuous from one approach embankment to the other. In this case, bridge deck, piers, abutments, embankments and ground must all be considered as a single compliant system. The load distribution in integral bridges depends on the relative stiffnesses of all the components and all the material, and structural stiffnesses must, therefore, estimated as realistically as possible. There will be thermal movement between the abutments and the approach road, but integral bridges are likely to be most useful in circumstances where the pavement does not have to be interrupted by mechanical expansion joints. Integral bridges are generally designed with the stiffnesses and flexibilities spread throughout the structure–soil system. Expansion and contraction of an integral bridge will cause the abutment to move. This movement will be resisted by earth pressure behind the abutments, and possibly by friction underneath. Abutments should normally be designed to minimize such resistance to movement, as these horizontal forces have to be carried as compression or tension by the deck. Even small abutments are likely to be sufficient to resist the horizontal traffic loads.

The analysis of the interaction of an integral bridge with its environment has been presented by Hambly (1991) with the aid of a global model. Figure 3.80(a) shows a longitudinal section of half of the length of the bridge with run-on slab while a plane frame model is shown in Figure 3.80(b). The deck is continuous over the piers (on a footing of 14×3m) and is built into the abutments (on a footing of 14×2m). The run-on slabs are modelled with pin-joints at the abutments. Each abutment is supported by two spring supports and restrained horizontally by lateral springs on the footing and at the centroid of passive earth pressure. Each pier is supported on two spring supports and restrained horizontally by a lateral spring and has a moment release at the top to simulate a bearing with dowel. The spring stiffnesses of the supports are calculated with the Equations (3.81)–(3.84). For example, under the action of short-term braking forces, the soil might have Young's modulus $E = 40$ MPa. With Poisson's ratio γ, of the order of 0.3 to 0.5 we can calculate for springs under the footings of the piers as:

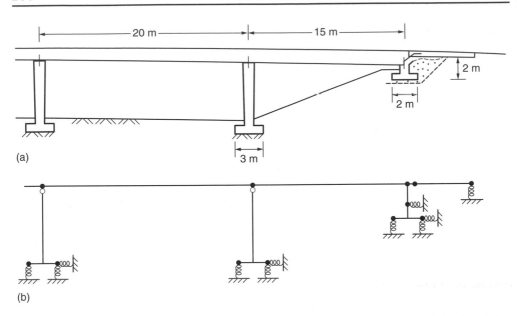

Figure 3.80 Longitudinal section and plane frame model of an integral bridge (after Hambly, 1991).

Two vertical springs	$K = 0.5 \times 1.5EA^{0.5} = 200$ MN/m
at spacing	$l = 0.82 \times 14^{0.25} \times 3^{0.75} = 3.6$ m
horizontal springs	$K_x = EA^{0.5} = 260$ MN/m

In the same way for springs under the footings of the abutments we have:

Two vertical springs	$K = 0.5 \times 1.5EA^{0.5} = 160$ MN/m
at spacing	$l = 0.82 \times 14^{0.25} \times 2^{0.75} = 2.7$ m
horizontal springs	$K_x = EA^{0.5} = 210$ MN/m

It is found from calculations under ultimate temperature loads that the displacement just exceeds 0.01 m and then the horizontal springs on the abutment footings are replaced by the limiting forces $F = 2.1$ MN resisting sliding. It is easy to check by using two vertical springs under each footing that no spring goes into tension under combined weight and temperature loading. It is shown that the total range of thermal movement is about 40 mm, or 20 mm at each abutment. No special treatment of shrinkage effects is required for the bridge (Nicholson, 1994).

More detailed treatment of integral bridges may be found in the publication by the Steel Construction Institute (Biddle *et al.*, 1997). It provides an introduction to the concepts relating to 'integral bridges' and illustrates ways in which the ordinary composite beam-and-slab deck bridge can be adapted to become an integral bridge. Worked examples in a companion publication (Way and Yandzio, 1997) illustrate many of the design aspects of integral bridges.

3.9 Structural dynamics

3.9.1 Equation of motion

The essential physical properties of a linearly elastic structural system subjected to external dynamic loading are its mass, stiffness properties and energy absorption capability or damping. The principle of dynamic analysis may be illustrated by considering a simple single-storey structure as shown in Figure 3.81. The structure is subjected to a time-varying force $f(t)$. k is the spring constant that relates the lateral storey deflection x to the storey shear force, and the dash pot relates the damping force to the velocity by a damping coefficient c. If the mass, m, is assumed to be concentrated at the beam, the structure becomes a single-degree-of-freedom (SDOF) system. The equation of motion of the system may be written as:

$$m\ddot{x} + c\dot{x} + kx = f(t) \qquad (3.85)$$

Solutions to Equation (3.85) can give an insight into the behaviour of the structure under dynamic situation.

3.9.2 Free vibration

In this case the system is set to motion and allowed to vibrate in the absence of applied force $f(t)$. Letting $f(t) = 0$, Equation (3.85) becomes:

$$m\ddot{x} + c\dot{x} + kx = 0 \qquad (3.86)$$

Dividing Equation (3.86) by the mass m, we have:

$$\ddot{x} + 2\xi\omega\dot{x} + \omega^2 x = 0 \qquad (3.87)$$

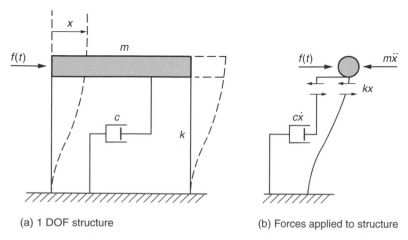

(a) 1 DOF structure   (b) Forces applied to structure

Figure 3.81 A single-storey structure – principle of dynamic analysis.

where:

$$2\xi\omega = c/m \text{ and } \omega^2 = k/m \qquad (3.88)$$

The solution to Equation (3.87) depends on whether the vibration is damped or undamped.

Case 1: Undamped free vibration

In this case, $c = 0$, and the solution to the equation of motion may be written as:

$$x = A\sin \omega t + B\cos \omega t \qquad (3.89)$$

where $\omega = \sqrt{(K/m)}$ is the circular frequency. A and B are constants that can be determined by the initial boundary conditions. In the absence of external forces and damping, the system will vibrate indefinitely in a repeated cycle of vibration with an amplitude of:

$$\sqrt{(A^2 + B^2)} \qquad (3.90)$$

and a natural frequency of:

$$f = \frac{\omega}{2\pi} \qquad (3.91)$$

The natural period is:

$$T = \frac{2\pi}{\omega} = \frac{1}{f} \qquad (3.92)$$

The undamped free vibration motion as described by Equation (3.89) is shown in Figure 3.82.

Case 2: Damped free vibration

If the system is not subjected to applied force and damping is present, the corresponding solution becomes:

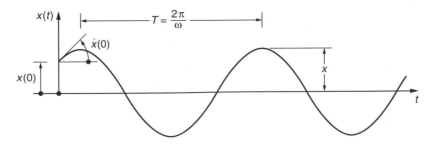

Figure 3.82 Undamped free vibration motion.

$$x = A \exp(\lambda_1 t) + B\exp(\lambda_2 t) \tag{3.93}$$

where:

$$\lambda_1 = \omega\left[-\xi + \sqrt{\xi^2 - 1}\right] \tag{3.94}$$

$$\lambda_2 = \omega\left[-\xi - \sqrt{\xi^2 - 1}\right] \tag{3.95}$$

The solution of Equation (3.93) changes its form with the value of ξ defined as:

$$\xi = \frac{c}{2\sqrt{mk}} \tag{3.96}$$

If $\xi^2 < 1$, the equation of motion becomes

$$x = \exp(-\xi\omega t)(A \cos \omega_d t + B \sin \omega_d t) \tag{3.97}$$

where ω_d is the damped angular frequency defined as:

$$\omega_d = \sqrt{(1 - \xi^2)}\omega \tag{3.98}$$

For most building structure ξ is very small (about 0.01) and therefore $\omega_d \approx \omega$. The system oscillates about the neutral position as the amplitude decays with time t. Figure 3.83 illustrates an example of such motion. The rate of decay is governed by the amount of damping present.

If the damping is large, then oscillation will be prevented. This happens when $\xi^2 > 1$ and the behaviour is generally referred to as over-damped. The motion of such behaviour is shown in Figure 3.84.

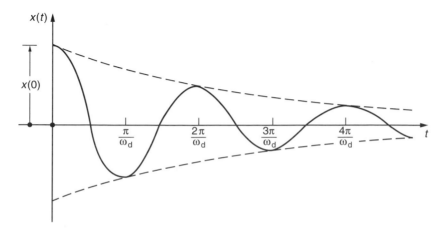

Figure 3.83 Damped free vibration motion.

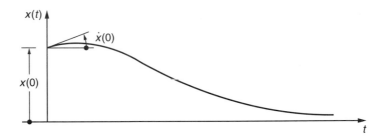

Figure 3.84 Damped free vibration motion – overdamped.

Damping with $\xi^2 = 1$ is called critical damping. This is the case where minimum damping is required to prevent oscillation and the critical damping coefficient is given as:

$$c_{cr} = 2\sqrt{(km)} \qquad\qquad (3.99)$$

The degree of damping in the structure is often expressed as a proportion of the critical damping value. Referring to Equations (3.98) and (3.99), we have:

$$\xi = \frac{c}{c_{cr}} \qquad\qquad (3.100)$$

ξ is called the damping ratio.

3.9.3 Forced vibration

If a structure is subjected to a sinusoidal motion such as a ground acceleration of $\ddot{x} = F \sin \theta_f t$, it will oscillate and after some time the motion of the structure will reach a steady state. For example the equation of motion due to the ground acceleration (from Equation 3.87) is:

$$\ddot{x} + 2\xi\omega\dot{x} + \omega^2 x = -F\sin \omega_f t \qquad\qquad (3.101)$$

The solution to the above equation consists of two parts; the complimentary solution given by Equation (3.89) and the particular solution. If the system is damped, oscillation corresponding to the complementary solution will decay with time. After some time the motion will reach a steady state, and the system will vibrate at a constant amplitude and frequency. This motion, which is called forced vibration, is described by the particular solution expressed as:

$$x = C_1\sin \omega_f t + C_2\cos \omega_f t \qquad\qquad (3.102)$$

It can be observed that the steady forced vibration occurs at the frequency of the excited force, ω_f, not the natural frequency of the structure, ω.

Substituting Equation (3.102) into (3.101), the displacement amplitude can be shown to be:

$$X = -\frac{F}{\omega^2} \frac{1}{\sqrt{\left[1-\left(\dfrac{\omega_f}{\omega}\right)^2\right]^2 + \left(\dfrac{2\xi\omega_f}{\omega}\right)^2}}$$ (3.103)

The term $-F/\omega^2$ is the static displacement caused by the force due to the inertia force. The ratio of the response amplitude relative to the static displacement $-F/\omega^2$ is called the dynamic displacement amplification factor, D, given as:

$$D = \frac{1}{\sqrt{\left[1-\left(\dfrac{\omega_t}{\omega}\right)^2\right]^2 + \left(\dfrac{2\xi\omega_f}{\omega}\right)^2}}$$ (3.104)

The variation of the magnification factor with the frequency ratio ω_f/ω and damping ratio ξ is shown in Figure 3.85.

When the dynamic force is applied at a frequency much lower than the natural frequency of the system ($\omega_f/\omega < 1$), the response is quasi-static. The response is proportional to the stiffness of the structure, and the displacement amplitude is close to the static deflection.

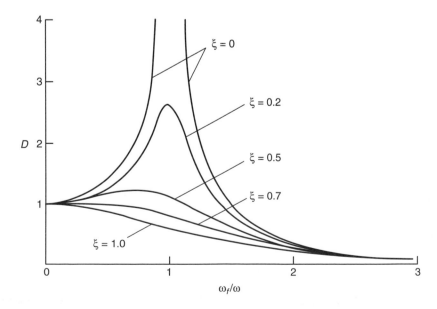

Figure 3.85 Variation of amplification factor.

When the force is applied at a frequency much higher than the natural frequency ($\omega_f/\omega < 1$), the response is proportional to the mass of the structure. The displacement amplitude is less then the static deflection ($D > 1$).

When the force is applied at a frequency close to the natural frequency, the displacement amplitude increases significantly. The condition at which ω_f/ω is called resonance.

Similarly, the ratio of the acceleration response relative to the ground acceleration may be expressed as:

$$D_a = \left|\frac{\ddot{x} + \ddot{x}_g}{\ddot{x}_g}\right| = \frac{\sqrt{1 + \left(\dfrac{2\xi\omega_f}{\omega}\right)^2}}{\sqrt{\left[1 - \left(\dfrac{\omega_f}{\omega}\right)^2\right]^2 + \left(\dfrac{2\xi\omega_f}{\omega}\right)}} \tag{3.105}$$

D_a is called the dynamic acceleration magnification factor.

3.9.4 Response to suddenly applied load

Consider the spring–mass damper system of which a load P_o is applied suddenly. The differential equation is given by:

$$M\ddot{x} + c\dot{x} + kx = P_o \tag{3.106}$$

If the system is started at rest, the equation of motion is:

$$x = \frac{P_0}{k}\left[1 - \exp(-\xi\omega t)\left\{\cos\omega_d t + \frac{\xi\omega}{\omega_d}\sin\omega_d t\right\}\right] \tag{3.107}$$

If the system is undamped, then $\xi = 0$ and $\omega_d = \omega$, we have:

$$x = \frac{P_o}{k}[1 - \cos\omega_d t] \tag{3.108}$$

The maximum displacement is $2(P_o/k)$ corresponding to $\cos\omega_d t = -1$. Since P_o/k is the maximum static displacement, the dynamic amplification factor is equal to two. The presence of damping would naturally reduce the dynamic amplification factor and the force in the system.

3.9.5 Response to time-varying loads

Some forces and ground motions that are encountered in practice are rather complex in nature. In general, numerical analysis is required to predict the response of such effects, and the finite element method is one of the most common techniques to be employed in solving such problems.

The evaluation of responses due to time-varying loads can be carried out using the piecewise exact method. In using this method, the loading history is divided into small time intervals. Between these points, it is assumed that the slope of the load curve remains constant. The entire load history is represented by piece-wise linear curve, and the error in this approach can be minimized by reducing the length of the time steps. Description of this procedure is given in Clough and Penzien (1993).

Other techniques employed include Fourier analysis of the force function followed by solution for Fourier components in the frequency domain. For random forces, random vibration theory and spectrum analysis may be used (Warburton, 1976; Dowrick, 1988).

3.9.6 Multiple degree systems

In multiple degree systems, an independent differential equation of motion can be written for each degree of freedom. The nodal equations of a multiple degree system of n degrees of freedom may be written as:

$$[M]\{\ddot{x}\} + [c]\{\dot{x}\} + [k]\{x\} = \{F(t)\} \tag{3.109}$$

where $[m]$ is a symmetrical $n \times n$ matrix of mass, $[c]$ is a symmetrical $n \times n$ matrix of damping coefficient and $\{F(t)\}$ is the force vector which is zero in the case of free vibration.

Consider a system under free vibration without damping, the general solution of Equation (3.109) is assumed in the form of:

$$\begin{Bmatrix} x_1 \\ x_2 \\ \vdots \\ x_n \end{Bmatrix} = \begin{bmatrix} \cos(\omega t - \phi) & 0 & 0 & 0 \\ 0 & \cos(\omega t - \phi) & 0 & 0 \\ \vdots & \vdots & \vdots & \vdots \\ 0 & 0 & 0 & \cos(\omega t - \phi) \end{bmatrix} \begin{Bmatrix} C_1 \\ C_2 \\ \vdots \\ C_n \end{Bmatrix} \tag{3.110}$$

where angular frequency ω and phase angle ϕ are common to all xs. In this assumed solution, ϕ and C_1, C_2, C_n are the constants to be determined from the initial boundary conditions of the motion and ω is a characteristic value (eigenvalue) of the system.

Substituting Equation (3.110) into Equation (3.109) yields:

$$\begin{bmatrix} k_{11} - m_{11}\omega^2 & k_{12} - m_{12}\omega^2 & \cdots & k_{1n} - m_{1n}\omega^2 \\ k_{21} - m_{21}\omega^2 & k_{22} - m_{22}\omega^2 & \cdots & k_{2n} - m_{2n}\omega^2 \\ \vdots & \vdots & \vdots & \vdots \\ k_{n1} - m_{n1}\omega^2 & k_{n2} - m_{n2}\omega^2 & \cdots & k_{nn} - m_{nn}\omega^2 \end{bmatrix} \begin{Bmatrix} C_1 \\ C_2 \\ \vdots \\ C_n \end{Bmatrix} \cos(\omega t - \phi) = \begin{Bmatrix} 0 \\ 0 \\ \vdots \\ 0 \end{Bmatrix} \tag{3.111}$$

or:

$$[[k] - \omega^2[m]]\,\{C\} = \{0\} \tag{3.112}$$

where $[k]$ and $[m]$ are the $n \times n$ matrices, ω^2 and $\cos(\omega t - \phi)$ are scalars, and $\{C\}$ is the amplitude vector. For non-trivial solution, $\cos(\omega t - \phi) = 0$ thus solution to Equation (3.112) requires the determinant of $[[k] - \omega^2[m]] = 0$. The expansion of the determinant yields a polynomial of n degree as a function of ω^2, the n roots of which are the *eigenvalues* $\omega_1, \omega_2, \omega_n$.

If the eigenvalue ω for a normal mode is substituted in Equation (3.112), the amplitude vector $\{C\}$ for that mode can be obtained. $\{C_1\}, \{C_2\}, \{C_3\}, \{C_n\}$ are therefore called the eigenvectors, the absolute values of which must be determined through initial boundary conditions. The resulting motion is a sum of n harmonic motions, each governed by the respective natural frequency ω, written as:

$$\{x\} = \sum_{i=1}^{n}\{C_i\}\cos(\omega_i t - \phi_i) \tag{3.113}$$

3.9.7 Distributed mass systems

Although many structures may be approximated by lumped mass systems, in practice all structures are distributed mass systems consisting of infinite number of particles. Consequently, if the motion is repetitive, the structure has infinite number of natural frequency and mode shapes. The analysis of a distributed parameter system is entirely equivalent to that of a discrete system once the mode shapes and frequencies have been determined, because in both cases the amplitudes of the modal response components are used as generalized coordinates in defining the response of the structure.

In principle an infinite number of these coordinates are available for a distributed-parameter system, but in practice only a few modes, usually those of lower frequencies, will provide significant contribution to the overall response. Thus the problem of distributed-parameter system can be converted to a discrete system form in which only a limited number of modal coordinates is used to describe the response.

Flexural vibration of beams

The motion of the distributed mass system is best illustrated by a classical example of a uniform beam of span length L and flexural rigidity EI and a self-weight of m per unit length, as shown in Figure 3.86(a). The beam is free to vibrate under its self-weight. From Figure 3.86(b), dynamic equilibrium of a small beam segment of length dx requires:

$$\frac{\partial V}{\partial x}dx = mdx\frac{\partial^2 y}{\partial t^2} \tag{3.114}$$

in which:

$$\frac{\partial^2 V}{\partial x^2} = \frac{M}{EI} \tag{3.115}$$

and:

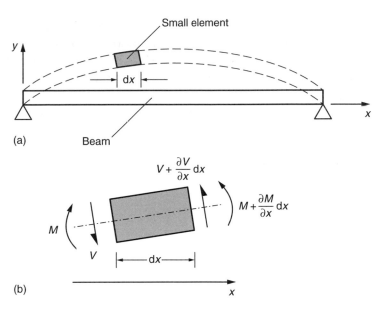

Figure 3.86 Motion of distributed mass system.

$$V = -\frac{\partial M}{\partial x}, \quad \frac{\partial V}{\partial x} = -\frac{\partial^2 M}{\partial x^2} \tag{3.116}$$

Substituting these equations into Equation (3.114) leads to the equation of motion of the flexural beam:

$$\frac{\partial^4 y}{\partial x^4} + \frac{m}{EI}\frac{\partial^2 y}{\partial t^2} = 0 \tag{3.117}$$

The above equation can be solved for beams with given sets of boundary conditions. The solution consists of a family of vibration mode with corresponding natural frequencies. Standard results are available in Figure 3.93 to compute the natural frequencies of uniform flexural beams of different supporting conditions. Methods are also available for dynamic analysis of continuous beams (Clough and Penzien, 1993).

Shear vibration of beams

Beams can deform by flexure or shear. Flexural deformation normally dominates the deformation of slender beams. Shear deformation is important for short beams or in higher modes of slender beams. Figure 3.94 gives the natural frequencies of uniform beams in shear, neglecting flexural deformation. The natural frequencies of these beams are inversely proportional to the beam length L rather than L^2, and the frequencies increase linearly with the mode number.

Combined shear and flexure

The transverse deformation of real beams is the sum of flexure and shear deformations. In general, numerical solutions are required to incorporate both the shear and flexural deformation in the prediction of natural frequency of beams. The following simplified formula may be used to estimate the beam's frequency in respect of beams with comparable shear and flexural deformations:

$$\frac{1}{f^2} = \frac{1}{f_f^2} + \frac{1}{f_s^2}$$
(3.118)

in which f is the fundamental frequency of the beam, and f_f and f_s are the fundamental frequencies predicted by the flexure and shear beam theory (Rutenburg, 1975).

3.9.8 Damping

Damping is found to increase with the increasing amplitude of vibration. It arises from the dissipation of energy during vibration. The mechanisms which contribute to energy dissipation are material damping, friction at interfaces between components and energy dissipation due to foundation interacting with soil, among others. Material damping arises from the friction at bolted connections and frictional interaction between structural and non-structural elements such as partitions and cladding.

The amount of damping in a building can never be predicted precisely, and design values are generally derived based on dynamic measurements of structures of a corresponding type. Damping can be measured based on the rate of decay of free vibration following an impact; by spectral methods based on analysis of response to wind loading; or by force excitation by mechanical vibrator at varying frequency to establish the shape of the steady state resonance curve. However, these methods may not be easily carried out if several modes of vibration close in frequency are presented.

3.9.9 Numerical analysis

Many less complex dynamic problems can be solved without much difficulty by hand methods. For more complex problems, such as determination of natural frequencies of complex structures, calculation of response due to time-varying loads and response spectrum analysis to determine seismic forces, may require numerical analysis. Finite element method has been shown to be a versatile technique for this purpose.

The global equations of an undamped motion, in matrix form, may be written as:

$$[M]\,\{\ddot{x}\} + [k]\{\dot{x}\} = F(t)$$
(3.119)

where:

$$[K] = \sum_{i=1}^{n}[K_1] \quad [M] = \sum_{i=1}^{n}[M_i] \quad [F] = \sum_{i=1}^{n}[f_i]$$
(3.120)

are the global stiffness, mass and force matrices, respectively. $[k_i]$, $[m_i]$ and $\{f_i\}$ are the stiffness, mass and force of the ith element, respectively. The elements are assembled using the direct stiffness method to obtain the global equations such that intermediate continuity of displacements is satisfied at common nodes and, in addition, interelement continuity of acceleration is also satisfied.

Equation (3.119) is the matrix equations discretized in space. To obtain solution of the equation, discretization in time is also necessary. The general method used is called direct integration. There are two methods for direct integration: implicit or explicit. The first, and simplest, is an explicit method known as the central difference method (Biggs, 1964). The second, more sophisticated but more versatile, is an implicit method known as the Newmark method (Newmark, 1959). Other integration methods are also available in the works by Bathe (1982).

The natural frequencies are determined by solving Equation (3.119) in the absence of force $F(t)$ as:

$$[M]\,\{\ddot{x}\} + [K]\{x\} = 0 \tag{3.121}$$

The standard solution for $x(t)$ is given by the harmonic equation in time:

$$\{x(t)\} = \{X\}e^{i\omega t} \tag{3.122}$$

where $\{X\}$ is the part of the nodal displacement matrix called natural modes which are assumed to be independent of time, i is the imaginary number and ω is the natural frequency.

Differentiating Equation (3.122) twice with respect to time, we have:

$$\ddot{x}(t) = \{X\}(-\omega^2)e^{i\omega t} \tag{3.123}$$

Substitution of Equations (3.122) and (3.123) into (3.121) yields:

$$e^{i\omega t}([K] - \omega^2[M])\{x\} = 0 \tag{3.124}$$

Since $e^{i\omega t}$ is not zero, we obtain:

$$([K] - \omega^2[M])\{X\} = 0 \tag{3.125}$$

Equation (3.125) is a set of linear homogeneous equations in terms of displacement mode $\{X\}$. It has a non-trivial solution if the determinant of the coefficient matrix $\{X\}$ is non-zero; that is:

$$[K] - \omega^2[M] = 0 \tag{3.126}$$

In general, Equation (3.126) is a set of n algebraic equations, where n is the number of degrees of freedom associated with the problem.

3.9.10 Seismic bridge behaviour

It is imperative for a bridge designer to understand the important principles of structural dynamics in order to apply the appropriate modelling and analysis tool to the bridge seismic response problems. A detailed treatment of this subject is given by Priestley *et al.* (1996); a brief account of the subject extracted from the book is given in the following sections.

The dynamic excitation and response of a bridge subjected to earthquake ground motion in the form of ground acceleration $\ddot{u}_g(t)$ can best be explained by means of a single-degree-of-freedom model of a bridge structure shown in Figure 3.87. This simplified model is applicable only if the bridge is straight consisting of a large number of equal spans and piers of equal height or stiffness as shown in Figure 3.87(a). It is assumed that all piers are exposed to the same coherent earthquake ground acceleration $\ddot{u}_g(t)$ perpendicular to the bridge axis and that the superstructure moves as a rigid body. In this model, the seismic mass m is assumed to be lumped at the top of the single-column bent at a height H above ground level. The cantilever stiffness of the assumed massless bridge pier can be expressed by k, which is the force required to produce a unit displacement at the centre of mass relative to the column base. Furthermore, if damping of the bridge system can be expressed in the form of viscous damping, the characteristic damping force required to resist a unit velocity at the point of mass can be expressed as c. The dynamic response of the bridge model can be expressed in the form:

$$u_t = u_s + u_g \tag{3.127}$$

in which u_t is the total displacement, u_s structural displacement of the cantilever pier and u_g the ground displacement relative to an absolute frame of reference.

Equilibrium of all forces acting on the system at this single displacement degree of freedom requires that the inertia force $f_i(t) = m\ddot{u}_t(t)$, the viscous damping force $f_d(t) = c\dot{u}_s(t)$ and the restoring force $f_s(t) = ku_s(t)$ can be combined in a single equilibrium equation at the unknown displacement degree of freedom in the form:

$$m(\ddot{u}_g + \ddot{u}_s) + c\dot{u}_s + ku_s = 0 \tag{3.128}$$

or:

$$m\ddot{u}_s + c\dot{u}_s + ku_s = -m\ddot{u}_g \tag{3.129}$$

which represents the equation of motion of the single-degree-of-freedom bridge model shown in Figure 3.87(c) under transverse earthquake ground accelerations.

Bridge dynamic response characteristics

Independent of the specific dynamic input, each bridge system is represented within the elastic range by dynamic response modes typically referred to as the natural modes of vibration, characterized by independent mode shape Φ_i with correspond-

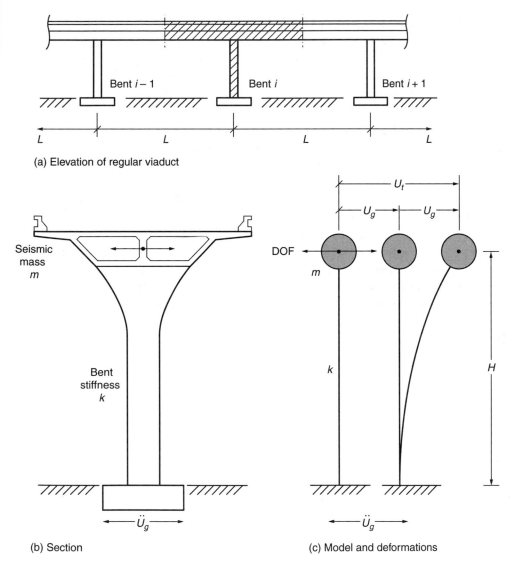

(a) Elevation of regular viaduct

(b) Section

(c) Model and deformations

Figure 3.87 Transverse dynamic bridge response model.

ing periods of vibration T_i. The number of characteristic mode shapes and vibration periods of a bridge model depend on the selected number of dynamic degrees of freedom defined during the analytical model discretization. In the case of a prototype bridge, there are an infinite number of vibration modes featured. But on the other hand, a selected finite number of degrees of freedom and associated modes of vibration are featured in the analytical model of a bridge. However, it is possible to capture the governing dynamic response of a bridge by the contribution of a limited number of vibration modes. A good indication of the dynamic response of a bridge can often be obtained from the fundamental or lowest mode of vibration. Single-degree-of-freedom models, therefore, serves as an invaluable tool for design office use.

Modelling of bridge structures

The mathematical formulation of the bridge behaviour satisfying a particular assessment or design requirement for a quantitative response determination should be simple. The model has to capture the physical and mechanical interactions of earthquake input and structure response. It should be capable of describing adequately the geometric domain, the seismic mass, the connection and boundary conditions, and the loading of the prototype. For each element in the model, member end forces and deformation relationships are defined by kinematic, constitutive and static relationships. Also, the correct seismic weight or mass associated with each degree-of-freedom also needs to be determined. It is essential to characterize the soil–structure interface and when the earthquake loading to the bridge is generated by ground motion. In the case of longer bridges consisting of a number of spans and frames attention should be paid to the necessity and validity of modelling the entire bridge in a global model. The model should be capable of capturing the effects of curves in plan and elevation, highly skew supports, ramps and interaction between frames in order to meet the seismic demand assessment. Some specific modelling issues in respect of bridge components are given below:

Superstructure

Most bridge structures can be considered as linear structures, where the span length L between bents is larger than the width B or depth D of the superstructure as shown in Figure 3.88. In many cases, the bridge superstructure can be assumed to move as a rigid body under seismic loads. The entire modelling problem is reduced to the stiffness modelling of the bents with geometric constraints simulating the rigid superstructure. In long and narrow bridges and interchange connectors, for example, the superstructure cannot be considered rigid. In such cases, the superstructure can modelled as a grillage of beam elements as shown in Figure 3.88(c) or a spine with beam elements following the centre of gravity of the cross section along the length of the bridge as shown in Figure 3.88(d). Equivalent member properties for the spine or beam elements representing the overall effective superstructure stiffness can be obtained by using the methods described in Section 3.6 on grillage analysis. For superstructure-carrying wide roadways, the spine model may be erroneous, particularly when combinations of earthquake forces with gravity loads and loads need to be investigated.

Single-column bents

In the seismic response analysis of bridges, the bents or supports are the critical structural elements that provide gravity-load and earthquake force transfer to the ground and ground motion input to the bridge superstructure. For a generic single-column bent as shown in Figure 3.89(a), different discretizations need to be considered, depend-

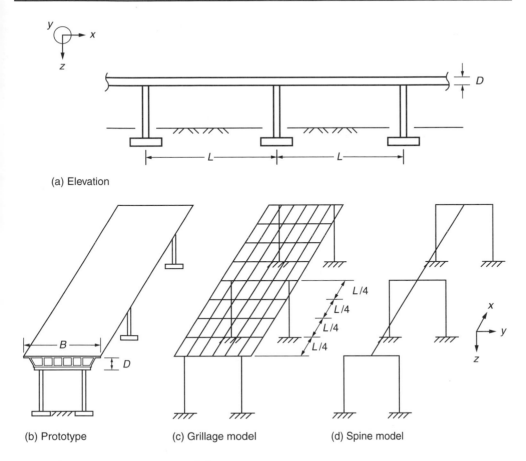

Figure 3.88 Superstructure models.

ing on the bent geometry and the expected seismic response. For essentially elastic response and a prismatic column, as shown in Figure 3.89(b), a single-column element connected at nodes 2 and 3 at the superstructure soffit and the top of the footing is sufficient to model the seismic response. For a tapered or flared single-column bents or bents where column mass effects may be important, a stepped discretization as shown in Figure 3.89(c) may be used. In the case of bridges with very tall or slender columns, the combined influence of gravity load and displacement can be significant even though the displacement may be small compared to the overall bridge dimensions. In these cases only a second-order or $P - \Delta$ effects need to be considered by evaluating equilibrium in the deformed configuration. $P - \Delta$ effects in the analytical modelling can be accounted by means of reduced capacity for lateral loads, as shown in Figure 3.90, and the yield and ultimate deformation limit states together with bending moment contributions from the lateral earthquake force F and the vertical load P due to gravity.

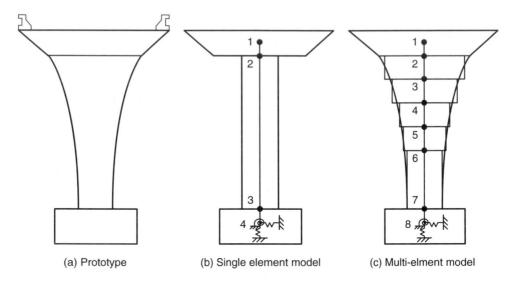

(a) Prototype   (b) Single element model   (c) Multi-elment model

Figure 3.89 Single-column-bent models.

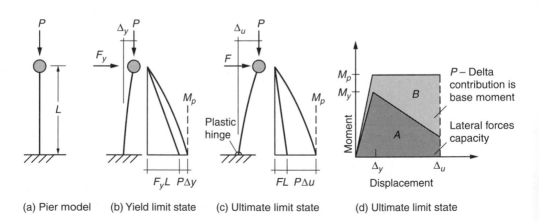

(a) Pier model   (b) Yield limit state   (c) Ultimate limit state   (d) Ultimate limit state

Figure 3.90 P − Δ effect on a bridge column.

Multicolumn bents

In the multicolumn bents the framing action or coupling between columns contributes to the seismic response in terms of stiffness, capacity, and axial load levels in the various frame members during cyclic earthquake loading. In the analytical model, all of these effects can be incorporated in a planar frame model along the bent axis, consisting of beam and column elements with appropriate effective member properties. Special elements for joints and boundary conditions may be provided as shown in Figure 3.91, if required.

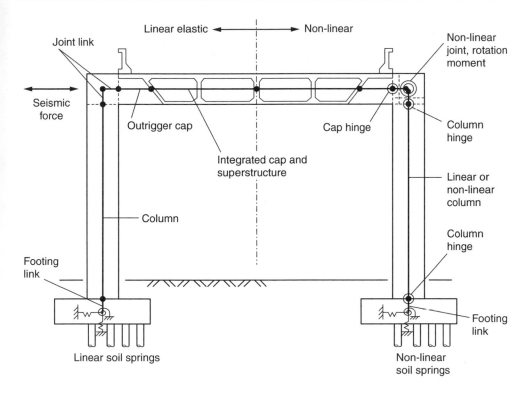

Figure 3.91 Multi-column-bent model.

Foundations

Foundations of bridge piers are the structural elements at or below ground level that support the pier and provide vertical, lateral and rotational resistance to gravity loads and seismic forces. This resistance depends upon the type and geometry of the foundation, the characteristics of the surrounding soil and the interaction between soil and structure. Spread or pile footings are considered to be rigid bodies. It is therefore possible to model at a single point with boundary springs at the bottom of the column or pier model at the end of the effective length extension link into the footing as shown in Figure 3.92. Only a vertical, a translational and a rotational boundary spring need to be defined for a two-dimensional column, bent or frame model. In a three-dimensional model, on the other hand, six springs, one for each possible degree-of-freedom at the column bas are required. For a pile shaft, the flexibility of the pile and the surrounding soil should be modelled in a seismic demand analysis.

Abutments

In the modelling of abutments, assumptions made for abutment stiffnesses and capacities as well as damping of the surrounding soil mass can have a significant effect on the analytical dynamic response characteristics. This is particularly true in shorter bridge structures. Abutments are, generally, massive structures; interact with large

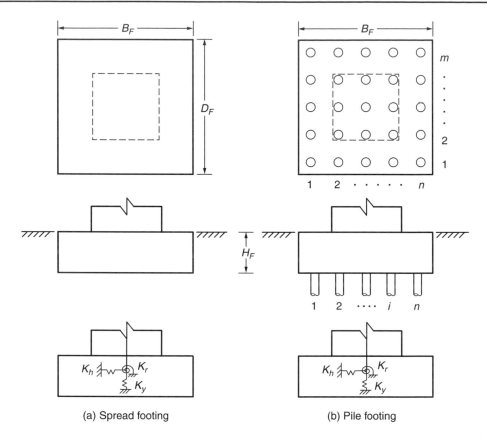

Figure 3.92 Pile and spread-footing models.

soil mass; exhibit significantly higher stiffness values and thus attract higher seismic forces. Abutment capacity and stiffness characteristics are often based on empirical relationships.

Methods of analysis

Tools for the analysis of seismic response quantification of bridges range from simple linear elastic static analyses in the form of hand calculations to three-dimensional dynamic non-linear response history analyses for global bridge systems with large number of degrees of freedom.

Static or quasistatic analysis

It is most convenient, in many cases of seismic analysis, to apply the seismic actions in the form of an equivalent static force to the bridge model, particularly when seismic force distributions or likely deformation modes can be estimated. The magnitude of the equivalent static force is either specified as absolute acceleration coefficient a_s or obtained from the design- or assessment-level earthquake in the form of an expected peak ground acceleration which then needs to be amplified by an appropriate

Frequencies and mode shapes of beams in flexural vibration

$f_n = \dfrac{k_n}{2\pi} \sqrt{\dfrac{EI}{mL^4}}$ HZ $n = 1, 2, 3...$	L = Length (m) EI = Flexural rigidity (Nm^2) M = Mass per unit length (kg/m)

Boundary conditions	K_n; $n = 1, 2, 3$	Mode shape y_n $(\dfrac{x}{L})$	A_n; $n = 1, 2, 3$
Pinned – Pinned 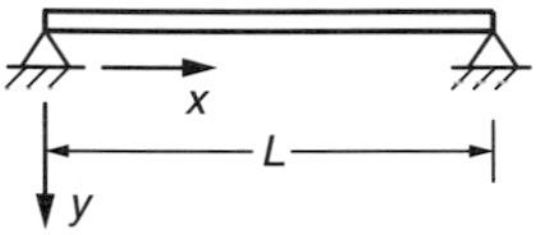	$(n\pi)^2$	$\sin \dfrac{n\pi x}{L}$	
Fixed – Fixed 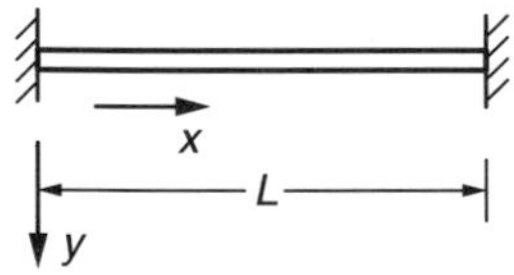	22.37 61.67 120.90 199.86 298.55 $(2n+1)\,\dfrac{\pi^2}{4}$; $n > 5$	$\cosh \dfrac{\sqrt{K_n}x}{L} - \cos \dfrac{\sqrt{K_n}x}{L}$ $- A_n \left(\sinh \dfrac{\sqrt{K_n}x}{L} - \sin \dfrac{\sqrt{K_n}x}{L} \right)$	0.98250 1.00078 0.99997 1.00000 0.99999 1.0; $n > 5$
Fixed – Pinned 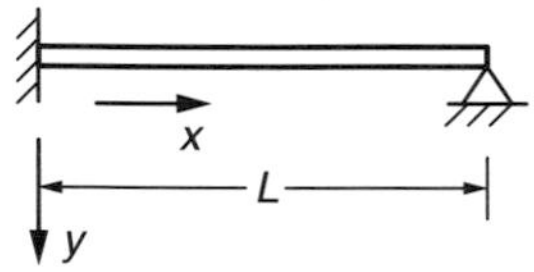	15.42 49.96 104.25 178.27 272.03 $(4n+1)^2\,\dfrac{\pi^2}{4}$; $n > 5$	$\cosh \dfrac{\sqrt{K_n}x}{L} - \cos \dfrac{\sqrt{K_n}x}{L}$ $- A_n \left(\sinh \dfrac{\sqrt{K_n}x}{L} - \sin \dfrac{\sqrt{K_n}x}{L} \right)$	1.00078 1.00000 1.0; $n > 3$
Cantilever 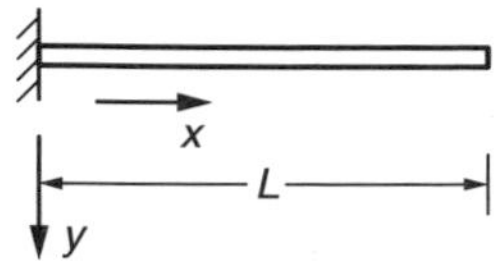	3.52 22.03 61.69 120.90 199.86 $(2n-1)^2\,\dfrac{\pi^2}{4}$; $n > 5$	$\cosh \dfrac{\sqrt{K_n}x}{L} - \cos \dfrac{\sqrt{K_n}x}{L}$ $- A_n \left(\sinh \dfrac{\sqrt{K_n}x}{L} - \sin \dfrac{\sqrt{K_n}x}{L} \right)$	0.73410 1.01847 0.99922 1.00003 1.0; $n > 4$

Figure 3.93.

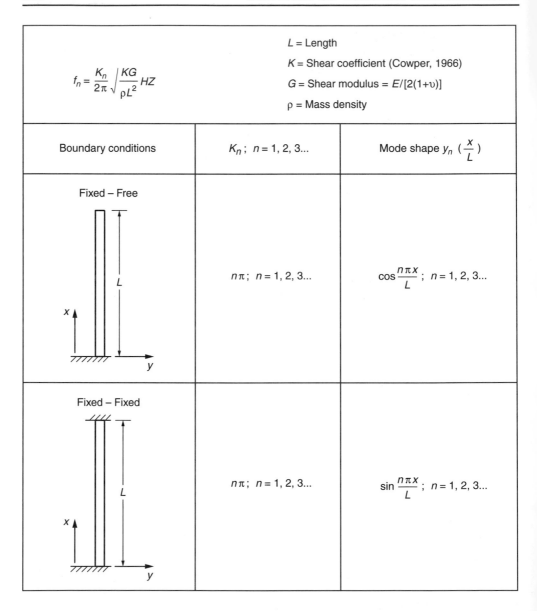

Figure 3.94 Frequencies and mode shapes of beams in shear vibration.

spectral acceleration response factor for specific site or soil characteristics. The acceleration coefficient a_s is then multiplied by the total seismic weight W_s to determine the earthquake force $E = W_s a_s$ lumped at the centroid of seismic mass, or distributed proportional to the expected fundamental mode shape. This force and distribution pattern is then applied to the structure either as static monotonic force or in the form of a quasi-static stepwise monotonic or cyclic force.

Response spectrum analyses

Modal spectral analyses are demand analysis tools to determine maximum response quantities from the spectrum of a given ground motion or from smoothed design spectra. Analysis models used for modal spectral analysis are linear elastic models based on effective stiffness properties and on assumed equivalent viscous damping ratios. Response spectrum analyses can be performed: (i) for bridge systems expected to perform essentially in the linear elastic range based on cracked or effective stiffness properties, (ii) for inelastic response of bridge systems where the equivalent response is linearized to the initial effective stiffness and subsequently modified by means of equal energy or equal displacement principles and (iii) for substitute structure analyses.

Time-history analyses

This analysis utilizes a particular earthquake ground motion input and provide bridge response quantification for this earthquake input in the form of time histories of the various response quantities. Similar to the static or quasistatic analyses, models with linear elastic materials behaviour, non-linear cyclic materials characteristics and geometric non-linearities as $P - \Delta$ or full non-linear geometry can be analysed. However, now in addition to two or three spatial dimensions, an additional dimension t, has to be accounted for. Step-by-step integration in the time domain or superposition of normalized modal time histories in the time domain or evaluation of frequency-dependent response contributions with transformation to and superposition in the time domain can be applied for the earthquake time-history analysis of bridge models. Of the three methods, stepwise linearized modal time-history analyses have been successfully employed in the non-linear dynamic response analysis of global bridge models.

Bibliography

Argyris JH. *Energy Theorems and Structural Analysis,* Butterworths, London, 1960.

Barkan DD. *Dynamics of Bases and Foundations* (translated from Russian by L Drashevska). McGraw-Hill, New York, 1962.

Bathe KJ. *Finite Element Procedures for Engineering Analysis.* Prentice Hall, Englewood Cliffs, NJ, 1982.

Biddle AR, Iles DC and Yandzio E. *Integral Steel Bridges: Design Guidance.* The Steel Construction Institute, UK, 1997.

Biggs JM. *Introduction to Structural Dynamic,* McGraw-Hill, New York, 1964.

British Standards Institution. BS 5400: Part 3: 1982, *Code of Practice for Design of Steel Bridges.* British Standards Institution, London, 1982.

Bushnell D. A computerised information retrieval systems. *SMCP Symposium,* 735–804.

Chatterjee S. Design of webs and stiffeners in plate and box girders. *Proceedings, International Conference on The Design of Steel Bridges,* Cardiff, UK, 189–214, 1980.

Chatterjee S and Dowling PJ. The design of box girder compression flanges. *Proceedings, International Conference on Steel Plated Structures,* London, July, 1976.

Chen WF. (Ed.) *The Civil Engineering Handbook,* CRC Press, London, 1995.

Cheung MS and Cheung YK. *Analysis of Curved Box Girder Bridges by Finite Strip Method.* IABSE, 31-I, 1–20, 1971.

Cheung MS, Cheung YK and Ghali A. Analysis of slabs and girder bridges by the finite strip method. *Building Science*, 5, 1970.

Clough RW and Penzien J. *Dynamics of Structures*, 2nd edn. McGraw-Hill, New York, 1993.

Committee of Inquiry into the Basis for Design and Method of Erection of Steel Box Girder Bridges, *Interim Design and Workmanship Rules*. Department of the Environment, London, 1973.

Cook RD. *Concepts and Applications of Finite Element Analysis*. John Wiley, New York, 1981.

Cusens AR and Pama RP. *Bridge Deck Analysis,* John Wiley, London, 1975.

Das BM. *Principles of Foundation Engineering*, 2nd edn. PWS-Kent, Boston, MA, 1990.

Desai CS and Abel JF. *Introduction to the Finite Element Method*. Von Nostrand Reinhold, New York, 1972.

Dowrick DJ. *Earthquake Resistant Design for Engineers and Architects*, 2nd edn. John Wiley, New York, 1988.

Evans HR. Longitudinally and transversely reinforced plate girders. In *Plated Structures – Stability and Strength* (Ed. R Narayanan), 1–37. Applied Science Publishers, London, 1983.

Evans HR and Shanmugam NE. The elastic analysis of cellular structures containing web openings. *Proc. Instn Civ Engrs*, Part 2, 67, Dec., 1035–1063, 1979a.

Evans HR and Shanmugam NE. An approximate grillage approach to the analysis of cellular structures. *Proceedings, Institution of Civil Engineers*, Part 2, 67, March, 133–154, 1979b.

Evans HR and Shanmugam NE. An experimental and theoretical study of the effects of web openings on the elastic behaviour of cellular structures. *Proceedings, Institution of Civil Engineers*, Part 2, 67, Sept., 653–676, 1979c.

Evans HR and Shanmugam NE. The elastic analysis of cellular structures containing web openings. *Proceedings, Institution of Civil Engineers*, Part 2, 67, Dec., 1035–1063, 1979d.

Evans HR and Shanmugam NE. An experimental study of the ultimate load behaviour of small-scale box girder models with web openings. *Journal Strain Analysis*, 16, 4, 251–259, 1981.

Evans HR and Taherian AR. The prediction of the shear lag effect in box girders. *Proceedings, Institution of Civil Engineers*, Part 2, 63, March, 69–92, 1977.

Faulkner D. A review of effective plating for use in the analysis of stiffened plating in bending and compression. *Journal of Ship Research*, 19, 1, 1–17, 1975.

Gorbunov-Possadov MI and Serebrajanyi V. Design of structures upon elastic foundations. *Proc. 5th Inter. Conf. Soil Mech. Found. Eng*, Paris, Vol. 1, 1961.

Hambly EC. *Bridge Deck Behaviour,* E & FN Spon, London, 1991.

Hambly EC and Burland JB. *Bridge Foundations and Substructures,* HMSO, London, 1979.

Hambly EC and Nicholson BA. Prestressed beam integral bridges. *The Structural Engineer,* 68, 23, 474–481, 1990.

Harding JE. The interaction of direct and shear stresses on plate panels. In *Plated Structures – Stability and Strength* (Ed. R Narayanan), 221–255. Applied Science Publishers, London, 1983.

Hoglund T. Strength of thin plate girders with circular or rectangular web holes without web stiffeners. *Proceedings, Colloquium, International Association of Bridge and Structural Engineering*, London, 1971.

Holtz RD and Kovacs WD. *An Introduction to Geotechnical Engineering,* Prentice-Hall, Englewood Cliffs, NJ, 1981.

Horne MR. Structural action in box girders. CIRIA Research Guide, Construction Industries Research and Information Association, London, 1975.

Horne MR and Narayanan R. Strength of axially loaded stiffened panels. *Memoires of the International Association of Bridge and Structural Engineering*, Zurich, 36-I, 125–157, 1976a.

Horne MR and Narayanan R. Ultimate capacity of stiffened plates used in box-girders. *Proceedings, Institution of Civil Engineers*, London, 61, 253–280, 1976b.

Horne MR and Narayanan R. Design of axially loaded stiffened plates. *Journal of the Structural Division, American Society of Civil Engineers*, 103, ST 11, Nov., 2243–2257, 1977.

Lambe TW and Whitman RV. *Soil Mechanics, SI version*. John Wiley, New York, 1979.

Lee SL, Karasushi P, Zakeria M and Chan KS. Uniformly loaded orthotropic rectangular plates supported at the corners. *Civil Engineering Transactions*, October, 1971.

Lightfoot E and Sawko F. Structural frame analysis by electronic computer : grid frameworks resolved by generalised slope deflection. *Engineering*, 187, 18–20, 1959.

Moffat KR and Dowling PJ. Shear lag in steel box girder bridges. *The Structural Engineer, London*, 53, October, 439–448, 1975.

Mulligan GP and Pekoz T. analysis of locally buckled thin-walled columns. *Proceedings of the Seventh Int. Natl. Specialty Conference on Cold-Formed Steel Structures*, St Louis, MO, 93–126, 1984.

Narayanan R. Ultimate shear capacity of plate girders with openings in webs. In *Plated Structures – Stability and Strength* (Ed. R Narayanan), 39–76. Applied Science Publishers, London, 1983.

Narayanan R and Der Avenessian NGV. Strength of webs containing circular cut-outs. *IABSE Proceedings*, P-64/83, August, 141–152 1983a.

Narayanan R and Der Avenessian, NGV. Equilibrium solution for predicting the strength of webs with rectangular holes. *Proc. Instn Civ Engrs*, Part 2, 75, June, 265–282, 1983b.

Narayanan R and Der Avenessian NGV. An equilibrium method for assessing the strength of plate girders with reinforced web openings. *Proc. Instn Civ Engrs*, Part 2, 77, June, 107–137, 1984.

Narayanan R and Shanmugam NE. An approximate analysis of stiffened flanges. *Memoires of the International Association of Bridge and Structural Engineering, Zurich*, P-24, 1–12, 1979a.

Narayanan R and Shanmugam NE. Effective widths of axially loaded plates. *Journal of Civil Engineering Design*, 1, 3, 253–272, 1979b.

Nath B. *Fundamentals of Finite Elements for Engineers*. Athlow Press, London, 1974.

Nethercot DA, Salter PR and Malik AS. *Design of Members Subjected to Bending and Torsion*. SCI Publication 57, The Steel Construction Institute, Ascot, UK, 1998.

Newmark NM. A method of computation for structural dynamics. *Journal of Engineering Mechanics, ASCE*, 85(EM3), 67–94, 1959.

Nicholson BA. Effects of temperature, shrinkage and creep on integral bridges. In *Continuous and Integral Bridges* (Ed. B Pritchard). E & FN Spon, London, 1994.

Oden JT. *Finite Elements of Nonlinear Continua*. McGraw-Hill, New York, 1972.

Owens GW and Knowles P. *Steel Designer's Manual*, Blackwell Science, Oxford, 1994.

Poulos HG and Davis EH. *Pile Foundation Analysis and Design*. John Wiley, New York, 1980.

Priestley MJN, Seible F and Calvi GM. *Seismic Design and Retrofit of Bridges*. John Wiley, New York, 1996.

Roberts TM. Patch loading on plate girders. *Plated Structures – Stability and Strength* (Ed. R Narayanan), 77–102. Applied Science Publishers, London, 1983.

Rockey KC, Evans HR and Porter DM. Test on longitudinally reinforced plate girders subjected to shear. *Proceedings, Conference on Stability of Steel Structures*, Liege, 1977.

Rockey KC, Evans HR and Porter DM. A design method for predicting the collapse behaviour of plate girders. *Proceedings, Institution of Engineers, London*, Part 2, 65, 85–112, 1978.

Rutenberg A. Approximate natural frequencies for coupled shear walls. *Earthquake Eng. Struct. Dynam.*, 4, 95–100, 1975.

Shanmugam NE and Evans HR. An experimental and theoretical study of the effects of web openings on the elastic behaviour of cellular structures. *Proc. Instn Civ Engrs*, Part 2, 67, Sept., 653–676, 1979.

Shanmugam NE and Evans HR. A grillage analysis of the nonlinear and ultimate behaviour of cellular structures under bending loads. *Proceedings, Institution of Engineers, London*, Part 2, 71, Sept., 705–719, 1981.

Shanmugam NE, Liew JYR and Lee SL. Thin-walled steel box-columns under biaxial loading. *Journal of Structural Engineering, ASCE*, 115, 11, 2706–2726, 1989.

Shanmugam NE, Rose H, Yu CH and Lee SL. Uniformly loaded rhombic orthotropic plates supported at corners. *Computers and Structures*, 30, 5, 1037–1045, 1988.

Shanmugam NE, Rose H, Yu CH and Lee SL, Corner supported isosceles triangular orthotropic plates. *Computers and Structures*, 32, 5, 963–972, 1989.

Smith CB. Some new types of orthotropic plates laminated of orthotropic material. *Journal of Applied Mechanics*, 20, 2, 1953.

Steel Construction Institute. *Steelwork Design Guide to BS 5950: Part 1: 1990*, Vol. 1: Section Properties, Member capacities, 5th edn. The Steel Construction Institute, Ascot, UK, 1997.

Timoshenko SP and Gere JM. *Theory of Elastic Stability,* McGraw-Hill Book Company, New York, 1961.

Timoshenko SP and Goodier JN. *Theory of Elasticity*, McGraw-Hill Kogokusha Ltd., 1970.

Timoshenko SP and Krieger SW. *Theory of Plates and Shells*, McGraw-Hill, 1959.

Tsai SW and Cheron T. *Anisotropic Plates* (translated from the Russian edition by SG Lekhnitskii). Gordon and Breach Science Publishers, New York, 1968.

Von Karman T, Sechler EE and Donnel LH. Strength of thin plates in compression. *Transactions of the American Society of Mechanical Engineers, Journal of Applied Mechanics*, 54, 2, 53, 1932.

Warburton GB. *The Dynamical Behaviour of Structures*, 2nd edn. Pergamon Press, Oxford, 1976.

Way JA and Yandzio E. *Steel Integral Bridges: Design of a Single-span Bridge – Worked Examples*. The Steel Construction Institute, UK, 1997.

West R. *Recommendations on the Use of Grillage Analysis for slab and Pseudo-slab Bridge Decks*. Cement and Concrete Association, London, 1971.

Winter G. Strength of Thin Steel Compression Flanges. *Transactions, American Society of Civil Engineers*, 112, 527, 1947.

Zienkiewicz OC and Cheung YK. *The Finite Element Method in Structural and Continuum Mechanics*. McGraw-Hill, London, 1967.

Pages 99–100 include extracts from the *Steel Designer's Manual, 1992,* edited by G Owens and P Knowles. Reproduced with permission of Blackwell Science Ltd.

4 Design of reinforced concrete bridges

P. JACKSON

4.1 Introduction

Most modern small bridges are of reinforced concrete construction and nearly all modern bridges contain some elements of reinforced concrete. In this chapter, the design of reinforced concrete bridge superstructures is considered and some aspects of the design criteria for reinforced concrete which are also relevant to other reinforced concrete elements are reviewed.

In situ reinforced concrete construction has the great advantage of simplicity; formwork is placed, reinforcement fixed and concrete poured and the structure is then complete. In modern practice, precast bridge elements are usually prestressed. For smaller elements, this is because pretensioning on long line beds is a convenient method of providing the steel. For larger structures, post-tensioning provides the most convenient way of fixing manageable sized elements together. The result is that purely reinforced concrete bridges are almost invariably cast *in situ*. An exception is very short span structures which can be most economically precast effectively complete as box culverts, leaving only parapets and, where required, wing walls to cast *in situ*. Precast reinforced beams were used in the past but are unusual in modern construction. This chapter is therefore primarily concerned with *in situ* construction.

In the following, the basic design approach for a simple solid slab bridge is considered and the particular features of various other types of bridge are then reviewed. Figure 4.1 shows the types of sections considered.

4.2 Solid slab bridges

4.2.1 Form

The solid slab is the simplest form of reinforced concrete bridge. Ease of construction resulting from the simplicity makes this the most economic type for short span structures. Solid slabs also have good distribution properties which makes them efficient at carrying concentrated movable loads such as wheel loads for highway bridges.

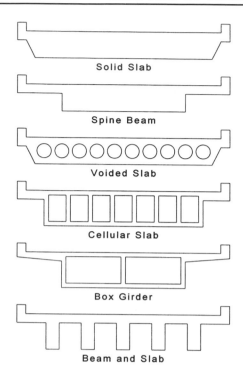

Figure 4.1 Types of section considered.

However, above a span of around 10 m the deadweight becomes excessive, making other forms of construction more economic.

Solid slab bridges can be simply supported on bearings or built into the abutments. For several decades up to the present, bridge engineers tended to be quite pedantic about providing for expansion and even bridges as short as 9 m span were often provided with bearings and expansion joints. However, bearings and expansion joints have proved to be amongst the most troublesome components of bridges. In particular, deterioration of substructures due to water leaking through expansion joints has been common especially in bridges carrying roads where de-icing salt is used.

Recently, the fashion has changed back to designing bridges that are cast integral with the abutments or bank seats (Department of Transport, 1995). Apart from the durability advantages, this can lead to saving in the deck due to the advantage of continuity. On short span bridges with relatively high abutment walls, being able to use the deck to prop the abutments can also lead to significant savings in the abutments. However, this normally depends on being able to build the deck before backfilling behind the abutments.

A feature of the design of integral bridges which has not always been appreciated is that, because the deck is not structurally isolated from the substructure, the stress state in the deck is dependent on the soil properties. This inevitably means that the analysis is less 'accurate' than in conventional structures. Neither the normal at rest pressure behind abutments nor the resistance to movement is ever very accurately known. It might be argued that because of this designs should be done for both upper and lower bounds to soil properties. In practice, this is not generally done and the design criteria used have sufficient reserve so that this does not lead to problems.

In the past, some *in situ* multi-span slab bridges were built which were simply supported. However, unlike in bridges built from precast beams, it is no more complicated to build a continuous bridge. Indeed, because of the absence of the troublesome and leak-prone expansion joints, it may actually be simpler. It is therefore only in exceptional circumstances (for example construction in areas subject to extreme differential settlement due to mining subsidence) that multiple simply supported spans are now used.

Making the deck continuous or building it into the abutments also leads to a significant reduction in the mid-span sagging moments in the slab. The advantage of this continuity in material terms is much greater than in bridges of prestressed beam construction where creep redistribution effects usually more than cancel out the saving in live load moments.

In multi-span slab bridges, various approaches are possible for the piers. Either leaf piers can be used or discrete columns. Unlike in beam bridges, the latter approach needs no separate transverse beam. The necessary increase in local transverse moment capacity can be achieved by simply providing additional transverse reinforcement in critical areas. This facility makes slab bridges particularly suitable for geometrically complicated viaducts such as arise in some interchanges in urban situations. Curved decks with varying skew angles and discrete piers in apparently random locations can readily be accommodated.

Whether discrete columns or leaf piers are used, they can either be provided with bearings or built into the deck. The major limitation on the latter approach is that, if the bridge is fixed in more than one position, the pier is subject to significant moments due to the thermal expansion and contraction of the deck. Unless the piers are very tall and slender, this usually precludes using the approach for more than one or two piers in a viaduct.

4.2.2 Design approach

The shape of the bridge is first decided guided by experience and typical span-to-depth ratios. A typical simply supported slab has a span-to-depth ratio of around 10–15 but continuous or integral bridges can be shallower. Because the concentrated live load (i.e. the wheel load) the deck has to carry does not reduce with span, the span-to-depth ratio of short span slabs tends to be towards the lower end of the range. However, deck slabs of bigger bridges often have greater span-to-depth ratios than slab bridges. This is economic because the deadweight of the slab, although an insignificant part of the load on the slab, is significant to the global design of the bridge.

There was a fashion for very shallow bridges in the 1960s and 1970s as they were considered to look more elegant. However, unless increasing the construction depth has major cost implications elsewhere (such as the need to raise embankments) it is likely to be more economic to use more than the minimum depth. The appearance disadvantage on short span bridges can be resolved by good detailing of the edges.

Bridge decks with short transverse cantilevers at the edges tend to look shallower than vertical sided bridges even if they are actually deeper.

Having decided the dimensions of the bridge, the design calculations then serve primarily to design the reinforcement. Most of the principles of the design calculations for a reinforced concrete slab type bridge superstructure also apply to other types of reinforced concrete decks. They also apply to the deck slabs of beam and slab type decks irrespective of the type of beams. The principles that apply to all the other types are therefore explained here.

Ultimate strength in flexure and torsion

Reinforced concrete is normally designed for ultimate strength in flexure first. This is partly because this is usually, although not invariably, the critical design criterion. Another reason is that reinforcement can be more readily designed directly for this criterion. For other criteria, such as crack width or service stresses, a design has to be assumed and then checked. This makes the design process iterative. A first estimate is required to start the iterative procedure and the ultimate strength design provides such an estimate.

Although other analytical methods give better estimates of strength, elastic analysis is usually used in design. This has to be used when checking serviceability criteria. Because of this, the use of more economic analyses at the ultimate limit state (such as yield-line analysis) invariably results in other criteria (such as cracking or stress limits) becoming critical.

Concrete slabs have to resist torsion as well as flexure. However, unlike in a beam, torsion and flexure in slabs are not separate phenomena. They interact in the same way that direct and shear stresses interact in plane stress situations. They can be considered in the same way, that is using Mohr's circle. Theoretically, it is most efficient to use orthogonal reinforcement placed in the directions of maximum and minimum principal moments. Since there is no torsion in these directions, torsion does not then have to be considered. However, it is not often practical to do this as the principal moment directions change with both position in the slab and load case.

In right slabs the torsional moments in the regions (the elements of the computer model where this is used) where the moments are maximum are relatively small and can often be ignored. In skew slabs, in contrast, the torsions can be significant. The usual approach is to design for an increased equivalent bending moment in the reinforcement directions. Wood (1968) has published the relevant equations for orthogonal steel and Armer (1968) for skew steel. Many of the computer programs commonly used for the analysis of bridge decks have post-processors that enable them to give these corrected moments, commonly known as 'Wood–Armer' moments, directly. To enable them to do this, it is necessary to specify the direction of the reinforcement.

When the reinforcement is very highly skewed, the Wood–Armer approach leads to excessive requirements for transverse steel. When assessing existing structures, this problem can be avoided by using alternative analytical approaches. However, in design it is usually preferable to avoid the problem by avoiding the use of very highly skewed reinforcement. The disadvantage of this is that it makes the reinforcement detailing of skew slab bridges more complicated. This arises because the main steel in the edges of the slab has to run parallel with the edges. Orthogonal steel can therefore only be achieved in the centre of the bridge either by fanning out the steel or by providing three layers in the edge regions. That is, one parallel to the edge in addition to the two orthogonal layers.

When torsion is considered, it will be found that there is a significant requirement for top steel in the obtuse corners even of simply supported slabs. It can be shown using other analytical methods (such as yield-line or torsionless grillage analysis) that equilibrium can be satisfied without resisting these moments. The top steel is therefore not strictly necessary for ultimate strength. However, the moments are real and have caused significant cracking in older slab structures which were built without this steel. It is therefore preferable to reinforce for them.

Ultimate strength in shear

Shear does not normally dictate the dimensions of the slab. However, codes allow slabs (unlike beams) which do not have shear reinforcement and it is economically desirable to avoid shear reinforcement if possible. Use of links is particularly inconvenient in very shallow slabs, such as in box culverts or the deck slabs of beam and slab bridges, and many codes do not allow them to be considered effective. The shear strength rules can therefore be critical in design.

Whereas flexural strength is well understood and test results (and also codes of practice from different countries and different periods) are consistent, shear strength is much less well understood and test results are very variable. The result is that the shear rules in codes of practice are essentially empirical and differ significantly between codes.

Modern codes, including both the new European Code (EC2: Part 2) (British Standards Institution, 2000) and the current British one (BS 5400: Part 4) (British Standards Institution, 1990) , generally consider that the allowable shear that can be taken on a concrete section varies with the main reinforcement percentage. Once links are required they have to comply with a minimum percentage area, usually the same as that for links in beams. Partly because of this sudden step in the requirement, it may pay to increase the area of main steel to avoid the requirement for links. In particular, this may often make curtailing the main steel uneconomic. The weight of the extra links required can easily exceed the weight of steel saved by curtailing. Even if it does not, curtailing may still increase cost as the links are more expensive per tonne because of the greater fixing cost. It is therefore often only for practical reasons that it is worth curtailing the main steel. For example, if the total length of a deck was around 14 m and the maximum standard length of bar was 12 m (the length of a standard articulated lorry trailer) it is convenient to use staggered 12 m bars. Alternate bars then stop 2 m from the end of the bridge.

The shear strength of concrete members without links does not increase with size as normal dimensional analysis would suggest; the stress at failure reduces as the depth of section increases (Bazant, 1984). This is also allowed for by many modern design codes. However, older codes did not consider either this effect or the effect of main steel area on shear strength. Even now, some codes (including the American AASHTO (1989) code) do not consider the effect of scale. AASHTO also only considers a very limited effect of reinforcement percentage.

Codes which did not consider steel area or scale effect often gave significantly higher allowable shear stresses than modern codes in deep sections with low main steel percentages. Test results suggest that these stresses cannot be justified. Even the current AASHTO can apparently be unsafe in some situations (Collins and Kuchma, 1999). However, fortunately, the maximum shear stresses normally arise close to the supports. In these situations, the actual shear strength increases due to so called 'short shear span enhancement'. This effect is not considered in the older codes and is underestimated by AASHTO. The theoretical worst case of these codes is therefore unlikely to arise in practice.

Most of the codes that have the low shear stresses from tests for members with low reinforcement percentage do allow some form of 'short shear span enhancement' to be considered in design. It arises because shear failures in reinforced concrete normally occur on planes inclined at a shallow angle (typically around $\tan^{-1} 0.4$) to the horizontal. Anything that forces the plane to be steeper results in a higher failure load. One disadvantage to the designer of being allowed to use this effect is that it makes it less obvious where the critical section for shear will be. A simpler approach which is sometimes used is not to consider short shear span enhancement directly but to check shear an effective depth from the support, rather than at the support.

In a simple beam or one-way spanning slab under uniform load, both the shear stress distribution and the shear span are well defined. However, in many slab situations they are not. Partly because of this, BS 5400 does not apparently allow short shear span enhancement to be considered in slabs. It may appear odd to say BS 5400 does not 'apparently' allow short shear span enhancement but this is because of the interrelationship between 'vertical' or 'flexural' shear and punching shear. This is not often appreciated but will be considered later.

An alternative to considering short shear span enhancement, as considered above, is to consider the ratio of the moment to the shear force. To make it dimensionally correct, the ratio of the moment to the shear force times effective depth is actually used. This approach is used in AASHTO (1989). In a typical simply supported test specimen loaded by a single concentrated load, the moment shear force ratio and the shear span-to-depth ratio are the same thing. However, in continuous beams they are not related and the moment to shear force ratio is low at points of contraflexure where shear planes are not forced to be steep. The American code avoids allowing excessive shear at these points by limiting the moment to shear force ratio that can be used in calculations.

In contrast to slabs, all beams are required to have links by modern codes. However, once the shear stress in a slab is sufficient for links to be required, the design approach is the same as for beams. Most codes design for shear using the 'addition principle'. In this the shear strength of the beam is assumed to be the sum of the strength without links and the strength due to the links. The latter is calculated using

45° truss analogy where the member is assumed to made up of the links acting as ties and concrete struts angled at 45°.

An alternative approach which can be used in EC2 (British Standards Institution, 2000) is the 'varying angle truss' approach. All the shear is assumed to be taken by the analogous truss but the angle of the truss can be varied to give the greatest strength. A flatter angle truss enables a given area of links to take a higher shear force but this implies a greater force in the main tension steel. This does not affect the maximum area of main steel required in a member but does alter the design of any curtailments.

The evidence indicates that the varying angle truss approach gives a more realistic representation of behaviour and more consistent safety margins than the 'addition principle'. The behaviour of beams with links is fundamentally different from those without. Unless they have short shear span-to-depth ratios, beams without links fail as soon as the shear crack appears. However, the links can only be sufficiently highly stressed to be effective after the concrete has cracked. The addition principle is therefore a purely empirical approach. It has no theoretical basis and is justified only by test results. Indeed, the code rules for the strength of beams without links are themselves purely empirical.

In contrast, the varying angle truss approach does have a theoretical basis (Nielsen, 1984). It is based on plastic theory. However, this is not perfect either because the behaviour is too brittle for plastic theory to be strictly valid. The beams therefore fail at lower loads than pure plastic theory suggests. In order to obtain good predictions for shear strength, it is necessary to multiply the normal concrete compressive strength by an empirical 'effectiveness factor' (Nielsen, 1984; Batchelor et al., 1986) to obtain the compressive strength to be used for the inclined concrete struts. It is also necessary to limit the angle of the struts. The limits in EC2 are $\tan^{-1} 0.4$ and 2.5 without curtailments in the main steel or $\tan^{-1} 0.5$ and 2.0 with curtailments.

Given the requisite area of main steel, the varying angle truss approach gives the greatest shear strength with the flattest truss and this is normally greater than that given by the addition approach. There is, however, an exception with small link areas. This arises because the addition principle implies that any amount of links would give an increase in strength. In contrast, with small areas of links, varying angle truss analogy gives a strength which is less than the strength without links. What happens in practice is that the section fails much as it would with no links and small areas do not really increase strength. This is why codes specify a minimum link area for slabs as mentioned above. Theoretically, it is not required if varying angle truss design is used.

The varying angle truss approach also automatically imposes an upper limit on the shear strength, irrespective of the amount of links, and gives a design criterion for curtailments. When the more conventional addition approach is used, more empirical rules are required to cover these aspects. However, the upper or 'web crushing' limit in shear is rarely critical in solid slab structures. It is more likely to be critical in box and other flanged sections where it pays to make the webs as thin as practical.

Ultimate strength in punching shear

Codes require a check on punching shear around concentrated loads or supports. In European practice, this is done on a section at 1.5 times effective depth, d, from the face of the loaded area. This section has no physical significance; it is simply the distance at which the shear stress calculated for test specimens at failure matched the limiting shear to the codes. The actual failure cone extends further from the loaded area. Because of this, although the check is done on a surface $1.5d$ from the load, you can only use main steel area for calculating the shear strength if the relevant bars extend to some $3d$ from the load.

The American code uses a critical section that is closer to the loaded area. This requires a different (higher) limiting stress and gives a greater sensitivity of predicted strength to loaded area which is probably more correct.

The punching check is generally considered to be an extra check in addition to the normal shear check. However, most bridge codes (including BS 5400) require the use of an elastic shear distribution for the normal or 'flexural' shear check in slabs. This gives high concentrated values near concentrated loads. In a case that is anywhere near the 'punching shear' limit, these are invariably outside the code rules. The implication is that the code does not require you to check the elastic stress so close to the loaded area. The physical explanation of why this works is that a failure so close to the load would attract some short shear span enhancement, hence the earlier comment that BS 5400 does not 'apparently' allow short shear span enhancement in slabs.

Punching shear is relatively rarely critical in bridge deck slabs except around discrete pier supports.

Service stresses

As well as ultimate strength in flexure, service stresses have to be considered in most design codes. It is arguable that a stress check is contrary to the principles of limit state design. Stress is not a fundamental design criterion and it is theoretically possible to design a perfectly acceptable bridge structure in which the reinforcement has yielded. However, checking such criteria as deflection and crack width for a structure in which the steel had yielded would be extremely difficult. Once a structure has gone outside the elastic range there is no guarantee that it will recover from deflection or crack opening caused by transient loads. It is therefore theoretically necessary to undertake a non-linear analysis of the complete load history to ensure the structure is satisfactory. This is not practical.

The usual approach is to impose limitations on steel stress in tension and concrete stress in compression to ensure the material stays within the linear range. This enables crack widths and deflections to be checked without considering non-linear behaviour.

Service stresses are normally checked against criteria related to the yield strength of reinforcement and crushing strength of concrete. Because structures do not recover from non-linearity, it is always necessary to check for the worst load case and combination. However, because the service stress check ensures that structures will recover from all loadings, other criteria may not have to be checked for all loads. For example, if crack widths or deflections that arise only fleetingly and occasionally are of no concern, these criteria do not have to be checked for infrequent loads such as abnormal vehicles.

The British design code (BS 5400: Part 4) allows the stress check to be avoided provided certain restrictions in the ultimate analysis are complied with. An explanation of this and a fuller discussion of the philosophy of stress checks in limit state codes are given by Jackson (1987).

Crack widths

Although it is a basic principle of reinforced concrete that the concrete cracks, wide cracks are considered undesirable partly for aesthetic reasons. It is often also assumed that concrete with wide cracks is less good at protecting reinforcement from corrosion. Accordingly narrower widths than would be needed for aesthetic reasons are often imposed. Research does not support this relationship (Beeby, 1978) and there does not appear to be any justification for limits more severe than around 0.3 mm.

Different codes have different formulae for the width of structural cracks and the background to some is given by Beeby (1979). The rules can give very different results and some involve considerable amounts of complicated calculation. However, there does not appear to be any real justification for this additional work. Many codes, including EC2, allow the crack width calculation to be avoided provided stress limits related to bar size or spacing are complied with. This is justified as excessive structural cracking has normally only arisen where either reinforcement has yielded or poor detailing has been used, such as very wide-spaced bars.

Excessively wide cracks were more common in side faces of very deep beams. To avoid this, side-face reinforcement should be provided in beams and slabs which are deeper than around 500 mm.

A particular problem can arise with concrete that is restrained when it is first cast. The concrete heats up as the cement hydrates and tends to crack as it cools down again. With sections in excess of around a metre thick, the problem can arise even without external restraint. This is because the cooling surface layers are restrained by the core which stays hot for longer.

Detailed rules on this 'early thermal cracking' and the reinforcement required to prevent excessive cracking due to it are given in BD27/87 (Department of Transport, 1987). This document has been found to give adequate control. However, it does suggest that the crack widths are proportional to the restrained strain. More recent work has shown that this is not correct. In the normal range of strains, the strain only controls the number of cracks. Because each crack arises when a force corresponding to the tensile strength of the concrete is released and transferred locally to the reinforcement, there is no reason for their width to increase greatly with strain. The section of EC2 dealing with water retaining structures uses this newer theory.

In practice, early thermal effects are rarely critical in decks except locally where components like string courses or concrete parapets are cast after the deck so that they are restrained by the deck. They are more frequently critical in walls.

Fatigue

Some codes, including EC2-2, require a check on fatigue for the reinforcement and even, in some cases, the concrete itself. Introducing such checks has been controversial not least because there do not appear to have been any fatigue failures in reinforced concrete structures.

EC2 and BS 5400 both enable a full Miner's curve fatigue assessment of reinforcement to be undertaken similar to that used for steelwork. However, a simpler approach of checking compliance with a live load stress range under service loads is also provided. The stress range currently specified in EC2-2 is extremely cautious, representing the extreme worst case. UK Highways Agency documents and the draft UK 'NAD' (National Application Document) (British Standards Institution, 2000) provide more realistic ranges. Only the stress range due to normal traffic load ('HA' in BS 5400, 'Frequent' in the Eurocode) is considered because the number of cycles of the rarer load is much lower.

These more realistic stress ranges, which vary with the loaded length, should avoid the need for full checks in most cases. However, due to the high live load ratio, they would have been likely to be restrictive in the deck slabs of beam and slab bridges. These are, however, exempted from the check by the aforementioned UK NAD. The justification for this is that tests show that the stress range in the reinforcement is much lower than the calculations suggest (Holowka and Csagoly, 1980).

Full fatigue checks may still be required for some railway structures and for cases where welded reinforcement is used.

Detailing checks

In addition to the global design issues considered above, codes cover such things as cover required (normally governed by durability), minimum and maximum bar spacing, bond and lap strength and many other aspects.

There are also minimum steel areas which serve a number of purposes. One of their objectives is to ensure ductile failure modes. With very low steel areas, the ultimate moment capacity with the concrete cracked and the steel yielding may actually be less than the moment required to cause the first crack. Theoretically, the steel area required to prevent this increases both as the effective depth to total depth ratio reduces and as the concrete tensile strength increases. In practice there is no control over the *maximum* concrete strength so it is not possible to write a rigorously correct rule. It can be shown that the minimum steel areas in many codes are not theoretically adequate but few resulting problems have been observed. One reason for this is because, due to the interaction between the code checks, a seriously inadequate 'minimum' steel area can only arise where the calculated applied moment is significantly less than the cracking moment. The section is therefore unlikely to crack. Although reinforced concrete is always designed to crack, many reinforced concrete structures do not crack; they actually work due to the tensile strength of concrete.

4.3 Voided slab bridges

Above a span of about 10 m, the dead weight of a solid slab bridge becomes excessive. For narrower bridges, significant weight saving can be achieved by using relatively long transverse cantilevers giving a bridge of 'spine beam' form as shown in Figure 4.1. This can extend the economic span range of this type of structure to around 16 m or more. Above this span, and earlier for wider bridges, a lighter form of construction is desirable.

One of the commonest ways of lightening a solid slab is to use void formers of some sort. The commonest form is circular polystyrene void formers. Although polystyrene appears to be impermeable it is actually only the much more expensive closed cell form which is. The voids should therefore be provided with drainage holes at their lower ends. It is also important to ensure that the voids and reinforcement are held firmly in position in the formwork during construction. This avoids problems that have occurred with the voids floating or with the links moving to touch the void formers, giving no cover.

Provided the void diameters are not more than around 60% of the slab thickness and nominal transverse steel is provided in the flanges, the bridge can be analysed much as a solid slab. That is, without considering either the reduced transverse shear stiffness or the local bending in the flanges. BS 5400 specifies a nominal steel area of the lesser of 1500 mm^2/m or 1% of minimum flange section in tensile regions and 1000 mm^2/m or 0.7% in compressive regions.

The section is designed longitudinally in both flexure and shear allowing for the voids. Links are always provided and these are designed as for a flanged beam with the minimum web thickness.

The shear stresses are likely to become excessive near supports, particularly if discrete piers are used. However, this problem can be avoided by simply stopping the voids off, leaving a solid section in these critical areas.

If more lightening is required, larger diameter voids or square voids forming a cellular deck can be used. These do then have to be considered in analysis. The longitudinal stiffness to be used for a cellular deck is calculated in the normal way, treating the section as a monolithic beam. Transversely, such a structure behaves quite differently under uniform and non-uniform bending. In the former the top and bottom flanges act compositely whereas in the latter they flex about their separate neutral axes as shown in Figure 4.2. This means the correct flexural inertia can be an order of magnitude greater for uniform than non-uniform bending. The behaviour can, however, be modelled in a conventional grillage model by using a shear deformable grillage. The composite flexural properties are used and the extra deformation under non-uniform bending is represented by calculating an equivalent shear stiffness.

Having obtained the moments and forces in the cellular structure, the reinforcement has to be designed. In addition to designing for the longitudinal and transverse moments on the complete section, local moments in the flanges have to be considered. These arise from the wheel loads applied to the deck slab and also form the transverse shear. This shear has to be transmitted across the voids by flexure in the

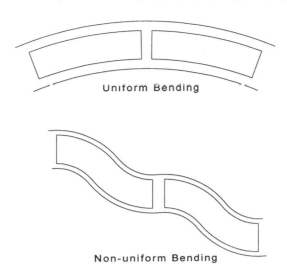

Figure 4.2 Transverse bending in cellular slab.

flanges, that is by the section acting like a vierendeel frame as shown in Figure 4.2. This aspect is considered by Clark (1983) who gives a method for checking decks with large circular voids.

4.4 Beam and slab bridges

In recent years, *in situ* beam and slab structures have been less popular than voided slab forms whilst precast beams have generally been prestressed. Reinforced beam and slab structures have therefore been less common. However, there is no fundamental reason why they should not be used and there are thousands of such structures in service.

One of the disadvantages of a beam and slab structure compared with a voided slab or cellular slab structure is that the distribution properties are relatively poor. In the UK at least, this is less of a disadvantage than it used to be. This is because the normal ('HA') traffic load has increased with each change of the loading specification, leaving the abnormal ('HB') load the same. However, reinforced concrete beam and slab bridges do not appear to have increased in popularity as a result. They are more popular in some other countries.

The relatively poor distribution properties of beam and slab bridges can be improved by providing one or more transverse beams or diaphragms within the span, rather than only at the piers. In bridges built with precast beams, forming these 'intermediate diaphragms' is extremely inconvenient and therefore expensive so they have become unusual. However, in an *in situ* structure which has to be built on falsework, it makes relatively little difference and is therefore more viable.

The beams for a beam and slab structure are designed for the moments and forces from the analysis. The analysis is now usually computerized in European practice although the AASHTO code (1989) encourages the use of a basically empirical approach.

Having obtained the forces, the design approach is the same as for slabs apart from the requirement for nominal links in all beams. Another factor is that if torsion is considered in the analysis the links have to be designed for torsion as well as for shear. It is, however, acceptable practice to use torsionless analysis at least for right decks.

Because the deck slab forms a large top flange, the beams of beam and slab bridges are more efficiently shaped for resisting sagging than hogging moments. It may therefore be advantageous to haunch them locally over the piers even in relatively short-span continuous bridges.

The biggest variation in practice in the design of beam and slab structures is in the reinforcement of the deck slab. A similar situation arises in the deck slabs of bridges where the main beams are steel or prestressed concrete and this aspect will now be considered.

Conventional practice in North America was to design only for the moments induced in the deck slab by its action in spanning between the beams supporting wheel loads (the 'local moments'). These moments were obtained from Westergaard (1930) albeit usually via tables given in AASHTO. British practice also uses elastic methods to obtain local moments, often Westergaard but also sometimes influence charts such as Pucher (1964). However, the so called 'global transverse moments', the moments induced in the deck slab by its action in distributing load between the beams, are considered. These moments, obtained from the global analysis of the bridge, are added to the local moments obtained from Westergaard or similar methods. Only 'co-existant' moments (the moments induced in the same part of the deck under the same load case) are considered and the worst global and local moments often do not coincide. However, this still has a significant effect. In bridges with very close spaced beams (admittedly rarely used in North America) the UK approach can give twice the design moments of the US approach.

Although the US approach may appear theoretically unsound (the global moments obviously do exist in American bridges) it has produced satisfactory designs. One reason for this is that the local strength of the deck slabs is actually much greater than conventional elastic analysis suggests. This has been extensively researched (Hewit and Batchelor, 1975; Kirkpatrick et al., 1984).

In Ontario (Ontario Ministry of Transportation and Communications, 1983) empirical rules have been developed which enable such slabs to be designed very simply. Although these were developed without major consideration of global effects, they have been shown to work well within the range of cases they apply to (Jackson and Cope, 1990). Similar rules have been developed in Northern Ireland (Kirkpatrick et al., 1984) and elsewhere but they have not been widely accepted in Europe.

4.5 Longer span structures

In modern practice, purely reinforced structures longer than about 20 m span are quite unusual; concrete bridges of this size are usually prestressed. However, there is no fundamental reason why such structures should not be built.

The longest span reinforced concrete girder bridges tend to be of box girder form. Although single cell box girders are a well-defined form of construction, there is no

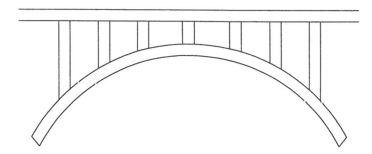

Figure 4.3 Open spandrel arch.

clear-cut distinction between a 'multi-cellular box girder' and a voided slab. How-ever, the voids in voided slab bridges are normally formed with polystyrene or other permanent void formers whereas box girders are usually formed with removable form-work. The formwork can only be removed if the section is deep enough for access, which effectively means around 1.2 m minimum depth. Permanent access to the voids is often provided. In older structures, this was often through manholes in the top slab. This means traffic management is required to gain access and also means there is the problem of water, and de-icing salt where this is used, leaking into the voids. The fashion has therefore changed to obtaining access from below.

In a continuous girder bridge the hogging moments, particularly the permanent load moments, over the piers are substantially greater than the sagging moments at mid-span. This, combined with the greater advantage of saving weight near mid-span, encourages longer span bridges to be haunched. Haunching frequently also helps with the clearance required for road, rail or river traffic under the bridge by allowing a shallower section elsewhere.

The longest span and most dramatic purely reinforced concrete bridges are open spandrel arches as shown in Figure 4.3. The true arch form suits reinforced concrete well as the compressive force in the arch rib increases its flexural strength. As a result, the form is quite efficient in terms of materials.

Because of the physical shape of the arch and the requirement for good ground conditions to resist the lateral thrust force from the arch, this form of construction is limited in its application. It is most suitable for crossing valleys in hilly country. The biggest problem is the cost of the falsework. Because of this, such bridges are often more expensive than structurally less efficient forms, such as prestressed can-tilever bridges, that can be built with less temporary works. However, they may still be economic in some circumstances, particularly in countries where the labour required to erect the falsework is relatively cheap. A further factor may be local avail-ability of the materials in countries where the prestressing equipment or structural steelwork required for other bridges of this span range would have to be imported.

The efficiency of arch structures, like other forms used for longer span bridges, arises because the shape is optimized for resisting the near-uniform forces arising from dead load which is the dominant load. The profile of the arch is arranged to minimize the bending moments in it. Theoretically, the optimum shape approximates to a catinary if the weight of the rib dominates or a parabola if the weight of the deck dominates but the exact shape is unlikely to be critical.

Arch structures can be so efficient at carrying dead weight that applying the usual dead load factor actually increases the live load capacity by increasing the axial force in, and hence flexural capacity of, the rib. The code's lesser load factor (normally 1.0) for 'relieving effects' should therefore be applied to dead weight. However, the letter of many codes only requires this to be applied in certain cases which are defined in such a way that it does not appear to apply here. This cannot be justified philosophically and the reduced factor should be used.

Because the geometry is optimized for a uniform load, loading the entire span is unlikely to be the critical load case unlike in a simple single-span beam bridge. It will normally be necessary to plot influence lines to determine the critical case. For uniform loads, this is often loading a half-span.

Steel arch bridges have been built in which the live load bending moments are taken primarily by the girders at deck level, enabling the arch ribs to be very slim in appearance. However, with concrete the usual approach is to build the arch rib first and then build the deck structure afterwards, possibly even after the falsework has been struck so that this does not have to be designed to take the full load. The deck structure is then much like a normal viaduct supported on piers from the arch rib and the rib has to take significant moments.

Bibliography

American Association of State Highways and Transportation Officials. *Standard Specification for Highway Bridges.* 14th edn. AASHTO, Washington, 1989.

Armer GST. Correspondence, *Concrete.* August, 319–320, 1968.

Batchelor BdeV, George HK and Campbell TI. Effectiveness factors for shear in concrete beams. *Journal of the structural division. American Society of Civil Engineers*, 112, ST6, June, 1464–1477, 1986.

Bazant ZP. Size effect in shear failure of longitudinally reinforced beams. *American Concrete Institution Journal*, September–October, 456–468, 1984.

Beeby AW. *Cracking and Corrosion*. Concrete in the Oceans. Technical Report 1. CIRIA, C&CA. DoE. 1978.

Beeby, AW. The prediction of crack widths in hardened concrete. *The Structural Engineer*, 57A, 1, January, 9–17, 1979.

British Standards Institution. *BS 5400 Steel, Concrete and Composite Bridges: Part 4. Code of practice for the design of concrete bridges*. BSI, London, 1990.

British Standards Institution. *Eurocode 2: Design of Concrete Structures: Part 2. Bridges (together with United Kingdom National Application Document)*. BSI. London, 2000.

Clark LA. *Concrete bridge design to BS 5400*. Construction Press, London 1983 (including supplement incorporating 1984 revisions 1985).

Collins MP and Kuchma D. How safe are our large, lightly reinforced concrete beams, slabs and footings? *American Concrete Institute Structural Journal*, 96, 4. July–August, 482–490, 1999.

Department of Transport. *BD 28/87. Early thermal cracking in concrete*. Department of Transport, London, 1987 (Amended 1989).

Department of Transport *BA 57/95 Design for durability*. Department of Transport, London, 1995.

Hewit BE and Batchelor B deV. Punching shear strength of restrained slabs. *Journal of the Structural Division. American Society of Civil Engineers*, 101, ST9, September, 1831–1852, 1975.

Holowka M and Csagoly P. *Testing of a Composite Prestressed Concrete Aashto Girder Bridge*. Ontario Ministry of Transport and Communications. Downsview. Ontario, Canada, Research Report 222, 1980.

Jackson PA. The stress limits for reinforced concrete in BS 5400. *The Structural Engineer*, 65A, 7, July, 259–263, 1987.

Jackson PA and Cope RJ. The behaviour of bridge deck slabs under full global load. *Developments in Short and Medium Span Bridge Engineering '90*, Montreal, Canada, August, 253–264, 1990.

Kirkpatrick J, Rankin GIB and Long AE. Strength evaluation of M-beam bridge deck slabs. *The Structural Engineer*, 62B, 3, 60–68, 1984.

Nielsen MP. *Limit analysis and concrete plasticity*. Prentice Hall, New Jersey, 205–219, 1984.

Ontario Ministry of Transportation and Communications. *Ontario Highway Bridge Design Code*. Downsview, Ontario, Canada, 1983.

Pucher A. *Influence Surfaces for Elastic Plates*. Springer, Wien, 1964.

Westergaard HM. Computation of stresses in bridge deck slabs due to wheel loads. *Public Roads*, 2, 1. March, 1–23, 1930.

Wood RH. The reinforcement of slabs in accordance with a pre-determined field of moments. *Concrete*. February, 69–76, 1968.

5 Design of prestressed concrete bridges

N.R. HEWSON

5.1 Introduction

Concrete is strong in compression but weak in tension; however, prestressing of the concrete can be used to ensure that it remains within its tensile and compressive capacity. Prestressing is applied as an external force to the concrete, by the use of wires, strands or bars and can increase the capacity of concrete alone many times. The development of prestressed concrete bridges has given the bridge engineer increased flexibility in the selection of bridge form and in the construction techniques available, resulting in prestressed concrete frequently being the material of choice for bridges with spans ranging from 25 m up to 450 m.

This chapter reviews the design and construction of the different types of prestressed concrete bridges and assumes that the reader has a basic understanding of prestressed concrete design which can be applied to the specific application of bridgeworks.

5.2 Principle of prestressing

Prestressing of concrete members is achieved by force transfer between the prestressing tendon and the concrete. The tendons are placed within the concrete member as internal tendons, or alongside the concrete as external tendons. For post-tensioning the tendon is pulled using a jack and the force is transferred directly on to the hardened concrete, while for pretensioning the tendon is jacked against an anchor frame prior to concreting and the force released from the anchor frame to the concrete when it has obtained sufficient strength. In this way an external compressive force is applied to the concrete which can be used to counter the tensile stresses generated under the bending moments and shear forces present.

The concrete can be either fully prestressed, which ensures that the longitudinal stresses are always in compression, or partially prestressed, which allows some tension to occur under certain loading conditions.

5.3 Materials

5.3.1 Concrete

The required strength of the concrete is determined by the compressive stresses gen-erated in the concrete by the prestress and applied forces. A minimum strength of f_{cu} equal to 45 N/mm^2 is typical for prestressed concrete; however, it is becoming more common to use higher strengths, with f_{cu} up to 60 N/mm^2 not unusual, while even higher values have been achieved on specific projects. The rate of gain of strength of the concrete is important as this governs the time at which the prestress can be applied. At the time of transfer of the prestress force to the concrete it is normally required that the concrete strength be at least a minimum of 30 N/mm^2, although this can vary depending on the tendon and anchor arrangement and the magnitude of load applied.

To minimize creep and shrinkage losses in the prestressing, the cement content and water/cement ratio of the concrete should be kept to a minimum, compatible with the high concrete strengths required.

5.3.2 Prestressing steel

High tensile steel is used, as wire, strand or bars, with nominal tensile strengths vary-ing between 1570 N/mm^2 and 1860 N/mm^2 for wire or strand and between 1000 N/mm^2 and 1080 N/mm^2 for bars.

After the load is applied to the prestressing steel, stress relaxation occurs which results in a reduction of the force in the tendon. The magnitude of relaxation varies depending on the steel characteristics and the initial stress levels applied, with typical values of between 2.5–3.5% when a stress of $0.70f_{pu}$ is applied, which reduces to 1% when the initial stress is $0.50f_{pu}$ or less.

Typical material properties for tendons are shown in Table 5.1.

5.3.3 Cement grout

Used to fill the void around post-tensioned tendons and their ducts, a water/cement ratio of between 0.35 and 0.40 is typically used, with admixtures occasionally added to improve flow and to reduce shrinkage and the water/cement ratio.

5.3.4 Grease or wax grout

Used for external tendons to enable destressing or replacement, the grease or petroleum wax is injected into the duct at 80 or 90°C before cooling to give a soft, flexible filler.

5.4 Prestressing systems

5.4.1 Wires

Individual wires are sometimes used in pretensioned beams but have become less common in favour of strand, which has better bonding characteristics. Wire diameters are typically between 5 and 7 mm with a minimum tensile strength of 1570 N/mm^2 and carry forces up to 45 kN.

Table 5.1 Typical tendon properties

	Nominal diameter (mm)	Nominal area (mm²)	Nominal mass (kg/m)	Yield strength (N/mm²)	Tensile strength (N/mm²)	Minimum breaking load (kN)	Young's modulus (kN/mm²)	Relaxation after 1000 h at 20°C and 70% UTS
7-wire strand								
13 mm super	12.9	100	0.785	1580	1860	186.0	195	2.5%
15 mm super	15.7	150	1.18	1500	1770	265.0	195	2.5%
Stress bars								
20 mm	–	314	2.39	835	1030	325	170	3.5%
32 mm	–	804	6.66	835	1030	828	170	3.5%
50 mm	–	1963	16.02	835	1030	2022	170	3.5%
Cold-drawn wire								
7 mm	7 mm	38.5	302	1300	1570	60.4	205	2.5%
5 mm	5 mm	19.6	154	1390	1670	32.7	205	2.5%

Reference should be made to the manufacturers' literature for the properties of individual wires, strands or bars being used.

Figure 5.1 Prestressing strand, anchor and duct (courtesy of Freyssinet International).

5.4.2 Strands and tendons

A 7-wire strand with a tensile strength of 1860 N/mm^2 and either 13 or 15 mm diameter, is a very common form of prestressing and can be used either singularly for pretensioning or in bundles to form multistrand tendons for post-tensioning as shown in Figure 5.1. The most common post-tensioned tendon sizes utilize 7, 12, 19 or 27 strands to suit the standard anchor blocks available, but tendons can incorporate up to 55 strands for larger tendons. Stressing to 75% UTS gives a jacking force of 140 or 199 kN for the 13 or 15 mm diameter strand, respectively, while the larger multistrand tendon can carry forces upto 10 000 kN. During stressing, jacks are placed over the tendon, gripping each strand and pulling it until the required force is generated. Wedges are then placed around the strand and seated into the anchor block so that on the release of the jack the wedges grip the strand and transfer the force on to the anchor and into the concrete.

5.4.3 Bars

Individual bars can vary in diameter from 15 up to 75 mm and are used in post-tensioned construction with jacking forces ranging from 135 to 3000 kN. The bars

Figure 5.2 Prestressing bar and anchor (courtesy of McCalls Special Products Ltd).

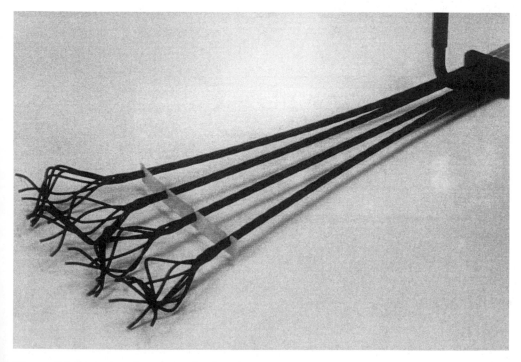

Figure 5.3 Strand with cast-in dead end anchorage (courtesy of VSL International).

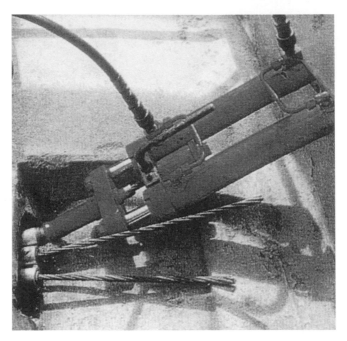

Figure 5.4 Single strand stressing jack (courtesy of VSL International).

Figure 5.5 Multistrand stressing jack (courtesy of Freyssinet International).

are placed into ducts which have been cast into the concrete between two anchor blocks on the concrete surface. The bars are pulled from one end by a stressing jack and held in place by a nut assembly which transfers the load from the bar to the anchor block and then into the concrete, see Figure 5.2.

5.4.4 Anchorages

At each end of the tendon the forces are transferred into the concrete by an anchorage system. For pretensioned tendons the anchorage is by bond of the bare strand cast into the concrete, while for post-tensioned tendons the anchorage can be either by anchorage blocks or by bond for some types of cast-in dead-end anchors. Figures 5.1 and 5.2 show typical live-end anchorage for strands and bars respectively, while Figure 5.3 is a typical cast-in dead-end anchorage for post-tensioned strand.

When external, post-tensioned tendons are used they should be removable and replaceable and the detail of the anchorage should be arranged to allow this. Where cement grout is used this can be achieved by providing a lining on the central hole of the anchor to ensure the grout around the tendon does not bond to the anchor, allowing the tendon to be cut and pulled out if necessary.

5.4.5 Stressing jacks

Typical jacks for single strand and multistrand tendons are shown in Figures 5.4 and 5.5, respectively, while Figure 5.6 shows a typical jack for bar tendons. For single strands, smaller tendons or smaller bars the jacks, weighing up to 250 kg, can be easily handled and manoeuvred into position, while for the larger tendons lifting frames or cranes are required to move the jacks which can weigh up to 2000 kg.

Designs should always take into account the access needed to set up and operate the jacks and associated equipment.

5.4.6 Tendon couplers

It is possible to couple tendons together to assist in the stage-by-stage construction method frequently adopted. Figure 5.7 shows the arrangement for strands, where a special anchor block is used to enable the tendon to be extended after it has been stressed. This arrangement can simplify the tendon layout and save the cost of a separate anchor.

Bars can also be joined together by a simple threaded coupler as shown in Figure 5.8.

5.4.7 Corrosion protection and ducting

Wires, bars and strand are generally used uncoated; however, to protect the tendons during storage or to reduce friction losses during stressing they may be coated with soluble oil which is washed off prior to grouting the duct. Galvanized bars and strand are also available, as are epoxy-coated strand in some countries, although these have yet to be commonly adopted in normal prestressing works.

Post-tensioned tendons are normally placed inside ducting to allow them to be stressed inside the hardened concrete and to provide protection to the tendons. For internal tendons the ducts have traditionally been manufactured using galvanized mild

Figure 5.6 Jack for prestressing bars.

steel strips; more recently plastic ducting systems are being adopted to provide a watertight barrier around the tendon, as protection against corrosion additional to the cement used to fill the ducts after the tendons have been stressed.

For external tendons high density polyethylene (HDPE) ducts are used with either cement grout or grease around the strands. The HDPE provides long-term durability while the advantage of using grease is that it allows the tendons to be more easily de-stressed and re-stressed or replaced, which is an important aspect for external tendons. Where cement grout is used re-stressing of the tendon is not possible, and removal of the external tendon involves cutting it up into short lengths

Figure 5.7 Tendon coupler (courtesy of VSL International).

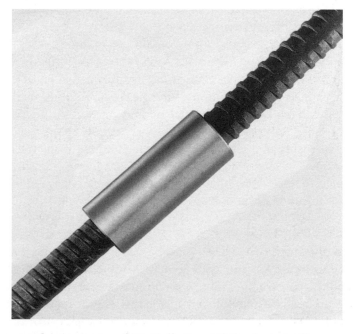

Figure 5.8 Bar coupler (courtesy of McCalls Special Products Ltd).

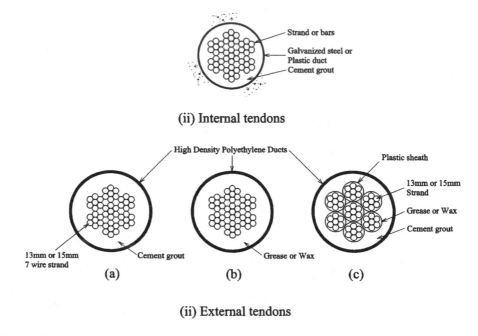

(ii) Internal tendons

(ii) External tendons

Figure 5.9 Tendon ducts and protection systems.

and pulling it out of the deviators and anchor blocks, which requires careful detailing of the duct arrangements. The HDPE ducts need to be strong enough to prevent deformation when pressurized during grouting and to resist the strand punching through at deviated positions, requiring a wall thickness generally greater than 6 mm. Figure 5.9 shows the different protection systems used.

Grouting of the ducts is normally carried out on site, although grease filling of external tendons can be done in the factory for single strands in individual sheaths or for smaller multistrand tendons. For cement grout, the mixing and pumping into the duct is carried out in a continuous operation with typical equipment shown in Figure 5.10. The grout is pumped in at the low end and steadily pushed through until the duct is full. Grout vents need to be placed at regular intervals along the duct and at all high and low points to ensure that all the air is expelled and the ducts filled. Discovery of poorly grouted ducts in several bridges in the UK has led to a review of procedures and to the Concrete Society Technical Report No. 47 being issued to give guidance on all aspects of grouting of tendons.

5.5 Prestress design

5.5.1 General approach

The choice of prestress type and arrangement is influenced by the structural form of the deck and the construction methods employed. Pretensioned strands or wires are used in precast beams with typical arrangements as shown in Figure 5.11(i), where the economies from construction repetition can offset the cost of providing a casting bed with anchor frames strong enough to resist the total prestress force.

Figure 5.10 Grouting equipment.

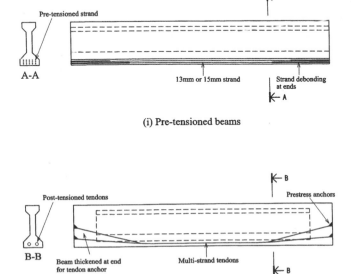

(i) Pre-tensioned beams

(ii) Post-tensioned beam

Figure 5.11 Typical prestress layout in beams.

Figure 5.12 Prestressing bars used for deck erection.

Prestressing bars generally have a higher cost per kN of force than strand, but are usually considered to be easier to handle and install and are often used where they simplify construction, such as during the erection of precast segmental decks shown in Figure 5.12 or during the launching of box girder decks. Post-tensioned tendons, such as the arrangements shown in Figure 5.13, have the advantages of being installed on site, with flexibility in the tendon layouts that can be achieved and in the forces that can be applied. Jacking systems used normally require a tendon length of more than 5 m, while for internal tendons above 120 m long the friction losses, even when

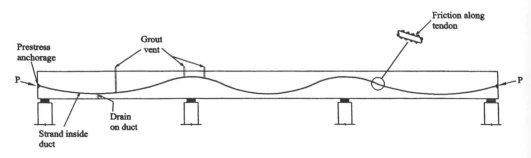

Figure 5.13 Typical post-tensioned tendon layouts.

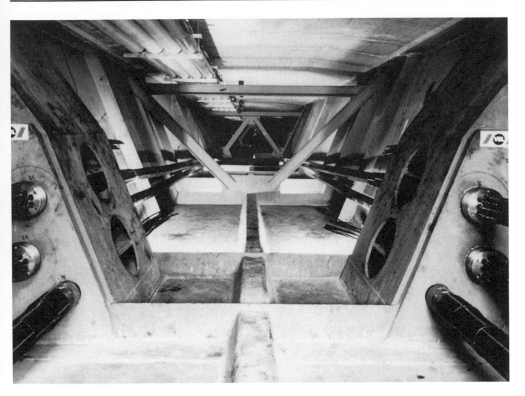

Figure 5.14 External tendons (courtesy of VSL International).

double-end stressed, become excessive. External tendons have much lower friction losses and have been installed with lengths over 300 m.

External tendons as shown in Figure 5.14 have been used extensively for bridge strengthening as well as for new construction, and offer a solution where the tendons can be easily inspected and replaced if necessary, something that is not easy to do where internal tendons are used. For new construction external tendons cost more per tonne than the equivalent internal tendons; but this can be offset against savings elsewhere, as with the tendons outside the concrete section the concrete sizes, particularly the webs, can be reduced in thickness, and reinforcement fixing and concrete placing is made easier. In segmental construction, using external tendons with span-by-span erection has resulted in very rapid erection of long viaducts to give overall cost savings in the project.

The prestress tendon layouts for most types of bridge structures are now well established, and any new design would normally start by considering a tendon arrangement similar to that used on previous structures of the same type, for example straight strand in the bottom of precast beams, cantilever tendons and continuity tendons for balanced cantilever construction or simple draped tendons for *in situ*, continuous decks.

For most structures, the prestress quantity is governed by the serviceability stress check, and the number of tendons required at each critical section such as mid-span and over supports can be derived by calculating the stresses on the section due to the applied loads and construction effects and then estimating the prestress required to keep these stresses within the allowable limits based on an initial estimate of the secondary moment. The problem is that at this stage the prestress secondary moment is not known, and therefore after estimating the prestress layout required it is necessary to calculate the actual secondary moments generated and to compare these with the values used in the initial estimate of prestress. It may take several iterations of estimating the secondary moment and adjusting the prestress arrangements before the required prestress stresses are obtained and the actual secondary moment matches the assumed secondary moment, to give the final design. After this the ultimate moment should be checked at the critical sections and if necessary additional tendons added to ensure adequate resistance can be mobilized, although this would require a recheck of the serviceability stresses to ensure that they are still within the acceptable limits.

For structures with external, unbonded tendons the quantity of prestress can be governed by the ultimate limit state, especially where the increase in stress in the tendon is small as described in Section 5.5.7 and no non-prestressed reinforcement is available as for precast segmental construction. In these cases the number of tendons required at critical sections can be determined from the ultimate moment check, and the serviceability check then carried out to ensure the stresses are within their acceptable limits.

The magnitude of the secondary moments set up can vary greatly for any type of structural form and will depend on the prestress layout chosen; however, for bridgeworks it is normally found that the secondary moment is sagging, reducing the design hogging moments over the piers and increasing the design sagging moments in mid-span. For concrete box girders with spans in the range of 35–40 m and with typical tendon layouts, the prestress secondary moments can be of the order of 5000 kN m at internal piers, while for heavily prestressed structures and for longer spans the secondary moments can be significantly higher.

Although not specified in the UK, many countries now recommend that when designing post-tensioned structures, allowance is made for the future installation of an additional 10% or more of prestress, which can be easily incorporated with external tendons. The need for this has come about because of excessive loss of prestress or deflections in some existing structures, or the need to upgrade or strengthen the structure to take heavier loads, and it would seem a sensible provision to make on all new post-tensioned bridges.

5.5.2 Primary and secondary effects

When the prestressing tendons apply load to the structure the resultant forces and moments generated can be considered as a combination of primary and secondary or parasitic effects.

The primary effect is the moment, shear and axial force generated by the direct application of the force in the tendon on the section being considered. Secondary effects occur when the structure is statically indeterminate and restraints on the structure prevent the prestressed member from freely deflecting when the prestress is applied.

For simply supported decks, unrestrained at the ends, the structure is free to deflect when prestressed and no secondary effects are set up. For continuous decks, when the prestress is applied the intermediate supports restrain the deck from vertical movement, and secondary moments and shears occur. These secondary effects can be derived using a number of different methods, the most common being the equivalent load method where the loads from the tendons are applied directly to the structural model, and the combined primary and secondary affects as shown in

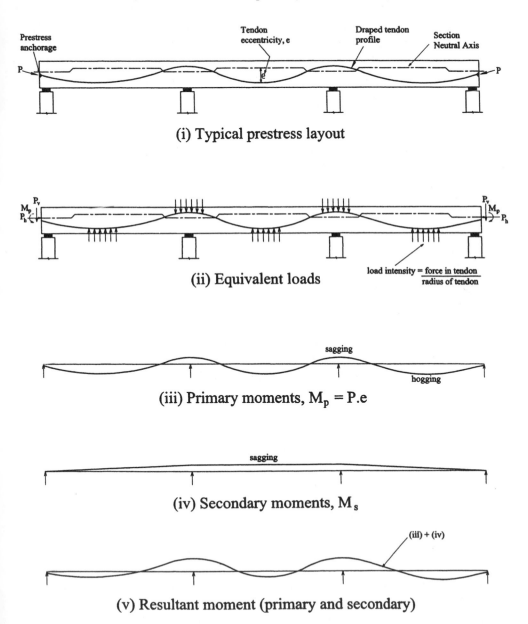

(i) Typical prestress layout

(ii) Equivalent loads

(iii) Primary moments, $M_p = P.e$

(iv) Secondary moments, M_s

(v) Resultant moment (primary and secondary)

Figure 5.15 Prestress in continuous decks.

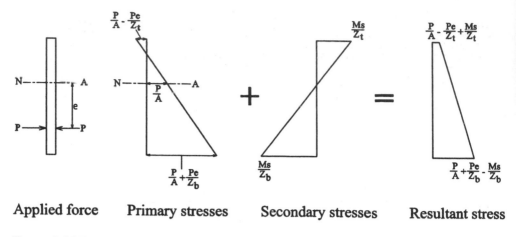

| Applied force | Primary stresses | Secondary stresses | Resultant stress |

Figure 5.16 Prestress stresses on section.

Figure 5.15(v) are derived directly from the analysis output. The forces and moments from the tendon P_h, P_v and M_p are applied at each anchor position, while along the structure equivalent loads are applied to the model wherever the tendons have an angle change. Care needs to be taken in setting up the model and applying the pre-stress load to ensure the structural behaviour is accurately represented. The member layout should follow the neutral axis of the structure, and friction losses in the tendons incorporated by applying appropriate forces and moments on the members along the model.

The primary and secondary moments both set up longitudinal bending stresses on the section, as indicated in Figure 5.16, and the resultant stress is taken into the serviceability stress analysis.

An alternative approach to derive the secondary effects is the influence coefficient method which takes into account changes in section and friction losses and is readily adaptable for use with spreadsheets. With the prestress loading on the structure as shown in Figure 5.17, and by using the theory of least work, the following equations can be derived:

$$f_{11}X_1 + f_{12}X_2 = -U_1$$
$$f_{21}X_1 + f_{22}X_2 = -U_2$$

where:

$$f_{11} = \int_s (m_1^2/EI)\,ds$$
$$f_{12} = \int_s (m_1 m_2/EI)\,ds = f_{21}$$
$$f_{22} = \int_s (m_2^2/EI)\,ds$$
$$U_1 = \int_s (m_1 M_p/EI)\,ds$$
$$U_2 = \int_s (m_2 M_p/EI)\,ds$$

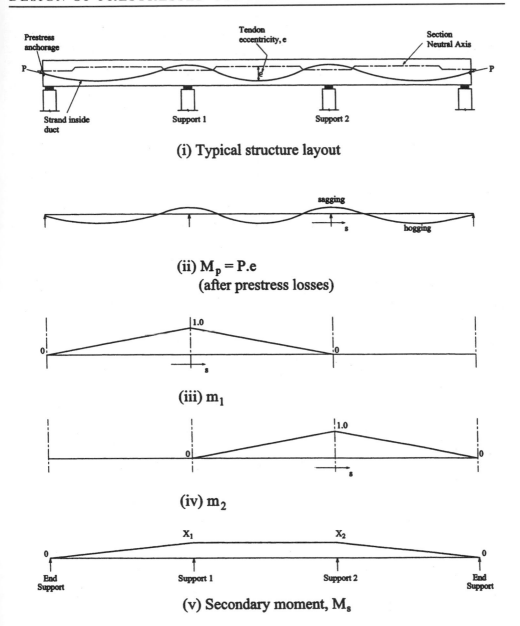

(i) Typical structure layout

(ii) $M_p = P.e$
(after prestress losses)

(iii) m_1

(iv) m_2

(v) Secondary moment, M_s

Figure 5.17 Secondary moments using the influence coefficient method.

By solving the equations, X_1 and X_2 can be determined which are the secondary moments generated at their respective support. This approach can be extended to derive secondary moments for structures with many more spans. The secondary shear forces can be determined by consideration of the changes in the secondary moments along the structure.

Where the deck is built integral with abutments or piers, the restraints will again set up secondary effects from the prestressing which must be taken into account in the design. An example of this is shown in Figure 5.18. The prestress tends to compress the portal beam, but this compression is resisted by the columns, which generates secondary moment and 'tensile' forces on the beam. Similarly the prestress tends to bend the portal beam under its primary effects, and this bending is again resisted by the columns.

Where post-tensioned tendons are used in curved decks, torsions are generated due to the secondary effects of the prestress. The intermediate supports along the structure resist the curved decks' tendency to twist, and where the prestress generates secondary moments along the deck and vertical reactions in the supports a complementary torsion is also present. An estimate of the torsion can be derived by computing the secondary moment diagram assuming the bridge is straight and dividing this by the horizontal radius of the deck. This M/R diagram is then applied as a load to the structure and the shears generated along the deck are the torsions that will be present in the original structure. Provided the tendons are symmetrical about the vertical axis of the section, the primary effects of the tendons are not affected by deck curvature; however, where this is not the case, such as with external tendons between anchor and deviator points, a transverse bending and further torsions can be generated.

5.5.3 Prestress force and losses

Tendons can be stressed up to 80% UTS, although some codes limit the jacking load to 75% UTS, and after the jack releases the strand to a maximum of 70 or 75% UTS, for post-tensioning and pretensioning, respectively. Prestressing bars are normally used straight and encounter little friction losses. When a multistrand tendon is pulled by the jack, the movement is resisted by friction along the duct, and the force F at any point along the tendon, x metres from the stressing anchor, is given by:

$$F = F_o e^{-(kx + \mu\theta)}$$

Typical values for the coefficients are shown in Table 5.2.

For external tendons, or pretensioned strands, $k = 0$.

When the tendon is released by the jack a length of reverse friction is set up as the strand pulls in and the wedges grip. Figure 5.19 gives a typical force profile along the tendon, while the typical pull-in at lock-off could be up to 7 mm.

Table 5.2 Typical values for the coefficients

	k	μ
Bare strand in metal ducts	0.001–0.002	0.2–0.3
Bare strand in UPVC ducts	0.001	0.14
Bare strand in HDPE ducts	0.001	0.15
Greased and plastic sheathed strand in polyethylene ducts	0.001	0.05–0.07

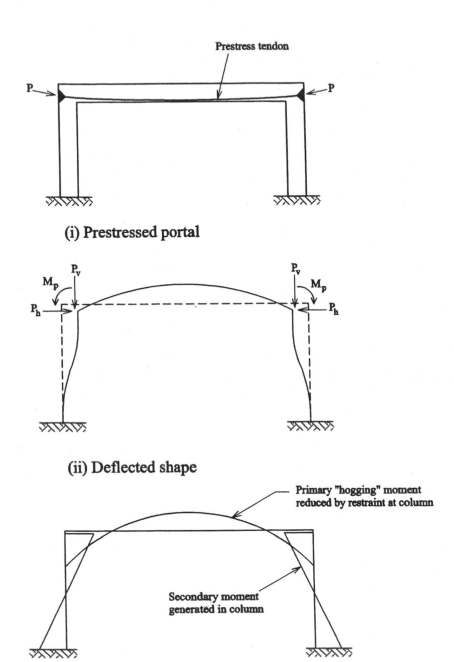

(i) Prestressed portal

(ii) Deflected shape

(iii) Prestress moments diagram

Figure 5.18 Prestress effect with integral bridge decks.

The force loss along the tendon is greatly dependent on the tendon profile, and for tendons longer than 40 m excessive losses can occur; however, this can be compensated for by double end stressing, where the tendon is jacked from both ends.

The extension of the tendon during stressing can be calculated from the force profile:

$$\text{Extension} = (\Sigma F d_x / A_t E_t) = (\text{Area under force profile}/A_t E_t)$$

Elastic shortening of the concrete causes losses in the tendons already stressed. The losses can be calculated by the following formula:

$$l_E = \sigma_c (E_t / E_c) \text{ N/mm}^2$$

where σ_c is taken as the increase in stress in the concrete adjacent to the tendon, occurring after the tendon has been stressed.

For pretensioning the full prestress force will generate elastic shortening of the concrete and give a constant loss in each tendon. For post-tensioning each tendon at a section will suffer a different loss depending upon when it is stressed in relation to the other tendons. For most designs it is sufficient to calculate an average loss and apply this to all the tendons with σ_c taken as half the total final stress in the concrete, averaged along the tendon length.

Relaxation of the tendon will reduce the prestress force over a period of time, as will creep and shrinkage of the concrete. Typical relaxation percentages are given in Table 5.1.

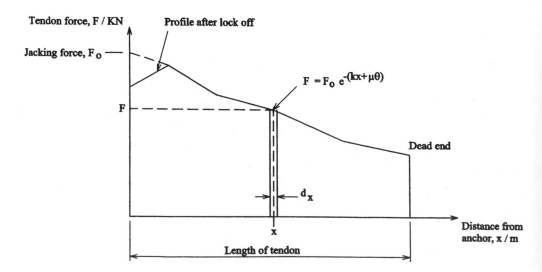

Figure 5.19 Typical tendon force profile.

$$\text{Creep losses, } l_c = \sigma_c(E_t/E_c) \, \phi \text{ N/mm}^2$$

$$\text{Shrinkage losses, } l_s = \Delta_{cs}E_t \text{ N/mm}^2$$

Derivations of $\varnothing$ and Δ_{cs} are given in BS 5400 Part 4, Appendix C, and σ_c is the total stress in the concrete adjacent to the tendon. Typical values for shrinkage and creep losses are given in Clauses 6.7.2.4 and 6.7.2.5 of BS 5400 Part 4.

External tendons are not bonded to the concrete and the elastic shortening, creep and shrinkage losses should be estimated by considering the total deformation between the anchorages or other fixed points as being uniformly distributed along the tendon length.

5.5.4 Serviceability limit state stress check

Longitudinal stresses in the prestressed deck have to be checked at each stage of construction, and throughout the structure's design life the stresses in the concrete under service loads have to be within allowable limits. Typical allowable stresses in the concrete under service load would be as follows:

Compressive stresses
$0.5f_{ci} \leqslant 0.4f_{cu}$ with triangular stress distribution, or
$0.4f_{ci} \leqslant 0.3\,f_{cu}$ with uniform stress distribution

Tensile stresses
0 N/mm^2 with permanent loads and normal live load
$0.45\sqrt{f_{cu}}$ full loading and other effects on pretensioned members
$0.36\sqrt{f_{cu}}$ full loading and other effects on post-tensioned members.

Tensile stresses during construction
1 N/mm^2 during erection with prestress and coexistent dead and temporary loads
$0.45\sqrt{f_{ci}}$ for pretensioned members under any loading
$0.36\sqrt{f_{ci}}$ for post-tensioned members under any loading.

The stress levels should be checked at all the critical stages in the structure's life, including:

(i)   at transfer of prestress to the concrete
(ii)   during construction, with temporary loads applied
(iii) at bridge opening, with full live load
(iv)   after long-term losses in the prestress, and full creep redistribution of moments have occurred.

With external tendons, to allow for replacement the design of the deck should be checked for the situation when any two tendons are removed. The live loading applied may be reduced to reflect traffic management systems that could be implemented.

5.5.5 Deflections

For normally proportioned prestressed concrete decks and stresses within the allowable limits, deflections are not critical and do not need to be checked, other than to confirm the precamber values to be catered for.

5.5.6 Vibrations and fatigue

The vibration from traffic or wind seldom creates a problem and for most prestressed concrete decks it is not necessary to consider vibrations further; however, if external tendons are used the vibrations of the individual tendons can give rise to fluctuations in stress levels in certain circumstances.

For either internal or external tendons, the fluctuation in the direct stress in the tendons due to live and other loading is very small and fatigue is not critical; however, where external tendons are used provisions have to be made to ensure that the tendons are not subjected to vibrations that are likely to give rise to fretting of the strand or bending stresses that could cause fatigue problems. This is normally done by ensuring the frequency of the free length of tendons, between anchors or deviator points, is not the same as the natural frequency of the deck or the traffic using the bridge. Approximate values for frequencies are as follows:

Frequency of the external tendon $= 1/2L_t\sqrt{(F/m_t)}$ Hz

Frequency of the deck $= (k_f^2/2\pi L_{sp}^2)\sqrt{(EI/m_d)}$ Hz

k_f depends on span continuity and is π for simply supported spans and between 3–4.5 for continuous deck.

Vibration frequency of live load is often taken as approximately 1–3 Hz for highway traffic, while for rail traffic the behaviour of track supports, track stiffness and natural frequencies of the train bogey leads to a more complex behaviour.

To prevent vibrations in highways bridges, it is normally sufficient to limit the free length of the tendon to 12 m or less, which should result in the tendon frequency being different from that of the deck and traffic.

5.5.7 Ultimate moment

As the bending moment on a prestressed beam increases the compression on one side goes up while on the other side the concrete goes into tension. When the tensile strength of the concrete is exceeded, cracking will occur and the load is transferred either to the prestressing tendons or to any non-prestressed reinforcement present. As the moment further increases, the crack in the concrete opens and propagates across the section with a pure couple set up between the compression in the concrete and tension in the tendon and non-prestressed reinforcement, with the maximum moment reached when either the concrete or tendon and reinforcement fail. To ensure that there is a sufficient factor of safety against failure the ultimate moment of resistance of any section along the deck must exceed the bending moment generated by the ultimate loads. It is normally sufficient to only check critical points for the deck, such as over the pier and at mid-span.

The ultimate moment of resistance is derived by considering a balance of tensile force in the tendon and compressive force in the concrete as shown in Figure 5.20 for internal tendons.

In the case of rectangular sections, or flanged sections with the neutral axis in the flange, the compressive stress in the concrete is taken as an average of $0.4f_{cu}$ over the depth of compression, while the concrete strain is limited to 0.0035. The increases in stress in the tendon, f_{pi}, and non-prestressed reinforcement, f_{si}, are derived by considering the increase in stress caused by the bending within the beam and can be obtained by reference to the tendon stress/strain curves similar to Figures 2 and 3 in BS 5400 Part 4, with $\gamma_m = 1.15$.

To determine the ultimate moment of resistance, an iterative approach can be used, with a first estimate of d_c taken, from which C and T can be calculated.

$$C = 0.4f_{cu}d_cb_s$$

$$T = A_t((f_p + f_{pi})/1.15) + A_p((f_s + f_{si})/1.15)$$

d is then adjusted and C and T recalculated until $C = T$, giving the ultimate moment of resistance $= TL$.

The American bridge design code (AASHTO) requires that the factored moment at a section, $M \leqslant \phi M_r$, where $\phi = 1.0$ for precast concrete and 0.95 for *in situ* concrete, while for rectangular sections or flanged sections with the neutral axis in the flange:

$$M_r = A_t(f_p + f_{pi}) \, d \, [1 - 0.6(A_t/bd)((f_p + f_{pi})/f'_c)]$$

For external tendons unbonded from the concrete over long lengths, the change in stress under ultimate loads is derived by considering the increase in tendon length between its fixed points as the deck deflects and then using this to estimate an average strain change. The increase in stress in the tendon is thus limited and ultimate failure is normally governed by the compression in the concrete which can give rise to a brittle failure mechanism. The estimate of the Ultimate Moment of Resistance follows the same procedure as for internal tendon above, although for long tendons installed over several spans $f_{pi} = 0$ N/mm^2 is normally taken. For short tendons anchored within one span, f_{pi} can be significant depending on the tendon layout and deck arrangement.

When the ultimate moment of resistance of a section with external tendons is being considered, as the deck deflects the tendon will remain a straight member between the nearest restraints, and at a point away from the anchors or deviators the tendon moves closer to the compression force with the lever arm, L, being reduced. This can become significant if the free length of external tendons is greater than a quarter of the span length and the tendons are not restrained near the critical points.

Similar to the serviceability stress check, with external tendons the section should have adequate capacity to carry a reduced live loading with any two of the tendons removed.

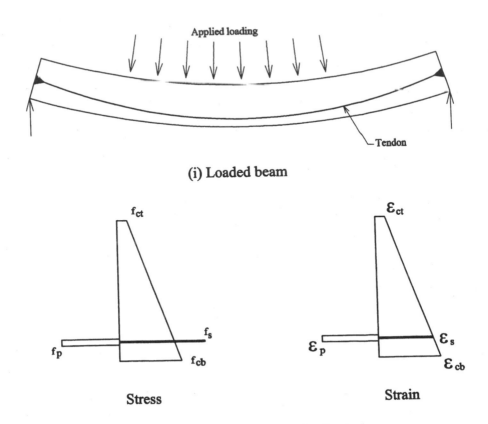

(i) Loaded beam

(ii) Initial stress and strain distribution

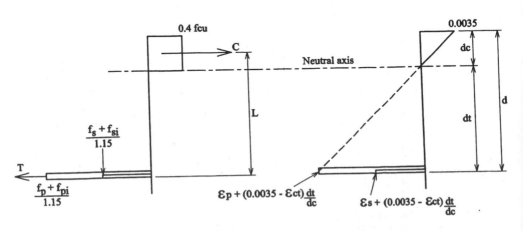

(iii) Stress and strain distribution at ultimate moment of resistance

Figure 5.20 Ultimate moment of resistance.

5.5.8 Ultimate shear

Shear need only be considered at the ultimate limit state, when the shear resistance must exceed the shear forces generated by the ultimate loads applied. The shear in the deck is checked adjacent to each pier and at regular intervals along the spans, and reinforcement, normally in the form of links, provided as required to give adequate resistance when combined with the concrete resistance. All the shear is assumed to act on the webs of a section and only the web resistance taken into account in the design. Where a deck is haunched, such as for long span box girders, the longitudinal bending in the deck gives rise to compression in the bottom slab which acts parallel to the soffit as shown in Figure 5.21. This compression has a vertical component that acts against the shear forces and can be deducted from the shear force, V, in the design.

The shear capacity at any section is the sum of V_c the ultimate shear resistance of the concrete and V_s the ultimate shear resistance of the reinforcement present.

With internal tendons, designing to BS 5400 Part 4, V_c is the lesser of V_{co}, the ultimate shear resistance of a section uncracked in flexure and V_{cr}, the ultimate shear resistance of a section cracked in flexure.

$$V_{co} = 0.67\, bh\sqrt{(f_t^2 + f_{cp}f_t)} + V_p$$

$$V_{cr} = 0.037\, bd\, \sqrt{f_{cu}} + (M_{cr}/M)V$$

$$M_{cr} = (0.37\sqrt{f_{cu}} + f_{pt})(I/y)$$

The values for V_p, f_{cp} and f_{pt} should be based on the prestressing forces after all losses have occurred and be multiplied by the partial safety factor, $\gamma_{fL} = 0.87$. Where the vertical component of prestress, V_p, resists shear at the section being considered, this should be ignored when calculating V_{cr}.

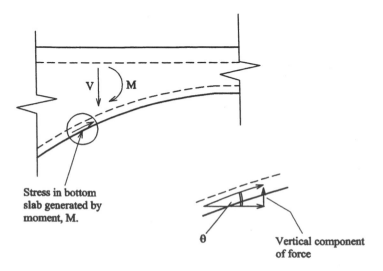

Figure 5.21 Resistance to shear in haunched decks.

BS 5400 Part 4 requires a minimum shear reinforcement to be provided such that:

$$(A_{sv}/S_v)(0.87f_{yv}/b)0.4 \text{ N/mm}^2$$

When V exceeds V_c shear reinforcement is required such that:

$$(A_{sv}/S_v) = (V + 0.4bd_t - V_c)/(0.87f_{yv}d_t)$$

Longitudinal reinforcement should be provided in the tensile zone such that:

$$A_{sl} \geqslant V/2(0.87f_{yl})$$

This quantity can include both the non-prestressed reinforcement and any bonded prestressing tendon not being utilized at the ultimate limit state for other purposes.

The maximum shear force V, is limited to $0.75\sqrt{f_{cu}}$ to prevent excessive principal stresses occurring in the concrete.

Where internal post-tensioned tendons are placed within ducts in the webs of the concrete, allowance has to be made for the reduction in effective web width. Prior to grouting the tendons, the full duct width should be deducted from b, while after grouting, two-thirds of the duct diameter should be deducted.

When checking the ultimate shear capacity at the ends of pretensioned beams, allowance must be made for the loss of prestress over its anchorage length, L_t as defined in Section 5.6.1. It is normally sufficient to consider the prestress force as varying linearly over this length and, by applying the formula above, derive V_c. If the shear resistance based on the beam being simply reinforced, and ignoring the pre-stress, is greater than that calculated assuming a prestressed beam, then the higher value can be taken.

AASHTO's approach to load factor design differs from BS 5400 Part 4 in that for shear, $V \leqslant \phi(V_c + V_s)$, with V_c as the lesser of:

(i) $bd(0.29\sqrt{f_c'} + 0.3f_{pc}) + V_p$

or

(ii) $0.05 \, bd \, \sqrt{f_c'} + V_d + (M_{cr}/M) \, V$

with $d \geqslant 0.8h$.

$$M_{cr} = (I/y)(0.5\sqrt{f_c'} + f_{pe} + f_d)$$

M and V are the factored moment and co-existent shear from the load combination causing maximum moment at the section.

$$V_s = (a_{sv}f_{yv}d/S_v) \geqslant 0.644\sqrt{f_c'} \, bd \text{ and } d \geqslant 0.8h$$

The strength reduction factor, ϕ, is taken as 0.9.

For external tendons, AASHTO treats shear design in the same way as for internal tendons when calculating V_c and the shear reinforcement to be provided, while in the UK, BD 58/94 requires that sections with unbonded external tendons are designed as reinforced concrete columns, subjected to an externally applied load, with:

$$v(V - 0.87V_p/bd)$$

where v should not exceed 5.8 N/mm^2.
   When:

$$v \leqslant \xi_s v_c, \quad A_{sv} \geqslant 0.4bs_v/0.87f_{yv}.$$

When:

$$v > \xi_s v_c, \quad A_{sv} \geqslant bs_v (v + 0.4 - (1 + (0.05(0.87P)/A_c))\xi_s v_c)/0.87f_{yv}$$

where v_c and ξ_s are given in BS 5400 Part 4, Tables 8 and 9, respectively.
   With external tendons the shear capacity should be adequate to carry a reduced live loading with any two tendons removed.

5.5.9 Ultimate torsion
Torsion does not normally govern the dimensions of the concrete members, but will require additional reinforcement in the section and is considered at the ultimate limit state.

In determining the torsional capacity it is necessary to calculate the torsional shear stresses generated by the ultimate loads and where these stresses exceed V_{tmin} as given in Table 10 of BS 5400 Part 4, then reinforcement needs to be provided by means of transverse links and longitudinal bars. The shear stresses generated by torsion and shear must be added together and the sum must not exceed $0.75\sqrt{f_{cu}}$, and be less than 5.8 N/mm^2.

For rectangular sections, the torsional shear stress, v_t, is given by:

$$v_t = (2T_u/h_{min}^2(h_{max} - h_{min}/3))$$

and torsional reinforcement provided such that:

$$A_{st}/S_v \geqslant T_u/(1.6x_iy_i(0.87f_{yv}))$$

$$A_{sl}/S_L \geqslant A_{st}(f_{yv}/f_{yl})/S_v$$

where T and I sections are used, the torsion is considered as acting on the individual rectangular elements, with the section divided up to maximize the sum of $(h_{max}h_{min}^3)$ of each rectangle. Each rectangle is then designed to carry a proportion of the torsion based on its value of $(h_{max}h^3_{min})$ in relation to the sum of the values for all the rectangles, reinforcement being determined as for normal rectangular sections and detailed to tie the individual rectangles together.

For box sections:

$$v_t = T_u/2h_{wo}A_o$$

and reinforcement provided such that:

$$A_{st}/S_v \geqslant (T_u/2A_o(0.87f_{yv}))$$

$$A_{sl}/S_L \geqslant A_{st}/S_v \, (f_{yv}/f_{yl})$$

Where a part of the section is in compression, this compressive force may be used to reduce the longitudinal reinforcement required.

This reduction in A_{sl}/S_L is given by $f_c h_{wo}/0.87f_{yl}$.

For precast segmental decks where no reinforcement passes through the joints, and they should be in compression at all times, the longitudinal torsional stresses need to be overcome by a residual longitudinal compression from the prestress equal to $T_u/2h_{wo}A_o$.

5.5.10 Longitudinal shear

As the bending moments change, the flow of stress through the section gives rise to longitudinal shear which needs to be checked at the slab/web interfaces, as indicated in Figure 5.22. The ultimate longitudinal shear force per unit length is given by:

$$V_L = V(A_L y/I)$$

when A_L and y refer to the area of concrete outside of the section being considered.

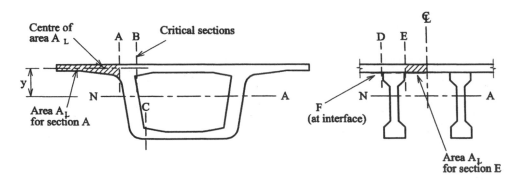

Figure 5.22 Longitudinal shear.

The concrete is able to provide some resistance to this shear, with reinforcement required to ensure that V_L is less than both:

$$\text{(a) } k_1 f_{cu} L_S$$

or:

$$\text{(b) } v_L L_S + 0.7 A_s f_y$$

Values for v_L and k_1 are given in BS 5400 Part 4, Table 31. For precast beam-and-slab construction, a minimum area of reinforcement of 0.15% of the contact area should be provided across the interface between the beam and slab. BS 5400 Part 4 allows reinforcement that is provided for other purposes to be utilized to resist the longitudinal shear.

5.5.11 Partial prestressing

Classified in BS 5400 Part 4 as Class 3, partial prestress is not currently adopted on bridgeworks in the UK; however, several other countries, including Denmark and Australia, have successfully adopted this approach into their bridge designs. For fully prestressed structures the design is based on the concrete being uncracked under service loading, either as Class 1 where the stresses are in compression under all loading conditions, or as Class 2 where under transient loading small tensions are allowed which are kept below the tensile strength of the concrete. For partially prestressed structures the philosophy is to allow the concrete to crack under service loading and to limit the crack widths to the normal allowable limit. This is often related to a hypothetical allowable tension in the concrete of up to 6 or 7 N/mm² when carrying out the longitudinal stress check.

The ultimate moment and torsion design is carried out in the normal manner, while for ultimate shear the concrete resistance is calculated as cracked in flexure, with:

$$V_{cr} = (1 - 0.55(f_{pe}/f_{pu}))v_c bd + M_o(V/M)$$

with $M_o = f_{pt}(I/y)$ and any vertical component of prestress ignored.

The values of f_{pe} and f_{pt} are based on the prestressing force after all losses have occurred and multiplied by the partial safety factor, $\Upsilon_{fl} = 0.87$, and for the purpose of this equation f_{pe} should be $\leqslant 0.6 f_{pu}$ while v_c is given in BS 5400 Part 4, Table 8 in which the value of A_s is the total area of prestressed or non-prestressed reinforcement in the tensile zone.

A fatigue check of the prestress is needed to ensure that the service stress fluctuation in the strand is not critical for the level of stress present. Normally checking to ensure the stress range is below 120 N/mm² is sufficient.

The advantages of partial prestressing include reduction in the quantity of prestress, full utilization of non-prestressed reinforcement for ultimate strength, smaller deflections due to prestress and reduced creep in the concrete. The disadvantages include the need for more non-prestressed reinforcement and a fear that the durability of the structure will be reduced in harsh environments, although this fear has not been borne out where partially prestressed structures have been used.

5.5.12 Precamber

Deflections of the concrete deck occur under the self-weight and from the weight of the permanently applied loads followed by further movements due to long-term creep of the concrete and losses in prestress. The deck must be cast and erected so that the theoretical profile is achieved upon completion. The adjustment made to the profile during casting to achieve the desired shape is called the precamber and this will be affected by the construction sequence and concrete properties.

The creep effects will cause the deck to change its profile over its life and it is normal to aim to achieve the desired alignment at time of bridge opening, although the long term changes in deflections should be checked to ensure they are not excessive.

5.5.13 Construction sequence and creep analysis

The way the bridge is built affects the moments and shears generated in the structure and this needs to be fully taken into account during the design. The structure should be checked for strength and stability, and serviceability stresses assessed at each stage of construction with the final moments and shears derived to reflect the construction sequence. For example, Figure 5.23 shows the dead load bending moments in a four-span deck constructed in stages, with the final moments after creep being between the as-built moments and the moment if the deck was built instantaneously.

When a concrete structures statical system is changed during construction, creep of the concrete will modify the as-built bending moments and shear forces towards the 'instantaneous' moment and shear distribution, the amount of the change being dependant on the creep factor, $\varnothing$, of the concrete.

$$\varnothing = \text{creep strain/elastic strain}$$

$\varnothing$ varies depending on the concrete constituents and details, environmental conditions and age of concrete, and this variation can be from 1.3 to 3 or more. For precast construction, $\varnothing$ is normally around 1.6, while for *in situ* construction it would normally be between 2.0 and 2.5. BS 5400 Part 4, Appendix C.2 gives the derivation of $\varnothing$ in more detail.

Where the change to the statical system is sudden, such as connecting cantilevers with a mid-span stitch, the modification to the moments is:

$$M_{final} = M_{as\text{-}built} + (1 - e^{-\varnothing})\,(M_{inst.} - M_{as\text{-}built})$$

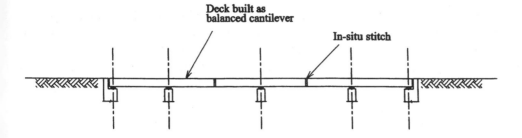

(i) Structure arrangement

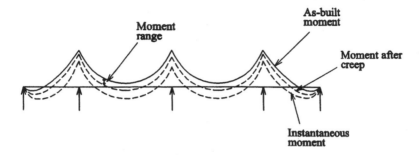

(ii) Dead load bending moment diagram

Figure 5.23 Creep redistribution of moments.

Where the change is gradual, such as the differential shrinkage between precast beams and *in situ* top slab, the modification to the moments becomes:

$$M_{final} = M_a \left((1 - e^{-\varnothing})/\varnothing \right)$$

Shears are modified in a similar way.

5.5.14 Temperature effects

Changes in effective temperature of the deck will cause it to expand or contract, while differential temperature gradients through the concrete result in stresses that need to be considered in the prestress design.

Under a uniform temperature change, the change in deck length is $\Delta_L = \Delta_t \propto L$.

The co-efficient of thermal expansions, $\propto$, is typically $12 \times 10^{-6}/°C$ for normal concrete or $9 \times 10^{-6}/°C$ with limestone aggregates.

Where a deck is free to expand or contract this overall change in effective temperature will not give rise to any forces in the structure, although the movement does need to be allowed for in the bearing and expansion joint design. When a restraint exists that restricts free movement, such as integral decks or multiple fixed piers, forces and stresses will be generated throughout the structure which must be taken into account in the prestress design.

Differential temperature effects are generated through the concrete section as the outer surface heats or cools more quickly than the rest. BS 5400 Part 2 Figure 9 defines a series of simplistic temperature gradients that can be applied to different

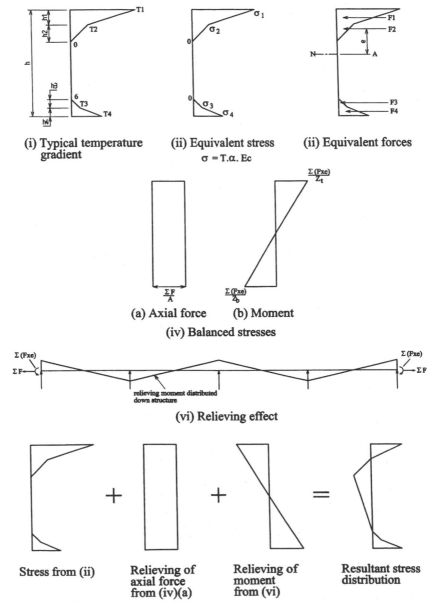

Figure 5.24 Differential temperature stresses.

structural types. Figure 5.24 illustrates how the temperature gradients generate stresses and forces within the deck. The equivalent forces generated by the temperature differential give rise to an out-of-balance axial force and moment. Provided the deck is not restrained, the force and moment will relieve at the ends and the stresses adjusted accordingly to give the resultant stress distribution across the section. These stresses are then catered for in the design when considering the serviceability stress check of the prestressed concrete.

The moment effect gives rise to shear forces in the spans and these should be considered during the ultimate shear design.

5.6 Design of details

5.6.1 Anchorage

With pretensioned strand, the force transfer into the concrete is achieved through bond between the two materials. At the end of the strand it slips into the concrete as the bond gradually builds up the force transferred, until the total force in the strand has been taken up by the concrete over a transmission length, L_t. With $F_o \leqslant 75\%$ UTS and $f_{ci} \geqslant 30$ N/mm^2, then this length can be defined as:

$$L_t = k_t D_t / \sqrt{f_{ci}}$$

k_t can be taken as 600 for plain, indented or crimped wire with a wave height of less than $0.15D_t$, 400 for crimped wire with a total wave height greater than $0.15D_t$, 240 for 7-wire standard and super-strand or 360 for 7-wire drawn or compacted strand.

When post-tensioned tendons are anchored, they apply a large concentrated force to the concrete which needs to be contained. Ciria Guide No. 1 (Clark, 1976) describes the behaviour and design of anchor blocks with a typical arrangement as shown in Figure 5.25. Behind each anchor, splitting forces and tensile stresses occur and reinforcement needs to be provided as follows.

Bursting reinforcement

This is provided as a spiral or series of links around each individual anchor, with:

$$A = F_{bst}/0.87f_y$$

F_{bst} depends on the end block arrangement and is given by:

y_{po}/y_o	0.3	0.4	0.5	0.6	0.7
F_{bst}/F_o	0.23	0.20	0.17	0.14	0.11

$0.87f_y$ should be replaced by a stress of 200 N/mm^2 or less to control cracking when the concrete cover is less than 50 mm.

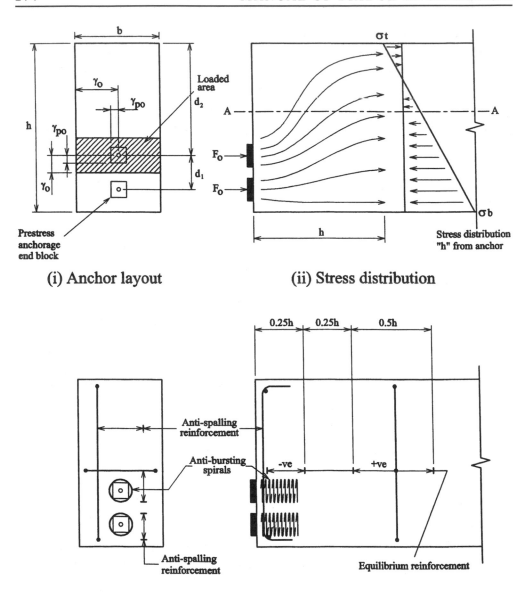

(i) Anchor layout   (ii) Stress distribution

(iii) Reinforcement design

Figure 5.25 End block design.

Bursting reinforcement should be placed between $0.2y_o$ and $2y_o$ from the anchor face and enclose a cylinder or prism with dimensions 50 mm larger than the face of the anchor block. The bursting forces in the two principal directions should be determined and sufficient reinforcement provided to suit. The loaded area used for determining y_o is taken as symmetrical about the anchor, extending to the nearest edge of the concrete or to the midpoint of any adjacent anchors.

Spalling reinforcement

This is provided to prevent spalling of the end face around the anchor with:

$$A = 0.04F_o/0.87f_y$$

The stress in the reinforcement $(0.87f_y)$ should be kept to below 200 N/mm² to control cracking while the bars should be placed as near to the end face as possible and anchored around the concrete edges.

Where the anchor is positioned non-symmetrically on an end face, additional spalling stresses are set up and additional reinforcement must be provided in the unsymmetrical face such that:

$$A^l = 0.2[(d_1 - d_2)/(d_1 + d_2)](f_o/f_y)$$

Equilibrium reinforcement

This is provided to maintain the overall equilibrium of the end block.

The force from the anchor block is assumed to have fully spread out at a distance h from the anchor face. By considering the concrete block with the anchor force on one side and the stress distribution on the other, the equilibrium of any horizontal plane, such as A–A in Figure 5.25(ii), is checked and vertical reinforcement provided to resist the out-of-balance moment, using a lever arm of $h/2$ to calculate the quantity of reinforcement needed. The reinforcement is then placed over the distance $0.25h$ or $0.5h$ as indicated in Figure 5.25(iii) depending on the direction of the out of balance moment.

The shear on each horizontal plane should also be checked to ensure that the shear stress does not exceed:

$$(2.25 + 0.65pf_y) \text{ N/mm}^2$$

where p is the ratio of the reinforcement crossing the plane and is equal to A_s/bh.

5.6.2 Anchor blisters

Where anchors are placed on blisters, or concrete blocks cast on the side of the concrete member, Figure 5.26(i) shows the additional reinforcement needed to tie the blister into the main body of concrete.

The bursting and spalling reinforcement quantities are calculated as for standard anchors, but in addition tie-back reinforcement is required to prevent cracks occurring behind the anchor due to the tensile forces generated to achieve strain compatibility in the concrete. It has been traditional to provide tie-back reinforcement to cater for 50% of F_o, although finite element analysis can show that significantly less reinforcement than this is needed in some cases.

Equilibrium effects occur as the force distributes across the section and into the webs and slabs, requiring reinforcement to be provided as described above.

5.6.3 Anchor pockets

Figure 5.26(ii) shows an arrangement where recesses or pockets are provided in the concrete to anchor the tendons. The anchor forces tend to cause cracking of the

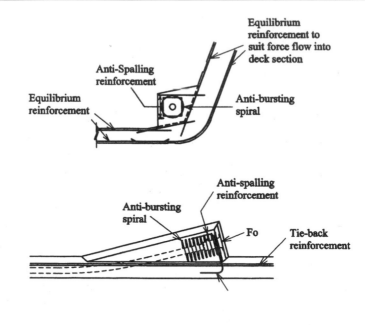

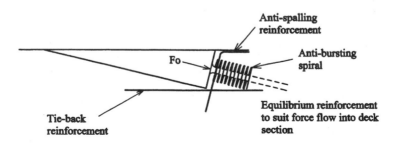

(i) Blister reinforcement

(ii) Anchor pocket reinforcement

Figure 5.26 Anchor reinforcement.

concrete behind the recesses and similar tie-back reinforcement should be provided as for the anchor blisters, as well as the normal bursting and spalling reinforcement, while equilibrium reinforcement is again needed in the webs and slabs to restrain the force as it spreads out.

5.6.4 Ducts

Duct sizes are governed by the practicalities of having space to thread the tendon through and to allow the grout to flow freely around the strands or bars, with the area of a duct normally not less than twice the tendon area.

With internal tendons, the ducts have to be supported and held in place during the concreting operation. This is normally done by installing additional reinforcement around the duct and fixing this to the normal reinforcement cage. The spacing of those supports will depend on the duct size and stiffness and is normally between 0.5 and 1.0 m.

To ease concrete placing and prevent delamination of the adjacent concrete or the tendon bursting out of the section, ducts cast into the concrete should have a minimum clear spacing of not less than the greater of the following:

(i)   duct internal diameter
(ii)   aggregate size + 5 mm
(iii)   35 mm.

The concrete cover to the deck should not be less than 50 mm although this depth may need to be increased when the ducts are curved.

With curved tendons, the strand bears against the edge of the duct and exerts a pressure on the concrete which causes splitting forces that need to be considered. The splitting force on the concrete depends on the force in the tendon and the radius of the curve. When the radii is large enough no additional reinforcement is needed as the tensile strength of the concrete is sufficient. Where the radius of the duct is significantly less than the value given in Tables 36 and 37 of BS 5400 Part 4 for the tendon force and cover or duct spacing, reinforcement should be provided to restrain the tendon as shown in Figure 5.27. Where a curved tendon exerts a force on the

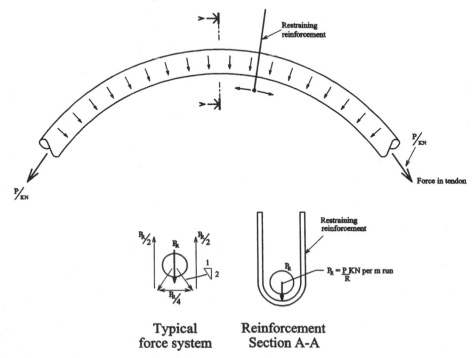

Typical
force system

Reinforcement
Section A-A

Figure 5.27 Restraint of curved tendon.

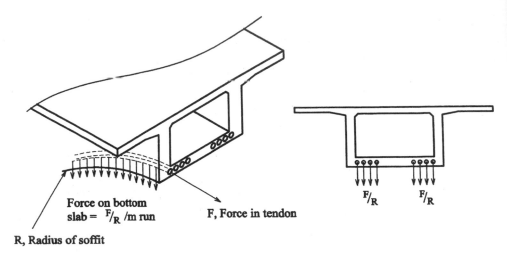

R, Radius of soffit

Figure 5.28 Force on curved soffit.

Table 5.3 Typical duct details

Typical tendon size	Duct size internal diameter	Typical tendon offset	Minimum radii for internal ducts	Minimum radii for external ducts
7×15 mm strand	65 mm	10 mm	3 m	2.5 m
19×15 mm strand	90 mm	17 mm	5 m	3.0 m
27×15 mm strand	110 mm	20 mm	7 m	5.0 m

deck section, it is sometimes necessary to provide additional reinforcement to counter the effects, such as where the tendons placed inside the bottom slab of a haunched box girder give a resultant downward force as indicated in Figure 5.28. This will generate bending and shears in the bottom slab and tension in the webs.

Where the ducts are curved, the tendon will move to the edge of the duct, resulting in the tendon centreline being offset from the duct centreline. This offset will vary depending on tendon and duct sizes and arrangement, and needs to be taken into account when considering the tendon eccentricity in the design.

With strands, the tendons can follow a profile with fairly tight curves, the minimum radii being governed by the ability of the duct to bend. Individual strands can be bent to radii as little as 0.5 m. Typical design offsets and duct radii for tendons are indicated in Table 5.3.

5.6.5 Diaphragms

Diaphragms are generally used in the deck at points of support to transfer the load from the webs into the substructure below. Typical diaphragm arrangements are shown in Figure 5.29. For the arrangement shown in Figure 5.29(i), a truss analogy is normally used to model the force transfer from the webs into the bearings. To ensure that all the load is picked up and transferred on to the truss at the top of the webs,

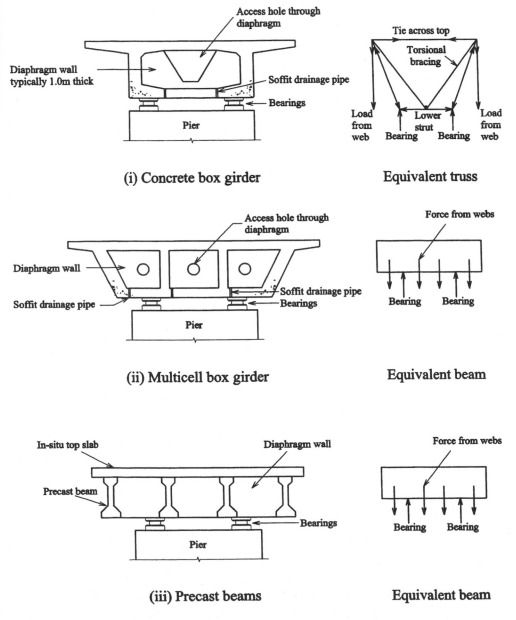

Figure 5.29 Diaphragm arrangements.

'hanging reinforcement' needs to be provided in the form of vertical bars in the web and diaphragm. The vertical force is then held by a 'tie' across the top of the diaphragm and a strut towards the bearing. Where significant torsion exists in the deck, it is also necessary to consider the horizontal forces present in the top and bottom slabs, and reinforcement provided to tie these together.

In Figure 5.29(ii) and 5.29(iii) the diaphragm will behave as a beam and bending moments and shears can be determined and reinforced against in the normal way. Where a web is directly over a bearing, then the load will go straight into the support, elsewhere 'hanging reinforcement' will be needed to take the force from the web up to the top of the diaphragm beam.

The concrete above the bearings needs to contain the high loads applied and can be designed in a similar manner to an anchor end block, with spalling and bursting reinforcement being provided accordingly.

5.6.6 Deviators
Where deviators are used to deflect external tendons to give the desired profile they can be subject to large forces that need to be tied into the deck section. Figure 5.30 shows several different types of deviator arrangements that can be used.
The force applied to the deviator is equal to the force in the tendon multiplied by its 'angle change' (in radians) and the deviator should be designed to cater for this load. In the UK, where BD 58/94 is used, the design is carried out at the ultimate limit state with the applied ultimate force based on the characteristic strength of the tendon.

5.7 Bridge construction and design
5.7.1 Deck form
Many factors affect the choice of the bridge type, the span arrangement and general layout, while prestressed concrete decks cover a wide range of construction forms with span lengths ranging from 25 m for single spans to over 400 m in cable-stayed bridges. Below spans of 25 m reinforced concrete is likely to be preferred, while for spans above 400 m steel or composite cable-stayed decks are used; however, between these span lengths prestressed concrete often gives an economic, aesthetic and simple solution.

The choice for the deck form depends largely on the individual site constraints and the advantages and disadvantages of each option need to be carefully considered to arrive at the optimum solution. Figure 5.31 indicates the typical span range for different deck types.

For decks with an overall length of 60 m or less, it is common to build the deck continuous and integral with the piers and abutments to reduce the number of expansion joints and bearings and eliminate the maintenance problems that can occur with these elements. For longer structures, the deck should still be made continuous for as long as practical depending on the structural form, with concrete box girders having been constructed up to 1.7 km between expansion joints. The need for bearings between the deck and substructure depends on the deck type and stiffness of

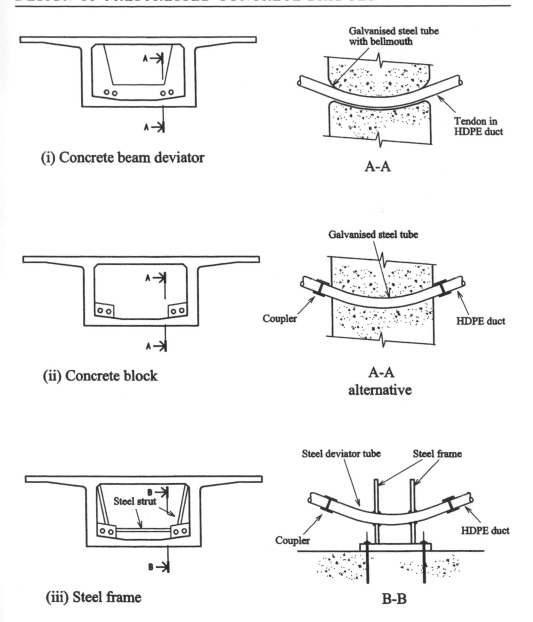

(i) Concrete beam deviator

A-A

(ii) Concrete block

A-A
alternative

(iii) Steel frame

B-B

Figure 5.30 Deviator arrangements.

the substructure. Precast segmental construction usually utilizes bearings to simplify erection while *in situ* deck construction can have bearings, or can easily be built into the substructure provided that deck movements do not generate excessive forces in the piers and foundations.

Span arrangements are frequently governed by the nature of the obstruction being bridged, but if possible the spans should be arranged to suit the type of bridge being constructed. With precast beams, it is preferred to keep a standard beam length

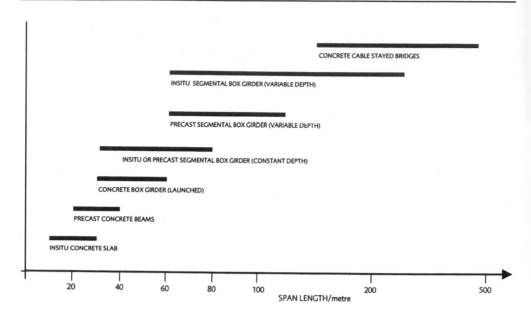

Figure 5.31 Typical span range for different deck types.

throughout to standardize construction equipment, while with *in situ*, precast segmental or launched continuous decks, the end spans should be arranged to be approximately 70% of the length of the internal spans to balance out the design bending moments. Where the deck is built by balanced cantilever, the end spans should be reduced to approximately 60% of the length of the internal spans to minimize any out-of-balance effects of the end balanced cantilever, although care should be taken to ensure that no uplift occurs at the abutment bearings, otherwise a tie-down arrangement would be needed.

5.7.2 Solid-slab bridges
In situ, solid-slab decks can be used for shorter spans where easy access is available, with the advantages including simple construction and formwork layout, while the disadvantages are that prestressing a solid concrete element is inefficient and dead load becomes excessive as spans get longer, and for these reasons prestressing of solid-slabs is not commonly seen for bridgeworks.

5.7.3 Voided-slab bridges
In situ, voided-slab decks were commonly used during the early period of prestressed concrete bridge development within the span range of 25–35 m and depth/span ratios of up to 1:20. The voids are likely to be shaped from polystyrene void forms, either round as shown in Figure 5.32 or rectangular with bevelled corners leaving a 150 mm thick concrete section above and below, while the concrete is cast, supported by falsework from the ground. The advantages include simple construction and form-

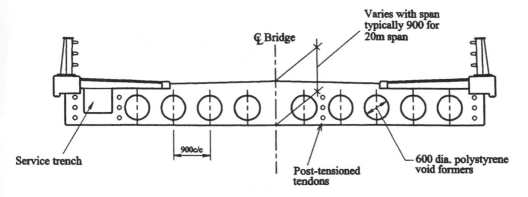

Figure 5.32 Typical voided deck slab arrangement.

work arrangement, while the disadvantages include needing to hold down the voids during concreting and a fairly heavy deck section, and this has more recently resulted in beam-and-slab or multicell boxes being used in preference.

Analysis of the voided-slab can easily be carried out with sufficient accuracy using a grillage model. It can be convenient to place the longitudinal grillage members to match the layout of the prestress tendons or groups of tendons, with transverse members placed over the supports and at suitable spacing between to give an evenly balanced grid pattern. The section properties of both the longitudinal and transverse members are based on the deck width to mid-way between the members, with the resultant voided section taken longitudinally and the section at the centre of the void taken for the full width transversely. This is normally found to give sufficiently accurate results for carrying out the design. When assigning the torsional stiffness to the grillage members it is sufficient to give both longitudinal and transverse members the same value of twice the 'I' value of the longitudinal members.

The longitudinal prestress design is carried out by considering the bending moments and shear forces from the imposed loads on the longitudinal grillage members and designing the prestress to balance the serviceability limit state stresses, with checks to ensure adequate ultimate limit state strength in the normal way.

5.7.4 Beam-and-slab bridges

Beam-and-slab bridges can utilize either precast or *in situ* concrete beams, with the deck slab cast *in situ*, see Figure 5.33. Spans normally range from 25 m up to 40 m although spans of up to 50 m can be found. Depth/span ratios would be approximately 1:16 with either post-tensioned or pretensioned strand positioned to take the sagging moments as indicated in Figure 5.11. Figure 5.34 shows some typical arrangements for the beams at piers, where the beams can be connected or the top slab made continuous over the piers to minimize the need for expansion joints in the road surface. The beams are normally connected at the support location by a transverse *in situ* concrete beam, or diaphragm, which provides rigidity to the arrangement and

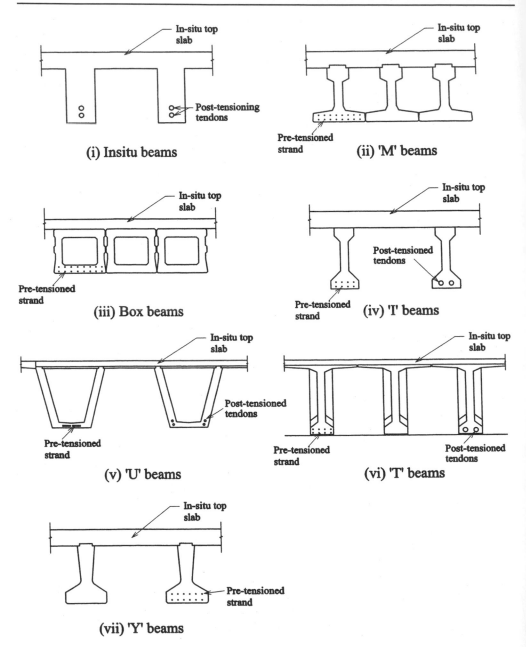

Figure 5.33 Typical prestressed beam-and-slab arrangements.

can be used to transfer load from the beams on to the supports. The advantages of beams include economics in repetitive precasting and rapid construction while the disadvantages include the need for good access and heavy lifting equipment for the longer spans.

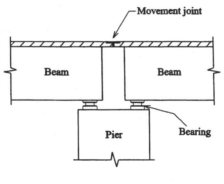

(i) Non-continuous

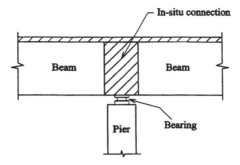

(ii) Continuous over support

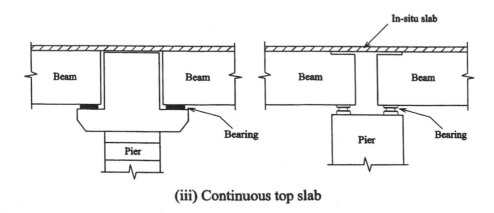

(iii) Continuous top slab

Figure 5.34 Beam continuity.

Where the beam-and-slab deck is cast *in situ,* traditional formwork and falsework arrangements can be used to support the concrete while it hardens. Post-tensioning tendons can then be threaded through the ducts cast into the concrete and stressed, lifting the deck off the formwork and allowing the falsework and formwork to be removed.

Precast beams are normally made in special casting yards which have a series of casting beds to form the beams. For pretensioned beams, jacking frames at either end of the casting bed provide anchorages for the strands to be stressed against. The reinforcement and shutters are assembled around the stressed strand, and the concrete poured. Although the side shutters can be stripped away after 1 or 2 days, the concrete is normally required to reach at least 35 N/mm^2 before the strand is released from the jacking frames and the force transferred into the beam. For post-tensioning, the beams are made with ducts and anchorages cast in, to allow the tendon to be threaded through and stressed after the concrete has sufficient strength. Figures 5.35 and 5.36 show a typical pretensioning and post-tensioning casting bed arrangement, respectively. For simply supported beams the applied moment reduces at the end of beams to zero at the supports if simply supported, and if a constant prestress force is applied along the beam's length this can generate high compression in the bottom and tension in the top over a significant length of the ends. To prevent this in post-tensioned beams, the tendons can be draped with the anchors raised up at the ends while with pretensioned beams some of the strands can be debonded by placing in a plastic tube over the ends, or alternatively some strands can be deflected up by means of an anchor frame at 1/4 or 1/3 points along the beam.

The placing of the beams into position is normally done by crane; however, for long, multispan bridges or where ground access is poor, an erection gantry can be used to lift the beams and carry them into place. Figures 5.37 and 5.38 show crane and gantry erection, respectively. After positioning the beams, the falsework is placed to allow the top slab to be cast.

Figure 5.35 Casting bed for pretensioned beam.

Figure 5.36 Casting bed for post-tensioned beam (courtesy of Hyder Consulting Ltd).

Figure 5.37 Crane erection of precast beams (courtesy of Hyder Consulting Ltd).

Figure 5.38 Gantry erection of precast beams.

Design of beam-and-slab decks involves the superimposing of many different effects to build up the overall design state. Dead load, superimposed dead load and live load effects, both longitudinally in the beams and transversely in the deck slab, can be derived from a grillage analysis which will give satisfactory results for most standard beam-and-slab arrangements. For the grillage analysis, the deck is modelled as a series of discrete members both longitudinally and transversely. Longitudinally it is simplest to provide a grillage member along the line of each beam web, with section properties to include a portion of the top slab up to midway to adjacent webs and similarly a portion of the bottom slab where U-beams or box-beams are used. Where box structures are present and are represented by individual grillage members for each web, the torsional stiffness of the box is calculated and a quarter of this is assigned to each web, with the remaining half of the torsional stiffness assigned to the transverse members. Transversely, grillage members would be positioned at each diaphragm location with section properties to match the diaphragm beam, and also at regular intervals between with section properties to match the length of the top and bottom slabs between, where appropriate. The location of the transverse members should result in a ratio of the spacing of longitudinal members to transverse member of between 1:1 and 1:2 to achieve reasonably accurate load distribution.

Where unusual beam-and-slab arrangements are used or secondary effects such as distortions or transverse bending in the beams become significant then three-dimensional finite element models should be used for deriving the load effects. Shell

elements can be used to build up the webs, slabs and diaphragms of the three-dimensional structure, and the dead, superimposed and live load applied to the model to
give the forces and moments at the critical sections for the beam design.

Full account of the construction and stressing sequence needs to be allowed for in
the design. When the beams are cast they are supported by the formwork along their
total length; however, on applying the prestress they tend to lift up over the length
of the beam and when removed to the storage yard they would be supported at the
ends of the beams only, giving a typical stress profile at mid-span as indicated in

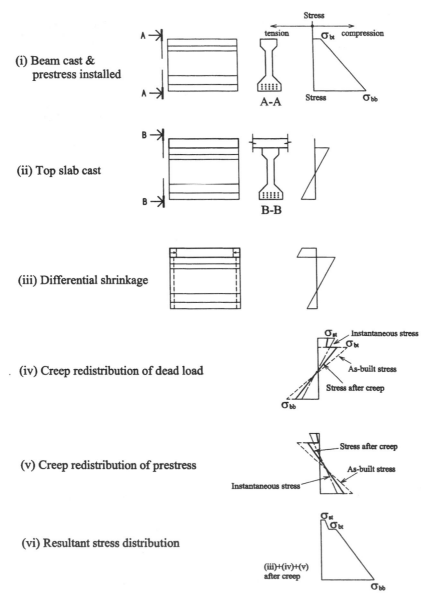

Figure 5.39 Longitudinal stress distribution in beams.

Figure 5.39(i). At this stage the beams are subject to their own dead weight and the applied prestress and the resulting stresses at the top and bottom of the beam can be accurately derived. The common problem at this stage is to ensure that the compression in the bottom and tension in the top of the beam do not exceed allowable limits. Once erected, the top slab is cast with the weight being carried by the beam section alone as indicated in Figure 5.39(ii). When the top slab concrete has hardened, creep on the concrete will redistribute the dead load stresses and prestress forces across the section. The superimposed and live loads are carried by the combined section of the beam and top slab.

The creep redistribution of the dead load and prestress stresses can be estimated by considering the as-built condition, where all the load is carried on the beam and the stresses are a combination of Figure 5.39(i) and (ii), and the theoretical 'instantaneous' condition where the full dead load and prestress is applied to the composite section. The stresses will creep from the as-built condition towards the instantaneous condition with the final stress at any level in the section being given by:

$$\sigma_{final} = \sigma_{as\text{-}built} + (1 - e^{-\phi})\,(\sigma_{inst} - \sigma_{as\text{-}built})$$

The loss of prestress forces due to relaxation in the strand, and concrete creep and shrinkage that occurs after casting of the top slab will effect the stresses in both the slab and beam, and these long term prestress losses should be considered in two phases. For the first phase before the deck slab is cast, the prestress losses should be estimated and the prestress force in the beam reduced accordingly. In the second phase, the remaining losses that occur after casting the slab should be estimated and then these applied as a 'tensile' force to the composite section at the position of the centroid of the strand or tendons.

When the top slab is cast, the concrete in the beam will already have completed a large proportion of its shrinkage resulting in differential shrinkage occurring between the slab and beam. The deck slab tries to shrink more than the beam and is restrained by it, creating tension in the slab and a combined compressive force and sagging moment in the beam. The equivalent differential shrinkage force, assuming the top slab was fully restrained, can be estimated from:

$$\text{force, } F_s = \Delta_s A_s E_c((1 - e^{-\phi})/\phi)$$

with $(1 - e^{-\phi})/\phi$ reducing the effect due to creep of the concrete.

The stresses in the section can be derived by adding the 'fully restrained' stress to the 'released' stress in the composite section, i.e.

(i)   tensile stress from force F_s acting on slab alone at the centroid of slab, plus
(ii) stresses generated from a compressive force F_s acting on the composite section, applied at the centroid of the slab

giving a stress profile similar to Figure 5.39(iii).

The stresses generated at the top of the slab, top of beam and bottom of beam must be checked at each stage of construction as well as in the permanent condition after opening and again after the long-term effects have occurred.

When the beams are fully connected longitudinally over the piers to make the deck continuous, the dead load and prestress effects will be redistributed due to creep of the concrete. The creep from the prestress load on the concrete can cause a hogging of the beam at mid-span and a sagging restraint over the pier, which can lead to excessive cracks in the soffit in this region if the effect is not fully considered and suitable reinforcement provided. The final moment at the pier due to the crept dead load and prestress can be derived by a similar approach to that described in Section 5.5.13, with:

$$M_{final} = M_{as\text{-}built} + (1 - e^{-\phi})(M_{inst} - M_{as\text{-}built}).$$

ϕ should be based on the residual creep left in the beam at the time of casting the connection and $M_{as\text{-}built}$ is equal to zero where the connections are made after erecting the beam and casting the top slab. The moment calculated as if the deck was built 'instantaneously' as continuous, M_{inst}, should include the prestress secondary moment as well as the prestress primary moment and dead load. The differential shrinkage between the top slab and beam will generate a secondary hogging moment in the deck at the pier after the connection has been made, which must also be considered in the design.

The design for ultimate moment, shear and torsion is carried out in the normal manner, while the check of the longitudinal shear along the interface between the slab and beam must include the forces generated by the differential shrinkage and creep effects.

During the lifting and transporting of the beams they can be subjected to additional loads resulting in a change to the bending moments which must be designed against. Care must be taken to minimize any impact or dynamic loading during handling, while lifting or temporary support positions should be located as near to the permanent support position as possible. Where long, slender beams are used, transverse instability can occur either during lifting, or when placing the concrete for the top slab, and suitable restraint will need to be provided.

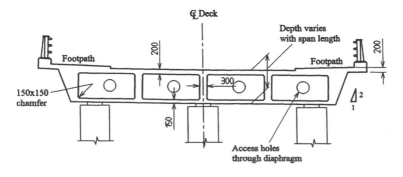

Figure 5.40 Typical multicell box girder arrangement.

5.7.5 In situ multicell box girder decks

Multicell box girder bridges are similar to voided slabs with the voids occupying a larger proportion of the deck section with a typical arrangement as Figure 5.40, which would have transverse diaphragms at each pier position. Concreting and formwork construction normally dictate a minimum depth of 1200 mm, and with depth/span ratios of up to 1:25, the span lengths are normally greater than 30 m and can extend to 50 m. The advantages include efficient use of concrete and prestress and simple construction where easy access is available. The disadvantages include the need for extensive labour activities on site, and the falsework can become complex if access is difficult.

The formwork for concreting can be supported either from falsework off the ground or from a gantry spanning between the piers. It simplifies the formwork to cast the deck section in several stages, with the bottom slab, outer webs and diaphragms being cast first, followed by the interior webs and top slab soon after. Figure 5.41 shows a multicell box girder being prepared to cast the first stage.

Where thin webs and slabs are being used, the box section can significantly deform and distort under the imposed loading, and the deck analysis can most accurately be carried out by using a three-dimensional finite element model with shell elements used for each web and slab section. The moments and shears in each element in each direction can be taken directly from the model and used for both the longitudinal prestress design and the transverse reinforcement design. For the longitudinal design the deck section is divided up into a series of I beams consisting of a web and asso-

Figure 5.41 Multicell box girder deck under construction (courtesy of Hyder Consulting Ltd).

ciated top and bottom slabs up to the mid point of each cell. Prestress layouts are then designed for each web to balance the stresses derived from the finite element analysis and to give adequate ultimate strength.

5.7.6 In situ single cell box girder bridges

In situ single cell box girders such as the typical section shown in Figure 5.42 are used for a wide range of spans from 40 up to 270 m when haunched. The advantages are the efficient use of concrete and prestress and the flexibility in span arrangements, while the disadvantages include the labour intensive activities on site and the long construction times needed for the larger structures. For spans up to 60 m a constant depth section with depth/span ratio of 1:20 would be used, with the concrete cast in sections span-by-span supported by falsework from the ground as Figure 5.43 or with a truss spanning between the piers. Above 60 m spans it is common to use a haunched section cast within a travelling form as a balanced cantilever about a pier as shown in Figure 5.44, with segments between 3 and 5 m long and a typical casting cycle resulting in a segment being cast every 7 or 8 days. At mid-span, the two opposing cantilevers are joined to make the deck continuous. Depth-to-span ratios are typically 1:16 at the pier, reducing to 1:45 at mid-span.

Traditionally, for a single cell box girder the longitudinal and transverse designs have been considered separately although it is now becoming more common to establish three-dimensional finite element models that combine the effects.

The deck can be modelled longitudinally as a two-dimensional or three-dimensional frame by standard structural analysis software, and the moments and shears derived for dead load, superimposed dead load and live load as well as for the secondary effects such as support settlement. Creep effects of the dead load and the prestress and the temperature effects can be determined as described earlier in the chapter.

Transversely, the box is subjected to bending and shear due to the dead and super-imposed dead loads and the live loads on the top slab. A three-dimensional finite

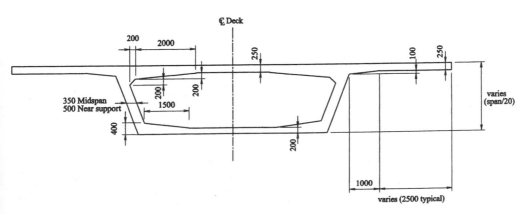

Figure 5.42 Typical single-cell box girder section.

Figure 5.43 In situ box girder, construction on falsework (courtesy of Hyder Consulting Ltd).

Figure 5.44 In situ balanced cantilever construction (courtesy of Hyder Consulting Ltd).

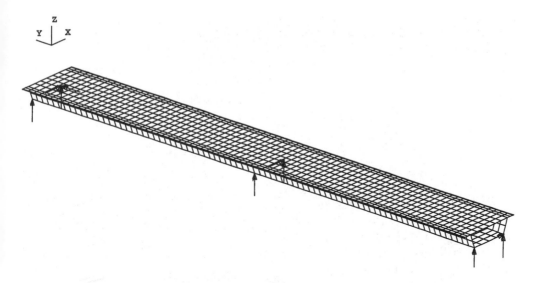

Figure 5.45 Typical finite element model of box girder.

element model with shell elements should be set up, similar to Figure 5.45, covering sufficient length of the box to allow the longitudinal distribution of the live load and the three-dimensional behaviour of the box to be accurately modelled. Once the moments and shears around the box have been calculated, reinforcement can be provided as necessary.

Under torsional loading, concrete box girders warp with their cross-sections undergoing out-of-plane displacements. The sections undergo torsional and distortional warping while the applied loads can also cause distortion to the cross-section. These effects are fully described in the C&CA Technical Report on concrete box beams (Maisel and Roll, 1974). If a full three-dimensional finite element model is set up over the full length of the deck, the warping effects can be directly modelled, along with both the longitudinal and transverse effects.

For normal proportioned box girders, the prestress tends to balance the dead and live load moments, and shear lag is not a problem and need not be considered further. For wider or slimmer decks with unusual proportions shear lag can be significant and needs to be allowed for in the design. In these cases, the shear lag will modify the stress distribution across the section and peak stresses will need to be determined for use in the serviceability longitudinal stress check; however, shear lag need not be considered in the ultimate limit state. The shear lag effect can be determined from a full three-dimensional finite element model of the structure, although care is needed in modelling the prestress forces to give an accurate representation of their effect.

5.7.7 Precast segmental box girders
Precast single-cell box girders are found to be very economic for long bridge lengths due to the savings associated with maximizing repetition in factory conditions. Span lengths vary typically from 40 m up to 150 m, above which the segment weights become

Figure 5.46 Precast segmental erection by crane.

Figure 5.47 Precast segmental erection by gantry.

excessive. Up to spans of 70 m a constant depth section would be used with a depth:span ratio of 1:20, and segments could be positioned by crane if access is suitable or by erection gantry as shown in Figure 5.46 and 5.47, respectively. Above 60 m spans the deck is more likely to be haunched with depth-to-span ratios of 1:16 at the pier, reducing to 1:35 at mid-span. The advantages include rapid construction with minimal on-site work, while the disadvantages include the costs of setting up the casting yard and the special erection equipment needed.

The segments are made in purpose-built casting yards in specially designed form-work, most commonly using the short bed method shown in Figure 5.48, where the segment is cast against its neighbour, and when hardened it is moved into the coun-tercast position to be cast against, after which it is moved to a storage area. In the alternative long bed method, the segments are cast against each other in a long line, the formwork moving down the line for each segment casting. After casting a series of segments they are all removed from the bed to the storage area. By casting direct-ly against each other either in the short bed or long bed method, a perfect fit is achieved when the segments are erected, and by carefully controlling the relative position of the segments when they are being cast, complex horizontal and vertical alignments can be achieved.

Before the segment being placed is positioned against the already erected portion of the deck the end faces are lightly sandblasted to remove any deleterious material and to provide a good key for the epoxy glue which is spread evenly over

Figure 5.48 Precast segment casting bed (courtesy of Hyder Consulting Ltd).

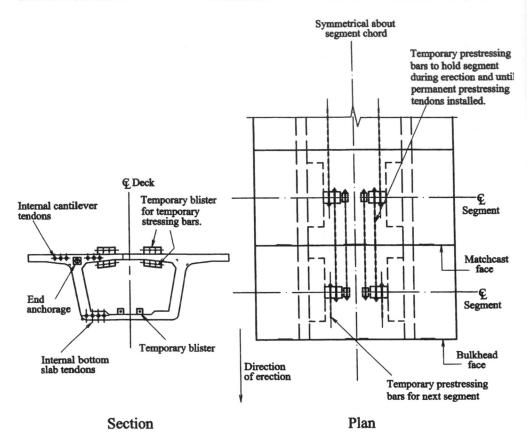

Section   **Plan**

Figure 5.49 Typical temporary prestress for precast segmental erection.

the matching faces. Temporary prestress is then applied to hold the segment in place as indicated in Figure 5.49. The temporary prestress should apply an average compressive stress of between 0.2 N/mm^2 and 0.3 N/mm^2 with a minimum local stress of 0.15 N/mm^2 and a flexural stress difference less than 0.5 N/mm^2 over the segment joint being epoxied, to ensure that the joint thickness is constant through-out, normally between 1 and 3 mm. The epoxy helps to lubricate the joint during erection, allowing the segment to slide into position, and once hardened the epoxy seals the joint and provides structural continuity between the segments.

Several bridges have been built with dry joints, where no epoxy is used between the segments and the design relies on shear friction to transfer the loads in both the permanent and temporary conditions. To date dry joints have been used only in tropical climates where freeze–thaw cycles do not occur and their effectiveness in more aggressive climates has yet to be proved.

Figure 5.50 indicates three alternative systems for segment erection. The most common method of erecting the segments is by the balanced cantilever technique, either with a gantry as shown in Figure 5.47, a crane or by a special lifting frame fixed to the deck. The first segment over the pier is lifted into position and placed

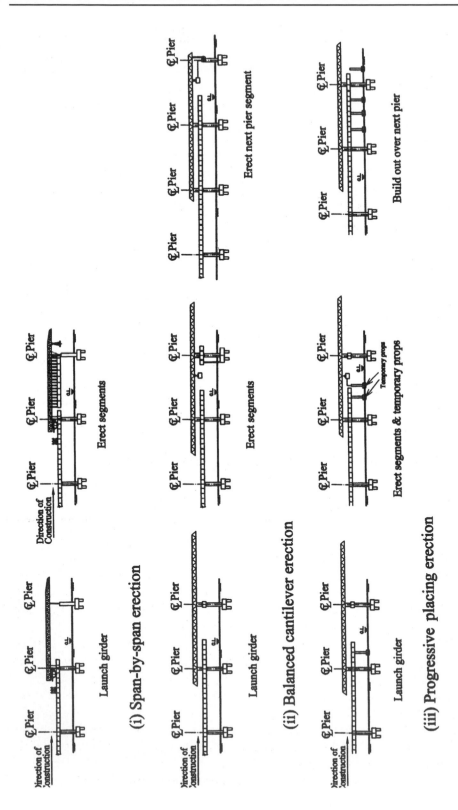

Figure 5.50 Precast segmental deck erection methods.

on temporary supports to allow it to be correctly aligned. Subsequent segments are transported into position and lifted either side of the pier segment, being fixed to it by epoxy and temporary prestress. Further segments are positioned with erection progressing out from both sides of the pier in a balanced manner. When the cantilever reaches mid-span it is connected up to the opposing cantilever from the adjacent pier by an *in situ* concrete stitch and prestressing tendons are installed along the length of the span and across the stitch to complete the connection.

For spans of less than 50 m, an alternative erection method is to place the segments span by span. In this technique a gantry, either overhead as shown in Figure 5.51 or underslung, is used to support a complete span of segments which are then pulled together by applying the permanent prestress, allowing the gantry to release the span on to the bearings and to launch itself forward ready to erect the next span.

A third method is progressive placing, where erection is started at one end of the bridge and continued out progressively until the other end is reached. This method usually needs temporary props along the span to support the deck until the next pier is reached, such as is shown in Figure 5.52.

With match cast segments erected with epoxy in the joints the design of the concrete box in its completed condition is similar to that for an *in situ* box girder. The presence of the joints creates a discontinuity in the longitudinal reinforcement and the joints should be kept in compression under all loading conditions. The epoxy creates a bond of greater strength than the concrete itself and the ultimate shear checks can be carried out as if the deck is monolithic; however, where wide, cast

Figure 5.51 Span-by-span segments erection (courtesy of Hyder Consulting Ltd).

Figure 5.52 Progressive placing of segments.

in situ concrete joints are incorporated into the structure, the shear force under ultimate loads should not be greater than:

$$0.7(\tan \alpha)\, 0.87P$$

where tan α depends on the interface and for roughened faces may be taken as 0.7 during erection and 1.4 after completion of the bridge, and P is the horizontal component of the prestress force after all losses.

Shear keys are provided on the surface of the joint to help align the segments during erection and to transfer the shear forces before the epoxy has set. Either large or small multiple shear keys can be used as shown in Figure 5.53, and both are designed to prevent a shear failure under the temporary erection loads. Considering the shear friction concept as given in AASHTO, the capacity of the shear key can be estimated using a coefficient of friction of 1.4, resulting in the shear resistance being:

$$V_k = 1.4[A_{sk}f_{pk} + A_r(0.87f_y)].$$

For the large shear keys these can be designed as a reinforced concrete corbel where appropriate reinforcement is provided, while for the small shear key, reinforcement is not normally required.

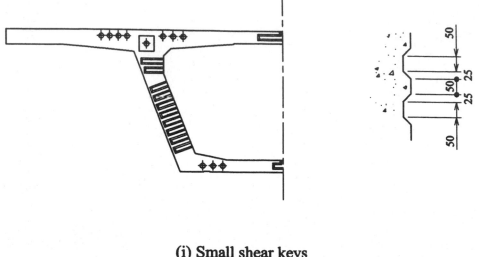

(i) Small shear keys

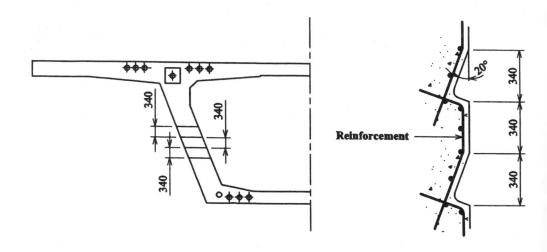

(ii) Large shear keys

Figure 5.53 Shear key arrangement.

5.7.8 Incrementally launched box girder bridges

Where the bridge alignment is straight or on a constant radius curve, either vertical or horizontal, launched single cell box girders can be used to overcome access problems or to avoid obstructions at ground level. Generally used for spans up to 60 m, the technique has been used for longer spans up to 100 m with the help of

Figure 5.54 Incrementally-launched box girder (courtesy of Freyssinet International).

temporary piers placed to reduce the effective span length during launching. Deck depth must be constant with the ratio to the launched span generally 1:16 or less.

Cast in segments behind the abutment, the deck is pushed or pulled out over the piers as shown in Figure 5.54. A specially prepared casting area is located behind the abutment with sections to assemble the reinforcement, to concrete, and to launch with a typical layout shown in Figure 5.55. Segment lengths are normally standardized for a bridge and are usually 20–30 m long.

The first segment is cast and moved forward on temporary bearings. The second segment is then cast against the first and both are moved forward by a further increment, with subsequent segments cast and the deck moved until it reaches the opposite abutments and its final position.

The area immediately behind the casting bay is utilized for steel fixing and placing the prestressing ducts, which can progress at the same time as other operations. When the deck is launched the steel cage is attached to the concrete and pulled into position ready for concreting. The formwork system needs to be specifically designed to allow it to be lowered, leaving the deck on temporary supports ready for launching. Temporary bearings are used on each pier and in the casting area to launch the deck and consist of a steel plate, surfaced with stainless steel, on laminated rubber pads, with a teflon pad fed in between the bearing and concrete deck to provide a

Figure 5.55 Casting area layout for launched box girder (courtesy of Hyder Consulting Ltd).

low-friction sliding surface. The launching devices can be fixed to the abutments which provide the resistance against the thrust needed. A typical launching device as shown in Figure 5.56 would jack up the deck a small amount to grip the structure and then push or pull the deck forward a small amount before dropping down to release the structure and to move back before starting another stroke.

As the deck moves out the pushing force has to increase to overcome the frictional force on the temporary bearing, which can be between 2–6% of the vertical load. Greater pushing forces are needed where a deck is being launched up a slope, while if going down a slope a braking device would be needed. The launching force is normally resisted by the abutment which has to be designed to prevent sliding or over-turning. Additional resistance can be mobilized by providing the casting area with a ground slab as a working platform and by connecting this slab with the abutment. To keep the deck correctly aligned during the launching operations, guides are fixed to the piers. When the launching is complete, the deck is jacked up and the temporary bearings replaced by the permanent bearings

As the deck is launched over a pier, large cantilever moments occur until the next pier is reached and to reduce these moments a temporary lightweight steel launching nose is fixed to the front of the box. The effectiveness of the launching nose is governed by its length and stiffness, and the optimum arrangement needs to be chosen to balance the cost of the nose against the cost of catering for additional

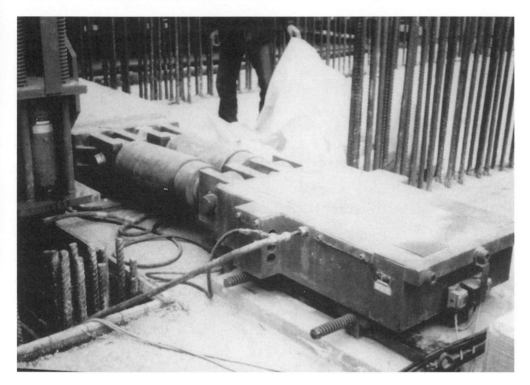

Figure 5.56 Launching device (courtesy of Hyder Consulting Ltd).

moments in the deck. Typically the launching nose length would be about 60% of the span length, and would have a stiffness (*EI*) of about 10–15% of the concrete deck. An alternative to a launching nose is to utilize a temporary tower and stay-cables over the front portion of the deck in order to reduce bending moments. The tension in the stays is adjusted as the deck passes over a pier to control the moments and forces imposed on the structure.

As the deck moves over the piers, each section is subject to a change in moment and shear. At some stage, each section will be over a pier and subject to hogging moments, or in mid-span and subject to sagging moments, and the prestress design needs to cater for the full range of moments, indicated in Figure 5.57. During launching, a first stage prestress would be arranged to give a constant compressive stress across the section, normally of the order of 5 N/mm^2. This prestress would normally consist of straight tendons placed in the top and bottom slabs, anchored on the construction joint between segments and coupled or lapped to extend through each new length of deck cast. After completion of launching, a second stage prestress is installed with a draped profile along the spans to balance the bending moment profile generated by the deck in its final position.

The full length of box section also needs to be strong enough to resist the shear forces and the temporary bearing load under the webs, as the deck passes over the

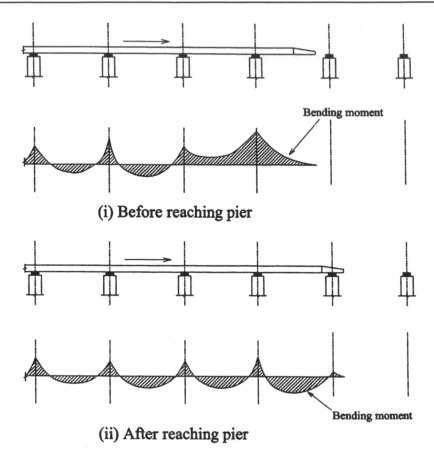

(i) Before reaching pier

(ii) After reaching pier

Figure 5.57 Bending moment range in deck during launching.

piers. The webs are normally kept a constant thickness and the web/bottoms slab corners need to be stiffened throughout.

Unevenness of the concrete surface and differential settlement of the piers and temporary supports generate additional moments and shears in the deck during the launching operation which need to be allowed for in the design.

As the deck is launched, the friction in the temporary bearings applies a load to the top of the piers. The temporary bearings are aligned parallel to the deck rather than horizontal like the permanent bearings and this induces a further horizontal load on to the pier. The piers have to be designed to resist these horizontal loads in combination with the co-existent vertical loads, although the effects can be reduced by providing stays or guys to the top of the piers.

5.7.9 Concrete cable-stayed bridges

Providing stay cables to support the deck can result in a slimmer deck section or in longer spans being achieved. The deck elements can be either a concrete box or a beam-and-slab form with typical arrangements being shown in Figure 5.58. The deck

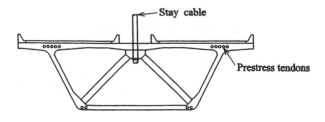

(i) Box girder with single plane of stays

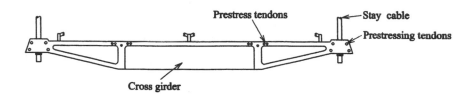

(ii) Semi-box girder with twin plane of stays

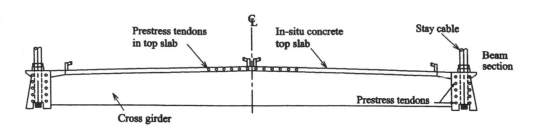

(iii) Beam-and-slab with twin plane of stays

Figure 5.58 Types of cable-stayed decks.

can be made up of precast elements, transferred to site and lifted into position, or *in situ* segments cast inside a travelling form. Either method normally involves building the deck out from the pylons in a balanced manner, installing the stay cables as the construction proceeds. The advantages of a concrete deck are that the horizontal component of the stay force causes compression in the deck which is readily taken by the concrete, while the mass and stiffness of the deck reduces its susceptibility to vibrations or aerodynamic movements.

Figure 5.59 Vasco da Gama Bridge under construction.

In Figure 5.58(i) the box girder has inherent torsional strength which makes it suitable for use with a single plane of stays, while the concrete cross-section is well suited to precasting to enable fast and simple erection. This form of deck was used for the Brotonne Bridge in France, which had a main span of 320 m and box depth of 3.8 m, and also for the Sunshine Skyway Bridge in Florida with a main span of 366m and box depth of 4.47 m.

Where twin planes of stays are used then either a 'semi-box' as shown in Figure 5.58(ii) or a beam-and-slab as shown in Figure 5.58(iii) are commonly used. A semi-box arrangement was used for the Pasco-Kennewick Bridge, USA with a main span of 299m and deck depth of 2.13 m, while typical examples of the beam-and-slab type are the Vasco da Gama Bridge, Portugal shown in Figure 5.59, with a main span of 420 m and deck depth of 2.6 m and the River Dee Crossing, UK with a main span of 194 m.

5.7.10 Extra-dosed bridges

A variation on the cable-stayed arrangement, the 'extra-dosed' bridge combines a stiff concrete deck with 'shallow' cable stays anchored to a reduced height pylon and

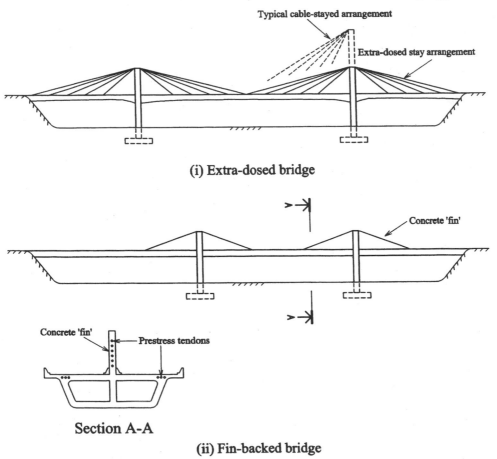

Figure 5.60 Extra-dosed and fin-back bridge arrangements.

Figure 5.61 Bubiyan Bridge under construction (courtesy of Bouygues).

has proved economic in the 100–200 m span range. The relative stiffness of the deck gives a structural behaviour more like an externally prestressed deck than a cable-stayed structure and the design is carried out on that basis. A typical arrangement is shown in Figure 5.60(i).

5.7.11 Fin-backed bridges
These are similar to the extra-dosed bridge, with the stays encased in a concrete wall or 'fin' as shown in Figure 5.60(ii). The fin is an extension of the deck section and stiffens the deck adjacent to the pier while also providing protection to the stay arrangement.

5.7.12 Truss bridges
Although not commonly used because of the complexity in casting and joining the struts and chords of the truss, several precast segmental viaducts and cable-stayed decks have utilized trusses for the 'webs' in order to reduce weight, with a typical arrangement such as that used on Bubiyan Bridge, Kuwait shown in Figure 5.61.

Bibliography
AASHTO. Standard Specification for Highway Bridges. America Association of State Highway and Transportation Officials.
BD58/94. The design of concrete highway bridges and structures with external and unbonded pre-stressing. Design Manual for Roads and Bridges. Nov, 94.

British Standards Institution. BS5400. *Steel, Concrete and Composite Bridges*, Part 2, 1978: Specification for Loads. British Standards Institution, London, 1978.

British Standards Institution. BS5400 *Steel, Concrete and Composite Bridges,* Part. 4, 1990. Code of practice for the design of concrete bridges. British Standards Institution, London, 1990.

Clark JL. A Guide to the design of anchor blocks for post-tensioned prestress concrete. CIRIA, 1976.

Concrete Society Technical Report No. 47, Durable Bonded Post-Tensioned Concrete Bridges.

Hambly EC. *Bridge Deck Behaviour*. Wiley, New York, 1979.

Maisel BI and Roll F. Methods of analysis and design of concrete box beams with side cantilevers. Cement and Concrete Association, 1974.

Podolny W and Muller JM. *Construction and Design of Prestressed Concrete Segmental Bridges*. Wiley, New York, 1982.

Appendix I. Definitions

Bonded tendons	Where the tendon is bonded to the concrete member, by being cast-in as for pretensioning or through grout for post-tensioning.
External tendons	Where the tendon is placed outside the concrete section.
Internal tendons	Where the tendon is placed within the concrete section.
Post-tensioning	Where the tendon is stressed after the concrete has hardened, with load transfer during the stressing operation.
Prestressing tendons	A wire, strand, a bar or bundle of strands used to prestress the concrete.
Pretensioning	Where the tendon is stressed before placing the concrete, with load transfer occurring after the concrete has set.
Unbonded tendons	Where the tendon is not connected to the concrete other than at the anchor positions, allowing the tendons and concrete to act independently along the tendon length.

Appendix II. Symbols and notation used

A or A^1	$=$	Area of reinforcement to be provided.
A_c	$=$	Area of concrete section.
A_L	$=$	Area of concrete considered for longitudinal shear check.
A_o	$=$	Area enclosed by median wall lines around the box.
A_p	$=$	Area of non-prestressed reinforcement.
A_r	$=$	Area of reinforcement across failure plane.
A_s	$=$	Area of slab.
A_{sk}	$=$	Area of shear key.
A_{sl}	$=$	Area of longitudinal reinforcement.
A_{st}	$=$	Area of leg of link around section.
A_{sv}	$=$	Area of shear reinforcement.
A_t	$=$	Cross-sectional area of tendon.
b	$=$	Breadth of member or web.
b_s	$=$	Width of slab.
C	$=$	Compressive force generated in concrete at ultimate moment capacity.

d	=	Distance from tendons to compression face.
d_c	=	Depth of compression in concrete at ultimate moment capacity.
d_t	=	Distance from reinforcement to compression face.
d_1	=	Larger dimension from line of action of anchor force to the boundary on non-symmetrical prism.
d_2	=	Smaller dimension from line of action of anchor force to the boundary on non-symmetrical prism.
D_t	=	Nominal diameter of tendon.
E_c	=	28 day secant modulus of elasticity of concrete.
E_t	=	Modulus of Elasticity of tendon.
f_c	=	Stress in the concrete at point considered.
f_c'	=	28 day cylinder strength of concrete.
f_{ci}	=	Cube strength of concrete at age being considered.
f_{cp}	=	Compressive stress of prestress at centroid.
f_{cu}	=	28 day cube strength of concrete.
f_d	=	Stress due to unfactored dead load at tensile face of section subject to M_{cr}.
f_p	=	Initial stress in tendon.
f_{pc}	=	Stress due to prestress only at the centroid of the tendons.
f_{pe}	=	The effective prestress after all losses.
f_{pi}	=	Stress increase in tendon.
f_{pk}	=	Average compressive stress over shear keys.
f_{pt}	=	Stress due to prestress only at the tensile face.
f_{pu}	=	Characteristic strength of tendon.
f_s	=	Initial stress in non-prestressed reinforcement.
f_{si}	=	Increase in stress in non-prestressed reinforcement.
f_t	=	Tensile strength of concrete, taken as $0.24\sqrt{f_{cu}}$
f_y	=	Characteristic strength of reinforcement.
f_{yl}	=	Characteristic strength of longitudinal reinforcement.
f_{yv}	=	Characteristic strength of link reinforcement.
F	=	Force in tendon at point being considered.
F_{bst}	=	Anchor bursting force.
F_o	=	Force applied by the jack at the anchor.
F_s	=	Force generated in top slab due to differential shrinkage.
h	=	Overall depth of member.
h_{max}	=	Larger dimension of the section.
h_{min}	=	Smaller dimension of the section.
h_{wo}	=	Web or slab thickness.
I	=	Second moment of area of section.
k	=	Wobble coefficient.
k_i	=	Concrete bond coefficient.
k_t	=	Coefficient dependent on type of tendon.
K_L	=	Longitudinal shear coefficient.
l	=	Lever arm at ultimate moment.

l_c	=	Losses of stress in tendon due to creep of concrete.
l_E	=	Losses of stress in tendon due to elastic shortening of concrete.
l_s	=	Losses of stress in tendon due to shrinkage of concrete.
l_t	=	Transmission length for anchorage of pretensioned strand.
L	=	Length of deck.
L_s	=	Width of longitudinal shear failure plane.
L_{sp}	=	Length of span of deck.
L_T	=	Free length of tendon.
M	=	Moment at section due to ultimate loads.
M_a	=	The moment generated by the change applied 'instantaneously'.
$M_{as\text{-}built}$	=	Moment as constructed.
M_{cr}	=	Cracking moment of section.
M_{final}	=	Final moment after creep effects.
$M_{inst.}$	=	Moment if the structure is built instantaneously.
M_o	=	Moment necessary to produce zero stress in the concrete at the tensile face.
M_p	=	Primary moment from prestress on section.
M_r	=	Ultimate Moment of Resistance at a section.
M_s	=	Prestress secondary moment.
m_t	=	Mass per metre run of tendon.
m_d	=	Mass per metre run of deck
m_1	=	Moment due to unit restraint moment applied at pier 1.
m_2	=	Moment due to unit restraint moment applied at pier 2.
$N\text{-}A$	=	Neutral axis of section.
P	=	Total, unfactored prestress force acting on section.
P_h	=	Horizontal force from prestress tendon.
P_v	=	Vertical force from prestress tendon.
p	=	Ratio of reinforcement, A_r/bh.
S_v	=	Spacing of link reinforcement.
S_L	=	Spacing of longitudinal reinforcement.
T	=	Tensile force generated at ultimate moment.
T_s	=	Torsional moment due to serviceability loads.
T_u	=	Torsional moment due to ultimate loads.
v	=	Shear stress in the concrete due to ultimate loads.
v_c	=	Ultimate shear stress allowed in concrete.
v_L	=	Ultimate longitudinal shear stress in the concrete.
v_t	=	Torsional shear stress.
V	=	Shear force due to ultimate loads.
V_c	=	Ultimate shear resistance of concrete at section.
V_{cr}	=	Ultimate shear resistance of concrete cracked in flexure.
V_{co}	=	Ultimate shear resistance of concrete uncracked in flexure.
V_k	=	Ultimate shear resistance of shear key.
V_s	=	Ultimate shear resistance provided by the reinforcement.
V_d	=	Shear force at section due to unfactored dead load.

V_L	=	Longitudinal shear force per unit length.
V_p	=	Vertical component of prestress.
x	=	Distance of point being considered from the tendon anchor.
x_i	=	Smaller centreline dimension of torsion link.
y or $y.$	=	Distance in section from N–A to point being considered.
y_i	=	Larger centreline dimension of torsion link.
y_o	=	Half length of side of anchor block.
y_{po}	=	Half length of side of loaded area.
y_t	=	Distance in section from NA to tensile face.
z_t	=	Elastic sectional modulus referred to top face (I/y).
z_b	=	Elastic sectional modulus referred to bottom face (I/y).
θ	=	Total angle change (in radians) in the tendon over distance x.
μ	=	Friction co-efficient.
σ_c	=	Stress in concrete adjacent to prestress tendon.
σ_{final}	=	Final stresses in section.
$\sigma_{as\text{-}built}$	=	Stress in section due to construction sequence.
$\sigma_{inst.}$	=	Stress in section if built instantaneously.
ϕ	=	Creep factor.
Δ_l	=	Change in deck length due to Δ_t.
Δ_t	=	Change in effective temperature.
Δ_{cs}	=	Shrinkage strain deformation of the concrete.
Δ_s	=	Differential shrinkage strain between *in situ* slab and precast beam.
ξ_s	=	Depth factor for shear, given in Table 9 of BS 5400 Part 4.
$\propto$	=	Coefficient of thermal expansion of the concrete/°C.
γ_m	=	Partial safety factor for strength.
ϕ	=	AASHTO strength reduction factor.
ε_{ct}	=	Strain in concrete at top fibre.
ε_{cb}	=	Strain in concrete at bottom fibre.
ε_p	=	Strain in prestress tendon.
ε_s	=	Strain in non-prestressed reinforcement.

6 Design of steel bridges

G.A.R. PARKE AND J.E. HARDING

6.1 Introduction

Structural steel is an extremely versatile material eminently suited for the construction of all forms of bridges. The material, which has a high strength-to-weight ratio, can be used to bridge a range of spans from short through to very long (15–1500 m) supporting the imposed loads with the minimum of dead weight.

Steel bridges normally result in light superstructures which in turn lead to smaller, economical foundations. They are normally prefabricated in sections in a factory environment under strict quality control, transported to site in manageable units and bolted together *in situ* to form the complete bridge structure. Using this construction method the erection of a steel bridge is usually rapid, resulting in minimal disruption to traffic, a very important factor if traffic delays, be it road or rail, are properly costed in the construction project.

Rolled steel sections, the largest manufactured in the UK being a 914 × 419 × 388 kg/m universal beam, are economical for short span highway bridges and, when designed to act compositely with the concrete deck, are capable of spanning 25–30 m. For spans in the range of about 25–100 m plate girders, again acting compositely with the deck, provide an economical solution. In order to optimize the concrete deck, which has to distribute wheel load transversely across the bridge, it is usual to arrange for a plate girder spacing of around 3 m. For longer spans exceeding 100 m box girders are the favoured choice. Although box girders have a higher fabrication cost than plate girders, box girders have substantially greater torsional stiffness and if carefully profiled good aerodynamic stability. For very long spans, in excess of 250 m stiffened steel box girders with an integral orthotropically stiffened steel top plate, forming the primary support for the running surface, provide a very economical lightweight solution. Figure 6.1 gives cross-sections taken through typical bridge structures using hot rolled, plate girder and box girder sections.

6.2 Truss bridges

Lattice truss structures have been used very successfully for both railway and highway bridges throughout the last 150 years. There are three main truss configurations

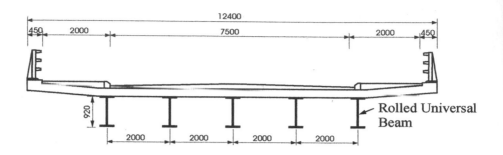

UNIVERSAL BEAM COMPOSITE BRIDGE

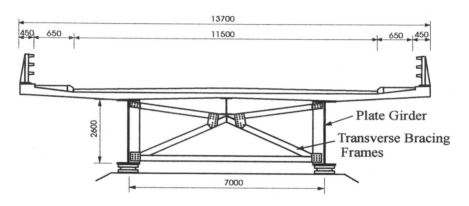

PLATE GIRDER COMPOSITE BRIDGE

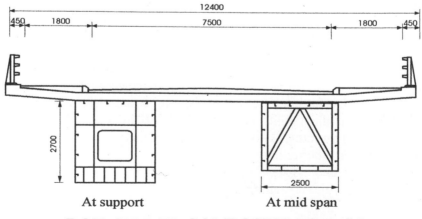

At support   At mid span

BOX GIRDER COMPOSITE BRIDGE

Figure 6.1 Cross-sections through typical composite bridges.

in use today, namely the Warren truss, the Modified Warren truss and the Pratt truss all of which can be used as an underslung truss, a semi-through truss, or a through truss bridge. Figure 6.2 gives details of the three truss types together with sections showing the differences between an underslung, semi-through and through trusses. Figure 6.3 shows a section through a typical through truss highway bridge and gives details of the terminology used in truss bridge design.

With an underslung truss, as shown in Figure 6.2, the live loading due to either the passage of vehicles or trains is carried directly by the top chord of the truss. Underslung trusses are used almost exclusively for rail bridges in situations where the depth of construction or clearance under the bridge is not critical. In semi-through trusses vehicles or trains pass through the truss bridge, but due to the height of the vehicles or trains relative to the upper chord, the transient live load projects above the top chord members. Consequently, in semi-through trusses it is not possible for the top chords to be braced laterally and these chord members must obtain lateral

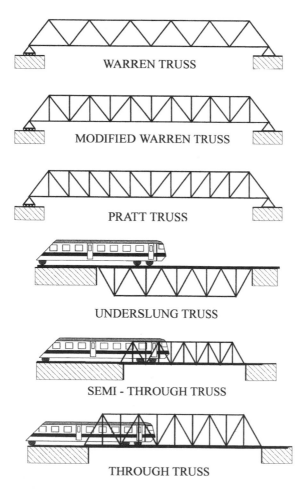

Figure 6.2 Typical truss bridges.

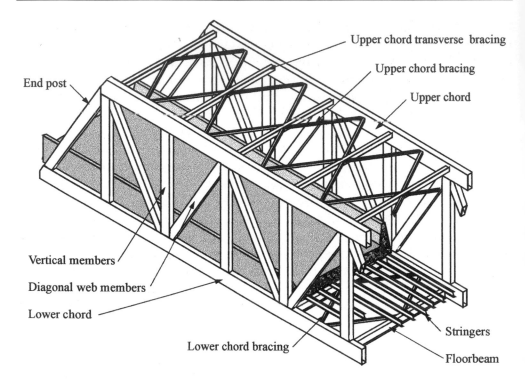

Figure 6.3 Truss bridge terminology (Chen and Duan, 1999).

stability from 'U' frame action described later in the sequel. In through truss bridges as shown in Figure 6.2, the vehicles or trains pass through the centre of the truss bridge and the clearance between the live load and the top chords is such that it is possible to brace laterally the top chord members.

Provided a truss bridge is detailed so that the live loading is effectively applied at the nodes of the structure, the members in the truss will carry primarily only axial tension or axial compression forces. The global bending moment acting on the bridge can be resolved into a couple formed from compression forces acting in the top chord and axial tension forces acting in the bottom chord. Similarly, the global shear force acting on the truss bridge is carried by the diagonal web members again either in axial tension or compression depending on the truss configuration. In a Warren truss the diagonal web members carry compression and tension alternately along the bridge. However, in a Pratt truss all of the internal diagonal web members carry tension only, while the shorter vertical web members are loaded in compression.

6.2.1 Analysis

Due to the way in which truss bridges transmit the imposed loads to the foundations, via axial tension and compression member forces, it is acceptable to analyse these structures as pin jointed assemblies either as a two dimensional plane truss, or prefer-ably, as a three-dimensional space truss. This type of analysis assumes that member connections are pinned and consequently it is not possible for any of the truss members to attract moment or torsion forces. A two-dimensional plane truss analysis can be undertaken by hand either by using equilibrium equations to resolve the forces at each joint in turn, or by using the 'method of sections' to freebody parts of the bridge truss, again using equilibrium equations to determine member forces.

The stiffness method can also be used to determine first, node displacements followed by member forces. Nowadays, truss bridges do not have pinned joints, the connections being either welded or bolted, however analysing the structure as a two or three dimensional pin jointed assembly will permit an accurate assessment of member axial loads but will over-predict the truss node displacements. In order to obtain a more realistic prediction of node displacements together with an assessment of the secondary bending and torsion moments, which will be small but nevertheless present, because the joints are not pinned in reality, it is necessary to analyse the truss as a three-dimensional space frame with six degrees of freedom at each node. The secondary moments and torsions acting on the structure can influence the fatigue life of the bridge especially if the truss is continuous, spanning over several supports. Secondary forces and hence stresses can be minimized by ensuring that the neutral axis of all members meeting at a node intersect at a single point in three-dimensional space.

Members

Several different member types are suitable for use for the chords and web members of truss bridges. Rolled 'H' sections and rolled square hollow sections are suitable for the tension and compression chords and also for the web members of short span highway trusses (30 to 50 m), where as for longer highway truss bridges or trusses supporting railway loading, larger fabricated sections such as a 'top hat' section or box section will be required for the chords as shown in Figure 6.4. Open 'top hat' sections will permit easy connection to 'H' section web members if the two section types being connected together are carefully sized so that the flanges of the 'H' section web members pass inside the 'top hat' chord members. Top hat sections will require lacing and battening across the open side to ensure that the section is fully effective. Box sections form very efficient compression chord members having both an improved lateral stiffness and aerodynamic stability when compared to 'H' and 'top hat' sections, however connecting web members to box sections chords with very limited internal access may prove awkward.

Compression members

Compression chord members, irrespective of the cross-section type, should be kept as short as practicable to maximize axial load capacity. Careful attention must be given to determine the appropriate effective lengths for buckling, both in the plane of the truss and normal to the plane of the truss. It is most likely that, due to the lateral bracing running between the top chords and the differences in the radius of gyration about the two major axes of the member, the flexural buckling capacity for a chord will be different for in and out of plane buckling. Effective lengths for in- and out-of-plane flexural buckling can be determined by undertaking an elastic critical

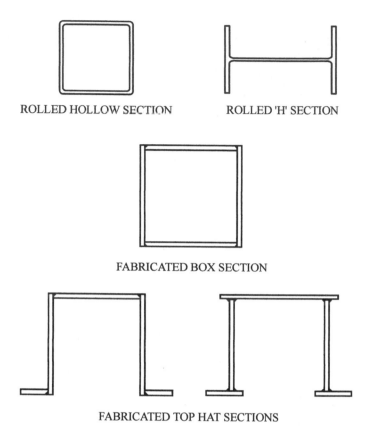

ROLLED HOLLOW SECTION   ROLLED 'H' SECTION

FABRICATED BOX SECTION

FABRICATED TOP HAT SECTIONS

Figure 6.4 Typical chord members (ESDEP 1985).

buckling analysis of the whole truss, however, guidance on choosing the appropriate effective lengths can be obtained from Table 6.11, BS 5400 Part 3 (1982) given below in Table 6.1.

For simply supported underslung truss bridges, where the top compression chord is continuously supported by a steel or reinforced concrete deck, the chord members can be considered to be effectively restrained laterally throughout their length if the connection between the deck and the chord member is capable of resisting a lateral force, distributed uniformly along the member length, of 2.5% of the maximum force in the chord. Where the deck provides adequate lateral stability to the chord members their effective length for buckling in the plane of the deck may be taken as zero. Where lateral restraint is provided by discrete cross-members, provided the cross-members have sufficient stiffness and are also capable of carrying at least 2.5% of the maximum force in the chord they are restraining, then the effective flexural buckling length of the chord member in the lateral plane is equal to the spacing of the cross-members.

For chord or web compression members whose cross-section is not doubly symmetric, such as 'top hat' sections which have only one plane of symmetry, failure can

Table 6.1 Effective length le for compression members in trusses (BS 5400 Part 3 1982)

| Member | Effective length l_e | | |
| | Buckling in plane of truss | Buckling normal to plane of truss when: | |
		compression chord is effectively braced by lateral system	compression chord is unbraced
Chord	0.85 × distance between intersections with web members	0.85 × distance between intersections with lateral bracing members or rigidly connected cross beams	See clause 12.5.1 BS 5400 Part 3 (1982)
Web member			
Single triangulated system	0.70 × Distance between intersections with chords	0.85 × distance between intersections with chords	Distance between intersections with chords
Multiple intersection system with adequate connections at all points of intersection	0.85 × greatest distance between any two successive intersections	0.70 × distance between intersections with chords	0.85 × distance between intersections with chords

occur by flexural buckling about one of the principal axes, or by torsional buckling involving twisting about the shear centre, or by flexural–torsional buckling, which is a combination of both flexure and torsion as shown in Figure 6.5. Open sections which have their shear centre and centroid coincident, together with closed sections, are not subject to either torsional or flexural torsional buckling.

Unbraced compression chords

In semi-through truss bridges where vertical clearance requirements for the live load prevent the incorporation of a lateral bracing system for the top chords, the lateral restraint to the top chord is provided by 'U' frames, formed from transverse members and vertical web members as shown in Figure 6.6.

In these circumstances the top chord compression members are considered to behave like a column braced at intervals by elastic springs whose stiffness is equivalent to the stiffness of the transverse 'U' frames. The buckling behaviour of a pin-ended column, restrained laterally at intervals by a series of springs, depends on the spring stiffness. If the spring stiffness is low, column buckling will take the shape of a single half-wave over the full length of the member as shown in Figure 6.7. Alternatively, if the spring stiffness is significant, then nodes will form at the intermediate restraint points, and the buckled shape will exhibit a shorter wave length indicating a substantial increase in buckling capacity.

Figure 6.5 Flexural torsional buckling of top hat compression members (Galambos, 1998).

Early investigations into the behaviour of a column transversely braced at intervals by elastic springs was reported by Engesser (1885). Engesser assumed that the elastic supports were equally spaced and that each restraint had the same stiffness. Also that the compression chord was straight, of uniform cross-section, equally stressed throughout its length and pinned, but rigidly held in position, at both ends. Engesser indicated that the stiffness required from a transverse 'U' frame in order to ensure that the complete truss compression chord supports an axial force of P_c is given by:

$$C_{req} = \frac{P_c^2 l}{4EI} \qquad (6.1)$$

where l is the panel length (Figure 6.6) and EI the flexural rigidity of the chord. Due to the simplifying assumptions made by Engesser the expression given in Equation (6.1) is not suitable for short span semi-through truss bridges which have only a small number of panels.

Engesser's investigation has been re-appraised by Hu (1952) who also studied the behaviour of elastically restrained compression chords in semi-through truss bridges. Hu stated that a suitable value for the transverse spring stiffness is given by:

$$C = \frac{E}{h^2\left[\left(\dfrac{h}{3I_C}\right)+\left(\dfrac{b}{2I_B}\right)\right]} \qquad (6.2)$$

where the symbols correspond to the 'U' framework shown in Figure 6.8. It is important to appreciate that this equation neglects the contributions from the torsional stiffness of the compression chords and in addition does not include the transverse bending stiffness of the diagonal web members.

For design purposes the effective length l_e for a compression chord restrained laterally by 'U' frames is given in BS 5400 Part 3 (1982) clause 12.5.1 as:

$$l_e = 2.5k_3 \left(EI_C \, a\delta\right)^{2.5} \text{ but } l_e \text{ must not be less than the term } a \qquad (6.3)$$

where a is the distance between the 'U' frames as shown in Figure 6.7. The term I_C represents the maximum second moment of area of the chord about the weak Y–Y axis and the parameter k_3 depends on the degree of rotational restraint on plan, given

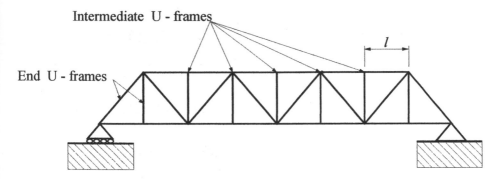

Figure 6.6 'U' frames in a typical truss bridge (BS 5400 Part 3 1982).

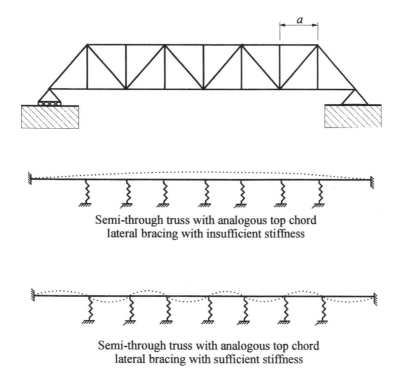

Semi-through truss with analogous top chord
lateral bracing with insufficient stiffness

Semi-through truss with analogous top chord
lateral bracing with sufficient stiffness

Figure 6.7 Top chord restraints for semi-through truss bridges.

to the ends of the chord member. If the compression chord is free to rotate in plan at both ends k_3 must be taken as 1.0, however if the compression chord is partially restrained against rotation in plan, at the points of support, the value of k_3 may be reduced to 0.85. The term δ (used in Equation 6.3) is given by:

$$\delta = \frac{d_1^3}{3EI_1} + \frac{usd_2^2}{EI_2} + fd_2^2 \qquad (6.4)$$

where the parameters d_1, d_2, I_1, I_2, and s are shown in Figure 6.9.

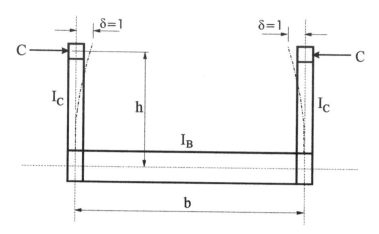

Figure 6.8 'U' framework for semi-through truss (Hu, 1952; Galambos, 1998).

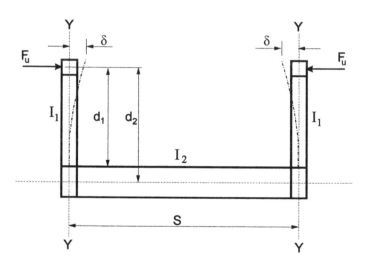

Figure 6.9 'U' framework for semi-through trusses (BS 5400 Part 3 1982).

For a semi-through truss bridge the value of u can be taken as 0.5. The parameter f represents the flexibility of the connection between the cross-member and the verticals of the U frame, expressed in radians per unit moment, and may be taken as 0.5×10^{-10} radians/N mm when the cross-member is bolted through unstiffened end plates or cleats, 0.2×10^{-10} radians/N mm when the cross-member is bolted through stiffened end plates or, 0.1×10^{-10} radians/N mm when the cross-member is fully welded or bolted through stiffened end plates to a stiffened part of the vertical member or stiffened section of the chord. Examples of typical cross-member to vertical connections and their respective flexibility are given in BS 5400 Part 3 (1982) and shown in Figure 6.10.

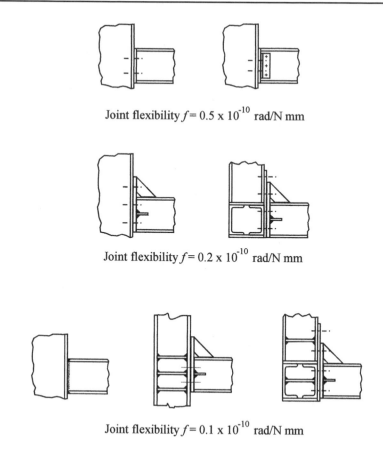

Joint flexibility $f = 0.5 \times 10^{-10}$ rad/N mm

Joint flexibility $f = 0.2 \times 10^{-10}$ rad/N mm

Joint flexibility $f = 0.1 \times 10^{-10}$ rad/N mm

Figure 6.10 Joint flexibilities (BS 5400 Part 3 1982).

6.3 Plate and box girder bridges

6.3.1 Introduction

As discussed previously in this chapter, one of the most common forms of steel (or composite) bridge, the plate girder, is comprised of steel plated elements, often of relatively slender construction. These elements are found in the webs and flanges of plate girders (generally fabricated by welding together of individual plates) and also in the stiffeners although the latter are normally made from hot rolled sections. The box girder, somewhat less common, is found in longer span bridges either as a composite construction or, for very long spans, as an all steel structure with stiffened steel decks. Such long span girders may have additional support provided by cables such as found in cable stayed or suspension bridges (see Chapter 11). Very long span bridges can have extremely complex cross-sections of aerodynamic shape with complex stiffening arrangements. Because of the slender nature of the individual plate components, they are prone to local buckling. In addition, other buckling modes might occur such as lateral torsional buckling of a plate girder between points of lateral flange restraint. This is discussed later in this chapter. In order to design the plated elements of a plate or box girder it is desirable to have a degree of understanding of the behaviour of plates and the functions and effects of stiffening under various types of loading and the next section of this chapter will explain the buckling behaviour of plates including the effects of initial imperfections and boundary conditions representative of those found in bridge structures.

6.3.2 The behaviour of plates

Plates in compression

The reference case for an initial understanding of plate behaviour is a simply supported square plate loaded by uniform in-plane compression along two opposite edges. Such a plate is shown in Figure 6.11.

For a plate to be defined as slender, the in-plane dimensions of the plate, a and b, are significantly greater than the plate thickness, t. The dimension b is normally the dimension transverse to the main component of in-plane loading or in the case of a laterally loaded plate, loading applied transverse to the surface of the plate, it is normally taken as the smaller of the two dimensions. The dimensionless slenderness b/t is the most important slenderness in terms of the buckling capacity (and behaviour) of the plate and, while there are differences, it could be considered equivalent to the slenderness l/r for a column. The buckling behaviour of a plate is in many ways similar to a column in that a perfect elastic plate loaded by increasing compression will carry all applied stresses by in-plane response with no out-of-plane deflection until the plate reaches the critical buckling stress (equivalent to the Euler stress). At this stress level the plate will suddenly buckle out-of-plane, but unlike the column which is essentially in a state of neutral equilibrium after elastic buckling, the plate has a significant postbuckling reserve and the buckling resistance rises as the lateral deflections increase. This is shown in the plot of stress against lateral deflection shown in Figure 6.12.

Prior to buckling, the in-plane stresses in the plate will be dependent on the in-plane boundary conditions along the edges. If the longitudinal edges are free to move, they will move outwards because of Poisson expansion. If prevented from moving, the

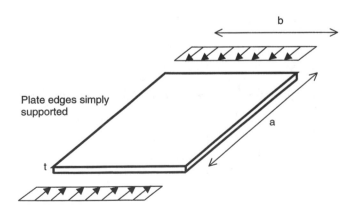

Figure 6.11 A square plate loaded by uni-axial compression.

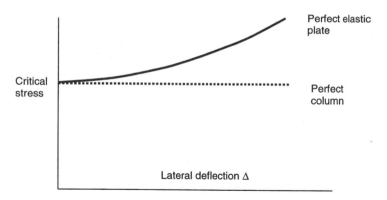

Figure 6.12 Postcritical reserve of a perfect elastic plate.

resistance to displacement will cause transverse compressive stresses to be set up at the boundary. After buckling, the stress states will, because of the out-of-plane action, be dependent on both the in-plane and out-of-plane boundary conditions at the edges.

The critical buckling stress of the plate is a function of the plate slenderness (b/t), the plate aspect ratio (a/b), the boundary conditions and the nature of the loading. The general expression for the critical buckling stress of a plate is:

$$\sigma_{cr} = k\,\frac{\pi^2 E t^2}{12(1-v^2)b^2} \tag{6.5}$$

where E and v are the material Young's modulus and Poisson's ratio, respectively, and k is the buckling coefficient which has a numerical value which is dependent on the loading and boundary conditions. For the uniaxially compressed plate with simply supported boundaries, k is generally taken to be four although the actual value of k varies with aspect ratio as shown in Figure 6.13.

This variation occurs because a perfect plate under in-plane compression will buckle into m square half waves if the plate aspect ratio is an integer $(m = a/b)$. This

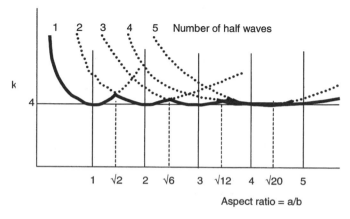

Figure 6.13 Variation of k with aspect ratio.

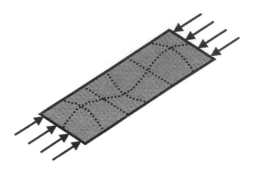

Figure 6.14 Buckling of a plate with a 3:1 aspect ratio.

corresponds to the lowest energy mode and for other aspect ratios (e.g. 1.5) the plate will in theory have a higher buckling stress. In practice this increase has no real significance because imperfections will interfere with this perfect behaviour. Buckling of a plate with an aspect ratio of 3:1 is shown in Figure 6.14.

Table 6.2 shows values of k for boundary conditions relevant to bridge design.

The free edge cases with their much lower values of k (for example 0.425 for one edge free and one edge simply supported) have a particular relevance for the behaviour of some stiffener cross-sections. This will be discussed later.

After critical buckling of the perfect plate, lateral deflections increase with the applied loading (in a square half wave for the case being considered) and because the plate does not form a developable surface, membrane tensions are set up which resist the growth of deflection and cause the postbuckling reserve mentioned above. Bending moments and twists are also set up in the plate due to out-of-plane action and there is a coupling between in-plane and out-of-plane behaviour. The form and magnitude of the in-plane tensions are dependent on the transverse boundary condition along the unloaded edge. A Poisson effect will cause compressive stresses to develop near the plate corners while the deflection of the plate will cause tension stresses to develop near the centre. If the boundary is completely unrestrained, however, the stresses can not anchor at the boundary and any development of transverse stresses has to self-equilibrate within the plate and hence the postbuckling reserve is modest. If the longitudinal plate edge is fully restrained against transverse movement, large values of edge stress can develop and a significant net boundary tension can occur with benefit to the postbuckling reserve. An intermediate condition exists, thought to approximate to the condition of an internal compression panel of a multi-stiffened flange of a box girder, where the edge of the panel is allowed to move but constrained to remain straight by adjacent panels. In this case boundary stresses are possible but the central tensions have to self-equilibrate with the corner compressions because the boundary is unable to carry a net force. This boundary stress state is shown in Figure 6.15.

The applied compression may also vary from the uniform condition shown in Figure 6.11 as the plate buckles, depending on the in-plane boundary conditions on the loaded edges. The uniform stress condition is not maintained, for example, if the loaded edge remains essentially straight either because of the presence of an edge

Table 6.2 Buckling coefficients for different boundary conditions

Longitudinal edge support conditions	Diagrammatic representation	*K* value
Both edges simply supported		4.00
One edge simply supported, the second fixed		5.42
Both edges fixed		6.97
One edge simply supported, the second free		0.425
One edge fixed, the second free		1.277

b

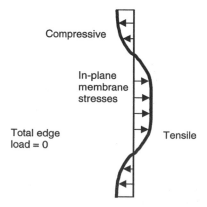

Compressive

In-plane membrane stresses

Total edge load = 0

Tensile

Figure 6.15 Transverse boundary membrane stresses for a constrained plate edge.

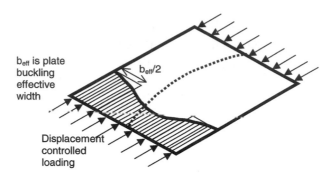

Figure 6.16 Distribution of stresses in a panel due to buckling.

beam or in practice because of the presence of longitudinally adjacent panels. In this case, the buckling of the panel causes a decrease of the stress at the centre of the edge as shown in Figure 6.16.

This stress reduction occurs because of a reduction in in-plane stiffness along the centre line of the plate caused by the lateral deflection. It is an out-of-plane effect and should not be confused with the reduction in in-plane stresses at the centre of a flange caused by the in-plane behaviour associated with shear lag. This will be discussed later in this chapter. The reduction in stress caused by buckling action gives rise to a concept often used in design called the buckling effective width. In this simple representation, the actual stress distribution in the plate is replaced by an idealized distribution which is of uniform magnitude (equal to the maximum of the actual distribution) but with a width, less that the actual width of the plate, known as the effective width. This width is evaluated so that the total force carried by the plate is the same as for the actual response. Figure 6.16 shows this idealization. A commonly used design approach defines failure as the point at which the stress for the effective width reaches the yield stress.

Figure 6.17 shows the form of the effective width for a plate girder and simple box girder that might be used in a design approach.

So far the discussion has related to the behaviour of a perfect plate. In reality, no structural element or system is perfect. In particular, for the type of element being

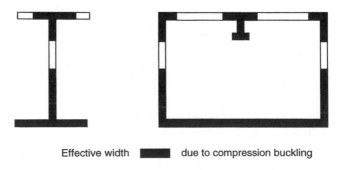

Effective width ▬▬▬ due to compression buckling

Figure 6.17 The effective elements in a plate girder and box girder due to compressive buckling (thicknesses exaggerated).

considered for bridge design, the most important imperfections can be considered to be of two forms. Geometric imperfections occur because of handling and fabrication processes. Because of the nature of plate buckling, the most important imperfections are out-of-plane defections of the plate, which may or may not be sympathetic with the buckling mode. If they are sympathetic they will tend to lower the initial buckling resistance, if not sympathetic they may even enhance it by resisting the development of buckling in the lowest critical mode. The second form of imperfection present in fabricated girders is a residual stress, caused primarily by the welding process. This is not dissimilar to the residual stresses caused by hot rolling, for example in an H-section column member but is somewhat different in form. As a plate edge is welded, for example to an adjacent plate or to a stiffener, the metal has to cool from a molten state at the location of the weld. This differential cooling introduces in-plane stresses, which are compressive near the centre of the plate and tensile around the weld location, which is last to cool. If the plate is slender, the compressive stress is non-linear due to out-of-plane deflection. The tensile stress will be at yield because of the material having been in a molten state. The state of residual stress is often idealized by a tensile block of width ηt at yield with a uniform compressive stress over the remainder of the plate width. This is not unlike the effective width idealization. The tensile and compressive stress blocks have to be self-equilibrating because no external forces are acting on the plate at this stage. Hence:

$$(b - 2\eta t)\sigma_{RC} = 2\eta t\sigma_y \tag{6.6}$$

The build-up of compressive residual stresses actually causes some additional out-of-plane imperfection in the plate panel which tends to be in the critical buckling mode shape.

For practical fabrication the geometric imperfection is around 1/200th of the plate width for a typical mild steel plate. Design codes often provide an equation which relates the fabrication tolerance to the plate slenderness and yield strength. For example that provided within BS 5400 Part 3 is given below. The buckling strength curves provided within the code then relate to this level of imperfection.

$$\Delta_0 \leqslant 0.145\beta t \tag{6.7}$$

where Δ_0 is the maximum imperfection value and β is the slenderness parameter $\beta = b/t\sqrt{(\sigma_y/E)}$. The residual stress level σ_{RC} is typically around 10% of the yield stress for bridge fabrication although it can be as high as 30% for heavy welding such as sometimes used in heavy steelwork such as ships. The effects of geometric imperfections and residual stresses are, however, not directly summative (with a significant imperfection present a small residual stress level will only produce a modest additional reduction in strength) and hence many codes allow for residual stresses by providing a sensibly conservative fabrication tolerance requirement as residual stresses are not easily monitored.

The effect of geometric imperfections on the elastic response of plates, as for columns, is to initiate a gradual loss of plate stiffness from the onset of loading. This is shown in Figure 6.18 where the gradual increase in the out-of-plane deflection can be seen as the load is increased. As with a column, because the postcritical response is stable, the elastic buckling curve eventually becomes asymptotic to the critical buckling curve as the deflections become large. The level at which this occurs depends on the magnitude of the initial imperfection.

The stress–strain or end shortening response of the plate is also affected by the level of the geometric imperfection. This is shown in Figure 6.19 where it can be seen that the bi-linear response of the perfect plate changes to a gradual non-linear response. In this case all curves start from zero strain or displacement because the initial imperfection is orthogonal to the end shortening.

Residual stresses do not directly affect the elastic response because there is no limit to the stress level that can be attained.

While the elastic buckling response and in particular the elastic critical buckling value provide a reference for design (indeed codes have in the past applied imperfection and material factors to the critical buckling stress to obtain a design resistance value), with limit state design it is the ultimate collapse load of the plate panel that is of importance to the designer. This ultimate performance is affected by the non-linear response of the material, the degree of which varies depending on the plate slenderness. Three ranges of slenderness can be defined depending on the relationship between the yield stress and the critical buckling stress. Stocky plates can be defined as those where $\sigma_{cr} \gg \sigma_y$, slender plates where $\sigma_y \gg \sigma_{cr}$ and plates of intermediate slenderness where $\sigma_{cr} \approx \sigma_y$. The effect of geometric imperfections and residual stresses are different depending on within which range the slenderness falls.

For a stocky plate, failure occurs as an in-plane phenomenon with material yield controlling collapse. Out-of-plane deflection is negligible and hence elastic critical

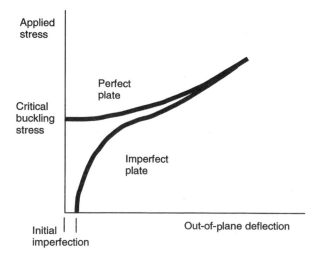

Figure 6.18 Effect of geometric imperfection on elastic buckling response.

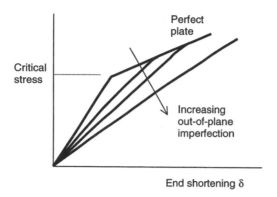

Figure 6.19 Load–end shortening response of an elastic plate.

buckling is irrelevant to the response. Geometric imperfections of normal magnitudes therefore have no influence on the response and the collapse stress–strain curve follows that of the material response, typically approximating to a bi-linear curve with a plateau as shown in Figure 6.20. Residual stresses on the other hand do influence the response because at a stress approximating to $\sigma_y - \sigma_{cr}$ the central area of the panel yields and the in-plane stiffness reduces.

However, because of the stable nature of the stocky plate, as the load increases the stresses redistribute from the centre to the edge strips and the plate eventually reaches the same value of collapse load. This is also shown in Figure 6.20. Hence the design value for such a plate is equal to the yield stress. Even when present, work hardening of the steel is ignored in the design because it occurs generally at high values of strain and hence affects collapse behaviour at a late stage, particularly if buckling is taking place. In any event, the precise nature of the material stress–strain response would not generally be known at the design stage.

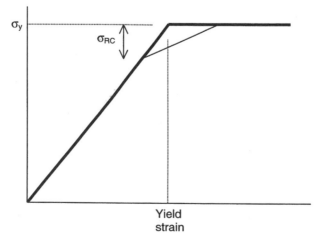

Figure 6.20 Stress–strain response of a stocky plate.

For a slender plate, the behaviour is dominated by the out-of-plane response with buckling and hence elastic behaviour being dominant. As in-plane stress levels are relatively low and yield occurs as an out-of-plane extreme fibre phenomenon particularly concentrated at the central zone of the plate, residual stresses are of little importance. While geometric imperfections will affect the behaviour and reduce the plate in-plane stiffness, because of the postbuckling stability, the sensitivity of the collapse behaviour to imperfection levels will be modest and ultimate load will be dictated by interaction between buckling defections and out-of-plane yielding which will gradually reduce the plate stiffness until it eventually reaches zero. Figure 6.21 demonstrates the resulting stress–strain behaviour and it can be seen that the collapse stress is higher than the critical buckling stress because of the postcritical reserve.

A plate of intermediate slenderness, as for a column where the Euler stress and yield stress are approximately equal, is in the region of maximum imperfection sensitivity. This is because both buckling and yield interact in determining the collapse behaviour and ultimate load of the panel. However, plates are only moderately sensitive to imperfections because of their inherent stability. This sensitivity is a function both of the type of loading and also the boundary conditions. For example, fully supported laterally loaded plates are insensitive while plates with free edges loaded in compression are relatively sensitive. Plates in shear and in-plane bending, relevant to the design of the webs of bridge girders are also relatively insensitive for reasons that will be described later.

Bearing in mind the description of the above slenderness zones, the strength slenderness curve for a plate is shown in Figure 6.22. The imperfection sensitive area can be seen for intermediate slenderness and the postbuckling strength can be seen for high slenderness. For mild steel the zone limits occur around a b/t of 30–40 and around 80.

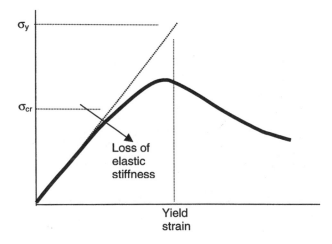

Figure 6.21 Stress–strain response of a slender plate.

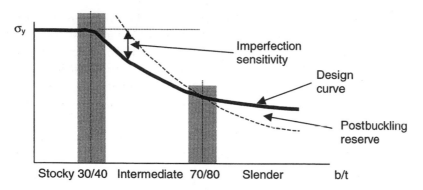

Figure 6.22 Strength slenderness curve for a plate panel in compression.

The point of maximum imperfection sensitivity can be defined for a particular yield stress by equating the yield stress to the critical buckling stress.

For a mild steel yield stress of 245 N/mm² and an E value of 205 000 N/mm² this gives:

$$\sigma_y = 245 = \frac{k\pi^2 E}{12(1-v^2)}\left(\frac{t}{b}\right)^2 = 4\frac{\pi^2 205000}{12(1-0.3^2)}\left(\frac{t}{b}\right) \tag{6.8}$$

which gives a b/t value of 55.

In BS 5400 Part 3 the curves are presented as non-dimensionalized with respect to a yield stress value of 355 (see Figure 6.23) which is the standard value adopted

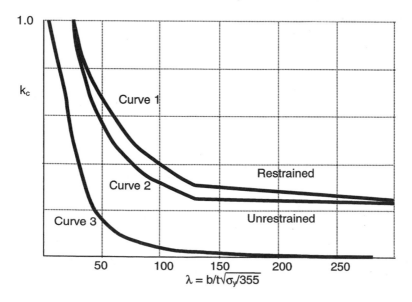

Figure 6.23 Compression panel ultimate strength curves in BS 5400 Part 3.

in the code. Curves for two different boundary conditions are presented labelled as unrestrained and restrained although the latter corresponds more closely to the constrained condition described above. The use of these two curves is described later in the context of the design of box girder compression flange panels. The ordinate k_c is multiplied by the yield stress to obtain the panel ultimate collapse stress.

In Eurocode 3 the equivalent curve is given in terms of a non-dimensionalized slenderness $\bar{\lambda}_p$, where:

$$\bar{\lambda}_p = \sqrt{\frac{\sigma_y}{\sigma_{cr}}} \qquad (6.9)$$

This curve is shown in Figure 6.24.

Approximate expressions are given in Part 1.5 of the Eurocode which define the buckling effective width factor ρ:

$$\rho = 1 \text{ when } \bar{\lambda}_p \leqslant 0.673 \qquad (6.10)$$

$$\rho = (\bar{\lambda}_p - 0.22)/\bar{\lambda}_p^2 \text{ when } \bar{\lambda}_p > 0.673 \qquad (6.11)$$

It was mentioned above that the critical buckling coefficient for a plate with a free edge with the other edges simply supported is very low ($k = 0.425$). The buckling behaviour of such a plate is also significantly more unstable than for a four-side supported plate, as it does not have the postbuckling reserve of the latter. During buckling such a plate has a tendency to shed a high proportion of its direct stress and, if acting as a stiffener to a stiffened plate, the inertia of the section also reduces as the stiffening element twists towards the attached plating. For this reason BS 5400

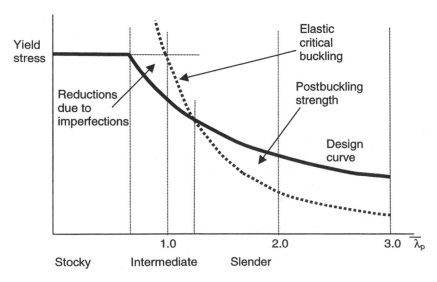

Figure 6.24 Compression panel ultimate strength curves in Eurocode 3.

applies strict limits to the slenderness of outstand elements (flats, angles, tees and bulb flats) acting as stiffeners in plated structures. These limits were based on elastic critical buckling behaviour of the elements factored by an additional safety factor to allow for non-linear behaviour.

The buckling behaviour of such elements depends on their outstand slenderness and in particular on whether they are buckling elastically or plastically but in either case the mode of local buckling is not unlike the lateral torsional buckling that can affect a beam member. If elastic, the critical mode is with the outstand in a half-sine wave between supports while if plastic, buckling occurs as a short wave S-shaped buckle which can exhibit well-defined yield lines. These buckling modes are shown in Figure 6.25.

Plates in shear

Figure 6.26 shows a plate panel loaded by in-plane shear along its edges. A perfect elastic plate will, as for compression, carry load by in-plane action alone prior to reaching its elastic critical stress. The plate will distort in-plane with stretching of one diagonal and shortening of the other creating orthogonal tension and compression stresses. It is the compressive component of the stress state that causes buckling and hence the buckling mode is of a diagonal form as shown in the figure.

The single buckle shown occurs in a square plate but, like a plate in compression, the number of waves varies with the aspect ratio. Unlike the compression case,

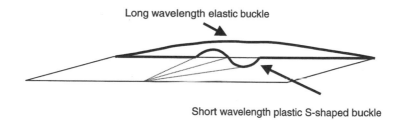

Figure 6.25 Buckling modes for a plate with one unsupported edge.

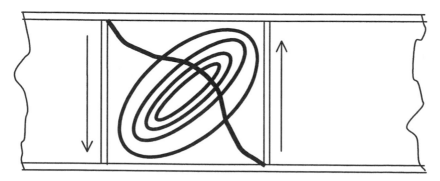

Figure 6.26 A plate in shear showing the diagonal buckling mode.

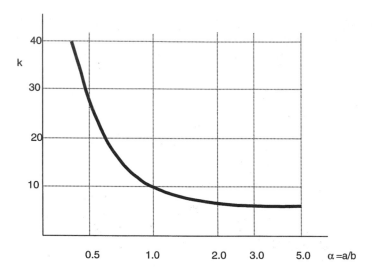

Figure 6.27 Buckling coefficient for a plate in shear.

however, the critical buckling coefficient and hence load is significantly affected by aspect ratio. The critical buckling expression can be approximated by the equation $k = 5.34 + 4(b/a)^2$ for $b/a \leqslant 1$ (Figure 6.27). This gives a value of k of 9.34 for a square plate. The true expression is discontinuous with wave number.

The higher value compared with the value of four for compression reflects the presence of the tensile stresses inherent in the loading. The consequence of this is that plates can be designed with a higher slenderness in shear compared with compression and still be economic. This can be seen by comparing the slenderness values at which the critical stress and yield stress are equal (the point of maximum imperfection sensitivity). The yield stress in shear is $\tau_y = \sigma_y/\sqrt{3}$.

This can be appreciated from the governing Mises yield equation:

$$\sigma_y^2 = \sigma_x^2 + \sigma_y^2 - \sigma_x\sigma_y + 3\tau^2 \tag{6.12}$$

Therefore by equating the shear yield stress to the critical buckling stress with a k value of 9.34 for the same material values as used previously:

$$\frac{245}{\sqrt{3}} = 9.35\frac{\pi^2 205000}{12 \times 0.91}\left(\frac{t}{b}\right)^2$$

This gives $b/t = 111$ compared with the value of 55 for compression. A typical range of slenderness for the design of a panel in shear is $60 \leqslant b/t \leqslant 180$ with the stockier panels being used in the presence of combined compression and the more slender panels being used in the presence of tension.

A perfect elastic plate in shear undergoes two phases of behaviour. The first, prebuckling, corresponds to in-plane response where diagonal tensile and compressive contributions are equal. The second occurs after buckling where the compressive stresses stabilize and begin to fall and the tensile stresses continue to increase, in principle up to the yield value. This postbuckling behaviour can be thought of as a band of diagonal tension, the vertically resolved component of which gives the panel additional shear capability. This behaviour leads to a simplified design process for webs called tension field design. The diagonal band is shown Figure 6.28. Where the panel is adjacent to a flange or other panel the tensile stresses can anchor off the boundary increasing this contribution to the capacity. A restrained panel will have a strength close to the shear yield strength, even for very slender panels and will be effectively imperfection insensitive. In fact the tension field response means that all shear panels have a low imperfection sensitivity.

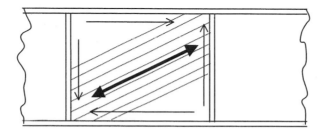

Figure 6.28 Tension field band in a postbuckled shear panel.

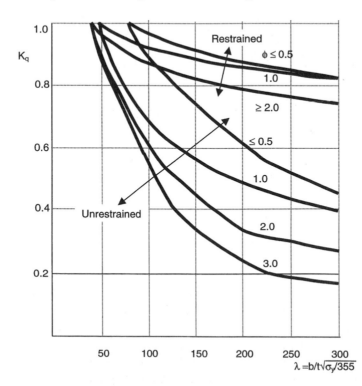

Figure 6.29 Shear panel ultimate strength curves in BS 5400 Part 3.

Figure 6.29 shows the shear strength curves in BS 5400 Part 3, which illustrates the low sensitivity of strength to slenderness when the panel is restrained. Because the critical buckling coefficient is significantly affected by the aspect ratio, the latter also has an important effect on the collapse stress and the code therefore presents different curves for different aspect ratios.

This contrasts with the code curves for compression, which were independent of aspect ratio. The value k_q is multiplied by the shear yield stress to obtain the collapse shear stress in the code, which is assumed to act uniformly down the panel. However, as will be seen later, an idealized tension field approach rather than a panel strength approach is used in the code where the panel extends for the full depth of the web.

Plates under in-plane bending

The buckling coefficient for a plate in bending is very significantly influenced by the fact that half (in linear response) of the load is applied in tension. Figure 6.30 shown the resulting buckling mode for such a plate and it can be envisaged that the plate would buckle at a stress equivalent to the buckling stress of a stockier compression plate, i.e. a buckling width appreciable less than the breadth of the actual plate.

In fact the k value for a plate in equal bending is 23.9 and, like the compression plate, has curves with minima which are independent of the aspect ratio although these minima occur at aspect ratios somewhat less than in compression.

Again because of the presence of the tension, plate behaviour in bending is somewhat less sensitive to imperfection and collapse loads, because of the variation of k with aspect ratio, are also insensitive to aspect ratio. One further important phenomenon occurs in a plate whose edges are effectively held straight during the buckling response. As for shear, as the plate deflects, the compressive stresses tend to stabilize and decline while the tensile stresses continue to increase. The neutral axis therefore shifts towards the compression side of the panel as the moment increases. While only stocky panels will reach the in-plane plastic moment capacity

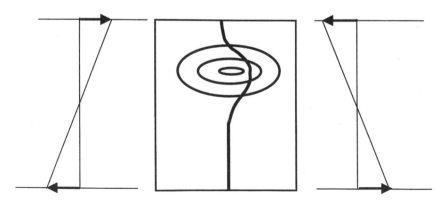

Figure 6.30 Buckling mode for a plate loaded by in-plane bending.

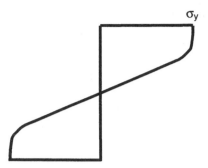

Figure 6.31 Redistribution of bending stresses in a stocky panel.

M_P, panels with intermediate slenderness, as high as b/t values of 100 and beyond, will exceed the in-plane yield moment value M_y. This redistribution of stress towards a plastic type distribution can be seen in Figure 6.31.

In BS 5400 Part 3, the strength of a panel in bending is referenced to the yield stress value as for compression by the ultimate load coefficient K_b shown in Figure 6.32. This is referenced to a linear stress distribution. In order to reflect the above redistribution and in order to provide strengths above M_y, the k_b values increase above 1.0 to reflect the section shape factor.

In Eurocode 3 Part 1.5, the same governing equations are used as for compression (Equations 6.10 and 6.11), with λ being defined in terms of the bending critical stress with a k factor of 23.9. The value of ρ produced defines an effective width for the compression zone of the web which is distributed 0.4/0.6 between the effective area of the web near the flange and the area next to the tensile zone when evaluating an effective section. The tensile zone is taken to be fully effective. The basis of the effective section approach within Eurocode 3 Part 1.5 is described later.

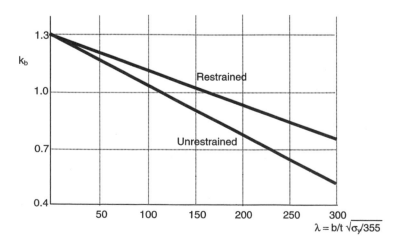

Figure 6.32 In-plane bending panel ultimate strength curves in BS 5400 Part 3.

Panels under lateral loading

The concept of critical buckling has no relevance to panels loaded exclusively by lateral rather than in-plane loading. As the lateral load is applied, the panel will increasingly deflect out-of-plane and out-of-plane moments will increase. As large deflection behaviour occurs, membrane tensions will begin to resist increased deflection and may become very large, particularly if they can anchor off the boundaries. Behaviour will be limited by these membrane tensions reaching yield and hence the panel collapse load is imperfection insensitive. Of course, the significant deflections caused by the lateral loading could have a major effect on the stability of the plate under any coincident in-plane destabilizing stress components as they would essentially have the effect of large imperfections.

However, while relevant to, for example, the design of ship hulls, the only significant design case where lateral loading could be relevant to the element design in a bridge is for wheel loading on an orthotropic bridge deck. Even in this case BS 5400 ignores the lateral load in the context of the design of the flange plate panels, only making an allowance for it in terms of an added in-plane stress contributing to the panel stress for considering yield in terms of serviceability.

Plates under combined in-plane stresses

The influence of plate slenderness and aspect ratio on the behaviour of a plate panel loaded in a combined stress state, e.g. uni-axial compression and shear, reflects the influence they exert on the individual load cases. In addition, there is an interaction between the individual buckling modes that might enhance or detract from the collapse loads under the individual stress components essentially depending on whether the modes from the individual components are in sympathy. The elastic critical buckling interaction between bi-axial compression, bending and shear is given below:

$$\sqrt{\left(\frac{\sigma_{cx}}{\sigma_{cx,cr}}\right)^2 + \left(\frac{\sigma_{cy}}{\sigma_{cy,cr}}\right)^2 + \left(\frac{\sigma_b}{\sigma_{b,cr}}\right)^2 + \left(\frac{\tau}{\tau_{cr}}\right)^2} = 1 \qquad (6.13)$$

where σ_{cx}, σ_{cy}, σ_b and τ are the applied compressive stress in the x-direction, the y-direction, the applied bending stress and the applied shear, respectively, and the cr values are the corresponding critical stresses for the panel for each individual type of loading.

This shows that compression has the greatest influence on interactive buckling while bending and shear have a lower influence. It also shows the effect of bi-axial compression in reducing the elastic critical buckling load.

No exact interaction equation exists for the ultimate collapse strength of a panel under a combined stress state. Figure 6.33 illustrates the behaviour by showing the collapse stress interaction diagram for combined compression/tension and shear. For a stocky panel which exhibits no buckling the response follows the Mises yield curve given by Equation (6.12) and is circular in form. It can be seen that the tensile

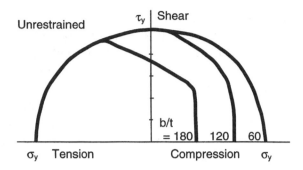

Figure 6.33 Collapse interaction diagram for applied tension/compression and shear.

stresses have a beneficial effect on shear buckling for high levels of shear and low levels of tension while for high levels of tension buckling has no influence even for modest levels of shear even for slender panels and the yield surface is again reached. For slender panels and high levels of compression it can also be seen that shear has little influence until the levels of shear approach around 60% of the shear yield value.

Figure 6.34 shows the corresponding diagram for bending moment and shear. There is little influence of shear on buckling behaviour for shcar values less than about 50% of shear yield and little influence of bending on buckling behaviour for bending values less than about 40% of M_u. In this diagram M_u is the collapse moment in the absence of shear.

An empirical curve fitting is used in BS 5400 Part 3 to represent these interaction diagrams to relate the strength under individual stresses to the combined stress state. This interaction equation, based on that for critical buckling, is defined in Equation (6.14). The equation is:

$$\sqrt{\left(\frac{\sigma_{cx}}{k_c\sigma_y}\right)^2+\left(\frac{\sigma_{cy}}{k_c\sigma_y}\right)^2+\left(\frac{\sigma_b}{k_c\sigma_y}\right)^2+\left(\frac{\tau}{k_q\tau_y}\right)^2}=1 \qquad (6.14)$$

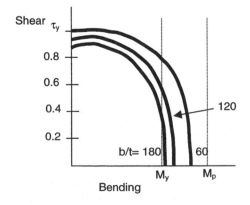

Figure 6.34 Collapse interaction diagram for applied bending and shear.

where k_c are the appropriate values using the value of b transverse to the applied direct stress. Curve 3 of Figure 6.23 is used with the other dimension of the panel if in either case it leads to a higher value of k_c. The code indicates that an individual square term should be taken as negative enhancing the buckling capacity if one of the stresses is tensile but it would seem to be logical in this case to remove the k_c term from the appropriate expression as it is potentially non-conservative otherwise.

The code also includes a parameter that allows for stress shedding in this equation. This is discussed later in the context of longitudinally stiffened web design.

6.3.3 Design of plate girders

General design considerations

Plate girders are fabricated by welding flanges to a web plate as shown in Figure 6.35. The flanges are generally significantly thicker than the web because of the lower buckling capability of one edge unsupported plates as described above. Occasionally the unsupported edges of the flange are themselves stiffened by an outstand but this is not often done in bridge design.

If the girder is to be used compositely with a concrete deck, the top flange will generally be narrower and will carry shear connectors for composite action with the concrete slab. In such a case the width of the flange has to be sufficient for the construction condition carrying the wet concrete before composite action is achieved.

Cross-girders may be welded or bolted between adjacent girders, for example for a rail bridge where they support the deck that carries the track. In all cases the main girders need to be appropriately designed to resist lateral torsional buckling. In the former case this can only occur during construction as the concrete deck, once acting compositely, will provide full restraint to the steel top flange. In such a case restraint is normally proved by cross-bracing at appropriate intervals bolted to outstands welded to the girders. Where the girders are connected by cross-beams near the bottom of the main girders, U-frame action whereby the cross-beams provide a moment restraint stiffness can be used to prevent torsional buckling of the girders. The design of lateral torsional buckling restraint is illustrated in Figure 6.36. Plan bracing will also be needed between adjacent girders.

If adjacent spans are continuous rather than simply supported, cross-bracing may be needed near the supports to prevent lateral torsional buckling of the bottom flange.

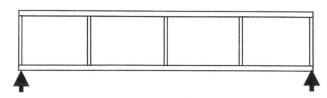

Figure 6.35 Typical form of a plate girder showing possible stiffener arrangements for a transversely stiffened girder.

Figure 6.36 Lateral torsional buckling restraint provided by cross-bracing and U-frame action.

Lateral torsional buckling occurs because the low torsional and transverse stiffnesses of the girder, compared to the main vertical stiffness, allow failure by sideways and twisting deformation of the girder even when loaded vertically. The section will be particularly prone to this form of failure if the vertical loading is applied to the compression flange providing an increasing eccentricity of load as the girder deflects sideways. It is prevented through the measures discussed above by restraining sideways movement of the compression flange either directly through bracing or indirectly through the U-frame action.

Because they can be fabricated from plates of any width and thickness, plate girders can be used for much longer spans than using beams from hot rolled sections with their restricted availability. Indeed spans of over 200 m are possible, often with haunches provided near continuous supports to increase moment capacity. Fabrication can be particularly effective where stiffening is kept to a minimum and welding carried out automatically.

Girders are generally relatively deep to provide the moment resistance and webs therefore generally need only to be thin to resist the applied shear. This leads to relatively thin web design. In contrast there is little benefit in having thin flanges so these are designed to reduce buckling problems. This tends to lead to webs which require stiffening, at least over the supports to prevent crippling caused by the high point loads and to allow the shear in the web to be transferred as compression down the bearing stiffeners into the support. Transverse stiffeners are also often provided for longer spans to enhance the buckling capacity of the web, principally by increasing its shear capability. In a minority of cases longitudinal web stiffeners are also provided to increase the web buckling capacity by reducing the slenderness of the panels within the depth of the girder as shown in Figure 6.37.

However, this should be considered carefully as it significantly increases fabrication complexity both by preventing the use of automatic welding and also by introducing

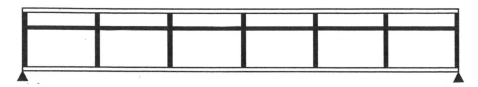

Figure 6.37 Plate girder stiffened by longitudinal and transverse stiffeners.

Figure 6.38 Curtailment of transverse web stiffeners to improve fatigue response.

a significant number of complex cutting and welding operations at the connection between transverse and longitudinal stiffeners. It is worth noting that intermediate transverse stiffeners are often curtailed short of the tension flange in order to provide better fatigue resistance. This is shown in Figure 6.38. Such curtailment does not affect the buckling enhancement of the stiffening as the latter still provides out-of-plane bending support to the web plate.

In longer span girders it is possible to vary the cross-section in the longitudinal direction to match more closely the variation of moment and/or shear along the span. Flange thickness can be varied with full strength butt welding providing a smooth external flange surface. Flange width can also be varied. The possibility of a haunched girder has already been mentioned. Rather than vary the web thickness, it is more normal to vary transverse stiffener spacing or introduce a longitudinal stiffener over part of the span, for example in the compression region over a continuous support. With modern fabrication it is also possible to use different yield strength steels for different sections along the bridge to achieve a variation in capability. In all cases the benefit of better matching the resistance of the structure to the applied loading needs to be weighed against any increased fabrication complexity and hence cost that might ensue.

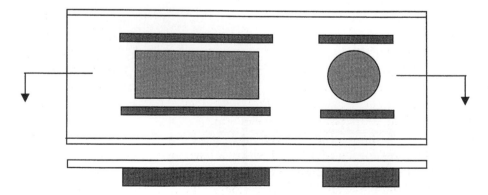

Figure 6.39 Plate girder with openings in the web.

Where holes are required in the girder web to accommodate services, it is necessary to provide some form of stiffening around the openings. A rule of thumb design approach is to replace the material removed from the depth by the area of the stiffeners but the designer must ensure that the remaining web section is capable of carrying the applied shear. Figure 6.39 illustrates such openings.

It is possible to define a range of practical dimensional proportions that are typical of bridge construction. Plate girders with spans up to around a 1000 m have been used as suspension structures but one notorious example, the Tacoma Narrows bridge with a span of 853 m, failed disastrously, because of its flexibility to wind excitation, in 1940. Box girders are now generally accepted as more appropriate for the longer spans because of their inherent torsional stiffness. More modest plate girder suspended spans of up to 400 m are not unusual. Cable stayed composite plate girder bridges have been constructed with spans up to about 500 m. Composite plate girders without additional support are used for many bridge structures over a range of more modest spans as a competitor to prestressed concrete structures.

Overall girder depths range between one-tenth to one-twentieth of the span with the larger values used for longer spans. Flange widths will tend to be between one-third and one-fifth of the girder depth. As has been noted previously flange plates are usually designed to be stocky to preclude loss of strength by buckling. They would normally be designed as at least semi-compact, a definition for slenderness used in Eurocode 3 although not in BS 5400 Part 3, an older code, which restricts section classifications to compact and non-compact. In terms of the Eurocode this means applying a limit of 14 $t_f \varepsilon$ to the flange outstand width where t_f is the flange thickness and ε is $(235/\sigma_y)^{0.5}$. This can be compared with a limit of $7\sqrt{(355/\sigma_y)}$ for a compact section in BS 5400 Part 3 and a limit for any flange outstand of $12\sqrt{(355/\sigma_y)}$ in the latter.

The terms compact, semi-compact and non-compact define the moment that a girder can carry prior to buckling and also its ability to redistribute moment along the span prior to ultimate collapse. In modern codes there are four classes of section. Class 1 sections, generally called plastic, are able to reach their plastic moment M_p value and to attain sufficient rotation prior to buckling to allow redistribution of moments and hence the use of plastic collapse analysis. Class 2 sections, compact, are also able to reach the M_p value but have limited rotational capacity. Class 3 sections, semi-compact, are able to reach their first yield moment M_y (a moment which at least achieves σ_y in the extreme fibre of the girder) prior to buckling. Class 4 sections, slender girders (the equivalent of non-compact in BS 5400), will not reach M_y and require design rules which either define a limit stress lower than the yield stress or which define a reduced effective section either through an effective width concept or by using an artificial reduced thickness. Figure 6.40 illustrates the response of girders with the four classes of cross-section.

In all cases both the webs and flanges of a girder have to adhere to the individual classification limits for a girder to fall within a particular higher class.

Because of the range of stiffening options a web might have a depth to thickness ratio ranging from between 80 and 500. Longitudinal stiffening can be considered for webs with slenderness ratios larger than about 200.

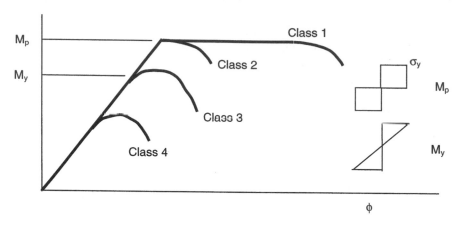

Figure 6.40 Moment rotation behaviour of girders with different classes of cross-section.

Plate girder design rules

This section describes the principles behind the rules of BS 5400 Part 3 and Eurocode 3 Part 1.1 for the design of plate girders presenting the main elements of the design process. Alternative rules for girder design are presented in Eurocode 3 Part 1.5 which are specifically aimed at more complex stiffened cross-sections but which are applicable to both plate and box girder design. These are presented in the section on box girder design rules.

In all following sections where strength is defined, the design must incorporate appropriate safety factors. With the partial safety factor format used in modern limit state design codes these are applied both to the load and resistance side of the equation. Reference should be made to the codes to establish appropriate values for these. The equations given below are also only a guide and should not be used out of the context of the code. The descriptions are intended to outline the key factors in the design process but should not be used in isolation from the full code requirements.

A simple and often effective design basis for a girder is to design the flange to carry all the moment and the web to carry all the shear. Even where a more complex basis is used for the final design, this can be a very effective method for initial sizing. This approach recognizes the inherent capabilities of the two elements. This approach is permitted by Eurocode 3 Part 1.1. The required area of the flange can be readily obtained for a given cross-section by dividing the moment by the distance between the flange centre lines (allowing of course for appropriate safety factors). As this involves knowing the flange thickness, a minor iteration might be required. The design of the web will depend on its slenderness, although if only transversely stiffened the shear stress can be assumed to be uniform down the depth.

In Eurocode 3 Part 1.1, web buckling is only considered where the web slenderness, d/t_w exceeds 69ε. Prior to this value the full yield stress may be used. Beyond this value Eurocode 3 Part 1.1 allows two design methods for plate girder design. These are the simple postcritical method and the tension field approach. The latter has already been mentioned in the context of panels in shear earlier in this chapter.

For the latter design method to be used within Eurocode 3 Part 1.1 transverse stiffeners must be present with spacing between the web depth and three times the web
depth. BS 5400 places no such restriction on the equivalent design approach. Because
the tension field approach allows for the full postcritical behaviour of the girder, higher collapse loads are provided than by the simple postcritical method.

Simple postcritical method within Eurocode 3 Part 1.1
In this approach the shear buckling resistance ($V_{ba.Rd}$ in Eurocode 3) can be determined simply from three equations which are applicable depending on the slenderness of the web. These equations provide a design shear stress (τ_{ba}) which can be
multiplied by the web area to provide the shear strength.
    Figure 6.41 shows the design curve for τ_{ba} as a function of $\bar{\lambda}_w = \sqrt{(\tau_y/\tau_{cr})}$. Equa-
tions are provided in the code which, for stiffener spacings of greater than the web
depth, correspond to that given previously for critical shear buckling.
    Equations for zones AB, BC and CD are given in the code which are termed stocky
or thick, intermediate and slender or thin, respectively.

Tension field method within Eurocode 3 Part 1.1
While tension field behaviour has been described in the context of an individual shear
panel the behaviour of a girder web exhibits additional features. The diagonal tension field band that occurs after the critical buckling stress is reached anchors off top
and bottom flanges and also off the transverse stiffeners on either side of the web
panel being considered. The degree of anchorage is dependent on the transverse stiffness of the flanges as well as the adequacy of the stiffener design. A third additional
component of resistance comes from the bending deformation of the flange (out-of-
plane) which must occur for the girder to fail in shear. This third component is
normally evaluated by assuming a plastic collapse mechanism to occur in the flanges
for an effective section of the flange plus an effective depth of web.

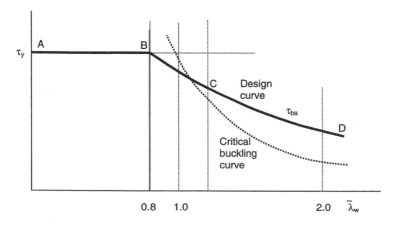

*Figure 6.41 Simple postcritical design curve for shear buckling according to Eurocode 3
Part 1.1.*

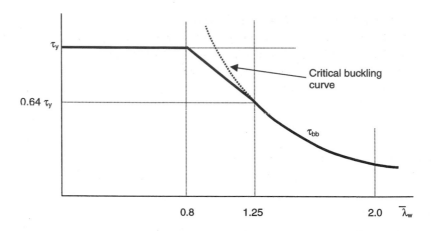

Figure 6.42 Strength corresponding to web buckling not allowing for tension field.

The equation representing the capacity of the web in shear in the code is:

$$V_{bb.Rd} = dt_w\tau_{bb} + 0.9gt_w\sigma_{bb}\sin\phi \qquad (6.15)$$

where τ_{bb} is the shear strength of the panel and the second term provides the tension field contribution.

If the web is stocky, $\bar{\lambda}_w < 0.8$, the web will reach shear yield. If of higher slenderness, the web will reach its elastic buckling stress according to equations which are reproduced in Figure 6.42.

It can be seen that Figures 6.42 and 6.41 are similar in form but the former provides lower strengths for thin webs because it makes no allowance for postcritical reserve which in this method is evaluated as a separate tension field action.

Calculation of the tension field effect is relatively complex. The angle of the tension field band ϕ is unknown and in principle could be established by an iterative procedure which maximizes the resistance. However, studies have shown that if a value of $\phi = \theta/1.5$ is used where θ is the angle of the web panel diagonal ($\tan^{-1}(d/a)$ where d is the web panel depth and a is the transverse stiffener spacing) a reasonable and conservative value is obtained for the tension field strength contribution. σ_{bb} is the level of tension field stress in the tension band which is the material yield stress reduced to allow for the shear stress present τ_{bb}. Using the Mises yield criterion produces:

$$\sigma_{bb} = 0.5(f_y^2 - 3\tau_{bb}^2 + (1.5\tau_{bb}\sin 2\phi)^2) - 1.5\tau_{bb}\sin 2\phi \qquad (6.16)$$

where f_y is the yield strength of the web material and τ_{bb} is the above elastic critical buckling stress.

Finally the width g of the tension field band is evaluated by a consideration of the flange plastic mechanism. It is a function of the positioning of the flange hinges which is itself dependent on the plastic moment of resistance of the flange reduced to allow for coexistent axial force in each flange due to the bending moment in the girder. The full equation is to be found in the code.

Interaction between shear and bending in Eurocode 3 Part 1.1
The presence of bending stresses in the girder web will reduce its shear carrying capacity. These influence both elastic buckling capacity and the extent of the tension field as well as the flange plastic resistance as mentioned above.

For the simple postcritical method, interaction between moment and shear can be considered as limited as shown in Figure 6.43.

For moments less than the moment resistance of the flanges alone (the simple design basis mentioned at the start of this section), there is no need to allow for interaction and the full shear resistance calculated from the simple postcritical method can be used. If flanges are not semi-compact there would be a need to define an effective width to allow for buckling prior to the calculation of M_f which is the moment provided by the two flanges alone.

Using the simplified approach of ignoring the web contribution to moment resistance and designing the flange as semi-compact in combination with the simple postcritical method for shear therefore leads to an effective and simple, if slightly conservative, design approach. If the web is designed to carry moment then, when the applied shear is less than half the shear resistance produced by the simple postcritical method, there is no need to reduce the moment capacity of the girder including the web. However, for the region of the interaction curve where the shear is higher, the relationship:

$$M_{Sd} \leqslant M_{f.RD} + (M_{pl.Rd} - M_{f.Rd}) \left[1 - (2V_{Sd}/V_{ba.Rd} - 1)^2\right] \qquad (6.17)$$

should be used to define the interaction. This gives the design moment capacity as a function of the design shear value, the simple postcritical shear resistance and the moment capacities of the girder including the web and the flanges.

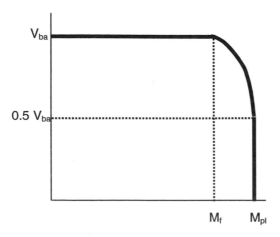

Figure 6.43 Interaction between bending moment and shear force for the simple post-critical method of Eurocode 3 Part 1.1.

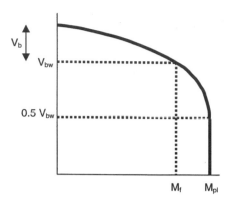

Figure 6.44 Interaction between bending moment and shear force for the tension field method of Eurocode 3 Part 1.1.

If the web is more slender than given by $d/t_w \leqslant 124\varepsilon$ the moment capacity needs to be reduced to allow for web buckling due to the in-plane compression.

For the tension field approach there is a greater interaction between shear and bending in the girder. For moments less than the moment resistance of the flanges alone, a shear capacity including the web tension field but excluding the plastic moment flange contribution can be carried. This can be evaluated by giving the flanges a zero plastic moment of resistance in the shear capacity equation. It can be seen from Figure 6.44 that this is conservative.

At the other end of the interaction shear may be ignored as long as the shear force is less than half the tension field shear capacity of the girder ignoring the flange out-of-plane bending contribution. The governing equation for intermediate interaction is:

$$M_{Sd} \leqslant M_{f.RD} + (M_{pl.Rd} - M_{f.Rd})\left[1 - (2V_{Sd}/V_{bw.Rd} - 1)^2\right] \tag{6.18}$$

Bending resistance of plate girders in BS 5400 Part 3

The bending resistance of a girder in BS 5400 depends on whether the girder is designed as a compact section. For the section to be compact both the flange and web must satisfy the code limits.

For the flange:

$$\frac{b_{f0}}{t_{f0}} \leqslant 7\sqrt{\frac{355}{\sigma_y}} \tag{6.19}$$

where the subscript zero refers to the flange outstand.

For the web, the depth of the web compression zone must satisfy the following slenderness requirement:

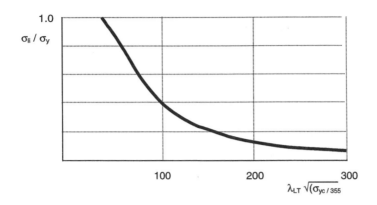

Figure 6.45 Allowable stress as a function of lateral torsional buckling slenderness.

$$y_c \leqslant 28t_w \sqrt{\frac{355}{\sigma_{yw}}} \qquad (6.20)$$

In this case the design moment can be evaluated from the yield stress and the plastic section modulus of the girder. The yield stress should be reduced to allow for lateral torsional buckling where relevant. The code provides an equation linking the lateral torsional buckling slenderness of the girder to the allowable stress shown in Figure 6.45.

The lateral torsional buckling slenderness is a function of the unrestrained length of girder and is defined for a range of section shapes within the code.

The design resistance (with allowance for safety factors) is M_D:

$$M_D = Z_{pe}\sigma_{lc} \qquad (6.21)$$

For compact sections, the value of σ_{lc} in Equation (6.21) is equal to σ_{li} from Figure 6.45.

For a non-compact section, the width of the flange outstand must still be limited as given by Equation (6.22):

$$\frac{b_{f0}}{t_{f0}} \leqslant 12 \sqrt{\frac{355}{\sigma_y}} \qquad (6.22)$$

The section is governed by elastic rather than plastic behaviour and the limit stress is reduced below yield even if lateral torsional buckling does not occur. The governing equations are:

$$M_D \leqslant Z_{xc}\sigma_{lc} \leqslant Z_{xt}\sigma_{yt} \qquad (6.23)$$

where Z_{xc} and Z_{xt} are the elastic section moduli for compressive and tensile extreme fibres allowing for a reduced web thickness for slender webs as defined below. σ_{lc} is the lesser of σ_{yc} or $D\sigma_{li}/2y_t$ where D is the full girder depth and y_t is the depth of the web tension zone.

The full thickness of the web can be considered to act if:

$$\frac{Y_c}{t_w}\sqrt{\frac{\sigma_{yw}}{355}} \leq 68 \qquad\qquad (6.24)$$

For higher web slendernesses, a reduced web thickness should be used in evaluating the elastic section moduli. This is evaluated as a linear interpolation between the above and the slenderness value for an effective web thickness of zero given by Equation (6.25).

$$t_{we} = 0 \text{ if } \frac{Y_c}{t_w}\sqrt{\frac{\sigma_{yw}}{355}} \geq 228 \qquad\qquad (6.25)$$

Shear resistance of plate girders in BS 5400 Part 3

If the web of the girder is only stiffened with transverse stiffeners and there are no longitudinal stiffeners, the shear resistance is calculated using a tension field approach which is very similar to that adopted by Eurocode 3 above. If the girder has longitudinal stiffeners a completely different approach is adopted for the web design. This is described later in this chapter in the context of box girder bridge design.

As for the Eurocode, the three components of response, critical buckling shear, web tension field and flange contribution are allowed for in the design process.

The code adds the three components which are presented as a design shear stress in the form of a series of graphs. It is possible to use related equations.

The shear resistance V_D is related to the design shear stress (with allowance for safety factors) by the equation:

$$V_D = t_w \tau_l d_w \qquad\qquad (6.26)$$

where d_w is the web depth (allowing for the presence of any openings) and τ_l is the shear strength defined by the graphs in the code. An example of one of the code graphs is given in Figure 6.46.

The limiting stress is given as a function of the web slenderness $\lambda = (d_w/t_w)\sqrt{(\sigma_{yw}/355)}$ and a parameter m_{fw} that allows for the transverse bending strength of the flange contributing to shear capacity.

$$m_{fw} = \frac{\sigma_{yf}b_{fe}t_f^2}{2\sigma_{yw}d_w^2 t_w} \qquad\qquad (6.27)$$

where b_{fe} is an effective width of flange equal to half the girder width but $\leq 10t_f\sqrt{(355/\sigma_{yf})}$.

The graph of Figure 6.46 corresponds to an m_{fw} value of 0, which corresponds to a zero flange contribution. This has relevance to the interaction between shear and

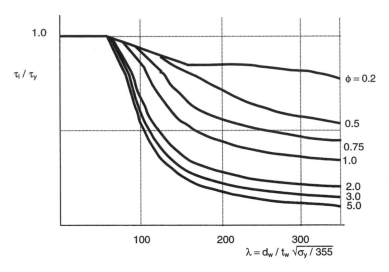

Figure 6.46 Limiting shear strength in BS 5400 Part 3 for $m_{fw} = 0$.

bending described in the next section. A series of graphs is presented in the code for a range of values of m_{fw} and the actual value should be found for design by linear interpolation between them.

Interaction between shear and bending in BS 5400 Part 3

The basis of this interaction is not dissimilar to that in the Eurocode. It is, however, in the form of a multi-linear interaction diagram, rather than the curves of the Eurocode. The diagram is presented in the code as four linear equations represented by Figure 6.47.

As for the Eurocode, significant levels of bending and shear are allowed to co-exist with no interaction. In this context BS 5400 caters for slightly less interaction than

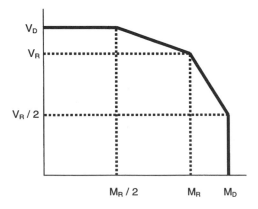

Figure 6.47 Interaction between bending moment and shear force for the tension field method of BS 5400 Part 3.

the Eurocode. If the applied shear is less than half the shear capacity of the web, not allowing for the flange moment contribution – the graph corresponding to $m_{fw} = 0$, the girder can carry its full moment capacity (flange and web). If the shear is equal to the web shear capacity, the moment capacity is that of the flanges alone. There is a linear interpolation between these two values. If the applied moment is equal to the bending resistance of the flanges alone, the shear capacity is limited to the shear capacity of the web without the flange moment contribution ($m_{fw} = 0$) and if less than half this value the full shear capacity is available. Again there is linear interpolation between these two limits as shown. One of the limits therefore corresponds to the case where the flange carries all the bending and the web all the shear.

Web crippling

Occasionally bridge girders are subjected to significant local concentrated loading that can induce high localized compressive stresses in the web. A good example of this occurs if the girder is launched over a rolling support. In this case the web may be subjected to very severe local patch loading at sections which are not strengthened by transverse stiffeners. In such circumstances it is necessary to check the web for local web crippling, the resistance to which is a function of the load length, the spacing of the stiffeners and the depth of the panel (in some circumstances). Different buckling behaviour occurs depending on the relationship of these parameters. Both the Eurocode and BS 5400 have sections dealing with this form of buckling phenomenon.

Transverse web stiffeners

Intermediate transverse stiffeners at regular spacing along the web of a plate girder are used to increase the buckling capacity of the web. Stiffeners provided at support points require a different design approach and these are considered in the next section. The design of intermediate transverse stiffeners in box girders adheres to similar requirements. In general transverse stiffener cross-sections are made of hot rolled sections and are either flats or bulb flats, angles or T-sections. The latter are used where the depth of the stiffener needs to be great enough to permit longitudinal stiffeners to pass through cut-outs in them. Longitudinal stiffeners would normally be either bulb flats or angles and would be welded to the transverse stiffeners. It is an important principle that the main longitudinal load carrying elements, e.g. the longitudinal stiffeners, are continuous to avoid eccentricities in the longitudinal load path.

As mentioned above the Eurocode requires stiffeners to be spaced at between one and three times the web depth. While BS 5400 Part 3 imposes no such restriction it is a sensible range for normal design application. If the stiffener spacing was any closer it is likely that increased fabrication cost would more than offset the savings in web material achieved through using a thinner web and if more widely spaced the enhancement to web buckling capacity would be limited.

Transverse stiffeners have two main functions in terms of the web. Firstly they will increase the out-of-plane buckling resistance of the web by acting as a nodal line preventing out-of-plane deformation. Secondly they provide anchorage for tension field forces thereby enhancing ultimate web shear strength. They can also carry limited direct loading from the deck into the web (not relevant for composite plate girders) and can also help to prevent the flange buckling into the web.

The Eurocode and BS 5400 differ in their design approach. The Eurocode provides an inertia requirement (a stiffness) to prevent buckling of the web distorting the stiffener out of the plane of the web and a column (strut) requirement to ensure that the stiffener can carry the resolved tension field force. In contrast, BS 5400 converts the web buckling loading, via a critical buckling equivalence, into an equivalent compressive loading and bases the stiffener design on the ultimate collapse of an eccentrically loaded strut. There is evidence that the latter approach is over-conservative in its representation of a lateral beam load as a compressive column load because of the $P - \Delta$ effect the latter introduces in the non-linear response of the column. Non-linear analyses carried out by the second author of this chapter suggests that a simple beam model with a beam spanning between the flanges, loaded by a uniform load which is a function of the in-plane stress state in the web, provides an accurate design alternative. This leads to the use of a beam column model when the other load components are included.

Looking first at the Eurocode 3 Part 1.1 design method, the equations for the inertia of the stiffener are given in terms of the web depth, web thickness and stiffener spacing and are also dependent on the panel aspect ratio. The strut approach uses the resolved tension field force N_S acting at the centre line of the web. It is therefore an eccentric load if the stiffener is only on one side of the web plate as is the norm for intermediate stiffeners. The stiffener and an associated width of web plate ($15\varepsilon t_w$ on either side of the stiffener section) are designed as a strut section using the column rules in the code. The load applied to this section is:

$$N_S = V_{sd} - dt_w \tau_{bb} \qquad (6.28)$$

where τ_{bb} is the shear buckling resistance without any allowance for tension field and V_{sd} is the design value of the shear force.

In BS 5400 Part 3, the effective strut is defined as the stiffener section comprised of the stiffener together with a total effective width of web equal to 32 times the web thickness. A strut equation combining moment and axial load is provided and used in combination with the column design curve in the code which is a function of the l/r of the effective section where l is the length of the stiffener.

There are a number of load components applied to the stiffener which include the resolved tension field force, the axial force representing, through equivalence, the destabilizing influence of web panel buckling, any moment applied through U-frame action and compressive loads from direct loading to the flange or through cross-frames and due to any curvature of the flange. The approach is too complex to present in detail here, but the first two load actions will be described.

The tension field force, which acts at the mid-plane of the web, is defined as F_{tw} which is the smaller of:

$$F_{tw} = (\tau - \tau_0)t_w a \text{ or } F_{tw} = (\tau - \tau_0)t_w l_s \qquad (6.29)$$

where l_s is the length of the transverse stiffener, τ is the average shear stress present in the web and τ_0 is a reference shear buckling stress:

$$\tau_0 = 3.6E\left[1+\left(\frac{b}{a}\right)^2\right]\left(\frac{t_w}{b}\right)^2\sqrt{1-\frac{\sigma_1}{2.9E}\left(\frac{b}{t_w}\right)^2} \tag{6.30}$$

but is equal to zero if the square root term is negative.

A is the panel length, b is the panel width (web depth between flanges) and σ_1 is the average longitudinal stress in the panel (+ve if compressive). The equation references web panel width because the same transverse stiffener design equations are used in the design of transverse stiffeners in longitudinally stiffened webs.

Again the approach is not dissimilar to Eurocode where the force results from the increased web capacity above a certain critical buckling stress.

The axial force representing the destabilizing action, acting at the centroid of the effective strut, is given by:

$$F_{wi} = \frac{l_s^2}{a_{max}}t_w k_s \sigma_R \tag{6.31}$$

where a_{max} is the maximum transverse stiffener spacing to satisfy the web design (can be taken as a), k_s is a numerical parameter which is a function of the strut slenderness with a maximum value of 0.4 for very slender struts, and σ_R represents the destabilizing stresses present in the web (with an allowance for web longitudinal stiffeners if present).

$$\sigma_R = \tau_R + \left(1+\frac{\Sigma A_s}{I_s t_w}\right)\left(\sigma_1 + \frac{\sigma_b}{6}\right) \tag{6.32}$$

where ΣA_s is the total area of longitudinal web stiffening, τ_R is the lower of τ or τ_0, σ_1 is the average value of web longitudinal stress (compression +ve) and σ_b is the maximum stress due to bending alone.

A simpler design requirement for the effective area of the stiffener is provided in the code if the only load component is the destabilizing web action (i.e. the last component above).

End post or load bearing support stiffener design

End posts have to transfer significant shear loads from the web into the support bearings. They also have to anchor any tension field forces in the absence of an adjacent web panel. Because of the magnitude of the compressive load carried, stiffeners would normally be located on both sides of the web to avoid eccentricity. A grouping of four stiffeners, two on either side and close to the bearing on both sides of the web may be used to provide a strong H section column over the bearing. Possible arrangements of bearing stiffeners are shown in Figure 6.48.

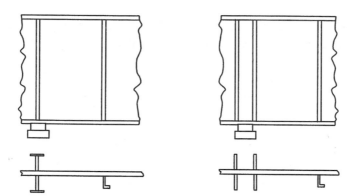

Figure 6.48 Possible arrangements for end posts.

BS 5400 defines a range of forces and moments which the end post section have to resist. These include direct loads, the destabilizing force from the web buckling and a moment produced by the eccentric tension field force. The code provides strength, stiffness and yield requirements. Again the requirements are too complex to be presented here in detail.

In addition to providing a similar approach including definition of a tension field behaviour that allows the geometry of the mechanism to be defined, the Eurocode suggests the option of designing the final web panel by the simple postcritical procedure even if the remaining panels are designed using the tension field approach so that the tension field components are not present. In order that the end panel shear strength is the same as the remainder of the adjacent girder so that it does not present a weak link in the load transfer path, the end panel is closed up with a closer stiffener spacing than elsewhere.

6.3.4 Design of box girders

General design considerations

Box girders, while more expensive to fabricate than plate girders because of their complexity, have a number of significant advantages, particularly for longer spans. Firstly because of the shape of the box, the top flange itself can act as the decking without the need for a concrete deck. They can also be designed with an aerodynamic shape, again making them ideal for long spans. Their high torsional rigidity again helps in the context of long spans, but also in providing flexibility for bearing arrangements and where a structure curved in plan is required.

For intermediate spans a box can be used compositely, either with a full width top flange or with two narrow flanges. In all cases, shear connection must be present over the full flange width. Depending on the width of slab needed, the former type can have transverse cross-girders connected to the boxes. Cantilever cross-girders may also be used to increase the bridge width. The cross-girders provide additional support to the concrete flange allowing it to span in two directions. An example of each type of structure is shown in Figure 6.49.

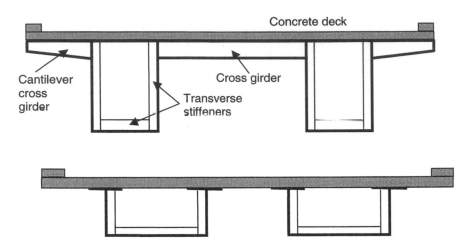

Figure 6.49 Composite box girders with and without a full width top flange.

Figure 6.49 shows bridges with two boxes. It is possible to design bridges with a number of smaller boxes and a composite deck, but the benefits of using boxes compared with plate girders are then reduced.

For the longest spans a single box, either cable stayed or with suspension support, is used. There are now many quite significant examples of box girder suspension bridges with spans of 1000 m or more. The Humber bridge in the UK, for example, has a span of 1410 m while the Akashi Kaikyo bridge in Japan has a main span of 3910 m. All longer span bridges of this type have an orthotropic steel deck in place of concrete to minimize weight. The decks are very slender, multiply-stiffened and aerodynamically shaped to minimize wind induced drag and possible oscillations. An example of such a cross-section is shown in Figure 6.50.

The construction of the very large spans is carried out by cantilever balancing construction away from a pier. New sections are bolted to the existing structure lifted from the water by a temporary gantry sitting on the existing deck. This is illustrated in Figure 6.51. The two halves of the bridge are joined together at mid span.

Figure 6.50 An idealized example of a slender box girder cross-section.

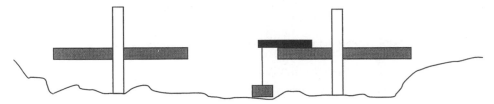

Figure 6.51 Diagrammatic representation of construction process.

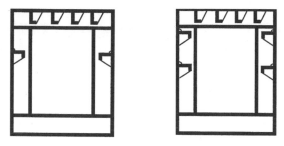

Figure 6.52 Simple longitudinally stiffened box cross-sections (diagrammatic representation).

The webs of box girders are stiffened with intermediate transverse stiffeners and, unless stocky, with one or more longitudinal stiffeners. A sensible initial design assumption for one stiffener is to place it at one-third of the depth away from the compression flange while two can be placed at mid-depth and one-sixth of the depth from the compression flange. These locations can be adjusted to optimize the performance of the web plate panels. For a large box the webs can have multiple longitudinal stiffeners generally equally spaced down the inclined web. Cross-sections for simple horizontally stiffened girders are shown in Figure 6.52. For this type of construction the breadth to depth ratio would tend to be in the range from 3:1 to 1:2 depending on the bridge configuration.

While longitudinal stiffeners are not required in the tension flange for the main loading components, they are often provided for robustness unless the flange is particularly stocky.

If the box has a wide flange and is non composite, the top flange will be fabricated as an orthotropic steel deck. The deck has to carry wheel loads as well as the main in-plane girder loading and therefore has to have closely spaced longitudinal stiffeners. These are of bulb flat or angle or closed section. The latter can be trough or V-shaped stiffeners, which have the advantage of multiple support for the flange plate as well as high torsional rigidity when welded to the deck. Figure 6.53 shows the range of stiffener sections that may be used in box girders.

All may be used for flange longitudinal stiffeners although T-sections tend to be used for transverse stiffeners and flats suffer from the disadvantage that they have very low torsional stiffness. In all cases, longitudinal stiffeners should be threaded through and welded to cut-outs in the transverse stiffeners so as to maintain the integrity of the longitudinal load path. This is particularly important for compressive stress where eccentricities should be avoided. Figure 6.54 shows the configuration of a typical orthotropic deck, fabricated using open section stiffeners.

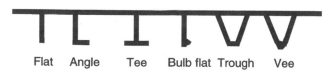

Flat   Angle   Tee   Bulb flat   Trough   Vee

Figure 6.53 Range of stiffener cross-sections used in stiffened plating.

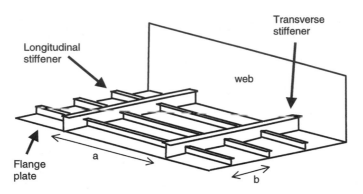

Figure 6.54 Arrangement of stiffeners in a typical orthotropic deck.

The buckling modes for such a deck include buckling of the sub-panels between the longitudinal stiffeners, buckling of the stiffened flange between cross-girders and buckling of the overall flange with a longitudinal wavelength longer than the transverse stiffener spacing. In the latter case the cross-girders will deform transverse to the flange. These buckling modes are shown diagrammatically in Figure 6.55. All buckling modes can occur either upward or downward. In the case of modes involving the stiffeners this will mean that both the plate and the stiffener outstand could be in local bending compression (in addition to the direct compressive stress) so failure can occur by buckling of either element or by yield in tension. As will be seen later this leads to both deformation directions being checked in the design process.

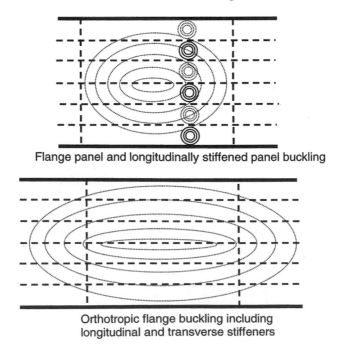

Figure 6.55 Possible buckling modes for a stiffened flange.

Figure 6.56 Buckling of plating and longitudinal stiffeners.

It also leads to the possible local buckling of the plates of the stiffener sections themselves although this is normally eliminated from design by making these sub-elements stocky. This approach is taken in BS 5400 Part 3 where slenderness limits are defined for the elements of the stiffener cross-sections, which provide a factor of safety against elastic critical buckling to allow for non-linear effects.

Figure 6.56 shows the buckling of the panels between longitudinal stiffeners and the buckling of longitudinal stiffeners between cross-frames. In can be seen that in both cases adjacent elements will deform in opposing directions because of continuity. The weakest mode will determine the failure capacity. The torsional rigidity of the stiffeners will determine whether they remain straight or whether they will twist to remain orthogonal to the plating or longitudinal stiffeners. The former has been shown in the figure.

Buckling of the webs is similar to the case of plate girders although the increased use of longitudinal stiffeners adds complexity. The buckling restraint provided by the flanges is also significantly lower, particularly where these are longitudinally stiffened and hence very slender. For any girder where longitudinal stiffeners are present in either web or flange, BS 5400 does not allow the use of the tension field approach, resorting to a more complex approach that considers the strength of each sub-panel in detail. This is described later. The reasons for this are two-fold. Firstly, the large shear deformations needed to achieve full tension field action could build in significant eccentricities and destabilize a slender stiffened compression flange. Secondly, the tension field model for a longitudinally stiffened web is complex with the diagonal buckles rotating from the sub-panel buckles through the longitudinal stiffeners themselves. This form of behaviour was precluded from the code.

The final main elements that are needed in a box girder are the intermediate cross-frames or diaphragms and the bearing diaphragms. Intermediate cross-frames are needed at regular intervals to maintain cross-sectional shape. They also have a role in supporting the flange longitudinal stiffeners providing nodal lines to limit the buckling effective length of the longitudinally stiffened flange. The location of these cross-frames or transverse stiffeners coincides with transverse stiffeners in the webs and tension flange so that a welded ring exists which strengthens the cross-section. As these cross-frames are normally placed around 1.5–3 times the width of the flange apart, additional web transverse stiffeners are normally placed between them to enhance web buckling capacity.

At less regular intervals the cross-frames may be braced to add additional strength to the cross-section to prevent distortion of the cross-sectional shape. This is also shown in Figure 6.57. A more complex cross-frame arrangement combined with cross-girders and web stiffeners is shown in Figure 6.58.

For larger more slender boxes, plated intermediate diaphragms may be used to fulfil this function. Such diaphragms are not dissimilar to bearing diaphragms but

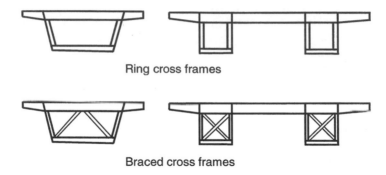

Ring cross frames

Braced cross frames

Figure 6.57 Intermediate ring frames and braced cross-frames.

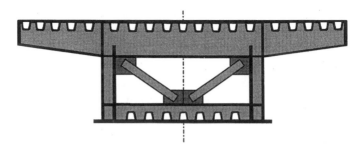

Figure 6.58 Example of a braced cross-section.

are generally lighter with some stiffening when slender but without the load distribution stiffening needed in a bearing diaphragm because of the high compression above the bearings.

There is a substantial range of possible geometries for bearing diaphragms depending on the size and shape of the cross-section but also dependent on the location of the bearings relative to the centre line of the box and the webs. A simple bearing diaphragm for a small stocky box is shown in Figure 6.59. The full depth stiffeners prevent the plated diaphragm buckling from the compression above the bearings and also help the shear distribute from the webs into the bearings. Because of the stocky nature of the diaphragm there is no need for horizontal stiffeners because of the limited horizontal stress resulting from transverse bending.

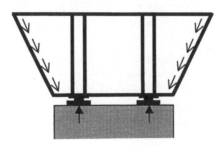

Figure 6.59 A simple plated bearing diaphragm for a stocky box section.

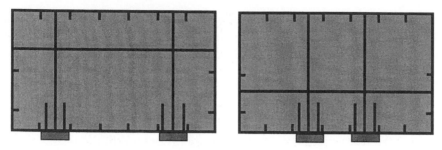

Figure 6.60 Diaphragms illustrating different stiffener arrangements because of bearing location.

Figure 6.60 shows two diaphragms illustrating the possible effect of bearing location. In the diaphragm on the left, the bearings are located near the webs introducing transverse bending putting the top of the diaphragm in horizontal compression. The horizontal stiffener has been added to prevent buckling of the central plate caused by this stress combined with the vertical compression and shear. In the diaphragm on the right, the reverse is true with the bearings close together and the horizontal compression from transverse bending present in the bottom of the diaphragm. The horizontal stiffener is therefore placed in this critical area. Both diaphragms have stub bearing stiffeners to prevent local plate crippling as well as full depth stiffeners for shear transfer.

Bearing location is often determined by the location and size of the piers which are themselves influenced by the general environment of the bridge, the nature of the ground, its height, etc.

Figure 6.61 shows a complex multi-stiffened diaphragm, which has widely spaced bearings, a wide slender box and an access hole for maintenance and inspection purposes. There is a complex series of stiffeners around the large opening to prevent the centre of the diaphragm plate from buckling. The design of such a diaphragm is not unlike a multi-stiffened web as it is essentially a deep beam with high levels of shear, bending and transverse compression. The stress state in each panel has to be considered separately to prevent individual panel buckling.

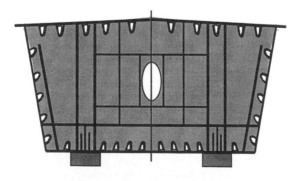

Figure 6.61 Example of a complex multi-stiffened diaphragm.

It makes sense in design to keep diaphragms relatively stocky to avoid complexity. Weight is not a significant issue for a support diaphragm and fabrication costs can be disproportionate if slenderness is not required because of the size of the box.

Box girder design rules

The Eurocode rules for plated structures are published as a European prestandard at the current time (Eurocode 3 part 1.5). While having been in existence for a number of years, the rules within BS 5400 Part 3 are probably still the most comprehensive available in the context of complex stiffened plated elements.

This section deals in some detail with the principles behind both the Eurocode Part 1.5 rules and the BS 5400 Part 3 rules, presenting some of the main design equations. However, both codes are complex in this area, particularly the British Standard, with many minor rules providing important additional requirements. In some areas these are important in providing the global conditions under which the main rules apply. In design, it is therefore important to use the source documents in order to ensure the overall integrity of the design process. The totality of the design rules is too complex to present here. Other design codes deal with the area of box girder design, e.g. in the USA, but in general these follow similar principles and often relate back to the same source research and development material.

The overall approach in the two codes is somewhat different. BS 5400 Part 3 essentially designs webs and flanges separately (while allowing for contributions to the web shear strength from the flange, the possibility of stress shedding from the web to the flange and the influence of boundary conditions). Eurocode 3 Part 1.5 defines an effective section for the girder (box or plate girder) for bending which is then checked against the applied bending moment. Clauses provide effective sections for the compression flange and for stiffeners in the compression zone of the web. A separate check is carried out for web shear capacity (again allowing for a flange contribution). The two governing equations are defined below:

For design for bending:

$$\eta_1 = \frac{\sigma_{x,Ed}}{f_{yd}} = \frac{M_{Sd}}{f_{yd}W_{eff}} \leqslant 1.0 \qquad (6.33)$$

For design for shear:

$$\eta_3 = \frac{\tau_{Ed}}{\chi_v \dfrac{f_{ywd}}{\sqrt{3}}} = \frac{V_{Sd}}{\chi_v \left(\dfrac{f_{ywd}}{\sqrt{3}}\right)bt} \leqslant 0 \qquad (6.34)$$

where W_{eff} is the modulus for the effective section for bending incorporating the effective sections for the elements in compression mentioned above and χ_v is the reduction factor for shear. These are fully described later. More complex expressions are given in the Eurocode for use in the presence of axial force and transverse stresses.

The above equations need to be checked in an interaction equation if the value of η_3 exceeds 0.5. The interaction equation is given in Equation (6.35):

$$\eta_1 + \left[1 - \frac{M_{f,Rd}}{M_{pl,Rd}}\right](2\eta_3 - 1)^2 \leqslant 1.0 \qquad (6.35)$$

where $M_{f.Rd}$ is the design plastic moment resistance of the cross-section consisting only of the flanges (basing it on the smaller flange size) and $M_{pl.Rd}$ is the plastic resistance of the whole cross-section not allowing for any reduction due to slenderness. In Equation (6.35), η_1 may be evaluated using gross section properties.

Stiffened compression flanges in BS 5400

The main elements of a stiffened compression flange are shown in Figure 6.54 and associated buckling modes are shown in Figures 6.55 and 6.56.

As indicated previously a number of buckling modes have, in principle, to be considered:

- buckling of the sub-panels between the stiffeners
- buckling of the longitudinally stiffened panel between the cross-frames
- orthotropic panel buckling of the entire stiffened flange including the cross-frames
- local buckling of the elements of the stiffeners themselves.

BS 5400 Part 3 deals with these in different ways. Buckling modes 3 and 4 are not considered by the designer directly. They are precluded by placing design requirements on the individual stiffeners. In the case of the transverse stiffeners, requirements for the stiffness and strength of the stiffeners are introduced in the code while in the case of the stiffener cross-section, local buckling slenderness limits are placed on all the individual components of the outstands.

The panel buckling strength curves of Figure 6.23 are used in the code to define a buckling effective width to incorporate into the behaviour of the longitudinal stiffeners. The main buckling check in the code therefore deals with the buckling of the longitudinally stiffened panels. This is not, however, dealt with as a stiffened plate but as a series of isolated struts spanning between the cross-girders. The struts have a plate effective width of K_c times the width of the plate panel b. This is illustrated in Figure 6.62.

Representation of the stiffened plate in this way ignores the transverse continuity with the control of flange deformations exerted by the webs. This can be significantly conservative in the case of slender flanges with a significant number of stiffeners but the approach is often used because of its relative simplicity. Other methods are available which consider the buckling of the stiffened orthotropic plate. These tend to be based on an elastic orthotropic plate buckling theory with ultimate collapse related to simple yield requirements.

The two buckling directions for the effective strut section acting as a column in compression with an effective length equal to the cross-girder spacing are dealt with by two column equations in the code.

The first equation deals with the stiffener outstand in compression as shown in Figure 6.63. Again it should be emphasized that factors of safety have not been included.

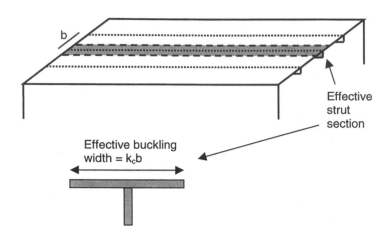

Figure 6.62 Representation of a longitudinally stiffened flange as a series of isolated struts.

$$\sigma_a + 2.5\tau_1 k_{s1} \leqslant k_{l1}\sigma_{ys} \qquad (6.36)$$

where σ_a is the longitudinal compressive stress at the centroid of the effective section of the strut (positive), τ_1 is in-plane shear stress in the flange plate due to torsion (positive) and σ_{ys} is the design yield stress of the stiffener material.

K_{s1} is a factor which reduces the effect of shear on the buckling of the strut for low column slenderness and k_{l1} is the column buckling coefficient. Values of k_{s1} and k_{l1} are given in Figure 6.64.

The abscissa in Figure 6.64 is the column slenderness where r_{se} is the radius of gyration of the effective strut section. λ for buckling in this direction is related to the yield stress of the stiffener. η is an effective imperfection factor for the column which for buckling with the stiffener outstand in compression is given by:

$$\eta = y_0 \Delta / r_{se}^2 \qquad (6.37)$$

Δ is made up from a number of components of imperfection. The first is a simple geometric out-of-straightness as a function of the column length, the second allows for the eccentricity in the axial load caused by variation in stress down the outstand caused by global bending of the box and the third relates to any vertical curvature of the bridge over the length $l(e_f)$:

$$\Delta = \frac{l}{625} + \frac{r_{se}^2}{2y_{Bs}} + \frac{e_f}{2} \qquad (6.38)$$

Figure 6.63 Column buckling producing additional compression in the stiffener outstand.

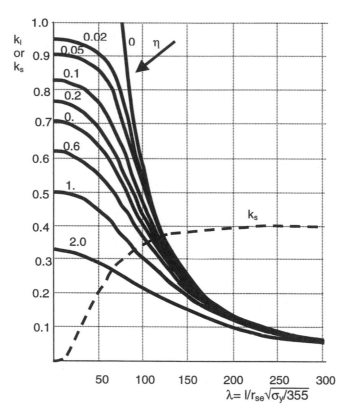

Figure 6.64 Factors for the design of longitudinal flange stiffeners.

The distance y_{Bs} is from the neutral axis of the box to the centroid of the effective strut section and is illustrated in Figure 6.65. It is, in fact, illogical to use the value of y_{bs} in the same way for both directions of flange buckling.

Equation (6.36) thereby allows this direction of strut buckling to be checked. As a direct equivalent, Equation (6.39) provides the basis for checking buckling of the effective strut section with the plate, rather than the outstand, in additional compression as shown in Figure 6.66.

$$\sigma_a + 2.5\sigma_1 k_{s2} \leqslant k_{l2}\sigma_{ye} \tag{6.39}$$

where σ_{ye} is the design yield stress of the plate material reduced to allow for coexistent shear stress:

$$\sigma_{ye} = \sqrt{(\sigma_{yf}^2 - 3\tau^2)} \tag{6.40}$$

where τ is the in-plane torsional shear stress plus half the complementary shear stress at the flange edge due to vertical shear of the girder. An averaging across the flange width is assumed for the latter. The process is directly equivalent to that defined above but with appropriate terms in the equations.

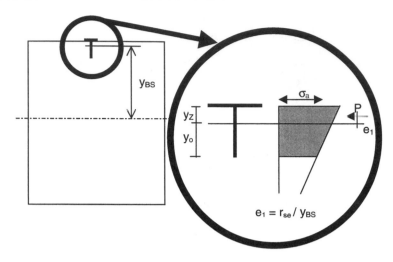

Figure 6.65 Definition of y_{BS}.

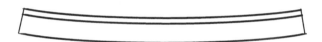

Figure 6.66 Column buckling producing additional compression in the plate.

The imperfection factor of Equation (6.37) for use in Figure 6.64 becomes:

$$\eta = y_z \Delta / r_{se}^2 \tag{6.41}$$

where y_z is the distance from the mid-plane of the flange plate to the centroid of the effective strut section. λ in Figure 6.64 for establishing k_{l2} is now related to the plate yield stress.

In addition to the column bucking requirements, BS 5400 also has clauses relating to yielding of the section.

The design of transverse stiffeners for the stiffened flange, unlike the web, is based on a beam model with the beam spanning between the webs (length B). Stiffness and strength requirements are provided for the transverse members to ensure that they act as nodal lines for the longitudinal stiffener buckling.

The strength requirement is based on an effective section of the stiffener (generally of T-section) plus an associated width of flange plate in the longitudinal direction of half the transverse stiffener spacing for a non-composite top deck or one-seventh of the main web spacing for the bottom flange in the hogging moment region of a continuous span. The former value allows for transverse in-plane flange compression.

The loading applied to the beam is a uniformly distributed load along the beam span which equals the total axial compressive load carried by the flange/unit flange width divided by 200 distributed along the length of the beam. This is illustrated in

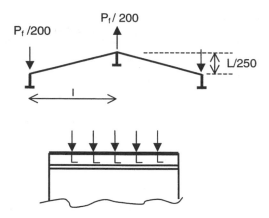

Figure 6.67 Loading applied to transverse flange stiffener.

Figure 6.67. Other loading is added to this, which includes any directly applied loads, the effect of any bridge vertical curvature and also restraint against distortion and effect of concrete shrinkage (if present). The code gives details of how to evaluate these other components.

The standard beam strength equations, Equations (6.21) and (6.23), depending on whether the stiffener section is compact, should be used. As a hot rolled section will almost certainly be used for the stiffener, it is likely that the compact requirements will be met and it is sensible to design the beam as a plastic member in this context.

The stiffness requirement for the flange transverse stiffeners in BS 5400 is based on satisfying the equation:

$$I_{be} \geqslant \frac{9\sigma_f^2 a B^4 A_f^2}{16KE^2 I_f} \tag{6.42}$$

where I_{be} is the required effective inertia of the transverse member with an effective flange width equal to one-quarter of the web spacing (not greater than the transverse stiffener spacing), σ_f is the average longitudinal compressive stress in the flange, A_f is the total flange area including longitudinal stiffeners/B, I_f is the inertia of the stiffened flange about its centroidal axis and K is a factor which varies depending on the location of the cross-girder and the type of stiffeners. For example where the beam is spanning between two box webs and the flange longitudinal stiffeners are of open cross-section the value of K is 24.

Other values exist for parameters in Equation (6.42); for example the value of K will be different if the beam is a cantilever beyond the box. Again, the code provision is relatively complex.

As mentioned above, there is no need to check the buckling of individual stiffener plate components because the code limits their slenderness to ensure they do not buckle. This is because buckling of free edge elements, as discussed in terms of plate behaviour earlier, is a relatively unstable phenomenon with the possibility of significant loss of stiffness. Examples of the slenderness limits applied are, for flat stiffeners:

$$(h_s/t_s)\sqrt{(\sigma_y/355)} \leqslant 10 \qquad\qquad (6.43)$$

where h_s/t_s is the outstand slenderness and for angle stiffeners:

$$(b_s/t_s)\sqrt{(\sigma_y/355)} \leqslant 11 \qquad\qquad (6.44)$$

and:

$$(h_s/t_s)\sqrt{(\sigma_y/355)} \leqslant 7 \qquad\qquad (6.45)$$

where b_s is the stiffener flange width and h_s is the stiffener depth.

   In presenting these design rules no mention has been made of the phenomenon of shear lag. BS 5400, in contrast to other codes, allows shear lag to be ignored in the collapse limit state. There is a need to account for it in particular circumstances for the serviceability limit state, notably in the context of cable supported structures.

   Shear lag occurs as a non-linear distribution of direct stress across the flange caused by the in-plane shear flexibility of the flange plating. It is an elastic phenomenon and should not be confused with the non-linear stress distribution caused by out-of-plane buckling. BS 5400 assumes that at the collapse limit state the shear lag induced stress variation redistributes itself through plastic redistribution although research shows there is a limit to this redistribution and other codes do allow for shear lag at collapse in certain circumstances. The appearance of the stress variation is demonstrated in Figure 6.68.

   The phenomenon of shear lag comes about because the direct stresses in a flange are introduced through shear along the web–flange boundary due to vertical bending of the web. This stress has to be transferred across the width of the flange through the in-plane shear stiffness of the flange plate. For a typical beam with low width to thickness ratio there are no significant stress variations across the width but for a flange that is wide compared with the bridge span, a significant variation can occur with central stresses being as low as 20% or less of the edge compression. In effect, plane sections are no longer remaining plane and simple bending theory no longer applies.

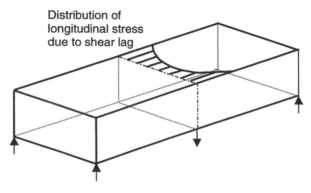

Distribution of
longitudinal stress
due to shear lag

Figure 6.68 Non-linear flange stress distribution caused by shear lag.

BS 5400 Part 3 provides tables of shear lag effective widths for use in serviceability calculations and one is reproduced below to illustrate the magnitude of the effect and its dependence on a number of parameters. The table presented is for a simply supported girder for a stiffened flange between two webs.

In Table 6.3, α is the degree of longitudinal stiffening equal to the total area of the flange stiffeners divided by the total area of the flange plate.

It can be seen that the shear lag effective width Ψ is heavily influenced by the parameter $B/2L$, which is the half width of the flange divided by the bridge span. For a continuous bridge, L would normally be defined as the distance between points of contraflecture but this is allowed for in the code by providing appropriate values of Ψ in a separate table in terms of L (defined still as the beam span). The effective width of the flange is ΨB, which can also be used to define the ratio between centre and edge flange stresses.

The influence of location along the span is relatively modest for very wide bridges where the stress has little opportunity to continue to develop across the width but is very significant for longer spans where stress becomes more uniform away from the supports. The stiffening ratio has a moderate influence as the lumped stiffener mass reduces the degree to which the shear stiffness can redistribute the stress and hence high stiffening ratios result in lower values of effective width.

Stiffened compression flanges in Eurocode 3 Part 1.5

There is a requirement to allow for shear lag at ultimate collapse in Eurocode 3 Part 1.5 (a separate requirement is also specified for serviceability considerations). The shear lag reduction factor β^k multiplies by an effective section which allows for buckling to give an effective flange area for compression flanges.

$$A_{eff} = A_{c,eff}\beta^k \text{ but } A_{eff} \geqslant \beta A_{c,eff} \tag{6.46}$$

where $A_{c,eff}$ is the plate buckling effective area and β is the shear lag effective width factor. The values of β and κ are given in the code but for a simply supported span and a stiffened compression flange between two webs, κ is given by Equation (6.47):

Table 6.3 Shear lag factors from BS 5400 Part 3

$B/2L$	Mid-span $\alpha = 0$	Mid-span $\alpha = 1$	Quarter-span $\alpha = 0$	Quarter-span $\alpha = 1$	Support $\alpha = 0$	Support $\alpha = 1$
0.00	1.00	1.00	1.00	1.00	1.00	1.00
0.05	0.98	0.97	0.98	0.96	0.84	0.77
0.10	0.95	0.89	0.93	0.86	0.70	0.60
0.20	0.81	0.67	0.77	0.62	0.52	0.38
0.30	0.66	0.47	0.61	0.44	0.40	0.28
0.40	0.50	0.35	0.46	0.32	0.32	0.22
0.50	0.38	0.28	0.36	0.25	0.27	0.18
0.75	0.22	0.17	0.20	0.16	0.17	0.12
1.00	0.16	0.12	0.15	0.11	0.12	0.09

$$\kappa = \alpha_0 b_0 / L_e \tag{6.47}$$

where:

$$\alpha_0 = \left(1 + \frac{A_{sl}}{b_0 t}\right)^{0.5} \tag{6.48}$$

where b_0 is the half flange width, A_{sl} is the area of stiffening within this width (hence $A_{sl}/b_0 t$ is the flange stiffening ratio) and L_e is the girder span.

The shear lag effective breadth b_{eff} is equal to the flange width times β. For a simply supported span the values of β are given by Table 6.4.

The factor ρ for plate buckling has been defined in Equations (6.10) and (6.11). If the flange has no longitudinal stiffeners, the appropriate value $A_{c,eff}$ can then be defined:

$$A_{c,eff} = \rho A_c \tag{6.49}$$

This equation is also used to define the sub-panel effective width to be used in association with the stiffeners (see below):

$$\rho_{pan} = \rho$$

In the presence of longitudinal flange stiffeners, $A_{c,eff}$ is given by:

$$A_{c,eff} = \rho_c A_c \tag{6.50}$$

where A_c is the reduced area of the stiffened compression flange.

A_c is the effective area of the stiffeners together with the total sub-panel area reduced to allow for plate buckling:

$$A_c = A_{sl,eff} + \Sigma_c(\rho_{pan} b_{c,pan} t) \tag{6.51}$$

$A_{sl,eff}$ is reduced in accordance with Equations (6.9)–(6.11) if local buckling affects the plates of the stiffener cross-section. σ_{cr} in Equation (6.9) is evaluated for the relevant boundary conditions (i.e. a plate with a free edge if an outstand element).

Table 6.4 Values of b in Eurocode 3 Part 1.5 for a simply supported span

κ	β
≤ 0.02	1.0
0.02–0.07	$1/(1 + 6.4\kappa^2)$
> 0.07	$1/(5.9\kappa)$

ρ_c allows for overall buckling of the longitudinally stiffened flange as a plate with a smeared stiffness which allows for the stiffeners and buckling of a stiffener and effective width of plate as a strut between cross-frames. It is hence more sophisticated than BS 5400 in that it allows for the orthotropic behaviour while BS 5400 Part 3 presumes the flange strength is based on that of the individual effective strut.

The final equation for ρ_c is:

$$\rho_c = (\rho - \chi_c)\xi(2 - \xi) + \chi_c \qquad (6.52)$$

where ρ is the reduction factor to allow for buckling of the plate of smeared stiffness and χ_c is the reduction factor for strut buckling.

$$\xi = (\sigma_{cr,p}/\sigma_{cr,c}) - 1 \qquad (6.53)$$

where $\sigma_{cr,p}$ is the elastic critical buckling stress of the smeared longitudinally stiffened plate and $\sigma_{cr,c}$ is the elastic critical buckling stress of the equivalent plate with longitudinal edge support removed.

$\sigma_{cr,p}$ is evaluated from Equation (6.5) where b is the total flange width, t the flange plate thickness and k is a buckling coefficient from orthotropic plate theory:

$$k \text{ is equal to } k_{\sigma,p} = \frac{((1+\alpha^2)^2 + \gamma)}{\alpha^2(1+\delta)} \text{ if } \alpha < (1+\gamma)^{0.25} \qquad (6.54)$$

or:

$$k_{\sigma,p} = \frac{2\left(1+\sqrt{1+\gamma}\right)}{1+\delta} \text{ if } \alpha > (1+\gamma)^{0.25} \qquad (6.55)$$

where $\gamma = (I_x/I_p) > 50$ is the second moment of area of the stiffened panel divided by the bending stiffness of the plate $(bt^3)/(12(1-v^2))$ (b is the width of the stiffened plate), $\delta = A_{sl}/A_p$ is the gross area of all longitudinal stiffeners divided by the gross area of the plate bt and $\alpha = a/b > 1$.

ρ is evaluated from Equation (6.10) or (6.11) with $\bar{\lambda}_p$ given by:

$$\bar{\lambda}_p = \sqrt{(\beta_a \sigma_y/\sigma_{cr,p})} \qquad (6.56)$$

where β_a is the effective area of the stiffened plate allowing for buckling (A_{eff}) of the sub-panels divided by the gross area of the stiffened plate (A).

$\sigma_{cr,c}$ is evaluated from Equation (6.57) where I_x is as defined above, a is the transverse stiffener spacing and A is the gross area of the stiffened plate.

$$\sigma_{cr,c} = (\pi^2 EI_x)/(Aa^2) \qquad (6.57)$$

χ_c the reduction factor for strut buckling is given by the column buckling expression in Eurocode 3 Part 1.1 with α replaced by $\alpha_e = \alpha_0 + (0.09/(i/e))$ where $i = \sqrt{(I_x/A)}$ is the radius of gyration of the stiffened plate, e is the larger distance from the centroid of the stiffened plate to the centroid of the stiffener section or the centre of the plating and α_0 is 0.34 for hollow section stiffeners or 0.49 for open section stiffeners. This gives an appropriate imperfection factor for an eccentricity stiffened plate, a bow of $a/500$ (in contrast to the values used for columns in Part 1.1). The appropriate column slenderness to use in evaluating the column strength is:

$$\bar{\lambda}_c = \sqrt{(\beta_A \sigma_y / \sigma_{cr,c})} \tag{6.58}$$

$$\chi_c = 1/(\phi + (\phi^2 - \bar{\lambda}^2)^{0.5}) \text{ but} \leqslant 1 \tag{6.59}$$

where:

$$\phi = 0.5(1 + \alpha(\bar{\lambda}c - 0.2) + \bar{\lambda}_c^2) \tag{6.60}$$

Stiffened webs in BS 5400

The design of transversely stiffened webs for girders with no longitudinal stiffeners has been dealt with in the context of plate girders and would be applied in the same way for a box. However, BS 5400 requires the use of a different method for the design of webs when either the web or flange has longitudinal stiffeners for reasons defined previously. For the latter case the basis of the web design considers the ultimate capacity of the individual web panels in the cross-section (a single panel if there are no longitudinal stiffeners present in the web). The shear capacity of the web, in the presence of other in-plane stresses, is the summation of the individual panel capacities in a vertical section (Figure 6.69).

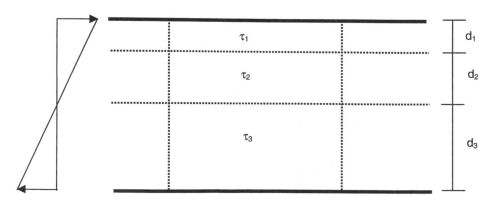

Figure 6.69 Basis for web design for a girder with longitudinal stiffeners.

The total web shear strength is given by the equation:

$$V_D = (\tau_1 d_1 + \tau_2 d_2 + \tau_3 d_3 + ... + ... \tau_n d_n) t_w \qquad (6.61)$$

where n is the number of panels through the web depth. The values of the individual shear strengths are evaluated as the capacity of each panel in shear allowing for the other direct stresses present.

The distribution of direct stresses for which the individual panel capacity is to be assessed is based on a simple linear sub-division of the total stresses evaluated for the web based on simple elastic bending theory. Only the ratio of the individual stresses is initially needed, i.e. the ratio between bending stress, uniform direct compression or tension and shear (which is assumed to be uniform down the web depth at this stage) is needed. A representation of these stresses is shown in Figure 6.70.

It can be seen from Figure 6.70 that the direct stresses in the top panel are mainly compressive with a small bending component, in the centre panel they are largely bending with a small compressive component and in the bottom panel they are an intermediate level of tension and bending. The slope of the bending stress in all three panels is identical to the slope of the original bending stress distribution while the combination of compression/tension and bending at each of the panel boundaries adds up to the direct stress level at that location in the original distribution.

The stresses, of course, have to be evaluated on an initial trial of a web thickness and stiffener locations (and a trial flange geometry) as with all designs. Once the web panels have been checked the thickness of the web or the locations of transverse and

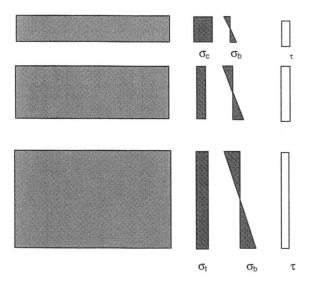

Figure 6.70 Individual stress states for each of the web panels.

longitudinal stiffeners can be adjusted to optimize the design. There is also a need to iterate between flange and web designs. Because of the relative complexity of the calculations, a desk top computer program should be used.

To evaluate the collapse stress combination of each panel, the strengths of the panels under the individual stress components is determined using the curves of Figures 6.23, 6.29 and 6.32 for the cases of compression, shear and bending, respectively. Values of k_{c1}, k_{q1}, k_{b1}, k_{c2}, k_{q2}, k_{b2}, k_{q3} and k_{b3} are therefore obtained. While the code does not make this clear, it is logical to take the value of k_{c3} as 1 because the panel is in tension. If a lower value is taken it is potentially non-conservative because the effect is beneficial as a negative term in the interaction equation which is then artificially large.

In evaluating these coefficients, an assumption has to be made about the boundary condition for each panel (the degree of in-plane restraint). The code defines internal panels as being restrained and panels abutting a flange as being unrestrained unless the latter satisfy conditions relating to the slenderness of the edge panel or the bending restraint provided by the flange to the web (in order to anchor transverse in-plane boundary stresses). The edge panel either has to have a slenderness λ less than 24 or a value of m_{fw} greater than (including factors of safety):

$$[\sigma_{yf}^2/(\sigma_{yf}^2 - \sigma_f^2)] \times (0.00025(\lambda - 24)) \tag{6.62}$$

for $24 \leqslant \lambda \leqslant 84$, where $\lambda = b/t_w \sqrt{(\sigma_{yw}/355)}$ and $m_{fw} = (\sigma_{yf}b_{fe}t^2f/2\sigma_{yw}b^2t_w)$, a measure of the out-of-plane bending stiffness of the flange compared with the thickness of the web. b_{fe} is an effective width of flange used as the section which restrains transverse stress $= 10t\sqrt{(355/\sigma_{yf})}$ (but not greater than half the web spacing).

Each panel is then checked in the interaction equation presented previously as Equation (6.14) (which also allows for any vertical compression). The form given in the code is:

$$m_c + m_b + 3m_q \leqslant 1 \tag{6.63}$$

where:

$$m_c = \sigma_1/(\sigma_{yw}k_1(1 - \rho)) \tag{6.64}$$

$$m_b = (\sigma_b/(\sigma_{yw}k_b(1 - \rho)))^2 \tag{6.65}$$

$$m_q = (\tau/(\sigma_{yw}k_q))^2 \tag{6.66}$$

$k_1 = k_c$ for the appropriate curve. The value of three in Equation (6.63) comes from the use of σ_{yw} in Equation (6.65) rather than τ_y in Equation (6.14) ($\sigma_{yw} = \sqrt{(3\tau_y)}$ from the Mises yield criterion).

The function $1 - \rho$ in Equations (6.64) and (6.65) allows for the differential benefit of stress redistribution in the context of restrained and unrestrained panels. The code allows shedding of up to 60% of direct stresses from any or all web panels into the flanges (which can carry them more efficiently) as long as overall axial and moment equilibrium are maintained. The stresses used in Equations (6.64)–(6.66) are the reduced levels of stress and hence benefit ensues from the interaction equation for the particular panel being considered. $\rho = 0$ for restrained panels and hence has no effect but equals the proportion of stress shed for an unrestrained panel eliminating any benefit in stress shedding for that panel in terms of the buckling criterion (the reduced value of stress is used in the yield criterion within the rules). This limitation was introduced because of concerns about the strain discontinuities required to introduce this stress differentiation in the context of an unrestrained panel with its less stable postpeak behaviour. If stress shedding is required it is worth considering the provision of a narrow edge panel in order to ensure it is restrained (if not already satisfying the restraint requirements) as this can improve the economy of the design. This stress shedding concept is an interesting one in that it goes some way towards the simple design process allowed in Eurocode 3 Part 1.1 for plate girders whereby webs carry the shear and the flanges carry all the bending moments. With reduced in-plane stress levels, the web panels can carry higher shear stresses leading to a higher shear capacity for the girder but the flange of course has to be designed for the increased bending moment in order to maintain equilibrium.

Once each of the panels, with or without stress shedding, has been checked in the interaction equation design iterations will be required. These depend on the outcome of the panel buckling factors. For example, if the value from Equation (6.63) is greater than one in all cases the whole of the web is under designed and either the web thickness (or web depth) or the number of web longitudinal stiffeners must be increased. In certain circumstances, if the over-stress is modest, the transverse stiffener spacing can be reduced which can have an effect on the shear buckling resistance. Clearly if all are more than marginally less than one, the web is over designed and savings can be made using the converse of the above. If the result varies between panels (i.e. some factors greater than one and others less, it is possible to improve the efficiency of the design by moving the location of the longitudinal stiffeners to improve the balance of the resistance of each of the panels. Panels with factors greater than one should have their breadth reduced, etc. Clearly, as with all design work, there is a practical limit to the value of endless iteration.

Design of web stiffeners in BS 5400 Part 3

The design of transverse web stiffeners has been dealt with in the context of plate girders above. The only significant difference is the inclusion of the longitudinal stiffener area (when present) which has been incorporated into Equation (6.32).

The design of longitudinal web stiffeners is based on similar principles. The longitudinal stiffener is designed as a column spanning between transverse web stiffeners with an effective section comprised of the stiffener plus an effective width of web in the transverse direction equal to $32(1 - \rho)$ times the web thickness (as for transverse web stiffeners other than for the inclusion of the stress shedding parameter).

The loading on the stiffener relates to the axial load present in the web together with an allowance for shear. The design equation is:

$$\sigma_{se} \leq \sigma_{ls} \qquad (6.67)$$

where σ_{ls} equals the column strength obtained from a column design curve in the code.

The slenderness of the column $\lambda = a/r_{se}\sqrt{(\sigma_{ys}/355)}$, where r_{se} is the radius of gyration of the effective stiffener section.

$$\sigma_{se} = k'_s \sigma_1 + \left(2.5\tau + \frac{a^2}{b^2}\sigma_2\right)\frac{bt_w k_s}{A_{se}} \qquad (6.68)$$

where $k'_s = 1$ for continuous longitudinal stiffeners, σ_1 is the longitudinal stress at the stiffener, σ_2 is a transverse compression stress when present and τ is the average web shear stress at the location of the stiffener. K_s is a reduction factor equivalent to that shown in Figure 6.64 and used in Equation (6.36). A_{se} is the area of the effective stiffener cross-section.

Stiffened webs in Eurocode 3 Part 1.5

The design of webs in Eurocode 3 Part 1.5 can follow the process defined for compression flanges above although it is somewhat more complex because the stress gradient down the web has to be allowed for and buckling factors, used to define the effective section, only apply in the compression zone of the web. The modified equations are not presented here but are available in the Eurocode.

A simpler method is available for webs with only one or two longitudinal stiffeners in the compression zone, the common case for all but very large boxes, which avoids calculating $\sigma_{cr,p}$ for the actual geometry of the web. The basis of the method is to define a gross area for a fictitious column which is the gross area of the stiffener A_{sl} together with associated plating. If a sub-panel is fully in compression, one-half of the sub-panel width is used, if stresses change to tension in the sub-panel, one-third of the compression width should be used. This gross area is then reduced to allow for buckling by evaluating reduction factors for the plate elements using Equation (6.10) or (6.11) with $\bar{\lambda}_p$ replaced by:

$$\bar{\lambda}_{p,red} = \bar{\lambda}_p\sqrt{(\sigma_{comm,Ed}/\sigma_y)} \qquad (6.69)$$

where $\sigma_{comm,Ed}$ is the maximum compressive stress in the gross cross-section of the plate.

The critical buckling stress $\sigma_{cr,p}$ is then given by:

$$\sigma_{cr,p} = \frac{1.05E\sqrt{I_{sl}t^3 b}}{Ab_1 b_2} \quad \text{if } a \geq a_c \qquad (6.70)$$

or:

$$\sigma_{cr,p} = \frac{\pi^2 EI_{sl}}{Aa^2} + \frac{Et^3 ba^2}{4\pi^2(1-v^2)Ab_1^2 b_2^2} \quad \text{if } a \geqslant a_c \qquad (6.71)$$

where A is the gross area of the effective column, I_{sl} is the second moment of area of the gross column cross-section about its centroidal axis, b_1 and b_2 are the distances from the centre line of the stiffener to the compression and tension flange and $b = b_1 + b_2$.

If there are two stiffeners in the compression zone, they should first be considered separately and then a single stiffener with an A and I_{sl} located at the position of the resultant axial force in the stiffeners should be used to evaluate $\sigma_{cr,p}$.

Using this elastic buckling stress in Equation (6.56) with A and A_{eff} relating to the compressed part of the web to give $\bar{\lambda}_p$ and then Equation (6.10) or (6.11) to give ρ gives the first reduction factor of Equation (6.52).

The column buckling stress $\sigma_{cr,c}$ may be evaluated allowing for the stress gradient by the use of an effective length factor given in Part 2 of Eurocode or Equation (6.57) may be conservatively used. Equations (6.59), (6.52) and (6.53) are then used as for compression flanges to define ρ_c.

If $\rho_c\sigma_y$ is greater than the average stress in the stiffened column ($\sigma_{c,Ed}$), the reduction factor ρ_c is applied to the area of the fictitious column with plate widths reduced to allow for plate buckling as defined above, otherwise the effective area is further reduced by multiplying the effective area by $\sigma_y/\sigma_{c,Ed}$.

The reduction factor for shear χ_v for use in Equation (6.34) is comprised of contribution from web and flanges:

$$\chi_v = \chi_w + \chi_f \leqslant \eta \qquad (6.72)$$

The contribution χ_w depends on whether end posts are rigid. A rigid end post essentially requires two pairs of transverse stiffeners on either side of the web a distance at least one tenth of the web depth apart with the first stiffener above the centre line of the bearing and the second outside the span of the beam. The section modulus of each stiffener should be at least $4h_w t^2$, where h_w is the web depth for bending out of the plane of the web or if a flat stiffener it should have an area of $4h_w t^2 e$, where e is the stiffener spacing.

The value of χ_w is given by Table 6.5.

η has the value of 1.2 for normal strength steels multiplied by a yield stress factor.

Table 6.5 Values of χ_w

$\bar{\lambda}_w$	Rigid end post	Non-rigid end post
$<0.83/\eta$	η	η
$0.83/\eta \leqslant \bar{\lambda}_w$	$0.83/\bar{\lambda}_w$	$0.83/\bar{\lambda}_w$
$\geqslant 1.08$	$1.37/(0.7 + \bar{\lambda}_w)$	$0.83/\bar{\lambda}_w$

$\bar{\lambda}_w$ depends on the type of web stiffening present but for a web with intermediate transverse stiffeners and/or longitudinal stiffeners the expression for $\bar{\lambda}_w$ is:

$$\bar{\lambda}_w = \frac{b_w}{37.4t\varepsilon\sqrt{k_\tau}} \tag{6.73}$$

where:

$$k_\tau = 5.34 + 4.00(h_w/a)^2 + k_{\tau st} \text{ when } a/h_w \geqslant 1 \tag{6.74}$$

$$k_\tau = 4.00 + 5.34(a/h_w)^2 + k_{\tau st} \text{ when } a/h_w < 1 \tag{6.75}$$

and:

$$k_{\tau st} = 9\left(\frac{h_w}{a}\right)^2\left(\frac{I_{sl}}{t^3 h_w}\right)^{0.75} \text{ but } \geqslant \frac{2.1}{t}\left(\frac{I_{sl}}{h_w}\right)^{\frac{1}{3}} \tag{6.76}$$

where I_{sl} is the second moment of area of the longitudinal stiffener about the web (the sum of the web stiffener contributions where there are more than one).

The contribution from the flanges χ_f is reduced to allow for applied bending and is given by:

$$\chi_f = \frac{b_f t_f^2 \sigma_{yf}^{\sqrt{3}}}{cth_w \sigma_{yw}}\left[1-\left(\frac{M_{Sd}}{M_{f.Rd}}\right)^2\right] \tag{6.77}$$

where:

$$c = \left[0.25 + \frac{1.6b_f t_f^2 \sigma_{yf}}{th_w^2 \sigma_{yw}}\right]a \tag{6.78}$$

If an axial force is present in the girder there is a further reduction.

If the intermediate transverse stiffeners are required to act as rigid supports to internal panels they should have a stiffness of:

$$I_{st} \geqslant 1.5h_w^3 t^3/a^2 \text{ if } a/h_w < \sqrt{2} \tag{6.79}$$

$$I_{st} \geqslant 0.75h_w t^3 \text{ if } a/h_w \geqslant \sqrt{2} \tag{6.80}$$

They should also be designed for an axial force which is defined in the code.

Design of stiffeners in Eurocode 3 Part 1.5

In Eurocode 3 Part 1.5, as for BS 5400, transverse stiffeners in either the compression flange or web are designed to provide support for longitudinal stiffeners. Transverse stiffeners must satisfy stiffness and strength requirements. The effective section of the transverse stiffener includes an effective width of plate equal to $30\varepsilon t$. This is almost identical to the effective section used by BS 5400 with the exception of the yield stress multiplier. All definitions below relate to this effective section.

However, unlike the web requirements of BS 5400, the transverse stiffener is treated as a beam both for stiffness and strength requirements with a span equal to the web depth between flanges or webs as appropriate and an imperfection equal to the distance between transverse stiffeners a divided by 300. The transverse stiffener may be loaded by a resolved component of the in-plane compressive stress which is then checked against yield in the stiffener or a deflection increment limit of $b/300$. Alternatively, these may be satisfied by providing a second moment of area for the effective stiffener cross-section of I_{st}:

$$I_{st} = \frac{\sigma_m}{E} \left(\frac{b}{\pi}\right)^4 \left[1 + w_0 \frac{300}{b} u\right] \qquad (6.81)$$

where:

$$\sigma_m = \frac{\sigma_{cr,c}}{\sigma_{cr,p}} \frac{N_{Sd}}{b} \left[\frac{2}{a}\right] \qquad (6.82)$$

$$u = \frac{\pi E e_{max}}{\sigma_y 300b} \geq 1.0 \qquad (6.83)$$

e_{max} is the extreme fibre distance from the stiffener centroid and N_{Sd} is the largest design compressive force of an adjacent panel.

The code provides requirements for longitudinal stiffeners to prevent torsional buckling of the outstands.

Diaphragm design

The principles behind the design of diaphragms have been described earlier. In BS 5400, the design is based around the checking of the conditions of the sub-panels in a similar way to multiply stiffened webs in association with the design of the individual stiffeners again in line with the principles applied for web stiffeners above. The details within the code are too complex to present in summary form.

6.4 Connections

6.4.1 Introduction

The design of the connections forms an important part of the overall design of a bridge structure. Early in the project advice should be sought from steelwork fabricators as to which factors are significant when minimizing the cost of the connections. However, cost is not the dominant factor for the bridge designer and careful attention must also be given to the strength and fatigue behaviour of the chosen connections. The bridge designer also has to consider other important factors such as the optimum location of the joint, which connections are to be welded, which are to be bolted and, following on from this, which connections are to be made in the fabricator's workshop and which connections are to be undertaken on site. When considering the location of joints, it is desirable to position the joints at points of contra-flexure or areas of low stress. Also consideration must be given to optimizing the total number of joints taking into account transportation to site, the weight and corresponding ease of erection of a particular section of the structure. It is generally uneconomic to transport very large sections of steelwork which because of their size, will require a police escort in the UK, and may also demand the 'one off' hire of special lifting equipment in order to erect the part. In general most 'shop' joints effected in the fabricators workshop are welded. In the workshop it is easier to obtain the correct 'fit up' of the various parts for welding and also to make allowances and, if necessary remedy, any distortions resulting from welding. In bridgework *in situ* joints can be welded, bolted or made using a combination of welding and bolting. However, most *in situ* connections are bolted with extensive use being made of high strength friction grip (HSFG) bolts.

Advantages and disadvantages of bolted and welded in situ connections

Due to restrictions imposed by the need to transport to site individual sections of the structure shorter than approximately 25 m, *in situ* connections are nearly always required to construct the complete bridge. In order to assist the bridge designer in making the correct choice of connection, Table 6.6 summarizes the advantages and disadvantages of both bolted and welded *in situ* connections.

6.4.2 Welded connections

The welding process

The process of welding requires the formation of a molten metal weld pool which is used to fuse together the faces of the parts to be joined. Arc welding is the predominant method used for welding structural steelwork and with this process a weld pool is formed by passing a high current through an electrode which when placed close to the earthed work piece allows a plasma arc to form. The high current passing through the electrode and subsequently through the arc into the earthed work piece causes the tip of the electrode to melt and allows the transfer of molten metal from the electrode to the work piece, to form the weld bead. Several arc welding techniques are available for use by the fabricator and the major methods are briefly outlined below.

Manual metal-arc process (MMA)

This process is the oldest of all of the arc processes and is widely used by fabricators for attaching stiffeners, etc. The process is a manual operation with the quality of the finished weld depending to a large extent on the skill of the welder. The electrode consists of a steel wire coated with a flux formed from cellulose, silicates, iron oxide, etc. The primary purpose of the flux is to shield the weld pool to prevent the molten

Table 6.6 Advantages and disadvantages of bolted and welded in situ connections

Bolted joints	Welded joints
Advantages	
• Generally cheaper.	• Good aesthetics forming a visually unobtrusive connection.
• Connections are almost self-aligning and relatively quick to make.	• Minimizes girder weight with no deductions required for holes in the tension flange.
• Only semi-skilled labour required.	• Full strength connection can be achieved by using full penetration butt welds.
• Allows adjustment to vertical and longitudinal alignment if normal clearance holes are used.	
• Easy to inspect.	
• Normally does not govern bridge fatigue life.	
• Not weather sensitive during 'bolting up'.	
Disadvantages	
• May be visually unacceptable.	• Expensive unless in large numbers.
• Deduction for holes in tension flange may lead to the use of a larger section.	• Weather sensitive, must be protected from moisture.
• Can be prone to corrosion if not properly protected.	• Preheat of steelwork is usually necessary requiring a tent/shelter to prevent heat loss and protection against inclement weather.
	• Expensive to inspect for defects.
	• If defects are present it is expensive and time consuming to repair with subsequent risk of delay to project.
	• May govern the fatigue life of the structure. Skilled operatives required, working to an approved specification.
	• Temporary cleats needed to assist in preweld alignment. Cleats normally have to be carefully removed and parent material restored to original condition to prevent fatigue problems.

metal oxidizing and in addition taking into solution nitrogen and other unwanted gasses. Some of these gasses if present, are released back into the air as the weld cools and in the process cause porosity in the weld metal. The flux on the electrode also helps to stabilize the plasma arc and can be used to control the deposition rate of the molten metal. It is most important that the electrodes chosen are of the correct grade of steel, at least equivalent to that of the parent parts to be joined, and that they are completely dry and corrosion free before use. Eurocode 3 (1992) specifies that the filler metal in the electrodes must have mechanical properties, namely yield strength, ultimate tensile strength, elongation at failure and a minimum Charpy V-notch energy value, equal or better than the values specified for the steel grade being welded. The electrodes used in the manual metal arc process are of a relatively short length and hence for long continuous welds there will be numerous stop–start positions, each the potential source of an imperfection.

Submerged arc welding (SAW)

Unlike the manual metal arc process the submerged arc method is a fully automatic method, employing a travelling gantry or robotic unit. The electrode is a continuous bare steel wire which is unwound from a storage drum as required, and fed, via the welding nozzle, into a bed of flux which has been deposited on the surfaces to be welded. The flux, which completely encapsulates the plasma arc, can also be used to enhance the composition of the weld metal. The submerged arc process can only be successfully used in the 'down hand' position and is generally the preferred method for making long welds, such as the web to flange welds, required for plate girder fabrication.

Gas shielded processes

These processes, like the submerged arc process, use a continuous bare wire electrode. However, the weld pool is protected from the atmosphere not by a flux, but by a gas which is also fed via the welding nozzle, around the plasma arc and weld pool. The shielding gas used may be inert argon or non-reactive carbon dioxide and the processes are termed metal inert gas (MIG) and metal active gas (MAG), respectively. These processes are very versatile and can be used in both the 'down hand' and 'overhead' positions but must be sheltered from draughts which tend to disturb the gas flow.

Welded connection design

There are two types of structural weld in common use namely butt welds and the fillet welds. A butt weld is normally made within the cross-section of the abutting plates, whereas a fillet weld is a weld of approximately triangular cross-section applied to the surface of the plates to be joined. When plates greater than about 5 mm thick are to be butt welded together, the plate edges will have to be prepared before welding in order to obtain a full penetration weld. Figure 6.71 shows typical butt welds together with the required edge preparation. A full strength penetration butt weld in structural steel, made correctly with the appropriate electrodes, is considered to be a strong as the parent steel plates, and consequently no strength calculations are necessary, unless fatigue is a consideration.

However, fillet welds do not require any edge preparation for the parent plates and consequently are usually cheaper to undertake than butt welds. Fillet welds are usually specified by requiring a minimum leg length for the weld. These welds are considered to carry all of the loads applied to them, through shear which acts on the effective throat of the fillet weld. Figure 6.72, taken from Figures 53 and 54 of BS 5400 Part 3 (1982), shows the method for calculating the effective throat area which

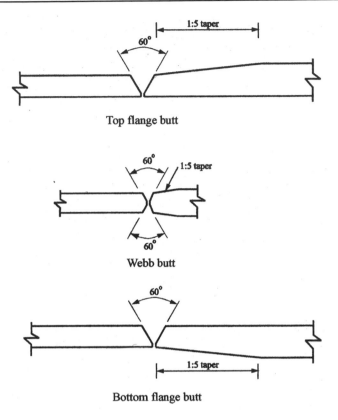

Top flange butt

Webb butt

Bottom flange butt

Figure 6.71 Typical butt welds showing the required edge preparation (Needham, 1983).

is described in Clause 14.6.3.9 of BS 5400 Part 3 (1982) as the height of a triangle that can be inscribed within the weld cross-section and measured perpendicular to its outer side.

The UK Code of Practice BS 5400 Part 3 suggests two alternative procedures for assessing the stress in a fillet weld. The first procedure requires that the vector addition of all of the shear stresses acting on the weld should not exceed the shear stress capacity of the weld τ_D given by:

$$\tau_D = \frac{k(\sigma_y + 455)}{\gamma_m \gamma_{f3} 2\sqrt{3}} \quad \text{Clause 14.6.3.11.1, BS 5400 Part 3 (1982).} \quad (6.84)$$

where σ_y is the nominal yield stress of the weaker part joined, k is eqial to 0.9 for side fillets, or equal to 1.4 for end fillets in end connections, and equal to 1.0 for all other welds. γ_m is the material partial safety factor given in Table 2 of BS 5400 Part 3 (1982) for the ultimate limit state as equal to 1.10, and γ_{f3} is a loading partial safety factor taken to be equal to 1.1 for the ultimate limit state as given in Clause 4.3.3. of BS 5400 Part 3 (1982).

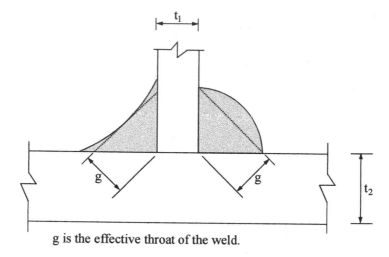

g is the effective throat of the weld.

If either t_1 or t_2 is greater than 4mm, g has to be at least 3mm.

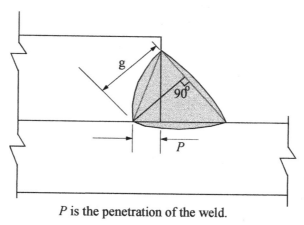

P is the penetration of the weld.

g is the throat of the weld.

Figure 6.72 Effective throat thickness of fillet welds (BS 5400 Part 3 1982).

The second method of assessment states that the following stress condition should be satisfied:

$$\sqrt{\sigma^2 + 3\left(\tau_1^2 + \tau_2^2\right)} \leq \frac{k(\sigma_y + 455)}{2\gamma_m\gamma_{f3}} \quad \text{Clause 14.6.3.11.2. BS 5400 Part 3 (1982)} \quad (6.85)$$

where σ is the stress normal to a section through the throat of the weld, τ_1 is the shear stress acting at right angles to the length of the weld on a section through the throat, τ_2 is the shear stress acting along the length of the weld on a section through the throat. The terms σ_y, k, γ_m and γ_{f3} are defined as given above.

By permitting the parameter k to be equal 1.4, both methods allow for the fact that end fillet welds have proved to be stronger than side fillets. The latter procedure, where it is required to calculate the stresses acting on the weld throat, is more time consuming but less conservative than the former method.

6.4.3 High strength friction grip bolted connections

It is generally a requirement that for bridgework all bolted structural connections are made using high strength friction grip (HSFG) bolts. These bolts, which are tightened up to achieve a specified shank tension, act by clamping together the plates to be joined, so that under normal loading conditions, the applied forces are transferred through the connection by friction acting at the plate interfaces. High strength friction grip bolts are manufactured to the requirements given in BS 4395 (1969) which covers three different grades of bolts namely:

Part 1   General grade – with a strength grade for the bolts of about 8.8 together with grade 10 nuts.

Part 2   Higher grade – with a strength grade 10.9 for the bolts together with grade 12 nuts.

Part 3   Higher grade – with a waisted shank – with a strength grade 10.9 for the bolts together with grade 12 nuts.

For bridgework the majority of HSFG bolts used are general grade.

Bolt tension

The reliable control of bolt tension is essential for ensuring predictable performance in HSFG bolted joints. The Code of Practice BS 4604 (1970) which is published in two parts corresponding to Parts 1 and 2 of BS 4395 (1969), outlines the procedures acceptable for tightening these bolts. The Code of Practice gives two main procedures for tightening namely, the 'Part Turn Method', used for General grade bolts, and the 'Torque Control Method', used for higher grade bolts. In the 'Part Turn Method' the nut is rotated a specific number of turns from the snug position which tensions the bolt. The 'Torque Control Method' requires the use of a manually operated torque wrench or power driven wrench to achieve the required bolt tension. There are other methods of obtaining the required bolt tension the most frequently used method employing load indicator washers. This method relies on the plastic compression of nibs, which are incorporated into the washer, down to a predetermined gap, giving a reliable and relatively easily inspected bolt assembly. Another propriety system available for use is the 'Tension Controlled Bolt' (TCB). These HSFG bolts have a special spline section at the end of the bolt which shears off at a predetermined torque. Figure 6.73 shows a typical tension control bolt, tightened using a special shear wrench which simultaneously tightens the nut whilst holding on to the spline end. Figure 6.74 also shows a typical splice connection in a plate girder made using tension control bolts.

Figure 6.73 A tension control bolt.

Figure 6.74 Typical splice connection in a plate girder – using tension control bolts.

HSFG bolt connection design

High strength friction grip bolted connections designed to comply with BS 5400 Part 3 (1982) are not permitted to slip at the serviceability limit state. At the ultimate limit state, the connection is allowed to slip and the strength of the joint is governed by the bearing or shear capacity of the bolts, whichever is the lower. Clause 14.5.4.2 of BS 5400 Part 3 (1982) states that the design capacity of a HSFG bolt at the serviceability limit state is the friction capacity P_D given by:

$$P_D = k_h \frac{F_v \mu N}{\gamma_m \gamma_{f3}} \tag{6.86}$$

where γ_m is the material partial safety factor given in Table 2(b) of BS 5400 Part 3 (1982) for the serviceability limit state as equal to 1.2, F_v is the prestress load equal to the proof load in the absence of external applied tensions, μ is the slip factor given in Clause 14.5.4.4. of BS 5400 Part 3 (1982), N is the number of friction interfaces equal to 1.0 for a single lap joint and equal to 2.0 for a double lap joint, $k_h = 1.0$ where the holes in all the plies are of normal size as specified in BS 4604 (1970), and γ_{f3} is a loading partial safety factor taken to be equal to 1.0 for the serviceability limit state as given in Clause 4.3.3. of BS 5400 Part 3 (1982).

At the ultimate limit state Clause 14.5.4.1.1. of BS 5400 Part 3 (1982) states that the design capacity is the greater of the following.

The friction capacity using:

$$P_D = k_h \frac{F_v \mu N}{\gamma_m \gamma_{f3}} \tag{6.87}$$

where $\gamma_m = 1.3$ Table 2(a) BS 5400 Part 3 (1982), or the lesser of the shear capacity or the bearing capacity given by Clause 14.5.3.4. and Clause 14.5.3.6. of BS 5400 Part 3 (1982), respectively.

Clause 14.5.3.4 states that τ the average shear stress given by $\tau = (V/nA_{eq})$ must comply with the following condition:

$$\tau = \frac{V}{nA_{eq}} \leqslant \frac{\sigma_q}{\gamma_m \gamma_{f3} \sqrt{2}} \tag{6.88}$$

where V is the load on the bolt and A_{eq} is the cross-sectional area of the unthreaded shank, provided the shear plane or planes pass through the unthreaded part, or the tensile stress area of the bolt if any shear plane passes through the threaded part. The parameter n is the number of shear planes resisting the applied shear; σ_q is the yield stress of the bolts and the partial safety factors γ_m and γ_{f3} are both equal to 1.1 as given in Table 2(a) and Clause 4.3.3. BS 5400 Part 3 (1982), respectively.

The bearing capacity is given by Clause 14.5.3.6 BS 5400 Part 3 (1982) which states that σ_b, the bearing pressure between a fastener and each of the parts, given by $\sigma_b = V/A_{eb}$ must comply with the following condition:

$$\frac{V}{A_{eb}} \leqslant \frac{k_1 k_2 k_3 k_4 \sigma_y}{\gamma_m \gamma_{f3}} \tag{6.89}$$

where A_{eb} is the product of the shank diameter of the bolt and the thickness of each connected part loaded in the same direction, irrespective of the location of the thread, $k_1 = 1.0$ for HSFG bolts, k_2 varies with edge distance and is equal to 2.5 when the edge distance, measured from the centre of the hole, is greater than three times the hole diameter, k_3 depends on whether the part being checked is enclosed on both faces when $k_3 = 1.2$ or $k_3 = 0.95$ in all other cases, k_4 depends on bolt tension and is equal to 1.5 when fasteners are HSFG bolts acting in friction or equal to 1.0 for all other cases. Further details concerning this parameter are given by Needham (1984). σ_y is the yield stress of the bolt or plate, whichever is the lesser, $\gamma_m = 1.05$ Table 2(a) BS 5400 Part 3 (1982), and $\gamma_{f3} = 1.1$ Clause 4.3.3. BS 5400 Part 3 (1982).

Other considerations

Lack of fit

In a HSFG bolted connection good 'fit up' of the parent and cover plates is essential if a load bearing connection is to be achieved. If this is not the case, and the joint faces are not flat and/or not properly aligned, then some or possibly all of the prestress in the bolts will be used to bring the plates into contact with little or no bolt tension available to induce the friction resistance in the faying surfaces. CIRIA has published a report on the lack of fit in steel structures which offers valuable advice on how to circumvent such problems (CIRIA, 1981).

Relaxation of bolt tensions

Research has shown that the resistance to slip along the faying surfaces is a result of shearing of the contact interfaces of microscopic protrusions on the surface of the plates. This resistance can be enhanced with surface contamination and roughness. In a HSFG bolted joint these surface effects, together with a small reduction in joint plate thickness caused by in-plane plate tensile stresses due to the applied load, can cause a loss of bolt pretension. The loss of bolt tension due to plate thinning has been shown to increase rapidly when the in-plane tensile stress becomes high enough, which when combined with the normal bolt pressure, causes local yielding of the plate material (Cullimore and Eckhart, 1974). The loss of bolt tension from this cause will therefore increase with the ratio of bolt prestress to the yield stress of the parent plates, and with other parameters remaining constant, the slip factor will decrease as the bolt tension increases.

Slip factor

Clause 14.5.4.4. of BS 5400 Part 3 (1982) gives values of the slip factor μ which can be used when assessing the friction capacity of a HSFG bolted connection. Generally, for UK bridge construction, the faying surfaces are grit blasted and then masked until the erection of the steelwork, allowing the use of a slip factor μ equal to 0.5. As mentioned previously, the slip load of a connection will be improved by increasing the friction coefficient of the faying surfaces. If this is achieved by increasing the roughness of the faying surfaces this will generally result in offsetting the accompanying loss of bolt tension. Where difficulties arise in assessing a suitable value for μ the characteristic value of the slip factor should be determined in accordance with the procedure given in BS 4604 (1970).

6.5 Fatigue

6.5.1 Introduction

Fatigue is the mechanism caused by cyclic loading which permits crack growth in a member finally resulting in failure of the element. The process is strongly influenced by the magnitude of the applied stress range to which the element is subject to and also by the presence and acuity of stress concentrations occurring within the element.

6.5.2 Fatigue fracture surface

A fatigue fracture surface normally exhibits two distinct regions, typically a smooth flat area and a rougher area which forms the remaining part of the fracture cross-section as shown in Figure 6.75. The smooth region frequently exhibits concentric rings or 'beach marks' which surround the fracture point, and in addition radial lines which tend to point towards the fracture nucleus. The smooth area is the region of the fracture surface over which the fatigue crack grew, initially in a slow stable manner. The rougher region is the final rupture area through which the fatigue crack progressed rapidly in an unstable manner.

The relative size of the two regions gives a good indication of the stress level causing failure. If the rough area is large, compared to the smooth area, then the stress level was high indicating that only a small part of the overall cross-section, represented by the smooth region, could be lost due to the crack propagation. If the smooth area is large, relative to the rough area, then a large percentage of the cross-section could be lost due to propagation of the fatigue crack before failure occurred indicating a low stress level acting on the cross-section.

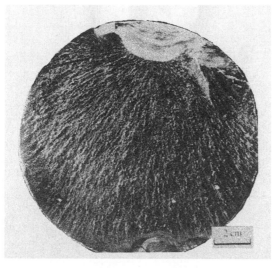

Figure 6.75 Typical fatigue failure in a steel component (ESDEP 1998).

6.5.3 Stress concentrations

In welded steel bridges the fatigue performance of the entire structure is usually governed by the fatigue strength of the individual welds. Fatigue cracks will initiate and grow from both load-bearing and non-load-bearing welds. This is because the welding process causes metallurgical discontinuities together with physical changes in shape, both of which cause local stress concentrations. Stress concentrations are also produced by notches, holes and any abrupt change of shape in a member which interrupts a smooth stress path. The stress concentration factor, which is used to define the magnitude of the stress concentration, is defined by the relationship $K_t = S_p/S_{net}$ where S_p is the peak stress and S_{net} the average stress on the net cross-section as shown in Figure 6.76.

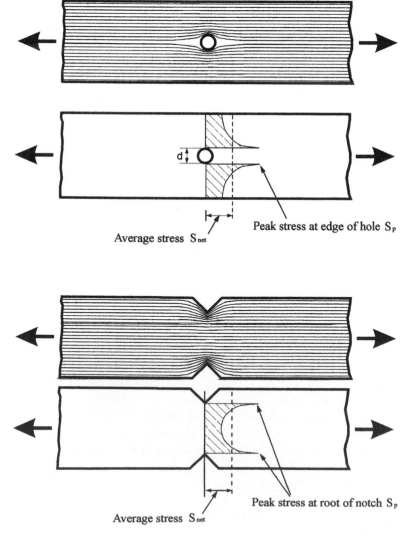

Figure 6.76 Stress concentrations (Gurney, 1979).

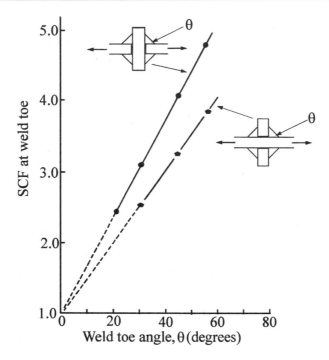

Figure 6.77 The influence of weld toe angle on the stress concentration at the weld toe, obtained using finite element analysis (Gurney, 1976).

Where stress concentrations arise from a sudden local change in shape it is not necessary for the change in shape to involve a reduction in the cross-sectional area of the member, an abrupt increase in the cross-sectional area can also induce stress concentrations.

For welded joints, where the weld is transverse to the direction of stress, the weld toe angle has a major influence on the stress concentration factor (SCF) with the SCF increasing as the toe angle increases (Figure 6.77). The weld size also has an influence on the SCF and in non load carrying joints the SCF tends to increase with increasing weld size, although for load carrying welds also transverse to the stress direction, there is a dramatic decrease in SCF with increase in weld size (Figure 6.78).

It is important to appreciate that fatigue cracks and failure can occur in members which are nominally in compression if they are welded to the structure or contain a welded detail. During the welding process shrinkage of the weld induces large tensile residual stresses into the weld regions. The large tensile residual stresses occurring in the weld vicinity must be added to the fluctuating compressive stress acting in the member resulting in a tensile stress variation in the welded regions causing crack propagation.

6.5.4 Fatigue assessment to BS 5400 Part 10

Part 10 of BS 5400 (1980) covers the method of fatigue assessments of parts of bridges which are subject to repeated fluctuations of stress. The Code of Practice gives

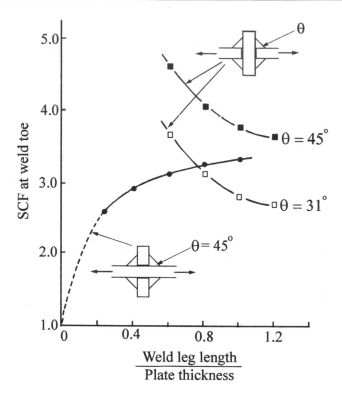

Figure 6.78 The influence of weld size on the stress concentration at the toe of transverse fillet welds (Gurney, 1976).

general guidance on the factors influencing fatigue behaviour and describes the loadings which have to be considered for fatigue assessment, the ensuing stress calculations and methods of calculating fatigue damage for a particular detail.

Loadings for fatigue assessment

In order to simplify and represent the wide spectrum of loads which cause fatigue damage to bridge structures a 'standard fatigue vehicle' is used to load the bridge in order to calculate the stress range occurring at a particular detail in the structure. The dimensions of the standard fatigue vehicle are shown in Figure 6.79.

The vehicle has four standard 80 kN axles and represents the most damaging group of commercial vehicles found on trunk roads in the United Kingdom. As every commercial vehicle which passes over a bridge causes a small proportion of the total fatigue damage, an assessment of the total number of commercial vehicles using the structure must be made. Table 1 of BS 5400 Part 10 (1980) reproduced here in Table 6.7, gives the annual flow of commercial vehicles in millions, for different road types and lane classifications. These values are used to sum the total damage occurring to a bridge structure.

Table 6.7 Annual flow of commercial vehicles ($n_c \times 10^6$) (BS 5400 Part 10 1980)

Category of road		Number of lanes per carriageway	Number of millions of vehicles per lane, per year (n_c)	
Type	Carriage layout		Each slow lane	Each adjacent lane
Motorway	Dual	3	2.0	1.5
Motorway	Dual	2		
All purpose	Dual	3	1.5	1.0
All purpose	Dual	2		
Slip road	Single	2		
All purpose	Single	3		
All purpose	Single (10 m*)	2	1.0	
Slip road	Single	1		Not applicable
All purpose	Single (7.3 m*)	2	0.5	Not applicable

*The number of vehicles in each lane of a single carriageway between 7.3 and 10 m wide should be obtained by linear interpolation.

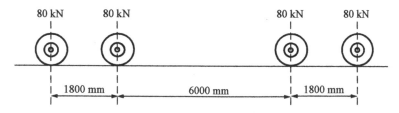

Axle arrangement of standard fatigue vehicle

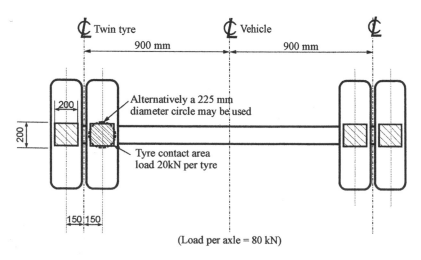

(Load per axle = 80 kN)

Figure 6.79 Standard fatigue vehicle with standard axle (BS 5400 Part 10 1980).

Application of loading

The standard fatigue vehicle is considered to be traversed along the slow lane and adjacent lanes only of the bridge. It is usual to bar commercial vehicles from travelling in the fast lane. Only relatively light vans and cars weighing less than 1500 kg should be permitted in this lane and these vehicles cause no fatigue damage to normal bridge structures. Only one standard fatigue vehicle is assumed to be on the structure at any one time and each of the loaded lanes should be traversed separately (Clause 7.2.3.5 BS 5400 Part 10). The fatigue vehicle is traversed along each of the loaded lanes so that the mean centre line of travel is along a path parallel to and within 300 mm of the lane centre line.

Stress calculation

The stress range resulting from the standard fatigue vehicle traversing each slow and adjacent lane in turn has to be calculated for each element being assessed. The stress range required for the assessment of a plate or element is the greatest algebraic difference between principal stresses occurring on principal planes not more than 45° apart in any one stress cycle. For welds, the stress range is the algebraic or vector difference between the greatest and least vector sum of stresses in any one stress cycle Figures 6.80 and 6.81 show how to calculate the principal stress in the parent metal adjacent to a potential crack location.

The maximum and minimum stress values required to determine the stress range at a particular point in the structure must be calculated using elastic theory taking into account all axial, bending and shearing stress resulting from the standard fatigue vehicle loading. No plastic redistribution is permitted. Additional effects which have to be included in the stress calculation are given in Clause 6.1.5 of Part 10 BS 5400 (1980). For the assessment of non-welded details if the stress range is entirely compressive the effects of fatigue may be ignored. If the non-welded detail is subject to stress reversal, with the stress fluctuating from tension to compression, the effective stress range to be used in the fatigue assessment should be obtained by adding 60%

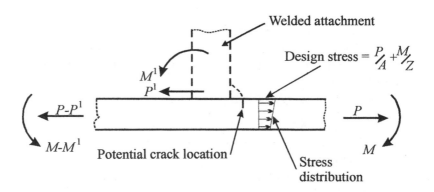

Figure 6.80 Reference stress in parent metal (BS 5400 Part 10 1980).

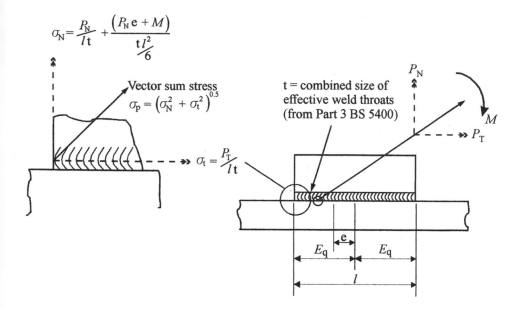

Figure 6.81 Reference stress in weld throat (BS 5400 Part 10 1980).

of the range from zero stress to maximum compressive stress to that part of the range from zero stress to maximum tensile stress.

Classification of details

In order to undertake a fatigue assessment of a construction detail it is necessary to classify the detail into a particular strength group. These strength groups have been obtained from constant amplitude fatigue tests undertaken on a wide range of samples containing different welded detail types. Table 17 of BS 5400 Part 10 (1980) gives a variety of construction details and indicates the prerequisites required to classify the relevant detail. The main features that affect the detail classification can be listed as follows:

(i)   the location of the anticipated crack initiation site which must be defined relative to the direction of the fluctuating stress
(ii) the geometrical arrangement and proportions of the detail. Important features are the weld shape, the size of component and its proximity relative to unwedded or free edges
(iii) the methods of manufacture and inspection.

There are several positions at which potential fatigue cracks may occur in welded details. The fatigue crack may initiate in either the throat of the weld or in the parent metal of the component parts joined together. In the latter case the fatigue crack can initiate at the toe of the weld, or at the end of the weld and even at a change in direction of the weld. The particular location depending on the direction of the fluctuating stress.

Classification of typical details

Figure 6.82 shows some of the most commonly used welded details in bridge structures and gives their classification details.

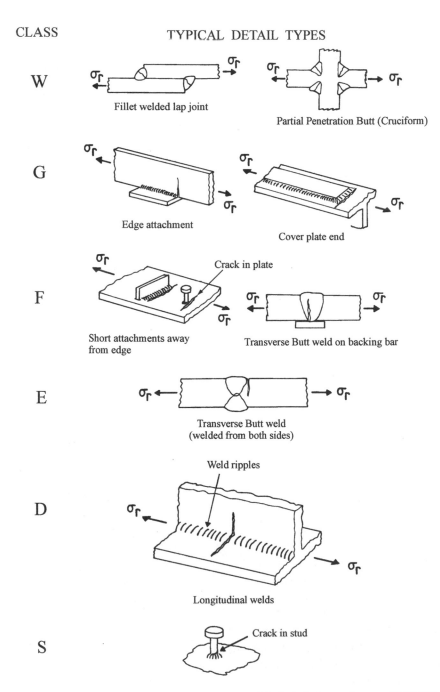

Figure 6.82 Typical details of weld arrangements with their fatigue classes (ESDEP, 1985).

Methods of assessment

Three methods of fatigue assessment are outlined in BS 5400 Part 10 (1980). The accuracy of the assessment increases with the complexity of the method chosen. Guidance will be given for the first two methods only.

Simplified procedure

This is the easiest method to adopt, but also the most conservative. It is the most suitable method for initial design purposes and may only be used if the detail being checked can be classified under the headings given in Table 17 of the Code. In addition, the design life for the bridge structure must be 120 years and the assumed fatigue loading is the standard load spectrum with the annual flow of commercial vehicles in accordance with Table 1, Part 10. Figure 6.83 gives the basic steps in the simplified assessment procedure.

Figure 8 of the Code shows the relationship between the limiting stress range σ_H and the span of the bridge for each detail classification and for four different road categories. The stress range occurring at the detail under assessment must be less than or equal to the limiting stress range σ_H given in the figure for the detail to have 120 year design life.

Single vehicle method

This procedure gives a more precise assessment than the method described previously. It may be used where the standard design life of 120 years or the annual flow given in Table 6.7 are not applicable. However, the procedure may only be used if the detail under assessment can be classified using Table 17 of the Code, but is not a Class S detail and the fatigue loading is the standard load spectrum. Figure 6.84 outlines the basic steps in this method of assessment.

Figure 9 of the Code used in this assessment procedure shows how to derive the stress range σ_v and the effective annual flow of commercial vehicles if required for the damage calculation. If the highest peak and lowest trough occur with the vehicles in the same lane, Case 1 of Figure 9 can be used in conjunction with Table 1 of the Code to determine the number of millions of vehicles per lane per year.

Figure 10 of BS 5400 Part 10 gives the relationship between the stress range σ_v and the lifetime damage factor for one million cycles per annum for 120 years. The relationship gives an assessment of the cumulative fatigue damage caused by the design load spectrum represented by the passage of a standard fatigue vehicle. The damage chart of Figure 10 (BS 5400 Part 10) is calculated assuming an influence line base length of 25 m. To allow for influence line base lengths of less than 25m and for the effects of multiple vehicles the adjustment factor K_F is obtained from Figure 11 (BS 5400 Part 10). To allow for the effects of the number of vehicles other than the 120 million assumed the lifetime damage factor d_{120} (Figure 10 BS 5400 Part 10) is multiplied by n_c the number of vehicles in millions per year traversing the relevant lane of the bridge.

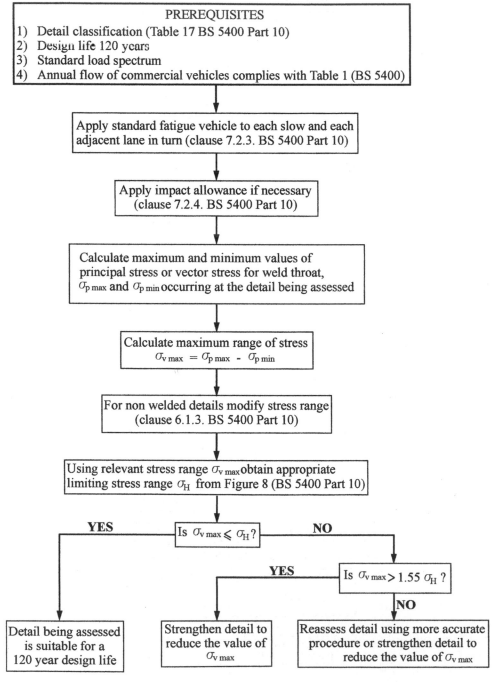

Figure 6.83 Flow chart for the simple assessment procedure to comply with BS 5400 Part 10 1980.

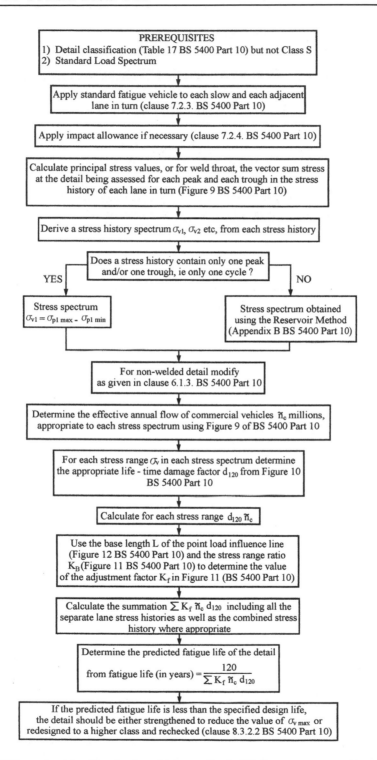

Figure 6.84 Flow chart for the single vehicle method of fatigue assessment to comply with BS 5400 Part 10 1980.

6.5.5 General design requirements

The effects of fatigue damage can be minimized by careful design of the structure. The following points should be given special consideration:

(i)   Avoid physical discontinuities and abrupt change in rigidity in the structure.
(ii) Design weld details on the basis of providing the easiest path for stress flow through the weld.
(iii) Where possible use full strength penetration butt welds in preference to load-carrying fillet welds. If load-carrying fillet welds are unavoidable make the weld size large enough to ensure that failure does not occur through the weld material.
(iv) Minimize the number and severity of joints. As far as possible welded joints should be positioned at points of low fatigue stress.
(v)   Specify automatic shop welding in preference to manual shop welding and site welding.

To ensure that the principles previously described are applied during construction, it is very important that the designer communicates effectively with the fabricator. This is to make sure that the fabricator understands the reasons for the chosen design details, which if made differently from that specified, may result in unacceptably low fatigue strengths.

Bibliography

Design rules

British Standard BS 4395, *Specification for High Strength Friction Grip Bolts and Associated Nuts and Washers for Structural Engineering*. British Standards Institution, London, 1969.
British Standard BS 4604, *Specification for the Use of High Strength Friction Grip Bolts in Structural Steelwork*. British Standards Institution, London, 1970.
British Standard BS 5400, Part 10, *Steel, Concrete and Composite Bridges, Code of Practice for Fatigue*. British Standards Institution. British Standards Institution, London, 1980.
British Standards Institution. BS 5400, Part 3, *Steel, Concrete and Composite Bridges, Code of Practice for the Design of Steel Bridges*. British Standards Institution, London, 1982.
Dowling PJ, Harding JE and Bjorhovde R (Eds). *Constructional Steel Design – An International Guide*. Elsevier Applied Science Publishers, London, 1992.
Eurocode 3, Design of steel structures, ENV1993-1-1: Part 1.1, General rules and rules for buildings, CEN,1992.
Eurocode 3, Design of steel structures, ENV1993-1-5: Part 1.5, General rules: Supplementary rules for planar plated structures without transverse loading, CEN,1997.
Eurocode 3, Design of steel structures, ENV1993-2, Steel bridges, CEN,1997.
European Recommendations for the Design of Longitudinally Stiffened Webs and Stiffened Compression Flanges, Publication 60, ECCS, 1990.
Galambos TV (Ed.). *Guide to Stability Design Criteria for Metal Structures*, 4th edn. John Wiley, New York, 1988.
Wolchuk R and Mayrbourl RM. Proposed design specification for steel box girder bridges, Rep. No. FHWA-TS 80-205, US Department of Transportation, Federal Highway Administration, Washington, DC, 1980.

Trusses

Chen W and Duan L. *Bridge Engineering Handbook*. CRC Press, London, 1999.
Engesser F. Die Sicherung offener Brücken gegen Ausknicken. *Centralbl Bauverwaltung*, 415, 1884; 93, 1885.

ESDEP. *European Steel Design Education Programme*. The Steel Construction Institute, UK, 1985
Hu LS. The instability of top chords of pony truss bridges. Dissertation, University of Michigan, Ann Arbor, MI, 1952.

Plate behaviour

Brush DO and Almroth BO. *Buckling of Bars, Plates and Shells*. McGraw-Hill, New York, 1975.
Bulson PS. *The Stability of Flat Plates*. Chatto and Windus, London, 1970.
Dowling PJ, Harding JE and Agelides N. *Collapse of Box Girder Stiffened Webs. Instability and Plastic Collapse of Steel Structures*. Granada Publishing, London, 1983.
Dwight JB and Little GH. Stiffened steel compression flanges – a simpler approach. *Structural Engineer*, 54, 1976.
Harding JE. The interaction of imperfection wavelength and buckling mode in plated structures. *Schweizer Ingenieur und Architect.*, No 1–2, January 1985.
Harding JE. Non-linear analysis of components of steel plated structures. *1st European Conference on Steel Structures. Eurosteel '95*, Athens, May, 1995.
Harding JE and Hobbs RE. The ultimate behaviour of box girder web panels. *The Structural Engineer*, 57B, 3, September, 1979.
Harding JE, Hobbs RE and Neal BG. The elastic-plastic analysis of imperfect square plates under in-plane loading. *Proc. Instn. Civ. Engrs*, Part 2, 63, March, 1977
Jetteur P. *et al*. Interaction of shear lag with plate buckling in longitudinally stiffened compression flanges. *Acta Technica CSAV*, 3, 1984.
Naryanan R (Ed.). *Plated Structures: Stability and Strength*. Applied Science Publishers, London, 1983.
Structural Stability Research Council. *Stability of Metal Structures, A World View*, 2nd edn. Structural Stability Research Council, USA, 1991.
Szilard R. *Theory and Analysis of Plates*. Prentice-Hall, Englewood Cliffs, New Jersey, 1974.
Timoshenko S and Winowski-Krieger S. *Theory of Plates and Shells*. McGraw-Hill, 1959.

Plate and box girder design

Basler K. Strength of plate girders in shear. *ASCE J. Struct. Div.*, 87, 1961.
Cooper PB. Strength of longitudinally stiffened plate girders. *Proc ASCE J. of Struct. Div.*, ST2, 1967.
Dalton DC and Richmond B. Twisting of thin walled box girders. *Proceedings of the Institution of Civil Engineers*, January, 1968.
Dowling PJ and Chatterjee S. Design of box girder compression flanges. *2nd Int. Colloq. Stab., European Convention for Constructional Steelwork*, Brussels, 1997.
Dubas P and Gehri E. Behaviour and design of steel plated structures. Technical Committee 8 Group 8.3, ECCS-CECM-EKS, No 44, 1986.
Evans HR and Tang KH. The influence of longitudinal web stiffeners upon the collapse behaviour of plate girders. *Journal of Constructional Steel Research*, 4, 1984.
Harding JE and Dowling PJ. The basis of the new proposed design rules for the strength of web plates and other panels subject to complex edge loading. *Stability Problems in Engineering Structures and Components*, Applied Science Publishers Ltd., 1978.
Harding JE, Hindi W and Rahal K. 'The Analysis and Design of Stiffened Plate Bridge Components. SSRC Annual Technical Session, St Louis, USA, April, 1990.
Hindi W and Harding J E. Behaviour of transverse stiffeners of orthogonally stiffened compression flanges. *First World Conference on Constructional Steel Design*, Acapulco, Mexico, December 1992.
Horne MR, CIRIA Guide 3, Structural action in steel box girders, Construction Industry Research and Information Association, London, 1977.
Inquiry into the basis of design and method of erection of steel box girder bridges, Report of the Committee and Appendices, HMSO, London, 1973.

Isles DC. Design guide for composite box girder bridges. The Steel Construction Institute, Ascot, 1994.

Jetteur P. A new design method for stiffened compression flange of box girders. In *Thin Walled Structures*, Granada, London, 1983.

Moffatt KR and Dowling PJ. Shear lag in steel box girder bridges. *Structural Engineering*, 53, 1975.

Naryanan R and Rockey KC. Ultimate load capacity of plate girders with webs containing circular cut-outs. *Proc. Inst. Civ. Engrs*, Part 2, 1981.

Porter DM, Rockey KC and Evans HR. The collapse behaviour of plate girders in shear. *Structural Engineer*, 53, 1975.

Rahal K and Harding JE. Transversely stiffened girder webs subjected to shear loading in Part 1: Behaviour. *Proc. Instn Civil Eng*, Part 2,89, March, 1990.

Rahal K and Harding JE Transversely stiffened girder webs subjected to shear loading in Part 2: Stiffener Design. *Proc. Instn Civil Eng*, Part 2,89, March, 1990.

Roberts TM and Rockey KC. A mechanism solution for predicting the collapse loads of slender plate girders when subjected to in-plane patch loading. *Proc. Inst. Civ. Engrs*, Part 2, 1979.

Rahal K and Harding JE. Transversely stiffened girder webs subject to combined inplane loading. *Proc Instn Civil Eng*, Part 2, 91, June, 1991.

Rahal K and Harding J E Design of transverse web stiffeners in plate and box girder bridges. *First World Conference on Constructional Steel Design*, Acapulco, Mexico, December 1992.

Roberts TM. Patch loading on plate girders. In *Plated Structures: Stability and Strength* (Ed. R Naryanan). Applied Science, London, 1983.

Rockey KC and Evans HR (Eds). Design of steel bridges. *Proc. Int. Conf.*, *University College, Cardiff*. Granada, London,1980.

Institution of Civil Engineers. Steel box girder bridges. *Proc. Int. Conf.*, Institution of Civil Engineers, London,1973.

Connections

Blodgett OS. *Design of Welded Structures*. James F Lincoln Arc Welding Foundation, Cleveland, 1966.

CIRIA. Lack of fit in steel structures, Report 87, 1981.

Cullimore MSG and Eckhart JB. The distribution of clamping pressure on friction grip bolted joints. *The Structural Engineer*, 52, 5, 1974.

Needham FH. Site Connections to BS 5400 Part3. *The Structural Engineer*, 61A, 3, 1983.

Needham FH. Discussion – site connections to BS 5400 Part3. *The Structural Engineer*, 62A, 3, 93–100, 1984.

Owens GW and Cheal BD. Structural Steelwork Connections Butterworths, London, 1989.

Pratt JL. Introduction to the welding of structural steelwork. The Steel Construction Institute, Ascot, Berks, UK, 1989.

Fatigue

ESDEP. European Steel Design Education Programme. The Steel Construction Institute, UK, 1998.

Gurney TR. Finite element analysis of some joints with the welds transverse to the direction of stress. *Weld. Res. Int.*, 6, 4, 40–72, 1976.

Gurney T. *Fatigue of Welded Structures*. Cambridge University Press, Cambridge, 1979.

Gurney T. *Fatigue of Steel Bridge Decks*. HMSO, London, 1992.

Maddox SJ. *Fatigue Strength of Welded Structures*. Abington Publishing, 1991.

7 Composite construction

D. COLLINGS

7.1 Introduction

Composite bridges are structures that combine materials like steel concrete, timber or masonry in any combination. In common usage today composite construction is normally taken to mean either steel and *in situ* concrete construction or precast concrete and *in situ* concrete bridges. Composites are also a term used to describe modern materials such as glass or carbon reinforced plastics, etc. These materials are becoming more common but are beyond the scope of this chapter.

Composite structures are a common and economical form of construction used in a wide variety of structural types. This chapter initially reviews the forms of structure in which composite construction is used, then each of the more common forms of composite construction is considered in more detail. Compliance with codes (British Standards Institute, 1978, 1982a, b, 1992; Eurocode; AASHTO, 1996) and regulations is necessary in the design of a structure but is not sufficient for the design of an efficient, elegant and economic structure. An understanding of the behaviour, what physically happens and how failure occurs is vital to any designer. Without this understanding the mathematical equations are a meaningless set of abstract equations. One aim of this chapter is to give an understanding of the behaviour of composite structures.

7.2 Materials

The behaviour of the composite structure is heavily influenced by the properties of its component materials. A brief summary of the key properties particularly relevant to composite action is summarized in Table 7.1, however, the reader wanting to understand composite bridges should first have a good understanding of the properties and design methods for the individual materials as set out in other chapters of this book. In particular the reader should note the differences between materials as it is the exploitation of these different properties that makes composite construction economic. For example, the use of a concrete slab on a steel girder uses the strength of concrete in compression and the high tensile strength of steel to overall advantage.

Table 7.1 Comparison of material properties

Material	Strength	Elastic modulus	Other properties
Steel	$f_y = 250\text{–}450 \text{ N/mm}^2$ similar in tension and compression	$E = 200 \text{ kN/mm}^2$	Tensile range may be affected by fatigue and compression range by buckling
Concrete	$f_{cu} = 30\text{–}60 \text{ N/mm}^2$ $f_{ct} = 36 \text{ N/mm}^2$	$E = 30 \text{ kN/mm}^2$	Creep and shrinkage effectively lower the elastic modulus over time.
Masonry			
Timber	$f_t = 12 \text{ N/mm}^2$ $fc = 10 \text{ N/mm}^2$	$E_{long} = 10 \text{ kN/mm}^2$ $E_{trans} = 1 \text{ kN/mm}^2$	Creep and shrinkage effectively lower the elastic modulus over time.

7.3 Basic concepts

There are two primary effects to consider when looking at the basic behaviour of a composite structure:

- the differences between the materials
- the connection of the two materials.

7.3.1 The modular ratio

Differences between the strength and stiffness of the materials acting compositely affect the distribution of load in the structure. Stronger, stiffer materials like steel attract proportionally more load than materials such as concrete or timber. In order to take such differences into account it is common practice to transform the properties of one material into that of another by the use of the modular ratio.

At working or serviceability loads the structure is likely to be within the elastic limit and the modular ratio is the ratio of the elastic modulus of the materials. For a steel–concrete composite the modular ratio is:

$$m = E_s/E_c \tag{7.1}$$

The value of this ratio varies from 7 to 15 depending on whether the short-term or long-term creep affected properties of concrete are utilized.

At ultimate loads the modular ratio is the ratio of the material strengths. This ratio is dependent on the grade of steel and concrete utilized. For design, the different material factors need to be considered to ensure a safe structure. For a steel concrete composite:

$$m_u = 0.95f_{fy}/0.4f_{cu} \tag{7.2}$$

7.3.2 Interface connection

The connection of the two parts of the composite structure is of vital importance. If there is no connection then the two parts will behave independently. If adequately connected the two parts act as one whole structure, potentially greatly increasing the structure's efficiency.

Imagine a small bridge consisting of two timber planks placed one on another, spanning a small stream. If the interface between the two were smooth and no connecting devices were provided the planks would act independently. There would be significant movement at the interface and each plank would, for all practical purposes, carry its own weight and half of the imposed loads. If the planks were subsequently nailed together such that there could be no movement at the interface between them then the two parts would act compositely and the structure would have an increased section for resisting the loads and could carry about twice the load of the non composite planks. The deflections of the composite structure would also be smaller by a factor of approximately four, the composite whole being substantially stronger and stiffer than the sum of the parts. A large part of the criteria in the following sections is aimed at ensuring this connection between parts is adequate.

7.4 Structural forms

Most common composite structures are either precast prestressed concrete beams with an *in situ* concrete slab or steel girders with a concrete slab. Composite structures are very versatile and can be used for a considerable range of structures from foundations, substructures and superstructures through a range of forms from beams, columns, towers and arches and for a diverse range of bridge structures from tunnels, viaducts, elegant footbridges and major cable stayed bridges.

7.4.1 Foundations

Piles formed from driven cylindrical steel casings filled with concrete are used for carrying loads into the ground. Pile sizes range from micro piles 200–300 mm in diameter which use high strength steel pipe and cement grout through to large 2 m diameter caissons formed from a cylindrical steel shell with a concrete core (Biddle). More commonly, composite piles are formed from concrete bored piles with a steel beam embedded within it. This latter form of pile is often used in contiguous or secant pile retaining walls subject to relatively large bending moments from the retained soil (Hubbard).

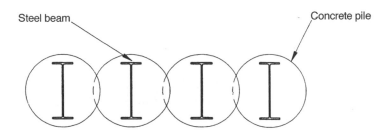

Figure 7.1 Composite steel–concrete secant pile wall.

7.4.2 Substructures

Where high compression loads are required to be carried, a composite steel–concrete column may be economic (Kerensky and Dallard, 1968), particularly where the size needs to be minimized. For the support of small or medium span structures, a concrete-filled rolled hollow section (to 500 mm) or pipe section (to 2200 mm) may be used.

7.4.3 Towers

For the towers of cable stay bridges or suspension bridges, an outer fabricated stiffened steel plate structure with a concrete core could be used (Aparicio and Casas, 1997) or a fabricated steel core surrounded by reinforced concrete (Figure 7.2). The force regime in the towers of cable stay bridges is extremely complex with the stay anchorage's imposing both horizontal and vertical loads. A fabricated steel core carrying the horizontal tensile component and local anchorage loads with shear connectors transferring load to an outer concrete skin primarily carrying global compression and bending provides an economical form and has been used on many of the worlds largest structures (Virlogeux *et al.*, 1994). Where an outer steel structure is used it serves a dual function of carrying the loads induced by bending and axial loads as well as acting as a permanent form work system.

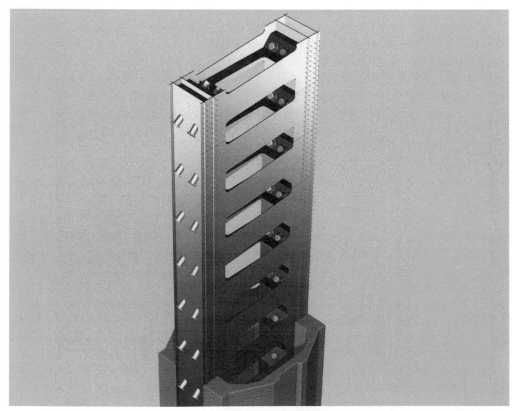

Figure 7.2 Steel–concrete transfer structure at tower cable anchorage of a cable stay structure.

7.4.4 Tunnels

The requirement for water tightness makes steel–concrete composite structures ideal for immersed tube tunnels subject to significant hydrostatic pressures, an impermeable non structural steel skin protects the main structural concrete core. Recently, research has been carried out on the use of sandwich plate construction (Narayanan *et al.*) where steel plates with a series of shear connectors form a shutter lining and provide an impermeable skin. The connectors ensure the steel skin and concrete act structurally reducing the amount of conventional reinforcing bar required (Figure 7.3).

7.4.5 Simply supported beam and slab bridges

For modest simply supported spans, a composite beam and slab structure can give an optimal structural solution. The structure may be formed from a number of possible material combinations:

Precast concrete beams

A bridge made of precast prestressed concrete beams forming the web and tensile flange with an *in situ* top slab constituting the compression flange uses these two types of concrete efficiently. This kind of structure usually avoids the extensive falsework

Figure 7.3 Detail of a composite sandwich panel construction for an immersed tube tunnel.

normally associated with *in situ* concrete construction. For short spans in the 3–15 m range, the beams are often encased in concrete to form a slab. While this form may be relatively heavy it is comparatively simple to construct and the large contact area between the beams and concrete infill ensures low stresses at the interface. Hence additional reinforcement is not required. However, the most common method is to place a thin *in situ* slab over the beams, this form of construction is useful for spans up to approximately 30 m. Beyond 40 m the weight of the beams means cranage costs will be relatively significant and this form of construction is far less economic.

Steel beams

For small spans – often the transverse beams of railway bridges – filler joist construction may be used. Here the beam is encased entirely in concrete (which is a relatively inefficient in use of material in bending but is simple to construct) and again the stresses at the interface between steel and concrete are relatively low such that no shear connectors are required for full composite action.

A series of rolled steel beams forming the webs and tensile flanges, with a concrete top slab forming the compression flange, utilizes the properties of the two different materials in a much more efficient manner on simply supported spans of 10–25 m. Above 25 m the symmetrical rolled section is usually replaced by a purpose fabricated asymmetric plate girder.

7.4.6 Integral bridges

Integral bridges, where the superstructure is connected to the sub structure, are preferable to those using joints and bearings. The integral bridge is more robust particularly with regard accidental impact, and requires less maintenance of expansion joints and bearings. Essentially, the design of the composite integral structure is similar to a conventional bridge but the effects of soil pressures and movements need to be considered more carefully (Hambley, 1997).

7.4.7 Continuous beams

The use of a continuous beam (Figure 7.4) is often preferable to a series of simply supported spans as the number of bearings and expansion joints is reduced. Both are potential long-term maintenance liabilities. However, in this form of structure the continuity moments over the supports are of opposite sign to those at mid span and the composite section is not as structurally efficient as the concrete slab is now in the tensile zone. For composite precast prestressed beams, heavily reinforced *in situ* concrete stitch sections are required to create the continuity. For composite steel and concrete construction, cracking of the top slab will occur causing some redistribution of moment unless prestress is added to counteract this (this is common in many European bridges).

7.4.8 Box girders

For longer spans, or for bridges with significant horizontal curvature, or simply for aesthetic reasons, steel–concrete composite box girders may be used. The boxes may

Figure 7.4 Continuous steel–concrete bridge spanning the M5 motorway (Collings, 1994).

be complete steel boxes (with torsional stiffness) and an overlay slab (Dickson), or an open (often trapezoidal) box (Figure 7.5) where the concrete slab closes the top of the box. The use of the open steel box section allows the reintroduction of a bottom concrete compression flange at hogging moment regions by in-filling over supports giving a doubly composite section. In a number of large structures constructed segmentally in short sections both top and bottom flanges are precast concrete with the web a fabricated steel section (Lecroq, 1988). Folded plate webs have also been used with this form of structure to reduce the amount of stiffening required in the steel web, as have webs formed of trusses.

7.4.9 Trusses

In its simplest form a composite truss may consist simply of a steel truss with a concrete top slab forming the main compression chord (Steel Construction Institute, 1992). However, other forms utilizing composite members in the truss are also used. Steel sections embedded in a concrete member (Cracknell, 1963) or a tubular steel section filled with concrete may be used. For both forms, the behaviour and design of the joints is a key issue in design. Trusses usually have a relatively low span to depth ratio (10–16 as opposed to the 20–30 normal for beams) and are relatively stiff. They are more common for rail bridges where the live loads are relatively large.

7.4.10 Arches

The arch form is ideal for a material like masonry or concrete whose strength is primarily compressive. However, the requirement for extensive centring, falsework or other temporary works often makes this form of structure uneconomic. Steel

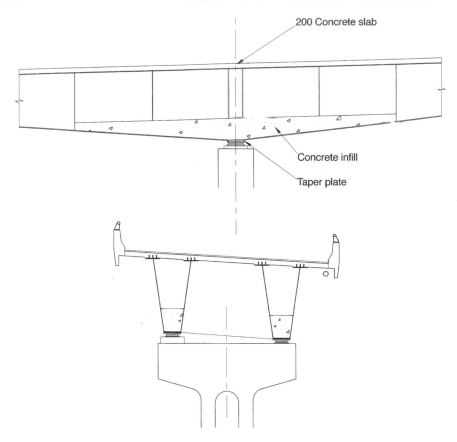

Figure 7.5 Trapezoidal box section near a support showing a doubly composite section (details courtesy Robert Benaim & Associates).

Figure 7.6 Falsework supporting an in situ concrete cantilever slab on a steel–concrete girder bridge.

arches reduce the temporary works but require extensive bracing to prevent buckling instability, thereby increasing their material content. The use of a composite structure with a steel framework being erected first and the concrete being cast around
(or inside) it provides a structure utilizing the properties of concrete and steel to maximum advantage. Composite arches of over 400 m have been constructed using this
technique (Zuou and Zhu, 1997).

7.4.11 Slabs

Often *in situ* slabs are utilized in composite structures requiring significant formwork
(Figure 7.6) or falsework, particularly at the edge where the slab cantilevers beyond
the beams.

The use of composite slabs using either steel or concrete permanent formwork
(Figure 7.7) often provides a convenient solution requiring no falsework with the flexibility of monolithic *in situ* construction. In the UK concrete permanent form work
is commonly used on bridges being constructed over or adjacent to major roads, railways or rivers where access to the soffit is limited (Dickson; Collings, 1994).

Occasionally it is possible to cast the slab onto the steelwork prior to erection
(Institute of Civil Engineers, 1997). However, other forms of construction are also
popular. Precast slabs are economic in major viaduct and cable stay structures (Ito
et al., 1991) where otherwise large areas of *in situ* concrete would be on the critical
path of construction. The precast elements are specially fabricated for the individual bridge to suit its geometry. The connection between precast elements and the
beams of the bridge needs to be carefully designed and detailed to ensure composite action. This is particularly important in cable stay bridges where the slab not only
carries traffic loads but also carries a significant proportion of the high longitudinal
compression induced by the stay cables.

There has been extensive research into the behaviour of slabs on composite bridges
(Kirkpatric *et al.*, 1986; Csagoly and Lybas, 1989). This research indicates that for

Figure 7.7 Permanent formwork on a steel girder flange.

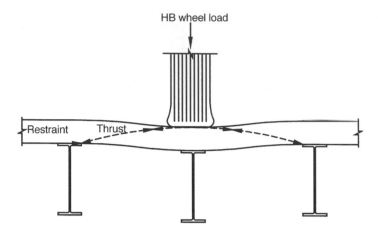

Figure 7.8 Arching action in a concrete slab.

restrained internal panels the slab carries the load primarily by arching action (Figure 7.8) with the slab behaving as a membrane rather than a flexural element. The use of membrane action for internal restrained slab panels can lead to significant reductions in reinforcement quantities.

7.5 Precast concrete composites

Concrete to concrete precast composite construction is a relatively good point to start in looking at composite action as the elastic moduli and strengths of the two composite elements are similar and the complications of modular ratios can often be ignored.

Consider the behaviour of a simply supported precast prestressed beam with an *in situ* concrete slab cast on top. Initially the beam will be subject only to its own self-weight and the initial prestress. When the slab is placed the stresses in the beam change, with the compression at the bottom of the beam reducing. As other loads such as surfacing and live loads are added the fully composite section resists the load. A compression develops in the slab with the compression at the bottom of the beam reducing further. At these normal working loads the concrete is uncracked and the shear stresses at the beam slab interface are low. There is no relative slip between the slab and beam. The stresses in the section are the sum of the stresses induced by each stage of construction. The sum of these stresses should be less than the allowable limiting value at the stage considered.

$$f_n > M_1/Z_1 + M_2/Z_2 + M_n/Z_n \qquad (7.3)$$

The force transfer at the interface for composite sections is related to the rate of change of the force in the slab (usually simply the rate of change in moment). At the serviceability limit state the shear flow q_n at stage n is:

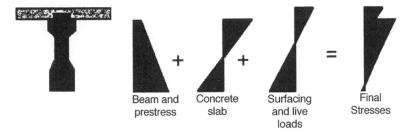

Beam and   Concrete   Surfacing   Final
prestress   slab   and live   Stresses
                     loads

Figure 7.9 Build up of stresses at various stages of construction – precast concrete composite construction.

$$q_n = (\delta M_n/\delta x) \int ty\,\delta y \qquad (7.4)$$

or:

$$q_n = VAy/I_n \qquad (7.5)$$

From Equation (7.5) it can be seen that the longitudinal shear is a function of the vertical shear force. For uniform loads the longitudinal shear will vary linearly with the maximum at the support and a zero requirement at mid span. For concentrated loads, particularly those acting near supports, the local shear flows may be large (Figure 7.10).

As more load is applied to the structure the force in the slab increases and the stress in the lower part of the beam drops until, at a tensile stress of approximately $0.1f_{cu}$, the concrete will crack. The extent of cracking will increase with load until at the ultimate limit the steel in the beam (prestressing steel and any other reinforcement) is at yield. At this point the moment of resistance is:

$$M_{us} = (zA_sf_y)/\gamma \qquad (7.6)$$

The concrete in the slab may also be approaching failure. Thus in order to ensure a ductile failure the section should be designed such that $M_{us} < M_{uc}$:

$$M_{uc} = (zA_cf_{cu})/\gamma \qquad (7.7)$$

As $A_c = bt$ at the ultimate limit state, the maximum change in force in the slab over a length from the support to mid span is:

$$\Delta F = 0.4f_{cu}bt \qquad (7.8)$$

and the shear flow at this stage is:

$$q_u = (2\Delta F)/L = (0.8f_{cu}bt)/L \qquad (7.9)$$

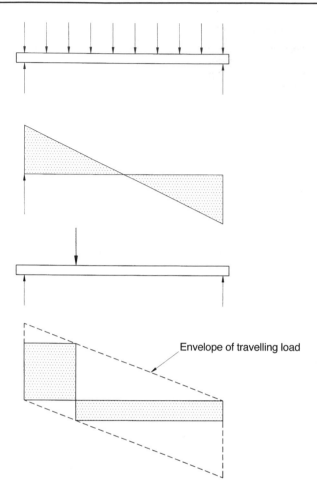

Figure 7.10 Shear diagrams for uniform and concentrated loads.

If the section is capable of significant plastic deformation, the shear flow may be considered to be uniform. However, for most sections it should generally follow the shape of the shear diagram. Typically codes allow a 10–20% variation from the elastic shear distribution. At the ultimate limit state the longitudinal shear plane (the interface between beam and slab) must have sufficient capacity to resist the shear flow. The capacity of the shear plane may be estimated as the lesser of:

$$q_p = vL_s + 0.7A_e f_y \qquad (7.10)$$

or:

$$q_p = KL_s f_{cy} \qquad (7.11)$$

Table 7.2 Ultimate interface shear stresses and coefficient k for composite members

Type of shear plane	Longitudinal shear stress v_l for 40 N/mm² concrete	Interface coefficient, k
Monolithic construction	1.25 N/mm²	0.15
Prepared surface	0.80 N/mm²	0.15
As cast surface	0.50 N/mm²	0.09

From the above behaviour it should be noted that for the composite section to function two key criteria should be met:

- that at working loads slip should not occur
- at the ultimate condition the interface must have sufficient strength.

For design purposes a simplified approach is adopted by current codes. The section is designed at the ultimate limit state ignoring creep, shrinkage and other secondary effects. However, the design shear flow used is the elastic value given by Equation (7.5), using the ultimate shear force. To ensure interface slip is small the values of v and k in Equations (7.10) and (7.11) are relatively conservative. The coefficients are dependent on the roughness of the interface (Table 7.2). To ensure a robust structure and that there is not a pulling apart of the components a minimum area of reinforcement A_e of 0.15% is normally recommended to pass through the shear plane.

The vertical shear capacity of the section at the ultimate limit state should be larger than the flexural capacity (again to ensure a ductile failure). The shear capacity of the beam may be conservatively assumed to be carried entirely by the precast beam assuming it to be cracked. If there is significant prestress and the section uncracked a method superimposing the principal stresses at the composite beam neutral axis may be used to derive a less conservative shear capacity for the composite section (Clarke, 1980).

7.5.1 Creep and shrinkage

The above example, a simply supported span with a relatively compact section, neglected the effects of creep and shrinkage in the concrete. In some cases, particularly slender construction or some continuous beams, differential creep and shrinkage between beam and slab has a significant effect and should not be neglected.

Shrinkage

The shrinkage deformation of concrete is dependent on the environment (particularly the humidity), the composition of the concrete (the water–cement ratio and cement content), the size of the member, the amount of reinforcement and the age of the concrete. Typical variations of shrinkage with time are shown in Figure 7.11 with the range indicating the difference between a dry and a moist environment. For concrete to concrete composite sections the amount of differential shrinkage is important. The slab is cast after the beam and has a different rate of shrinkage impos-

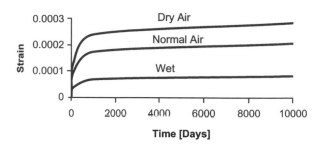

Figure 7.11 Typical shrinkage curve for concrete (British Standards Institute, 1978, 1982a, b, 1992).

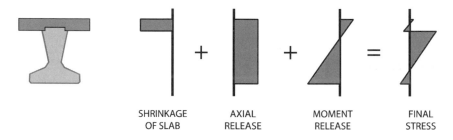

Figure 7.12 Stresses induced in the composite section due to shrinkage.

ing a force across the interface (Figure 7.12). Because the slab is not at the neutral axis of the section a moment is induced. These additional forces and moments may need to be considered in design.

Creep
Creep is a time-dependent deformation of the concrete that like shrinkage is dependent on environmental conditions, the composition of the concrete and the thickness of the section. It is also dependent on the concrete maturity at loading and the levels of applied stress or strain. A typical variation of strain with time under a 1 or 5 N/mm^2 stress is shown in Figure 7.13. The effect of creep is significant as it can reduce the effect of the shrinkage strains and redistribute load from the beam to the composite section.

$$M = (1/(1 + \phi))M_{beam} + (\phi/(1 + \phi))M_{composites} \qquad (7.12)$$

where ϕ is a creep coefficient which varies from approximately 1 to 3. Where the effects of creep are large, eventually the composite section will carry all of the applied loads.

7.5.2 Continuous construction
For simply supported structures the effects of creep and shrinkage are such that often they may be neglected. For continuous concrete composite structures the effects of shrinkage and creep are often significant and need to be considered in more detail.

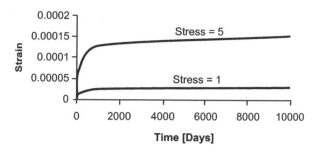

Figure 7.13 Typical creep strain curve for concrete (British Standards Institute, 1978, 1982a, b, 1992).

Methods of forming a continuous bridge from precast beams vary considerably. Three common types are shown in Figure 7.14. The method in Figure 7.14(b) provides continuity of the slab only, improving durability by removing the expansion joint. Many spans can be joined together forming a long continuous deck. Design of the beam is essentially as a simply supported span. The only additional complexity is in the design of the slab. At the joint the slab must be designed to carry normal live loads, longitudinal tensions or compressions from the bearings and accommodate the rotational movements of the two adjacent spans. This is achieved by ensuring the slab is debonded from the beam over a sufficiently large length (usually about five times the slab depth).

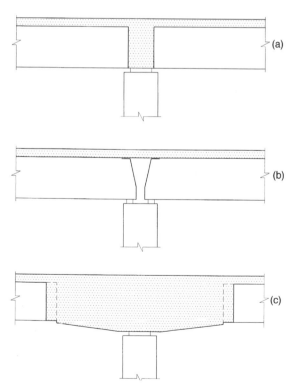

Figure 7.14 Continuity joints in precast beam bridges.

By placing a reinforced concrete diaphragm at the supports as shown in Figure 7.14(a) the precast elements can be made continuous, avoiding joints and reducing the number of bearings. This form of structure would normally be constructed by placing the beams on temporary bearings and then casting the top slab on the simply supported beams. The diaphragm would then be cast forming continuity for live loads. At normal working loads the diaphragm is subject to a hogging moment from live loads, as is the part of the precast beam adjacent to it. The design and detailing of the precast beam is complicated by the reversals of moment, often leading to the need for extensive debonding or deviation of the prestressing wires, increasing the cost of the unit. The precast beams are prestressed to counteract the main dead and live loads, consequently the stress diagram is likely to be more compressive on the lower flange. The creep effect will be larger here and the result is a tendency for the beam to curve upwards with time, causing a long term sagging effect at the diaphragm that requires reinforcement in the bottom of the diaphragm. At the ultimate limit state there is likely to be significant cracking at the diaphragm and a redistribution of moment from the support to mid-span. The bridge capacity may not be significantly greater than that of a simply supported bridge, and consequently this type of structure is rarely economic.

The use of a larger *in situ* concrete section at the support, with the precast beam starting near the point of contra-flexure as shown in Figure 7.14(c) may aid simplification of beam design as the reversal of moments will be lower. The potential span of the bridge is also increased. However, the increased *in situ* concrete means falsework is required as is a significant temporary support for the precast beam. Considerable care is required in the design of the connection between *in situ* and precast elements as the shear transfer from the edge beam can be particularly problematic.

Construction of precast concrete composite bridges is usually carried out by crane erecting beams. Using permanent formwork, an *in situ* slab is then constructed. For larger structures with multiple spans, the use of a purpose-built gantry is often the quickest method of erection.

7.6 Steel–concrete composite beams

The properties of steel and concrete are very different and an understanding of the behaviour of the structure and the interface between materials is vital.

Consider the idealized behaviour of a simply supported composite bridge comprising a steel beam with a concrete top slab. Initially, on completion of the slab construction, the unpropped steel section only is stressed (ignoring at this stage the effects of shrinkage) and there is no force transfer at the steel–concrete interface. For loads added after this stage the composite section carries them. Stresses in the beam increase, stresses in the slab occur and there is a force transfer at the steel–concrete interface. The force at the interface is the rate of change of force along the slab (see Equation (7.4)). The load is increased to cause yielding of the bottom flange (to ensure ductile behaviour this must occur prior to concrete crushing). Any further increase of load increases the zone of yield in the beam until the section has become fully plastic. Because of the redistribution of stresses, the composite section

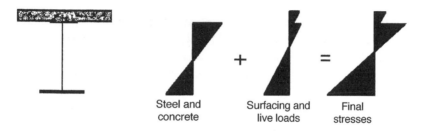

Steel and concrete   +   Surfacing and live loads   =   Final stresses

Figure 7.15 Build up of stresses at various stages of construction, steel–concrete composite construction.

is now carrying all the load including the steel and concrete self-weight originally carried entirely on the non composite steel beam. The development of stresses is shown in Figure 7.15.

The above behaviour implicitly assumes the interface connection does not fail or deform significantly prior to failure of the composite section. The connectors at the interface have their own load deflection behaviour. Any significant movement or slip at the interface will reduce the capacity of the structure.

Where the applied moments cause tension in the concrete, cracking modifies the behaviour slightly. Consider a cantilever beam with a reinforced concrete top slab. Initially, as before, the steel beam is carrying all of the self-weight. For loads added after this stage the composite steel–concrete section carries the load until the tensile capacity of the concrete is exceeded. When cracking occurs the section properties change from those of the steel–concrete structure to those of the steel beam and the slab reinforcement only. As the load is increased the steel yields and this ultimately results in the formation of a plastic section. Unless there is a significant proportion of steel (4% or so) the tensile capacity of the slab will be less than its compressive capacity. Consequently, in the cracked section, the force is lower and the rate of change in force is less and fewer connectors are required in this area. However, at the boundary of the cracked to uncracked section of slab there is a more significant change in force and more connectors are required at this location. Moment rotation curves for the section with the slab in compression and tension are shown in Figure 7.16 outlining this difference in behaviour.

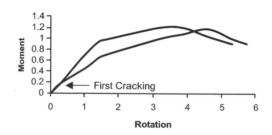

Figure 7.16 Moment–rotation curves for steel–concrete composite construction.

In summary, there are a number of important issues to be noted:

- below the elastic limit the force at the steel–concrete interface is proportional to the load applied after the composite connection was made
- at failure the force at the steel–concrete interface is proportional to the total load on the bridge. The force at the interface is the rate of change of force along the slab
- the ultimate strength of the connectors at the interface must be greater than the forces applied at the interface
- the slip at the interface must be small
- additional connectors may be required at changes of section.

7.6.1 Section properties

In steel–concrete composite construction the strength and stiffness of the two materials are very different. In the analysis and design of a structure it is normal to transform the concrete to an equivalent steel area using the appropriate modular ratio (see Equations (7.1) and (7.2)).

The compactness of the section is also important in determining the section properties to be used. The definition of compact, semi compact or slender varies slightly with the design code used. The rules shown in Table 7.3 are a reasonable guide.

Compact sections can be designed assuming fully plastic section properties. Semi-compact sections use the full elastic section properties. Slender sections need to take into account the effective reductions in web or flange areas that may occur due to out of plane buckling and shear lag. The differing modes of failure are illustrated by their idealized stress–strain curves as shown in Figure 7.17.

Table 7.3 Shape limitations for beams

Section type	Flange out stands in compression	Depth of web in compression	Spacing of shear connectors
Compact	$7t_f$	$28t_w$	$12t_f$
Semi compact	$12t_f$	$68t_w$	600 mm
Slender	varies with shear lag, dependent on b/L ratio and span configuration (see Figure 7.29)	from $68t_w$ to $228t_w$ the effective web thickness reduces to zero at $228t_w$	600 mm

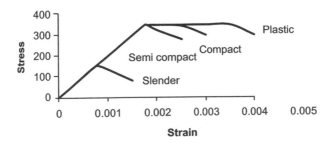

Figure 7.17 Typical stress–strain curves for steel–concrete composite construction.

7.6.2 Interface loads

The force transfer at the interface for the steel–concrete composite section is related to the rate of change of force in the slab (or beam). At the serviceability limit state the shear flow q_n at stage n is identical to Equations (7.4) and (7.5) for the concrete–concrete composite. Similarly at the ultimate limit state the maximum change in force in the slab over a length from the support or point of contraflexure to mid span is as Equation (7.9). In order to simplify design Equation (7.5) is used at both serviceability and ultimate.

$$q_n = VAy/I_n \qquad (7.5)$$

The design of the composite section is normally carried out at ultimate conditions for a compact section and at the serviceability limit state for a non-compact section. For continuous bridges the concrete over the support may be cracked and the designer has two options. The first option is to ignore the concrete in tension and use the appropriate section properties based on the beam and slab reinforcement only. Secondly, the uncracked section properties may be assumed along the whole beam. The first method is likely to give a lower connector requirement because of the reduced slab area, however, there will be an increased connector requirement at the point at which the section is assumed to change from uncracked to cracked.

7.6.3 Construction

For steel–concrete composite bridges the methods and sequences of construction are vitally important. As the concrete is placed, significant stresses are set up in the steelwork. Two basic construction assumptions can be made, that the section is propped or unpropped.

Propped construction

Propping the steelwork prior to concreting can aid slender or non compact sections. The majority of load is carried by the composite section immediately the props are removed. For most medium span bridges, the cost of propping is likely to be larger than any saving in steelwork (from reduced bracing and top flange requirements). Consequently it is not often used. On larger span bridges, the potential saving resulting from the use of propping may be more significant and is often worth investigation.

Unpropped construction

For unpropped construction, the bridge is built in stages. The steel section initially carries the self-weight of steel and concrete with the composite section carrying subsequently applied loads. For non-compact sections the stresses induced at each stage of construction should be summed. The sum of these stresses should be less than the allowable for the stage considered.

$$f_n > M_1/Z_1 + M_2/Z_2 + M_n/Z_n \qquad (7.3)$$

and for the forces at the steel–concrete interface:

$$q_n = q_1 + q_2 + q_n \cdots \qquad (7.13)$$

where q_1, q_2, etc. are calculated from Equation (7.5) using the properties appropriate to each stage.

The ultimate moment capacity of the non-compact section, M_u (for the tension flange) is based on the elastic section modulus Z_E.

$$M_u = (Z_E f_y)/\gamma \qquad (7.14)$$

For compact or semi-compact sections, the effect of construction sequence induced stresses may be offset by redistribution of stresses in the partially or fully plastic section. The entire load may be assumed to act on the cross-section appropriate to the stage under consideration:

$$f_n > (M_1 + M_2 + M_n)/Z_n \qquad (7.15)$$

and for the forces at the steel–concrete interface:

$$q_n = ((V_1 + V_2 + V_n)Ay)/I \qquad (7.16)$$

The ultimate moment capacity of the section, M_u (for the tension flange), is based on the plastic section modulus Z_p.

$$M_u = (Z_p f_y)/\gamma \qquad (7.17)$$

7.6.4 Shear connectors

Shear connectors are devices for ensuring force transfer at the steel–concrete interface. They carry the shear and any coexistent tension between the materials. Without connectors slip would occur at low stresses. Connectors are of two basic forms, flexible or rigid (Figure 7.18).

- Flexible connectors such as headed studs behave in a ductile manner allowing significant movement or slip at the ultimate limit state (Figure 7.19). At the serviceability limit state the loads on the connectors should be limited to approximately half the connectors' static strength to limit slip.
- Rigid connectors such as fabricated steel blocks behave in a more brittle fashion. Failure is either by fracture of the weld connecting the device to the beam or by local crushing of the concrete.

Typical nominal static strengths P_u for various connector types for grade 40 N/mm² concrete are given in Table 7.4.

Where the connectors are subject to tensile loads in addition to the interface shear flow then the nominal static strength should be reduced to:

$$P'_u = P_u - (T_u/\sqrt{3}) \qquad (7.18)$$

However, if the tension on the connector is larger than 20% of its nominal static strength then a more positive purpose designed connection should be considered (Johnson and Buckby, 1986).

Figure 7.18 Typical shear connector types for steel–concrete composite construction.

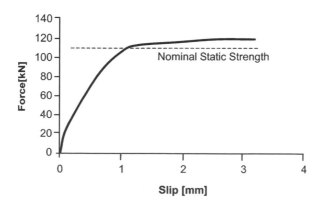

Figure 7.19 Typical load–slip curve for connectors in composite construction.

Table 7.4 Nominal static strengths of shear connectors

Type of connector	Connector material	Nominal static strength per connector for $f_{cu} = 40\text{N/mm}^2$
Headed studs,130 mm height	Steel with a characteristic strength yield stress of 385 N/mm^2 and minimum elongation of 18%	
Diameter		
19		109 kN
22		139 kN
25		168 kN
50 × 40 × 200 mm bar with hoops	Steel with a characteristic strength yield stress of 250 N/mm^2	963 kN
Channels	Steel with a characteristic strength yield stress of 250 N/mm^2	
127 × 64 × 14.9 kg × 150 mm		419 kN
102 × 51 × 10.4 kg × 150 mm		364 kN

7.6.5 Shear connector design

At the serviceability limit state there should be sufficient shear connectors to prevent excessive slip. The number of connectors required per unit length is:

$$N^o = q_n/0.55P_u \tag{7.19}$$

At the ultimate limit state the number of connectors required is:

$$N^o = (\gamma q_n)/0.8P_u \tag{7.20}$$

At the ultimate limit state the failure of shear planes other than at the interface $(x\text{–}x)$
may need to be investigated. These shear planes will be around the connector $(y\text{–}y)$
or through the slab $(z\text{–}z)$. Where haunches are used, a check on shear planes $(h\text{–}h)$
through the haunch may be required (Figure 7.20). Equations (7.10) and (7.11) are
relevant with v having a value of 0.9 and k a value of 0.15.

$$q_p = 0.9L_s + 0.7A_e f_y \tag{7.11}$$

$$q_p = 0.15L_s f_{cu} \tag{7.12}$$

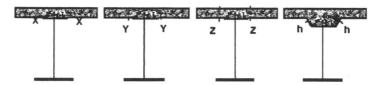

Figure 7.20 Typical shear planes in a steel–concrete composite structure.

For the majority of steel composite beams the amount of vertical shear carried into the composite flange is minimal and the steel webs of the beam can be assumed to carry all the shear. The shear capacity V_u will be given by:

$$V_u = (td\tau_n)/\gamma \tag{7.21}$$

where τ_n is the allowable shear stress:

$$\tau_n = (\beta f_y)/(\gamma/\sqrt{3}) \tag{7.22}$$

For slender or non-compact sections the shear capacity will depend on the slenderness of the web, the spacing between web stiffeners and the relative flange stiffness. For compact sections or webs with a slenderness (depth to thickness ratio) of 55 or less, Equation (7.20) may be used assuming $\beta = 1$. Above this slenderness the aspect ratio of the panel in shear (length to depth ratio) is important. The closer the stiffeners the higher the shear capacity. This is shown in Figure 7.21 for aspect ratios of 1 and 3 assuming a flexible flange.

The inertia of the web makes a contribution to the bending strength of the beam. The flange stiffness can improve the shear capacity of the composite section by spreading the area of web over which tension field action occurs. The increase in shear capacity with flange stiffness k_f is shown in Figure 7.22. For composite sections the top flange stiffness will be relatively large and the flange stiffness criteria governed by the lower non-composite flange.

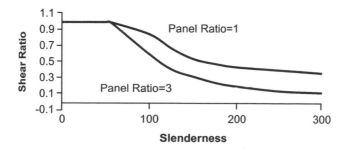

Figure 7.21 Variation in shear ratio with web slenderness and stiffener spacing.

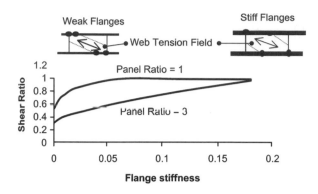

Figure 7.22 Effect of flange stiffness on shear capacity.

Because of the influence of the web on bending capacity and the flange on shear capacity the composite beam will not be able to carry its full moment and shear capacities simultaneously (Rockey and Evans, 1981). An interaction curve between bending and shear can be drawn based on the following limits:

- when the beam is designed with a shear capacity V_R based on a flexible flange ($k_f = 0$) then the beam can withstand a moment M_R
- when the beam is designed with a shear capacity V_u based on a stiff flange ($k_f > 0$) then the beam can withstand a moment of $0.5M_R$
- when the beam is designed for its full bending capacity M_u it can withstand a shear of $0.5V_R$.

M_R is the section capacity ignoring the web.

$$M_R = (DA_F f_y)/\gamma \tag{7.23}$$

The general interaction curve is shown in Figure 7.23(a). For compact sections with stocky webs and beams with small flanges $V_R \Rightarrow V_u$. For beams with slender webs $M_R \Rightarrow M_u$. The curve can then be simplified as shown in 7.23(b) and (c), respectively.

7.7 Construction methods

For steel–concrete composite bridges the methods and sequences of construction are vitally important. Prior to the concrete being placed the steelwork is relatively slender and often requires bracing to ensure stability. As the concrete is placed significant stresses are set into the steelwork, the sequence of placing of the concrete has an influence on the final behaviour of the structure.

For continuous bridges the sequence of construction will have an effect on the distribution of loads. Concreting spans adjacent to those already composite will induce additional stresses along the shear interface. It is common practice to construct mid span sections prior to support areas as this reduces tensile stresses in the slab.

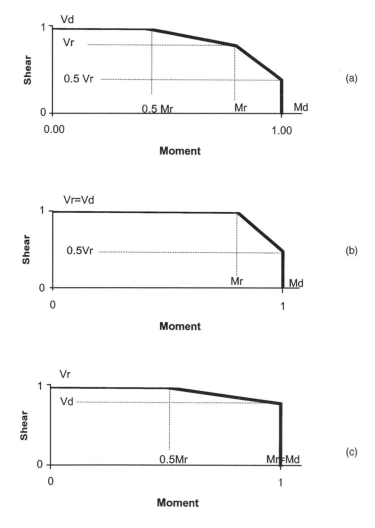

Figure 7.23 Typical shear–bending interaction curves. (a) General interaction curve. (b) Simplified curve for beams where $V_R \Rightarrow V_u$. (c) Simplified curve for beams where $M_R \Rightarrow M_u$.

7.7.1 Crane erection

The majority of small to medium span composite bridges are built using cranes. For bridges with good access and space to manoeuvre a crane, this method is simple and very economical. There are, however, a number of conditions that need to be addressed. Firstly, components must be of a size that is transportable to the site. In the UK this generally means a maximum length of 27.4 m and a width or height of 4.3 m. Where access roads do not permit this size, or larger units are required, the partial assembly of components on site may be necessary. There are many types of crane available from mobile cranes hired in for a specific lift to crawler cranes for smaller but more numerous lifts or for lifts requiring the crane to move under load.

For any lift, the crane's foundation must be adequately considered and ground improvements or special foundations provided, if necessary. This is particularly so for larger mobile units where significant loads may be placed on the outriggers.

Lifting points in the form of lugs or cleats will need to be attached to the beam prior to lifting. Normally these would be designed to be left in place and covered by the concrete slab. During lifting the beam must be stable. For long girders a bow-string brace may be required to stabilize the top flange. However, if beams can be lifted in pairs with the bracing required for concreting in place, the crane erection stage may not be a critical condition.

Joints in the girder will be placed by the designer to occur near points of contraflexure to minimize forces in the joint with due regard to transportation and lifting. Bolted joints using High Strength Friction Grip (HSFG) bolts are relatively quick and simple to build on site. They can be installed while the beam is supported by the crane. Site-welded joints are sometimes specified, however, temporary support towers may be required to hold the girder during welding. The welding process is likely to take longer than bolting. It is also relatively costly due to the high initial costs of testing and quality control and the need for cleaning and the application of a full paint system. In general, site welding should be avoided unless there is a sufficient number of joints to justify it. Aesthetic considerations are sometimes used as a justification for site welds, but in the author's opinion falsely as a well-positioned and detailed bolted joint is likely to look more functional.

7.7.2 Launched bridges

Incremental launching (Figure 7.24) consists of erecting the steelwork at a single location behind one of the bridge abutments. Once a section of structure is erected the whole structure is moved forward by the length of that section and the next section is then erected. For a composite bridge either the steelwork only or the steelwork and the concrete deck are constructed and launched. Launching of the steelwork only minimizes the moments and shears on the girders and means the rollers and jacking equipment will be comparatively modest. However, the deck slab will have to be cast by conventional suspended falsework after the launch. Launching the entire composite section minimizes the construction time as no major concreting operations are needed after erection. However, the heavier deck does require larger jacks and rollers (or slide plates) and greater moments and shears during launching will require additional stiffening of the girder webs. Sub-structures may need bracing to limit movement under launching.

The appropriateness of incremental launching as a method of construction depends primarily on the geometry of the structure. For launching to be economical and practical the structure is likely to be a viaduct (probably over 200 m long), have relatively consistent spans (of 40–60 m) and be of constant horizontal and vertical curvature. During launching the girders will be fitted with a launching nose. The purpose of the nose is to reduce the large cantilever moments that would otherwise occur in the deck prior to each pier being reached. The optimum length of the launching nose is usually that at which the cantilever moments induced in the girder adjacent to the

Figure 7.24 Launching of a large composite bridge (photograph courtesy of Robert Benaim & Associates).

nose are of a similar magnitude to those hogging moments that occur elsewhere. The launching nose will be profiled to overcome the tip deflection of the cantilever such that it can land on the adjacent pier.

7.7.3 Rolling in
A variation of the incremental launch is to construct the bridge adjacent to its final location and then roll it laterally to its correct location. This method is used extensively for replacing existing bridges under railways (Atkins and Wigley, 1988).

7.7.4 Cantilever construction
Cantilever construction is used extensively for the construction of composite cable stay structures (Institute of Civil Engineering, 1997) and is appropriate for trusses (Cracknell, 1963) and arches (Zuou and Zhu, 1997). It is not common in the erection of girder bridges because of the significant number of joints required.

7.7.5 Bracing of steelwork
For composite bridges constructed from steel beams or girders with concrete slabs cast on top, the construction stage is a critical area that needs careful consideration. The steelwork is relatively slender and requires bracing to ensure stability during construction. The bracing must be designed to ensure all likely buckling modes are suppressed. This includes instability of the girders between bracing points and overall instability of the whole bridge.

For an individual beam or girder, the main mode of instability is lateral torsional buckling (Timoshenko and Gere, 1961). The beam undergoes a simultaneous lateral movement and rotation. The tendency for a beam to buckle is influenced by a number of factors including: the nature of the external loads, the beam shape, the bracing type, strength and stiffness, fabrication tolerances, the length between effective bracing, residual stresses (from rolling or welding) and the stress in the compression flange. Bracing either in the form of plan bracing, anchoring to a rigid object, transverse bracing or a combination of methods can be used to suppress lateral torsional buckling. Common forms of bracing are outlined in Figure 7.25.

In order for the bracing to function it must have sufficient strength to resist the applied loads. The destabilizing forces generated by the compression flange are generally considered to be approximately 2.5% of the sum of the forces in the flanges being connected, or half of this value when considered with wind or other significant transverse effects. The force can vary significantly, the more flexible the bracing the larger the destabilizing force to be resisted (Wang and Nethercot, 1989) (Figure 7.26).

Typically bracing is required at 12–15 times the flange width. The determination of limiting stresses is based on a slenderness parameter λ using an effective length concept. The limiting compressive stress is determined from Figure 7.27. The curve is based on a best fit of experimental data (Rockey and Evans, 1981). Beams with a λ value of 45 or less can generate the full beam capacity, beams with a λ value greater than 45 are prone to instability and stresses must be limited.

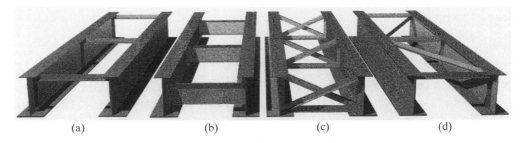

(a)   (b)   (c)   (d)

Figure 7.25 Bracing types for steel–concrete composite construction. (a) Ineffective bracing. (b) U frame bracing. (c) Cross-bracing. (d) Plan bracing.

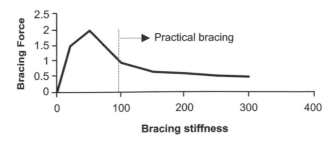

Figure 7.26 Variation in brace force with brace stiffness.

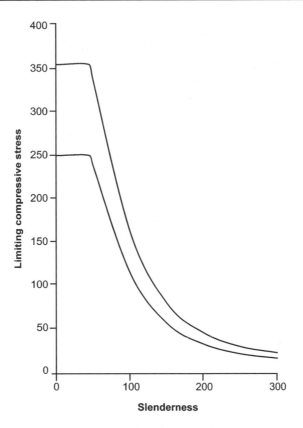

Figure 7.27 Limiting compressive stress for a steel girder.

$$\lambda = (l_b/r_y)nv \qquad (7.24)$$

where r_y is the radius of gyration.

Here n is a parameter related to the form of the external loads and the shape of the moment diagram between bracing positions. It varies from $n \Rightarrow 0.65$, where there is a change in sign of the moment to $n \Rightarrow 1.0$ for a constant moment (Figure 7.28a).

The parameter v is dependent on the shape of the beam in particular the relative sizes of the compression and tension flanges, given by Figure 7.28(b).

$$i = I_c/(I_c + I_t) \qquad (7.25)$$

For torsional cross-bracing shown in Figure 7.25(c) the bracing may be considered as rigid. For U frame braces (Jeffers, 1990) shown in Figure 7.25(b), plan bracing (7.25d) should also be used unless the U frame is sufficiently rigid, $l_b \ll l_e$, where:

$$l_e = 2.5(EI_y l_b \delta)^{0.25} \qquad (7.26)$$

δ is the deflection caused by a pair of unit forces on the U frame and l_b is the distance between rigid bracing.

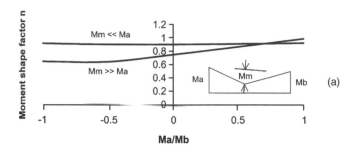

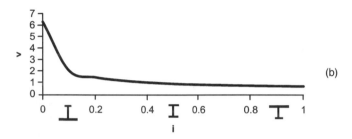

Figure 7.28 (a) Variation in n with moment. (b) Variation in v with beam shape.

There are two primary ways of reducing instability:

- increasing the lateral strength and stiffness by the use of plan bracing or by increasing the width of the compression flange (increasing i)
- increasing the torsional strength and stiffness by the use of transverse X or U bracing that connects the tension and compression flanges of adjacent beams. It should be noted that the connection of compressive flanges on adjacent beams (Figure 7.25(a)) will not provide stability unless the bracing has sufficient bending strength to act as a U frame.

For larger bridges or slender footbridges the use of cross-bracing alone may not be sufficient to suppress overall instability of the pair of girders. The effective length of the beam pair is determined from Equation (7.27) using the elastic critical buckling stress approach.

$$\lambda = \sqrt{(\pi^2 E/\sigma_{ci})} \qquad (7.27)$$

For I girders the torsional stiffness is generally small and the critical stress can be estimated by considering the warping strength of the girder pair.

$$M_{ci} = (\pi/l)^2 E(IK)^{0.5} \qquad (7.28)$$

$$\sigma_{ci} = M_{ci}/2Z \qquad (7.29)$$

where K is the warping constant (Roark and Young, 1985):

$$K = (D^3 B^2 t)/24 \qquad (7.30)$$

Where the overall slenderness of the section is critical, either plan bracing is required or a more stable box section should be used. Alternatively, adjacent pairs of girders should be connected. When more than two girders are connected the designer should be aware that the bracing will participate in the transverse spreading of load in the final condition. For this participating bracing the forces in the final condition may govern their design and the detailing of the connections with the slab and main girders must also consider the effects of fatigue, particularly at the area adjacent to web, flange and bracing connections.

In hogging regions of continuous girders the lower flange is in compression and nominally unrestrained for the full range of dead and live loads and bracing is often provided in this area to restrain the flange. The deck slab in this area provides restraint in the form of an inverted U frame, and means the critical buckling mode involves some distortion and bending. The pure lateral torsional buckling mode given by codes is conservative. Research on buckling in hogging regions (Weston et al., 1991) indicates that for many practical beam layouts the bracing is not required. Where the inverted U frame method is used, the effects of the additional tensile force on the connector capacity should be considered (see Equation 7.18).

7.8 Steel–concrete composite box girders

The box form is inherently stiffer and more stable than the plate girder form due to the fact that closed sections have a considerably greater torsional stiffness. Consequently, the box form is more suitable for slender structures, longer spans or curved bridges. For short spans the difference between plate girders and box girders is not significant and the simpler fabrication details of the plate girder make it more economic. For longer spans, where the high slenderness λ and low critical stresses require larger sections, box sections become the more economic form.

Composite box girders are generally of three types: a closed steel box of relatively compact proportions where plate stiffening and diaphragms are relatively minimal; an open-topped steel trough section that is made into a closed box by the concrete top slab or multi-celled boxes. The third form is rarely used as it is usually more expensive to fabricate and transport and usually less efficient structurally than the other two forms.

The relative compactness of the small closed box means that the plate thickness is high and requires less stiffening and consequently less welding per tonne fabricated. However, the closed form means that either welding has to be carried out inside the box or that the fillet welds normally associated with plate to plate connections have to be substituted by partial penetration welds from outside. The closed box type is commonly used in place of I girders on bridges of high curvature or where aesthetics are an important criterion. The design of this form of box is essentially similar to that of the composite I girder. For larger closed boxes the top flange of the box section may be affected by shear lag. Consequently, the distribution of forces on the shear connectors is not uniform but varies across the flange reducing as the distance from

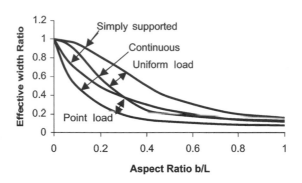

Figure 7.29 Variation in effective width of a flange allowing for shear lag effects.

the web increases. For design purposes an effective width concept is used. The effective width of the flange assumed to act with the web depends on the span-to-width ratio of the flange, the amount of stiffening, the type of loading (uniform or concentrated), the type of support restraints and the position on the span (Figure 7.29).

Generally, shear connectors should be concentrated adjacent to webs to be most effective and those outside the effective flange placed at nominal spacing merely to tie the steel flange to the concrete. The actual distribution of shear flow across the flange can be calculated from Equation (7.31). Where cross-girders or internal diaphragms stiffen the top flange, additional connectors will be required in transverse bands to maintain composite action.

$$q' = (q/N^o)[K(1 - x/b)^2 + 0.15] \qquad (7.31)$$

where q' is the force on a connector at a distance x from a web and b is the distance between webs. K is a coefficient that depends upon the number of connectors within a 200 mm band adjacent to the web. This is shown in Figure 7.30, where N^o is the total number of connectors.

The top flange of the box girder will act as permanent formwork for the concrete and must carry the weight of wet concrete. Once hardened the flange may be assumed to act with the slab for local wheel loads. Additional connectors will be required for this purpose.

The open form of box girder consisting of steel webs and a bottom flange has only small top flanges sufficient for stability during concreting but not closing the section (Figure 7.31). The advantages of this form are that access to all parts of the section for welding is available and that the web can be inclined allowing a larger span in the transverse direction of the slab. In some situations a single box may be sufficient thereby minimizing web material. Another significant advantage of this form in continuous construction is that the section can be made doubly composite (Figure 7.5). In the doubly composite section concrete is placed in the steel box adjacent to the supports (where the hogging moments occur) prior to placing of the top slab. The use of the bottom concrete in compression reduces the size of the steel to only that required for construction.

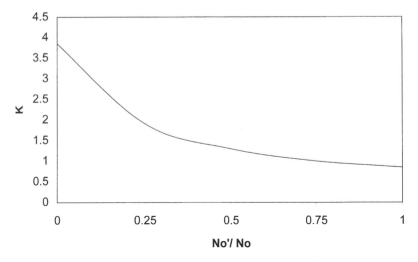

Figure 7.30 Coefficient K for connectors on box girders.

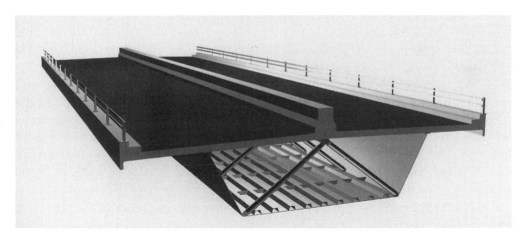

Figure 7.31 Large open topped steel trapezoidal box.

A disadvantage of the open box is that the high torsional stiffness of a closed section is not present during construction until the concrete slab has gained strength. For bridges with horizontal curvature, the torsional stiffness can be improved by the use of plan bracing. The plan bracing can be considered as an equivalent plate for calculating torsional properties.

$$t_{eff} = (A \sin^2\theta \cos \theta)/B \qquad (7.32)$$

7.8.1 Torsional properties
The use of a box form will aid the distribution of eccentric loads. The stresses induced by the loads will depend upon the magnitude of the load and its eccentricity, the box

geometry and the number and stiffness of diaphragms. In most cases the load causing the maximum torsional effect is not associated with the maximum load and the design of the main plate sizes should not be significantly influenced by torsion.

A circular section subject to a torsional load will resist load by a direct shear flow around the section. A square section will carry the load in a similar manner, but there will be a tendency for the section to distort. If the section is rectangular or trapezoidal, there will also be a tendency to warp under torsion (Chapman *et al.*, 1971).

Torsional warping
For a box section the torsional constant J can be approximated by the following:

$$J = (4A_o^2)/(\Sigma(B/t)) \tag{7.33}$$

where A_o is the area enclosed by the box, and B and t are the width and thickness of the section. Because the torsional behaviour is primarily a shear flow the effective thickness of the concrete elements could be based on transformed sections using a modular ratio based on the shear modulus.

$$m_J = G_S/G_c \tag{7.34}$$

However, for most practical design situations, given the uncertainty of creep and shrinkage on the section and the effects of concrete cracking and connector slip, the use of the modular ratio used for bending effects is usually satisfactory. The torsional warping stress can be estimated using the following equation:

$$f_{TW} = \beta(TD/J) \tag{7.35}$$

P varies with the section. It is one for a rectangular box, reducing as the top width and any deck cantilevers increase, it is zero for a triangular-shaped box.

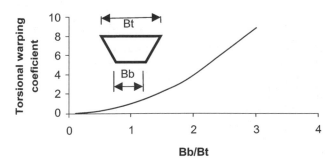

Figure 7.32 Variation of torsional warping with box geometry.

Distortional warping

The amount of distortion depends upon the shape of the box, the transverse stiffness of the web and flanges, the spacing and stiffness of diaphragms and on whether the torsional loads are point loads or uniformly spread. Two types of stresses are generated by distortional warping, a warping stress (f_{DW}) similar in nature to torsional warping and a transverse bending stress (f_{WB}).

$$f_{DW} = \beta(T_y L_D / BI) \qquad (7.36)$$

The variation in β is dependent on the transverse rigidity of the section and the diaphragm and can be computed using a beam on elastic foundation analogy (Wright *et al.*, 1968). The distortional stress can again be computed from the beam on elastic foundation model. The distortion is a transverse moment and is greatest at intermediate stiffeners. It is important as it can cause fatigue problems in the steel plates and a tensile load in the shear connectors at the flange that will lower the connector capacity (see Equation 7.18). Typically diaphragms are required at a spacing of three times the box depth. For larger boxes closer spacing may be required simply to aid fabrication and transportation.

The construction methods for box girders are similar to those of plate girders. However, as the boxes tend to be larger they are often only part fabricated for transportation with more assembly of parts on site. Because of their size, lifting is also more expensive and lifting techniques involving strand jacking of large elements or the construction of whole spans by gantry often proves viable (Figure 7.33).

Figure 7.33 Span by span construction of a box girder viaduct (courtesy of Robert Benaim & Associates).

7.9 Steel–concrete composite columns

Axially loaded members may be columns, arches or members of a truss or frame. Usually the steel and concrete elements have the same centroid (unlike most bridge beams). Initially the load will be distributed in proportion to the relative stiffness (EA) of the component making up the section. Normally the primary steel load bearing components would be fabricated and fitted together before encasing or filling with concrete. If the loads at this stage are significant, a check should be made of the capacity of the non-composite section. However, normally the section can be designed at the ultimate limit state ignoring the build up of stresses in the section

For short members not prone to buckling, the ultimate squash load N_u is the sum of the capacity of the steel, concrete and steel reinforcement.

$$N_u = (A_s f_y/\gamma) + (A_r f_r/\gamma) + (0.85 A_{cf} f_{cu}/\gamma) \tag{7.37}$$

No more than 4% reinforcement by area should be used and reinforcement less than 0.3% ignored. The steel contribution factor should be between 0.2 and 0.9.

$$\delta = (A_s f_y/\gamma N_p) \tag{7.38}$$

For $\delta > 0.2$ the section should be designed as a concrete column, while for $\delta < 0.9$ the section should be designed as a steel strut. For some concrete-filled tubes the capacity of the concrete is enhanced by the effects of confinement. For sections using very high-strength concrete or structures in seismic areas, this effect may be significant. For most practical design cases the additional requirement of compactness and low applied moments are not met and the enhancement due to confinement neglected.

For practical design situations the capacity of the section is influenced by buckling instability and the load that can be carried by the section is lower than the squash load.

$$N_R = k_1 N_u \tag{7.39}$$

where k_1 is determined from a buckling curve (Figure 7.34) using the slenderness parameter:

$$\lambda' = \sqrt{(N_u/N_{cr})} \tag{7.40}$$

$$N_c = ((EI)_e \sigma^2)/l_e^2 \tag{7.41}$$

$$(EI)_e = E_c I_c + E_s I_s + E_r I_r \tag{7.42}$$

For most practical applications the column will be subject to bending effects. For columns with both bending and axial load effects, there will be an interaction between the two effects (Figure 7.35) (Bergmann *et al.*, 1995).

At point A in Figure 7.35 the entire section is in compression and the capacity can be calculated from Equation (7.39) assuming zero moment. At point C the axial

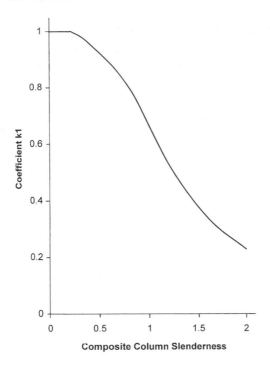

Figure 7.34 Buckling curve $k_1 - \lambda'$ for composite columns.

capacity is zero and the moment capacity is at its ultimate capacity. The maximum capacity of the section can be maintained under axial load to point C. At point C the capacity is limited to $k_2 N_u$. k_2 is dependent on the shape of the section and its slenderness. Approximately point C occurs at an axial load capacity equal to that of the concrete alone. The actual behaviour is more complex and is shown by the solid curve in the figure.

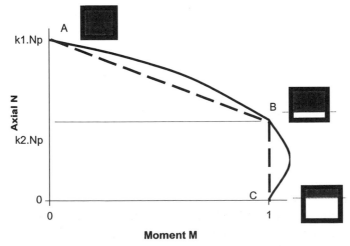

Figure 7.35 Axial load–bending interaction curve for composite columns.

7.9.1 Connections

Connections in composite columns and trusses are essentially similar to those commonly used in steel structures. The strength of the connection must be adequate to carry the proportion of load on the steel element of the composite section. Bolted connections should use HSFG bolts designed such that slip does not occur (this would transfer additional loads into the concrete). For concrete filled tubular sections the connections are usually made directly to the steel and provision must be made to distribute the loads to the concrete. Where the ultimate shear stress at the steel–concrete interface exceeds 0.4 N/mm^2, shear connectors should be provided to carry the load.

7.10 Prestressing of steel–concrete composite sections

Prestressing is commonly used in concrete to concrete composite sections, usually in the form of prestressed precast beams. The prestressing of steel–concrete composite sections is also feasible (Rosignoli, 1997) and can be used to control cracking or to increase the stress range of the structure. Prestress may be applied by the use of pre cambering and jacking, where the relative levels of supports are altered after the concrete deck has matured. This method is limited to continuous structures. The level of jacking required is likely to be approximately 0.1L, where L is the total length of the bridge. Consequently this method is generally limited to short or medium span structures.

Prestress may also be applied by stressing tendons or bars (Troitsky, 1990). Internal tendons directly in the concrete at areas of high tension can be used to eliminate cracking. External tendons may be used to prestress the whole section and redistribute forces in the structure. When prestress is used, the axial load effects will add additional loads at the interface. For prestressed steel, the anchorages and cable deflectors are significant items that can increase fabrication costs and offset the advantages of prestressing the section.

7.11 Fatigue

Fatigue is a phenomenon primarily affecting the steel elements of a composite structure. Fatigue is primarily influenced by the stress fluctuations in an element (the maximum range of stress, as opposed to the maximum stress), the structure geometry and the number of load cycles. The relationship between these variables is shown in Figure 7.36 where it can be seen that the allowable stress range for a bolted connection is significantly higher than for a welded connection detail. The stress range prior to significant damage varies with the number of load cycles. At high stress ranges the number of cycles to damage is low. There is a stress range below which an indefinitely large number of cycles can be sustained. For a bolted detail and a shear connector, the range is about 80 N/mm^2. For a welded detail it is nearly half this value. For design it is critically important to obtain a detail with the maximum fatigue resistance. This is achieved by avoiding sudden changes in stiffness or section thickness, partial penetration welds, intermittent welding or localized attachments.

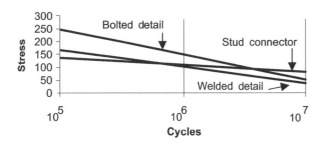

Figure 7.36 Typical stress range – number of cycles to failure for bolted and welded connections and stud shear connectors.

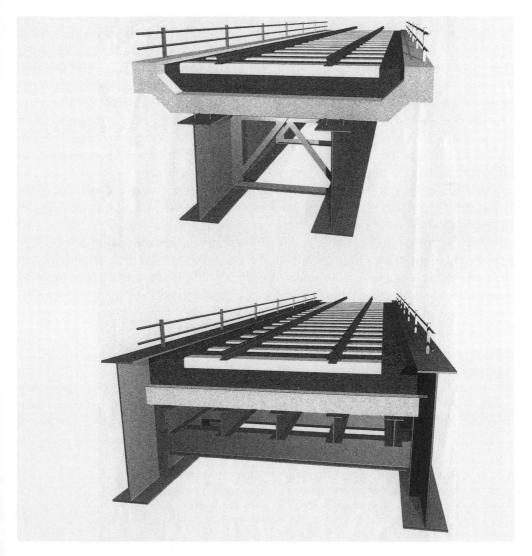

Figure 7.37 Examples of poor and good structural systems for fatigue.

Composite highway structures are not particularly sensitive to fatigue problems if properly detailed. The design of shear connectors near mid span may be governed by fatigue where static strengths dictate only minimum requirements. Participating permanent formwork, participating bracing and cruciform joints at cross-heads are also areas that are sensitive to poor fatigue details. Railway bridges with their higher ratio of live to permanent loads are more sensitive to fatigue. Figure 7.37 illustrates two composite bridge designs, the first with its numerous connection types, multiple member types and potential racking of girders is significantly more fatigue sensitive than the second example where loads are transferred directly to the main load bearing elements.

7.12 Steel–timber composites

Timber is in many ways a more complex material than either steel or concrete. It has similar strengths in compression and tension and it has a tendency to creep under sustained load. Being a natural material, it can have significant variations in properties and it is highly anisotropic, being significantly weaker when loaded across the grain rather than along it. However, it has a high strength to weight ratio. Studies in Canada (Bakht and Krisciunas, 1997) indicate that the lightness of a timber deck means that it is more economic than concrete for spans over approximately 50 m. This is particularly true in isolated areas where a good quality structural concrete is difficult to obtain.

Steel–timber bridges are not normally composite. Traditionally the steel beam spans longitudinally and the timber deck spans transversely. Because the stiffness of the timber is reduced by about one-sixtieth when loaded across the grain there is no advantage in composite action. Recently a number of steel timber composite bridges have been constructed, where the timber deck consists of a laminated construction spanning longitudinally between steel cross-frames. The deck is attached to the steel through shear connectors in pockets in the deck and the pockets are grouted with a fibre reinforced grout.

Bibliography

AASHTO. *Standard Specifications for Highway Bridges*, 16th edn, 1996.

Aparicio AC and Casas JR, The Alamilo cable-stayed bridge: special issues faced in the analysis and construction. *Proc ICE, Structures and buildings*, Nov., 1997.

Atkins FE and Wigley PJ. Railway underline bridges: developments within constraints of limited possession. *Proc. ICE*, Part 1, 84, Oct., 1988.

Bakht B and Krisciunas R. Testing of prototype steel wood composite bridge. *Structural Engineering International*, 1, 1997.

Bergmann R, Matsui C, Meinsma C and Dutta D. *Design Guide for Concrete Filled Hollow Section Columns under Static and Seismic Loading*, Verlag TUV Rheinland, 1995.

Biddle A. Steel bearing piles guide, The Steel Construction Institute.

British Standards Institution. BS 5400, *Steel, Concrete and Composite Bridges*, Part 5, *Design of Composite Bridges*. British Standards Institution, London, 1978.

British Standards Institution. BS 5400, *Steel, Concrete and Composite Bridges*, Part 3, *Design of Steel Bridges*. British Standards Institution, London, 1982a.

British Standards Institution. BS 5400, *Steel, Concrete and Composite Bridges*, Part 10, *Design of Bridges for Fatigue*. British Standards Institution, London, 1982b.

British Standards Institution. BS 5400, *Steel, Concrete and Composite Bridges*, Part 4, *Design of Concrete Bridges*. British Standards Institution, London, 1992.

Cartledge P (Ed.). *Proceedings of Conference on Steel Box Girder Bridges*. Institution of Civil Engineers, London, 1973.

Chapman JC, Dowling PJ, Lim PTK and Billington CJ. The structural behaviour of steel and concrete box girder bridges. *The Structural Engineer*, 49, Mar., 1971.

Clarke LA. *Concrete Bridge Design to BS 5400*. Construction Press, 1983.

Collings, D. M5 parallel widening. *New Steel Construction*, 2, Feb., 1994.

Cracknell DW. The Runnymede bridge, *Proc. ICE*, May–Aug. 1963.

Csagoly PF and Lybas JM. Advanced design methods for concrete bridge deck slabs. *Concrete International*, May, 1989.

Dickson DM. M25 orbital road, Poyle to M4: alternative steel viaducts. *Proc. ICE*, 82, Part 1, 309–326, April, 1987.

Eurocode. Common unified rules. No. 2, Concrete structures; No. 3, Steel structures; No 4, Composite steel and concrete structures; No. 5, timber structures. Commission of the European Communities.

Hambley EC. Integral bridges, *Proc. ICE, Transport*, Feb. 1997.

Hubbard HW, Pott D, Miller D and Burland JB. Design of the retaining walls for the M25 cut and cover tunnel at Bell Common. *Geotechnique*, 34, 4, 1984.

Institute of Civil Engineers. Second Severn crossing. *Proc ICE*, 120, Special issue 2, 1997.

Ito M, Fujino Y, Miyata T and Narital N (Eds). *Cable-stayed Bridges, Recent Developments and Their Future*. Elsevier Science, Oxford, 1991.

Jeffers E. U frame restraint against instability of steel in bridges. *The Structural Engineer*, 68, 18, Sept., 1990.

Johnson RP and Buckby RJ. *Composite Structures of Steel and Concrete*, Volume 2, *Bridges*. Collins, 1986.

Kerensky OA and Dallard NJ. The four level interchange at Almonsbury. *Proc. ICE*, 40, 295–322, July, 1968.

Kirkpatric J, Rankin GIB and Long AE, Strength evaluation of M beam bridge deck slabs. *The Structural Engineer*, 64B, Mar., 1986.

Lecroq P. Highway bridge decks in France. *ECCS/BCSA International Symposium on Steel Bridges*, Feb., 1988.

Narayanan R, Bowerman HG, Naji FJ, Roberts TM and Helou AJ, Application guidelines for steel–concrete sandwich construction: immersed tube tunnels. Technical Report 132, The Steel Construction Institute.

Roark RJ, Young WC. *Formulas for Stress and Strain*. McGraw Hill, New York, 1985.

Rockey KC and Evans HR (Eds). The Design of Steel Bridges. Granada, 1981.

Rosignoli M, Prestressed composite box girders for highway bridges. *Structural Engineering International*, 4, 1997.

Steel Construction Institute. Design of composite trusses, 1992.

Timoshenko SP and Gere JM. *Theory of Elastic Stability*, 2nd edn. McGraw Hill, New York, 1961.

Troitsky MS. *Prestressed Steel Bridges, Design and Theory*. Van Nostrand Reinfold, 1990.

Virlogeux M, Foucriat J and Lawniki J, Design of the Normandie bridge, Cable stayed and suspension bridges. *Proceedings International Conference*, AIPC-FIP, Deauville, October, 1994.

Wang YC and Nethercot DA. Ultimate strength analysis of three-dimensional braced I-beams. *Proc ICE*, Part 2, 87, March, 1989.

Weston G, Nethercot DA and Crisfield MA. Lateral buckling in continuous composite bridge girders. *The Structural Engineer*, 69, 5, Mar., 1991.

Wright RN, Abdel-Samad SR and Robinson AR. Beam on elastic foundation analogy for analysis of box girders. *Journal of the Structural Division, ASCE*, 94, July, 1968.

Zuou P and Zhu Z. Concrete filled tubular arch bridges in China. *Structural Engineering International*, 3, 1997.

8 Design of arch bridges

C. MELBOURNE

8.1 Introduction

The origins of the use of arches as a structural form in buildings can be traced back to antiquity (Van Beek, 1987). In trying to arrive at a suitable definition for an arch we may look no further than Hooke's anagram of 1675 which stated 'Ut pendet continuum flexilc, sic stabat continuum rigidum inversum' – 'as hangs the flexible line – so but inverted will stand the rigid arch'. This suggests that any given loading to a flexible cable if frozen and inverted will provide a purely compressive structure in equilibrium with the applied load. Clearly, any slight variation in the loading will result in a moment being induced in the arch. It is arriving at appropriate proportions of arch thickness to accommodate the range of eccentricities of the thrust line that is the challenge to the bridge engineer. Even in the Middle Ages it was appreciated that masonry arches behaved essentially as gravity structures, for which geometry and proportion dictated aesthetic appeal and stability. Compressive strength could be relied upon whilst tensile strength could not. Based upon experience many empirical relationships between the span and arch thickness were developed and applied successfully to produce many elegant structures throughout Europe.

The expansion of the railway and canal systems led to an explosion of bridge building. Brickwork arches became increasingly popular. With the construction of the Coalbrookdale bridge (1780) a new era of arch bridge construction began. By the end of the nineteenth century cast iron, wrought iron and finally steel became increasingly popular; only to be challenged by ferro cement (reinforced concrete) at the turn of the century.

During the nineteenth century analytical technique developed apace. In particular, Castigliano (1879) developed strain energy theorems which could be applied to arches provided they remained elastic. This condition could be satisfied provided the line of thrust lay within the middle third, thus ensuring that no tensile stresses were induced. The requirement to avoid tensile stresses only applied to masonry and cast iron; it did not apply to steel or reinforced concrete (or timber for that matter) as these materials were capable of resisting tensile stresses.

Twentieth century arch bridges have become increasingly sophisticated structures combining modern materials to create exciting functional urban sculptures.

8.2 Types of arch bridge

The relevant terms that are used to describe the various parts of an arch bridge are shown in Figure 8.1.

Arches may be grouped according to the following parameters:

- the materials of construction
- the structural articulation
- the shape of the arch.

Historically, arch bridges are associated with stone masonry. This gave way to brick-work in the nineteenth century. Because these were proportioned to minimize the possibility of tensile stress, they tend to be fairly massive structures.

By comparison the use of reinforced concrete and modern structural steel gives the opportunity for slender, elegant arches.

Nowadays, timber is restricted to small bridges occasionally in a truss form but more usually as laminated curved arches. Although timber has a high strength to density ratio parallel to the grain, it is anisotropic and strength properties perpendicular to the grain are relatively weak. This requires careful detailing of connections to ensure economic use of the material.

With regard to structural articulation the arch can be fixed or hinged. In the latter case either one, two or three hinges can be incorporated into the arch rib. Whilst the fixed arch has three redundancies, the introduction of each hinge reduced the indeterminacy by one until, with three hinges, the arch is statically determinate and hence, theoretically, free of the problems of secondary stresses. Figure 8.2 shows a range of possible arrangements. The articulation of the arch is not only dependent upon the number of hinges but is also fundamentally influenced by the position of the deck and the nature of the load transfer from the deck to the arch.

The traditional filled spandrel, where the vehicular loading is transferred through the backfill material onto the extrados of the arch, represents at first glance the simplest structural condition. As will be seen later this is not the case and has led to much research for the specific case of the masonry arch bridge in an attempt to improve our understanding of such structures.

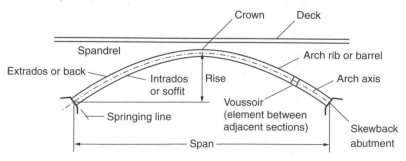

Figure 8.1 Arch bridge terms (O'Connor, 1971).

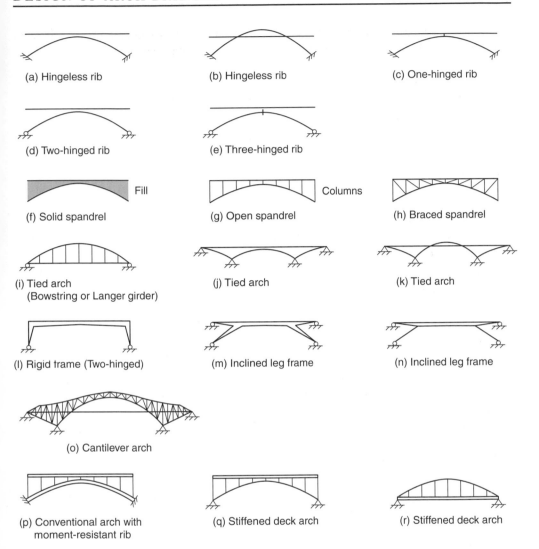

Figure 8.2 Types of arch bridge (O'Connor, 1971).

The spandrel may be open with columns and/or hinges used to transfer the deck loads to the arch. In an attempt to minimize the horizontal thrust on the abutments, the deck may be used to 'tie' the arch. Tied arches are particularly appropriate when deck construction depths are limited and large clear spans are needed (particularly if ground conditions are also difficult and would require extensive piling to resist the horizontal thrusts).

Returning to Hooke's anagram, the perfect shape for an arch would be an inverted catenary – this would only be the case for carrying its own self-weight. Vehicle loading and varying superincumbent deadload both induce bending moments. Consequently the arch has to have sufficient thickness to accommodate the 'wandering' thrust line.

For ease of setting out and construction simpler shapes are adopted nowadays – segmental or parabolic shapes are used. Although in situations where maximum widths of headroom have to be provided (say over a railway, road or canal) an elliptical shape may be required or its nearest 'easy' equivalent the three-centred arch.

It is worth commenting at this stage regarding the idealization of arch structures. Traditionally arches are perceived as being two-dimensional structures; this, of course is not true – but the extent to which it is not true should be of concern to the designer/assessor. Even in the case of a three-hinged arch which ostensibly is statically determinate the 'pins' are capable of transmitting shear even though they theoretically cannot transmit moments. In the case of non-uniform transverse distribution of loading the hinges will transmit a varying shear which will produce torsion in the arch. Moreover, in the case of skew arches or non-vertical ribs the structure has a much higher redundancy and hence will require greater attention to detail in regard to the releases which are engineered into the structure.

From an aesthetic point of view arches have a universal appeal. In spite of this, care must be taken as the impact of even the modest sized bridges is significant. Filled arches are invariably masonry (or widening of masonry) bridges. Cleanness of line, honesty of conception and the attention to detail are vital ingredients to a successful bridge. Certainly, simple stringcourses and copings are preferable to elaborate details which would be expensive and inappropriate for most modern bridges. Where stone is used it is important to be sensitive to the nature of the material. Modern quarrying techniques should be employed (laser cutting, diamond sawing, flame texturing and sandblasting) reserving traditional dressing to conservation schemes. If brickwork is used different textured or coloured bricks and mortar can be specified. Here stringcourses can be particularly useful to mask changes in bedding angle.

Historically abutments comprised either rock, or else were massive masonry supports relying on their weight to resist the thrust of the arch. In terms of structural honesty this is necessary as it is instinctive to expect such support.

Reinforced concrete and steel arches are altogether much lighter structures. 'The structure consists basically of the arch, the deck and usually some supports from the arch to the deck – in that order of importance. These elements should be expressed in both form and detail, and with due regard for their hierarchy' (Highways Agency, 1996).

It is important that the deck, if it rests on the crown of the arch, should not mask it in any way. Any support whether spandrel columns or hinges (in the case of the tied arch) should not be allowed to dominate. Preferably they should be recessed relative to the parapet and stringcourse.

Concrete arches can be either a full width curved slab or a series of ribs. Steel is almost always a series of ribs. Where ribs are used thought should be given (if they are going to be seen from underneath) to the chiaroscuro of the soffit.

The ratio of span to rise should generally be in the range 2:1 to 10:1. The flatter the arch the greater the horizontal thrust; this may affect the structural form selected i.e. whether or not a tie should be introduced, or the stiffness of the deck relative to the arch.

Figure 8.3 Monk New Bridge.

8.3 Typical structures

8.3.1 Monk New Bridge, Lancashire, UK

Monk New Bridge carries a re-aligned trunk road over a small river (Figure 8.3). The new bridge was constructed to replace an existing masonry arch bridge which was demolished as the road was realigned. The bridge was constructed using a mass concrete arch 725 mm thick. The square span is 9.9 m with a skew angle of 29°. The ground conditions were not conducive for an arch bridge but the original bridge on the site had been an arch, so to reduce the environmental impact, the new bridge took the same form. Consequently, piled abutments were used; these were tied together to resist the horizontal thrust from the arch barrel. The novel feature of the bridge was the use of through thickness inclusions in the mass concrete arch; the purpose of which was to act as crack inducers to convert the mass concrete continuum into voussoir blocks to replicate the articulation of the old stone arch bridges. Detailing the position of the inclusions required care. The bridge was subjected to test loading which demonstrated the low stress levels in the arch, its composite action with the backfill and that the forces were induced in the ties under live loading.

8.3.2 Chippingham Street Footbridge/Cycleway, Sheffield, UK

The Chippingham Street Bridge carries a footway/cycleway over a canal which passes through a steep-sided cutting 10.5 m deep and 52 m wide at the top (Figure 8.4). The ground conditions were good with 1.5 m of fill overlying sandstone and mudstone beds with a recommended safe bearing capacity of 600 kN/m². The choice of steel rolled sections together with the simplicity of line and form produced a structure which integrates well with its environment, offering an interesting focal point whilst reflecting the traditional aesthetics of canal structures.

Figure 8.4 Chippingham Street footbridge (Taylor, 1995).

To facilitate ease of construction a three-hinged approach was used comprising twin castellated universal column sections 458×305 bent to a circular radius of 20.5 m and braced transversely with angle sections. Vertical universal beam members 356×171 were used to support twin longitudinal beam sections of similar size. The deck comprised an *in situ* reinforced concrete slab. Lateral stability was achieved using diagonal bracing. Bolted connections were used throughout the steelwork. The hinges were achieved using 'pot' bearings at each of the springings and a bolted connection at the crown.

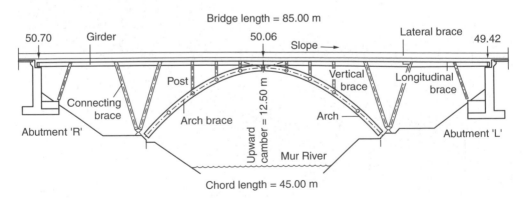

Figure 8.5 Mur River Bridge – elevation (Pischl and Schickhofer, 1993).

8.3.3 Mur River Bridge, Austria

The Mur River Bridge is a three-hinged parabolic timber arch bridge (Figure 8.5). The angled supports between the abutment and the arch springing point interact with the main girders to offer longitudinal stability. The vertical posts are rigidly connected to the main girders and arches using steel plates. The cross-section (Figure 8.6) and

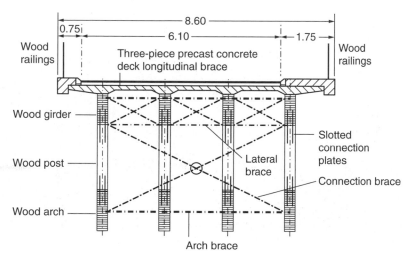

Figure 8.6 Mur River Bridge – construction (Pischl and Schickhofer, 1993).

Figure 8.7 Mur River Bridge – cross-section (Pischl and Schickhofer, 1993).

Figure 8.8 Mur River Bridge – bracing (Pischl and Schickhofer, 1993).

construction photograph (Figure 8.7) clearly show the attention to detail and structural complexity of the construction. The precast concrete deck units were placed on neoprene strips and fixed using glued threaded steel rods 30 mm in diameter. Longitudinal and transverse bracing was provided (Figure 8.8).

8.3.4 Wisconsin Avenue Viaduct, USA

The Wisconsin Avenue Viaduct in Milwaukee, Wisconsin is a 444 m long 11-span precast, prestressed concrete arch structure (Figure 8.9). The arches are provided by curved trough-shaped, precast post-tensioned arch segments that functioned as both load carrying structural members and as self-supporting permanent forms. The deck comprised pretension precast concrete beams that were framed into *in situ* concrete cross-beams. The use of precast concrete units mitigated against the construction constraints expected during severe weather conditions.

This structure is of particular interest as the alternative arch solutions were presented by the designers (Wanders *et al.*, 1995). Figure 8.10 shows the four alternative arch solutions which were considered. Alternative 1 comprised a deeply arched box girder. Although the temporary works would be reused it was thought that this might prove difficult.

The second alternative proposed a classical concrete arch with an open spandrel. The client had specified that deck replacement had to be possible without total demolition, This resulted in the provision of arch ribs and spandrel columns with longitudinal support girders below the deck. The designers were not convinced that a replacement deck could be installed economically.

Figure 8.9 Wisconsin Avenue viaduct (Wanders et al., 1995).

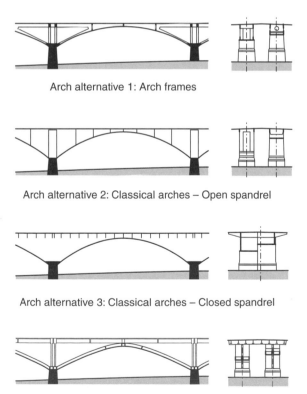

Arch alternative 1: Arch frames

Arch alternative 2: Classical arches – Open spandrel

Arch alternative 3: Classical arches – Closed spandrel

Arch alternative 4: Strutted arches

Comparative arch span units

Figure 8.10 Wisconsin Avenue schemes (Wanders et al., 1995).

A solid or filled spandrel arch design (alternative 3) offered a viable solution to the deck replacement issue. The spandrel walls would stiffen the arch rib and also provide support for the transverse deck ribs. The deck supports comprised equally spaced transverse beams and were thus fully replaceable. The forming system would be reusable. This proved to be the most expensive solution.

The adopted solution was alternative 4 which has already been described above. Because the arch ribs were precast minimal support from ground level was required. The segments were connected and prestressed at the crown then filled with concrete. The precast prestressed concrete deck beams were then installed and the deck, diaphragms and cross-beams finally cast using temporary supports of the extrados of the arch ribs.

8.3.5 Commercial Street Bridge, Sheffield, UK

The bridge carries an arterial tramway close to the Sheffield City centre (Figure 8.11). There were many constraints to the site which led to a tied arch solution that offered minimum depth of construction and an enhancing visual impact. All the main members are in steelwork. The segmental arch members are pinned at each end and are braced together over the central section. Each of the 1.6, deep steel edged beams tie their respective arch and are suspended by a series of 60 mm diameter solid high

Figure 8.11 Commercial Street Bridge (Wilson and Jones, 1995).

Figure 8.12 Commercial Street bridge (Wilson and Jones, 1995).

tensile steel bar hangers. The planes of the two arches are inclined towards each other at the crown with the result that the hangers appear to criss-cross from any viewpoint. The deck is formed using transverse rolled steel beams spanning between the steel edge beams and acting compositely with a 250 mm thick reinforced concrete deck slab.

There was a conscious effort by the designers to honestly express all the main features including the pins, hanger connections, bracing, etc. which produces a very satisfactory solution (Figure 8.12).

8.3.6 Hulme Arch Bridge, Manchester, UK

The bridge is 52 m span, supported by 22 diagonal spiral strand cables 51 mm in diameter and hung from a single parabolic arch (Figure 8.13). The arch is made from a tapering trapezoidal steel box section coated in aluminium paint. The trapezoidal section varies from 3 m wide and 0.7 m deep at the crown to 1.6 m wide and 1.5 m deep at the springings. The arch rises 25 m above the bridge deck and is connected to each of the springing foundations using 32 high tensile stainless steel 40 mm diameter bars. The bridge deck comprises a composite concrete slab cast on permanent

Figure 8.13 Hulme Arch Bridge (Hussain and Wilson, 1999).

formwork supporting 17 transverse girders spanning between steel edge beams. The support cables are attached to outrigger brackets.

Because the arch spans diagonally across the deck, the arch is subjected to not only the vertical arch forces but also significant out-of-plane bending moments and shears. This resulted in extensive stiffening and the crown section to be lightweight concrete filled. The lower sections of the arch were designed for sacrificial allowance of 0.6 mm per exposed surface and additionally, were shop-and-site-coated with a vapour corrosion inhibitor (VCI) using a micromist wet-fogging process.

8.3.7 Tyne Bridge, UK

The Tyne Bridge has served as a main route connection the Newcastle conurbation across the River Tyne since its construction in the 1920s. It comprises a two-pinned steel truss arch structure, spanning the entire width of the river (Figure 8.14). Most of the bridge deck is suspended from hangers attached to the lower members of the arch ribs, whilst the remaining deck is supported by spandrel columns of the arch ribs below. The main arches comprise two parallel steel arch ribs. Each rib spans 162.76 m between pinned skewbacks (the steel pins are 305 mm diameter) and rises 51.82 m to the highest point. The ribs have a box section made up of riveted steel angles and plates. The hangers are formed from two 305 × 102 mm channels connected using 254 × 12 mm plates riveted to their flanges.

Figure 8.14 Tyne Bridge (Lilley, 1995).

8.3.8 Runcorn–Widnes Bridge, UK

This bridge is a fine example of a truss cantilever arch (Figure 8.15). Its main span is 330 m (1082 ft) whilst the side spans are only 76 m (250 ft), a little smaller than is

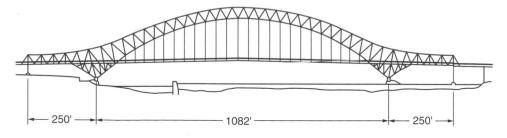

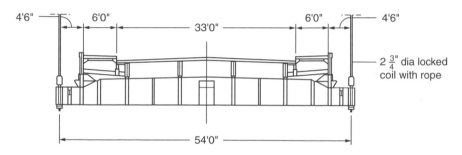

Figure 8.15 Runcorn–Widnes Bridge (O'Connor, 1971).

usual. The concrete deck is supported on stringers between the cross girders which
are supported by 70 mm (2.75 in) diameter locked coil hangers from the arches. Lateral bracing is provided in the deck and at the levels of the upper and lower chords
of the arch. There is no lateral bracing between the upper chords of the side spans
as they are in tension due to the cantilever action of the side spans which are held
down by two pairs of vertical links.

8.3.9 Port Mann Bridge, Canada

The Port Mann Bridge across the Fraser River near Vancouver is an interesting
variation of the tied arch (Figure 8.16). The central part of the main span is supported
by rigid hangers whilst the remaining parts are supported on spandrel columns secured
to the extrados of the arch, which descends below the deck to intermediate
piers. This offers some cantilever support to the orthotropic steel deck reducing its
deflections.

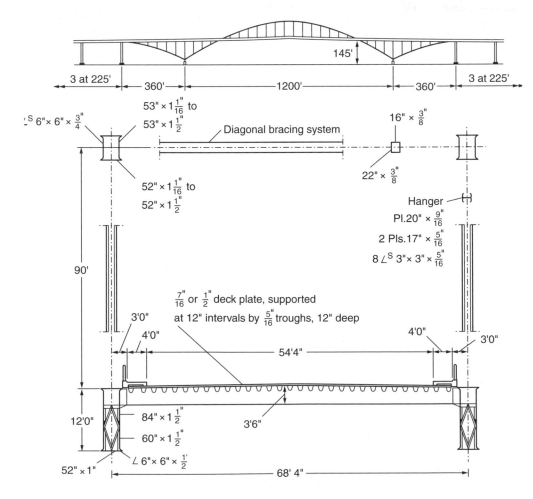

Figure 8.16 Port Mann (O'Connor, 1971).

Figure 8.17 Brno–Vienna Expressway (Strasky and Husty, 1999).

8.3.10 Brno-Vienna Expressway bridge at Rajhrad, Czech Republic

The bridge carries a local road across the Brno-Vienna expressway near the town of Rajhrad in the Czech Republic. The total length of the bridge is 110 m with a central arch span of 67.5 m (Figure 8.17).

At an early stage in the design it was decided that, for aesthetic reasons, an arch bridge was the preferred solution. The designers considered four options (Figure 8.18) (Strasky and Husty, 1999):

- two concrete arches supporting a double-T concrete deck on single spandrel columns
- a narrow steel box arch supporting a box deck on single spandrel columns
- a steel tube arch connected to a slender concrete deck with a tubular truss
- a steel tube arch supporting a channel deck with triangular steel struts.

Although the bridge is rectangular in plan, it crosses the expressway at a skew angle of about 30°. Computer graphics highlighted the visual disorder of the first and third options, so these were eliminated. The remaining options had similar cost estimates but the final option was considered more structurally efficient and easier to maintain.

The arch comprises a 900 mm diameter 30 mm thick steel tube filled with concrete. It has a radius of 74.75 m, a span of 67.5 m and a rise of 8.05 m. (Figure 8.19). The tube is internally stiffened every 2 m by a pair of ring diaphragms, which maintain the shape of the tube, transfer shear forces and act as bearing stiffness under the struts. The struts (at 6 m centres) are perpendicular to the arch which means that the deck

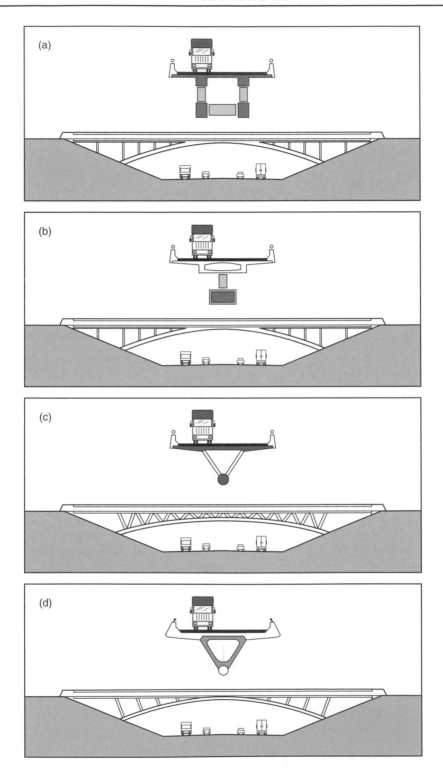

Figure 8.18 Brno–Vienna Expressway – schemes (Strasky and Husty, 1999).

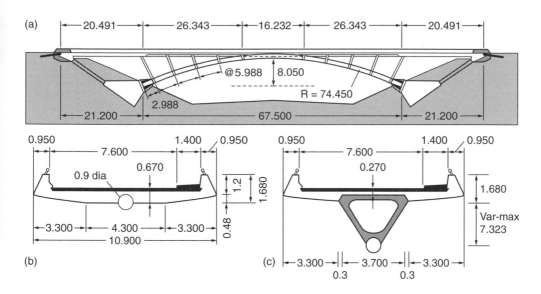

Figure 8.19 Brno–Vienna Expressway – structure (Strasky and Husty, 1999).

spans vary. They are welded to the arch at the locations of the internal diaphragms and are connected at the top by transverse beams that support the prestressed concrete deck. At midspan the deck is monolithically connected to the arch over a 5 m length. The concrete infill has a characteristic strength to 50 N/mm^2 and was pumped from the springings to the top of the arch. Prestressed concrete end struts support the side spans. They are connected to the arch abutments thus allowing longitudinal movement by rotation, whilst providing transverse stiffness through triangulation.

The short haunches are formed by welding two tapered channels to the top and bottom of the tube. Each end of the tube is welded to a steel base plate with longitudinal stiffness and openings for hydraulic jacks. The base plates are secured by four 100 mm diameter bolts to the footings which are essentially mass concrete. The bridge was analysed as a three-dimensional frame on spring supports.

8.4 Analysis

The simplest type of arch is the three-hinged arch. It is statically determinate and therefore can be analysed easily. It is usual to take a free-body diagram from a support to the internal hinge. Equilibrium of the free-body diagram together with the equilibrium of the arch as a whole enables the support reactions and horizontal hinge forces to be determined. The force at any other cross-section can be calculated by cutting the structure and considering equilibrium. It should be remembered that the shear is determined normal to the local axis of the arch which means that apart from at the crown, both the horizontal and vertical applied forces will contribute to the shear and axial arch rib forces.

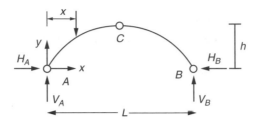

Figure 8.20 Three-hinged arch.

Influence lines can also be easily constructed to determine the approximate positions of loading to create maximum bending moments, reactions, etc. Different shapes, segmental, parabolic, etc. can be dealt with easily – the only complication being the geometry.

Consider a three-hinged parabolic arch span, L, and rise, h (Figure 8.20). The equation for the arch is $y = (4h/L^2)x(L - x)$.

In order to determine the influence line for the bending moment at the quarter span point, apply a unit load at x, as shown in Figure 8.20.

$$V_A = 1((L - x)/L) \text{ and } V_B = x/L$$

Taking moments at C for the free body BC.

$$V_B (L/2) - H_B h = 0, \text{ therefore } H_B = (L/2h)V_B = x/2h = H_A$$

When the unit load is applied at the quarter span.

$$\text{Bending moment } V_A (L/4) - H_A(3h/4) = (3L/32)$$

It can be shown that, as the unit load passes over to the side of the arch remote from the quarter span point being considered, the influence line bending moment linearly varies from $L/16$ at the crown to zero at the springing. Note that the bending moment is independent of h.

On the other hand the maximum horizontal reaction occurs when $x = L/2$; then $H_A = H_B = H_C = (L/4h)$ and consequently is dependent upon the span rise ratio.

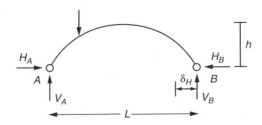

Figure 8.21 Two-hinged arch.

By removing one of the hinges the arch becomes statically indeterminate, having one redundancy. The two hinges are usually placed at the springings (Figure 8.21). From a practical point of view it may be considered that this is the most likely condition of an arch, given that most foundations will accommodate a little rotation.

For a linear elastic structure the strain energy U and complementary energy C are equal. It follows from Castigliano's theorems that the partial derivative of the strain energy, U, with respect to a force, is equal to the displacement in the direction of the force.

The strain energy of an arch rib is made up of three components:

(a) The strain energy due to bending $= U_B = \int_A^B (M^2 ds)/2EI$

(b) The strain energy due to axial thrust $= U_T = \int_A^B (T^2 ds)/2AE$

(c) The strain energy due to shearing force $= U_S = \int_A^B (S^2 ds)/2GA$

$$U = U_B + U_T + U_S$$

For an arch rib $\partial U/\partial H = \delta_H = 0$ for the special case where there is no movement of abutments. Hence:

$$\frac{\partial U}{\partial H} = \int_A^B \frac{M \frac{\partial m}{\partial H}}{EI} ds + \int_A^B \frac{T \frac{\partial T}{\partial H}}{AE} ds + \Delta \int_A^B \frac{S \frac{\partial S}{\partial H}}{GA} ds$$

where Δ = constant dependent on the shape of the cross-section.

At a preliminary stage or when checking computer analysis output, it is convenient to ignore the axial thrust and shearing force terms and assume no movement of abutments; in which case the above equation reduces to the much simpler:

$$\frac{\partial U}{\partial H} = 0 = \int_A^B M \frac{\partial m}{\partial H} \frac{ds}{EI}$$

The bending moment in any general arch may be considered as the sum of two loading cases (Figure 8.22).

- externally applied loading with the arch on a roller at one of the supports, and
- the arch with a roller at the same support and an unknown horizontal thrust applied to the roller.

Total bending moment at x is given by $(BM_x) = M_s - H_A y$, where M_s is statically determinate bending moment, therefore $\partial BM_x/\partial H_A = -y$. But the strain energy, $U_B = \int_A^B (M^2 ds)/2EI$.

$$\frac{\partial U_B}{\partial H_A} = \delta_A = \frac{\partial}{\partial H_A} \int_A^B \frac{M^2 ds}{2EI} = \int_A^B M \frac{\partial m}{\partial H_A} \frac{ds}{EI}$$

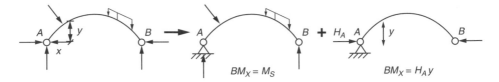

Figure 8.22 General case.

$$\delta_A = \int_A^B (M_s - H_A y)(-y)\frac{ds}{EI}$$

$$= -\int_A^B M_s y\frac{ds}{EI} + H_A\int_A^B y^2\frac{ds}{EI}$$

If $\delta_A = 0$ then:

$$H_A = \int_A^B M_s y\frac{ds}{EI} \bigg/ \int_A^B y^2\frac{ds}{EI}$$

If $\delta_A \neq 0$ or if there is a change in temperature, then bending moments are induced in the arch rib. Consider the two-hinged arch AB free to move horizontally at B (Figure 8.23). If the arch material has a coefficient of linear expansion of α per degree Celsius and is subjected to a rise in temperature of $t°C$ then the free expansion of the arch $= \alpha t L$, (which is very small compared with L). If the thrust H brings the arch back to its original position then $\partial U/\partial H = \alpha t L - 2kH = -\int M^2 y(ds/EI) + H\int y^2(ds/EI)$.

In the case of the thrust being resisted by a tie, then $\partial U/\partial H$ extension of the tie $= -(HL_t)/(A_t E_t)$, i.e.

$$\frac{\partial U}{\partial H} = \int M_s y\frac{ds}{EI} + H\int y^2\frac{ds}{EI} = -\frac{HL_t}{A_t E_t}$$

Therefore:

$$H = \frac{\int M_s y\dfrac{ds}{EI}}{\dfrac{L_t}{A_t E_t} + \int y^2\dfrac{ds}{EI}}$$

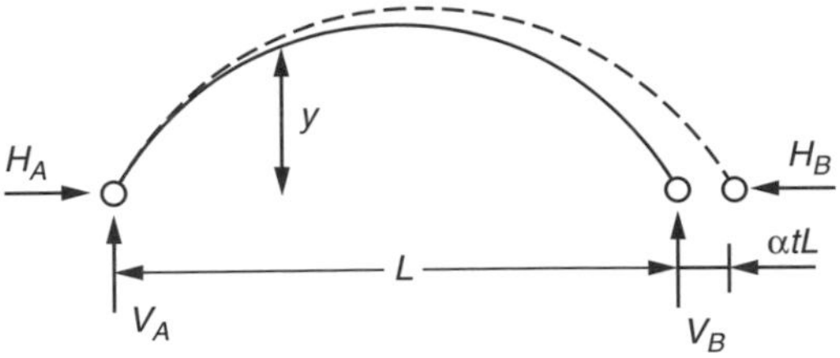

Figure 8.23 Thermal expansion.

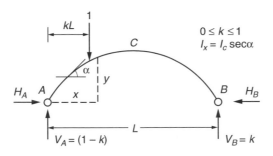

Figure 8.24 Two-hinged arch ($I_x = I_c\sec\alpha$).

It is interesting to note that if the tie and arch material both have the same coefficient of thermal expansion then no extra thrust is developed by changes in temperature.

Consider the specific case of a two-hinged parabolic arch with a secant variation of I (see Figure 8.24). We have that:

$$H = \frac{\int M_s y \dfrac{ds}{EI}}{\int y^2 \dfrac{ds}{EI}} = \frac{\int M_s y \dfrac{dx\sec\alpha}{EI_c\sec\alpha}}{\int y^2 \dfrac{dx\sec\alpha}{EI_c\sec\alpha}} = \frac{\int M_s y dx}{\int y^2 dx}$$

$$\int M_s y dx = \int_0^{KL} {}^1 (1-k)x\frac{4h}{L}\left(x - \frac{x^2}{L}\right)dx + \int_0^{(1-K)L} 1kx\frac{4h}{L}\left(x - \frac{x^2}{L}\right)dx$$

$$= (1-k)\frac{4h}{L}\left[\frac{x^3}{3} - \frac{x^4}{4L}\right]_0^{KL} + k\frac{4h}{L}\left[\frac{x^3}{3} - \frac{x^4}{4L}\right]_0^{(1-k)L}$$

$$= \frac{hL^3}{3}(k + 2k^3 + k^4)$$

$$\int y^2 dx = \int_0^L \left\{\frac{4h}{L}\left(x - \frac{x^2}{L}\right)\right\}^2 dx = 16\frac{h^2}{h^2}\left[\frac{x^3}{3} - \frac{x^4}{2L} + \frac{x^5}{5L^2}\right]_0^L = \frac{8h^2L}{15}$$

Therefore $H = ((hL^2/3)(k - 2k^3 + k^4))/(8((h^2L)/15)) = ((5/8)(L/h))(k - 2k^3 + k^d)$

The expression for H may be used to obtain the thrust due to a uniformly distributed load w per unit length/by considering a length δkL and integrating, thus:

$$\text{Thrust due to the element} = \delta H = 5L/8h(k - 2k^3 + k^4)w\delta kL$$

$$\text{Integrating } H = (5L^2w/8h)\int_{k1}^{k2}(k - 2k^3 + k^4)\delta k$$

Note that when $k_1 = 0$ and $k = 1$, i.e. the udl is applied across the entire span then $H = wL^2/8h$.

The conventional first-order theory may be unsafe for some geometries. It must be remembered that arches are mainly compressive structures and as such are susceptible to buckling. Consequently, global and member stability should be considered.

In plane buckling is possible in modern arches of slender proportions. Closed form solutions for the problem have been derived which are of the general form:

$$\text{Critical load is proportional to } EI/L^2$$

For a parabolic arch, span L rise R with a uniformly distributed dead load, if the dead load moments are zero the horizontal reaction, H, is given by

$$H = wL^2/8R$$

Therefore:

$$H_{cr} = (W_{cr}/8R)L^2 = C_1(EI/L^2)$$

Figure 8.25 plots values of C, against R/L for parabolic arches with 0, 1, 2 or 3 hinges and for the two cases, where I is constant and secondly where I is proportion to the secant inclination of the arch.

The critical load is significantly affected by the number of hinges and the stiffness of the springing.

Although imperfections in the initial shape may not be significant in terms of global stability, they may be critical for local buckling of members or plates and thus lead to collapse.

It is important to check stability at every stage (including especially construction) and at every level – from the smallest element to the whole structure.

8.5 Design

It is beyond the scope of this chapter to cover all aspects of design given the range of combinations of material and structural arrangements. Masonry arches will be considered in some detail later whilst in the present section the criteria for 'modern' materials are considered.

The economic viability of the scheme often depends on the suitability of the site's geology and soil. Clearly, the ideal site is a valley with sound rock valley walls but most engineers have to tussle with poor ground and therein lies the challenge of the design. The span/rise ratio has a fundamental influence of the magnitude of the horizontal thrust. If this becomes prohibitively large then the horizontal thrust must be

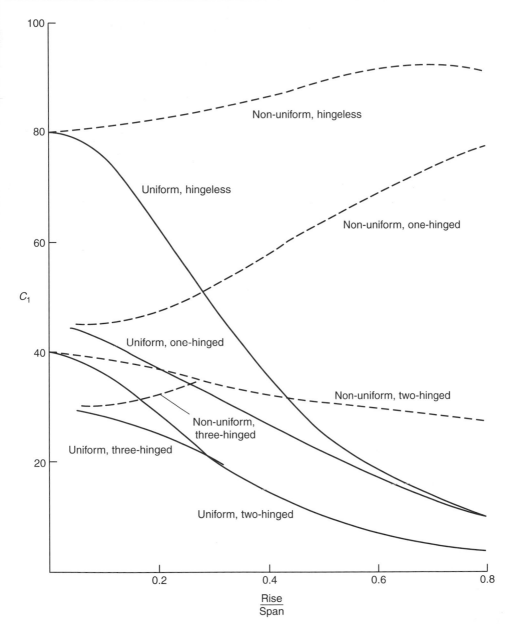

Figure 8.25 Coefficients for in-plane buckling of parabolic arch (O'Connor, 1971).

resisted by tying the abutments together in some way or using the deck to resist the arch thrust.

   If the deck is positioned high relative to the springings then this allows a smaller spandrel rise ratio which reduces the horizontal thrust (it also increases the relative vertical forces which improves the frictional forces available to resist the horizontal thrust).

A 'self-straining arch' (e.g. a bowstring arch) ensures that the abutment/pier are only called upon to resist vertical and braking forces.

The dominance of compressive forces implies the efficacy of concrete as a construction material. Formwork can be expensive but modern proprietary systems are now relatively economical. The segments can be precast, prestressed boxes (as in the case of Gladeville Bridge) or solid reinforced concrete. The need to reduce self-weight is important for larger spans but not so for small to medium spans where the extra dead weight helps to pre-compress the arch and thus enhances its performance as an arch.

For modest spans prefabricated steel sections can be used to great effect. Modern automated fabrication machinery allows complicated shapes to be produced economically and may include concrete 'filling' to act compositely with the steel should plate buckling be a concern.

Orthotropic decks can be an advantage when dead weight is a problem, although attention must be given to the aerodynamic stability of the deck and its interaction with the arch (and hinges if used).

Truss arches must be designed with particular sensitivity to aesthetic impact of the members. Merely considering elevations may not reveal the 'business' of the structure – packages now exist to allow a three-dimensional inspection of the structure. It is strongly recommended that such packages be used in all but the simplest frames. However, the weight advantage may become decisive in long span bridges. Truss bridges are usually relatively easy to erect but can suffer from the penalty of high maintenance costs and relatively low allowable stresses on account of the lateral stability requirements of the design.

8.6 Masonry arch construction

It is very important at the outset to dispel the idea that all masonry arches are of similar construction. Nothing could be further from the truth. Whether designing a new bridge, assessing an existing bridge or formulating a repair and maintenance strategy, it is vital that a holistic view is taken of the interaction between each of the elements of the bridge and that it is a three-dimensional structure.

Figure 8.26 shows a typical arch bridge construction. Where practicable the bridge foundations were taken down to rock. This was not always possible and so timber piles or grids; even faggots were used. Of course only the successful solutions have survived; no doubt there were many failures!

The barrel may take various shapes; semi-circular; parabolic; segmental; elliptical; gothic pointed etc and may comprise dressed stone, random rubble, brickwork or mass concrete. The backfill over the arch may be contained by spandrel walls which extend beyond the abutments to provide wing walls. The backfill may be anything from ash and rubble through to concrete. Clay was sometimes used as a waterproofing membrane over the extrados of the barrel. To lighten the structure and also to eliminate the horizontal soil pressures on the external spandrel walls, internal spandrel walls were used. Historically, this form of construction was used on some bridges with spans greater than about 12 m. The proportions of these internal spandrel walls

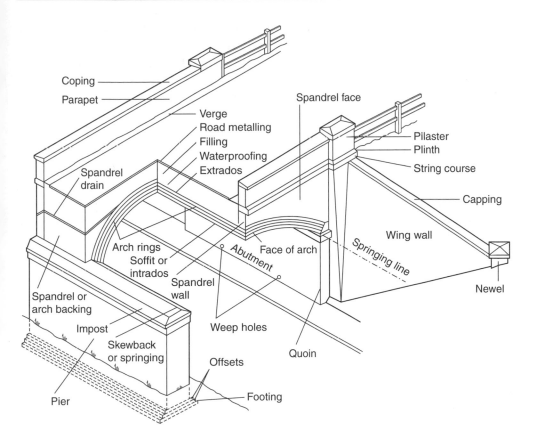

Figure 8.26 Typical masonry arch bridge (Sowden, 1990).

depended upon the nature of the available masonry and whether or not the over-spans took the form of stone, slabs or arches. Significantly, there are usually no external indications of the form of internal construction. Even the barrel thickness cannot be relied upon corresponding to that shown on the elevations. The latter was frequently proportioned to comply with the aesthetic demands of the client.

Alternatively, internal arches may be provided which span longitudinally and spring from the extrados of the main arch barrels. These may be totally internal, or extend-ed through the external spandrel walls to provide an aesthetic feature and, in the case of river bridges, an escape route for flood water.

An extension of this form of construction takes the form of a series of smaller arches supported by piers resting on the extrados of the main span. This form is particularly prevalent in China.

8.7 Assessment of masonry arch bridges

It is vital that any assessment takes a holistic approach; the materials, form of con-struction, loading etc should all be taken into account. All too often the assessment focuses upon the barrel with lesser regard to its interaction with the other elements of the bridge – when they, themselves, may be critical. In the UK, the current method of determining the load carrying capacity is embodied in the Department of Transport Department Standard BD 21/97 (Department of Transport, 1997) and Advice Note BA 16/97 (Department of Transport, 1997). The first step in the assessment is to apply the so-called MEXE method. If this yields a capacity which is too low or the type and nature of the bridge excludes its application, alternative methods of analysis and assessment should be used.

The MEXE method evolved from work undertaken in the 1930s which included both field and laboratory tests to calibrate theoretical work. During World War II, this research was used to develop a quick field method to classify bridges according to their capacity to carry military vehicles; this was subsequently adapted for civil use.

The method comprises the calculation of a provisional axle load (PAL) using either a nomogram given in the Advice Notice BA16/97 or the equation:

$$\text{PAL} = 740\,(d + h)^2/L^{1.3} \text{ tonnes (or 70 tonnes, whichever is less)}$$

where L is the span, d is the thickness of the arch barrel adjacent to the keystone and h is the average depth of fill at the quarter points of the transverse road profile, between the road surface and the arch barrel at the crown, including road surfacing. The limits set for $(d + h)$ are 0.25 and 1.8 m and for the span, L, 1.5 and 18 m.

The precise origin of the nomogram and its associated equation is not known, but it is attributed to the above mentioned research work started in the 1930s and is probably based upon the elastic analysis of a two-pinned parabolic arch with a span/rise ratio of four, subjected to a point live load applied at the crown. The induced stresses were limited to 0.7 N/mm^2 in tension and 1.4 N/mm^2 in compression. The results of a parametric study using the above criteria, modified by field observations led to the relationship given above.

Having determined the PAL it is modified by the following factors which take account of any departure from the 'standard' arch barrel geometry and condition. The span/rise factor (F_{sr}) assumes that steeper profiled arches are stronger than the flatter ones, with a ratio of 4 being taken as the optimum and hence a factor of 1. This reduces for higher span/rise ratios (Table 8.1).

The profile factor (F_p) takes account of the different shape of the arch; so far there has been an assumption that the arch is parabolic. This would have a rise at the quarter point (r_q) equal to 0.75 times the rise at the crown (r_c). For values of r_q/r_c less than or equal to 0.75 the profile factor F_p is taken to be unity. For ratios greater than 0.75, i.e. for profiles which may be elliptical or semi-circular etc the profile factor may be calculated from the expression:

Table 8.1 Span/rise factor (F_{sr})

Span/rise ratio	<4	5	6	7	8
F_{sr}	1.0	0.85	0.75	0.67	0.61

$$F_p = 2.3[(r_c - r_q)/r_c]^{0.6}$$

The material factor (F_m) is determined using an expression which incorporates a barrel factor, F_b, and fill factor, F_f, together with d and h.

$$F_m = ((F_b d) + (F_f h))/(d + h)$$

The norm for the barrel factor is taken as good condition limestone or building brick whilst the fill factor ranges from 0.5 for a weak fill through to 1.0 for concrete backfill. Typical values are presented in Tables 8.2 and 8.3.

It is recognized that the strength of masonry is influenced by the size and condition of the mortar joints. As the width of the joint increases the strength of the masonry decreases. Similarly, the poor condition of the mortar and incompleteness of the perpendicular bed both adversely affect the strength. The joint factor, F_j, is determined using the formula:

Joint factor (F_j) = width factor (F_w) × depth factor (F_d) × mortar factor (F_{mo})

Typical values for each of the factors are presented in Tables 8.4–8.6. Finally, a condition factor F_{cm} is applied (see Table 8.7 for typical values). It is important in

Table 8.2 Barrel factor (F_b)

Arch barrel	Barrel factor
Granite and Whinstone whether random or courses and all built-in course masonry except limestone, and all large shaped voussoirs.	1.5
Concrete or engineering bricks and similar sized masonry (but not limestone).	1.2
Limestone, whether random or courses, good random masonry and building bricks, all in good condition.	1.0
Masonry of any kind in poor condition (many voussoirs flaking or badly spalling, shearing, etc.). Some discretion is permitted if the dilapidation is only moderate.	0.7

Table 8.3 Fill factor (F_f)

Filling	Fill factor
Concrete	1.0
Grouted materials (other than those with a clay content)	0.9
Well-compacted materials	0.7
Weak materials evidenced by tracking of the carriageway surface	0.5

Table 8.4 Width factor (F_w)

Width of joint	Width factor
Less than 6 mm	1.0
Between 6 and 12.5 mm	0.9
Greater than 12.5 mm	0.8

Table 8.5 Mortar factor (F_{mo})

Condition of mortar	Mortar factor
Good	1.0
Loose or friable	0.9

Table 8.6 Depth factor (F_d)

Condition of joint	Depth factor
Good, fully filled	1.0
Unpointed joints, pointing in poor condition and joints with up to 12.5 mm from the edge insufficiently filled	0.9*
Joints with from 12.5 mm to one the nth of the thickness of the barrel insufficiently filled	0.8*
Joints insufficiently filled with more than one-tenth of the thickness of the barrel	At the engineer's discretion

*Interpolation between these values is permitted.

Table 8.7 Recommended condition factors (F_{cm})

Defect	Recommended condition factor F_c
Longitudinal cracks due to settlement of one edge to bridge	#0.4 (crack spacing #1 m) 0.4–0.6 (crack spacing > 1 m)
Transverse cracks or deformation of arch due to partial failure of arch or movement of abutments	0.6–0.8
Diagonal cracks	0.3–0.7
Cracks in the spandrel walls near the quarter points	0.8

Note. Where F_{cm} is less than 0.4, immediate consideration should be given to the repair or reconstruction of the bridge.

arriving at a value (between 0 and 1.0) that defects are not double-counted. For example friable mortar and weak backfill have already been considered.

The modifying factors are applied to the provisional axle load to give a modified axle load:

$$\text{Modified axle load (MAL)} = F_{sr} F_p F_m F_j F_{cm} \text{ (PAL)}$$

There is an assumption that the load is applied through a double-axled bogie and that the surface vertical alignment is such that there is no lift-off of either axle. If this is not the case then further modification is necessary.

There is the further assumption that the bridge comprises a single span between stable abutments. Clearly this is not always the case and further modifications should be made depending upon the situation. Values which were suggested at the time of the original development of the MEXE method were

- 0.9 for arches supported by one abutment and pier, while
- 0.8 should be used for arches supported on two piers.

8.8 Alternative methods to the modified MEXE method

There are many situations where the modified MEXE method cannot be used or where the bridge fails according to the method. If such cases exist it may be deemed necessary to consider alternative methods of assessment. Several approaches have been developed in recent years.

The masonry arch can be modelled as the pinned elastic arch and analysed using a suitable frame analysis or FE program. The analysis should be undertaken in two stages; under dead load and under live loads. The ultimate live load can then be determined by applying appropriate modifying factors usually equal to the product of the joint and conditions factors used in the MEXE method, but may also need adjustment for the fill (although this should be considered when modelling the barrel and backfill).

The arch ring idealization should comprise at least 12 elements with pinned support assumed at the springings. There should be several nodes in the region of the 1/4 to 1/3 span region as experimental and analytical studies have shown this region to be the most vulnerable load point for the arch. The load should be applied at the road level and dispersed through the backfill and arch material at slopes of two vertical to one horizontal. The analysis is usually based upon a 1 m width of arch so an estimation of the effective width in relation to the axle width is necessary. The guidance, based upon field observation, is to assume an effective transverse width, w, of the arch barrel equal to:

$$w = (h + 1.5) \text{ m}$$

where h is the fill depth, in metres, at the point under consideration.

The load may comprise more than one axle in which case the influence line (Figure 8.27), may be used to determine the first estimate of the worst position.

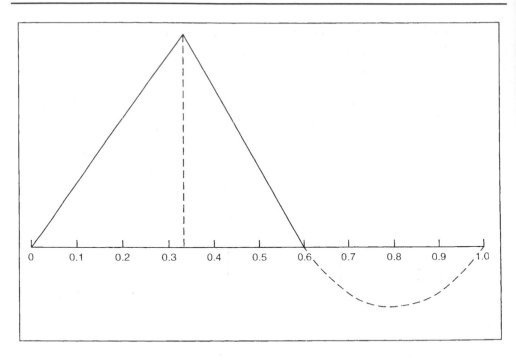

Influence Line for Determining Critical Load Position

Figure 8.27 Influence line (Highways Agency, 1998).

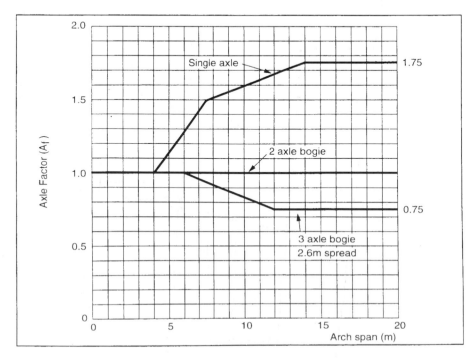

Figure 8.28 Axle factor, A_f (Highways Agency, 1998).

It may be appropriate, in the first instance, to undertake the analysis using a single axle. In the case of EC and C&U vehicles this can be used to derive the allowable multi-axle loads using Figure 8.28 and multiplying the single axle loading by the axle factor A_f.

The maximum compressive strength should be determined from laboratory tests if possible. In the absence of any actual strength tests data, conservative estimates may be made using available data; Figures 8.29 and 8.30 give examples of characteristic strengths.

The ultimate load capacity of the arch is assumed to have been reached when the total dead weight and live load compressive stress at any section, calculated using the full depth of section equals the ultimate compressive strength of the masonry. The corresponding load is the ultimate load. This must be modified to give an allowable load. The computer idealization has absorbed most of the modification factors that the MEXE method required with the exception of the joint factor, F_j, and the condition factor, F_{cm} both of which should be applied to the maximum axle failure load. To obtain allowable axle/vehicle load a partial factor γ_{FL} is applied such that:

$$\text{Allowable single axle load} \times \gamma_{FL} = \text{theoretical maximum single axle failure load}$$
$$\times F_j \times F_{cm}$$

where γ_{FL} is = 3.4.

If more than one axle is being considered then $\gamma_{FL} = 3.4$ should be used for the critical axle and a value of 1.9 for the remaining axles.

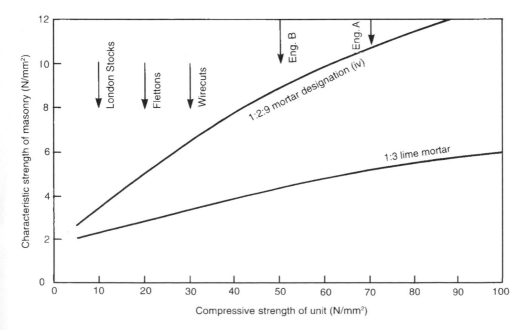

Figure 8.29 Characteristic strength – brickwork (Page, 1983).

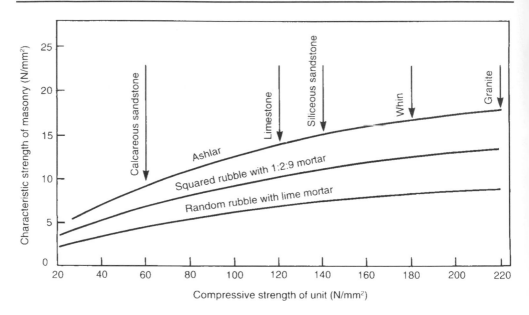

Figure 8.30 Characteristic strength – stonework (Page, 1983).

Table 8.8 Material parameters (unit, mortar and masonry) to be considered in a masonry arch bridge analysis

Compressive strength	Elastic modulus
Flexural strength	Shear modulus
Tensile strength	Poisson's ratio
Shear strength	Fracture energy
Bond strength (at mortar/unit interface)	Area of adhesion
Friction coefficient (at mortar/unit interface)	

A value of $\gamma_{FL} = 2.0$ may be used for situations where the axle loads are known with accuracy as is likely to be the case for abnormal loads.

It is important before embarking on any structural modelling that the analyst has an appreciation and understanding of the behaviour of the structure that is to be modelled and the experimental data which are going to be used to validate the computer models. Table 8.8 presents some of the parameters which must be considered. What is important is the realization that the construction of masonry arch bridges can be diverse and it is this diversity which can present the analyst at the macroscopic level with many problems. At the microscopic level different problems arise. These will be considered in the next section.

8.9 The influence of masonry materials

Masonry (whether comprising bricks or stone blocks) may be considered to be a composite where units are held together (or kept apart depending on one's percep-

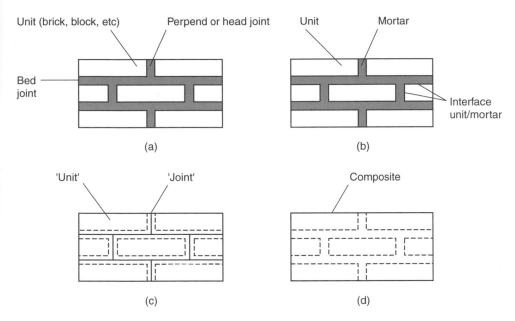

Figure 8.31 Modelling strategies for masonry (Lourenco, 1996).

tion) by mortar joints. In any event, the properties of the unit, mortar and the unit/mortar interface must all be robustly represented in the model. This may be in a form which allows the computer program to decide the nature and extent of the response or it may be greatly influenced by predetermined constraints. It is this range of interventionist modelling (this also applied to experimental strategies) which can influence the model response and hence the predicted behaviour and carrying capacity.

Figure 8.31 shows a range of possible modelling strategies that may be used for masonry structures. Ideally, the analyst would wish to represent each physical unit with elements possessing the mechanical properties of the unit material (Figure 8.31a). Similarly, the mortar would be appropriately represented. The two materials may then joined at the interface (Figure 8.31b). This level of 'accuracy' may be misplaced given the variability in the constituent materials' properties, which are not insignificantly influenced by the workmanship.

An alternative strategy replaced the mortar and units by an equivalent material and the interface by an equivalent interface. This retains the ability to locate and trace the development of crack patterns which can be correlated with actual structural behaviour (Figure 8.31c).

Finally a general composite may be provided which has the overall properties of the brickwork or stonework which is 'smeared' over the structural component. Crack criteria can be set but these may result in a crack pattern which is oblivious of the masonry bonding and so will need some post-processed interpretation (Figure 8.31d).

At a microscopic level both the units (whether brick, concrete or rock) and the mortar display softening that is a deterioration of mechanical performance under monotonic or cyclic increase in deformation. Such behaviour is usually attributed to the nature of and defects within ceramic materials and the complexity of their interaction. By the very nature of the construction process, masonry arches contain many flaws. As suggested above, bricks contain flaws that are consequential upon the manufacturing process and which in engineering quality bricks are stable at low stress levels. Similarly, the stone blocks may contain flaws initially formed during the geological process that created the rock from which the blocks were quarried. The latter process can reveal and/or aggravate defects.

Initially, the micro-cracks are stable and propagation is only caused by increased stress levels. (A durable unit will not be one where cyclic loading within a known stress regime does not cause crack propagation. Although it sounds quite straightforward to quantify, the situation is significantly complicated by the interaction between the mortar and the units. Deterioration of any of the mechanical properties by interface bond, may change the 'internal' response to a given 'external' stress regime and thus allow crack propagation. This can be an important consideration when assessing arches containing macroscopic defects like ring separation in multi-ring arch bridges). The effect of the presence of cracks is to cause stress redistribution with the crack surface attracting zero normal stress thus locally releasing the strain energy. (This phenomenon is utilized in the construction of mass concrete arches containing inclusion (Melbourne and Njumbe, 1998) and 'protecting' the distressed material.)

Typical stress displacement graphs for ceramic materials are shown in Figure 8.32, in each case there is a well-defined softening which under soft loading systems would result in 'brittle' failure. Some of the most advanced work (Rots, 1997; Lorenco, 1996) attempts to incorporate parameters like the fracture energy, G_f^I and the compressive fracture energy G_c as material properties. Clearly, in compression the behaviour of the masonry is significantly influenced by the unit-mortar interface performance under shear loading (Figure 8.32). This leads to the definition of mode II fracture energy G_f^{II} as the integral of the $\tau - \delta$ graph in the absence of a stress normal to the bedding plane.

Notwithstanding the above there is a strong correlation between the strength of the masonry and the strength of its constituent parts.

Although the unit-mortar interface may be considered a 'weakening' feature of masonry, it must be remembered that it is this very 'weakness' which provides masonry with the ability to articulate and accommodate movement without loss of structural integrity.

Van der Pluijm (1992) carried out a series of controlled laboratory tests on small solid clay and calcium silicate units. Mode I fracture energies G_f^I were obtained which ranged from 0.005 to 0.02 Nmm/mm^2 for tensile bond strengths ranging from 0.3 to 0.9 N/mm^2. Interestingly, it was observed that the bonded area was concentrated in the inner part of the specimen; this was probably a consequence of shrinkage resulting from surface moisture loss. The effect of this was to reduce the bonded area for the small square interfaces to about one third of the total area with the implication that the bonded area for a long wall would be about 60%. This of course makes no allowance for the effects of workmanship and presupposes that the mortar-unit is

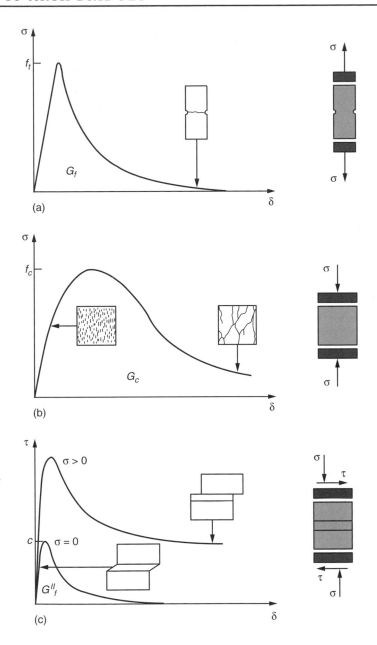

Figure 8.32 Typical behaviour of quasi-brittle materials under uniaxial, tensile, compressive and shear loading (Lourenco, 1996).

engaged under optimum conditions if the unit is neither totally dry or saturated (Hendry, 1990a) suggests about 75% full saturation) and there is 100% contact at the time of laying.

Mode II failure is notoriously difficult to simulate under experimental conditions (Riddington, 1997). Van der Pluijm (1993) studied the behaviour of clay and calcium

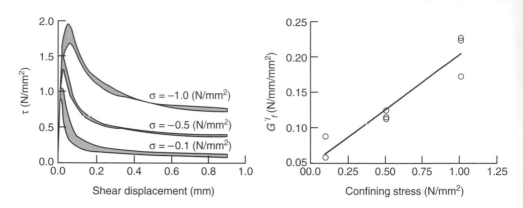

Figure 8.33 Typical shear bond behaviour of joints for clay bricks (van der Pluijm, 1993).

silicate units in a specially designed test apparatus which allowed shear stress measurements under confining compressive stresses of 0.1, 0.5 and 1.0 N/mm². The work raises several interesting observations which indicate depending on the nature of the units and mortar.

Figure 8.33 shows a plot of the shear stress against displacement. The area under the curve is defined as the mode II fracture energy G_f^{II} with values ranging from 0.01 to 0.25 N mm/mm² for initial cohesion, c, values ranging from 0.1 to 1.8 N mm/mm², respectively. The test also allowed the initial friction angle associated with a Coulomb friction model which ranged from 0.7 to 1.2 for the unit/mortar combinations tested. Significantly the residual internal friction angle remained approximately constant and equal to 0.75. The dilatancy angle, , was recorded as one unit sheared over another. What is important is its dependency upon the confining pressure, the unit surface roughness and the amount of slip. At low counting pressures the tangent ranged from 0.2 to 0.7 depending on the unit surface roughness. As the confining pressure increased the tangent decreased, an obvious consequence of local crushing of the sand particles as they attempted to pass over each other. Similarly, at large amounts of slip the shearing surfaces 'smoothed' with the tangent decreasing to zero.

Currently there does not appear to be any reliable test data on Mode III behaviour of masonry. In the absence of these data it has to be assumed that extrapolation of the Mode II data are the best that can be offered. The dependency confining pressure, unit roughness and amount of slippage can be assumed, although the contact area and shear distribution is open to some speculation.

Having considered in some detail the nature of the mortar/unit interface it is now necessary to consider the overall behaviour of the composite. There is the traditional dilemma between standardized test data and 'real' structure performance, for example in a typical series of tests the strength of two-units was about 11% larger than that of the one-unit wide specimens. Under uniform compressive loading failure usually occurs by the development of tensile cracks parallel to the axis of loading (see Figure 8.34). The degree to which the development of such cracks is prevented due to constraint will positively influence the carrying capacity. Similarly it is the lack of constraint of the unit of the mortar which results in the masonry having a lower strength

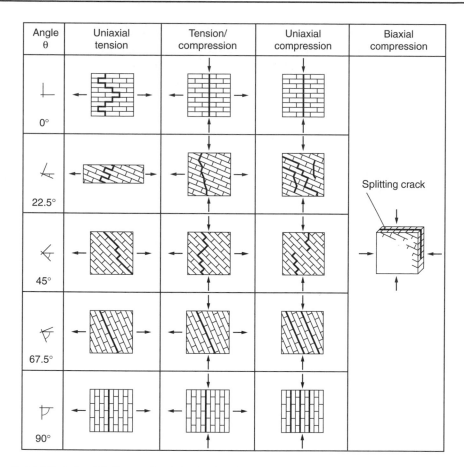

Angle θ	Uniaxial tension	Tension/ compression	Uniaxial compression	Biaxial compression

Figure 8.34 Models of failure of solid clay brickwork under biased loading (Dhanaseker et al., 1985).

than the unit. The converse is true for the mortar which is constrained by the units and consequently achieves a much higher strength although its relatively high poisson ratio leads to a propensity to tensile cracking of the units as stated above. If follows that the thickness of the mortar beds (and the volume of mortar per cubic metre of masonry in the case of random rubble masonry) has a significant effect upon strength.

Figure 8.34 shows the modes of failure of solid clay units masonry under biaxial loading. Although the behaviour of a masonry panel subjected to uniaxial loading perpendicular or parallel to the bedding plane may be well understood and well documented, the response is anisotropic and highly dependent on the bonding and unit/mortar strengths which makes the extrapolation of a continuous failure surface somewhat circumspect. Page (1981, 1983) demonstrated the strength enhancement for biaxial loading compared with uniaxial loading. It is believed that the failure mechanism is the same, namely tensile cracking resulting from the tensile stresses induced at a flaw as the compressive stress flows round it. It is important to notice that in the case of biaxial compression the expected mode of failure is splitting in the plane of the applied loads.

The above has used bonded brickwork to illustrate the relative significance strength parameters have upon behaviour. Additionally it must be remembered that masonry can exist in several types including brickwork, dressed stone and random rubble. Test data usually related to small parallel bedded specimens, as the masonry in the barrel is curved it is inevitable that in the case of brickwork the mortar joints are wedge-shaped because the bricks are a parallel pipe. Dressed stone, on the other hand may be cut to give wedge-shaped stone units (voussoirs), which will have a uniform bed thickness (historically some were so accurately cut that no mortar was required). Finally random rubble masonry will have varying thickness of both mortar and stone. Some care should be exercised when extrapolating small unit tests to large bridge structures. The Eurocode EC 6 has attempted to standardize procedures by designating the reference unit dimensions as 200×200 mm (height and width) in parallel to the introduction of a shape factor to accommodate national variations. Although EC 6 excludes bridges it may be used as a guide in the absence of any specific code.

BS 5628 gives information on the mean compressive strength at 28 days for a range of mortars although for assessment purposes it will probably be quite difficult to obtain *in situ* data. It is recognized that the mortar will usually be weaker and less stiff than the units, hence under compressive loading the masonry ultimately fails by cracking of the units. The thicker the mortar joint, therefore, the greater the Poisson's strains and the lower the failure load (all other things being equal). For example in brickwork variations between 5and 20 mm thick bedding joints results in approximately $\pm 10\%$ variation in brickwork uniaxial compressive strength compared with that of brickwork with a 10 mm thick bedding joints (Hendry, 1990b). The standard test data relates to axially loaded specimens. In an arch the masonry is subjected to eccentric loading; although this will result in higher stresses at the intrados or extrados it is recognized that stresses higher than those permitted for axial loading can be tolerated. A figure of 20% has been suggested (Hendry, 1990a) if the eccentricity of the thrust exceeds 0.15 of the arch thickness and is not greater than 0.4 of the arch thickness. Additionally, if the strength has been determined using small diameter 'radial' cores it is important to appreciate that not only are the results affected by size but that they are also perpendicular to the thrust plane in the arch barrel; because masonry is anisotropic this needs to be taken into account.

To determine the mortar strength in existing masonry is very difficult. It is virtually impossible to recover a competent specimen of sufficient size to test from the bridge. For old bridges lime mortar was invariably used with an estimated strength of 0.5–1.0 N/mm^2 even cement mortars are unlikely to excess 2.5 N/mm^2. Fortunately, it has to be shown in tests on masonry piers that weak mortars make little difference to masonry strengths (Hendry, 1990a).

In most research relating to the stress–strain relationship, compressive load has been applied perpendicular to the bedding joint. For this load case it is generally accepted that the relationship is closely represented by the parabola:

$$f/f_{max} = 2(\varepsilon - \varepsilon^1) - (\varepsilon - \varepsilon^1)^2$$

where f_{max} and ε^1 are, respectively, the stress and strain at the maximum point of the stress–strain curve. The initial tangent modulus is given by:

$$\varepsilon = 2f_{max}/\varepsilon^1$$

and the second modulus is 75% of this.

Approximate values for brickwork and thin jointed, coursed stone masonry are given by $900f_k$ or $400\text{–}600f_m$, where f_k and f_m are the characteristic and mean strengths, respectively (Hendry 1990). Random rubble may be less than half this value.

There is no guidance for the Young's modulus parallel to the bedding plane and certainly no help when considering the general case.

It will be appreciated that for multi-ring brickwork arches where there are no headers provided, there is a propensity for the rings to separate. Apart from mortar washout the mechanical bond between the rings can be broken by the development of excessive longitudinal shear caused by the thrust moving from one ring to the next or by the stresses induced by compatibility. If the compressive strength of the mortar is expected to be greater than 1.5 N/mm^2 then a shear strength of 0.35 N/mm^2 may be assumed otherwise assume a shear strength of 0.15 N/mm^2.

8.10 Analysis of masonry arch bridges

Having considered the behaviour of masonry as a material subjected to idealized loading conditions it is necessary to marry our knowledge of the behaviour of masonry arch bridges to the available mathematical models. It must be remembered that most masonry arch bridges were conceived as gravity structures for which mass and geometry were the design criteria. Certainly, the proportions passed down from antiquity had no thought of stress criteria and were probably based upon bitter experience.

Barlow had demonstrated in 1846 that there was no unique thrust line associated with a stable arch but that there were many possibilities. Navier (1826) had shown that for linear elastic materials, where plane sections remained plane, tension could be avoided by ensuring that the thrust line lay within the middle third of the section. Combining these two pieces of work led to the well-known middle third rule of design which aimed to eliminate tension in the masonry and thus avoid any cracking. It also afforded the analysts the luxury of modelling the arch as an elastic continuum.

Castigilano (1879) applied the theorems of minimum strain energy to the arch. (These have already been used to analyse the two-hinged elastic arches above). The position of the thrust line was determined and that it lay within the middle third; if this criterion was violated the tensile zone was 'removed' and the calculation iterated until no tension was present at any point in the arch. (This forms the basis of the CTAP assessment program.)

The main advantage of an elastic analysis is that stress levels and deflections can be calculated – how meaningful they are is open to much debate but it has to be conceded that they provide a 'feel' for the problem. However, it is universally accepted that masonry arch bridges crack – even before the centring is removed. This is a very important observation which should not be obscured by the sophistication of some of the currently available modelling software.

The simplest form of modelling is to idealize the arch barrel to a two-dimensional plane frame made up of at least 12 line elements. The work can be carried out using the computer program MINIPONT (1975) but any other suitable frame-analysis or finite element program could be used. Depending on the level of sophistication required to solve the problem, this may suffice; however, where the structure is more complex a more robust analysis may be justified. In which case the MINIPONT analysis can be used to identify regions of interest. There are many FE analysis packages currently available including ABAQUS, ANSYS, DIANA and LUSAS. Each is constantly evolving and have their advantages and limitations. As outlined in the section on masonry properties the analyst is faced with the dilemma between using a smear model which averages the unit/mortar composite and the expense of modelling the unit and mortar separately and the uncertainty of replication of the unit/mortar interface. Currently, only DIANA has had coding developed specifically for masonry structures, Rots (1997), Lourenzo (1996), although to date it has only been applied to buildings. The other FE packages listed require the designation of interface models drawn from a library developed mainly for concrete and so should be viewed in that light. As shown in the masonry section above the mode of shear failure at the interface is dependent upon several factors, all of which can be swamped by workmanship factors and the condition of the materials (whether wet or dry, subject to chemical attack or just simply the mortar has been washed away!). What is important is that the theoretical justification of contact elements, 'solid' elements, etc. should be investigated thoroughly in order to understand the limitations of the model. As mentioned earlier a simple model can be used to determine the regions of particular interest. This allows for some economy in computer time by subsequently concentrating the closely meshed sections into the regions identified using the simple model. It is often the case that a standard FE package does not possess a 'ready-made' gap element which has the desired properties. It is then up to the ingenuity of the analyst to assemble elements which in combination replicate the interface. Care must be taken in setting the limits to bond strength, tensile strength of the mortar, shear strength of the interface etc. If these are set too high not only does the analyst overestimate the strength but also convergence may be a problem to a built in 'brittleness' into the system. For example the shear bond strength for a 1:2:9 lime mortar would normally be about $0.2 N/mm^2$, even this will not be achieved if there is deterioration or poor workmanship present.

While a computer is commonly used for the analysis of arch bridges, it is useful to undertake some hand calculations to develop a feel for structural behaviour and to check the computer output. For this the 'closed' solutions offered earlier may be used in conjunction with influence lines to determine carrying capacity.

An alternative approach recognizes the particulate nature of masonry and the observed collapse mechanisms. The first scientific discourse on the failure of an arch which refers to the formations of mechanisms is reported by Couplet (1730) and illustrated by Frezier's (1737–1739) based on engravings of tests undertaken by Danyzy

(1732). Prior to this any reference to failure modes were purely descriptive. Throughout the nineteenth century and the early part of the twentieth century analysis and design of masonry arch bridges depended upon elastic analysis and the middle third rule. It was from the work of Pippard (1962) that the modern mechanism method evolved. He idealized the arch to a two-dimensional arch made up of blocks (voussoirs) which possessed the following properties:

- no tension
- the constituent elements are rigid
- infinite compressive strength
- sufficient inter-surface friction to prevent sliding.

As an elastic continuum it would have possessed three redundancies. Consequently, in order to become a mechanism four releases are required and they are provided in the present model by four hinges. (They may be provided by a combination of rollers, and hinges. Additionally, material failure is possible).

The simplest model considers only the dead weight of the different elements. The collapse load W (Figure 8.35) can be determined from statics by taking moments about each of the hinges in turn and solving for the V_a, V_d and H. This method considers a kinematically admissible mechanism and is therefore an upper bound solution. Consequently, various positions of the hinges must be considered and values for W calculated until the minimum value is determined. This can be done in a systematic way as can be shown with reference to Figure 8.36. OA represents a reference line (usually the intrados) of the arch. The dashed line represents the thrust line. The vertical eccentricity of the thrust line is ε_O and ε_A at O and A, respectively. Taking moments about the thrust line at A, equilibrium is maintained if:

$$(y_A + \varepsilon_A - \varepsilon_0)H + \left[x_n \sum_{r=1}^{n} W_r - \sum_{r=1}^{n} x_r w_r \right] = V x_A$$

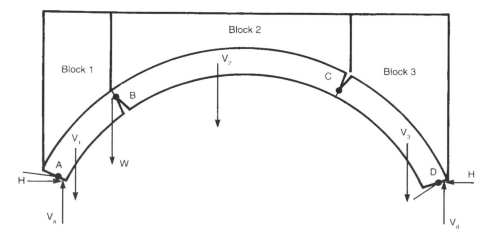

Figure 8.35 Simple mechanism (Page, 1993).

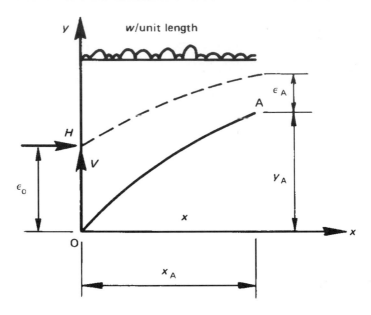

Figure 8.36 General thrustline (Heyman, 1982).

This can be rearranged to express in general terms the vertical eccentricity:

$$\varepsilon = \varepsilon_0 - y + \frac{V}{H}x - \frac{1}{H}\left[x_n\sum_{r=1}^{n}w_r - \sum_{r=1}^{n}x_r w_r\right]$$

By taking moments about the assumed hinged positions, V, H and W can be determined. Then, using the above equations the eccentricity at any point on the arch can be checked that it is within the limit sets. Care must be taken when solving these equations by hand as small differences of large numbers are usually involved, so rounding errors can have a significant effect on the result. Because the method gives an 'upper bound' solution it is necessary to vary the position of the live loading and hinges in order to determine the minimum value for the carrying capacity. This can be a tedious procedure, although several computer programs are now available which carry out this type of analysis.

In order to check any computer output it is useful to apply a quick approximate analysis. Heyman (1982) presents a simple idealized solution. Figure 8.37 shows the dimensions of the arch and defines the dimensionless parameters which are used to determine the value of P. It is assumed that hinges occur at each of the springings, directly under the load (with no dispersion through the fill or arch) and at the crown. It is also assumed that the backfill and masonry have the same unit weight . This yields the expression:

$$P = 16\frac{W_2 x_2\left\{\alpha + \left(1 - \frac{1}{4}k\right)\tau\right\} - \left(W_1 x_1 + \frac{1}{4}W_2\right)\left\{(1-\alpha) - \left(1 + \frac{1}{4}k\right)\tau\right\}}{(3 - 2\alpha) - (2 + k)\tau}$$

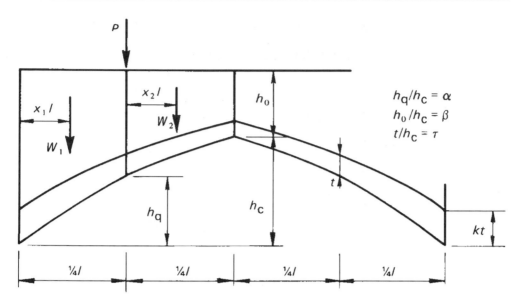

Figure 8.37 Heyman 'quick' method (Heyman, 1982).

The above equation gives the load which will just cause collapse; it also implies that when $(3 - 2\alpha) = (2 + k)\tau$ then P is infinite – this corresponds to a geometry which produces a 'perfect' structure that will be discussed later. (A 'perfect' structure may be defined as one where the thrust line cannot move out of the arch whatever the value of P. In reality failure can occur by crushing of the 'real' material). If it is assumed that the intrados comprises a series of straight lines then:

$$P = \frac{\gamma L h_c}{6} \left[\frac{(1+3\beta-\alpha)\left\{\alpha + \left(1-\frac{1}{4}k\right)\tau\right\} - (6+9\beta-5\alpha)\left\{(1-\alpha) - \left(1+\frac{1}{4}k\right)\tau\right\}}{(3-2\alpha) - (2+k)\tau} \right]$$

Heyman (1982) presents tables for quick calculations based on the above equation with a further constraint that $k = 1$ which is safe and only as a small effect for arches with span/rise ratios of four or more.

So far the barrel has been modelled as a two-dimensional arch comprising discrete rigid voussoirs; the reality is far more complex and requires a more sophisticated model which takes a more holistic approach to bridge behaviour. Early attempts to model other effects (Melbourne and Walker, 1989) demonstrated the need for such an approach.

An extension of the work of Pippard and Heyman has been the development of the 'rigid block' method of analysis (Gilbert, 1993). The method recognizes the formulation of the governing equations as being in a form which can be solved using standard linear programming techniques like the Simplex method.

Generalization of the model needs to take into account the effects of sliding between voussoirs, ring separation, soil pressures, load dispersion, local crushing, spandrel wall stiffening and skew.

In earlier discussion it was pointed out that a two-dimensional arch has three redundancies and hence requires four released to produce a mechanism. So far releases have been assumed to be hinges however they could be a roller bearing – or more realistically a frictional slider between blocks. Drucker (1953) showed that an interface that modelled frictional sliding which embraced the principle of normality by incorporating dilatancy, will produce an upper bound of the 'exact' solution. It has been shown that provided the coefficient of friction between the blocks is greater than 0.4, slip is unlikely in the radial directions, (Given that the axial thrust is much larger that the radial shear).

The same cannot be said of longitudinal shear. This, coupled with the propensity for multi-ring brickwork arches to suffer from mortar 'wash-out' present real difficulties for the assessing engineer. Here a judgement has to be made. Although the 'mathematical' model maintains a theoretical relationship between the surfaces this may not exist in reality and so a more pragmatic interpretation of the computer output may be necessary.

Pressures at the extrados present difficulty both for the analysis for design and for the assessment of existing structures. The former is slightly easier as the designer has some control of the specification of the material and method of construction. The latter situation is far more difficult as the nature of the backfill may not be known, let alone the pressure distribution. The important thing is to ensure that the free-body idealization takes into account the true stress/pressure distributions at the boundaries. This is reinforced by a holistic idealization of the bridge. In the mechanism, idealization should be taken account of the pressure distribution on each slice/block. The initial stressed state will invariably not be known so some realistic assumption must be made. As the arch deforms to create the failure mechanism the soil/backfill will be subjected to significant (albeit local) strains which may create large local pressures on the extrados. Significantly, the free body may not 'see' these.

So far the arch has been considered as a two-dimensional structure – this is clearly not the case. Even square span bridges behave as complex three-dimensional structures due to eccentric loading and the interaction between the elements of the bridge. Before considering the available analytical techniques it is informative to consider the forms of construction of the arch barrel. Figure 8.38 shows the development of three arrangements for a 45° skew segmental circular arch. The simplest form comprises units which are laid parallel to the abutment. This may be acceptable for small angles of skew but are clearly dependent upon the bed shear capacity to transfer load as the skew increases. Additionally, if bricks or square cut stone is used the evaluation of the arch barrel will be stepped and may need extra work to achieve an acceptable appearance.

The second method is the helicoidal or English method. It ensures that the bed at the crown is perpendicular to the longitudinal axis of the bridge and by inference the principal axis of the stress at the crown due to crown loading. There is an inferred

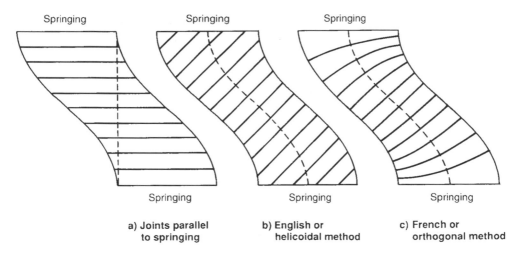

a) Joints parallel
to springing

b) English or
helicoidal method

c) French or
orthogonal method

Figure 8.38 Projection of the soffit of a 45° skew angle (Page, 1993).

assumption that (particularly if the arch is filled) the stress perpendicular to the axis reduces (or is better distributed closer to the abutment). The geometrical constraints are such that, for the beds to remain parallel, then their orientation causes the beds to 'roll-over' and thus rest on the springing at an angle.

Finally, the orthogonal or French method attempts to keep the bed orthogonal with the local edge of the arch and thus follow the perceived line of principal stress onto the abutment. Although this may appear to be desirable it does mean that all the units will have slightly different shapes and thus such construction is usually confined to stone arches (and the availability of highly skilled masons). Brickwork cannot be used as the bedding planes are not parallel.

The analysis of a skewed arch presents many difficulties. There is no universally accepted method but several methods have emerged which can be used to inform any assessment decisions or new design.

The simplest two-dimensional model was the four-hinged mechanism model. Even as a rib this becomes a more complex problem once skew is introduced. This is because in three-dimensional space the rib has six redundancies and therefore requires seven 'releases' in order to become a mechanism. Additionally, torsion will be induced on the rib and if the 'no-sliding' criterion is to be maintained, then rotation about the axis of the rib can only be accommodated by the existence of pins (in addition to the 'traditional' hinges) which will allow a kinematically admissible mechanism to form.

It has been observed in large-scale tests that even when full width line loading is applied parallel to the abutments the collapse mechanism which develops has more than four 'hinges' and none may be parallel to the abutments. This observation helps the analyst to idealize the structure. If Figure 8.39 represents a kinematically admissible mechanism then the following observations can be made:

- four or more 'fracture lines' are required
- elements of the arch barrel between 'fracture lines' are assumed to be rigid

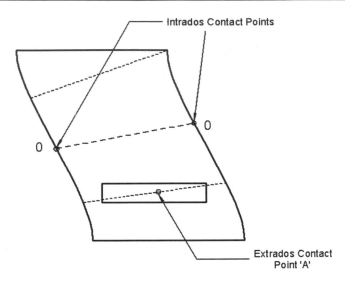

Figure 8.39 Skew bridge fracture lines.

- if the fracture line is not parallel to the bedding plane and the arch barrel is moving upwards, then the axis of rotation is determined by the intrados edge contact points on each elevation of the arch barrel (Figure 8.39 axis 00)
- if the fracture line is not parallel to the bedding plane and the arch barrel is moving downwards, then there is a single point of contact on the extrados (Figure 8.39 point *A*) this point can act as a universal joint thus allowing three rotational releases
- the fracture lines usually follow the bedding and perpendicular mortar joints unless the units are relatively weak or there are local high stress levels in which case the fracture line may go through the unit
- sliding is possible
- 'normal' shear and torsional rotation
- in the special case of a multi-ring brickwork arch barrel ring separation is possible and given the likelihood of concentrated local load transfer at fracture lines this will probably occur
- the abutments may not be rigid and can accommodate some movement, thus offering 'release' to the arch barrel
- given the concentrated local load transfer it is likely that local material failure will occur. This may manifest itself as local spalling, mortar failure or unit failure (splitting).

Whether the abutment moves as a result of the twist is important with regard to the soil pressures on the extrados particularly in the vicinity of springing. If the abutments are fixed and the arch barrel accommodates any rotation purely by forming a hinge at the point where the intrados (or extrados) meets the skewbacks, then the local soil strains will be small and the changes in pressure correspondingly small. On the other hand, if the abutments move (either by spreading or rotating) and thus contribute to the 'release' of the arch barrel, then the local soil strains may be significant, and consequently, the changes in soil pressure.

Notwithstanding the above, it must be remembered that the backfill is confined by the spandrel walls and the extent and stiffness of that confinement will influence the longitudinal stresses/pressures that the backfill will exert on the extrados. Additionally, if the thrust causes the abutments to spread then the changes in the backfill pressures at the springing will be much larger than those resulting from pure rotation of the barrel at the springing.

Dependent on the nature of the barrel material there may be a possibility of local crushing in the vicinity of the hinges. From a compatibility point of view it would be logical to assume that hinging occurs at the limit of the rigid block; (it being assumed that the material is either rigid or plastic). By removing the rigid material criterion, the problem becomes non-linear, but this can be overcome by introducing an iterative procedure in the algorithm.

There has been a tacit assumption that the backfill material is homogeneous – this is not the case. Not only might the extrados backfill contain/comprise internal spandrel walls and overslabbing but also the spandrel wing walls themselves will provide significant stiffening and resistance to movement. The final surface whether a road blacktop or a railway (or a canal for that matter) will all influence the local dispersal to the extrados – which may have a greater influence than the crushing strength of the barrel masonry! The most conservative assumption is that axles manifest their effect as a point/line load on the extrados – this is perfectly acceptable for quick hand calculations to check computer output. The more accepted dispersal is two vertical for one horizontal, i.e. the load is applied to an area of width equal to the depth to the extrados.

The influence of the spandrel and wing wall is significant – even when spandrel wall separation has occurred. This was recognized in the nineteenth century and is reflected in the design recommendations which passed to later generations (Alexander, 1927). It was suggested that the spandrel/wing walls 'stretch horizontally such a distance, that the friction of their base shall at least be equal to the excess of the horizontal thrust of the loaded half arch over that of the unloaded half … from two-thirds to one-half of the horizontal thrust of the arch at the crown'.

Having considered the behaviour of single span bridges, attention will now be turned to multi-span masonry arch bridges. Most of these bridges in the UK were constructed in the nineteenth century, for the railways. Bridges containing many spans often included a substantial pier at intervals to prevent the entire viaduct from collapsing in the event of one of the spans failing. This is a clear indication that our forefathers had an understanding on the interaction between the spans. Nowadays the assessment of the multi-span masonry arch bridges may be undertaken using an FE model and limitations on stress levels applied; thus, by an interactive procedure successfully eliminating tensile stressed material, an assessment of carrying capacity can be made.

If the piers and abutments are substantial, it is likely that the critical loading relates to the carrying capacity of the weakest single span. However, it is vital that a holistic approach is taken to identify possible failure mechanisms. If more than one span is considered, then it follows that more than four releases will be needed to convert a section of the viaduct into a mechanism. For example if two adjacent spans are assessed, then seven releases will be required to create a mechanism. Large-scale laboratory tests have confirmed this (Melbourne *et al.*, 1997). The applications of mechanism methods of analysis are also appropriate and certainly are a useful tool when used in conjunction with FE analysis. Programs now exist (Gilbert, 1998) which can deal with the multi-ring brickwork arches at the same time as multi-span construction.

From the limited test data available, several observations can be made:

- In the absence of robust piers, the failure mechanism involved both the loaded span and one or other of the adjacent spans. Figure 8.40 shows a typical result; note that the crown of the loaded span moves downwards whilst the crown of the adjacent span moves upwards. It is the resistance to this upward movement which is significant in determining the carrying capacity of the loaded span and hence the viaduct.
- The ultimate failure load is less than the comparable single span arch bridge.

Figure 8.40(a) Multi-span bridge.

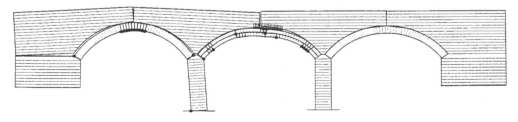

Figure 8.40(b) Multi-span bridge.

- The critical loading position is in the vicinity of the crown (rather than the quarter span) as this offers the greatest horizontal loading to the top of the intermediate pier and hence the greatest 'springing' loading.
- The spandrel walls have a significant effect upon the carrying capacity. In viaducts they often represent (particularly over the piers) a significant proportion of the overall width of the bridge and consequently (if only included as dead load) can make a crucial contribution to the stability of the arch barrel.
- Ring separation in multi-ring brickwork multi-span bridges can be just as significant as its influence on single span bridges.
- In addition to the above, skewed multi-span bridges appear to be weaker than their equivalent square span structure (for brickwork). They produce complex three-dimensional failure mechanisms which initially encourage the barrel to carry the load square to the abutments (i.e. the shortest span). This, between adjacent spans, causes torsion to develop in the piers (and non-uniform load distribution across the width of the abutment) even with fill width uniform loading. A recent test (Melbourne, 1998) on a two span multi-ring brickwork arch bridge with a 45° skew, showed that although the strain and deflection contours initially indicated that non-parallel hinges were forming due to the full width knife edge loading parallel to the abutments, the arch barrel articulation became more aligned to parallel hinge formation once the pier had become 'torsionally' released. The further significance of this observation relates to the fact that as the loading increased the mode of response of the structure could be different depending upon the most efficient mechanism.

8.11 Design of masonry arch bridges

In considering the design of masonry arch bridges it is important to remember that there are over 40 000 examples in the UK alone. Most have already exceeded the Department of Transport's design life of 120 years and therefore must be considered as an archive of good practice and proportion. Traditionally, the span to rise ratio varies from 2:1 (semi-circular) to 10:1 but is usually in the range of 3:1 to 6:1 with the 'ideal' taken as 4:1. Clearly the shape of the intrados is of great importance as this dictates of the complexity of the temporary support or centring. Nowadays, this usually takes the form of a proprietary metal system of supports upon which timber transverse joints are secured and a plywood facing attached. Care must be taken to limit the centring deflections during construction. It is good practice to provide a method statement which requires the barrel to be constructed at an even rate from each springing. The centring should not be removed until the spandrel walls (if the bridge is to be a filled spandrel bridge) have been taken to the string course and the wing walls extended sufficiently to resist the horizontal thrust and also contribute to the stability of the abutment. It is important to take a holistic approach and consider the interaction between the several elements of the bridge. It should not be forgotten that small settlements or rotations of the piers and abutments offer 'releases' to the structure and thus can contribute to the formation of a mechanism.

The initial sizing of the structure can be undertaken using any of a number of empirical equations.

Rankine advocated that the barrel thickness, d, for a segmental arch of radius R should be:

$$d - 0.19\sqrt{R} \text{ (single span)}$$

$$d = 0.226\sqrt{R} \text{ (multi-span)}$$

Heinzerling suggested:

$$d = 0.4 + 0.025R \text{ (dressed stone)}$$

$$d = 0.4 + 0.028R \text{ (brickwork)}$$

$$d = 0.4 + 0.032R \text{ (random rubble stone)}$$

Trautwine suggested (for span = L):

$$d = 1.0[0.061 + 0.138\sqrt{(R + 0.5L)}] \text{ (first class cut stone)}$$

$$d = 1.13[0.061 + 0.138\sqrt{(R + 0.5L)}] \text{ (second class cut stone)}$$

$$d = 1.13[0.061 + 0.138\sqrt{(R + 0.5L)}] \text{ (brickwork)}$$

Rennie and Stephenson related arch barrel thickness to span and radius, respectively.

$$d = span/30 \rightarrow span/33 \text{ (Rennie)}$$

$$d = R/26 \rightarrow R/30 \text{ (Stephenson)}$$

Historically, the addition of haunches to the arch ring at the abutments has been considered good practice (and in keeping with the usual assumption for two-hinged arch analysis that the second moment of area of the barrel varies as the secant of the tangent to the centreline).

The suggested ratio of springing to crown thickness varied from 1.2 to 2. In stonework and concrete this is relatively easy to achieve but in brickwork the haunching is usually concrete and there is a reliance upon the bond between it and the brickwork.

Abutment sizes for a 'gravity' solution (as opposed to a reinforced concrete or masonry solution) have been suggested by Baker (1909) where the abutment thickness, t, in metres is given by:

$$t = 0.012 (5L + 4H) + 0.3$$

where L is the span (metres) and H is the height from the top of the foundation to the springing line (metres).

Both the haunching rules and Trautwine's rule for abutment sizes are shown in Figure 8.41.

In the case of multi-span bridges, the thickness or piers vary. Rankine (1904) suggests a range from 1/10 to 1/4 of the span, the latter offering sufficient support to cater for the removal of one of the spans. Historically, the most common thickness for intermediate piers is from 1/6 to 1/7 of the span. These suggested thicknesses make no reference to the height of the pier which is an important parameter when considering any out of balance forces between adjacent spans that have to be accommodated by the pier. Of equal importance, then, is the slenderness ratio of the pier. Recent laboratory tests have demonstrated that even with a slenderness (height/thickness) ratio of 3.4, the failure mechanism involves at least two span. It is therefore suggested that a slenderness ratio of less than two is required to ensure the independent behaviour of each span; otherwise any analysis must consider the whole structure. In any case, the stability of the piers must be verified.

Modern limit state codes require both serviceability and ultimate limit states to be considered. In the case of the masonry arch serviceability criteria are difficult to quantify. Generally, serviceability criteria limit deflection for aesthetic or functional reasons and cracking for durability considerations. For masonry arches the stress levels are generally very low and hence deformations are small. Additionally, any cracking which might occur in response to de-centring and 'bedding-in' of the structure is usually only an indication that it is working and should cause no concern. On the other hand, it is important that the structure is designed to avoid proliferation of micro-cracking under repeated loading (fatigue); particularly if this could lead to

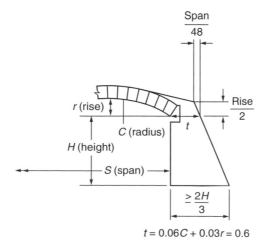

Figure 8.41 Empirical rule for abutment sizes.

ring separation in multi-ring brickwork arch barrels. It has been observed in field and laboratory tests that the first hinge forms at approximately half the ultimate load carrying capacity of each bridge. Using this observation and in the absence of any more reliable criteria, it is suggested that for HA type loading a partial load factor γ_{FL} of 3.4 should be used for the critical axle (knife-edge load) and $\gamma_{FL} = 1.9$ for the other live loads. Whilst the HB type abnormal loading $\gamma_{FL} - 2.0$ may be used. These loadings should be applied in conjunction with a partial load factor for all the dead load and superimposed dead load of 1.2 when they are adversely affecting the structure and 1.0 when they are resisting hinge formation.

Initial design procedure:

1. Choose span and rise.
2. Select materials to be used.
3. Determine trial section using a selection of empirical equations.
4. Ignoring horizontal soil pressures, calculate the required arch barrel thickness using a simple 'block' mechanism for the ultimate load condition. This can be assumed to comprise appropriately factored HA loading.
5. Check the compressive stress based on 0.1 (arch thickness) or 100 mm whichever is the greater.

$$\text{Compressive stress} \leqslant \lambda_u f$$

where λ_u is a coefficient = 0.35 for concrete grades 15 and 20, 0.4 for concrete grades 25 and above, and 0.44 for masonry. f is the characteristic cube strength of concrete, f_{cu}, or the compressive strength of masonry f_K as appropriate (allowance for γ_M has been made).

6. Check that the radial shear at the crown is less than 0.4 (horizontal reactions).
7. In the case of multi-ring brickwork arches check that the horizontal mid-depth shear stress at the crown is less than 0.15 N/mm^2.
8. Check abutment stability and stress levels.
9. Check foundation stability and stress levels.

Example
Consider an 8 m span 2 m-rise segmental brickwork arch bridge with a total available construction depth of 0.9 m at the crown. Following the initial design procedure:

1. Span = 8 m; rise = 2 m; intrados radius = 5 m
2. Empirical equations:

$$\text{Rankine} = 0.19\sqrt{R} = 0.19\sqrt{5} = 0.42 \text{ m}$$

$$\text{Heintzerling} = 0.4 + 0.028R = 0.54 \text{ m}$$

$$\text{Trautwine} = 1.33(0.061 + 0.138\sqrt{(R + 0.5L)})$$
$$= 1.33(0.061 + 0.138\sqrt{(5 + 4)}) = 0.631 \text{ m}$$

It is therefore likely that the arch ring thickness will be about 0.5 m – it follows that a construction depth of 0.9 m is probably enough.

A simple 'block' mechanism analysis using a 1 m 'slice' of the bridge can be undertaken to determine the barrel thickness. The minimum lane width of 2.5 m is used with a $\gamma_{FL} = 3.4$ for the KEL and $\gamma_{FL} = 1.9$ for the udl.

$$\text{The KEL} = (3.4 \times 120)/2.5 = 163.2 \text{ kN/m width}$$

$$\text{udl} = 1.9 \times (336/2.5)(1/\text{load length} = 4 \text{ m})^{0.67} = 101.3 \text{ kN/m run}$$

The hinges are assumed to occur at A, B, C and D. It is also assumed that the soil offers no horizontal support and that γ_{FL} for the self-weight on the loaded side of the arch is 1.2 while on the unloaded side of the arch it is assumed to be unity, as shown in Figure 8.42.

3. By taking moments about C, B and A for the free bodies to the right of each of the hinges three independent equations can be written and solved or the unknowns V_D, H and t:

$$t = 0.373 \text{ m}$$

$$H = 447 \text{ kN}$$

$$V_D = 231 \text{ kN}$$

Making allowance for the potential loss of mortar depth to the intrados and rounding to the nearest half brick – assume a barrel thickness of 450 mm.

4. At the crown the compressive force $= H = 447$ kN so the characteristic compressive strength required $(447 \times 10^3)/(0.44 \times 100 \times 10^3) = 10.2 \text{ N/mm}^2$

5. At the crown the vertical shear V_D – (self-weight to the right of C) $= 331 - 115.5 = 115.5$ kN (this is less than $0.4H = 0.4 \times 447 = 178.8$ kN).

6. At the mid-depth of the barrel at the crown:

longitudinal shear stress $=$ (vertical shear/I(width))$A\bar{y}$

$$((115.5 \times 10^3) \times 1000 \times 225 \times 112.5)/(((1000 \times 450^3)/12) \times 1000)$$
$$= 0.39 \text{ N/mm}^2 > 0.15 \text{ N/mm}^2$$

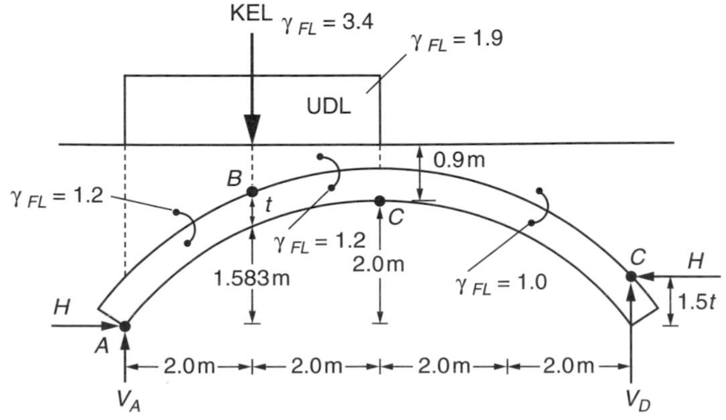

Figure 8.42 Preliminary design example.

Headers should be provided or some form of mechanical bond introduced. If neither of these is practical the arch should be thickened to reduce the stress.
Additionally, the longitudinal shear stress should be checked at the interface
between the brickwork rings as the thrust line moves from one ring to the next for
example between A and B.

7. If the abutment height at the springing to foundation is 3 m and the base is 2 m
   thick (i.e. 2/3 height), it can be shown that the abutment is unstable and will rotate
   about the rear face. The stabilizing effects of the wingwalls may be considered at
   this stage to check that stability is achievable.
8. The foundation stability and stress levels would then be checked.

Having determined the initial proportions some consideration must be given to the
practicalities of construction and the specification prior to undertaking a more detailed
analysis for the different load cases using a more accurate analytical model based
upon a realistic representation of the material properties, including foundations and
soil properties.

Depending upon the sophistication of the analysis, the output may indicate the
stress levels which can be compared at critical section with the allowable stress. For
simpler analysis using a frame analysis program like MINIPONT the nodal forces
must be assessed using the standard equation:

$$\sigma = P/A \pm My/I$$

$$\left(\begin{array}{l}\text{stress, } \sigma, \text{ at a distance} \\ y \text{ from the neutral axis}\end{array}\right) = \left(\frac{\text{axial force}}{\text{cross} - \text{sectional area}}\right) \pm \left(\frac{\text{bending moment} \times y}{\text{second moment of area, } I}\right)$$

If tensile stresses are induced then that part of the section should be ignored and
the analysis iterated until the designer is satisfied that all tensile stresses have been
eliminated.

The design of spandrel walls, wing walls and parapets requires careful consideration as historically it is recognized that they usually show signs of deterioration first.
Spandrel and wing walls may be designed as gravity structures. Tensile stresses should
be avoided. It is important that these are checked at each section change and at
regular heights up the wall. Sliding along bed joints should be checked. The interaction between the arch barrel and the spandrel wall should be designed to minimize the risk of spandrel wall separation.

The forces on the walls may be reduced by using reinforced soil or foamed concrete as backfill. The walls may be connected to the soil reinforcement. It is preferable to ensure that the backfill is less stiff than the barrel to ensure that the arch
behaves as a particulate structure and can articulate in response to movement without creating 'hard' (or 'soft') spots of interaction with the backfill.

Parapets should be designed to contain vehicles. In order to comply with the
current design standards (Department of Transport, BD52 and BD37), it is inevitable
that either a conventional parapet is provided or some form of masonry grouted –

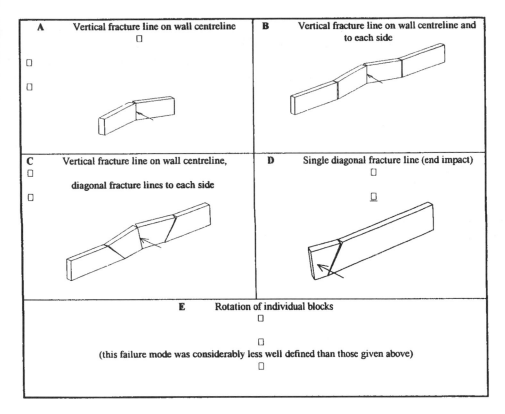

Figure 8.43 Parapet failure modes (Hobbs et al., 1998).

cavity wall has to be adopted. The cavity reinforcement and ties should be austenitic stainless steel. The reinforcement should be secured to a torsion beam or slab which spans between abutments. Recent research (County Surveyors' Society, Hobb *et al.*, 1995) has shown that unreinforced parapets have a greater containment capacity than had previously been thought. The impact resistance of plain masonry parapets depends on two principal factors. These are the internal forces which are governed by wall mass and geometry and on the bond conditions between the masonry units and the base of the parapet. Figure 8.43 shows the principal failure modes. It is unlikely that a simple 'equivalent static load' will predict the actual behaviour, only a dynamic analysis can do this. However, it is reassuring that masonry parapets with a thickness of 400 mm or more and a minimum length of 10 m have the capacity to contain a 1.5 tonne vehicle, travelling at 60 mph (100 kph) impacting at an angle of 20° irrespective of mortar strength. (Middleton, 1995).

8.12 Specification

Unreinforced arches may comprise brickwork, stone, reconstituted stone, precast and *in situ* concrete. Additionally there are proprietary systems which would need independent certification by the Bridge Authority.

Class A or B Engineering brick with a durability designation of FL complying with BS3921 'Specifications for clay and bricks' should be used. Calcium silicate bricks are not recommended. Concrete bricks and precast blocks should have a minimum characteristic strength of $30N/mm^2$ with a minimum cement content of $350 \, kg/m^3$ and manufactured using dense natural aggregates. Reconstituted stone blocks should be independently certified.

Natural stone should comply with BS5390 'Code of practice for stone masonry'. Stone should be selected on the basis of proven durability and weather resistance.

Mortar for masonry should be in accordance with Table 8.9. Portland cement based mortars will be stronger and more durable, but less accommodating of movement than lime mortars.

Table 8.9 Arch bridge faults and repair/strengthening techniques

Fault	Repair/strengthening
Deteriorated pointing	Repoint
Deterioration of arch ring material	Saddle Sprayed contents to soffit Prefabricated liner to soffit Grout arch ring
Arch ring thickness assessed to be inadequate to carry required traffic loads	Saddle Sprayed contents to soffit Prefabricated liner to soffit
Internal deterioration of mortar, e.g. separation between rings of a multi-ring brick arch	Grout arch ring Stitching
Foundation movement	Mini-pile Grout piers and abutments Underpin
Scour of foundations	Underpin Invert slab Cut water
Outward movement of spandrel walls	Tie bars Replace fill with concrete Take down and rebuild Grout fill if it is suitable
Separation of arch ring beneath spandrel wall from rest of arch ring	Stitch (short tie bars spanning the crack)
Weak fill	Replace fill with concrete Grout fill if it is suitable
Water leakage through arch ring	Waterproof road surface Waterproof arch ring extrados + improve drainage

Brickwork should be detailed in accordance with BS5628: Part 3. Clay bricks should not be used within 14 days of filling in order for the irreversible expansion due to moisture rake-up to be complete. Masonry in the arch should be flush jointed. Bucket-handle or other finishes may be considered for spandrel and wing walled masonry. It is difficult to achieve continuous flush mortar joints in the intrados in which case a neoprene rubber strip may be inserted during brickwork construction. This can be removed after the centring has been struck and the joint gun pointed using similar mortar.

The timing for removal of the centring should be such that the barrel is not required to carry stresses of 1/3 of its strength at the time of loading but not within 7 days of completion of the arch. Historically it is advised that the spandrel walls should be taken up to the stringcourse prior to the removal of the centring (Alexander, 1927).

Concrete should be designed using 20 or 40 mm durable natural aggregates with a minimum cement content of 325 or 295 kg/m^3, respectively and a minimum free water/cement ratio of 0.5. The concrete should be air-entrained to guard against freeze–thaw action (4.5% for 40 mm aggregate concrete and 5.5% for 20-mm aggregate concrete). In order to reduce shrinkage, 40 mm aggregate concrete is preferred. The concrete should be well compacted and cured. Specialist cements may be used where ground conditions or climate demand or shrinkage is to be minimized.

Care should be exercised to accommodate movement. This is equally true for all materials and should include allowances for:

- seasonal thermal movements
- movement of foundation
- creep of highly stressed elements (e.g. prestressed masonry)
- relative stiffness
- long-term irreversible expansion of clay brickwork due to moisture
- seasonal moisture movement (usually very small).

Joints in brickwork are usually required at 10–15 m centres. It is important that if joints are provided that they are able to work! All too often the three-dimensional nature of the structure is overlooked and restraints are present which prevents the efficient functioning of the joint.

In situ concrete arch barrels have been constructed using planar crack inducers to 'convert' what would otherwise be a brittle elastic continuum into a series of *in situ* voussoirs. The crack inducers may take the form of plastic sheets in which holes have been cut to ensure good load transfer and prevention of slip (Melbourne and Njumbe, 1998). It is important to have a continuous strip at the intrados (which should be filleted) to prevent stress concentration and the opportunity for local spalling (Figure 8.44).

Attention should be paid to brickwork bonding in the arch ring. Individual rings may be built in English or Flemish bond but where this results in no headers then the desired monolithic behaviour could be affected by ring separation. The bonding should therefore accommodate some bonding between rings every fourth course. This may be a header or some form of durable mechanical connection. Bricks (units) can

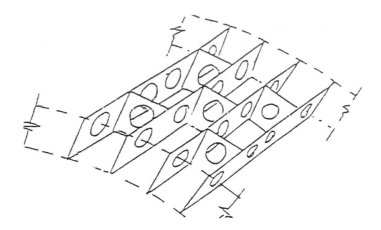

Figure 8.44 Planar inclusions.

be manufactured with a taper which matches the arch curvature. It is preferable to use larger mortar joints and accept lower compressive strength than to provide a multi-ring arch without headers. Concrete haunching may be provided.

Alternatively, the arch may be divided by voussoir-shaped blocks. These may act as ring separation 'arrestors' but they are no substitute for the provision of headers throughout.

The extrados and all buried structures in contact with the backfill should be protected by a waterproofing system which is suitably protected – preferably by a permeable layer.

It is vital to the long life of the bridge that any water which enters the backfill is efficiently removed and disposed into an appropriate drainage system.

8.13 Defects and rehabilitation

Most masonry arches are over 100 years old and have survived the vicissitudes of the UK climate and continue to carry loads very different to those envisaged by their constructors. It is inevitable that such old structures often show signs of distress, the causes of which can usually be attributed to substructure changes in loading regime and/or material deterioration.

Unlike trabeate construction, arches impose significant horizontal thrusts onto the abutments and piers. These may result in horizontal movements which cause the masonry barrel to crack. The situation is further complicated by non-uniform distribution of the thrust with the consequence that the abutments and piers are subjected to torsion across the width of the bridge. (This may be brought about by eccentric loading of the span and/or skew.)

As with most structures the preferred founding material is rock. Unfortunately this is not always the case, consequently many bridges are founded on 'soil'.

Historically, timber piles, grillages or faggots have been used to improve the support conditions. Bridges which were given inadequate foundations have long since collapsed, so bridges which have developed defects in recent times have done so as a result of changes. It is not doubted that well constructed masonry is one of the most durable materials. However, there are conditions which accelerate material deterioration and lead to problems, the main agent being water. In new construction low porosity bricks should be used. In existing structures the bricks may be showing signs of failure,e.g. spalling, cracking, friability. This may be local in which case there may be a local cause that can be remedied or it may be more general and thus require a more drastic solution.

The two consequences of water in masonry are the aggravation of freeze–thaw effects and the enhancement of chemical activity. The former is a particular problem in temperate climates due to the frequency of freeze–thaw cycles during winter months (which may be several in a single 24-hour period). As water freezes it expands so if the water is present near the surface of the masonry the incremental detcrioration may take place. It is therefore important that water should be excluded from the bridge. Although this may be possible in the case of highway bridges where the highway construction is impervious, it may not be possible in the case of railway bridges as the installation of a waterproofing membrane may be prohibitively expensive. Historically, puddle clay was used as a waterproofing membrane laid directly onto the extrados. (This should be remembered when modelling the backfill material properties of soil pressure profiles.)

In addition to aggravating the freeze–thaw effects, water can transport sulphates (and chlorides) from the backfill, de-icing salts or the masonry itself; the effects of which are not only to cause mortar wash-out but also to facilitate chemical attack and deterioration of the bricks/stone and mortar. The mortar is usually more vulnerable than the units. Certainly, the appearance of a well-pointed intrados should not lull the assessing engineer into a false sense of security as the underlying mortar will be weak, friable, perhaps non-existent! Additionally, if the pointing has been undertaken using a modern cement-base mortar this will produce 'hard-spots' in the intrados which will aggravate stress concentrations and freeze–thaw effects potentially causing spalling. Under no circumstances should an impervious membrane be applied to the intrados – the masonry must be allowed to 'breathe'. In any case any re-pointing of the masonry must be undertaken using a mortar compatible with that in the original structure – which is probably a lime mortar.

It should be remembered that bricks fresh from the kiln expand irreversibly (often referred to as 'moisture expansion'). This can be a very beneficial phenomenon when patch repairing as the slight expansion can lock the patchwork into the original structure and thus enable the new brickwork to carry some dead load as well as the live loading. Additionally, as the historical brickwork will be constructed using lime mortar it will be quite accommodating of the patchwork and finally, the creep characteristic and natural particulate nature of the brickwork will allow stress concentrations to be relieved by redistribution. Some attempt should be made to achieve compatibility between 'old' and 'new' with regard to Young's modulus, coefficients of thermal expansion, etc.

If colour matching is an important issue then sufficient of the existing structure should be cleaned to establish brick and mortar colour before the repairs are undertaken and trial panels made to ensure a good match.

The deterioration of stonework is related primarily to the chemical nature of the material and its physical condition, in particular its porosity. Notwithstanding the natural variability of stone, the natural durability of stone depends upon the susceptibility of the stone's constituents to acid attack from the carbonic, sulphurous or nitrous acid rain which results from the carbon dioxide, sulphurous oxide and nitrogen in the atmosphere. The subsequent leaching which takes place increases the porosity of the stone whilst reducing its cohesiveness.

Leaching is more severe in stone with fine pores as this increases the capilliarity and hence the ease with which moisture can be held by surface tension. (This also holds for susceptibility to freeze–thaw effects.) Porosity can be measured using the saturation coefficient which is determined by comparing the amount of water absorbed naturally with that resulting from vacuum impregnation.

It should be remembered that when replacing defective stone, it may have taken well in excess of 100 years for the stone to deteriorate and therefore is perfectly adequate. Additionally, the chemical and physical state of the original stone should be carefully established to ensure that any replacement stone does not interact unfavourably with it. Particular care should be taken with regard to bedding planes and their orientation to exposed faces and setting (whenever possible) perpendicular to the principal stress. The damp surface of the masonry is also a breeding ground for bacteria which actively produce sulphates and nitrates. Additionally, algae will also develop in a few hours given suitable conditions of dampness, warmth and light. Temporary dryness may cause their death, however, the resulting humus offers a suitable host for lichen whose fungal hyphae penetrate the stone generating organic acids. On dark, damp surfaces fungal slime may develop. Having established that the physical, chemical and biological state of the stone are all significant in its durability, it is obvious that great care must be taken when repairing or cleaning the masonry.

It is important to keep the structure as dry as possible. High pressure water jetting can achieve the best results under normal circumstances. Biocidal treatment should be considered post-cleaning to control bacteria/fungal attack. Under no circumstances should surface sealants be used particularly if water penetration is expected from the buried/rear face (i.e. extrados of the arch, back of a retaining wall).

It was generally accepted that some movement would take place during the construction of a masonry arch bridge and that the construction sequence was very important in limiting and controlling any movement. '... the centre of an arch is always struck when the superstructure is only partly built, else the superstructure would crack, due to the after-subsidence, ... the earth must be filled behind the spandrels before the remainder of the superstructure is built' (Alexander, 1927).

Given that most masonry arch bridges are old structures it may be reasonable to assume that all dead load settlement will have taken place. If the structure is showing signs of distress associated with foundation movement, then serious thought should be given to the reasons. Founding strata may not be uniform which may have result-

ed in differential settlement or spread of the abutments with consequential distortion of the spandrel walls and arch barrel; the effect of which over the passage of
time may have been cracking and load path adjustment. This may have aggravated
the condition thus causing further deterioration. Add to this the prevailing climate
eg prolonged rain, drought, coupled with long-term deterioration of the fabric of the
bridge and it is possible to see how defects can worsen.

The effect of spreading and/or differential settlement is to cause cracking in the
spandrel walls and (more particularly) in the arch barrel with the consequential reduction in carrying capacity – albeit empirically assessed (Table 8.7, above).

Figure 8.45 shows the different modes of spandrel wall failure. The interaction
between the barrel and the spandrel walls and the lateral soil pressures are important. With heavier traffic which often runs close to the spandrel walls, many filled
masonry arch bridges show signs of distress. The longitudinally stiff spandrel wall and
more flexible arch barrel crack at the interface and separate. The precise mechanism
which governs this type of defect is not well understood. It is certainly aggravated

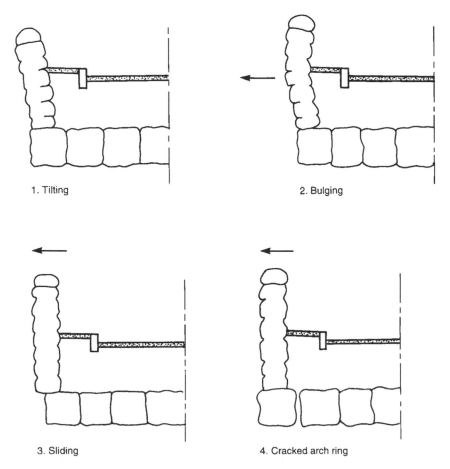

Figure 8.45 Spandrel wall failure (Page, 1993).

by the ingress of water from the fill above which washes out the mortar and causes further distress if it freezes in the cracks.

One significant problem which is specific to brickwork arch bridges is ring separation. This occurs in multi-ring arch barrels where the mechanical bond between adjacent rings of brickwork is lost with the resulting loss of load carrying capacity. For example the shear bond strength for a 1:2:9 lime mortar would normally be about 0.2N/mm^2, even this will not be achieved if there is deterioration or poor workmanship present.

It is very unlikely that the backfill over the arch fails in the structural sense. It is usually wet. This is independent of whether or not the road construction is supposed to be impervious or the backfill to be provided with drainage. Notwithstanding the effect that water has on the properties of the backfill, the presence of water saturating the masonry, washing out the mortar, and generally causing long-term deterioration of the fabric of the bridge is undesirable.

Although any change in vertical alignment of rail track is routinely adjusted, the same cannot be said of the maintenance of road surfaces which frequently develop potholes. These can increase the dynamic effects of tyres hitting the pothole and thus accelerate the deterioration of the bridge.

This is aggrevated by the heavier vehicles which are now using our roads and the increase in the volume of such traffic.

Having considered the range of defects which masonry arch bridges develop and their causes, it is essential to review the range of repair and strengthening techniques. Table 8.10 presents a list of common faults and their associated repair/strengthening techniques.

Table 8.10 Mortars

Location/element	Mortar designation
Work below or within 150 mm of finished ground level	(i) or (ii)
Work 150 mm or above finished ground level:	
• abutments, spandrel/wing walls, piers and parapets	(i) or (ii)
• unreinforced arch ring	(iii)
• reinforced/prestressed brickwork	(i) or (ii)

Notes.
1. Mortar designations correspond to proportions by volume of Portland cement:hydrated lime:sand as follows (BS 5628:Part 3):
   (i)   1:0–1.4:3.
   (ii) 1:2:4-42
   (iii) 1:1:5–6.
2. Alternative mortar mixes such as cement:sand with plasticizer may be suitable for some uses (see BS 5268: Parts 1 and 2).
3. Where FN classification bricks are used or when sulfates are naturally present in soil or groundwater in sufficient quantities to be damaging the sulfate-resisting Portland cement based mortars should be used.

The MEXE method of assessment quantifies the effect of loss of mortar with the implication that reinstate of the lost material will reinstate load carrying capacity. Although this is the desired outcome care must be taken in undertaking the work.

Firstly, it is important to determine the extent of the material loss and the nature of the original mortar. (This is because the extrados may have been repointed several times during its life). It is particularly important nor to use a mortar which is too stiff as this could lead to local splitting of the units and spalling.

Additionally, care should be taken not to remove so much material that the units become loose. This may be exacerbated by poor initial workmanship which may have left incompletely filled joints hence producing voids which have acted as a conduit for water to flow through the arch barrel.

The most effective way of removing the mortar is by jetting. A skilled and conscientious workforce is essential.

Hand pointing may be acceptable for small areas but repointing can only be effective if it is consistent and complete. This is best achieved using some form of mechanical pointing. A pumphole mortar mix, usually containing PFA and non-ionic wetting agent, is used and delivered through a nozzle at pressure of $0.1–0.3 N/mm^2$. The joints are filled from the inner voids outward. It should be remembered that in removing the friable mortar the units are vulnerable to being dislodged. Depending on the viscosity of the mortar and the rate of stiffening, considerable forces can be exerted between the unit. For example, a brick presenting a full face of 215×104 mm, if pointed to full depth at $0.2 N/mm^2$ would have an out of balance force of 44.7 kN which would probably be greater that the shearing resistance available at the mortar bed between the rings. The result could be delamination of the bottom ring, ring separation and hence a reduction in load carrying capacity. The residual compressive pressure helps the inner part of the arch barrel to play an active role in carrying the self-weight of the structure. Creep allows long tem re-distribution of 'hard-spot' stresses.

In situations where running water/surface water is present quickset mortar can be used. Care should be taken in the planning of the work to ensure that the future presence of the water is taken into account so drainage points should be incorporated into the joints to allow it to escape. Otherwise, by sealing the intrados, untold damage could be done to the rest of the arch barrel.

If it is felt that significant amounts of mortar have been lost from within the arch then repointing will not be sufficient to rehabilitate the structure. In such cases it is usual to offer grouting as a viable option. Pressure or vacuum grouting are the usual methods adopted. Selection of grout material is important. In the past, hydraulic limes and pozzolans have been used but nowadays thixotropic grouts and low viscosity grouts are available. As with mechanical pointing, the effect of the internally induced stress must be considered and the grouting sequence adjusted accordingly. Suitable arrangement must be made for the displaced air and water to escape as the grout is injected (otherwise voids are left and brick pressure forces the grout back out). It is inevitable that some grout escapes and this should be cleaned off immediately. It is unlikely that pre-pointing will be able to contain the grout. At best there will be some leakage and at worst the pointing will be 'blown' out. Grouting usually proceeds from the lowest to highest point. If it is judged that the pressure required to force the grout into small voids or fissures would cause damage then vacuum injections may be the solution. A vacuum is first established within the section to be treated. Grout is then introduced and 'sucked' into the voids to balance the vacuum. A low viscosity grout is usually used. The method for the arch barrel can only be used if the extrados of the barrel is in good condition and can provide an effective seal when the vacuum is established.

It is important to realize that the purpose of the grout is to reinstate a pliable 'glue' between the units. If a 'rigid' material is used then the arch barrel will become a brittle continuum highly redundant and susceptible to secondary stresses.

Where it is felt that the arch barrel is unable to carry the required loading one method of rehabilitating the bridge is to provide support from underneath. This is only possible if the resultant loss of headroom and the change in appearance is acceptable. It may be achieved by spraying concrete onto the intrados or by providing a structural liner. In both cases it is vital that adequate provision is made for any water to escape after installation.

Whichever method is used it is vital that there are adequate foundations available. They should be checked for the change in load path. Clearly, if the distress in the arch barrel has been brought about by movement of the abutments, then providing an 'underarch' may actually aggravate the situation rather than ease it.

Sprayed concrete (Gunite, UK; Shotcrete, USA) has been used in many situations. It is particularly suited to providing a relieving arch. The concrete is sprayed at high velocity onto the prepared intrados using either a dry or wet process. It is usual to provide a grid of pins to which is attached a steel rebar mesh. The pins and mesh also act as a guide to thickness. The maximum aggregate size is usually 10 mm (although 20 mm can be accommodated). Both processes can produce dense good quality concrete (typically 30–50 N/mm^2 28-day strength) with a satisfactory bond to a prepared intrados. A composite process has been developed in Hungary which claims to give better control on concrete quality and also a reduction in rebound loss.

Admixtures can be used to reduce shrinkage cracking although a recent survey (Ashurst, 1992) reported visible cracking in all the bridges surveyed.

To reduce visible impact, the edge of the relieving arch may be recessed (into the shadows) or cosmetically treated to blend with the bridge elevation.

An alternative relieving arch can be provided using a series of curved universal beams or columns bent about their major axis to 'fit' the intrados of the arch. It is usually necessary to provide 'needles' to support the steel ribs. If the headroom is sufficient, permanent shutters in the form of corrugated steel sheeting or grp may be used to provide an annulus that can be filled with pumped concrete.

Determining the distribution of load between the original arch and the relieving arch is somewhat circumspect and so it is best to assume that eventually the relieving arch will be carrying all the load.

Recently techniques using near surface reinforcement have to been used with some success. These involve the use of steel reinforcement or steel strip. The former is installed in purpose cut grooves using compatible adhesive whilst the latter can be bolted to the intrados (the length of the bolts can be such that they penetrate to within 100 mm of the extrados). The analysis of the strengthened arch barrel can be undertaken using FE software (Sumon, 2000). Care has to be taken when modelling the interface between the reinforcement and the adhesive and the adhesive and the masonry. As discussed earlier the situation is further complicated by the need to represent the mortar/unit interface and, in the case of multi-ring brickwork arch barrels, the inter-ring properties. All this is before considering spandrel wall and skew effects and soil-structure interaction. A simple mechanism model should be used to inform any FE modelling. This can be achieved by modifying the mechanism model to include a moment of resistance equivalent to that of the reinforced masonry at hinges where the reinforcement is in tension.

It should be noted that by including transverse reinforcement the load distribution can be improved. Additionally, by extending the reinforcement down into the abutments extra strengthening can be achieved. In the case of multi-ring brickwork arches some form of stitching should be undertaken to ensure that ring separation does not occur (it is likely that the installation of the intrados reinforcement will have adversely affected the mortar and the units and the bond between them – grouting may be required).

If there is sufficient depth of cover over the arch and an acceptable traffic management programme can be agreed with all interested parties, then the provision of a saddle or relieving slab may offer the best long-term solution. Again there are some pitfalls which must be avoided. It is important to determine why the bridge is showing signs of distress or failing in load carrying capacity assessment. If these are the result of abutment failure or spandrel wall instability then providing a saddle might make things worse. Additionally, some thought should be given to the consequences of relieving the arch barrel if long-term stress with the future possibility of units falling from 'soft spots' in the intrados. These considerations will influence whether or not the saddle is bonded to abutments and extrados (the latter may be so rough that bonding is inevitable unless some form of compressible layer is installed). The saddle usually comprises a reinforced concrete layer cast onto the extrados. Sometimes curved steel universal beam or column sections are used. If the cover is large or abnormal loading conditions have to be accommodated, a relieving slab deck may be provided which spans over the entire arch barrel onto new supports. In all cases careful attention to detailing of waterproofing is important to ensure ingress of water to the old arch barrel is minimized and any water in the backfill is efficiently drained away.

The barrel may become delaminated or in the particular case of multi-ring brickwork arch barrels, ring separated. Grouting or local patching may be possible but in the case of brickwork stitching may be the best option to reinstate the mechanical bond between the rings and hence ensure that the total thickness of the barrel is acting as one. It is best that the pins are installed radially and at close enough centres to ensure good load distribution. Saturated stitching should be avoided unless it is accepted that a potentially brittle continuum may result.

Stitching can be used more generally to strengthen structural elements and stabilize suscept foundations. The latter usually take the form of mini-piles. As many masonry arch bridges were constructed using timber piles or grillages problems have arisen due to rotting of the timber or settlement due to increased loading, etc. Such structures can be rehabilitated by a system of mini-piles installed by drilling through the existing abutment/pier. The reader is directed to specialist texts for details of the installation (which vary depending upon the site conditions). It is important to recognize that the new foundation will change the load path through the existing structure and that there will be a period of load transfer from the original to the new support. This may induce increased stresses in other elements of the bridge.

The installation of mini piles is usually associated with stitching of the piers, abutment, barrel and walls. It should be appreciated that 'blanket' stitching changes the masonry structure from a gravity, particulate material which can accommodate small settlement, thermal and load induced movements, without distress to a brittle continuum with all consequential material behavioural problems. Careful thought should be given to strategies which will mechanically connect the units together in such a way as to maintain their particulate articulation. In the case of multi-ring brickwork arch barrels radial pins reinstate through composite thickness behaviour of the brickwork by ensuring mechanical connection between the rings whilst at the same time allowing particulate behaviour between the pins. If the stitching is criss-crossed through the arch barrel this particulate behaviour is prevented. Additionally, the percussive nature of the installation necessitates high quality grouting to reinstate the units strength and their interface bond and to produce a waterproof structure.

Many of the problems with masonry arch bridges are associated with the stability of the spandrel and wing walls (and their interaction with the arch barrel). The reasons for their distress are multifarious and include excessive backfill pressures, diurnal and seasonal movements, settlement, vehicles being allowed to run close to the inner face, accidental impact, deterioration, etc. It may be that reconstruction is necessary but this will only be done in extreme cases. More usually tie-bars are used. These can be installed by drilling horizontally through the bridge. Typically, the tie bar would be 32 mm diameter and the spreader (or pattress) plates 600×600 mm. It is usual for the tie-bars to be installed with only a nominal tensioning – the logic being that the tie is there to prevent further movement rather than to pull the wall back to some predetermined position. Laboratory tests have shown that the ties have very little effect upon the barrel behaviour until the barrel has formed a mechanism and enhanced soil interaction has been mobilized. At which stage the ties (by holding the walls together) confined the backfill and hence stiffened it – it was only at this stage that the tie force increased. The test was undertaken over a short period and therefore was not subject to long term effects (Melbourne et al., 1995). Tie-bars can be installed within the arch barrel itself. The dimensions of the pattress plates, in this case, are usually determined by the barrel thickness. At least five tie-bars are provided (i.e. at the crown, quarter points and springings). Their effect is to offer some transverse strengthening which enhances the transverse arching contribution to load distribution; it also improves the longitudinal material properties by confining the material.

If the road surface is rutted or the track vertical alignment is a constant cause for concern then it may be that backfill is failing and needs improvement or replacement. Soil improvement techniques can be applied but usually (given the age of most of the arch bridge stock) the problem relates to ingress of water or changes in loading patterns. In any case, an investigation of the causes should be undertaken. As with all rehabilitation work the presence and requirements of statutory undertakers and other users of the bridge must be consulted and agreement reached regarding the proposed works. Grouting can cause particular problems for statutory undertakers.

If it is deemed necessary to remove the backfill, then saddling, relieving slabs etc are the likely solutions. The reinstatement of the backfill may then take the form of weak concrete or reinforced soil, i.e. methods of relieving the spandrel and wing walls of any soil pressures.

It is very important when considering repairing a masonry arch bridge that a holistic approach is taken. The engineer should consider the bridge as a gravity, particulate structure the many elements of which interact with each other and the soil upon which they rest. To change the nature of that interaction by intervention must be robustly justified.

Although the MEXE method is generally accepted as a good starting point for any assessment (or even sizing a new bridge) there are areas of uncertainty where further guidance may be sought.

It is vital that the true dimensions of the bridge are considered and that material properties are determined. The whole basis of the MEXE method is that 'the arch is assumed to be parabolic in shape with a span/rise ratio of four, soundly built in good quality brickwork/stonework, with well pointed joints, to be free from cracks, and to have adequate abutments'. It is clear from studying the background to MEXE other conditions were assumed. These include a secant variation of the second moment of area between the crown and the springing. Additionally, the stress levels are such that the arch barrel can be assumed to comply with linear elastic theory, and that the material is isotropic. Even if all these conditions are fulfilled it can be shown that the extension of Pippard's work was flawed by allowing dead load and live load stresses to be added together to give a range of permissible axles. This is because their effects are in opposite senses and so where the depth of fill (h) exceeds the arch barrel thick (d) then there is a real chance that the dead load conditions alone will cause a violation of the tensile stress criterion. It would therefore be expedient initially to limit MEXE to an $h \leq d$ criterion. Frequently haunching is provided which complies with the secant condition otherwise the barrel thickness must be checked at the crown, quarter span and springing under the carriageway to ensure that the correct figures are being used.

Additionally, the interaction between fill and arch profile is ignored (probably because the initial work assumed that the backfill contribution to stability was confined to vertical dead load only and that any horizontal active or passive contribution was ignored). Clearly arches with a smaller span to rise ratio will derive a greater contribution from the active and passive soil pressures than a flat arch.

In the case of flatter arches greater attention should be paid to the stability of the abutments and piers, particularly if the arch is being checked for an abnormal load or a change in the normal loading regime.

It should also be remembered that the profile factor F_p is based upon a parabola being the best shape for crown loading. Crown loading is assumed to give the highest stress levels, which is a logical approach for a permissible stress criteria method. Collapse is more likely to occur by the formation of a mechanism for which the worst loading case will be 1/4 to 1/3 span axle loading and so it is important not to mix factors and criteria between the two approaches.

Similarly, it is important when applying each of the factors that 'double counting' of the defects, etc. does not take place. For example, in the material factor if the brickwork of the arch barrel is deemed to be in poor condition and the factor F_b is taken to be 0.7 then the condition factor F_{CM} should not include an allowance for the conditions of the brickwork – it should be based on other deficiencies. Some guidance regarding the condition factor F_{CM} is given in BA16/97 with the statement 'where the condition factor is less than 0.4 immediate consideration should be given to the repair or reconstruction of the bridge. If the bridge has been the subject of regular inspections, the records of which are available, then it is important to determine whether or not the defects are the subject of continual deterioration, historic and now stable, or new. This should have great bearing upon the condition factor and the recommended course of action.

Defects do make modelling difficult if the application of the condition factor results in remedial action being needed to make the bridge safe. Longitudinal cracks are fairly easy to deal with as a two-dimensional idealization is the usual starting point and any contribution of spandrel walls (as usual) can be ignored. Very weak material or mis-shaped arch barrels can be accommodated in the input files. Things become more difficult when ring separation or diagonal cracking or non-uniform spandrel wall movements are present. In these cases a more sophisticated model is required, probably three-dimensional. All the above are further complicated if the bridge is skewed.

Bibliography

Alexander T and Thomason AW. *The Scientific Design of Masonry Arches*. The University Press, Dublin, 1909.

Ashurst D. An assessment of repair and strengthening techniques for brick and stone masonry arch bridges. Department of Transport. TRRL Contractor Report 284. Transport Research Laboratory. Crowthorne, 1992.

British Standards Institute. BSI BS 6779. *Highway Parapets for Bridges and other Structures*. British Standards Institute, London.

Barlow WH. On the existence of the line of equal horizontal thrust. *Proceedings of Institution of Civil Engineers*, 5, 1846.

Bridle RJ and Hughes TG. An energy method for arch bridge analysis. *Proceedings of Institution of Civil Engineers*, Part 2, September, 375–385, 1990.

Castigliano CAP. *Elastic Stresses in Structures* (translation by ES Andrews). Scott Grennwood, London 1879, 1919.

Chettoe CS and Henderson W. Masonry arch bridges: a study. *Proceedings of Institution of Civil Engineers*, 7, August, 723–774, 1957.

Choo BS, Coutie MG and Gong NG. Finite Element analysis of masonry arch bridges using tapered beam elements. *Proceedings of Institution of Civil Engineers*, Part 2, 91, December, 755–770, 1991.

County Surveyors' Society Bridges Group Report, No ENG/1-95, 1995. The Assessment and Design of Unreinforced Masonry Vehicle Parapets.

Couplet P. De la poussée des voftes, *Histoire l'Academie Royale des Sciences*, 79 and 117, 1730.

Cox D and Halsall R. *Brickwork Arch Bridges*. Brick Development Association, 1996.

Danyzy AAH, 1732. Methode générale pour déterminer la résistance qu'il opposer à la poussée des votes, Histoire de la Société Royale des Sciences établie à Montpellier, 2, 40, 1732.

Department of Transport. *The Design of Highway Bridge Parapets.* Department of Transport Standard BD52/93. Department of Transport, London, 1993.

Department of Transport. *The Assessment of Highway Bridges and Structures*. Department of Transport, Departmental Standard BD 21/97. Department of Transport, London, 1997a.

Department of Transport *The Assessment of Highway Bridges and Structures*. Department of Transport, Advice Note BA 16/97. Department of Transport, London, 1997b.

Department of Transport, 1987. The Assessment of Highway Bridges and Structures: Bridge Census and Sample Survey. Department of Transport, London.

Dhanasekar M, Page AW and Kleeman PW. The failure of brick masonry under biaxial stresses. *Proc. ICE*, Part 2, 79, 292, 1985.

Drucker DC. Coulomb friction, plasticity and limit loads. *Transactions Am. Soc. Mech. Engrs*, 76, 71–74, 1953.

Frazier. *La théorie et la pratique de la coupe des pierres* – 3 vols, Strasbourg and Paris, 1737–1739.

Gilbert M. On the analysis of multi-ring brickwork arch bridges. In *Arch Bridges* (Ed. A Sinopoli). Balkema 109–118, 1998.

Gilbert M. The behaviour of masonry arch bridges containing defects. PhD thesis. University of Manchester, UK, 1993.

Harvey WJ. Application of the mechanism analysis to masonry arches. *The Structural Engineer*, 66, 5, 1 Mar, 77-84, 1988.

Hendry AW. Masonry Properties for Assessing Arch Bridges. *Department of Transport. TRRL Contractor Report 244,* Transport Research Laboratory. Crowthorne, 1990a.

Hendry AW. Structural Masonry. Macmillan, 1990b.

Heyman J. *The Masonry Arch*. Ellis Horwood Ltd, Chichester, 1982.

Highways Agency. *The Appearance of Bridges and other Highway Structures*. HMSO, London, 1998.

Hobbs B, Gilbert M and Molyneaux T. Effects of vehicle impact loading on masonry arch parapets. In *Arch Bridges* (Ed. A Sinopoli) Balkema, 281–287, 1998.

Hodgson JA. The behaviour of skewed masonry arch bridges. PhD thesis. University of Salford, UK, 1996.

Hooke R. *A Description of Helioscopes and some other Instruments*. London, 1675.

Lilley DM. Research studies into the effects of transient loading on the Tyne Bridge. In *Arch Bridges* (Ed. C Melbourne), 55–64. Thomas Telford, London, 1995.

Lourenco PB. *Computational Strategies for Masonry Structures*. Delft University Press, Delft, 1996.

MEXE. Military load classification of civil bridges by the reconnaissance and correlation methods (SOLOG Study B38). Military Engineering Experimental Establishment, Christchurch, 1963.

Melbourne C. The collapse behaviour of a multi-span skewed brickwork arch bridge. In *Arch Bridges* (Ed. A Sinopoli). Balkema, 289–294, 1998.

Melbourne C, Gilbert M and Wagstaff M. The collapse behaviour of multi-span brickwork arch bridges. *The Structural Engineer*, 75, 17, 297–305, 1997.

Melbourne C and Njumbe S. Mass concrete arches. In *Arch Bridges* (Ed. A Sinopoli). Balkema, 331–340, 1998.

Melbourne C and Walker SJ. Load test to collapse on a full scale modes six meter span brick arch bridge. Dept of Transport TRRL Contractor Report 189. Transport Research Laboratory. Crowthorne, 1989.

Middleton WG. Research project into the upgrading of unreinforced masonry parapers. In *Arch Bridges* (Ed. Melbourne C), 519–528, 1995.

Navier LMH. Resumé des leHons donnees B l'Ecole des ponts et Chaussees sur l'application de la mechanique a l'establissement des construction et des machines, Part 1, 1826.

O'Connor C. *Design of Bridge Superstructures*. Wiley, Chichester, 1971.

Page AW. The strength of brick masonry under biaxial compression-tension. *Inst J. Masonry Constr.*, 3, 1, 26–31, 1983.

Pippard AJS and Baker J. The Voussoir Arch. In *The Analysis of Engineering Structures. London* (Eds Pippard and Baker), Chapter 16, 385–403. Edward Arnold, London, 1962.

Pischl R and Schickhofer G. The Mur River Wooden Bridge, Austria. *Struct. Eng. Inst.*, 4/93, 1993.

Pluijm R Vander. Structural behaviour of bed joints. *Proc 6th North Am. Masonry Conf. Philadephia*, 125–136, 1993.

Potts JG. Structural masonry. CUR Report 171, Balkema, 1997.

Riddington J and Jukes P. A review of masonry joint shear strength test methods. *J. Br. Mas. Soc, Masonry Inst.*, 11, 2, 37–43, 1997.

Smith FW, Harvey, WJ and Vardy AE. Three hinge analysis of masonry arches. *The Structural Engineer*, 68, No. 11, Jun, 203–213, 1990.

Sowden AM. *The Maintenance of Brick and Stone Masonry Structures*. E & FN Spon, London, 1990.

Strasky J and Husty I. Arch bridge crossing the Brno-Vienna Expressway. *Proc Inst Civ Engnr*, 132, Nov., 156–165, 1999.

Taylor JL. Chippingham Street footbridge/cycleway, Sheffield, UK. In *Arch Bridges* (Ed C Melbourne), 49–54. Thomas Telford, London, 1995.

Timoshenko SP and Gere JM. *Theory of Elastic Stability*. McGraw-Hill, London, 1961.

Van Beek GW. Arches and Vaults in the Ancient New East. *Sci. Am.*, July, 78–85, 1987.

Wanders SP, Manday MA, Redfield C and Strasky J. Wisconsin Avenue Viaduct. In *Arch Bridges* (Ed C Melbourne). Thomas Telford, London, 1995.

Wilson M and Jones H. Commercial Street Bridge, Sheffield, UK. In *Arch Bridges* (Ed C Melbourne), 75–86. Thomas Telford, London, 1995.

9　Seismic response and design

A.S. ELNASHAI

9.1 Introduction

An efficient transportation system plays a vital role in the development of a modern society, mainly due to the inter-reliance of various industries and the increased trend for outsourcing of various necessary ingredients within a single activity. Hence, transportation networks are referred to as lifelines, the integrity of which has to be protected alongside water supply, electricity and gas networks. Whilst roads are a most important component of transportation networks, bridges are both more important and sensitive to damage from natural disasters, since roads are more easily repairable and may be also readily by-passed. The closure of a bridge that represents the only or most important link between two areas separated by water or some geological feature (e.g. gorges) would potentially cause very severe consequences on industry, commerce and society as a whole. Recent examples abound as to the effects of earthquake damage to bridges, as discussed in subsequent sections. Two examples are quoted herein of the consequences of the closure of the Oakland Bay Bridge on traffic between San Francisco and Oakland (Loma Prieta, 1989) and the closure of several of the crossings between Kobe and Port Island (Hyogo-ken Nanbu, 1995), amongst several others. Not only did such closures affect the communities in the immediate vicinity of the bridge, but it had also knock-on effects on many other communities due to loss of business and delays in delivery of essential goods. In Table 9.1, estimates of economic loss due to bridge damage in three major earthquakes are give. These do not include indirect loss due to business interruption and lost revenue. They, however, serve to confirm the economic significance of bridge damage.

Table 9.1 Economic loss due to bridge damage in recent earthquakes

Earthquake	Cost (billion $)
Loma Prieta 1989	1.7
Northridge 1994	0.3
Kobe 1995	6.5

If the economic loss due to closure of a main arterial bridge is assessed alongside the cost of seismic retrofitting of the structure, the case of assessment and re-design of bridge structures in seismic areas will be immediately apparent. To emphasize this point, the effect of the San Fernando earthquake of 1971 is considered. Many of the cases of collapse of spans were attributed to the short seating length allowed at seismic joins. The cost of design and installation of restrainers (assuming that other failure modes would not be triggered) would have been a very small fraction of the direct cost of repair, and an even smaller proportion of the total cost including business interruption and loss of revenue. It is therefore of priority to re-assess bridge structures in areas subjected to seismic hazard with a view to minimize earthquake damage.

One of the serious problems facing the earthquake engineering community in reducing the earthquake risk to bridges is that whereas a feel for vulnerable parts in buildings and frequently encountered failure modes are common knowledge, engineers in general are less familiar with bridge structures. Therefore, increasing the awareness of bridge designers and earthquake engineers of the observed repetitive damage patterns following damaging earthquakes is a worthy cause. Moreover, adopting an appropriate concept for the bridge, including sub- and super-structures, is certainly an effective means of reducing drastically complications that may arise at the detailed design stage. Consequently, presenting conceptual design issues in a transparent manner and exploring their relationship with anticipated seismic behaviour would lead to reduced earthquake damage to bridges. Finally, there are considerable discrepancies between the leading international codes for seismic design of bridges. Hence, comparative assessment of codes for the design of bridges under earthquake motion, highlighting the differences and their origin, is important step towards improving the understanding of the engineering community of the seismic design procedure.

The objectives of this chapter are limited, and the audience well defined. It is intended for the non-specialist but the well aware of issues of dynamic response of structures, inelastic deformation and energy absorption and generalities of earthquake ground motion causes and effects. The scope is confined to fixed reinforced concrete bridges comprising decks, which may be steel or composite, supported on reinforced concrete piers. Other configurations are beyond the scope of this short chapter, and the reader is referred to the extensive literature.

9.2 Modes of failure in previous earthquakes

9.2.1 World-wide bridge damage observations

One of the earliest detailed and pictorial accounts of bridge failure is due to Milne (1892), where a description of the effects of the 'Great Earthquake of Japan, 1891' was given. This earthquake hit on 28 October 1891 and affected the prefectures of Gifu and Aichi and was felt over an area of approximately 90 000 square miles. The earthquake caused severe damage and some cases of collapse to bridges, such as the collapse of the masonry piers of the Kisogawa bridge and the total collapse of the Nagara Gawa steel truss bridge (Figure 9.1). Several other bridges suffered varying degrees

Figure 9.1 Damage to the Nagara Gawa Railway Bridge, Gifu (Japan) earthquake of 28 October 1891.

of damage. Since then, many earthquakes caused severe damage to bridges leading to very serious consequences, usually more economic than human. Among such earthquakes in Japan are the Great Kanto earthquake of 1923 (m=7.9), Nanki in 1946 (m_l=8.1), Fukui in 1948 (m_l=7.3), Imaichi in 1949 (m_l=6.4 and 6.7; multiple event), Tokachi-oki in 1952 (m_l=8.1), Northern Miyagi in 1962 (m_l=6.5), Niigata in 1964 (m_l=7.5), Ebino in 1968 (m_l=6.1), Tokachi-oki in 1968 (m_l=7.9), Miyagi-ken-oki in 1978 (m_l=7.4), Nihon-kai-chubu in 1983 (m_l=7.7), Kushiro-oki in 1993 (m_l=7.8), Hokkaido Nansei-oki in 1993 (m_l=7.8) and Hyogo-ken Nanbu in 1995 (m_l=6.9).

The Hyogo-ken Nanbu earthquake was particularly damaging to bridge structures in the area, knocking-off vital services such as the Shinkanzen line, Route 3 of the Hanshin Expressway (Figure 9.3) and affecting very seriously all other lines in the Kobe area (Priestley *et al.*, 1995, Elnashai *et al.*, 1995, Kawashima *et al.*, 1995).

In the USA, widespread damage was reported to bridges in the earthquakes of San Fernando of 1971 (m_s=6.6; Figure 9.3), Whittier Narrows of 1987 (m_s=6.3), Loma Prieta of 1989 (m_s=7.1) and Northridge of 1994 (m_s=6.9). In particular, the Northridge-inflicted damage on bridges was one of the major sources of financial loss due to the severe disruption of transportation lifelines (Broderick *et al.*, 1994; EERI, 1994).

Figure 9.2 Collapse of the Hanshin Expressway Elevated Road, Hyogo-ken Nanbu (Japan) earthquake of 17 January 1995.

Figure 9.3 Aerial view of the Golden State-Foothill Interchange, San Fernando (USA) earthquake of 9 February 1971 (aerial view).

Damage to bridges was also reported from other areas of the world, such as the Philippines earthquake of 1990 ($m_s=7.8$), the Talamanca earthquake of 1991 in Costa Rica and the Guam earthquake of 1993 ($m_s=8.1$).

9.2.2 Observed damage patterns

An appreciation of the possible areas of vulnerability in bridges is essential for the development of conceptual designs in seismic areas. Hence, below is a brief account of the most commonly observed failure patterns from various earthquakes.

Foundation soil

Soil lateral spreading or liquefaction imposes large deformation demand on bridge components (Figure 9.4), such as piles, abutment walls and simply supported deck spans. Some bridges founded on soft ground in the Kobe area suffered damage to piles due to negative skin friction resulting from soil failure. Also, approach structures and abutments have suffered substantial movement due to soil slumping.

Liquefaction was widespread in the Niigata earthquake, especially in the alluvial plains of the Shinano and Agano rivers. This caused significant damage due to large movement of pier and abutment foundations. Also, railway and highway bridges were affected by large ground displacement in the Costa Rica earthquake, where caisson and pier movements of 2.0 and 0.8 m, respectively, were observed.

Foundations and piles

Footings and piles are sometimes under-designed for earthquake loading, since the overstrength of piers they support would not have been taken into account. In

Figure 9.4 Soil spreading under a bridge pier, causing footing rotation (view looking down).

the Kanto earthquake, titling of foundations of mass concrete was observed, thus indicating inadequate consideration of over-turning. In Kobe, several investigated cases showed damage to footings, which cracked mainly in shear. Also, several piles were also damaged. It is relatively difficult to ascertain the cause of failure of sub-grade structures, but it is likely that such failures are due to unconservative estimates of the actions transmitted from the piers to the foundations. Also, the point of contraflexure of the pile–footing–pier system is often misplaced, hence the critical sections are not treated as such.

Sub-structure

Probably the most commonly observed failure is to the piers of bridges. Three modes of failure are possible, and their combinations; namely flexure, shear and axial distress. The single pier sub-structure of the Hanshin Expressway collapsed in the Hyogo-ken Nanbu earthquake due to the failure of gas welds on main reinforcing bars (Figure 9.2). Several cases of symmetric buckling of reinforcement and compressive failure of piers may be, at least in part, attributable to high vertical earthquake forces both in Kobe and Northridge. The three piers supporting the I10 (Santa Monica freeway) collector-distributor 36 suffered varying degrees of shear failure due to the short shear span that resulted from on-site modification of the original design.

Inadequate confinement in many piers caused premature failure of RC piers in recent earthquakes. Also, misinterpretation of the deformed profile of piers, with respect to the connection between pier and deck and pier and foundation, lead to critical sections developing at points of reduced longitudinal reinforcement. Several cases of damage in Kobe are attributed to the latter effect.

Figure 9.5 Total collapse of steel box girder pier due to weld failure.

Figure 9.6 Damage to sub-structure due to the Loma Prieta (USA) earthquake of 17 October 1989. Left, the Cypress Viaduct (1880); right, failure of the piers of the Struve Slough Bridge.

Unzipping of corner welds in steel box piers in Kobe lead to complete collapse of a number of piers which then were squashed by the weight of the heavy deck (Figure 9.5). Also, a number of columns suffered extensive local buckling in the Kobe area. In several cases, this coincided with the termination of concrete infilling, which is used to protect the piers from vehicle impact.

Multi-column sub-structures have also suffered damage in previous earthquakes. Examples of those are the Cyprus Viaduct in the Loma Prieta earthquake; frames collapsed along a distance of more than a mile due to shear failure of the RC section at the base of the top level column (Figure 9.6), and the Embarcadero RC elevated road that sustained heavy cracking especially at the beam-column connections. Piers of the Struve Slough bridge failed in the Loma Prieta earthquake also, and punched through the deck slab (Figure 9.6). There are many more cases of damage to piers in recent earthquakes, and more details are available in the literature (EERI, 1994; Broderick and Elnashai, 1995; Kawashima and Unjoh, 1997 amongst many others).

Super-structure

This includes pier–deck connections. Numerous cases of damage to or dislocation of bearings were observed in Kobe, with the ball or cylinder bearings sometimes found several metres away from their intended locations.

Figure 9.7 Damage by unseating at joints. Left, the Oakland Bay Bridge failed span in the Loma Prieta (USA) earthquake; left the Nishinomiyako Bridge collapse during the Hyogo-ken Nanbu (Japan) earthquake.

Damage to the super-structure is mainly not due to over-stressing, since decks are normally designed to remain near elastic in earthquakes. Many cases of collapse were observed in San Fernando, Loma Prieta and Northridge due to unseating at

Figure 9.8 Unseating of skew spans of the I5; Northridge (USA) earthquake.

Figure 9.9 Damage at abutment of the Tyler Street Bridge, San Fernando (USA) earthquake.

the seismic/expansion joints. This is due to defects in the bridge elsewhere, with the consequence of deck failure. Notable examples are the Oakland Bay Bridge steel truss (Figure 9.7) which lost a full span due to inadequate size of support angles and several high level crossings in the San Fernando valley. Also, a simply supported span of the Nishinomiyako bridge collapsed due to unseating (Figure 9.7). Where seismic restrainers are provided, damage to diaphragms occurred due to the very high local demands imposed at the restrainer anchorage point. Such effects are further aggravated in asymmetric or skew bridges, which are difficult to analyse at the design stage (Figure 9.8).

Impact damage between abutment and deck has also been observed, due to inadequate displacement tolerance there (Figure 9.9).

In a few cases, large inelastic deformation demands were imposed on decks, which were not designed for ductility. However, this is invariably due to the failure of an intermediate pier, thus imposing large flexural demand on a much-increased deck span.

9.2.3 Remarks

It is indeed noticeable that more recent earthquakes cause more damage to structures, notwithstanding the advancement in seismic design practice. This is due to the increased number of bridges of complex configurations coupled with the heightened consequences of bridge damage in developed societies.

The above quick over-view of damage patterns sets the scene for considerations of layout and configuration that go a long way towards ensuring the seismic safety of bridge structures. If the general guidelines presented below are adhered to, it is to be expected that final design confirmation by analysis would be straightforward and little additional provisions would be required, especially in areas of low to medium seismic hazard.

9.3 Conceptual design issues

9.3.1 Layout

Many design problems that would lead to unsatisfactory seismic performance and damage could have been anticipated by an earthquake engineering at the conceptual design phase. It is therefore of importance to discuss the commonly observed layout defects and their implications on seismic behaviour, following from the field observations above. Hereafter, a non-exhaustive review is given.

Plan layout

It is often necessary to construct skew or curved bridges. It should be noted though that curvature in bridges complicates the design and analysis and leads to difficulties in uniformly distributing ductility demand. It is often unconservative to analyse bridges with tight curvatures in 2D, hence a full 3D analysis is needed. Moreover, it is very difficult to quantify the degree of irregularity of a curved bridge, due to complicated mode shapes of response and interaction between location of piers, height of pier and mode shapes of most significance.

Skew bridges also pose design and analysis problems additional to those faced for straight bridges. Vibrations along the axis of a skew bridge cause torsional response that imposes large rotation demands on pier heads and out-of-plane movements on diaphragms. It was observed (Astaneh *et al.*, 1994) that diaphragms in steel decks would be subjected to lower demands if they were parallel to the axis of the pier, as opposed to normal to the axis of the connected girders.

In single pier bridges, an eccentricity between the deck axis (horizontal) and pier axis (vertical) would also lead to torsional response and non-uniform distribution of deformation demand. It is therefore concluded that the most effective layout is a straight bridge with the axes of piers coinciding with the centre line of the deck in single pier systems, or with the support frame axis normal to the deck axis in multi-column systems.

Layout in elevation

Contrary to common belief, the most regular bridge is not necessarily that with equal pier heights. This would be the case only in the special situation of full isolation of all piers and also abutments. The most regular response would be obtained from an elevation layout whereby the height of the pier is proportional to its distance from the abutment where fixity against out-of-plane displacement is available, taking into account the mode shape most likely to influence the response. In general, bridges are long period structures that are likely to be affected by higher modes. These notes are pertinent only to RC bridges of moderate lengths, as mentioned before, where the fundamental mode would be predominant. Therefore, if the height of the various piers follow a half sine wave spanning between two abutments where θ_y rotations are free and Z displacements are restrained, the most regular ductility demand will be imposed on the piers. In the case where the abutments allow lateral displacements, equal pier heights are the best solution. The relationship between pier stiffness and imposed displacement dictates the degree of regularity of the bridge. Conventionally, geometry considerations were used to quantify bridge regularity measures, or indices. In recent years, expressions including mode shapes, therefore the imposed demand, have been proposed (Calvi and Pavese, 1997). Further developments of regularity indices that taken into account the instantaneous supply and demand, in the inelastic range are currently underway. These are based on the concept that the pier causing the most irregular effect due to its stiffness is likely to be damaged first. By sustaining damage, its stiffness drops, hence the overall irregularity reduces.

Deck continuity

The majority of existing bridges have seismic or expansion joints. These were included in early design to reduce stresses from thermally induced deformations and/or simplify the analysis of the deck–pier system. As mentioned above, such joints have caused a large number of failures in previous earthquakes. Therefore, it is recommended that decks are designed as continuous structures wherever possible to eliminate problems with unseating and pounding. The latter problems are compounded when incoherent motion is imposed (incoherence is defined as significant differences in the input motion at different pier and abutment bases due to travelling wave and geometric incoherence of ground motion). The effect of incoherence on differential movement is negligible for most cases, compared to the effect of direct dynamic excitation (Monti *et al.*, 1996). However, for short span bridges with stiff abutment conditions, the incoherent case may become dominant, and conventional estimates of design forces may be unconservative (Tzanetos *et al.*, 1998).

Span length

In long span structures, a large axial force is imposed on piers due to their tributary part of the deck. Under earthquake motion, horizontal and vertical excitations are imposed, with the distinct possibility that vertical modes of vibration of the deck (acting as a continuous beam) would be excited. This will impose very high axial forces (and variations in axial forces) on piers, thus reducing their flexural and shear capacities. It is therefore advisable to use short span lengths by increasing the number of piers. Otherwise, analysis under vertical earthquake ground motion should be undertaken (Elnashai and Papazoglou, 1997) and the effect on axial forces in piers accounted for.

9.3.2 Foundation materials

Bridges are especially susceptible to damage from large imposed displacements. Therefore, it is important to found them on rock or stiff soil wherever possible. In cases where the surface soil is soft or potentially liquefiable, use of piled foundations is recommended. When piles are driven through sloping ground, stability issues should be considered, since slumping may impose very large lateral forces on piles and piers.

9.3.3 Foundation systems

Foundations are the first points of contact between the ground transmitting seismic waves and the structure. Therefore, the foundation system has a most marked effect on the response characteristics of the bridge. Various alternatives are shown in Figure 9.10 (Priestley *et al.*, 1995).

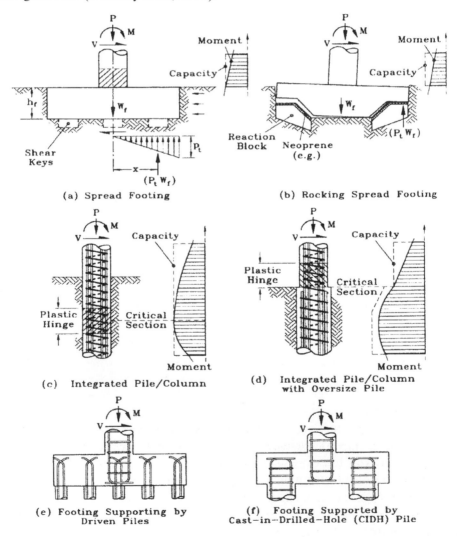

Figure 9.10 Options for foundation systems of RC bridges (Priestley et al., 1995).

The most commonly used foundation system is RC footings. These should be designed to resist the gravity loading in addition to the seismic forces appropriately scaled up by a factor to account for the force reduction used in the base shear calculation. Stability of the footing is provided by the contact with the foundation sub-strata due to gravity loading. Shear resistance is by friction on the horizontal plane under the footing and bearing on the vertical faces. Also, shear keys may be provided. Plastic hinges will form in the column base first, and their location may be controlled by detailing of longitudinal and transverse reinforcement. The damage location is easily accessible hence repair will not pose a problem. An alternative is shown in Figure 10(b), where the foundation is allowed to rock under earthquake motion, thus delimiting the column base force at the expense of increased top displacements.

In Figure 9.10(c) and (d) integral pier–piles systems are shown. These are rather economical and are used often in practice (e.g. the Jamuna bridge in Bangladesh, where concrete-filled steel circular tubes are used for piers and piles). In this case, the critical section is underground, hence is difficult to inspect and repair. To alleviate this shortcoming, the pile section may be increased, in order to impose plastic hinging in the pier, as shown in Figure 9.10(d).

Footings supported on piles are becoming widespread in practice. The most economical system (Priestley *et al.*, 1995) is the cast-in-drilled-hole RC pile, shown in Figure 9.10(f). In cases of footings supported on piles the objective is to keep the piles elastic and concentrate the inelasticity in the piers. Consequently, and due to cost considerations, the use of a small number of large diameter piles is recommended.

9.3.4 Foundation–pier connections

There are two options for this connection, namely pinned and fixed. In the pinned case, four bars are inserted at the centre of the pier and are anchored inside the footing, to resist nominal sliding shear. The pier section is cast discontinuous with the footing and rubber or felt sheets are inserted between the two. In this case, large moments will develop at the pier head acting as an inverted pendulum, but very low forces are exerted on the footing that requires only nominal reinforcement. The second option is full moment connection between footing and pier. In this case, a full design is required, since possible uplift forces may exist. Also, axial, bending and shear actions on the footing are much higher than in the pinned case.

9.3.5 Pier sections

There is a wide variety of RC and steel sections, as well as composite steel-concrete, used for bridge piers, some of which are shown in Figure 9.11 (Priestley *et al.*, 1995). The circular section is most desirable, especially in cases where the longitudinal and transverse demands are similar. One of its main advantages is that it provides uniform confinement (in contrast to rectangular sections) and adequately restrains the longitudinal bars from buckling. To provide torsional resistance and stability at the pier-deck intersection, a flare is commonly used, as shown in Figure 9.11 (section C–C). The shown additional bars are preferable to bending of the main bars into the

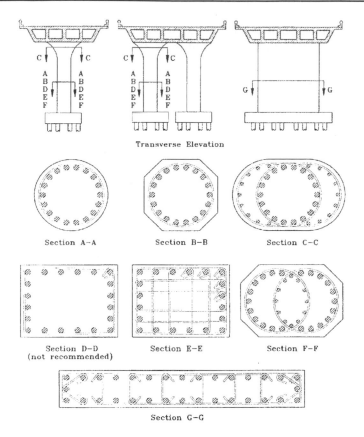

Figure 9.11 Options for pier sections and configuration (Priestley et al., 1995).

flare, for reasons of construction ease and also because of undesirable outward acting forces at the bending location.

The rectangular section D-D shown in Figure 9.11 has the disadvantage of inadequate protection of longitudinal bars against buckling. Also, the core is inadequately confined. These problems are mitigated in section E-E at the cost of a congested section and added workmanship. Section G-G, representing a shear wall type pier, is used where the longitudinal forces are carried mostly by the abutments, and only transverse stiffness and strength are required.

Hollow sections are used where the height of the piers is excessive hence the use of a solid section is not advisable (due to its high self-weight). It is most important though to check hollow sections against imploding due to the inward buckling of hoop inner layer of reinforcement.

Many bridges, especially in Japan, have a steel box girder for the deck. Buckling and unzipping problems have occurred in the past, hence these have to be checked carefully. Stiffeners are often used to enhance the buckling resistance of such members. The use of composite sections, constructed from concrete-filled steel tubes, is also popular, due to their high ductility capacity and ease of construction. The method of load transfer from deck to pier has an effect on the characteristics of the response, due to load sharing between the two materials. It is important that the steel tube does not separate from the concrete core, otherwise local buckling may occur. Sometimes, shear connectors are welded to the inside of the steel tube to ensure adequate interaction.

9.3.6 Lateral force resisting system

The frame action resisting earthquake motion may be either single column or multi-column structures, as shown in Figure 9.12 (Priestley *et al.*, 1995). Single column structures are easy to design and construct and are most suitable to situations where the demand along and across the bridge is similar. Also, since there is only one plastic hinge, response prediction is straightforward. It has several disadvantages though, such as low redundancy, high moment demand at the base, high seismic actions imposed on the foundation (due to the necessity of fixing the base) and high deck displacements.

Multi-column structures offer the option of fixed or pinned base solutions, hence drastic reduction in top section moments are availed of. Also, in general, displacements at the deck level are reduced, especially in the transverse direction. More-over, one of the most important advantages of this configuration is the degree of

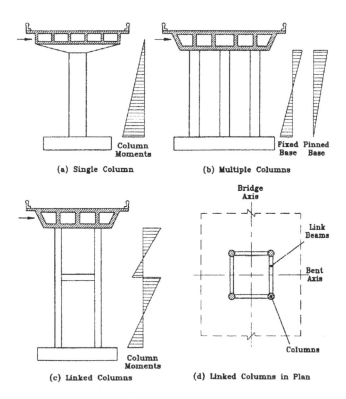

Figure 9.12 Options for lateral force resisting systems (Priestley et al., 1995).

redundancy, since re-distribution of action can occur between the various columns. Finally, load sharing between deck and columns, in monolithic structures, is better distributed than in the case of a single column system. The main disadvantages of this type of structure is that its response is more complicated, hence less predictable in the inelastic range and that there are more detailed connections than in the case of a single column. The demand imposed on columns will be non-uniform due to variations in axial forces and cap beam flexibility at various locations. As a consequence, the ductility demand in one column will be higher than that of the overall structure. In cases of very high piers, linking of the columns may be used, as shown in Figure 9.12(c). In this case, care should be exercised in shear design where the link creates a column of height shorter than about 6 times its plan dimension.

9.3.7 Connection between deck and piers

This may be either monolithic or bearing supported, as shown in Figure 9.13 (Priestley *et al.*, 1995) Monolithic construction is normally used for slender columns and small bridges (Priestley *et al.*, 1995). The energy absorption capacity of this configuration is in general larger than bearing supported systems (Elnashai and McClure, 1994) due to the double curvature of the former leading to potentially two plastic

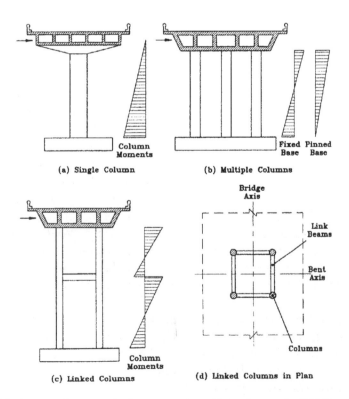

Figure 9.13 Options for pier-deck connection (Priestley et al., 1995).

hinging zones, instead of one in the case of bearings. For multi-column configurations, use of monolithic systems enables the stiffness in the longitudinal and transverse directions (when using circular and square columns) to be equal, thus eliminating the potential for a preferential response direction. Two more advantages of monolithic construction are that it allows consideration of using a fixed or pinned column-foundation condition (Section 9.3.4 above) and that it leads to higher redundancy of the lateral response system.

One of the major disadvantages of monolithic deck–pier systems is that large moments are transmitted to the deck that add to the moments from gravity forces thus creating critical conditions there. This is especially true for single column systems with wide decks, since the effective deck section resisting these high moments is relatively small. Another problem with this type of connection is the difficulties of ensuring that capacity design is respected, in terms of connection overstrength. Large diameter bars from the pier should be adequately anchored in the relatively shallow cap beam thus reinforcement congestion may become a problem.

Another problem is the imbalance between the longitudinal and transverse directions for single pier systems. In the transverse direction, the torsional stiffness of the deck only is acting against the pier behaving as a cantilever, thus the behaviour is close to such a condition. Longitudinally, the deck is monolithic with all piers, hence the restraint it applies on pier heads is much larger than in the transverse direction. Therefore, the pier is normally stiffer longitudinally than transversely. If a balanced design is sought, then a rectangular section will be required.

Thermal expansion imposes large displacement demand on monolithic systems longitudinally and therefore requires short spans between expansion joints. This is an unfavourable feature of monolithic pier–deck construction.

The second option is the provision of bearing support between pier and deck, allowing one or more translational and/or rotational degrees of freedom. A system by which rotations θ_z are allowed is in wide use. It is common also to have two shear keys in the transverse direction at a distance that would provide limited rotation θ_x restraint.

The most significant advantage of bearing supports is that the deck is not subjected to seismic forces, hence configurations not amenable to high moment resistance may be utilized. Also, the period of the bridge is elongated considerable as compared to monolithic bridges. This may be advantageous when the bridge is founded on rock or stiff soil, but is not suitable for soft sites. Another important advantage of bearing supports is that by suitable adjustment of bearing characteristics, stiff bearings may be placed on top of flexible piers and vice versa. Hence, a more uniform distribution of stiffness and strength than the case of monolithic structures would be easy to achieve.

Bearing supports have serious disadvantages, such as the effect of period elongation of the structure in areas of soft site conditions subjected to large distant earthquakes. Also, the bridge is subjected in general to larger displacements than its monolithic counterpart. For multi-column structures, the piers are placed in double curvature transversely whilst the longitudinal response is that of a cantilever. And the option of pinned pier–foundation condition is no longer available, hence footings are subjected to high seismic forces and are susceptible to uplift. Also, due to the existence of one potential plastic hinge in a single column structure, the ductility demand on the single pier is much higher than that of the overall structure (a global displacement ductility of 5 may correspond to a pier ductility demand of 14; Priestley *et al.*, 1995). Finally, bearing systems may undergo large inelastic displacements that are not restored at the end of the earthquake, hence impairing the use of the bridge. This drawback can be offset by suing self-restoring bearing systems.

9.3.8 Connection between deck and abutment

It is often the case that abutment design is not given due consideration leading to poor seismic response. This is partly due to the misconception that abutments are less sensitive to earthquake destruction than piers. Also, accurate representation of abutment seismic response is difficult, due to soil-structure interaction representation. However, site studies indicate that abutment failure is observed sufficiently frequently to warrant serious consideration in design. Moreover, studies on bridge models with various abutment conditions (Dodd *et al.*, 1996) concluded that boundary conditions at the abutment have a significant influence on all dynamic response characteristics of RC bridges.

A further point is the influence of abutment modelling and restraints under asynchronous motion that was studied by Tzanetos *et al.* (1998), as mentioned in earlier sections of this chapter. A number of analyses were undertaken for pinned and fixed abutment-deck conditions, which lead to the conclusion that the latter is more critical in the case of soft foundation material (low shear wave velocities). This was particularly noticeable and critical for the fixed abutment case, where the second mode contributed significantly for the asynchronous case, but not in the synchronous analysis. It is therefore important to ponder the effects of abutment configuration, stiffness in various planes and strength on the response of the bridge as a whole. The connection may be monolithic, bearing-supported or isolated, as described below.

Monolithic connections, such as those depicted in (a and b) of Figure 9.14 (Priestley *et al.*, 1995), are more commonly used for small bridges. They are designed to resist the total seismic force, whilst intermediate piers are designed for gravity loads only, but detailed for ductility as a precaution. In which case, the system shown in Figure 9.14(b) is more reliable than that of (a), since the latter relies at least in part on soil bearing resistance behind the abutment wall. This system will also have unequal stiffness in the push-pull longitudinal direction. The rigid (monolithic) abutment is designed for the peak ground acceleration without amplification, since it will move as a rigid body with the ground.

For large bridges, it is extremely difficult to resist the total seismic force at one location. Therefore, a bearing support may be employed, with all other vertical members sharing in the seismic resistance. Bearing supports have many configurations, two of which are shown in Figure 9.14(c) and (d). The stiffness in the negative longitudinal direction is provided by the wing wall, and the soil bearing stiffness (and strength) following the closure of the gap, which is designed for temperature changes.

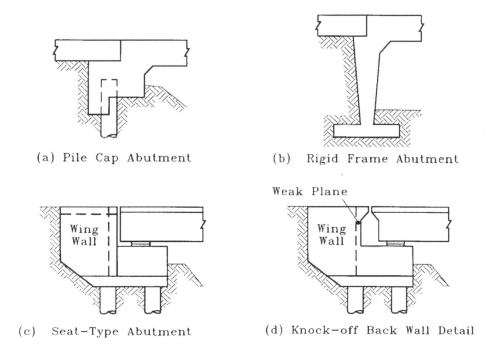

Figure 9.14 Options for abutment–deck connection (Priestley et al., 1995).

In the positive direction, stiffness depends on the bearing characteristics. Ball and cylinder bearings resting of horizontal planes or dish receptacles (for self-righting) are used in practice. The alternative shown in Figure 9.14(d) utilizes a knock-off detail, in which damage to the transverse wall is allowed in large earthquakes hence a large tolerance for displacements is afforded. For both configurations shown, the bearings may be substituted by isolators, which normally double up as dissipaters too, since the effect of period elongation of bridges on the seismic forces is less significant than the effect of damping by energy dissipation.

An issue of great significance is the transverse boundary conditions imposed on the deck by the abutment and the consequence of resulting forces imposed on the abutment. In general, it is more difficult to provide lateral stiffness and strength than it is for the longitudinal direction, as may be intimated from Figure 9.15(a) (Priestley *et al.*, 1995). Hence, buttresses, as shown in (b) may be utilized to increase resistance in both directions, which is also contributed to by the piles. With regard to the boundary conditions imposed on the deck, Figure 9.15(a) is examined. The moment constraint will be a function of the bearings used, their number and location. For a large number of bearings at some spacing resting on a high friction surface will provide a high degree of restraint, and vice versa. The force restraint depends on the existence and details of shear keys. Care should be exercised not to allow a very large difference in stiffness to develop between abutment and piers, since this will lead to a high concentration of force and deformation demand imposed on the former, which in turn will lead to severe distress.

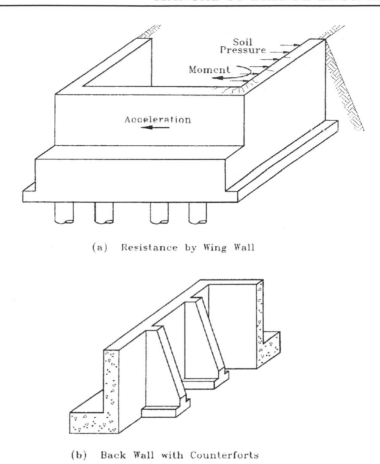

(a)   Resistance by Wing Wall

(b)   Back Wall with Counterforts

Figure 9.15 Mechanisms of resisting forces at the abutment (Priestley et al., 1995).

9.4 Brief review of seismic design codes

9.4.1 Historical note on bridge codes in Europe, the USA and Japan

The history of European practice in seismic design is rather recent, and cases of bridge damage in European earthquake are very few indeed. For Europe, the interest in seismic design of bridges arises from two main considerations, the potential for disastrous effects, and the export market. With regard to the former, several thousand bridges in Italy are potentially subjected to considerable earthquake risk, whilst major projects for bridge construction are underway in Greece, amongst several other European nations with a large bridge population and non-negligible exposure.

   Whereas national codes in Europe included seismic provisions for bridges, these were rather minimal considerations and intended to support European consultant working abroad. It was through the development of Eurocode 8 'Previsions for Earthquake Resistance of Structures' (CEN, 1994) that concerted efforts were dedicated to the development of seismic design guidelines for bridges in Europe. The first drafts, circulating in the late 1980s were quite brief, but unevenly so, and reflected the interest of the individuals involved in the drafting panels rather than the practice of design. With funding from national and European sources being directed towards research projects in the earthquake engineering field, extensive development work got underway in several academic and research institutions in Europe, including universities and research laboratories in Italy, Greece, Portugal, France, Belgium, Germany and the UK. Links with US, New Zealand and Japanese workers in the field also encouraged the development of a European approach to seismic bridge design. Subsequent versions of the Eurocode 8 draft, with part 2 dedicated to bridges, seemed more robust, comprehensive and reflecting a somewhat different approach to other codes. It is perhaps the relative freedom of drafting panels working on the draft, unhindered by practice (for better or for worse) that lead to the current format of EC8 Part 2.

The current version of EC8 part 2 is dated 1994. However, the bulk of the work was undertaken between 1997 and 1998. It therefore reflects a recent view of the European earthquake engineering community. The approach is that of utilizing ductility to reduce design forces, to detail potential plastic hinge zones for high levels of curvature ductility and to protect other bridge components by applying capacity design factors. It deals with both fixed and isolated bridges in steel and reinforced concrete. There are rather advanced topics in the code, such as models for spatial variability and conditions of acceptability of suites of artificial records for analysis. The code also attempts to give comprehensive guidance for modelling and analysis of bridges, statically and dynamically.

Eurocode 8 is currently in circulation as an ENV (voluntary European norm). The process of conversion from ENV to EN is currently underway for earlier parts, and Part 2 on bridges is soon to follow suit. It is therefore difficult to comment on EC8 part 2 at the moment since it is a moving target. Since the writer is involved with European networks working on bridge design developments, it is safe to speculate on the main developments. It is anticipated that the new draft will make use of recent development in multi-level earthquake design, thus covering various scenarios of action and attempting to meet different performance levels. It is also likely that the role of deformations will be promoted whilst force resistance will recede to be only a check. Improved criteria may be included to define regularity in a more representative fashion, by considering inelastic deformations. Finally the code will no doubt be affected by developments in other parts of the world.

The situation is fundamentally different in the USA. The Association of American State Highways and Transportation Officials (AASHTO) has been issuing seismic provisions for many years. However, the status of bridges in California, and the repeated relatively poor performance of bridges in Californian earthquakes, lead to the California Transportation Department (Caltrans) to develop its own design guidance). The Caltrans guidelines has been affected by the patterns and type of damage to bridges in earthquakes in the State. In the San Fernando earthquake of 1971, the interchange between the Golden State and Antelope Valley Freeways suffered total collapse. There were many other cases of heavy damage, partial and total collapse of bridges and elevated roads in the prosperous and heavily populated San Fernando Valley. This prompted major investment in research of design and strengthening of RC bridges in particular, and an extensive strengthening programme was initiated. The weakness of bridges in California was confirmed by the heavy damage sustained in the Loma Prieta earthquake of 1989. The catastrophic collapse of the Interstate 880 (Cypress Viaduct) caused heavy casualties as a result of the collapse of a mile-long stretch of the double deck RC structures (Elnashai *et al.*, 1989). Heavy damage was also inflicted on two major structures at the heart of the San Francisco network. The Embarcadero freeway suffered shear distress to many columns and girder-column connections alongside the terminal Separation. These were replaced at a cost of $120 million (1990 prices). Also, the Struve Slough bridge outside Watsonville collapsed completely with the piers punching through the deck. The State-funded research initiative for bridges was approved on 6 November 1989, less than one month after the earthquake.

Following the Governor of California's report on the earthquake 'Competing Against Time' published in May 1990, State and Federal funds were made available to a number of universities and research laboratories, in association with practising engineers. Interim guidance for repair and strengthening were issued and new design procedure, with further emphasis on ductility and capacity design principles. The next major disaster hit southern California in 1994. The Northridge earthquake of 17 January inflicted very heavy damage to bridges, intersections, over-crossings and even exit-entrance ramps (Broderick *et al.*, 1994, Broderick and Elnashai, 1995). It has shown that the battle against bridge damage in earthquakes was far from over. This sequence of events lead to the continuing state support for development of the Caltrans design and strengthening guidelines. It was therefore deemed appropriate that the subsequent comparison of bridge codes includes Caltrans guidelines as a US representative as opposed to AASHTO.

Following the Great Kanto earthquake of 1923, the first regulations for bridge design in Japan were introduced in 1926 (as reported by Kawashima *et al.*, 1998). This was updated in 1939, 1956 and 1964. These were rather simple guidelines mainly recommending the application of a lateral force of 20% of the weight of the structure, without reference to earthquake, structure or site characteristics. Moreover, the regulations pertained only to steel structures, since very few bridges were constructed in RC.

The first comprehensive seismic design code for bridges was issued in 1971, introducing design forces dependent on importance, hazard zone and site condition in the 'seismic coefficient method'. In a variant, termed 'modified seismic coefficient method' the structural response characteristics were also taken into account. And, due to the experience of the Niigata earthquake of 1964, assessment of liquefaction potential became part of the 1971 code.

Parts of the 1971 code that are common to the material sections (steel, concrete and composite) were further revised in 1980. This entailed mainly revision to the liquefaction potential assessment method and guidance on foundation design in liquefiable soils. A further, but more substantial, revision was introduced in 1990. The main revisions (Kawashima *et al.*, 1998) were:

- unification of the seismic coefficient and modified seismic coefficient methods
- checks on the dynamic strength and ductility of piers
- introduction of a frame analysis approach to evaluate force distribution in multi-span structures
- provision of response spectra for dynamic analysis.

Damage to bridge structures in the Hyogo-ken Nanbu earthquake of 17 January 1995 was significantly more than expected (Priestley *et al.*, 1995, EERI, 1996, Elnashai *et al.*, 1995 amongst others). Quick revision of the design guidance, especially for repair, was therefore published February 1995. These were applied to new structures as well, whilst the process of revision of the 1990 code was underway. The latter was issued in 1996 which included explicit reference to ductility-based design whilst retaining the old seismic coefficient method. The 1996 code is that currently in force.

In the writer's opinion, Japanese design practice for bridges has been traditionally based on strength concepts, thus ignoring the deformational capacity of structures. This leads usually to stiff and strong structures that exhibit very low levels of ductility. This could be a consequence of the cost of ductile detailing as opposed to that of additional concrete and reinforcement. The damage inflicted by the Hyogo-ken Nanbu earthquake in particular attests to this view. Contacts with Japanese experts working on code developments indicate that this approach is gradually changing, although the practice is still unaccustomed to applying large force reduction factors in design.

9.4.2 Comparison of provisions

A study comparing several codes from the USA and other countries has been undertaken (Rojahn *et al.*, 1997) In that study, the USA codes studied were AASHTO, ATC and Caltrans. Also included in the latter study were the codes of New Zealand, Japan and Eurocode 8. The tables of comparison are reproduced in Table 9.2 for a sub-set of codes. The Caltrans provisions are selected in preference to AASHTO, for the reasons outlined in the previous section. For brevity, only the provisions of the Japanese code and Eurocode 8 are given. Table 9.2 is, however, not strictly that given in the above-mentioned reference. Several clauses were altered to reflect new information available to the writer or different interpretation of code clauses. Also, in various locations, sources of additional information intended for use with the EC8 Part 2 on bridges are quoted, whereas in the original reference the authors dealt strictly with the provisions given in Part 2 of EC8.

It should be emphasized that all three codes discussed above are moving targets, since they are continuously under review and development. This is especially the case with the Japanese code, due to the time laps between issuing a new code version and its availability to an English speaking audience.

Table 9.2 Comparison of seismic codes

Provision	Caltrans	Eurocode 8	Japan
1. General			
a. Performance criteria	Implied. Structural integrity to be maintained and collapse during strong shaking to be prevented.	No collapse under safety-level event (ultimate limit state). No damage under frequent earthquakes (service limit state).	Faction to be maintained in small and moderate earthquakes. Collapse to be avoided during large earthquake.
b. Design philosophy	Adequate ductility capacity to be provided and failure of non-ductile elements and inaccessible to be prevented.	Sufficient strength of elastic structures in order to avoid damage. Brittle types of failure to be avoided in all structures.	Components to perform elastically under functional earthquake. Detailing of specific components to avoid damage.
c. Design approach	Single-level design. Desired performance at lower earthquake loads is implied.	Single-level design. Desired performance at lower earthquakes is implied (under review).	Utility level earthquake and working stress. Detailing to avoid collapse of girders checked. All under review.
2. Seismic loading			
	Elastic design spectra.	Normalized elastic response spectra with variable corner periods. All values set by national authorities (under review). Power spectrum representation permitted.	Elastic design spectra for magnitude 8± event.
a. Return period	Not defined. Maximum Credible Earthquake (MCE) with mean attenuation given by ARS curves.	475 years. Adjustment by Poisson distribution for different return periods and design life.	Not defined.
b. Geographic variation	Contour maps of peak rock acceleration developed for MCE and average attenuation as a function of distance from fault.	Set by national authorities. No agreed-upon maps for Europe.	Zone modification factor from map.
c. Importance considerations	Not addressed directly by design specifications.	Three categories based on national definitions.	Importance modification factor based on type of route or for particularly important bridges.

(*continued*)

Table 9.2 (Continued) Comparison of seismic codes

Provision	Caltrans	Eurocode 8	Japan
2. Seismic loading (continued)			
d. Site effects	Four soil profile types based on alluvial depth to bedrock.	Three (five in the future) soil types defined. Spatial variability must be considered for bridges with total length >600m and where abrupt change in soil type.	Three types of ground condition defined.
e. Damping	5% of critical.	Spectrum normalized to 5%. Scaling parameter given in an equation.	5% of critical implied.
f. Duration considerations	Not considered directly.	Specifications for artificial time histories.	Not considered directly.
3. Analysis			
a. Selection guidelines	Based on structure complexity only. Structure with $T > 3.0$ s treated as special case.	Three types of structure performance defined: ductile, limited ductility and elastic.	Based on structure complexity.
b. Equivalent static	Uniform lateral load preferred for hinge restrainer design.	Applicable for 'regular' bridges.	Reaction force method. Static frame method with static load.
c. Elastic dynamic	Multi-modal response spectrum analysis. 3-D lumped mass space frame with gross section properties.	Multi-modal response spectrum analysis. 3-D lumped mass space frame.	Considered special case.
d. Inelastic static	Not required.	Capacity design required for ductile structures.	Considered special case.
e. Inelastic dynamic	No guidelines.	Permitted but results may not relax requirements from elastic response spectrum analysis except for isolated bridges.	Considered special case.
f. Directional combinations	Case 1 $= L+0.3T$ Case 2 $= 0.3L+T$	Case 1 $= 0.3L+T+0.3V$ Case 2 $= L+0.3T+0.3V$ Case 3 $= 0.3L+0.3T+V$	Longitudinal and transverse loads examined separately.
g. Load combinations	VII $=$ DL + PS + b_E + EQ	$A=\gamma_D A_D+\gamma_L A_L+\gamma_E A_E$ (D:dead, L:live, E:eq.)	Earthquake combined with primary loads.

(continued)

Table 9.2 (Continued) Comparison of seismic codes

Provision	Caltrans	Eurocode 8	Japan
4. Seismic effects			
a. Design forces	Consider ductility and risk.	Consider ductility class and ground motion.	Elastic forces for functional earthquake. Forces adjusted for ductility when ductile behaviour is required.
(i) Ductile components	Component and period-dependent Z factors.	Elastic force reduction factor depends on structure type and component ($q = 1$–3.5)	Revise design coefficient considering ductility and magnitude $8\pm$ ground motion and check ultimate force capacity.
(ii) Non-ductile components	Capacity design or use $Z = 0.8$.	Capacity design with protection factors.	Force design using functional earthquake and ductility check with reduced ductility capacity.
b. Displacements	Use cracked stiffness	Use secant stiffness at yield for members with plastic hinges.	
c. Minimum seat width	Evaluated from seismic action and spans.	States that unseating should be checked.	
5. Concrete design			
a. Columns (flexure)	Ultimate strength design using nominal material strengths and a ϕ factor between 1.0 and 1.2 depending on axial stress.	Ultimate strength design with capacity-reduction factors.	Allowable stress design with 1.5 × normal design stresses.
b. Column (shear)	Demand based on capacity design (30%). Shear capacity based on contribution from concrete and transverse reinforcement. Concrete contribution at plastic hinges is reduced to zero for low axial loads.	Verification of capacity of diagonal compression, shear reinforcement, and sliding shear required.	Allowable stress design with 1.5 × normal design stresses.
c. Column spirals (confinement)	Three conditions given for outside plastic hinges, plastic hinges – small columns and plastic hinges – large columns.	Not required when axial load ratio < 0.08. Confinement derived from required curvature ductility.	Column design ductility factors are based on tests of typically reinforced Japanese bridge columns.
d. Piers	Nominal detailing with reduced ductility allowed.	Covered as tied columns.	As above

(continued)

Table 9.2 (Continued) Comparison of seismic codes

Provision	Caltrans	Eurocode 8	Japan
5. Concrete design (continued)			
e. Footings	Capacity design for seismic loading. Ultimate strength design with $\phi = 1.0$.	Detailing rules.	Allowable stress design with $1.5 \times$ normal stresses.
f. Superstructure and pier joints	Joint shear: $V_c = 12f'^{0.5}_c$ Spirals extended into cap.	Not covered.	Unknown.
g. Caps	Capacity design rule to ensure plastic hinge in columns not caps.	Capacity design in ductile structures.	Unknown.
h. Superstructure	Protected by capacity design factors.	Capacity design in ductile structures.	Unknown.
i. Shear keys	Abutment shear keys designed as sacrificial elements to protect stability of abutment.	Capacity design in ductile structures.	Unknown.
j. Column anchorage	ACI anchorage requirements	Covered in RC part (1.3)	Unknown.
6. Column splices	Not allowed in plastic hinge zones.	Not permitted in plastic hinge regions.	Unknown.
7. Steel design	No specification provisions for seismic design of steel members.	No specific provisions.	Unknown.
6. Foundation design			
a. Spread footings	Ultimate soil bearing capacity under seismic loading. Uplift allowed.	No specific provisions for spread footings in Part 2.	Unknown.
b. Pile footings	Designed for ultimate compressive capacity and 50% uplift capacity. Batter piles discouraged.	No specific provisions for pile footing in Part 2.	Unknown.
c. Liquefaction	Not addressed by specifications.	Covered in Part 5.	Method provided for determining liquefaction resistance factor $F_L = R/L$
7. Miscellaneous design			
a. Restrainers	Designed to remain elastic and restrict movement to allowable levels at expansion joints. Nominal restrainers required when ample seat width provided.	Required if minimum length without restrainers seat not met in ductile structures and for limited ductility and isolated bridges.	Standard devices available for tying superstructure and substructure together and for preventing the dislodging of movable bearings (stoppers).
b. Base isolation	Special case.	Specific chapters devoted to seismic isolation and elastomeric bearings.	Draft guidelines available.
c. Active/passive control	Not used.	Under development.	Draft guidelines available.

9.5 Closure

The above is a snap-shot of commonly observed bridge failure modes in previous earthquakes, given to highlight that there are repetitive patterns caused by inherent weaknesses. This was followed by a discussion of features of layout that are favourable to controlled and predictable seismic response of bridges. This favours relatively regular structures in plan and elevation. Finally, many of the options available for foundations through to the super-structure, and connections between various components, were presented and their likely effect on the response were discussed. Simple general guidelines adhered to during the conceptual design stage would facilitate considerably final design confirmation, and would increase seismic safety, especially in areas of low to moderate seismic hazard. Other alternative to the bridge configurations presented above, such as use of isolation and dissipation devices and active control mechanisms, are also feasible and have been gaining ground recently.

A brief review of seismic design codes and general practice in Europe, the USA and Japan indicate that the situation is far from uniform. In general, US and European practice utilize ductile response and capacity design principles more than Japanese practice. There are a number of issues that require further development and this are by-and-large underway. In all three cases considered, seismic design codes are undergoing radical revisions and hence the statements made above should be considered as time-qualified.

Bibliography

Astaneh A-A, Bolt B, McMullin K, Donikian R, Modjtahedi D and Cho S-W. Seismic performance of steel bridges during the 1994 Northridge earthquake. Report No. UCB/CE-Steel-94/01, UC Berkeley, 1994.

Broderick BM, Elnashai AS, Ambraseys NN, Barr J, Goodfellow RG and Higazy EM. The Northridge earthquake of 17 October 1994; observations, strong-motion and correlative response analyses, ESEE Report no. 94-4, Engineering Seismology and Earthquake Engineering Section, Imperial College, UK, 1995.

Broderick BM and Elnashai AS. Analysis of the failure of Interstate 10 ramp during the Northridge earthquake of 17 January 1994. *Earthquake Engineering and Structural Dynamics*, 24, 189–208, 1995.

Calvi GM and Pavese A. Conceptual design of isolation systems for bridge structures. *Journal of Earthquake Engineering*, 1, 1, 193–218, 1997.

CEN. Eurocode 8: Design provisions for earthquake resistance of structures, Part 2: Bridges, ENV 1998-2, European Committee for Standardization, Brussels, 1996.

Earthquake Engineering Research Institute. Northridge earthquake reconnaissance report, Earthquake Spectra Special Issue, Earthquake Engineering Research Institute, USA, 1994.

Elnashai AS and McClure DC. Effect of modelling assumptions and input motion characteristics on seismic design parameters of RC bridge piers. *Earthquake Engineering and Structural Dynamics*, 25, 435–463, 1995.

Elnashai AS, Bommer JJ, Baron IC, Lee D and Salama AI. Selected engineering seismology and structural engineering studies of the Hyogo-ken Nanbu (Great Hanshin) earthquake of 17 January 1995, ESEE Report no. 95-2, Engineering Seismology and Earthquake Engineering Section, Imperial College, UK, 1995.

Elnashai AS, Elghazouli AY and Bommer JJ. The Loma Prieta, Santa Cruz, California earthquake of 17 October 1989, ESEE Report no. 89-11, Engineering Seismology and Earthquake Engineering Section, Imperial College, UK, 1989.

Elnashai AS and Papazoglou A. Procedure and spectra for analysis of RC structures subjected to strong vertical earthquake loads. *Journal of Earthquake Engineering*, 1, 1, 121–156, 1997.

Dodd SG, Elnashai AS and Calvi GM, Effect of model conditions on the seismic response of large RC bridges, ESEE Report no. 96-5, Engineering Seismology and Earthquake Engineering Section, Imperial College, UK.

Kawashima K and Unjoh S. The damage to highway bridges in the 1995 Hyogo-ken Nanbu earthquake and its implications on Japanese seismic design. *Journal of Earthquake Engineering*, 1, 2, 505–542, 1997.

Milne J. The Great Japan earthquake of 1891, University of Tokyo report, 1892.

Monti G, Nuti C and Pinto P, Nonlinear response of bridges under multisupport excitation. *J. Struct. Eng. ASCE*, 122, 10, 1146–1159, 1996.

Priestley JMN, Seible F and Calvi GM, *Seismic Design and Retrofit of Bridges*. Wiley Interscience, New York, 1995.

Priestley JMN, Seible F and McRae G. The Kobe earthquake of January 17, 1995; Initial impressions, report no. SSRP-95-03, University of California at San Diego, 1995.

Rojahn C, Mayes R, Anderson DG, Clark J, Hom JH, Nutt RV and O'Rourke MJ. Seismic design criteria for bridges and other highway structures. Technical Report NCEER-97-0002, National Centre for Earthquake Engineering Research, New York, 1997.

Tzanetos N, Elnashai AS, Hamdan FH and Antoniou S. Inelastic dynamic response of RC bridges to non-synchronous earthquake input motion, ESEE Report No. 98-6, Engineering Seismology and Earthquake Engineering Section, Imperial College, London, 1998.

10 Cable stayed bridges

D.J. FARQUHAR

10.1 Introduction

The use of inclined stays as a tension support to the bridge deck was well known in the nineteenth century and there are many examples, particularly using the inclined stay to provide added stiffness to the primary draped cables of the suspension bridge. Unfortunately the concept was not well understood during the earlier part of the century as it was not possible to tension the stays and there was often inadequate resistance to wind induced oscillations. There were several notable collapses of such bridges, for example the bridge over the Tweed River at Dryburgh (Drewry, 1832), built in 1817, collapsed during a gale only 6 months after construction, in 1818. This caused the use of the stay concept to be abandoned in England. Nevertheless these ideas were adapted and improved by the American bridge engineer Roebling who used cable stays in conjunction with the draped suspension cable for the design of his bridges. The best known of Roebling's bridges is the Brooklyn Bridge, completed in 1883.

The modern concept of the cable stayed bridge was first proposed in the early 1950s for the reconstruction of a number of bridges over the River Rhine, these proving more economic, for moderate spans, than either the suspension or arch forms. It proved very difficult and expensive in the soil conditions of an alluvial flood plain to provide the gravity anchorages required for the cables of suspension bridges. Similarly for the arch structure, whether designed with the arch thrust carried at foundation level or as a tied arch, substantial foundations were required to carry the large spans. By comparison the cable stayed alternatives had light decks and the cable forces were part of a closed force system created by balancing the compression within the deck. Thus expensive external gravity anchorages were not required. The construction of the modern multi-stay cable stayed bridge can be seen as an extension, for larger spans, of the prestressed concrete, balanced cantilever form of construction. The tension cables in the cable stayed bridge are located outside the deck section, and the girder is no longer required to be of variable depth. However, the principle of the balanced cantilever modular erection sequence, where each deck unit is a constant length and erected with the supporting stays in each erection cycle, is retained.

The first modern cable stayed bridge was the Strømsmund Bridge (Wenk, 1954) in Sweden constructed by the firm Demag, with the assistance of the German engineer Dischinger, in 1955. At the same time Leonhardt designed both the North Bridge and the Theodor Heuss Bridge (Beyer and Tussing) across the Rhine at Dusseldorf. The North Bridge was constructed in 1958. The first modern cable stayed bridge constructed in the United Kingdom was the George Street Bridge over the Usk River (Brown, 1966) at Newport, South Wales, constructed in 1964. These structures were designed with twin vertical pylon and stay arrangements. The first structure with twin inclined planes connected from the edge of the deck to an A frame pylon was the Severins Bridge (Gregory and Freeman, 1960) crossing the River Rhine at Cologne, Germany. This bridge was also the first bridge designed as an asymmetrical two-span structure.

The economic advantages, proven in the construction of these early bridges, are valid to this day and have established the cable stayed bridge in its unique position as the preferred bridge concept for major crossings within an extensive range of spans.

10.2 Stay cable arrangement

Two basic arrangements have been developed for the layout of the stay cables:

- the fan stay system (including the modified fan stay system)
- the harp stay system.

These alternative stay cable arrangements are illustrated in Figure 10.1.

10.2.1 Fan cable system

The fan system was adopted for several of the early designs for the modern cable stay bridge, including the Strømsmund Bridge (Wenk, 1954), the method of supporting the stays on top of the pylon was taken from suspension bridge technology where the cable is laid within a tower top saddle. The floor of the saddle is machined to a radius so that each cable stay, anchored in the main span, can pass over the pylon and be anchored directly within the back or anchor span. This arrangement is structurally efficient with all the stays being located at their maximum eccentricity from the deck and applying minimum moment to the pylon.

The fan arrangement proved suitable for the moderate spans of the early cable stay designs, with a small number of stay cables or bundled cables supporting the deck. As larger spans became necessary the size of the limited number of cables increased, becoming uneconomically large and difficult to accommodate within the fan configuration. The anchorages were heavy and more complicated and the deck needed to be heavily strengthened at the termination point. There were also obvious difficulties with the corrosion protection of cables at the pylon head and with the replacement of individual stays in the event of any damage. A greater number of stays could be provided when the modified fan layout was introduced so that the stays are individually anchored near the top of the pylon. This is now the more commonly adopted system. In order to give sufficient room for anchoring the cable anchor points are spaced vertically at 1.5–2.5 m centres. Providing the anchor zone is maintained close to the pylon top there is little loss of structural efficiency as the behaviour of

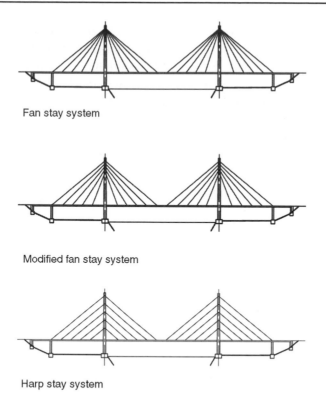

Fan stay system

Modified fan stay system

Harp stay system

Figure 10.1 Alternative stay cable arrangements.

the cable system will be dominated by the outer most cable attached to the top of the pylon and anchored at the supported end of the back span. The advantages of this arrangement are as follows.

- The large number of stays distribute the forces with greater uniformity through the deck section providing a continuous elastic support. Hence the deck section can be both lighter and simpler in its construction.
- As each stay supports a discrete deck module each module can be erected by the progressive cantilever method without resort to any additional temporary supports. Thus increased speed and efficiency of the deck erection is possible.
- The concentrated forces at each anchor point are much reduced.
- With the modified fan layout it is also possible to completely encapsulate each stay, thus giving a double protective system throughout its length and, should damage occur, replacement of the stay can be undertaken as a routine maintenance task.
- The large number of stays of varying length and natural frequency increases the potential damping of the structure.

10.2.2 Harp cable system
With the harp system the individual stays are anchored at equal spacing over the height of the pylon and are placed parallel to each other. This arrangement provides

a visual emphasis of the flow of forces from the back span to the main span and, in examples that are well proportioned, is aesthetically pleasing. However, the arrangement is not as structurally efficient as the fan layout and relies on the bending stiffness of the pylon and/or deck for equilibrium under non-symmetrical live loading. When loading one end only of the stay system the load may be divided into symmetrical and anti-symmetrical components of loading. The symmetrical loading will be resisted by the triangle of forces formed by the stays, pylon and deck but the anti-symmetrical loading can only be resisted by bending of the deck, the pylon or a combination of both depending on their relative stiffness. This disadvantage can be overcome by anchoring the back stay cables at approach pier locations so that any unbalanced load is resisted by the pier. An elegant example of this arrangement is the Knie Bridge over the River Rhine at Dusseldorf with its single pylon and 320m main span.

10.2.3 Multiple span bridges

The main concern with multiple span cable stayed bridges is the lack of longitudinal restraint to the top of the inner pylons, which cannot be directly anchored to an approach pier. Without providing additional longitudinal restraint a multiple span structure would be subject to large deformations under the action of live load. Increasing the stiffness of either the pylons or the deck can provide this additional restraint. Any increase in the deck stiffness will be accompanied by an unacceptable increase in the dead load and thus, the more practical approach is to stiffen the pylon. A typical example of the stiffened pylon is the A-frame braced pylon shown in Figure 10.2(a). However, such an arrangement will require a substantial increase in the pylon materials and require a much larger foundation. An alternative to relying upon the bending stiffness of the pylon is by the introduction of an auxiliary cable system to provide the required stability.

Two cable systems are illustrated, the first system, in Figure 10.2(b), connects the tops of the pylons and thus directly transfers any out of balance forces to the anchor stays in the end span. The second system, in Figure 10.2(c), connects the top of the internal pylons to the adjacent pylon at deck level so that any out of balance forces are resisted by the stiffness of the pylon below deck level. An example of this latter arrangement can be seen with the design of the Ting Kau Bridge, Hong Kong, which is a four span cable stayed bridge. The disadvantage of such an auxiliary cable system is that the individual cables are very long and the large sag will be visually dominant when compared with the neighbouring stay cables. Special measures may be necessary to limit the propagation of wind induced oscillations in these very long stays.

10.2.4 Number of cable planes

The cable layout may be arranged as either a single plane system or as a twin plane system. The twin plane system may either be formed as two vertical planes connected from the edge of the deck to two pylon legs located outside the deck cross-section or as twin inclined planes connected from the edge of the deck to either an A frame

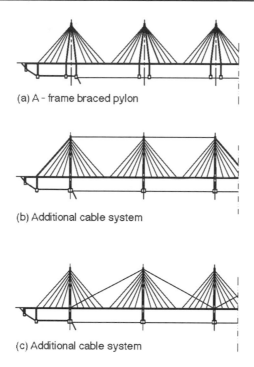

(a) A - frame braced pylon

(b) Additional cable system

(c) Additional cable system

Figure 10.2 Multiple span bridges.

or inverted Y frame pylon. The A frame pylon was first adopted for the Severins Bridge (Fischer, 1960). The vertical twin plane system, with its tensioned stay geometry, provides considerable rigidity between the deck and pylon when compared with the free hanging cables of the suspension bridge. Inclined stays further increase the stiffness and stability of the structure, with the stays and deck forming a transverse frame. Inclined stays are of particular benefit when adopted for the longer spans as they improve the torsion response of the structure to both eccentric live load and aerodynamic effects. When comparing the two alternative stay systems, one with two vertical stay planes and one with two inclined stay planes, supporting a deck which has low torsional stiffness the inclined stay system connected to an A shaped pylon will have approximately half the twist under eccentric loading. The inclined stay system is also aerodynamically superior reducing the magnitude of vortex shedding oscillations and increasing the critical wind speed of the structure. However, the geometry of the inclined stay planes must be carefully checked in relation to the traffic envelope and the clearances required may result in an increase in the overall width of the deck.

The single plane system creates a classic structural form avoiding the visual interference often associated with twin cable planes. However the single plane is not able to resist torsion loading from eccentric live loading and therefore this configuration requires the deck to be in the form of a strong torsion box. A deck section of this form is then likely to have excess resistance to the longitudinal bending of the deck, particularly when a multi-stay arrangement is used. The single pylon has to be located within the central median of the carriageway and as such an additional width of deck is required for the necessary clearances to traffic. Two outstanding examples of cable stayed bridges with a single stay plane are the Rama IX Bridge (Gregory and Freeman, 1987) and the Sunshine Skyway Bridge.

Figure 10.3 Rama IX bridge, Bangkok.

Figure 10.4 Sunshine Skyway Bridge, Florida (courtesy of Parsons Brinckerhoff).

Figure 10.5 Rama 8 bridge, Bangkok (courtesy of A Yee).

The Rama IX Bridge crosses the Chao Phraya River, Bangkok with a 450 m main span and an orthotropic steel box deck section which is 4 m deep. The deck section carries three lanes of traffic in each direction and is 33 m wide. The Sunshine Skyway Bridge crosses Tampa Bay, Florida with a 366 m main span and a 4.27 m deep trapezoidal concrete box deck section. The deck section carries two lanes of traffic in each direction and is 29 m wide.

The Rama 8 Bridge in Bangkok, now under construction, combines the use of twin and single plane arrangements in the single structure as incorporated into . This bridge has a single inverted Y pylon with twin inclined stay planes supporting a main span of 300 m, whereas the back span has a single stay plane anchored directly to a piled abutment.

10.2.5 Stay design
Many factors must be considered in the design of the stay system including the characteristic breaking strength and the effective stay modulus. The proportion of the breaking strength that can be realized depends on the relaxation of the stay under permanent loads. The irreversible strain arising from relaxation increases rapidly when the permanent load in the stay exceeds 50% of the breaking load. The Post-Tensioning Institute (PTI) Recommendations (1993) limit, for normal load combinations,

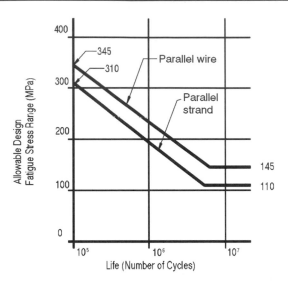

Figure 10.6 Wöhler endurance curve for parallel wire and parallel strand stays.

the maximum load in the stay to 45% of the stay breaking load and to 50% for exceptional load combinations.

The permissible load in the stay may also be limited by the fatigue performance of the stay under repeated live load cycles. However, this will depend on the magnitude of the live load cycles as a proportion of the permanent loads. Thus it is only the most heavily loaded stays of a highway structure that are possibly limited by fatigue whereas the stays for a railway structure where the live load is dominant will be much more fatigue sensitive. Fatigue endurance of a component is usually given in a plot of the stress range ($\Delta\sigma$) against the number of load cycles (N) known as the Wöhler curve. When the Wöhler curve is represented on a log-log scale the plot is represented by a series of straight lines. The PTI recommendations relate the design limit to the test acceptance criteria of the stay material, such as strand, bar or wire, and the test criteria of the assembled stay. The recommendations for parallel wire strand (PWS) and parallel strand are given in Figure 10.6.

The fatigue endurance of the assembled stay will not only result from variations in the applied tension but also be influenced by any secondary bending in the stay arising from either wind or traffic induced vibrations. The response to these factors is extremely complex and varies according to the manufacturing characteristics of the stay and its anchor. Because of this the PTI Recommendations propose that at least three representative samples of the stay assembly to be used in a project be fatigue tested. Testing is usually undertaken to two million cycles. The stress range depends on the generic type of stay being used but the upper limit of the stress range is always taken as 45% of the breaking load. Acceptance of the test is based upon a limit to the number of individual wires in the stay that may break and that the tensile test, undertaken after the fatigue test, achieves at least 95% of the guaranteed breaking load of the stay.

10.2.6 Stay types

Problems arose with the stays of early cable stay bridges as a result of deficiencies with the anchorage design, steel material problems and inadequate corrosion resistance. The development of modern stay systems has largely overcome these problems providing designs that minimize bending of the stay at the anchorage face and incorporate a double corrosion protection system throughout. Available stay systems include:

- locked coil (prefabricated)
- helical or spiral strands (prefabricated)
- bar bundles
- parallel wire strand (PWS)
- new PWS (prefabricated)
- parallel strand.

Locked coil stays have been incorporated into many of the earliest cable stay bridges. The stays are factory produced on planetary stranding machines, each layer being applied in a single pass through the machine and contra-laid between each layer. The core of the stay is composed of conventional round steel wires whilst the final layers comprise Z-shaped steel wires which lock together creating an extremely compact stay cross-section. A typical example of a locked coil stay is illustrated in Figure 10.7.

Modern locked coil stays provide all the wires in a finally galvanized condition and will achieve a tensile strength of up to 1770 N/mm^2. The stays are commonly anchored by zinc filled sockets although sometimes stays that are sheathed with a polyethylene protection have their sockets filled with epoxy resin. The largest locked coil stays manufactured to date are the 167 mm diameter stays supplied for the Rama IX Bridge over the Chao Phraya River, Bangkok.

Helical or spiral strands, which are illustrated in Figure 10.8, are also factory fabricated on a planetary stranding machine similar to the locked coil stay but are

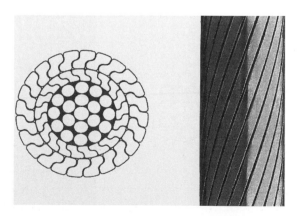

Figure 10.7 Locked coil cable (courtesy of Bridon International Ltd).

entirely manufactured from finally galvanized round steel wires. The wires are usually of 5mm diameter with a tensile strength of either 1570 or 1770 N/mm^2. The largest spiral strand stays manufactured to date are 164 mm diameter, as supplied for the Queen Elizabeth II Bridge over the River Thames at Dartford.

Bar bundles contain up to 10 No threaded steel bars with a tensile strength of 1230 N/mm^2 coupled together in 12-m lengths. The bars are conventionally placed within a steel tube and protected with a cement grout. The use of couplers connecting the bars will give a much reduced fatigue resistance when compared with the equivalent wire or strand systems. Coupled bar systems are thus rarely used where significant variations in the stay load are likely to occur. Tests have also been undertaken to assess the effectiveness of cement grout as a protective medium. These tests concluded that transverse and longitudinal cracking of the grout rapidly develops due to temperature and live load strains and wind vibration effects. Therefore cement grout encapsulation of the stay cannot be relied upon to give effective protection against corrosion.

Parallel wire strand (PWS) most commonly comprises 7 mm diameter finally galvanized round steel wires with a tensile strength of 1570 N/mm^2. PWS stays may either be prefabricated or assembled on site, the wires being installed without a lay or helix within a polythene tube and injected with cement grout or wax. When manufactured without a lay the prefabricated stays were difficult to handle and coil on to the reel and the system suffers from the doubts associated with grouted stays. The new PWS system, as illustrated in Figure 10.9, was developed with a tensile strength up to 1770 N/mm^2. The stay is prefabricated and with a long lay helix to improve coiling on to the reel. The largest stays can contain up to 400 wires and a coating of high density polyethylene (HPDE) is applied in the factory using the continuous extrusion process. The stays can be socketed using a patented system such as BBR's DINA or HiAm anchorages. The individual wires within these anchorages incorporate button heads transferring the full load to the anchor. The socket is then filled with a proprietary epoxy compound that is claimed to enhance the fatigue resistance of the stay.

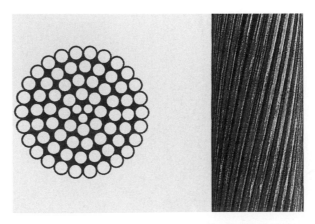

Figure 10.8 Spiral strand cable (courtesy of Bridon International Ltd).

Stay cables may also be manufactured from bundles of 15.7 mm diameter seven wire strands, which is usually galvanized and have a tensile strength of 1770 N/mm^2, to give a characteristic breaking load per strand of 265 kN. The strand bundle can typically comprise up to 110 strands and anchoring is by means of a prestressing anchor head with individual strands gripped by wedges. In order to provide adequate fatigue resistance it is essential that rotation of the strands at the face of the wedge grips, due to changes in stay load or wind oscillation, does not occur. Dampers located some distance in front of the anchor face prevent this by restraining the strands. An outer polyethylene tube covers the strand bundle providing protection against impact damage and preventing dynamic oscillation of the individual strands. Early designs filled the tube with cement grout or wax as corrosion protection but in later designs each strand is manufactured with a continuously extruded HPDE coating over a corrosion inhibiting grease. With this level of protection further cement or wax injection is unnecessary. European and Japanese practice has been to use galvanized strands but some American bridges have been constructed using epoxy coated strand. The risk with epoxy coated strand is that small pinholes or minor damage can propagate corrosion, forming a notch in the strand. This creates a local stress concentration that can lead to the premature failure of the strand. The use of Galfan, a licensed zinc and aluminium mixture, gives two to three times the protection for the equivalent weight of zinc coating, but the fatigue properties of the strand are reduced.

The outer covering to the stay has conventionally been manufactured from steel or polyethylene pipe. Where polyethylene pipe is used UV resistance was achieved by the use of carbon black pigment in the material. The design temperature differential between the stays and the deck or pylon can vary considerably. The PTI Recommendations (1993) notes that values of 9 and 22°C have been used for white painted or taped stays and black stays, respectively. The use of black stays is not preferred in tropical zones where there is a high solar gain. Early attempts to wrap the stays, as in the case of the Pasco-Kennewick Bridge over the Columbia River USA where a white plastic wrapping was used, were unsuccessful as the

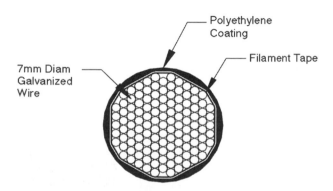

Figure 10.9 New parallel wire strand (PWS) system.

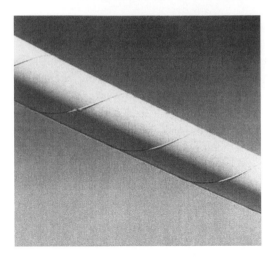

Figure 10.10 Outer sheath for parallel strand stay with helical ribs (courtesy of BBRV Systems Ltd).

coating deteriorated within a few years. Later coverings using Tedlar, as in the case of the Second Severn Bridge (Mizon *et al.*, 1997), have been more successful. More recently a range of light coloured polyethylene pipe has been developed with a high UV resistance. The pipe is manufactured in a bi-extrusion process where a thin coating of light coloured polyethylene is extruded over a black pipe core. These pipes can also be manufactured with fins or helical ribbing to give improved aerodynamic performance and thus minimize the effects of wind /rain induced vibrations of the stay, as shown in Figure 10.10.

10.2.7 Stay behaviour

The behaviour of the stay under load must be represented in the analysis of the structure. The modulus of the stay under load is a characteristic of the stay manufacture but will also be a non-linear variation with respect to both stay length and axial tension. When comparing the modulus of various types of stay that are manufactured, parallel wire strand (PWS) achieves the highest modulus at 205 kN/mm^2. This is close to the modulus of the steel wire itself. Seven wire strand achieves a modulus of some 195 kN/mm^2 while locked coil will be approximately 155 kN/mm^2. The modulus of helical strands will be within the range 155–175 kN/mm^2. The modulus of both locked coil and helical strands is variable depending on the lay angle, the galvanizing and the stay diameter. Factory produced locked coil and helical strand cable, which are prestressed as part of the manufacturing process, will give the stays a predictable elongation in service. Prestressing, where the cable is supported in a bed and preloaded, is the method used to remove the non-elastic stretch, resulting from the initial compaction of the strand. Prestressing is not required for PWS or parallel strand stays.

The non-linear behaviour of the stay may be represented by an equivalent modulus taking into account the sag in the loaded stay. The variation in the

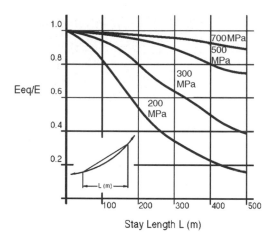

Based on E = 205 000 N/mm^2

Figure 10.11 Equivalent modulus of elasticity of stay cable.

equivalent modulus of elasticity of the stay (E_{eq}) is given in Figure 10.11 and may be expressed as:

$$E_{eq} = \frac{E}{1 + (\gamma^2 \times L^2 \times E/12 \times \sigma^3)}$$

where E is the guaranteed modulus of the straight stay, L = the horizontal length of the stay, γ = the specific weight of the stay and σ = the tensile stress in the stay

Allowance must be made during the erection of the deck for the dead load extension of the stay. In the case of a prefabricated stay this is achieved in manufacture, by reducing the stay length to compensate for this extension. In the case of the stay fabricated on site, using a strand system, the calculated extension of the stay is taken up within the anchorage. The length of the stays will also vary with changes in temperature and it is therefore necessary to measure the temperature of the stay and deck at the time of stay installation and calibrate the stay load accordingly.

10.3 Pylon

The pylon is the main feature that expresses the visual form of any cable stayed bridge giving an opportunity to impart a distinctive style to the design. The design of the pylon must also adapt to the various stay cable layouts, accommodate the topography and geology of the bridge site and carry the forces economically.

The primary function of the pylon is to transmit the forces arising from anchoring the stays and these forces will dominate the design of the pylon. The pylon should ideally carry these forces by axial compression where possible such that any eccentricity of loading is minimized.

10.3.1 Steel pylons

Early cable stay pylon designs were predominantly constructed as steel boxes and bridges such as the Strömsmund Bridge (Wenk, 1954) took the form of a steel portal frame, which was intended to provide transverse restraint to the stay system. However this restraint is largely unnecessary as sufficient transverse restraint can be provided within the stay system itself. When a single mast supports each stay plane any lateral displacement at the top of the mast is accompanied by a rotation of the stay plane. This rotation of the stay plane ensures, for the simple fan layout of stays attached to the top of the mast, that the resultant reaction from the main span and back span stay cables will pass through the pylon foot. The weight of the pylon will remain vertical but the reaction from the stays will be dominant. Thus the effective length of the mast in buckling will not be that of a simple cantilever, twice times the height (2*H*), but equal to the height (*H*). This effect is illustrated in Figure 10.12. With the harp layout of stays the loads will be applied at intervals down the pylon but similar principles apply. An example of the harp layout is the Theodore Heuss Bridge (Beyer and Tussing, 1955) where a slender strut supports each cable plane. When considering the buckling behaviour of such a pylon allowance must be made for constructional inaccuracies, typically for the stay reaction an eccentricity of 100 mm is adopted.

In the longitudinal direction the main span and back span stay cables will restrain the pylon against buckling providing the deck, to which the stays are anchored, is adequately restrained against longitudinal movement. The pylon behaviour when the deck is allowed to float is illustrated in Figure 10.13. When the deck is unrestrained any disturbing forces can only be resisted by the pylon acting as a cantilever with maximum bending at the base. Thus the effective length of the mast in buckling will be twice the height (2*H*). When the deck is effectively restrained at either an abutment or at one of the pylons the top of the mast is held in position by the stays and the effective length of the mast in buckling will be approximately 0.7 times the height. Connecting the back stays to an independent gravity anchorage is an equally effective solution. Early cable stay pylon designs, such as for the Strömsmund Bridge (Wenk, 1954), incorporated a pin at the pylon foot so as to ensure that the mast did not have to be designed for large bending moments. Later designs have adopted a fixed end cantilever mast, which is simpler and also more stable during erection.

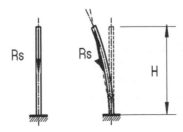

Figure 10.12 Transverse deflection of a simple pylon.

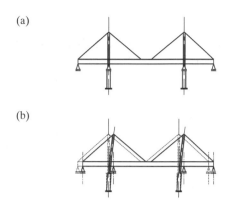

Figure 10.13 Longitudinal restraint of the pylon by the anchor stays. (a) Deck restrained by substructure. (b) Deck unrestrained.

Nevertheless the effect of a frictionless bearing can still be achieved with a fixed end mast providing the member is sufficiently slender, such that the maximum axial load approaches the buckling capacity of the mast in free cantilever. As the axial load increases in relation to the buckling capacity the location of the maximum moment will move up from the base and eventually the base moment will tend to zero. In this condition the pylon will offer no resistance against a longitudinal displacement at the pylon top. In practice the adoption of a particular slenderness may be limited by the need to maintain stability of the pylon during construction, when restraint from the stays is not available.

For the single mast pylon supporting a single plane of stays, two methods have been used to connect the mast at its base. It is either constructed encastre into a transverse girder forming part of the deck. In this case a bearing is required on top of the pylon foundation immediately beneath the mast and two further bearings at each end of the transverse girder so as to provide the necessary torsional restraint to the deck and pylon forces. Alternatively the mast can pass directly through the deck to sit upon the pylon foundation. In this case only bearings at each end of the transverse girder are required and these only have to provide the resistance to the deck forces. The latter method is more efficient and has been universally adopted in more recent designs. An example of the use of this method of support is in the design of the Rama IX Bridge.

10.3.2 Concrete pylons

Concrete is very efficient when supporting loads in axial compression. Advances in concrete construction and modern formwork technology has made the use of concrete increasingly competitive for pylon construction, despite the much greater self weight, when compared with a steel alternative. Concrete has proved particularly adaptable to the more complex forms of pylon. Many varied types of pylon have been developed to support both the vertical and inclined stay layouts. These include H frame, A frame and inverted Y frame pylons as illustrated in Figure 10.14.

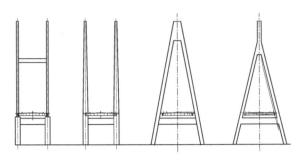

Figure 10.14 H frame, A frame and inverted Y frame pylons.

With the H frame pylon the stay anchors are normally located above the level of a crossbeam. With the modified fan arrangement of stays this crossbeam location would be between mid height and two-thirds of the pylon height above the deck. When the harp arrangement of stays is adopted the anchors are distributed over the full height of the pylon above the deck. Therefore, as in the case of the Øresund Bridge between Denmark and Sweden, the crossbeam is only provided below the deck level.

The deck section located at the pylon is usually the most highly stressed section, combining maximum negative moment and maximum thrust. When the stay is connected directly between the pylon leg and an edge stiffening girder within the deck section it is necessary to inset the pylon legs into the deck. In addition to this practical detailing problem, a notch is created giving a zone of concentrated stress at an already highly stressed section. There are several geometrical configurations that overcome this problem, the pylon can be widened and the stays connected to the deck by means of an out-stand bracket on the deck or the pylon leg can be sloped outward at its base. The leg may be inclined over the entire height of the pylon, in which case the pylon must be designed for a small eccentricity arising from the stay cable reactions. Alternatively the upper section of the leg can be maintained in a vertical plane and the pylon inclined only from below the level of the bottom anchorage. Locating the crossbeam at this change of direction conveniently ensures that the stay force reaction is transmitted as a direct thrust. Examples of this pylon geometry can be seen in the Annacis bridge over the Fraser River, Canada and the Vasco da Gama bridge (Capra and Leveille, 1998) over the Tagus River, Portugal as illustrated in Figure 10.15.

The A frame pylon is suitable for inclined stay arrangements and was first adopted in steel construction for the Severins Bridge (Fischer, 1960). A variation of the A frame is the inverted Y frame where the vertical leg, containing the stay anchors, extends above the bifurcation point. An example of the inverted Y frame is the pylon for the Normandie Bridge over the River Seine, France and the Rama 8 Bridge, Bangkok, Thailand.

10.3.3 Pylon geometry
Excessive land take, due to the wide pylon footprint, can occur with the inverted Y frame form of pylon, when a high clearance to the deck is necessary. This has been

Figure 10.15 Vasco da Gama bridge, Portugal (courtesy of A Yee).

overcome by breaking the pylon legs at or just below the deck to produce inward lean-
ing legs supported by a foundation. This diamond configuration is shown in Figure
10.16. However this modified arrangement is considerably less stiff when resisting
transverse wind or seismic forces and this can result in a significant increase in the
deflection of the pylon. This deflection can be mitigated only with a considerable
increase in the stiffness of the lower section of the pylon leg below the deck. Nevertheless, this arrangement was favoured for the pylon of the Tatara Bridge in Japan,
currently the world's longest cable stayed span at 890 m. This configuration was also
adopted, in a distinctive tandem configuration, for the twin cable stayed crossing of

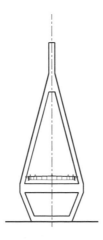

Figure 10.16 Diamond pylon.

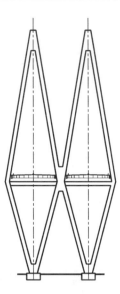

Figure 10.17. Twin diamond pylon, Houston Ship Channel, USA.

the Houston Ship Channel (Svensson, 1999). By connecting the twin diamonds and tying them together at deck level a strong truss was created which transmits the transverse wind loads to the foundations, see Figure 10.17.

It is also possible to incline the pylon in the longitudinal direction and many visually exciting structures have incorporated such a pylon as an architectural feature. However the resultant inclined thrust from the pylon must be carried by the foundation and a significant horizontal component will be developed. When the structure is founded in rock these horizontal reactions can be easily resisted with only a small displacement of the foundation. However, in typical estuarine soil conditions the foundation costs may represent a significant proportion of the overall project cost.

10.3.4 Stay connection

In early designs the connection between the stays and the pylon was formed in the same manner as for suspension bridges where the cables are laid in a saddle and carried through the pylon. The evolution of the modified fan and harp arrangements with stays anchored over a portion of the pylon leg has led to the use of separate stays for the main span and back span.

The most direct form of anchoring is to attach the stay socket or anchorage plate to the wall of the pylon. In this layout the hollow pylon shaft gives access to the stay anchors for stressing during erection and inspection or replacement in service. In the case of the concrete pylon the horizontal component of the cable forces will tend to split the shaft vertically and transverse prestressing is required to resist these forces. A typical layout of the prestressing, as adopted for the Helgeland Bridge (Svensson and Jordet, 1996), in Norway is shown in Figure 10.18.

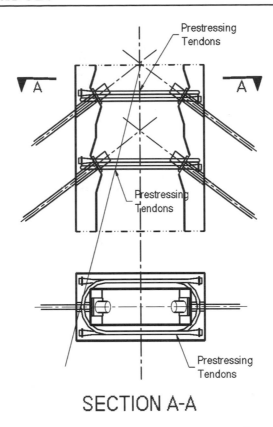

SECTION A-A

Figure 10.18 Prestressing layout for stay connection to concrete pylon.

An alternative arrangement, producing a more slender pylon, allows the main span and back span stays to cross so that they are anchored in rebates on the reverse sides of the pylon as illustrated in Figure 10.19. The horizontal component of the cable forces will place the pylon into compression. However, the two stays cannot be in the same plane and the pylon must be designed for the resulting torsion arising from this eccentric loading. This is a detailing problem that can be overcome by dividing the stay on one side of the pylon into a pair of stays and creating a balanced anchor layout as shown in Figure 10.20.

For the design of the Annacis Bridge over the Fraser River a different approach was employed where a steel fabrication was placed within the hollow pylon shaft at the level of each stay anchor to transfer the horizontal component of the stay force. The vertical component and any differential horizontal component was then transferred through a corbel cast on the inside of the shaft wall as shown in Figure 10.21.

All the above methods of connecting the stay to the pylon rely on the accurate placement of the steel formers and any anchor prestress and reinforcement within the concrete walls if the stay geometry and the strength of the connection intended in the design is to be realized. The complexity of the required details will often slow the progress of the erection throughout this critical zone of the pylon construction. In

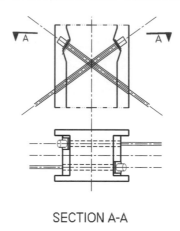

SECTION A-A

Figure 10.19 Alternative layout for stay connection to concrete pylon.

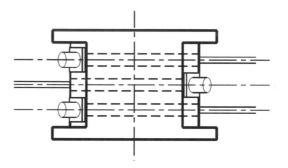

Figure 10.20 Alternative connection with balanced stay layout.

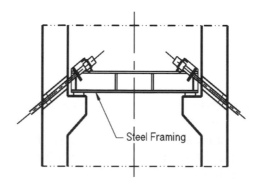

Figure 10.21 Stay connection with steel frame in concrete pylon.

order to mitigate these problems steel fabricated anchorage modules have been manufactured such that the required stay geometry is completely defined. This module can then be incorporated into the concrete shaft during its construction. Adequate shear connection, usually in the form of shear studs, is provided so that

Figure 10.22 Fabricated anchor modules, Ting Kau Bridge, Hong Kong (courtesy of Flint & Neill).

the concentrated anchorage forces in the fabrication can be transferred to the concrete shaft. An example of this form of pylon construction is the Normandie Bridge where the fabricated anchorage module is located centrally within the concrete shaft. A similar concept was incorporated into the pylon of the Ting Kau Bridge, Hong Kong. However, here the fabricated anchor modules were connected on the outside of the concrete core as shown in Figure 10.22.

It is essential that any eccentricity of the stay anchor within the pylon is accurately modelled as part of the analysis of the structure. When the inclination of the back stay and main stay cables are identical, with both anchors at the same level, the axes of the stay, and hence the stay forces, will intersect on the pylon centreline. However the inclination of the back stay and main span are rarely identical and hence the anchors must be located at different levels if the same intersection line is required. Alternatively, when the levels of the two anchors are maintained at the same level the vertical resultant of the stay forces will be slightly eccentric to the pylon. It is usually preferable to simplify the detailing of the anchor zone and accept the small eccentricity to the pylon in the design, as shown in Figure 10.23.

10.4 Deck
Rather than being merely supported by the cable, as in the suspension bridge form, the deck of the cable stayed bridge is an integral part of the structure resisting both

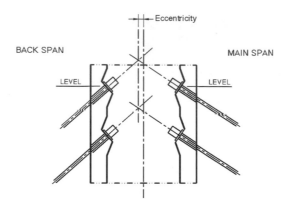

Figure 10.23 Stay anchor pylon geometry.

bending and an axial force derived from the horizontal component of the stay force. The economic solution for the suspension bridge is a minimum weight deck section. With the cable stayed bridge the greater participation of the deck in the overall structural behaviour gives the opportunity to consider alternative deck forms particularly with concrete being an efficient material when used for compression members. As such, cable stayed bridge deck forms have been developed as:

- a steel section incorporating an orthotropic road deck;
- a concrete section;
- a composite steel and concrete section.

10.4.1 Steel deck section

The early designs for cable stayed bridges had relatively few stays requiring the use of a light yet rigid deck capable of spanning between the wide stay anchor spacing. This structural arrangement ideally suited the all steel construction, incorporating an orthotropic road deck. The introduction of the multi-stay cable systems and the high fabrication costs have made the steel deck solution less economic for moderate spans. Nevertheless, for the longest spans where the reduction in deck weight is an economic imperative, steel construction has been retained for structures such as the Normandie Bridge and the Tatara Bridge.

The orthotropic road deck consists of a thin surfacing material laid on either a 12 or 13 mm steel plate stiffened longitudinally. The first designs incorporated longitudinal stiffeners of an open bulb flat section at approximately 300 mm spacing. Later the designs have adopted the closed trough stiffener, fabricated from folded plate, giving greater torsional rigidity to the deck plate system. The deck stiffeners are supported by transverse floor beams at 3–5 m spacing.

The design of the deck cross-section is dominated by the arrangement of the stays. Where only one central plane of stays is adopted the torsional resistance of the deck section is the only means of carrying any eccentric loading and therefore a strong

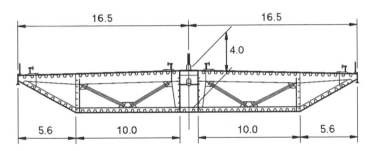

Figure 10.24 Multi-cell steel torsion box deck – Rama IX Bridge, Bangkok.

torsion box must be provided. The Erskine Bridge (Kerensky *et al.*, 1972) over the River Clyde in Scotland is an example of a steel box with only one central stay. In this case the girder has to span in excess of 100 m between the cable stayed anchor points. A common arrangement, originated by German designers, is to divide the cross-section of the torsion box into three or five cells. The central cell is the same width as the pylon such that the cable stays may be readily anchored within it. Examples of this arrangement are the Rama IX Bridge, Bangkok, divided into three cells and the Oberkasseler Bridge over the River Rhine, Germany, which is divided into five cells. The cross-section of the deck of the Rama IX Bridge is given in Figure 10.24.

With highway structures of moderate span and where two planes of stay cable are used the rigidity of the torsion box deck cross-section was unnecessary and it was possible to simplify the section to that of twin longitudinal girders. Early designs, such as the Knie Bridge over the River Rhine at Dusseldorf, Germany with a main span of 320 m combined longitudinal steel plate girders with an orthotropic deck. This bridge is illustrated in Figure 10.25.

When two planes of stays are connected to each edge of the deck eccentric loading of the deck can be carried by the cable system. The strong torsion box can still be beneficial by improving the distribution of the loads between the two cable planes, such as where both road and rail are carried, and the heavy rail loading has to be located eccentric to the deck section. The torsion box can also be more easily adapted into a streamlined shape, essential for reducing wind drag and for improving aerodynamic stability in very long spans.

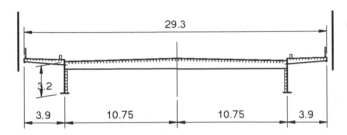

Figure 10.25 Twin girder steel deck – Knie Bridge, Germany.

10.4.2 Composite deck sections

The orthotropic deck, with its high manufacturing costs, has a cost penalty when compared with an equivalent concrete slab road deck. The cost advantage of the composite concrete slab over the steel deck is to limited extent balanced by the larger area of cable stays and the additional pylon and foundation costs required to support the heavier dead loads. German codes, at the time of the early cable stayed designs, did not permit tensile stresses in a concrete road slab and the provision of post-tensioning near the middle of the span was not considered economic. In a typical composite deck section the tensile stresses within the concrete deck slab are generally small. When the deck is spanning transversely between two planes of cables at the deck edge, the slab will act as the top flange of a simply supported beam and will remain in compression. Longitudinally, any tensile stresses due to bending are low because the neutral axis of the section is usually located close to the underside of the slab and will, for the majority of the span be less than the compression component from the stays. Even at the centre of the span where the compressive component reduces to zero it is possible by adjusting the forces in the stays to introduce a sagging moment, so maintaining compression in the slab.

When considering the cantilever erection of the composite deck two sequences are possible.

- The relatively light steel framing can be erected first, allowing the stay to be attached and partially stressed before forming the concrete deck, either as *in situ* concrete or as precast panel units. An early use of an *in situ* concrete deck was for the Second Hooghly River Bridge, India. The use of precast deck panels minimizes the amount of work during the critical erection phase and reduces the effects of shrinkage and creep that must be allowed for in the design. The precast deck panels sit on a neoprene strip on the edge of each beam and are made structurally continuous by overlapping hairpin reinforcement within an *in situ* concrete strip located over the top flange of the beams. This method of construction was first adopted for the deck of the Annacis Bridge over the Fraser River, Canada. More recently it was the method preferred by the contractor for the construction of the Houston Ship Channel Bridge (Svensson, 1999).
- Alternatively the concrete deck may be cast in the assembly yard, with each deck module, using a match casting process along side the neighbouring module. This ensures that lapping reinforcement is accurately aligned across the joint between each module. The procedure also reduces the effect of shrinkage and creep to be allowed for in the design and confines *in situ* concrete construction to the transverse stitch joint. Examples of structures incorporating this method of construction include the Kap Shui Mun Bridge, part of the Lantau Link, Hong Kong and the Second Severn Crossing (Mizon *et al.*, 1997), United Kingdom.

A further example of a composite deck section is that proposed for the Industrial Ring Road crossings of the Chao Phraya River, Bangkok, which is illustrated in Figure 10.26.

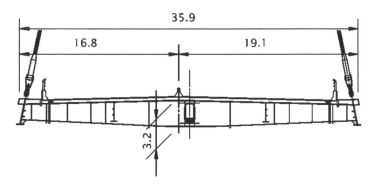

Figure 10.26 Twin girder composite deck – Industrial Ring Road Bridges, Bangkok.

10.4.3 Concrete deck section

Concrete deck sections were subject to developments similar to those for steel deck sections. The torsion box deck sections are commonly used in conjunction with a single central plane of stays. The first of such designs was the Brotonne Bridge crossing the River Seine near Rouen, France. A similar design, using precast segmental units for the deck, was adopted for the Sunshine Skyway Bridge, USA, as illustrated in Figure 10.27. In common with a number of major bridge projects in the USA this design was selected after a process of competitive pricing between alternative designs in steel and concrete.

   Concrete designs, in common with the evolution of the composite deck section, have developed a simplified deck form. Examples of this deck construction are the Dames Point Bridge over the St Johns River in Florida, USA, which is illustrated in Figure 10.28 and the Helgeland Bridge (Svensson and Jordet, 1996), Norway. In a typical arrangement the transverse floor beams are at 3–5 m centres supporting an *in situ* concrete road deck. The transverse beams on the Dames Point Bridge took the form of precast Tee beams. The longitudinal beams are located at each edge of the deck centrally beneath the cable planes and incorporate the stay anchors.

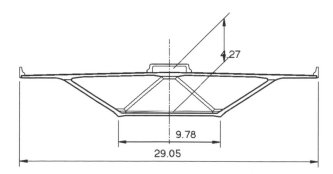

Figure 10.27 Concrete torsion box deck – Sunshine Skyway Bridge, USA.

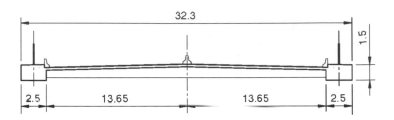

Figure 10.28 Twin beam concrete deck – Dames Point Bridge, USA.

Erection is by casting the deck in segments as a free cantilever using a form traveller. The stay is initially stressed against the form sufficient to minimize deflection during the placing of the concrete. The design of the Helgeland Bridge allowed for the precasting of a portion of the longitudinal edge girder, including the stay anchor tube to act as a compression strut. This element was bolted to the formwork before being incorporated within the deck concrete, see Figure 10.29. This system improves the accuracy of the stay anchor geometry and allows stressing of the stay anchor at an earlier stage in the erection cycle. Thus the stay anchor for this bridge is stressed in three stages, prior to concreting, immediately after concreting and when the transfer strength of the *in situ* concrete was achieved.

The Helgeland Bridge, which is illustrated in Figure 10.30, has an exceptionally narrow deck at 11.95 m and therefore the design of the *in situ* concrete crossbeams is not be a critical part of the design. With other bridges, where the width of the deck increases, this form of concrete deck construction becomes less economic. In the case

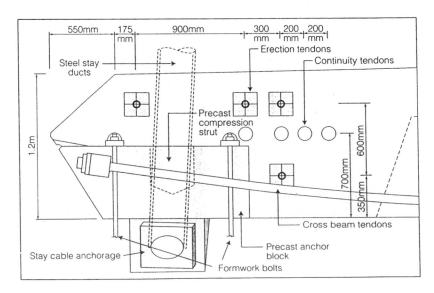

Figure 10.29 Precast anchorage and edge beam – Helgeland Bridge (Svensson and Jordet, 1996).

Figure 10.30 Helgeland Bridge, Norway (courtesy of E Jordet).

of bridge deck widths suitable for dual three lane highways, that is at least 30 m wide, the weight of the transverse floor beams becomes dominant and significant additional stay area must be provided to carry the extra dead load. One method of mitigating this problem, whilst still maintaining the simplicity of the concrete edge beam with its cast in stay anchor, is to incorporate a lighter composite plate girder as the transverse floor beam. This is the design philosophy adopted for the deck of the Vasco da Gama Bridge (Capra and Leveille, 1998) over the Tagus River, Portugal. The deck section for this bridge is illustrated in Figure 10.31.

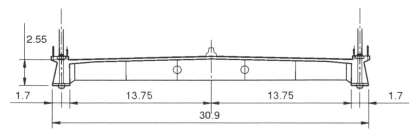

Figure 10.31 Combined concrete and steel deck section – Vasco da Gama Bridge, Portugal.

10.4.4 Design principles

As can be seen above there are a number of possible deck forms and when they are used in combination with an appropriate stay arrangement they can provide an economic design within their respective span ranges.

The deck is required to perform a number of particular functions within the overall structural system:

- to distribute the applied loading to the stay anchor points. With dead load and uniform live load this will take the form of transverse distribution to each stay plane and between each stay anchor. For a deck supported by a multi-stay system concentrated live loads will be distributed between a number of stays as a beam with elastic supports. It should be noted that that the moments in the deck will decrease as its longitudinal stiffness decreases
- to act as part of the global system with the stays and pylon. The primary force is the thrust component from the stay force and the deck must be sufficiently stiff to resist buckling from this compression. The capacity of the deck must be checked using a non-linear, second order analysis. In addition the deck may be required to resist bending moments induced by unbalanced loading within a harp cable stay system as described separately
- to resist transverse wind or seismic loads and transmit these loads to the substructure.

Thus in the case where the stays are widely spaced the design of the deck is dominated by longitudinal bending and, where there is only a single central stay plane, torsion effects may dominate. For the multi-stay systems the design of the deck is dominated by transverse bending combined with thrust and longitudinal bending derived from the global system.

The requirements for the deck design vary between the main span and the back span. The main span must be designed to give an acceptable aerodynamic performance and, for economy, the self weight of the deck section should be a minimum consistent with the choice of material. The following figures give a comparative estimate of deck weight:

Steel deck   2.5–3.5 kN/m^2
Composite deck   6.5–8.5 kN/m^2
Concrete deck   10.0–15 kN/m^2

The back span, through the stays, must stabilize the pylon when unbalanced by live loading within the main span and transmit any resulting uplift to the approach span piers. With any bridge constructed with an all steel deck section within both the main and back spans a means of transferring significant uplift must be provided. This has conventionally been achieved by the use of a pendel linkage that transfers the uplift force but still allows longitudinal translation to take place in response to temperature changes, an example of this arrangement is shown in Figure 10.32. A similar arrangement was used within the back spans of the Queen Elizabeth II Bridge over the River Thames at Dartford.

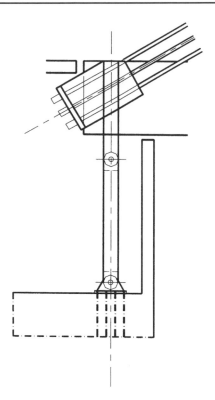

Figure 10.32 Pendel linkage over anchor pier.

It is clear that the back span deck does not have to comply with the criteria of minimum self-weight. Provided separate access for the erection of the back span is available, such that the main and back spans do not have to be erected in balanced cantilever, there is no reason why the back span should not be constructed from a different material to that in the main span. This arrangement was fully exploited for the construction of the Normandie Bridge over the River Seine. In this case the central 624 m of the 856 m main span was an all steel construction whilst the remainder of the main span and the back spans were of concrete construction. The Normandie deck was encastre with both the pylons and thus any strains due to temperature variation were accommodated by flexure within the lower section of the pylons and by variation in curvature of the deck profile.

Where the back span is over land it can be constructed in advance of the main span as a continuation of any approach viaduct. This can provide both programme and construction advantages during erection. The back spans can be constructed in parallel with the pylon, thus providing a high level access platform that is available at the commencement of cantilever erection within the main span.

10.5 Preliminary design

The cable stay bridge, incorporating multiple stays, is a highly redundant structure where the deck acts as a continuous beam with a number of elastic supports with varying stiffness. The deck and pylon of the cable stay bridge are both in compression and therefore bending moments in these elements will be increased, due to second order effects, arising from the deflection of the structure (the PΔ effect). With most cable stay structures these secondary moments will not exceed 10% but the application of these moments will be non-linear. This means that the use of influence lines, which rely on the principles of linear superposition, can only be used as an approximate method of determining the stay loads.

Non-linear material properties will also influence the design. Apart from the behaviour of the stays under load, discussed separately, all concrete and concrete and steel composite decks will be subject to the effects of creep and shrinkage during both construction and the service life of the completed structure. It can therefore be seen that a preliminary design by manual calculation should be considered as the first stage in an interactive design process, providing a basis for a more rigorous analysis.

10.5.1 Back span to main span ratio

When establishing the conceptual arrangement of the bridge it is important that the ratio between the back span and the main span be less than 0.5 in order to give a clear visual emphasis to the main span. This ratio is equally as important structurally as it influences the uplift forces at the anchor pier and the range of load within the back stay cables supporting the top of the pylon. The back stay cables have the largest stress amplitude and may therefore be critical when considering the fatigue endurance of the stays. Live load located within the main span will increase the anchor forces within the back stays and live load within the back span will decrease the anchor forces. Where there are no intermediate piers supporting the back span and there are no physical constraints imposed by the terrain, the foundations or any other requirements dictating the location of the abutment pier this ratio can be determined by the balance of the live load moments in the main span and the back span. Leonhardt and Zellner (1980) has determined the back span to main span ratio with respect to these parameters. For a highway structure where the live loading is typically 0.25 of the dead load the theoretical ratio is 0.38. However this calculation ignores the bending stiffness of the deck. When this stiffness is taken into consideration the optimum length of the back span is more likely to be between 0.4 and 0.45 of the main span.

On the other hand the optimum ratio for a structure carrying a heavy railway loading, where the live loading may be equivalent to the dead load, would be 0.18. Again when the bending stiffness of the deck is taken into consideration the length of the back span is more likely to be between 0.2 and 0.25 of the main span. Live load bending moments in the deck are likely to be a maximum near the end of the side span. Significant rotations are therefore possible at any expansion joint that is located at the anchor pier. Whilst this may be acceptable for a highway bridge a train crossing such an expansion joint at speed could experience an unacceptable level of acceleration caused by the sudden change in gradient across the joint. It is preferable in this case to provide continuity over the anchor pier by extending the deck into a small approach span sufficient to attenuate the live load rotations within the suspended back span.

To summarize it can be seen that the optimum main span-to-side span ratio is sensitive to the proportion of live load to dead load. Of course this will vary depending on whether a concrete, steel or composite deck is being proposed. For a bridge with a steel superstructure the length of the back span will be smaller than for a similar span bridge with a concrete superstructure.

Where there is ready access to the land beneath the back span it is possible to construct this part of the superstructure as an extension of any approach viaduct rather than as a balanced cantilever from the pylon. The back span will then be supported by intermediate piers that provide a direct anchorage for the back stays. This is of particular benefit in bridges with the harp arrangement of stays, as the additional stiffness will reduce the moment in the pylon normally associated with this stay arrangement. By constructing the back spans early in the construction schedule, independent of the pylon construction, early direct access to the pylon is possible and a platform is available from which the fabrication and erection of the stays and cantilever erection of the main span can proceed. This will usually be of benefit to the overall construction schedule.

10.5.2 Stay spacing

The spacing of the stay anchors along the deck should be compatible with the capacity of the longitudinal girders and limit the stay size so that the breaking load is less than 25–30 MN. The capacity of the longitudinal girders is likely to be critical when considering the case of an accidental severance of a stay (stay out condition). The spacing should also be small enough so that the deck may be erected by the free cantilevering method without the need for auxiliary stays or supports. These requirements will effectively limit the spacing within the range of 5–15 m. The heavier concrete construction will require the smaller stay spacing whilst the larger stay spacing is more suitable for steel or steel composite construction.

10.5.3 Deck stiffness

The deflection of the longitudinal girders is primarily determined by the stay layout. It is reasonable therefore that the depth of the longitudinal girders should be kept to a minimum, subject to sufficient area and stiffness being provided to carry the large compressive forces without buckling. When checking the longitudinal girders for the stay out condition the PTI Recommendations (1993) stipulate that the structure provide for the replacement of any individual stay with a controlled reduction of the live load during any stay exchange. The structure must also be capable of withstanding the accidental loss of any individual stay without structural instability occurring.

10.5.4 Pylon height

The height of the pylon will determine the overall stiffness of the structure. As the stay angle (α) increases the required stay size will decrease as will the size of the pylon. However the deflection of the deck will increase as each stay becomes longer. Both the weight of the stay and the deflection of the deck become a minimum when the expression $1/(\sin \alpha \times \cos \alpha)$ is also a minimum. Therefore the most efficient stay is

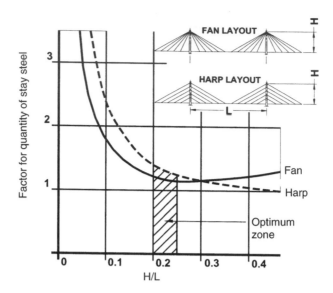

Figure 10.33 Optimum pylon height.

that with a stay inclination of 45°. In practice the efficiency of the stay is not significantly impaired when the stay inclination is varied within reasonable limits, which may be taken as 25–65°. The stay inclined at 25° will be the outer stay connecting the anchor pier and the deck panel adjacent to the centre of the main span to the top of the pylon. The stay inclined at 65° will be that located nearest the pylon. This implies an optimum ratio of pylon height above the deck (h) to main span (l) is between 0.2 and 0.25 as illustrated in Figure 10.33.

10.5.5 Preliminary stay forces

The main span stay forces resist the dead loads such that there is no deflection of the deck or pylon and the vertical components due to these loads are therefore known. By assuming that the live loads act in a similar manner an initial approximation of the stay force can be determined by considering the structure as a simple truss ignoring the bending stiffness of both the pylon and the deck. Thus the main span stay forces (P_{mi}) at an angle (α_i), as shown in Figure 10.34 can be determined from the expression:

$$P_{mi} = (W_{DL} + W_{LL})/\sin \alpha_i$$

and the horizontal stay component (F_h) is given by:

$$F_h = (W_{DL} + W_{LL})/\tan \alpha_i$$

The back stay anchoring forces can be calculated assuming the horizontal component of the main span and back span stay forces are balanced at the pylon.

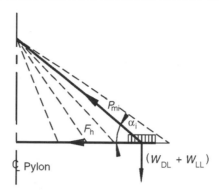

Figure 10.34 Main span–stay force diagram.

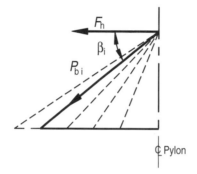

Figure 10.35 Back span–stay force diagram.

Ignoring the bending stiffness of the pylon will be a valid assumption as the bending stiffness of the pylon is usually small when compared to the axial stiffness of the stays. The back stay forces (P_{bi}) at an angle (β_i), as shown in Figure 10.35 can therefore be determined from the expression:

$$P_{bi} = F_h/\cos \beta_i$$

10.5.6 Deck form

The selection of the deck form will usually be based on an economic evaluation of the possible alternatives. The primary factors influencing the choice of deck will be the length of the main span and deck width. Other factors such as the cost of foundations, the local availability of materials or labour skills and the competitive conditions at the time of tendering may also have influence over the costs. A study by Svensson (1995) has undertaken an economic comparison of the various types of deck sections within the span range of 200–1000 m. The study concluded that a concrete deck section is the most economic deck section within the span range 200–400 m and the composite deck above 400 m. However the difference in cost is marginal within the span range 300–450 m and local factors are often decisive in the final choice. The study also does not consider the influence of any variation in the width of deck. The use of concrete construction in wide decks where there are six or more traffic lanes requires substantial crossbeams and the additional weight of these will penalize those spans near the upper end of the economic range. One solution, as used for the Vasco de Gama Bridge, was to design the crossbeams as steel plate girders, which are composite with the deck slab. Additional economy can be achieved, within the span range 350–600 m by using a hybrid combination of concrete back span and composite main span. Above 600 m a hybrid combination is still economic, but with the back span as concrete and the main span in an all steel construction. In the case of the Normandie Bridge the back span concrete deck was also extended more than 100 m into the main span.

10.5.7 Deck design

For the design of the deck it is possible, by tuning the loads in the stays, to minimize the moments in the deck, under the applied dead load, to the small local moments arising from the span between stays. For the design of the steel deck this dead load condition is applied following the structural completion of the deck and the application of all superimposed dead load. In the case of a concrete or composite deck the time dependant properties of the concrete must also be considered. Usually structural completion is assumed to be at time infinity ($t = \infty$), when all creep and shrinkage related strains have ceased.

Reducing the dead load moments in the deck to purely local effects will not provide the optimum solution. This is because the balance between positive (sagging) and negative (hogging) live load moments at any section along the girder will not be equal. As noted previously the magnitude of the moments will depend on the ratio of span lengths in the main span and the back span. The limiting stresses at the top and bottom of the section will not be the same as an allowance must be made for the effect of local bending within the road deck. The differences in material properties, as in the case of composite construction, will also effect the location of the neutral axis of the section. In most cases the properties of the deck section will be more favourable when resisting positive moments. The composite road slab can be kept in compression near the centre of the main span, where the normal forces are small, by inducing a positive dead load moment.

Taking these factors into account an optimum distribution of dead load moments can be determined. The design of the deck is then undertaken using an iterative process. For initial design purposes a minimum deck section can be assumed and using this section the dead weights and section properties are calculated. The structural system is then analysed incorporating a preliminary distribution of dead load moment and the sections checked against the distribution of total moments and normal forces. The sections where stresses exceed the permissible limits are then modified, the dead weights and section properties are re-calculated and the distribution of dead load moment is adjusted for the revised section. The structural system is then re-analysed and the cycle repeated until convergence is achieved.

10.5.8 Deck erection calculations

The common method of deck erection is by the successive cantilever method. With this method it is necessary to determine the stresses in the structure at each stage during the cantilever erection to ensure that stresses do not exceed the design limits. Erection by the successive cantilever method will give a quite different distribution of stay loads than when the entire bridge is loaded instantaneously with the weight of the superstructure. An instantaneous application of the deck load would also cause considerable extension of the stays giving rise to severe changes in geometry within the structure. In reality this does not occur as the stay extension is either absorbed within the anchor head at the time of jacking or, is deducted from the stay length during manufacture.

If the deck is of either concrete or composite construction the stay distribution will be further modified by the strains due to the effect of creep and shrinkage. It is therefore necessary, for calculation purposes, to also define the elapsed time from the initial casting of the various concrete pours at each construction stage.

The stay forces that are compatible with the final distribution of dead load moment and the defined structure geometry, at time infinity ($t = \infty$), are known. However, the initial stay forces introduced at each stage of the erection are not. Two methods are commonly used to determine these initial stay forces. The first method requires that the completed structure be dismantled stage by stage as follows:

(i)   using a model that represents the completed structure remove, where applicable, any element of the creep and shrinkage arising from time infinity to the time at the end of construction

(ii)   remove the superimposed dead load such as road surfacing, parapets, etc.

(iii)   apply any forces representing the formwork or other equipment required for connecting the two cantilevers at the centre of the main span

(iv)   using a model representing the completed cantilever, import the cable forces from (iii) and apply a moment at the tip of the cantilever that will give the optimum distribution of dead load moments

(v)   remove half the forces representing the formwork or other equipment necessary to connect the two cantilevers at the centre of the main span

(vi)   apply the weight of the lifting gantry or travelling formwork (bridge-builder) to the end of the cantilever (a typical arrangement is shown Figure 10.36)

(vii)   remove the last stay

(viii) remove the weight of the last segment

(ix)   the lifting gantry or bridge-builder is moved back along the cantilever so as to be in position for constructing the previous segment. This stay and segment are then removed in sequence

(x)   step (ix) is then repeated until the entire cantilever has been dismantled.

At each stage prior to the stay removal the initial stay force can be determined. The vertical component of the initial stay forces should be limited to less than the weight of the segment plus the weight of the lifting gantry.

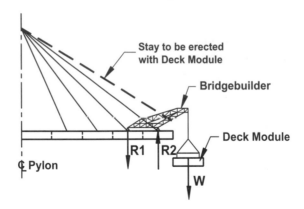

Figure 10.36 Typical cantilever erection using bridge-builder.

The alternative method of erection is to balance the vertical component of the stay force supporting each segment against the weight of each segment. Following the establishment of structural continuity at the centre of the main span each stay is then adjusted so as to ensure that the optimum distribution of dead load moment is achieved. This method does entail additional stressing of the stays but ensures that variations in the creep and shrinkage strains arising during the cantilever erection are discounted.

10.5.9 Static analysis
For the final analysis the most common approach is to model either a half or the entire structure as a space frame. The pylon, deck and the stays will usually be represented within the space frame model by 'bar' elements. The stays can be represented with a small inertia and a modified moduius of elasticity that will mimic the sag behaviour of the stay. In addition to carrying out the analysis of the completed structure

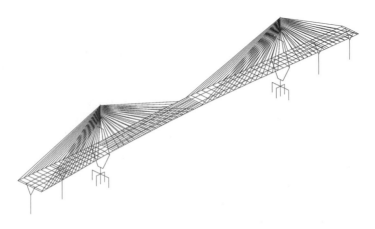

Figure 10.37 Typical space frame model.

the model can be used in the stage by stage erection analysis. An example of a typical space frame model is illustrated in Figure 10.37. The use of such a space frame model will ensure that the interaction between the deck, stays, pylon and piers and their foundations is accurately represented, both in the longitudinal and transverse directions. There are several computer packages commercially available that incorporate the facility to consider the non-linear behaviour of a structure and are suitable for the analysis of the cable stayed bridge.

10.5.10 Dynamic analysis

Dynamic analysis is the determination of the frequencies and the modes of vibration of the structure. This information is utilized for the following aspects of the design:

- the seismic analysis of the structure
- response of the structure to turbulent wind
- the physiological effect of vibrations.

Seismic analysis

Cable stayed bridges have the capacity to absorb large deformations without sustaining serious damage during an earthquake. Their flexibility is characterized by their low natural frequencies that are typically within the range 0.5–1.0 Hz. The dynamic analysis makes it possible to model the interaction between superstructure, substructure and the founding medium and assess the response of this model to the excitation produced during an earthquake. There are three methods of determining the seismic response, the equivalent lateral force method, the multi-modal response spectrum analysis and the time history analysis.

The equivalent lateral force method can be useful in preliminary design and gives a horizontal shear at each foundation proportional to the weight of the structure and the imposed ground motion. This method is limited in its application as it can only assess the effect on the structure from the fundamental mode of vibration and ignores the higher modes.

The multi-modal analysis determines the forces in each orthogonal direction for each modal frequency using a response spectrum. The response spectrum is a means of characterizing the earthquake over a large range of frequencies and levels of structural damping. Since the maximum response in each mode would not necessarily occur at the same instant of time, it would be over conservative to add the separate maximum modal responses. Based on the assumption that the peak modal response occurs randomly the orthogonal forces are combined either by the square root of the sum of the squares (SRSS) method or, more accurately by the combined quadratic combination (CQC) method. However, because this method relies on the principles of superposition it does not strictly apply to non-linear structures.

Both the equivalent lateral force method and the multi-modal response spectrum analysis assume that all the bridge foundations experience the same ground motion. With a large span structure there can be considerable discrepancy in ground motion at the different supports. This is known as asynchronous support motion. Where the structure is located within a high risk seismic zone a non-linear time history analysis is preferred. With this analysis a number of simulated earthquakes, given in the form of an acceleration versus time history, are applied to each foundation with an appropriate phase difference.

For cable stayed bridges located in seismically active areas it is beneficial to design the deck to 'float' past the pylon, in the longitudinal direction. Thus energy is dissipated within the cable system rather than resisted by the pylon. Provision is required at the bearing locations for a self-centring capability either through the shearing of flexible rubber pads or by an independently sprung buffer system.

Response to turbulent wind

The response to turbulent wind is divided into a mean wind component and fluctuating wind components. The fluctuating wind components can be separated into two responses. The 'broad-band' response is evaluated over the full range of turbulent frequencies and can normally be treated as a quasi-static load. However, the 'narrow band' response can produce a significant forced oscillation at one or more of the natural frequencies of the bridge structure.

Because the loading from turbulent wind is random the dynamic response of the structure has to be expressed in terms of probabilities. A method of analysis was developed by Davenport (1961, 1962, 1964) in which he suggested that the fluctuating velocity could be regarded as being formed from a large number of harmonic components and thus could be represented by the sum of a Fourier series. The analysis determines when the root mean square (rms) components of the dynamic response would be exceeded with a 50% probability. The total effect of turbulent wind may be determined by summing the mean load effect together with the factored 'broad band' and 'narrow band' effects.

The physiological effect of vibrations

The human response to vibration depends on a number of factors including the acceleration and frequency of vibration, the duration of the event and the direction of motion relative to the body. Whether it be from traffic or vortex shedding response to wind the vibration of bridges is predominately vertical. Because users are in the open and aware of the presence of wind or traffic increased levels of vibration are permissible relative to buildings. Irwin (1978) suggested a base curve for acceptable human response to the vibration of bridges subject to traffic loading and a further curve giving the acceptable limits for rare events such as storms. These curves are given in Figure 10.38. Where the evaluation of traffic induced vibration is to be investigated then a surface roughness must be established for the road and the passage of the vehicle evaluated at various speeds.

10.6 Aerodynamic behaviour

The behaviour of a long span structure in response to the wind is a key part of the design process. In order to determine the aerodynamic behaviour of the structure it

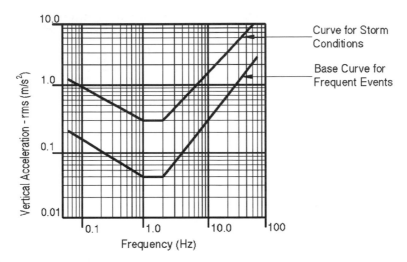

Figure 10.38 Acceptable bridge vibrations (vertical motion).

is necessary to have an understanding of the wind regime at the bridge site. The aerodynamic response of a long span structure may be sensitive to inclined wind and this effect is dependent upon the ground topography at the bridge site.

10.6.1 Definition of wind

Wind is normally defined by the wind speed and wind turbulence. The characteristic or reference wind speed is averaged over a specific time period, usually defined as the mean hourly wind speed. The mean hourly wind speed increases with height above the ground surface reaching a maximum at the gradient height. The gradient height is the height at which the wind is no longer disturbed by the earth's surface. The air flow below the gradient height is within the zone known as the boundary layer and this layer can vary in height from 500 to 3000 m depending on the type of terrain. The gradient height within a large city centre will be much higher than over the sea where the surface roughness is much reduced.

Various gust wind speeds that are averaged over different periods are also considered in the design. The 3-second gust wind speed is often used for the design of components less than 20 m in length. When checking a structure against aerodynamic effects the time taken for the motion to build up to an amplitude that is unacceptable is required. This build up in amplitude is highly variable but it is generally assumed that wind gusts with duration of at least 10 cycles of the critical vibration mode are required (Wyatt and Scruton, 1981).

The turbulence intensity depends on both the roughness of the terrain and the height of the deck above the ground.

10.6.2 Aerodynamic motion

The flow of air over the slender deck of the cable stayed bridge induces vertical bending and torsional/rotational oscillations of the deck section. There are several forms of such aerodynamic motion.

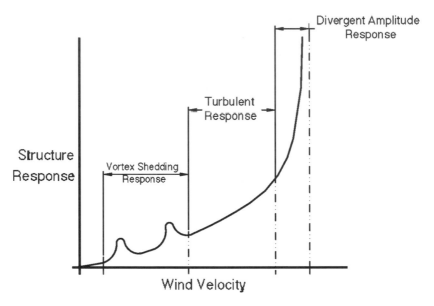

Figure 10.39 Aerodynamic response of a structure.

Vortex shedding response

The periodic shedding of vortices alternately from the upper and lower surfaces of the deck causes a periodic fluctuation of the aerodynamic forces on the structure. Over one or more limited ranges of wind speed the shedding of the vortices can be resonant with the natural frequency of the structure in the bending or torsional modes. For the most pronounced effect it is generally necessary for the wind flow to be perpendicular to the bridge cross section. The magnitude of the response will increase with increasing wind speed at which the resonant behaviour occurs, as shown in Figure 10.39. The response will decrease with increasing structural damping and with increasing turbulence intensity.

Turbulent or buffeting response

The forces developed by the wind fluctuate over a bridge deck within a wide range of frequencies. Only the lowest modes of vibration of the largest structures are affected at the high frequency end of the spectrum and turbulence arises from these fluctuating components of the wind acting about the three axes of the structure. If there is sufficient energy within the turbulent frequency bands that are resonant with the structure there may be a significant forced oscillation of the structure. Turbulence has the greatest impact with bluff deck cross sections and on the more flexible suspension structures. For such structures the total equivalent wind effect, including the dynamic enhancement can be twice the static force (Blom-Bakke *et al.*, 1993). The greatest response to turbulence for cable stayed structures is likely to be during cantilever erection when the deck can respond at the lower frequencies.

Divergent amplitude response

At the critical wind speed of the section there is a rapid increase in the response of the structure, as shown in Figure 10.39. This occurs when the sum of the equivalent aerodynamic damping, which can have either negative or positive sign, and the structural damping becomes negative resulting in zero total stiffness. At this point the amplitude can grow without limit exhibiting divergent behaviour. Thus the value of structural damping will increase the critical wind speed but will not decrease the magnitude of the oscillations. A divergent response can be either vertical or torsional, but for most practical deck sections the torsional behaviour will dominate. Usually the presence of turbulence will increase the critical wind speed.

Classical flutter

A very strong response of the divergent kind can arise in the event of a coupling of the vertical and torsional motion. In order that this form of response cannot occur it is necessary to ensure that the torsional and bending frequencies of the structure are separated.

When considering the effect of aerodynamic motion upon the design of the cable stayed bridge both divergent amplitude response and limited amplitude response must be considered. Divergent amplitude response will result in the failure of the structure and therefore an adequate margin of safety must be provided between the characteristic mean hourly wind speed and the critical wind speed. In the UK Design Rules for Aerodynamic Effects on Bridges (Department of Transport, 1993), a factor of 1.625 is adopted. Limited amplitude response, from either vortex shedding or turbulent wind must be considered in respect of both fatigue damage and any disturbing physiological effect upon the bridge user.

10.6.3 Example of aerodynamic motion

One interesting case study was the effect of the limited amplitude response on the Rama IX Bridge (Gregory and Freeman, 1987) in Bangkok. Wind-tunnel testing predicted that this structure could be subject to oscillations at low wind speed due to vortex shedding response. It is worth noting that this structure has one central stay plane and a light steel box deck that is comparatively wide, carrying three lanes of traffic in each direction. These oscillations where predicted in both the steel box pylon, due to transverse wind, and in the deck, again due transverse wind but with an upward component. The deck motion was predominantly torsional and, to a lesser extent, a vertical motion was also observed. These movements were successfully controlled by the introduction of tuned mass damping devices located at the top of each pylon and at eight locations for torsion and eight locations for bending within the steel box deck. The tuned mass damper consisted of a mass hanging from tension springs with 'viscodampers' attached. These dampers are cylinders, which are attached to the mass, and contain a viscous fluid with a plunger immersed in it. The plunger is attached to the bridge deck. The installation of tuned mass dampers on the Rama IX Bridge has successfully controlled any limited amplitude response to wind.

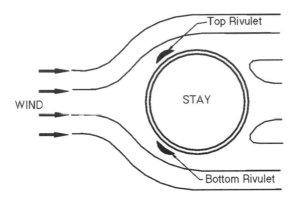

Figure 10.40 Wind–rain oscillation of stays.

10.6.4 Oscillation of stays

Another phenomenon that must be considered with cable stay bridge construction is the effect of stay oscillation. During cantilever erection a slender deck may be prone to wind induced movement. This can in turn excite the stays producing violent oscillations which have to be restrained with temporary straps. In service the stays may also be subject to oscillation, which can reach amplitudes of more than a metre. This behaviour was reported to have occurred with the Helgeland Bridge (Svensson, 1999). This behaviour is not solely due to wind but can be accentuated when wind and rain are combined as illustrated in Figure 10.40. The rain forms rivulets which run down the stay pipe such that the cross-section of the circular stay pipe is altered, sufficiently to induce lift and giving rise to an unstable divergent amplitude response. Rain–wind-induced oscillation of the stays usually occurs within the wind speed range 7–15 m/s. Below that range, the top rivulet does not form because the wind is insufficient to prevent it from running down the side of the stay. Above the range, the wind forces on the rivulets tend to blow them off the stay. In order for there to be a negligible risk of violent rain–wind oscillation it has been suggested that the Scruton number (S_c) for the stay should be at least 10. The Scruton number is defined as:

$$S_c = m \times \delta/\rho \times D^2$$

where m is the mass per unit length of the stay, δ is the damping ratio (typically in the range 0.005–0.01), ρ is the density of air (taken as 1.2 kg/m³) and D is the outer diameter of the stay.

Several methods have been employed to reduce or eliminate these oscillations. A simple method, that is incorporated in most stay systems and does not alter the appearance of the bridge, is with dampers installed between the steel anchorage pipes and the stay. Ribs, either parallel or helically wound, on the outer surface of the stay prevent the rivulets forming long continuous lengths and thus interfering with the aerodynamic behaviour. Helical ribbing is illustrated in Figure 10.10. As in the case of the Normandie and Helgeland Bridges, tuning ropes can be installed which connect the stays together and effectively reduce the fundamental period of the stay.

10.6.5 Wind tunnel testing

The UK Design Rules for Aerodynamic Effects on Bridges (Department of Transport, 1993) defines a long span structure requiring investigation in a wind tunnel as any structure with a span greater than 200 m. One of the most useful and economic tools for determining the aerodynamic behaviour of the bridge structure is through the testing of the sectional model in a wind tunnel. The sectional model consists of a representative section of the deck, geometrically and aerodynamically similar to the prototype. It is mounted in the wind tunnel in such a way as to measure the static and dynamic lift, drag and torsion produced by the wind. The model is usually located between the parallel walls of the wind tunnel so as to channel a two-dimensional wind over the model.

In early applications the sectional model was conceived as a substitute for the prototype bridge however the behaviour of the model cannot entirely represent the full scale bridge and considerable interpretation of the results are required.

In the fixed position, the sectional model can be instrumented to measure drag, lift and torsion. The bridge deck is rotated with respect to the air flow, simulating wind inclination, so as to measure the variation in the forces with a range of angles of attack, usually between $+10°$ and $-10°$. The force coefficients are usually specified as follows:

Drag $\qquad D = 1/2\rho V^2 B C_D$
Lift $\qquad L = 1/2\rho V^2 B C_L$
Torsion $\qquad T = 1/2\rho V^2 B C_T$

where ρ is the density of air, B is the width of the model deck, and C_D, C_L and C_T are dimensionless drag, lift and torsion coefficients, respectively.

When mounted on springs, with scaled mass per unit length, mass moment of inertia per unit length, structural damping and natural frequencies the sectional model can be used to investigate:

- the dynamic response to vortex shedding
- to determine the response to turbulence
- to ensure that the section is aerodynamically stable and has an acceptable factor of safety with respect to the design wind speed.

The structure response can be determined in either smooth or turbulent flow. However, turbulent conditions will usually reduce the response of the bridge deck to vortex shedding and also diminish the flutter behaviour of the section. Tests for these effects are therefore usually undertaken in low turbulence flow, sometimes called smooth flow.

Placing a uniform grid or spires into the wind tunnel simulates scaled atmospheric turbulence. The measured response to turbulent buffeting can be interpreted so as to provide estimates of the wind loading that can be used in the design of the bridge.

The typical geometric scales of the sectional model are 1:40 to 1:80 allowing the details of the deck to be modelled realistically. This is important as small changes,

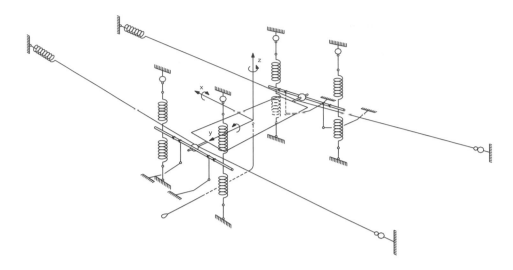

Figure 10.41 Sectional model in the wind tunnel.

particularly to the leading edge of the deck can have a significant influence on the aerodynamic behaviour.

A typical arrangement of a sectional model including the spring mounting system is shown in Figure 10.41.

Bibliography

Beyer E and Tussing F. Nordbrucke Dusseldorf. *Stahlbau*, 2, 3, and 4, 1955.

Blom-bakke L, Hellesland J and Vangnes A. Design and construction of the Askøy suspension bridge. *Proceedings of the Third Symposium on Straits Crossings*, Alesund, Norway, June, 75-82, 1994.

Brown CD. Design and construction of the George Street Bridge over the River Usk, at Newport, Monmouthshire. *Proceedings of the Institution of Civil Engineers*, 32, August, 552–561, 1966.

Capra A and Leveille A. Vasco da Gama Bridge, Portugal. *Journal of the Association for Bridge and Structural Engineering*, 8, 4, November, 261–262, 1998.

Davenport A.G. The application of statistical concepts to the wind loading of structures. *Proceedings of the Institution of Civil Engineers*, 19, August, 449-472, 1961.

Davenport AG. The response of slender, line-like structures to a gusty wind. *Proceedings of the Institution of Civil Engineers*, 23, November, 389–408, 1962.

Davenport AG. Note on the distribution of the largest value of a random function with application to gust loading. *Proceedings of the Institution of Civil Engineers*, 28, June, 187–196, 1964.

Department of Transport. Design Rules for aerodynamic effects on bridges. Department of Transport, UK BD 49/93, 1993.

Dewry CS. *A Memoir on Suspension Bridges*. Longman, Rees, Orme, Brown, Green and Longman, London, 25–26, 1832.

Fischer G. The Severin Bridge at Cologne (Germany). *Acier-Stahl-Steel*, 3, 97–107, 1960.

Gimsing NJ. *Cable Supported Bridges Concept & Design*, 2nd edn. John Wiley, Chichester, England, 1997

Gregory FH and Freeman RA. *The Bangkok Cable Stayed Bridge*. 3F Engineering Consultants, Bangkok, 1987.

Irwin AW. Human response to dynamic motion of structures. *The Structural Engineer*, 56A, September, 237–244, 1978.

Kerensky O A, Henderson W and Brown WC. The Erskine Bridge. *The Structural Engineer*, 50, April, 147–169, 1972.

Leonhardt F. and Zellner W. Cable-stayed bridges. *International Association for Bridge and Structural Engineering Surveys*, S-13/80, February, 1980.

Mizon DH, Smith N and Yeoman AJ. Second Severn Crossing – cable-stayed bridge. *Proceedings of the Institution of Civil Engineers*, 114, November, 49–63, 1997.

Podolny W Jr and Scalzi JB. *Construction and Design of Cable Stayed Bridges*, 2nd edn. Wiley, USA, 1986.

Post-Tensioning Institute Committee on Cable-stayed Bridges. Recommendations for Stay Cable Design, Testing and Installation. Post-Tensioning Institute Committee on Cable-stayed Bridges, August, 1993.

Svensson HC. The development of composite cable-stayed bridges. *Proceedings of Conference, Bridges into the 21st Century*, The Hong Kong Institution of Engineers, October, 45–54, 1995.

Svensson HC. The twin cable-stayed Houston Ship Channel Bridge. *The Structural Engineer*, 77/No5, March, 13–20, 1999.

Svensson HC and Jordet E. The concrete cable-stayed Helgeland Bridge in Norway. *Proceedings of the Institution of Civil Engineers*, 114, May, 54-63, 1996.

Troitsky MS. *Cable-stayed Bridges*, 2nd edn. BSP, Oxford, 1988.

Walther R, Houriet B, Walmar I and Moïa P. *Cable Stayed Bridges*. Thomas Telford, London, 1988.

Wenk H. The Strømsmund Bridge. *Stahlbau*, 23, 4, 73–76, 1954.

Wyatt TA and Scruton C. A brief survey of the aerodynamic stability problems of bridges. *Proceedings of Conference, Bridge Aerodynamics*, Institution of Civil Engineers, London, 21–31 March, 1981.

11 Suspension bridges

V. JONES AND J. HOWELLS

11.1 Introduction

11.1.1 Scope

The scope of this chapter is restricted to consideration of the classical three span suspension bridge configuration (Figure 11.1), with a stiffened load carrying deck structure supported by earth anchored cables. Bridges with unusual span or cable configurations, including bridges with multiple main spans, mono-cable bridges, self-anchored structures, and hybrid part suspension/cable stayed structures are not considered. Even with the above limitations, it is not possible in a short chapter to consider in detail many important aspects of suspension bridge design; in particular the analysis of cables and aerodynamic design requirements.

The principal structural elements of these classical suspension bridges are as follows:

- Two or more flexible main cables, which support the traffic carrying deck and transfer its loading by direct tension forces to the supporting towers and anchorages. These cables are formed from high strength steel wires, with a strength-to-weight ratio of around three times that of normal structural steels.
- A traffic carrying deck structure, supported from the main cables by hangers constructed from high strength wire ropes or strands.

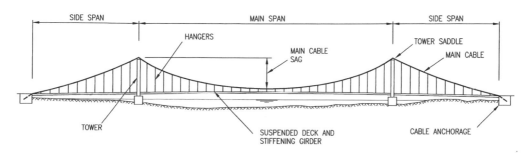

Figure 11.1.

- A longitudinal stiffening girder to distribute concentrated traffic loadings on the deck to the cable, thereby minimizing and controlling local deflections. This girder also provides the necessary flexural and torsional stiffness to prevent aerodynamically induced oscillations of the deck. The stiffening girder may be of truss or plate girders which, together with their associated lateral bracing systems, are structurally separate from the deck. Alternatively the stiffening girders and bracings may be integrated with the deck structure, with the logical development of this being the use of a box girder deck structure constructed from stiffened plates, and with a streamlined shape to minimize wind loading. The design of the deck and stiffening girder is one of the most critical aspects of the design of suspension bridges, as the deck dead load is entirely supported by the cable, towers, and anchorages. The most economical overall design will therefore generally result from the incorporation of the lightest practicable deck structure, as savings in deck weight will lead to further reductions in the weight and cost of the cables, towers, and anchorages.
- Towers to support the main cables at a level determined by the main span cable sag, combined with the clearance required above the waterway or other obstacle being crossed.
- Anchorages to secure the ends of the main cables against movement. These structures must resist large horizontal forces. In areas where ground conditions are poor, their cost may be very high, to some extent offsetting the economy of the main load carrying structure.

With cables constructed from very high strength steel loaded in direct tension as their primary load carrying members, suspension bridges are ideally suited to longer spans, and this is therefore the primary application for this type of structure. Although cable stayed structures have made considerable inroads into the lower end of the span range previously considered to be the domain of suspension bridges, the latter remain the unchallenged choice for spans over 1000 m.

Suspension bridges, when well designed and proportioned, are clearly the most aesthetically pleasing of all bridges. The simplicity of the structural arrangement, with the function of each part being clearly expressed, combines with the graceful curve of the main cables, the slender suspended deck and vertical towers, to produce a naturally attractive structure. This natural grace can also make a suspension bridge a suitable choice for relatively short span footbridges, in addition to its more usual function of economically spanning long distances.

11.1.2 Historical background

The method of crossing a gap using a suspended flexible rope as a support appears to have originated in ancient times, and was used by early civilisations in both Central and South America, and the Far East. This was no doubt due to the local availability of natural fibres, from which ropes could be produced of sufficient strength to enable modest suspended spans to be built.

In Europe and North America, the construction of the first suspension bridges dates from the Industrial revolution, when wrought iron bars became available, from which chain cables of pin connected eye bars could be manufactured. Chain cables were usually preferred for these early bridges because of their superior durability relative to the then currently available wire products. A notable example of these early bridges was the 176 m span Menai bridge, completed in 1826. The first major bridge supported by wire cables was the 273m span Grand Pont Suspendu across the Sarine valley at Fribourg in Switzerland, completed in 1834, which remained in service for almost 100 years.

By the mid nineteenth century, the USA had become the unchallenged centre of suspension bridge design and construction, retaining this position for the next 100 years. In 1849, the 308m span Wheeling bridge across the Ohio river was completed, becoming the world's longest span, but was destroyed by a severe gale after only 5 years service. The most famous suspension bridge engineer of this period was John Roebling, who was responsible for a series of major suspension bridges. The most notable of these are the 250m span Niagara bridge, completed in 1855, and the 322 m span Cincinatti-Covington bridge, completed in 1866. Roebling was also largely responsible for the design of the 486m span Brooklyn bridge, completed in 1883 after his death. For the construction of the Niagara bridge, Roebling originated the aerial *in situ* spinning method for the construction of parallel wire cables which, with progressive development and refinement, remains to this day the generally preferred method for cable construction.

For the next 50 years, American engineers constructed suspension bridges with steadily increasing span lengths, up to the 564 m span Ambassador bridge in Detroit, completed in 1927. However, within only 4 years, the 1006 m span George Washington bridge in New York was completed, almost doubling the record free span length, and this was soon followed by the 1280 m span Golden Gate bridge in San Francisco, completed in 1937.

Although many early nineteenth century suspension bridges had been damaged or destroyed by instability under the action of storm winds, the American bridges of the late nineteenth and early twentieth century, which had relatively heavy and deep stiffening trusses, did not experience such problems. As a result, the possibility of aerodynamically induced instability of these structures was largely ignored by suspension bridge designers and constructors. In parallel with this, the improved understanding of the behaviour of suspension bridge structures resulting from the development of the deflection theory, enabled designers to adopt progressively more slender deck structures.

These developments culminated in the construction of the 853 m span Tacoma Narrows bridge, completed in 1940. This bridge had an extremely low torsional stiffness, as its deck stiffening consisted only of 2.4 m deep plate girders without any bottom lateral bracing system. From the start, the structure oscillated in certain wind conditions, and within a few months of opening to traffic, its deck was completely destroyed by violent torsional oscillations induced by a wind speed of less than 20m/s. This disaster alerted designers to the importance of ensuring aerodynamic stability, and the analysis of this problem has become an essential part of the design process for suspension bridges.

Following World War II, the construction of long span suspension bridges continued to be American led, the most notable of these bridges being the Mackinac, Verrazano Narrows in the USA, and the Tagus bridge in Portugal. These were, however, the last long span bridges to be built by American engineers.

During the 1960s, American engineers also developed a new method for the construction of parallel wire cables, using preformed parallel wire strands (PPWS), although they were able to apply this technique only to the cables of the Newport and second Chesapeake Bay bridges, both with relatively modest spans. This method has, however, been developed and used by Japanese engineers for a number of long span bridges.

At around the same time as the last great American bridges were being constructed, the building of long span suspension bridges resumed in Europe. The first of these were the Tamar bridge in the United Kingdom and the Tan Carville bridge in France, with relatively modest spans of 335 m and 608 m, and with cables formed, respectively, from an assembly of locked coil and spiral strands.

These were soon followed by the 1006 m span Forth and 988 m span Severn bridges in the United Kingdom. These were the first European bridges to have parallel wire cables constructed by aerial spinning. The Forth bridge closely followed established American design and construction practices, but the Severn bridge represented a radical departure from these, with the use of a streamlined box girder for the deck structure rather than the conventional stiffening truss. In addition to a substantial saving in deck weight, the resultant reduction in wind load enabled further significant savings in the design of the tower structures. The main cables of the suspension system were of conventionally spun construction, but the hangers were arranged in an alternating inclined system as a means of improving the structural damping.

The streamlined box girder deck concept has been further developed in Europe with the subsequent construction of the 1074 m span Bosporus bridge in Turkey, the 1410 m span Humber bridge, which extended the record span length by a further 112 m, the 1090 m span Second Bosporus bridge, and finally the 1624 m Storebælt East bridge in Denmark.

From 1980, the Honshu-Shikoko bridge project in Japan has required the construction of more than ten major suspension bridges. The most significant of these is the Akashi Kaikyo bridge, which with a main span of 1990 m is currently the longest span in the world. Most of the Japanese bridges have truss stiffened deck structures, but a number of bridges have incorporated streamlined box girder decks. The 990 m span Kita and 1100 m span Minami Bisan Seto bridges are also notable in that they were the first very long span suspension bridges designed to carry both road and 'mine line' rail traffic. Many of these Japanese bridges, including all of those referred to above, have main cables formed from preformed parallel wire strands.

11.2 Cables as structural elements

11.2.1 Materials

The basic element for all cables is high strength steel wire, which for parallel wire cables will usually be between 5 and 5.5 mm diameter. Wire of this size is generally produced with a tensile strength of 1550 N/mm^2, and a 0.2% proof stress of around 1200 N/mm^2. Higher strength wire with a minimum tensile strength of 1800 N/mm^2 can be produced, and has been used in Japan for the cables of the Akashi Kaikyo bridge.

This wire is produced from steel rods with a much higher carbon content (around 0.8%) than the usual range of structural steels. The high tensile strength is obtained by cold drawing the wire through a series of dies to reduce its diameter from the initial rod size down to the final wire size. This process produces wire with consistent accurate dimensions, and a smooth flaw free surface. Production of high strength wire by this method does, however, have the comparative disadvantage that the resultant wire has low ductility, with the elongation at fracture being typically only about 4%.

For corrosion protection, the wire is galvanized after final drawing, the weight of zinc coating being typically 300 g/m^2. For storage and delivery, the final wire product is formed into coils of between 250 to 1200 kg, with an internal diameter of about 1500 mm, which is sufficient to ensure that the wire, when uncoiled, is sufficiently straight for subsequent cable spinning or strand manufacturing operations.

Because the construction of large suspension bridges occurs relatively infrequently, wire for the construction of parallel wire cables is produced specifically for each project. Long span bridges require very large quantities of wire, requiring an extended production period before commencement of cable work at site. As a result, storage for large quantities of wire coils in a controlled environment to prevent corrosion damage is generally necessary.

11.2.2 Cable types
Spiral bridge strands
Spiral strands are manufactured by the helical winding of multiple layers of round steel wires onto a straight centre core wire. Each layer of wires is laid with an opposite helix to the preceding layer, in a single pass through a stranding machine. The helical twist of the wires results in a decrease of the order of 15–25% in the stiffness of the strand relative to that of plain straight wire. The twisting of the wires also reduces the strand strength to around 90% of the sum of the individual wire breaking strengths. Figure 11.2 shows a typical arrangement of a spiral strand. For attachment to anchor points, spiral bridge strands are terminated by zinc-filled sockets.

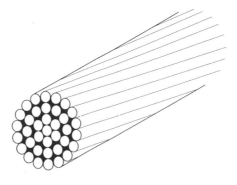

Figure 11.2

The twisting of the wires causes spiral strands to self-compact under axial load-ing, resulting in a significant non-elastic stretch occurring when the strand is first loaded. It is therefore essential to remove this as far as practicable during manufacture so that the strand behaves elastically when installed in the final cable. This is achieved by cyclic prestretching of the completed strand by applying an axial load between 10 and 50% of the breaking load until a stable modulus value has been reached. How-ever, despite this prestretching, some long term inelastic extension of spiral strands in service will generally occur, causing some permanent downward deflection of the bridge deck, as has occurred on the Lillebælt (Jensen and Petersen, 1994) bridge in Denmark.

Typical bridges using spiral stands for their main cables are:

Tancarville bridge (France)
Main span length: 608 m
Cable size: 56 no. 72 mm diameter strands
Year of completion: 1959

Lillebælt bridge (Denmark)
Main span length: 600 m
Cable size: 55 no. 68.7 mm diameter and 6 no. 41.4 mm diameter strands
Year of completion: 1969

Locked coil strands

Locked coil strands are also factory produced by the helical winding of multiple lay-ers of round steel wires onto a straight centre core wire. They differ from spiral bridge strands in that the final layers of wires are made up of interlocking Z-shaped wires, giving a larger proportion of wire area to strand 'cross-sectional' area, and more importantly, a smooth easily protected exterior surface, so that protective wrapping is not essential. Their stiffness and other characteristics are generally similar to those of spiral strands,

Typical bridges using locked coil strands for their main cables are:

Tamar bridge (UK)
Main span length: 335 m
Cable size: 31 no. 60 mm diameter strands
Year of completion: 1961

Rödenkirchen bridge (Germany)
Main span length: 378 m
Cable size: 37 no. 69 mm diameter strands
Year of completion: 1954, reconstructed and widened 1994

Askøy bridge (Norway)
Main span length: 850 m
Cable size: 21 no. 99 mm diameter strands
Year of completion: 1993

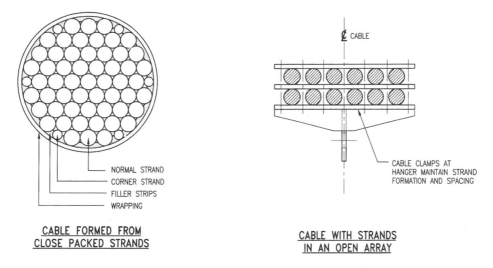

NORMAL STRAND
CORNER STRAND
FILLER STRIPS
WRAPPING

CABLE FORMED FROM
CLOSE PACKED STRANDS

CABLE

CABLE CLAMPS AT
HANGER MAINTAIN STRAND
FORMATION AND SPACING

CABLE WITH STRANDS
IN AN OPEN ARRAY

Figure 11.3.

Locked coil or spiral strands can be arranged either in a close packed hexagonal formation, or in an open rectangular array (Figure 11.3). The first of these has the advantage that, by the addition of aluminium or plastic spacers, the cross-section can be circularized, and then conventionally wrapped for corrosion protection. The open strand arrangement, which has been rarely used except on bridges in Norway, does not require the time consuming wrapping operation, and the cable bands can be simple fabricated structures. It has the disadvantages that the wind load on the cable is substantially increased, and inspection and maintenance of the inner strands is difficult.

Parallel wire cables
General

These are the most widely used type of cable, and consist of the required number of individual wires, laid straight and parallel throughout the complete cable length from anchorage to anchorage. Parallel wire cables are constructed by either aerial *in situ* spinning of the wires, or by the assembly of a number of preformed parallel wire strands. After completion of wire placement, both cable types are squeezed or compacted to form a single mass of parallel wires contained within a circular profile.

In situ spun cables

The aerial spinning method has been the preferred method for parallel wire cable construction since its original development in the mid nineteenth century, and spun *in situ* wires have been used to form the main cables of most long span suspension bridges. In aerial spinning, either two or four loops of wire are pulled by a travelling 'spinning' wheel from one anchorage to the other. After each trip of the wheel, the wires are adjusted to the required sag, this process continuing until all the wires in

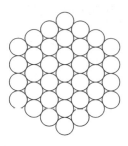

ARRANGEMENT OF 37 STRAND
PARALLEL WIRE CABLE BEFORE COMPACTION

Figure 11.4.

the cable have been assembled. The wires are anchored at both ends of the bridge by placing each loop of wire around a semi-circular shaped structure referred to as a strand shoe.

For convenience in construction, the individual wires are divided into groups, referred to as strands, each containing between 200 to around 550 wires. The number of wires in each strand must be a multiple of $2 \times 2 = 4$, the first factor of two arising from the looping of the wire round the strand anchors, and the second from the need for symmetry about the strand shoe centre line. The strands are usually arranged in a hexagonal formation with vertical sides, and cables are therefore generally made up from 19, 37, 61, or 91 strands arranged as shown in Figure 11.4. This arrangement of the strands in a vertical hexagon has the advantage that, during spinning, temporary spacers can be placed at intervals between the columns of strands. These spacers assist in keeping the strands of the part completed cable under control in the event of storm wind conditions occurring during construction. More importantly, they enable air to circulate between the strands, thus minimizing temperature differences, and keeping the cable at a more uniform temperature.

After the total number of wires required in the cable has been established, the number of strands can be derived, with the number being chosen so that the number of wires per strand is a multiple of 4, and is within the above limits. This will generally require that the cable will contain a few wires more than the theoretical minimum. For example, if (say) the theoretical requirement is for 14900 wires, the most suitable practical configuration would be a cable of 14948 wires (37 strands each of 404 wires).

Strands produced by spinning consist of two or more continuous wire lengths, each end of which is looped around the shaped anchor blocks (strand shoes). To achieve this, wire from a number of coils is joined together with swaged compression splices, a typical example being shown in Figure 11.5.

Prefabricated parallel wire strands (PPWS)
This alternative to aerial spinning was originally developed (Durkee, 1966) in the USA in the 1960s, but all the subsequent development and use of the method has been in

Figure 11.5.

Japan, and more recently in China. Instead of pulling individual wires across in groups of four or eight, bundles of wires are factory prefabricated into hexagonal shaped strands (PPWS), bound together at intervals by plastic tape, and fitted with sockets at each end. To generate the hexagonal shape of the strand, the number of wires in each PPWS must be one of the series of 'hexagonal' numbers, i.e. 37, 61, 91, 127, and so on. Early PPWS strands were each made up of 61 or 91 wires, but 127 wires per strand, arranged as in Figure 11.6, is now generally preferred. This gives a reasonable compromise between excessive reel weight and the erection of an unduly large number of strands.

The longest PPWS so far used are the 127 wire strands for the Akashi Kaikyo bridge, which have an average strand length of 4073 m, with each reeled strand weighing 94 tonnes. It can therefore be seen that the use of PPWS requires the transport and handling of large diameter heavy reels, as even for a 2000 m long stand, the reeled weight will be around 50 tonnes.

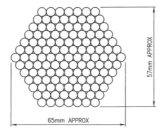

ARRANGEMENT OF 127 WIRE
PREFORMED PARALLEL WIRE STRAND

Figure 11.6.

Cables constructed using PPWS have a much larger number of small strands (e.g. 127 wires) than would be required in an equivalent spun cable. This has an impact on the design of the anchorages, as the size of the anchor face has to be significantly increased to accommodate the much larger number of strand terminations required. Except in Japan and China, PPWS cables have not thus far proved to be an economic alternative to spun cables.

11.2.3 Compaction

After completion of spinning or PPWS strand erection, parallel wire cables are squeezed together to form a single compact mass of wires within a nominally circular cross-section. The degree of compaction achieved in this operation is specified by the voids ratio, defined as the area of voids in the cable divided by the area enclosed by the circumscribing circle. A hexagonal arrangement of wires, with each wire in contact with six others, produces the closest packing of circular shapes, and in the absence of edge effects, results in a voids ratio of approximately 9.3%. Such a low void ratio is not achievable in actual parallel wire cables, experience having shown that the voids ratio after compaction will generally be slightly more than twice this value, at around 18–22%. This increase is due to edge effects, and the effect of wire crosses caused by the impossibility of achieving perfectly parallel wire alignments.

11.2.4 Corrosion protection

The main suspension cables are the primary structural elements of the bridge. Although limited local replacement of corroded outer wires has been carried out on some earlier bridges, the complete replacement of a large diameter compacted parallel wire cable has never been attempted, and is probably impracticable without also removing the deck structure. It is therefore essential to provide the main cables with corrosion protection which, with reasonable care and maintenance, will ensure that the cables will have a service life at least as long as the design life of the bridge.

Traditionally, parallel wire cables have been protected against corrosion by a three stage process. The primary protection is the galvanized zinc coating on the wires, with additional protection being provided by coating the exterior surface of the cable by a red lead and linseed oil paste contained by circumferential wrapping of the cable with closely spaced turns of galvanized low strength steel wire. Finally, a series of coats of paint are applied over the wrapping wires. This basic system of wrapping over an anti-corrosive paste represents the best system currently available for cable protection.

The use of lead based products is however no longer environmentally acceptable. Wrapping pastes based on metallic zinc powder, rather than red lead have therefore been employed on the recently constructed Høga Kusten (Sweden) and Storebælt East bridges. These alternative paste formulations have however proved to be somewhat more difficult to apply than the traditional red lead, and there is of course as yet no long-term experience of their corrosion protection effectiveness.

However, examination of existing bridge cables (Stahl and Gagnon, 1995) has shown that even when such a system is applied with attention to high standards, leakage of water into the cable frequently occurs with some resultant corrosion. Improvements to the corrosion protection provided by the wrapping process therefore continue to be sought (Furuya *et al.*, 2000). The use of interlocking S-shaped wires is one proposed method, so as to give improved resistance to water penetration. Another approach, which has been used on the Akashi bridge cables, is to circulate dry air into the cables in an attempt to maintain the relative humidity within the cable interior below that at which corrosion will occur.

Attempts were made in the early 1970s to develop alternative systems to wire wrapping over paste. On the Newport bridge, the cables were wrapped with multiple layers of glass fibre impregnated with acrylic resin, and on the second Chesapeake Bay bridge cables were wrapped with neoprene sheeting, applied over a liquid neoprene coating. Neither of these systems has been generally adopted, and all recent bridges have reverted to conventional circumferential wire wrapping.

11.2.5 Analysis and design of cables
Introduction
The cables are the primary load-carrying element of a suspension bridge, and an understanding of their behaviour under load is therefore essential. The case of a cable with self-weight loading only (the catenary) is first considered, followed by the case of uniformly distributed loading (parabolic cable). Finally the general case of arbitrary loading on a cable is discussed.

Cables are load-adaptive structures, which change their geometry to accommodate changes in both the magnitude and distribution of the applied loading. As a result of these geometrical changes with loading, the behaviour of cables is non-linear, with their stiffness increasing with increasing load. Explicit solutions are therefore not generally obtainable, and recourse must be made to iterative methods to derive the loaded geometry and internal forces.

Cables under self-weight loading only
The starting point for any analysis of cable structures is consideration of a flexible cable hanging between two fixed points. The term flexible is used here to mean that the cable has no bending stiffness, and can carry loading only by the development of tension forces directed along its local axis. The curve in which such a cable with uniform weight along its length hangs between two fixed points is called a catenary, and its shape is determined as follows.

The origin of co-ordinates x and y is chosen as a point directly below the lowest point of the cable profile as shown in Figure 11.7. Assuming that the cable is inextensible, the weight of the section OP is ws, where w is the weight per unit length of cable. Equilibrium of segment OP then requires that:

$$T\cos \psi = H \tag{11.1}$$

$$T\sin \psi = ws \tag{11.2}$$

As the cable is subjected to vertical loading only, the horizontal component of cable tension (H), is a constant. Defining therefore H/w as C, and dividing (11.2) by (11.1):

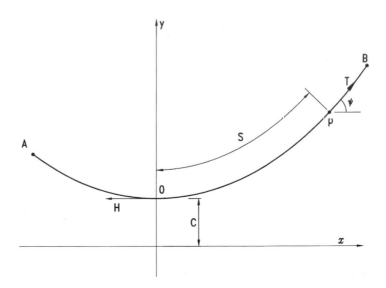

Figure 11.7.

$$s = C \tan \psi \qquad (11.3)$$

This is the intrinsic equation of the catenary, and the constant C is called the parameter of the catenary.

Equation (11.3) can be expressed in Cartesian co-ordinates by expressing $\tan \psi$ as (dy/dx) giving:

$$C \frac{dy}{dx} = s$$

Differentiating this equation, and noting that:

$$\frac{ds}{dx} = \sqrt{\left[1 + \left(\frac{dy}{dx}\right)^2\right]}$$

gives:

$$C\left(\frac{d^2y}{dx^2}\right) = \frac{ds}{dx} = \sqrt{\left[1 + \left(\frac{dy}{dx}\right)^2\right]}$$

Integrating this equation gives:

$$C \sinh^{-1}\left(\frac{dy}{dx}\right) = x + A$$

where A is a constant of integration. As the chosen origin is vertically below the lowest point of the curve, at $x = 0$, the slope $(dx/dy) = 0$, so that $A = 0$, and:

$$\frac{dy}{dx} = \sinh\left(\frac{x}{C}\right) \tag{11.4}$$

Integrating again gives:

$$y = C\cosh\left(\frac{x}{C}\right) + B$$

where B is another integration constant.

By making the y coordinate of point 0 equal to C, then for $x = 0, y = C$, and the Cartesian equation of the catenary is finally obtained as:

$$y = C\cosh\left(\frac{x}{C}\right) \tag{11.5}$$

The length of any part of the cable is obtained by combining equations (11.3) and (11.4):

$$s = C\sinh\left(\frac{x}{C}\right) \tag{11.6}$$

The tension in the cable at any point can be obtained by squaring and adding equations (11.1) and (11.2), giving:

$$T^2 = w^2(s^2 + C^2)$$

and using equations (11.5) and (11.6), this can be simplified to:

$$T = wy \tag{11.7}$$

The tension at any point is therefore directly proportional to its height above the origin, and at the lowest point is equal to H.

For a cable with its end supports at the same level, equations (11.5) or (11.6) can be used to determine the value of C, and hence the cable shape and tensions, provided either the cable length or sag are specified.

A more general case is where the supports are at different levels, and the length of the cable is known. In this case the value of C can be determined in relation to the cable length and end point geometry (Figure 11.8) as follows:

$$L = C\left[\sinh\left(\frac{x+k}{C}\right) - \sinh\left(\frac{x}{C}\right)\right]$$

$$b = C\left[\cosh\left(\frac{x+k}{C}\right) - \cosh\left(\frac{x}{C}\right)\right]$$

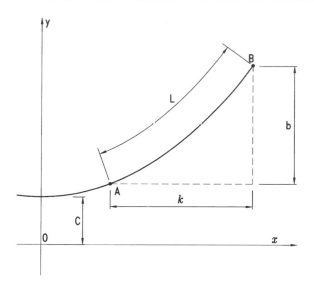

Figure 11.8.

By squaring and subtracting these equations, the following result can be obtained:

$$\sqrt{\left(L^2 - b^2\right)} = 2C \sinh\left(\frac{k}{2C}\right) \tag{11.8}$$

In this equation, the unknown C is not given explicitly in terms of the known dimensions (L, b, and k), and the solution must be obtained by an iterative procedure.

The catenary cable solution is principally required for the determination of the 'free cable' profile. This is the profile to which the free hanging cable must be constructed so that, after the inclusion of the cable deformation and extension resulting from the subsequent application of the deck dead and superimposed dead loading, the specified final bridge shape is obtained.

Cables with uniform applied loading

If the loading is uniformly distributed between the cable support points, rather than along the length of the cable, the resultant shape of the cable (Figure 11.9) can be obtained as follows:

$$T \sin \psi = w.x \tag{11.9}$$

$$T \cos \psi = H \tag{11.10}$$

Dividing (11.9) by (11.10) gives:

$$\tan \psi = \frac{wx}{H} = \frac{dy}{dx}$$

This equation can then be integrated to give the shape of the cable as:

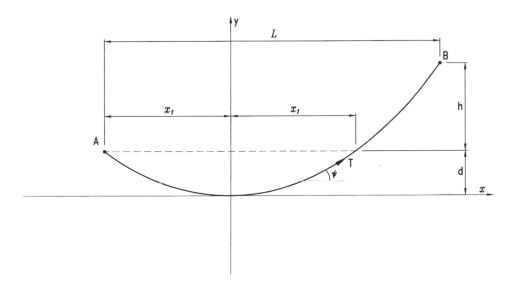

Figure 11.9.

$$y = \frac{1}{2}\left(\frac{wx^2}{H}\right) + A$$

If the origin is now taken at the lowest point of the cable; when $x = 0$, $y = 0$ the constant A will be zero, and the equation giving the cable shape becomes:

$$y = \frac{1}{2}\left(\frac{wx^2}{H}\right) \qquad (11.11)$$

A cable with a uniform load over its complete span therefore has a parabolic profile, with the tension in the cable being given by:

$$T = H\frac{ds}{dx} = H\sqrt{1 + \left(\frac{dy}{dx}\right)^2}$$

and since:

$$\frac{dy}{dx} = \left(\frac{w}{H}\right)x$$

the tension in the cable at any point is equal to:

$$T = H\sqrt{1 + \left(\frac{wx}{H}\right)^2} \qquad (11.12)$$

The value of H in terms of the geometry of the cable can be obtained from equation

(11) by inserting the values $y = d$ at $x = x_1$, and of $y = (d + h)$ at $x = (L - x_1)$, and solving the resulting quadratic equation for x_1. This gives:

$$x_1 = \left(\frac{L}{h}\right)\left(\sqrt{d(d + h)} - d\right) \qquad (11.13)$$

and:

$$H = \left(\frac{w(x_1)^2}{2d}\right)$$

For the simple case where the support points are at the same level, these expressions simplify to:

$$x_1 = \frac{L}{2}$$

and:

$$H = \frac{wL^2}{8d}$$

A uniform distribution of dead load along the span is a good approximation to the actual distribution for most suspension bridge decks. The resultant dead load suspension cable profile will therefore be somewhere between a parabola and catenary. Provided the deck weight per unit span length is significantly greater than that of the cables, which will be the case for most bridges, the resultant cable profile will be close to a parabola. In any case for the span/sag proportions normally adopted, the difference between a catenary and parabolic profile is quite small. As the parabola is analytically much simpler than the catenary, the assumption of a parabolic cable profile is therefore usual for preliminary design calculations.

Cables with an arbitrary loading distribution

The general case in which a cable is loaded by an arbitrary distribution of distributed and concentrated loads is not amenable to direct analytical solution. The cable profile and internal loads for such a case can however be obtained as follows. The cable length is divided into a series of short straight segments between points where concentrated loads are applied, and any distributed loads (such as self-weight) allocated to these points. By choosing a sufficiently small segment length the real cable geometry can be accurately represented.

The cable can then be made statically determinate by assuming a value of the horizontal tension component (H), and the resultant profile, including the effects of segment extension, can be calculated, working from one support. In general, the assumed value of H will not produce a cable profile terminating at the other support. From the magnitude and direction of the lack of fit, an improved value of H can be determined, and by a process of iteration the actual cable profile determined to any desired degree of accuracy.

Flexible cables support applied loads by adjusting their geometry to accord with the nature of the applied loading, with the elastic properties of the cable generally being of less significance. As a result their structural behaviour differs significantly from that of beam type structures. In the absence of other loads, the geometry of the cable will correspond to a funicular curve of its self-weight loading. The addition of applied loading will cause the cable to deflect, with this deflection being partly due to elastic extension of the cable and partly due to change in geometry. The relative importance of these effects depends on the uniformity of the applied loading.

The application of a uniformly distributed load over the whole span will result in an increase in H, and hence the cable tension (T) at all points. At each point, the change in T will be proportional to the applied load, and no geometrical change will be necessary to maintain equilibrium. However, the tension increase will result in a lengthening of the cable, so that all points will move downwards, with the maximum deflection being at midspan but with no change to the overall shape of the cable. The deflection of the cable is in this case entirely due to its elastic extension.

The effect of a concentrated or part span vertical load is rather different. In this case, because H must be constant, the tension is increased over the whole length of the cable, and to preserve vertical equilibrium, the cable profile flattens in the parts of the cable which do not have additional loading. As a result, the cable profile has to change, with those parts remote from the additional loading deflecting upwards as shown in Figure 11.10. Additional deflection will occur due to the cable extension, but this will be small relative to that due to the geometrical shape adjustment. In the limiting case of a concentrated or short length of distributed load, the maximum deflection will be almost entirely due to profile change rather than cable extension.

The stiffness of a cable against displacement by non-uniform loading is therefore related to its initial tension, as the higher the initial tension, the smaller the shape adjustment required to maintain equilibrium at each point. The resistance of a heavily loaded cable to displacement by additional loading is generally referred to as its gravity stiffness.

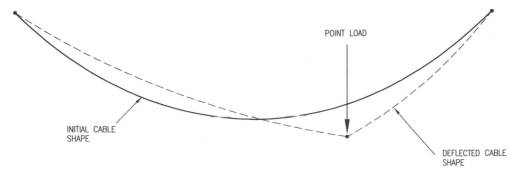

Figure 11.10.

A further important effect on the stiffness of the cable in the main span of a bridge is the extent to which its end supports at the tower tops move under applied loading. For most bridges, the longitudinal bending stiffness of the towers will be relatively small, and the movement of the tower top will be almost entirely controlled by the stiffness of the cable system. An increase in the loading of the main span cable will cause its horizontal tension component to increase, causing an imbalance at the tower, and the tower top will therefore move towards the main span. This movement will continue until a reduction in the side span cable sag, and the corresponding increase in its tension restores horizontal equilibrium. The inward movement of the tower top will of course produce additional deflection of the main span cable.

A short side span cable with a small sag will provide a more effective restraint to the tower top than a relatively more flexible long side span.

Cable design

The cables, towers, and anchorages are the primary load carrying system of the bridge, with the cable profile in each span taking the form of an equilibrium polygon for the loading of that span. At the design temperature, and with no applied live loading, the sags in each span are such that the horizontal components of their cable tensions are equal, and only vertical loading is applied to the towers. An increase in applied loading in one span will result in an increase in its cable tension, and an unbalanced horizontal force condition will therefore occur at the tower top, unless there is a corresponding change in the adjacent spans. Since friction prevents the cables from making the necessary adjustment by sliding through the tower saddles, the saddles must themselves move horizontally in a span-wise direction until the sags in each span have increased or decreased so that horizontal equilibrium is restored. In a similar manner, cable length change due to temperature variation, are also compensated by corresponding movements of the tower saddles.

With the saddles fixed to the tower tops, any movement of the saddles forces a corresponding movement of its tower top, causing a horizontal reaction generated by the tower resistance to longitudinal bending. This horizontal resistance slightly reduces the required saddle movement, as the main and side span horizontal components of cable tension are no longer exactly equalized. However, since the towers are relatively flexible, the effect of tower bending stiffness on saddle movement is small.

Ignoring the relatively small changes at the towers due to the effect of their longitudinal stiffness, the horizontal component of cable tension (H) is constant from one splay saddle to another across the whole length of the bridge for any vertical loading condition. The cable tension at any point is therefore proportional to the secant of the cable inclination to the horizontal at that point, and as the maximum cable angle is at the towers, the maximum cable tension occurs at this point.

Most suspension bridges have a constant cable cross-section throughout their length. However, for bridges with relatively short side spans, the cable angle on the side span side of the tower will be significantly larger than that on the main span side.

As noted above, the cable tension is directly proportional to this angle. For these bridges, it will often be advantageous to size the cable for the tension on the main span side, and to then provide additional strands in the side spans, running from anchor points on the tower saddles down to the anchorages.

During cable construction, individual wires and/or strands are adjusted so that in the completed cable, the total tensile load is distributed evenly between the wires. However, once the cable bands have been erected, these together with the saddles, form points of restraint at which additional secondary stresses are induced due to the erection of the deck and the subsequent application of traffic and environmental loads. These secondary stresses take the form of both bending stress in the wires and a redistribution of stress across the cable cross-section. Wyatt (1963) has derived a method by which these effects can be evaluated.

A further source of inequality in cable tension arises from the spreading out of the individual strands behind the splay saddles to their respective anchor points, so that they form a cone with its vertex just beyond the splay saddle. A direct tensile force applied by the side span cable does not therefore produce equal stresses in the strands behind the splay saddle, with the variation in stress being proportional to the square of the cosine of the strand angle from the centre line of the strand cone. This effect can be minimized by keeping the strand deviation angles to relatively small values.

When cable design has been based on allowable stress methods, the generally adopted allowable tensile stress has been 40% of wire breaking stress, and long experience has confirmed this to be satisfactory. Where limit state design is used, material factors (γ_{fL}) of 2.3 and 1.67 for the serviceability and ultimate limit states, respectively, have been widely adopted.

11.3 Structural form

11.3.1 General

In deciding the overall configuration of the bridge, the designer must make a number of decisions, the most important of these being:

- the length of the main span
- the main span cable sag-to-span ratio
- the ratio of main-to-side span length.

Associated with these decisions, the designer must determine:

- whether or not to support the sidespans from the cable
- the form of the deck structure
- the required clearance below the deck structure.

The determination of the optimum values of the above parameters for a particular bridge site is rather complex, with the only practicable method being the evaluation of the total cost of a series of designs, varying each of them within a range of values appropriate to the proposed bridge site.

11.3.2 Span length

The first decision to be made is the length of the main span. As the most suitable structural form for long spans, most suspension bridges will be used to provide a crossing of a navigable waterway. In these cases, the minimum main span length will be determined by the clear width required to ensure safe navigation of vessels using the waterway. This must include consideration of the most economical way of protecting the main tower foundations against ship impact, either by lengthening the span to position them in sufficiently shallow water depth to prevent the approach of large vessels, or the provision of protective works, such as a ship protection island. However, for really long spans the tower foundations may be sufficiently massive to resist accidental ship impacts without additional protective works being required. Additional span length may also be appropriate to locate the main tower foundations in areas with good foundation conditions, or on land to enable their rapid and economical construction.

The length of the side spans will generally be chosen to position the cable anchorages where they can be economically constructed. However, the restraint provided by the side span cables to the tower tops is an important determining factor in the vertical stiffness of the bridge main span. This therefore places an upper limit on side span lengths of around one-half of the main span, and as noted in earlier, a somewhat lower value will probably be found to be preferable. The minimum length of the side spans is determined by the resultant increase in cable slope adjacent to the towers which this cause. This results in increased cable tension and a larger overall cable, unless additional strands are provided in the side span only.

The side span length should preferably not exceed around 40% of the main span in order to provide an effective restraint to the tower top (as noted above). However, the side spans should not be less than 25% to 30% of the main span length to avoid an excessively high imbalance of cable tension at the towers. It should be noted that there is no need to have equal side span lengths.

11.3.3 Cable geometry

Associated with the choice of main span length, is the determination of the sag of this span. Almost all previously constructed suspension bridges have had main span sags within a range of from one-eighth to one-twelfth of the span length, and an appropriate value will generally lie between these limits. It is important to note that decrease in the main span (sag/span) ratio has the following effects:

- the tensions in the main cable will be increased, resulting in an increase in the quantity of cable wire, and hence its cost, and the time required for construction
- for a given navigation clearance, the required height and hence construction cost and time of the towers will be decreased
- the increased cable size and tensions will require enlargement of the anchorages with consequent increases in their cost and time for construction
- the global vertical stiffness of the bridge will be increased, giving reduced bearing and expansion joint movements, and hence reduced construction and maintenance costs.

As the length of the main span is increased, the self-weight of the cable becomes a more and more significant proportion of the total load to be carried. It would therefore be expected that there would be a trend towards the use of an increased sag ratio for larger spans. A study of the actual span/sag ratios of a representative sample of modern bridges confirms this general trend, although the Humber bridge is a notable exception.

Bridge	Main span length	Span/sag ratio
Akashi-Kaikyo	1990 m	9.9 to 1
Storebælt	1624 m	9.0 to 1
Humber	1410 m	12.2 to 1
Jiangyin (China)	1385 m	10.5 to 1
Tsing Ma (Hong Kong)	1377 m	11.0 to 1
Verrazano Narrows	1298 m	11.1 to 1
Second Bosporus	1090 m	12.0 to 1

Both transverse wind and non-symmetric traffic loading cause a tendency for large relative movements to occur between the deck and the cables in the centre of main span. The short relatively stiff hangers connecting the cable to the deck in this area resist these movements, resulting in fluctuating loads in these elements, which can cause fatigue damage. One way of dealing with this problem is to replace the hangers at the centre of the span by a completely rigid connection, a typical layout being as shown in Figure 11.11. Such a connection also has a beneficial effect on aerodynamic stability as the rigid connection to the deck resists asymmetric longitudinal movement of the cables, and therefore increases the torsional stiffness of the structural system.

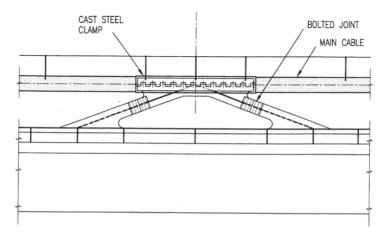

Figure 11.11.

11.4 Suspended deck structure

11.4.1 Choice of deck structure (box girder, plate girder or truss)

As noted in the introduction to this chapter, the design of the suspended deck structure is critical to the production of an economical overall bridge design. The most important design requirements for the deck structure are:

- low weight
- good aerodynamic characteristics
- high torsional stiffness.

A low weight is essential, as the whole of the deck weight is carried by the cable structural system (cables, towers, and anchorages). Any saving in deck weight will therefore produce further savings in these elements, with the benefit becoming more and more significant as the span length increases.

The long span and the relatively flexible structure of suspension bridges means that wind loading and aerodynamic stability are critical design issues. A good deck design must therefore have low wind resistance to transverse wind, and more importantly not be susceptible to the types of aerodynamically induced instability described in Section 11.9.

Because it is continuously supported by the cable system, the deck structure need have only very low longitudinal bending stiffness relative to its span length. It will therefore make an insignificant contribution to the overall bending stiffness of the structure, which will be dominated by that of the cable system. This is not however the case in torsion where the torsional stiffness of the deck can make a significant contribution to the overall structure stiffness, with consequent increase in the torsional natural frequency and improvement in aerodynamic properties. A high torsional stiffness is therefore an important requirement of the deck structure.

The requirement for a low deck weight means that a steel orthotropic deck will almost certainly be the preferred choice for the traffic platform, as the lower cost of a reinforced concrete slab will be more than counteracted by the additional costs of the supporting cable system.

To transfer the deck loads back to the hangers, and provide the required longitudinal bending, transverse, and torsional stiffness, the deck must be supported by a stiffening girder structure. Because of the importance of minimizing weight, this will invariably be a steel structure. The possible options for the stiffening girder structure (see Figure 11.12) are:

- open trusses
- plate girders
- an integrated box girder deck structure.

For many years, open truss stiffening girders were the preferred solution for providing the required flexural and torsional stiffness to suspension bridge decks. The stiffening girder structure then comprises two vertical trusses positioned at or near the deck edges, and connected by upper and lower plan bracing systems. The most efficient structural arrangement is obtained by integrating the bridge deck with the

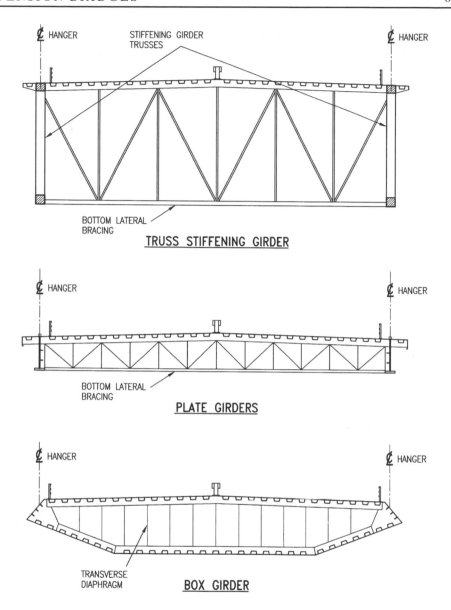

Figure 11.12.

truss top chord members so that it provides the upper bracing function by acting as a horizontal web plate between them. Transverse sway bracings between the trusses must be provided at regular intervals between the longitudinal trusses to maintain the cross-section shape. Although box girder decks are now generally the preferred solution, truss stiffened deck structures continue to be a competitive structural arrangement in two situations:

- where a bridge is required to carry traffic on two levels
- where a bridge is required to carry rail traffic.

A number of suspension bridges have been constructed with their decks stiffened by twin plate girders, but because of the poor aerodynamic characteristics of the resultant deck arrangement, this solution is suitable only for shorter span lengths where nowadays a cable stayed structure would be a more economic solution.

The streamlined box girder arrangement was first used for the Severn suspension bridge, completed in 1965. The streamlined box girder has a low self-weight due to the complete integration of the deck and stiffening girder structure, and an adequate torsional stiffness can be achieved by choosing a sufficiently large box cross-sectional area. Because of its relatively low exposed area and smooth external surfaces, maintenance of these structures is much simpler and of lower cost than for a comparable truss stiffened deck. A further advantage is that the reduced wind loads due to the low drag coefficient of the deck can produce significant savings in the main tower structure.

Exceptions to this are the Tsing Ma bridge (a requirement for two deck levels to include a railway), and the Akashi Kaikyo bridge, the longest span (1990 m) so far constructed. A truss-stiffening girder structure was chosen for the latter bridge because a span of around 1750 m appears to be the probable practicable limit for single box girder deck structures, at least for those locations where aerodynamic stability up to very high wind speeds (around 80 m/s) is required.

A streamlined box deck structure will generally consist of an orthotropic deck of conventional arrangement with longitudinal trough stiffeners supported at around 4 m centres by transverse bulkheads. In most suspension bridge decks of this type, the bulkheads have been have been stiffened plate diaphragms, but on the recently constructed Storebælt East bridge, transverse truss structures were found to be a more economical solution. The webs and bottom flanges of the box are formed from conventional stiffened plating.

Suspension bridge deck structures are supported at closely spaced regular intervals by the hangers, so that the dead load bending moments in the stiffening girder are generally insignificant. The actual hanger spacing will be chosen to suit the length of the modules into which the deck is divided for manufacture and erection, and in the case of truss stiffening girders the spacing of the truss nodal points.

The stiffening girder is required only to provide sufficient bending stiffness to keep local deformations under traffic loading within acceptable limits, and can therefore have a very high span-to-depth ratio. As the deflected shape of the stiffening girder is almost entirely controlled by the cable vertical stiffness, the lowest possible depth is desirable to minimize bending moments due to these imposed shape changes. The minimum depth will therefore generally be determined by the need to provide sufficient torsional stiffness to avoid aerodynamic instability. Truss stiffened decks typically have span-to-depth ratios in the range of 75–175, but with a box girder deck much higher ratios are possible, generally in the range 300–400.

The maximum bending moments and shears in the deck structure will arise from traffic and wind loads, combined with the effects of temperature changes. Maximum effects from traffic will arise from loading of a relatively short proportion of the span. The length of the influence line is a function of the intensity of loading due to the non-linearity of the structure response and it will be necessary to consider loading of a range of lengths. The situation is further complicated when the intensity of loading varies with loaded length, as in British traffic loading (BS5400 Part 2). Due to the non-linearity of the structure behaviour, it is also generally not possible to use superposition to derive the effects from various loading cases.

11.4.2 Structure articulation and bearings

Most suspension bridges have had separate deck structures in the main and side spans, with the ends of the main and side span stiffening girders being supported by bearings fixed to the tower, and with the main expansion joints for the bridge located there. However, experience has shown that the maintenance of bearings and movement joints constitutes a substantial part of the total maintenance costs, and the most recently designed bridges have their deck structures continuous between the anchorages, and supported vertically only by the main cables. Deck continuity also provides some improvement to the vertical and horizontal stiffness of the structure. A disadvantage is that the absence of rigid vertical supports at the towers removes the beneficial effect of torsional fixity at the towers on aerodynamic stability, and torsional deflections of the deck will also be increased.

Because of its continuity through the tower, the deck will experience increased bending moments in this area as it conforms to the angular rotation of the main cable at the saddles, and there will be increased deck longitudinal movements at the anchorages. These increased movements require the provision of a larger movement joint, and it may be necessary (as on the Storebælt East and Høga Kusten bridges) to provide a hydraulic damper or lock-up device. These resist loads due to traffic braking or wind gust loads, but permit slow longitudinal movements due to temperature changes.

At the ends of the stiffening girder, the supporting bearings have to accommodate large longitudinal movements and will also generally have to resist uplift forces in some loading conditions. Rocker bearings with pinned connections at each end have therefore been a natural choice for these bearings. If conventional free sliding spherical bearings are used, then either:

• the load distribution between the cable and the stiffening girder must be adjusted to ensure that the bearing reaction is positive for all load conditions; or
• an alternative means of resisting uplift forces must be incorporated.

11.5 Cable supports and attachments
11.5.1 Cable support at towers (tower saddles)

At the towers, the main cable must be supported by a saddle structure to transmit and distribute its vertical load into the supporting tower structure. The type of cable, in conjunction with the structural form of the tower determines the design of this saddle.

For cables of locked coil or spiral strands, the bottom of the saddle trough will consist of a series of stepped circular grooves so as to provide full bearing for the bottom layer of strands. The gaps between the strand layers are filled with shaped zinc

Figure 11.13.

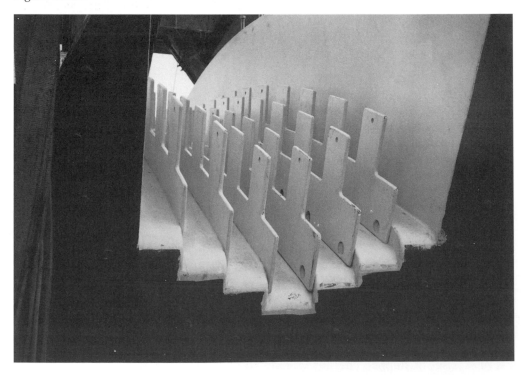

Figure 11.14.

or aluminium sections, a typical saddle for a cable composed of locked coil strands being as shown in Figure 11.13.

For parallel wire cables, the hexagonal arrangement of cable strands is placed in a curved trough structure, supported by a grid of longitudinal stiffeners and radial ribs which spread the cable loading into the supporting tower structure. A typical saddle arrangement for a parallel wire cable is shown in Figure 11.14.

The proportions of the trough grooves should be chosen to produce a cable shape in the saddle close to that of the compacted cable either side of the tower. This can be achieved by an approximately square arrangement of wires in each strand, with equal numbers of wire rows and columns. The groove width should then be determined to suit this arrangement, assuming alternating rows of N and $(N-1)$ wires as shown in Figure 11.15. The number of wires in each strand will however generally require that the top row of wires will be incomplete.

The height of the saddle side walls is determined by the efficiency of packing of the individual wires in the saddle. Although wires are packed as closely as practicable during placing in the saddle groove, edge effects and minor inaccuracies in wire packing cause the achievable voids ratio to be somewhat higher than the theoretical minimum value. A figure of between 14 and 16% is therefore a realistic value for a spun cable, with a rather higher value of between 18% and 20% for cables made up from PPWS.

To facilitate spinning, permanent steel separator plates are required between each column of strands, a typical arrangement being as shown in Figure 11.14. These support the strands laterally during spinning, and separate them for ease of strand movement through the saddle during the subsequent precise sag adjustment.

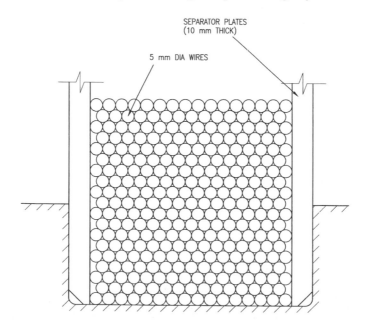

Figure 11.15.

Because the wires of parallel cables are laid individually during spinning, or in relatively small numbers for cable formed from PPWS, the cable as a whole does not experience significant bending during construction. The saddle radius of curvature can therefore be relatively small relative to the overall cable diameter, and is governed by the compressive forces between wires induced by their bending over the saddle. These compressive forces are determined by the cable tensile stress, the radius of the saddle, and the cable overall depth in the saddle. The radial compressive stresses are a maximum in the bottom wires of the cable, and their effect is to cause a reduction in the effective yield strength of the wire. Experience has shown that a cable centre-line radius of curvature of the order of nine to ten times the main cable diameter will generally be adequate.

Typical examples of tower saddle radii for some modern bridges are:

	Cable diameter (mm)	Saddle radius (mm)	Ratio
Second Bosporus Bridge	763	7400	9.7
Humber Bridge	684	6000	8.8
Tsing Ma Bridge	1095	12500	11.4
Storebælt East Bridge	827	7000	8.5

The maximum cable slopes at the ends of the saddle determine the required trough length. It is usual practice to dimension the saddle to the tangent points in the full dead load condition, and to provide sufficient saddle length beyond these points to accommodate the maximum cable rotations due to the most adverse combination of live load, or during erection of the deck. A short length of sharper radius curvature is normally provided at each end of the saddle trough to eliminate the possibility of a kink in the wires.

The cable produces a radial loading on the saddle trough, which is resisted by vertical reaction from the tower structure. Although the total forces are equal and opposite, the longitudinal distributions are different, and these differences in vertical forces produce longitudinal bending stresses in the saddle.

On many early suspension bridges, the tower saddles were made as a single large steel casting. It is now however more economical to manufacture the saddle from a combination of a cast steel trough section, supported by a grillage of welded steel plates.

For construction, the saddle must be offset from its final position towards the bridge side span to enable the cable to be built at its free cable geometry. This offset can be achieved in two ways.

1. Saddle permanently fixed to tower top. In this method, the saddle is fixed in its final position on the tower top. The required free cable offset of the saddle is obtained by deflecting the tower top towards the side span with some form of a pull back system. This method is most appropriate for steel towers, but has also been used for tall, relatively flexible concrete towers (e.g. the Storebælt East bridge).

2. Saddle initially offset to free cable position. In this method the tower is maintained vertical for cable construction, with the free cable position of the saddle being obtained by offsetting the saddle on the tower top by the required amount. A bearing is required between the saddle and the tower to enable the saddle to be moved into its permanent position on the tower top during construction of the bridge deck. This method is most appropriate for relatively stiff concrete towers. In some cases, a combination of saddle offset and pull back of the tower may be appropriate, this being the selected option for the Humber bridge concrete towers.

As noted earlier in the chapter, the maximum cable tension will generally occur on the side span side of the saddle, with the difference in cable tension being particularly significant for bridges with relatively short steep side spans. In this case it is advantageous to use additional strands in the side span cables, with these being terminated at and anchored to the tower saddles. A typical arrangement of saddle anchorages for the additional strands of a spun cable is shown in Figure 11.16.

The principal problem in the structural design of saddles is the evaluation of the lateral pressures developed by the cable wires, and the design of the saddles to resist these. These lateral pressures arise from the diagonal contact between the wire layers and are therefore related to the efficiency of wire packing achieved during construction. For perfect (60°) hexagonal packing of the wires, and ignoring friction between the wires, the ratio of horizontal pressure on the saddle walls to the

Figure 11.16.

vertical pressure at that level can be shown to be about 0.35. However, this quality of packing can not be generally achieved and actual pressures will be somewhat higher than this, although these are reduced to some extent by inter-wire friction effects.

11.5.2 Cable support and deviation at anchorage (splay saddles)

At the anchorage the cable has to be separated into its individual strands, so that they can be deviated laterally and vertically to their connection points on the anchorage structure. In order that they can be placed layer by layer during cable construction, the deviation of all strands should be downwards relative to the cable centreline at the bottom of the side span, resulting in an overall downwards deflection of the cable centre line. The structure providing these strand deviations is referred to as a splay saddle.

The principal component of the splay saddle is the trough containing the cable strands. To achieve the required vertical and lateral strand deviations, the saddle trough has to be curved in both elevation and plan as shown in Figure 11.17. The horizontal and vertical curvature of the saddle trough grooves requires careful coordination to ensure that the cable strands are correctly deviated without substantial changes in shape. The principal consideration is that horizontal curvature of the strands must only be carried out where they are in contact with the saddle, and there is a vertical force to maintain the wires in position. If strands are deflected horizontally only, wires drift outwards and pile up on the outside of the horizontal curve, resulting in a poor strand shape at the exit of the saddle.

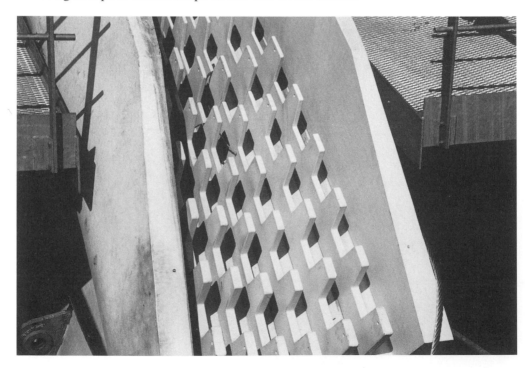

Figure 11.17.

As for the tower saddle, the splay saddle consists of a trough with a stepped transverse profile to suit the hexagonal arrangement of the cable strands, and curved vertically and horizontally to produce the required strand deviation angles.

To produce a resultant force directed through the centre of the saddle, the cable bearing pressure on the trough must be kept constant. This can be achieved by varying the trough longitudinal curvature so that the cable tension divided by the bend radius is kept constant as the strands peel off towards the rear of the saddle. A typical geometry is shown in Figure 11.18.

The splay saddle has to be able to move longitudinally, both during construction and also in service. During construction, the increase in cable tension from the cables only constructed condition to full dead load on the bridge will cause the length of the strands within the anchor chamber to extend. The saddle must therefore be ini-

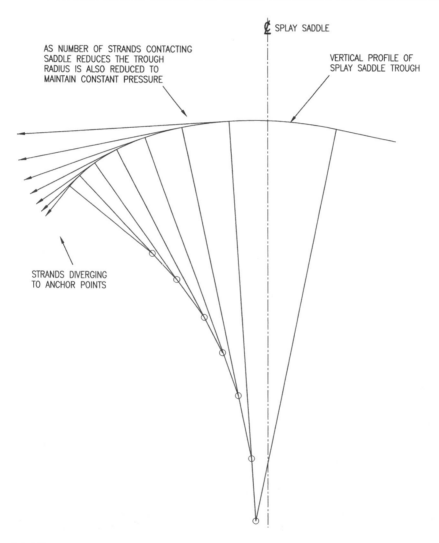

Figure 11.18.

tially positioned offset towards the strand anchor points, so that when the full dead load condition is reached, the strand extension causes the saddle to reach its final required position. In service, the saddle must be able to move to accommodate length changes in the strands within the anchor chamber due to fluctuations in live loading and strand temperature variations.

These movements can be provided for, by mounting the splay saddle on a sliding or rolling bearing. However, as the longitudinal movements required are rather small, a better solution is to design the saddle as shown in Figure 11.19, with the longitudinal movements being generated by the saddle articulating on a rocker bearing at its base.

The splay saddles are at the lowest point of the side span cables. As a result, any water leaking into the cable tends to run down to these low points, exiting at the rear of the saddle. Additionally, at the rear of the saddle, the divergence of the strands

Figure 11.19.

produces small horizontal and vertical gaps between them, which are extremely difficult to protect by painting. The splay saddle area of the cable is therefore particularly susceptible to corrosion. To provide the best possible protection against this:

- the vertical spaces between the strands at the rear of the saddle should be filled with solid steel spacer plates
- the splay saddle structure should be completely enclosed within the anchor chamber, which should be provided with a dehumidification system to maintain humidity levels below the threshold value for initiation of corrosion
- the side span cable entry point must be sealed to prevent leakage of water into the anchor chamber. This seal must be capable of accommodating any cable axial and rotational movements.

11.5.3 Hangers and cable bands

Hangers

The hangers connect the bridge deck and stiffening girder to the main cables. These hangers should be vertical and equally spaced along the span. The spacing of the hangers should be selected to:

- produce a hanger size which can be economically manufactured and erected
- conform to the length of the deck girder erection module (or a multiple of this).

Two basic types of hanger are possible, using either one or two parts of rope or strand.

Two part hangers

This type of hanger was used in most early long span suspension bridges. In this arrangement, the hanger is draped over the main cable, located in a groove in the associated cable band. The two bottom ends of the hanger are normally connected to the deck by simple sockets attached to the stiffening girder or deck structure, a typical arrangement being as shown in Figure 11.20. Hangers for this type of band generally need to be of steel wire rope, which has the necessary flexibility to conform to the relatively small bend radius around the cable band. However, locked coil ropes designed for enhanced flexibility have been used (e.g. replacement of the original hangers of the Rheinbrücke Köln Rödenkirchen).

Single part hangers

Single part hangers terminate on the underside of the cable, and are attached to the lower part of the associated cable band by a socket and pin connection. The connection of the bottom end of the hanger to the stiffening girder or deck structure can either be a simple bearing socket, or a socket and pin joint. Since it does not need to be sufficiently flexible to bend around the cable, this type of hanger can be of spiral strand or parallel wires, taking advantage of their superior strength and stiffness properties.

Figure 11.20.

Single part hangers have been used in an inclined configuration on a number of bridges (Severn, First Bosporus, Humber), with the objective of enhancing structural damping to improve aerodynamic performance. However, with this configuration, the hangers experience a significantly greater range of stress fluctuation due to live loading, and a conventional vertical hanger arrangement is now generally recognized to be preferable.

A particular problem in the design of hangers is the accommodation of relative horizontal movements between the top and bottom connections of the hanger, due to differential deflection of the cable and deck under wind or traffic wind loads, or from temperature effects. These movements cause bending of the hanger, which in the absence of preventative measures, is concentrated at the sudden increase of stiffness where the hanger enters its socket. The resultant high bending stresses at this point can produce premature fatigue failure of the hanger. The problem is usually

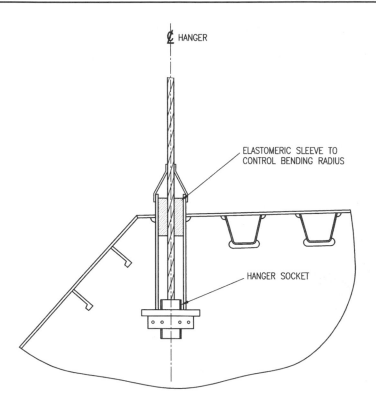

Figure 11.21.

most severe near the midspan areas of long spans, where significant differential deflections of deck and cable due to transverse wind can occur, and at the bottom end of the hanger where the environment is conducive to corrosion damage. For hangers connected to the stiffening girder by direct bearing of their bottom socket, a means of controlling the hanger bending curvature and moving this away from the socket entry point should be provided, a possible detail being as shown in Figure 11.21.

Spherical bearings can be used to connect hangers with a bottom pin connection, so that rotation in both horizontal axes is possible. The benefit of such a bearing can not however be fully realized as the moment to overcome bearing friction and initiate rotation will generally be sufficiently large to cause significant stress fluctuation in the hanger wires. However, the articulation provided does eliminate moments in the hanger due to geometrical errors causing misalignment of the cable and deck attachment points.

Finally, experience on previous bridges has shown that even well-designed and well-maintained hangers are likely to deteriorate sufficiently in service to the state where replacement will be required at some stage during the life of the bridge. The design of the hanger/cable–band/deck system should recognize this possibility, and be designed to permit the temporary removal of any hanger for replacement, without the need for traffic restrictions other than the closure of the immediately adjacent lane.

Cable bands

The cable bands connecting the hangers to the main cable each consist of two semi-circular half bands, clamped onto the cable by prestressed bolts to generate sufficient frictional force to resist the down-slope component of the hanger tension. Each half band is normally provided with interlocking and overlapping castellations to ensure accurate longitudinal alignment of the two parts, and to prevent wires being trapped during tightening of the bolts. The bore diameter of the band must be such as to allow the bands to be fitted on to the compacted cable, with at least a minimum engagement of the castellations.

For two part hangers, the cable bands are divided on their vertical centre line as shown in Figure 11.23. Cable bands for single part hangers are divided horizontally as shown in Figure 11.22, with the lower half being shaped into a gusset plate to enable the hanger to be attached by a pin connection to an open socket.

During tensioning of the clamping bolts, both types of cable band must deform slightly so that they contact and conform to the shape of the cable. A low wall thickness and ductile material properties are required to give this desired flexibility. The complex shape required, with the requirement for ductility, dictate that cast steel is the only practicable material for cable band manufacture.

Cable bands apply a concentrated vertical force to the cable, which produces a small angular change in the cable, with the longitudinal bending stiffness of the band causing this change to occur mainly at the ends. To distribute this angular change

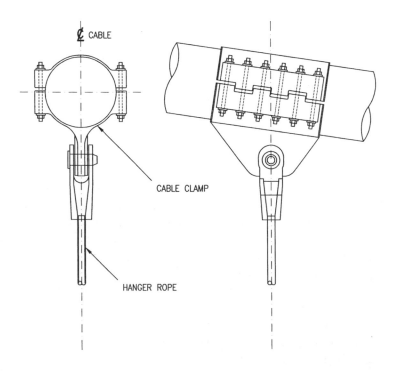

Figure 11.22.

Figure 11.23.

equally, the hanger axis should preferably be directed as near as practicable through the centre-line of the band.

The progressive increase in cable slope as the towers are approached produces a corresponding increase in the down-slope component of hanger tension, which must be resisted by friction between the cable band and the cable wires. For economy of manufacture and convenience in tightening, the same bolt diameter should be used for all cable bands, with the number of bolts increased as necessary to provide the additional required frictional resistance.

To produce good frictional resistance, a high standard of surface finish on the bore of the cable band is undesirable. It is therefore customary to specify that the bore is slightly roughened by final machining with shallow closely spaced circumferential cuts which are just visible to the naked eye, with a spacing of six cuts per centimetre being the usually selected value. For corrosion protection of the bore, zinc metal spray is the best choice, as it has a slightly rough texture and is compatible with the wire galvanizing.

The coefficient of friction which can be developed between the cable band and the cable wires can be assumed to be around 0.2, although little full scale data is available. A series of test was however carried out during construction of the Storebælt East bridge, by jacking apart two temporarily installed cable bands. These tests (Gimseng, 1998) indicated friction coefficients of only around 0.17 at the start of sliding, but it is probable that these rather low values resulted from the effective tension in the clamping bolts being somewhat less than specified.

An important factor in determining the clamping bolt requirements is the extent to which the clamping bolt tensions will reduce during the service life of the bridge. This relaxation of bolt tension is very rapid after the first tightening of the bolts, as the individual wires bed down under the sustained radial pressure from the bolt tensions. This is compensated for by re-tensioning the bolts several times during construction, with a final tightening operation being made as late as possible before bridge completion. However, measurements on earlier bridges have shown that relaxation of bolt tensions continues after completion, although at a reduced rate, and this can result in a significant long-term reduction in clamping force.

On most projects, the construction programme requires that the bore diameter of the cable bands has to be fixed, and manufacturing largely completed, prior to completion of cable construction and compaction. The compacted cable diameter must therefore be accurately predicted ahead of spinning or cable strand erection. It will be determined partly by the cable construction method, and partly by the efficiency of compacting and the compacting machine capability, and must therefore be estimated on the basis of the experience of previously constructed cables.

The determining factor in this is the voids ratio which is achieved in the compacted cable. A typical value for this is between 18 and 22%, with the lower value being typically achieved on cables constructed with PPWS with good compaction, and the higher value being typical of a spun cable with relatively inefficient compaction.

For cable bands with a typical ratio of diameter to wall thickness of around 25–30, the assessment of the clamping force required to prevent sliding can be based on the assumption that the band behaves as a flexible strap tensioned around the cable. With N pairs of screwed rods each tensioned to a load of P, the total tension force applied to each quadrant of the band will be (PN).

As the screwed rods are tensioned, the band will be pulled into tight contact with the cable around the complete periphery. Frictional losses due to the small relative circumferential movement between the band and the cable wires which this causes will tend to reduce the tension in the band to a minimum value (at 90° wrap angle) of:

$$NPe^{\frac{-\pi\mu}{2}}$$

where μ is the coefficient of friction between the wires and the cable band.

The effective total radial contact pressure will therefore be:

$$\frac{4NP}{\mu}\left[1-e^{\frac{-\pi\mu}{2}}\right]$$

and the resistance against sliding will be:

$$4NP\left[1-e^{\frac{-\pi\mu}{2}}\right]$$

To allow for the relaxation of screwed rod tension in service referred to above, the resistance value should be calculated assuming that the effective long-term clamping force will only be (say) 70% of the nominal value. Even so, re-tightening of the cable band bolts after some years of service has proved necessary on many bridges.

11.5.4 Cable end connections

At each anchorage, the individual strands from which the cable is formed must be terminated, and their load transmitted into the bridge anchorage. The way in which this is done depends on the type of cable construction.

For parallel wire cables constructed by *in situ* spinning, the wires are looped over a semi-circular strand shoe as shown in Figure 11.24. The wires are progressively laid into grooves with 60° tapered sides on each side of the anchor rods which transfer the strand load back to a base plate prestressed to the concrete face of the anchorage structure. These anchor rods provide a means of adjusting the position of the strand shoe during construction to enable the strand to be adjusted to the exact sag required. A radius of about 400 mm has been generally used for the strand shoes in previous bridges with spun cables made up of 5 mm diameter wires.

Figure 11.24.

The bending of the cable wires around the strand shoe during spinning induces a bending stress in each wire of:

$$\frac{Ed}{2R}$$

where E = wire modulus of elasticity, d = wire diameter, and R = strand shoe radius

For a typical wire diameter of 5 mm and strand shoe radius of 400 mm, the wire bending stresses around the strand shoe will therefore be ±1281 N/mm², comparable to the 0.2% proof stress of the wire material. Nevertheless, strand shoes of these proportions have given satisfactory performance on previous bridges, and this apparently high level of stress does not appear to have compromised the overall safety of the structures. This is probably because some plasticity and relaxation reduces the peak stresses.

For parallel wire cables constructed from PPWS, locked coil, or spiral strands, the individual strands are fitted with end sockets, normally secured to the strand by zinc socketing. The sockets are attached to the anchorage structure by either screwed rods, or by end bearing onto anchor beams. In the latter case, adjustment for slight variations in strand length is made by the use of shim plates between the socket and the anchor face.

11.6 Towers

11.6.1 Introduction

The primary function of the towers is to provide supports to the ends of the main span cable at a sufficient height to provide the required cable sag above the level of the stiffening girder. In addition, the tower generally has to provide support to the stiffening girders of the main and side spans. The primary loading on the tower is therefore the vertical load applied by the main cables at the tower top, together with wind loads from the cables and the bridge deck, and any loading applied by the stiffening girder.

As noted earlier in the chapter, an increase in the load on one of the spans supported by the tower generates an out of balance in the horizontal component (H) of the cable tension at the tower. Except for lightly loaded short bridges, this out-of-balance force is too large to resist by using stiff towers with fixed cable saddles. The tower design must therefore be such as to allow the saddles to move horizontally so that a balanced H condition is restored.

In most early suspension bridges this was achieved by designing the towers as stiff vertical cantilevers, with the required cable support movement being obtained by mounting the saddles on rolling or sliding bearings on the tower top. An alternative approach, used on some early bridges, was to design the tower with a hinged base, with the cable saddles fixed to the tower, so that the towers were free to rock in the plane of the main cables, offering no resistance to cable movements. The most notable example of this type is the Florianopolis bridge in Brazil.

All modern long span suspension bridges however have relatively flexible fixed base towers with cable saddles fixed to the tower tops, so that movement of the saddles due to varying traffic and temperature changes produces longitudinal bending of the tower legs.

11.6.2 Tower structural behaviour

The design loading for the tower will be that which produces the most adverse combination of vertical load from the cable, combined with the coincident longitudinal displacement of the tower top. This will generally occur with either both of the adjacent spans loaded with traffic, or the longer of these spans loaded.

Due to the tower longitudinal bending stiffness, longitudinal displacement of the tower top will induce a horizontal reaction there, with associated bending moments in the tower legs. However, the most significant effect of the horizontal displacement is the movement of the tower axis from the vertical, so that both the cable vertical load and the tower self-weight induce bending ($P\Delta$) stresses in the legs, in addition to their direct axial compressive stresses.

With a relatively stiff tower, the extent of the horizontal displacement of the tower top is reduced, and the $P\Delta$ stresses minimized, but at the expense of an increased horizontal out of balance load from the cable saddle. A stiffer tower also has a beneficial effect on the overall vertical stiffness of the cable and deck structure.

A more slender and flexible tower reduces the longitudinal out of balance cable force, but deflections and hence the $P\Delta$ stresses are increased, and the overall structural stiffness reduced. Although it is difficult to generalize, it seems probable however that overall, the most economical design will be achieved by using a tower of the highest practicable slenderness.

The limit on this is governed by the need for:

- an adequate margin against overall buckling of the tower
- sufficient strength and stiffness for the tower to be safely erected as a free standing vertical cantilever subject to wind loading.

Overall buckling of the tower is determined by its end support conditions. The base of the tower is clearly fixed in both position and direction. For the tower top, it is necessary to consider whether the cable system has sufficient stiffness to prevent longitudinal movement of the saddle other that that dictated by the applied loading. It can easily be demonstrated (Pugsley, 1968) that this will be the case, and that for longitudinal buckling, the tower behaves as a propped cantilever with an effective length of twice its height.

To prevent buckling and effectively resist wind loading in the transverse direction, the individual tower legs must be connected together. A diagonal cross-bracing system is the most efficient way of achieving this, although the bracing system must be interrupted at deck level to provide clearance for traffic, with lateral forces being transferred locally by bending and shear forces in the tower legs. However, for bridges with streamlined box girder decks, the resultant substantial reduction in wind loading enables the diagonal bracing system to be substituted by one or more cross-beams, so that the tower acts as a vertical vierendeel cantilever. Horizontal cross-members improve the appearance of towers, and in addition have lower future maintenance requirements than towers with diagonal bracing.

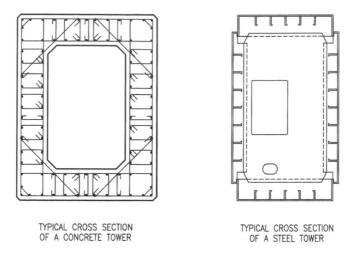

TYPICAL CROSS SECTION
OF A CONCRETE TOWER

TYPICAL CROSS SECTION
OF A STEEL TOWER

Figure 11.25.

11.6.3 Cross-section design
The predominant force carried by the tower in the as-built structure is the axial load from the cables, combined with some bending due to the eccentricity of its application. The tower cross-section must therefore be arranged to produce the most effective column section, with material placed at the maximum practical distance from the centroid. A natural choice for a steel tower is therefore a rectangular cross-section, made up of four stiffened plates. A similar hollow rectangular cross-section is also appropriate for a concrete tower. Typical cross-sections of steel and concrete towers are shown in Figure 11.25.

11.6.4 Tower material choice
Several factors influence the choice of material for the towers. The first of these is the relative cost of fabricated steel against concrete materials, including the cost of delivery to site. As the tower is predominantly loaded in compression, concrete is an obvious first choice material. However, the considerably increased self-weight over that of an equivalent steel structure may result in an uneconomic foundation requirement if ground conditions are relatively poor. This is particularly so for bridges in areas subject to moderate or severe earthquake conditions.

11.7 Anchorages
11.7.1 General
The anchorages resist the force from the main cables and transfer it to the ground. The precise direction of these forces is dictated by the side span cable geometry, but they will generally be predominantly horizontal with a smaller upwardly directed component. The requirement for the anchorages to resist a large, predominantly horizontal force makes their design difficult unless reasonably good ground conditions, and preferably sound rock, exist where the cable anchorages must be positioned.

As soon as each main cable enters the anchorage, it is divided into its constituent strands, and deflected downwards by the splay saddle as described in Section 11.5.2, with the individual strands being terminated and attached to the anchorage as described in Section 11.5.4. The termination points of the strands must be arranged in a closely spaced hexagonal or rectangular array so that the cable can be constructed by working upwards, progressively spinning or erecting strands. The strand terminations should be positioned sufficiently far back from the splay saddle so that the angular deviation of the outer strands from the cable centre line does not exceed about 10°.

From their anchor points, the strand forces have to be transmitted and distributed into the anchorage block. In early suspension bridges this was achieved by transmitting the cable force through a chain of embedded eye-bars to an anchorage girder at the back of the anchorage. The same result is nowadays more efficiently achieved by attaching the strand terminations to anchor slabs post tensioned against the face of the anchor block concrete (see Figure 11.24).

11.7.2 Geotechnical design

Where sound unfaulted rock exists at or very close to the surface, it is possible to transfer the cable forces directly to the rock. In anchorages of this type, an anchorage tunnel is driven to reach and penetrate far enough into sound unfaulted rock so that sufficient mass can be mobilized to resist the cable force with the required degree of safety, with the upper part of the tunnel forming the chamber in which the strands splay out to the anchor face. In most cases however, a 'gravity' type anchorage, in which the horizontal loads from the main cables are resisted principally by friction between the underside of the foundation and the supporting strata, will be required. Where the ground is suitable (e.g. weak rock) the base of the anchor block can be stepped to provide improved resistance by utilizing the passive pressure of the ground into which the block is embedded. The development of the necessary resistance against sliding by base friction requires the anchorage to have a very large mass, as can be deduced from consideration of the horizontal equilibrium of the anchor block.

$$C \cos \theta = \mu \times [M - C \sin \theta]$$

where C is the total cable force, M is the dead load of the anchor block, θ is the cable inclination to the horizontal and μ is the base coefficient of friction.

The required anchor block dead load is therefore:

$$M = C \times \left[\sin \theta + \frac{\cos \theta}{\mu} \right]$$

The inclination of the cable to the horizontal will generally be around 10–15°, and typical values of the base friction coefficient could be in the range 0.3–0.5. Taking the more adverse of these values, the required value of M would be $3.48C$, i.e. the anchorage dead load will need to be of the order of three-and-a-half times the total cable load. Gravity anchorages are therefore necessarily very large concrete structures

As the cable force is predominantly horizontal, it will produce a moment (Ch) about the front edge of the anchor block, where h is the height of the application of the cable force above base level. This will produce a base pressure distribution with a maximum at the front edge of the anchorage base, reducing progressively towards the rear. It is therefore advantageous to concentrate the mass of the anchorage towards its rear, producing an opposing pressure distribution to that from the cables, so that the resultant total pressure distribution is almost uniform. Consideration must also however be given to the construction condition, when the cable force is either absent or much reduced, and the anchorage base sized to ensure that allowable bearing pressures are not exceeded in this condition. Where ground conditions are rather poor, this may require a proportion of the anchorage dead load to be added concurrently with the construction of the cable and suspended deck structure.

The design of the anchor block foundation must of course include consideration of all possible soil failure modes, including slip circle failure of the ground initiated by the high vertical bearing stresses under its base.

11.7.3 Appearance

It is apparent from the above that gravity type anchorages for large bridges are necessarily massive structures with large horizontal and vertical dimensions, and very careful design is therefore required to produce an aesthetically acceptable solution. Possible ways in which anchorages can be given a more pleasing appearance are:

- placing as much as possible of the anchor block below ground level
- aligning the structure edges parallel and normal to the cable axis, so that the function of the anchorage in resisting the cable forces is logically expressed
- the use of suitable architectural features such as ribbing to break up the monotony of large concrete surfaces
- the use of an open structure, again with as much as possible of the required mass placed below ground.

Figure 11.26 (Storebælt East bridge) shows a particularly effective use of the last of these methods in a marine environment, in which the mass required has been concentrated mainly below water level, with an open frame upper structure arranged to minimize its visual impact.

11.8 Analysis of suspension bridges

11.8.1 General

The suspension bridge differs from other bridge forms in that it is a load adaptive structure in which the cable geometry has to vary for each load combination so as to produce equilibrium between the internal forces and the applied loading. An analysis of its structural behaviour must therefore take into account these displacements from the initial geometry, and standard methods of linear structural analysis can not be used.

Figure 11.26.

11.8.2 Classical theories

The first analysis of the complete suspension bridge system was made by Rankine, based on the assumption that the cable profile under dead load was parabolic, and that the stiffening girder was sufficiently stiff to distribute any imposed loading so that this profile remained parabolic. This in effect assumed that the loading in the hangers remained uniform for any given imposed loading. It was further assumed that the hanger loads were equal to the total load divided by the span length. With these assumptions, the increase in cable tension and the bending moments and shears in the stiffening girder could be derived. An improvement on this is the elastic theory, usually ascribed to Navier. This retains the assumptions of a parabolic cable profile and uniform hanger loading, but uses a strain energy method to derive a more rational hanger loading due to imposed loads.

Both the Rankine and elastic theories implicitly assume that the cable displacements under imposed load are small compared to the initial cable shape. This approximation introduces large errors in the stiffening girder bending moments, particularly for long spans. To eliminate these errors, the deflection theory was developed by Melan (1888). This method is based on combining the differential equations for both the cable and stiffening girder to derive an equation governing the behaviour of the complete system. Unfortunately, this equation can not be solved directly for practical cases, and recourse must be made to laborious numerical methods to derive solutions. To facilitate this, Bleich (1935) developed the linearized deflection theory, which assumes that the increase in cable tension due to imposed loading is small compared to that due to dead loading. An alternative approach was to use the relaxation method of Southwell to derive numerical solutions and in the final development of the classical theory, Crossthwaite (1947) finally developed this into a practical method which could take into account non-uniform stiffening girder properties, hanger extensions, and horizontal cable movements.

Although the required main cable size could be rapidly derived, a considerable amount of laborious hand calculation was required to derive bending moments in the deck stiffening girder. However, for preliminary designs approximate values could be quickly obtained using the methods of Hardesty and Wessman (1938).

11.8.3 Computer analysis

The widespread availability of finite element software for the three-dimensional large displacement analysis of geometrically non-linear structures has now rendered the above methods obsolete. It will generally be appropriate for global analysis to initially use a two-dimensional model of the main cable, hangers, towers and stiffening girder. This model can be used to determine the structure geometry for permanent loads, and to determine the deck girder bending moments, shear forces, and deflections due to traffic loads and overall temperature variation. Once initial sizing of members has been completed, the model can then be expanded into a full three-dimensional representation of the structure to analyse wind loading and differential temperature effects, torsional moments in the deck, together with hanger and bearing loads from asymmetric traffic loading.

For the global analysis, a typical three-dimensional analysis will generally model the stiffening girder as a six degree of freedom spine beam accurately representing the axial, bending, torsional, and shear stiffness properties of the actual structure. The connection between the hangers and the idealized stiffening girder is best represented by rigid elements. Constraint equations can be used, but are not recommended for large models. Particular care must be taken to correctly represent the torsional stiffness as any subsequent modal analysis used to determine natural frequencies and mode shapes for input to an aerodynamic stability analysis can be very sensitive to this property. Mode shapes and natural frequencies of the structure are generally determined by a linear eigenvalue analysis technique, using the structure equilibrium geometry and stiffness derived from the non-linear large deflection analysis of the structure under permanent loads.

Solution convergence can sometimes be difficult in the analysis of large displacement geometrically non-linear structures with large differences in member stiffness. Problems of this type can often be avoided by ascribing a small bending stiffness to the cables and hangers, rather than modelling them as tension only elements.

Critical loaded lengths for traffic loading must be derived by an iterative procedure, as due to the non-linear response of the structure, conventional influence lines will usually produce an underestimate of the required loaded length. This is a particular problem where the intensity of traffic loading is dependent on the loaded length, as is the case for British (e.g. BS 5400 Part 2) highway loading. However, influence lines generated by the application of a series of unit loads can be used to give a first approximation to the critical loaded lengths.

For relatively short spans, depending on the relative stiffness of the cables and deck, a sufficiently accurate global analysis for live loads can made using the permanent load geometry of the structure, and assuming linear elastic behaviour. This assumption of linearity must however be validated by comparison with results from a full non-linear analysis for a selection of typical load cases.

For critical areas of the structure, such as the hanger to deck connection, members adjacent to bearing positions, the tower and splay saddles, the global finite element model can not produce sufficiently detailed stress distributions. For these areas, additional conventional three-dimensional finite element models with boundary conditions derived from the global analysis will therefore be required.

11.9 Aerodynamics

11.9.1 General

The inherent flexibility of suspension bridges makes their design particularly sensitive to the effects of the natural wind. The aerodynamic actions and effects which must be considered by the designer can be summarized as follows:

- the mean (quasi-static) wind loading on the structure, principally due to the drag forces on the suspended deck, cables, and towers
- buffeting stresses induced in the structure by forced movements arising from the random turbulent wind fluctuations
- limited amplitude oscillations of the deck structure, either in vertical bending or torsion, caused by the periodic shedding of vortices
- aerodynamic stability, in which divergent oscillations of the suspended deck in either a torsional mode, or in a coupled bending torsional mode (classical flutter) can, if allowed to develop, rapidly increase in amplitude and destroy the structure
- vibration of hangers caused by the periodic shedding of vortices.
- the effects of wind strength and variation on the safety and comfort of traffic using the bridge.

A full consideration of these effects is beyond the scope of this chapter and only a brief summary of each is given below. For more detailed information, reference must be made to the extensive specialist literature on this subject (e.g. Larsen, 1992; Dyrbye and Hansen, 1997; Larsen and Esdahl, 1998).

11.9.2 Mean wind loading

This consists of the quasi-static load arising from the wind flow past the bridge, and is determined by its size and shape, the square of the mean wind speed, and its angle of inclination to the structure. The most important loading is that on the bridge deck and towers, on which the mean wind flow can produce horizontal (drag), vertical (lift), and torsional forces, with their relative magnitudes depending on the geometry of the cross-section. These mean wind forces are calculated in the usual way in terms of shape dependent non-dimensional coefficients, so that the horizontal (drag) force F_H is given by:

$$F_H = \frac{1}{2}\rho V^2 C_D A$$

where ρ is the air density, V is the mean wind velocity, C_D is the drag coefficient, and A is the exposed area.

Vertical (lift) and torsional forces can be similarly expressed in terms of coefficients C_L and C_M.

At the long spans for which suspension bridges are typically used, the lateral stiffness of the cable system will predominate over that of the stiffening girder. The horizontal wind loading on the suspended deck will therefore be carried largely by the cables to the tower tops, and is generally the determining load effect for the transverse design of the towers. An accurate assessment of this loading is required and, except for preliminary design, the calculation of quasi-static wind loading on the deck and towers should therefore be based on measured drag coefficients obtained from wind tunnel testing.

The importance of achieving low drag forces has been an important factor in the wide use of streamlined box girders for suspended deck structures as these can have drag coefficients of the order of 0.075 compared to typical values of around 0.25 for an open truss type stiffening girder.

11.9.3 Buffeting

This is the response of the bridge structure to the randomly fluctuating turbulent components of the wind. If the turbulent wind components contain sufficient energy at frequencies close to some of the natural frequencies of the structure, oscillations of the structure will develop. The response can occur in vertical, lateral, and torsional modes of vibration with the magnitude being dependent on:

- the cross-sectional shape of the structure
- the intensity of the wind turbulence
- the spatial distribution and correlation of the turbulence
- the natural frequencies and mode shapes of the structure.

Because the oscillatory response of the structure itself gives rise to additional motion induced wind forces, the growth of buffeting loads with increasing wind velocity is proportional to $1/2\rho(V)^N$, where N is significantly greater than two.

Buffeting induces fluctuating stresses in the structure which are additive to those of the mean wind and may cause vibrations which bridge users find unpleasant.

11.9.4 Vortex shedding

Vortices are shed periodically from the sides of all bluff obstacles to airflow, and cause alternating aerodynamic forces to be developed. These forces are strongest in the direction transverse to the airflow, but there are also much weaker alternating in-line forces. However, these rarely give rise to problems.

The frequency at which the vortices are shed is proportional to the velocity of the incident airflow and the transverse dimensions of the structure, with the factor of proportionality being called the Strouhal number S_t, so that:

$$f = S_t \left(\frac{V}{D} \right)$$

where f is the frequency of vortex shedding, D is a typical cross-wind dimension of the structure, and V is the velocity of the incident airflow.

For a circular cross-section, the value of the Strouhal number is around 0.2, whereas for rectangular cross-sections with a cross-wind dimensions substantially smaller than the downwind dimension, and therefore typical of bridge deck proportions, a value of around 0.1 would be typical.

If the frequency of vortex shedding is coincident with, or close to a transverse bending or torsion natural frequency of the structure, the resultant periodic cross-wind aerodynamic forces will excite resonant oscillations at that frequency. Because the vortex shedding frequency occurs only at a specific wind speed, these oscillations will occur only at that speed, or in a narrow range of wind speeds close to this. Vortex induced oscillations are most likely to occur when the incident airflow has a low intensity of turbulence, as turbulent variations in the wind speed disturb the regularity of vortex shedding and discourage the build up of oscillations.

The structure excitation due to vortex shedding is usually quite weak and a relatively small constant peak amplitude limited by the structure damping is therefore reached at which the structure continues to oscillate if the initial wind conditions persist. The oscillations do not usually cause levels of stress which would in themselves be sufficient to cause structural distress, but if the wind speed for onset is sufficiently low, leading to the frequent occurrence of oscillations, fatigue damage may result. In addition, the oscillations, whether vertical or torsional, can be alarming and unpleasant to bridge users, particularly to pedestrians.

The hangers to the suspended deck are particularly prone to vibrations induced by periodic vortex shedding. They will generally be of similar size and construction, and their tensions will not be significantly different. However their lengths will vary from only a few metres at mid span up to as much as several hundred metres adjacent to the towers, and they will therefore have a wide range of natural frequencies. Hangers constructed from wire strand or rope are light, very flexible and have low internal damping. There is therefore a high probability that at least some of the hangers will be susceptible to vortex shedding excitation at relatively low wind speeds. Possible measures to overcome this problem include the attachment of Stocksbridge dampers, linking the hangers with secondary stabilizing cables, and the use of viscous dampers at the deck connection.

11.9.5 Aerodynamic stability

The collapse of the Tacoma bridge in 1940 alerted bridge designers to the destructive power of aerodynamically induced instabilities of flexible structures. These are phenomena caused by interaction between the elastic properties of the structure, its deflections, and the resultant effects of these on the surrounding airflow. These instabilities develop when the mean wind speed reaches a critical value, and can cause oscillations with rapidly divergent amplitude, leading to partial or complete collapse of the structure. The structure must therefore be designed to have a critical speed for the onset of such instability with an adequate margin of safety over any probable wind speed that might occur during its life.

A prime consideration in the design of the suspended deck of the bridge is therefore the achievement of a critical speed for the onset of aerodynamically induced instability with an adequate margin of safety (generally taken as 1.3) over the highest predicted wind speed during the life of the bridge.

The instability can occur in a number of forms as follows, characterized by the causative aerodynamic action.

Static torsional divergence

The wind flow past the structure will in general produce a torsional moment on the deck cross-section, and this usually increases with increase in the incident angle of the wind. If this increase in applied torsional moment exceeds the resisting moment from the elasticity of the structure, the total stiffness against rotation can become zero, and torsional deformations will increase without limit to the point of failure. In practice, if the structure has sufficient stiffness for it to be stable in respect of other aeroelastic instabilities, then it will be acceptable in respect of static divergence.

Instability in transverse bending (galloping)

When a structure oscillates transversely in an airflow, the speed of the airflow relative to the structure is given by the vector addition of the incident airspeed and that of the transverse motion. As a result, the effective angle of incidence of the wind to the structure varies periodically. If the variation in lift coefficient of the cross-section with the angle of wind incidence is negative, then an upwards motion will produce a corresponding upwards force, reinforcing the motion, and galloping instability will occur, with increasing amplitudes of vibration.

Instability in torsion (stall flutter)

A similar effect to galloping can also occur in pure torsion. In this case, torsional oscillations lead to the alternating breakdown and separation of the flow from each side of the structure in turn. The separation of flow (stalling) produces a sudden reduction in transverse wind force, and the alternating nature of this can give rise to a periodic torsional moment on the cross-section reinforcing the initial motion, with continually increasing amplitudes of torsional oscillation.

Instability in coupled torsion and bending (classical flutter)

Instability can also occur when bending and torsional motions of the structure cause changes in the aerodynamic forces. These changes usually tend to damp any movement. However, if the bending and torsional natural frequencies are close together, torsional and transverse bending oscillations can become coupled together in such a way that the changes in aerodynamic forces reinforce the movements and energy is absorbed from the wind by the structure. If this is greater than that dissipated by the structural damping, which will generally be the case, the amplitude of the oscillations will increase and the resultant violent motions will result in failure of the structure. This mode of interdependent coupled unstable oscillations is known as classical flutter.

The prediction of the critical speed for all of the above types of instability requires a knowledge of the natural frequencies and mode shapes of the structure, and of the aerodynamic properties of the suspended structure cross-section. For practical reasons, even the slenderest 'streamlined' box girders are in fact relatively unstreamlined shapes, with sharp corners where flow separation readily occurs. Although considerable progress has been made in the field of computational aerodynamics in recent years, the aerodynamic properties of deck sections and the resultant critical speeds for instability can currently only be derived with certainty by wind tunnel testing of representative models as described in Section 11.9.9. However, empirical formulae have been developed which can give a rapid approximate assessment of the critical speeds, a typical example being the following formula, based on work by Selberg, for the critical speed for classical flutter:

$$V_f = 4 \times (f_T B) \left(1 - \frac{f_B}{f_T} \right) \left[\frac{mr}{\rho B^3} \right]^{1/2}$$

where V_f is the critical speed for the onset of flutter, f_T and f_B are the structure natural frequencies in torsion and bending, respectively, B is the width of the cross-section, m is the mass per unit length of the structure, r is the polar moment of inertia of the bridge cross-section, and ρ is the air density.

11.9.6 Effects of wind on traffic

The difficulty of driving in windy conditions is a familiar experience to all drivers on roads in exposed environments, and many accidents arise from impaired directional control, or are caused by the overturning of high-sided light vehicles. Long span bridges are often in locations exposed to severe winds and because there is usually no convenient alternative route in the event of the bridge becoming unusable due to an unsafe wind environment or blockage by an accident, the safety of bridge users requires careful consideration in design. To provide a level of protection which would enable the bridge to be usable in all but the most extreme conditions would require continuous edge barriers of sufficient height to produce adequately sheltered conditions on the roadway deck. However, there are considerable difficulties in the provision of such barriers as, the aerodynamic stability of the bridge may be unacceptably degraded, and the mean (quasi-static) wind forces will be significantly increased.

A particular problem for suspension bridges is the sheltering effect of the main towers, which during strong cross-winds, causes very rapid changes or even reversal in the wind loads on vehicles in this area. This problem can however be avoided with-out compromising the overall design of the bridge, by the provision of local barriers of increasing height and/or solidity as the towers are approached, eliminating the sudden changes in wind environment which would otherwise cause difficulty for drivers.

11.9.7 Aerodynamic design considerations

Although the overall configuration of the bridge is determined from functional and structural requirements, consideration of wind loading and aerodynamic stability is an essential part of the design process for the suspended deck of the bridge.

Drag reduction

As noted above, the use of streamlined box girders can reduce the horizontal (drag) forces on the bridge deck to a level at which the economic effect of the resultant load-ing on the towers is relatively insignificant.

Buffeting and vortex shedding

Solid web plate stiffening girders, in combination with a deck slab, have extremely poor aerodynamic properties and are highly sensitive to excitation by vortex shedding. This type of cross-section should therefore only be used in conjunction with very high structural stiffness, and even then may be unacceptable. Open truss deck stiffening girders generally do not develop significant aerodynamic forces transverse to the wind flow and the multiple small members shed small, poorly correlated vortices. Problems due to buffeting and vortex shedding are therefore rarely a problem with this type of structure, although achieved at the expense of high drag forces.

Vortex shedding can be a problem with box girders. The practical requirements of economically producing a section with a flat upper surface for traffic prevent an ideal fully streamlined cross-sectional shape, free from sharp corners at which flow separation occurs, from being adopted. Cross-sectional performance with respect to vortex shedding is generally improved by the incorporation of cantilevered footways at the mid height of the cross-section. If footways are not required, the incorpora-tion of non structural fairings or turning vanes at the edges of the deck can often eliminate vortex shedding problems.

Aerodynamic stability

The first requirement in achieving an acceptable margin for aerodynamic stability is the avoidance of cross-sectional shapes known to have poor aerodynamic stability characteristics, such as solid plate girders or rectangular box girders. The most impor-tant parameter under the control of the designer is however the torsional stiffness of the deck structure. As can be seen from Section 11.9.5, the ratio of torsional to bending natural frequency has a very significant effect on the critical speed for the onset of classical flutter. The global vertical bending stiffness of a suspension bridge is determined essentially by the main cables, with the contribution from the suspended deck generally being negligible, and the designer therefore can do little to change the bending frequency.

However, the suspended deck makes a much greater contribution to the overall torsional stiffness of the structure. The use of a high deck torsional stiffness is therefore the best way of achieving a good separation of the torsional and bending frequencies and hence increasing the critical flutter speed. Open truss suspended deck structures must therefore incorporate bottom lateral bracing so as to produce an effective closed torsion cell, with torsional stiffness comparable to that of a box girder. Further improvement to the performance of truss deck cross-sections can be achieved by incorporating longitudinal slots in the deck, which have been found to delay the onset of aerodynamic instability by permitting ventilation between the upper and lower surfaces. As noted previously, restraining the suspended deck torsionally at the towers (11.4.2), and a rigid connection between the main cable and deck at the centre of the main span (11.3.3) both have a beneficial effect on the aerodynamic stability of the structure.

11.9.8 Aerodynamic stability during construction

During erection, wind on the part completed suspended deck girder can present problems with respect to its stability and other wind load effects, as each stage of structural completion can have significantly reduced stiffness, and a different mass distribution from the final stage, and consequently different natural frequencies. The stiffness reduction arises largely because usually the part completed deck has a hogging curvature, so that the transverse joints between adjacent sections of deck can not be completed until erection of a substantial proportion of the deck has been completed.

As the suspended deck contributes significantly to the overall torsional stiffness of the structure, but much less so for vertical bending stiffness, the proportional reduction in torsional stiffness will be significantly greater than that in vertical bending. In addition, as the mass of the cables is concentrated at the edges of the deck, the addition of the deck girder produces a greater increase in vertical than rotational inertia. The effect of these changes is to reduce the ratio of torsional to bending frequencies to below that for the completed structure, with a corresponding reduction in critical flutter speed. As a result, it is found that as a general rule, when erection is commenced at the centre of the main span, working symmetrically outwards towards the towers, a minimum value of critical flutter speed occurs with 10–30% of the deck erected. Thereafter, as erection continues, the critical speed increases towards the completed structure value.

The bridge deck can also be vulnerable to buffeting and/or vortex excited oscillations during deck erection. As discussed in Section 11.10.6, the deck transverse joints can not usually be made rigid until a significant proportion of the deck has been erected, and the temporary connections required at these joints can provide only limited continuity of torsional deck stiffness. This overall reduction in stiffness and natural frequencies can result in a reduction in the critical wind speed for vortex shedding and an increased sensitivity to buffeting excitation at certain stages of deck completion, with considerably increased loads on the temporary connections between deck units.

11.9.9 Wind tunnel testing

The aerodynamic properties of a proposed structure and the critical speeds for the onset of unstable behaviour can only be determined in advance of actual construction by reference to testing of representative small scale models in a simulation of the natural wind, using a wind tunnel. The investigations can be carried out in a number of ways, as described below.

The most widely used of these techniques is sectional model testing, in which a geometrically scaled stiff model of the bridge cross-section is mounted either rigidly, or on springs and a damper simulating the elastic stiffness and structural damping in bending and torsion. A rigidly mounted model can be used to determine the static values of the drag (C_D), lift (C_L) and torsion moment (C_M) coefficients for the bridge cross-section, and by rotating the model relative to the flow direction, the variation of these with the flow angle of incidence. With a spring-mounted model, the aerodynamic stability of the cross-section can be investigated by measuring the response of the model as the wind speed is progressively increased in small increments. Tests can be carried out either in smooth (low turbulence) flow, or with the intensity of the natural wind simulated by the use of grids positioned upstream of the model.

Sectional model tests have the advantage that, as only a short section of the suspended deck is necessary, a relatively large model scale (say 1:50) can be used in which deck edge details, including parapets, which it is known can significantly affect aerodynamic behaviour, can be accurately represented. Further advantages are that, as the model is a simple rigid representation of the cross-section geometry, it can be inexpensively and quickly produced, and can be easily modified to investigate the effect of changes to the suspended structure layout. As a result the sectional model test is an indispensable tool for the investigation of aerodynamic performance.

An alternative approach is to wind tunnel test a full aeroelastic model of the proposed structure, simulating as accurately as practicable the geometry, stiffness and damping of the real structure. This has the advantage that three-dimensional effects, such as changes in the structure properties along the span, and variations in the ambient wind flow and its turbulence intensity due to topographical features can be investigated. Erection conditions, which are inherently three dimensional due to the incomplete state of the structure, can also be fully investigated in this way. However, even if a wind tunnel with a large working section is available, full aeroelastic models can only be tested at rather small scales (typically 1:100 to 1:300), so that accurate representation of cross-sectional features of the suspended deck is difficult. The construction and testing of a full three-dimensional aeroelastic model is significantly more time consuming and costly than for a sectional model, and changes to its properties are significantly more difficult and time consuming to incorporate. Testing of a full aeroelastic model is generally therefore only appropriate to finally confirm the aerodynamic behaviour and stability of an essentially finalized design, or where three-dimensional effects are especially significant.

11.10 Construction

Although suspension bridges can be constructed without any temporary intermediate supports, considerable temporary works are required for their construction. The most important aspects of the construction techniques required are briefly described below.

11.10.1 Towers

Steel towers

For short span bridges, the cross-section of the towers may be small enough to be fabricated and transported to site as a complete cross-section, requiring transverse site joints only. However, for long span bridges, the tower cross-section will be too large to permit this. The tower is then usually made up as an assembly of stiffened plates, with each plate forming one side of the tower leg cross-section as shown in Figure 11.25.

For land based towers, the base sections can be erected using large mobile or crawler cranes, with floating cranes being used for those in offshore locations. Except for bridges with relatively short towers, the towers are invariably too high for erection to be completed with such cranes. The usual construction method for the towers of earlier bridges was therefore to continue erection using climbing cranes, temporarily attached to the tower, and progressively moved upwards as erection proceeded. The lifting equipment and temporary supports for these cranes has to be specifically designed and manufactured to suit the tower being erected. Self-climbing tower cranes with sufficient capacity to erect steel tower sections are however now widely available, and will in many cases be a more appropriate choice for erection, as the only special equipment then required is the ties to attach the crane mast to the tower.

For suspension bridges with fixed saddles, the towers are necessarily very slender in the bridge axis direction. Although in the final bridge structure, the tower top is restrained longitudinally by the main cables, this support is however absent during construction and the towers must therefore be designed to be capable of free standing until the cable has been erected. The critical load case is with longitudinal wind loading, combined with any offset loading from a climbing crane. If the tower is pulled back to generate the free cable offset of the tower saddles, this will produce further tower bending moments, which will be additive to the above. These loads may produce tensile stresses on the windward face of the tower, which could be critical for transverse bolted joints if the tower joints have been designed to transmit compressive stresses mainly by direct bearing contact.

During construction, free-standing steel towers are much more flexible longitudinally than in the in-service condition, when they are effectively supported by the main cables. In this condition, steel towers are therefore susceptible to vortex induced vibrations. If analysis, supported by wind tunnel model testing predicts this, additional damping must be provided. This damping can be provided either by an external friction damper or, for taller towers, an internally mounted tuned mass damper.

Concrete towers

For concrete towers the possible options for construction of the tower legs are to continuously slipform, or to cast them in a series of lifts using jump shuttering. Slipforming is the quicker method and will give the shortest construction programme, but requires a continuous supply of concrete, with certainty of delivery in all weather conditions. Slipforming is therefore most suitable for construction of towers which are either on land, or close enough to the shore line to be accessible via a temporary bridge or causeway. At both the Humber and Tsing Ma bridges, one tower was on land with the other relatively close to the shore, and the towers of these bridges were therefore slipformed. However, for the Storebælt east bridge, where the towers are positioned around 2–3 km from the shore line, with the possibility of severe weather conditions interrupting concrete supplies, construction in 4.0 m lifts using jump shutters was chosen.

The construction of the portal frame members of concrete towers requires the use of a temporary girder to support the formwork and wet concrete, and spanning between the tower legs. For slipformed towers, the most convenient method is to complete slipforming of the tower legs, and to then construct the portal members from the uppermost downwards, and this procedure was adopted at Humber. At the Tsing Ma bridge, this was not possible, due to the possibility of typhoon wind conditions occurring. Additional protection against this was provided by installing steel cross-frames as tower leg construction proceeded, with these being subsequently incorporated into the permanent concrete cross-members.

Because of their greater mass and stiffness, vortex induced vibrations of concrete towers will generally be of very small amplitude, and special measures to control them are unlikely to be required.

11.10.2 Temporary footbridge

Irrespective of whether the main cables are constructed by spinning or by the erection of PPWS or other strand types, a working platform is required to give access to the complete length of the cable for this work. Working access is also required for the subsequent operations of compaction, cable band and hanger erection, deck erection, and finally for cable wrapping and painting.

This working platform is provided by the erection of a temporary footbridge (sometimes called a catwalk) positioned at a constant separation below the main cable construction (free cable) profile. The distance chosen must be such as to enable cable erection operations to be carried out at a convenient working height, but also to provide sufficient clearances for the later operation of the compacting and wrapping machines. A distance of around 1100 mm from the cable centre-line to the footbridge floor level will generally be adequate. The footbridge width must be sufficient to provide a walking access along both sides of the cable, and also provide clearance for the compacting and wrapping machines. A typical footbridge arrangement is shown in Figure 11.27.

Figure 11.27.

The footbridge is supported by a number of spiral strands or wire ropes, which are positioned at a transverse spacing usually between 500 and 600 mm below the working areas of the footbridge on either side of the cable centre-line, with a some-what larger gap along the cable centre-line to provide access for the erection of the hangers. The footbridge floor is usually made of welded steel wire mesh, typically made from 4 or 5 mm diameter galvanized wire. To provide a secure foothold, particularly on the steeper areas near the towers, the mesh must be fitted with trans-verse timber treads, with the spacing of these being varied to suit the local slope. U-frames fixed to the floor strands at intervals maintain the footbridge cross-section, and together with hand strands or ropes support wire mesh sides to provide a completely secure working environment.

The footbridges are very flexible structures and to stabilize them against unac-ceptable torsional displacements, they are linked together at intervals of around 200 m by temporary cross-bridges, which also provide convenient crossing points for per-sonnel. It is also usual to provide additional stiffness to the system by the provision of a storm system consisting of strands in an inverted catenary profile below the foot-bridge strands, and linked to them at intervals by vertical rope ties.

The footbridge strands can be either continuous over the tower, or terminated on each side and anchored to the tower structure. As noted previously, the footbridge is required as a working platform until after the deck has been erected and the bridge cables have deflected down to their full dead load profile. The footbridge strands must therefore be provided with adjustment at their anchor points, to enable the footbridge to be lowered to conform to the downward displacement of the cables.

The footbridge system is completed by two or more strands positioned above the floor strands, and linked to them by wire rope ties. These strands are linked together by cross-beams which provide support for an overhead tramway system which is required to pull the spinning wheel or haul out prefabricated cable strands.

11.10.3 Cable construction
Conventional aerial in situ spinning

This method, originally developed by John Roebling in the nineteenth century, and progressively improved since then, remains the preferred method for the construction of large parallel wire cables to this day. The construction of cables by *in situ* spinning consists of pulling two or four loops of wire across the temporary footbridge by means of an endless aerial ropeway, generally referred to as a 'tramway' (Figure 11.28). These loops of wires are wrapped around a 'spinning wheel' attached to the tramway rope. Two spinning wheels are attached to the tramway rope and are positioned so that, as a loaded spinning wheel makes an outward trip across the footbridge, the wheel from the previous trip returns empty to enable the next trip to be commenced with the minimum interruption to operations. Each spinning wheel is grooved to accommodate either two or four loops of wire, with two loops being the preferred choice for small to medium sized cables, and four for large cables. Cable spinning with two loops of wire is described below, but the same principles would apply for four loop spinning for larger cables.

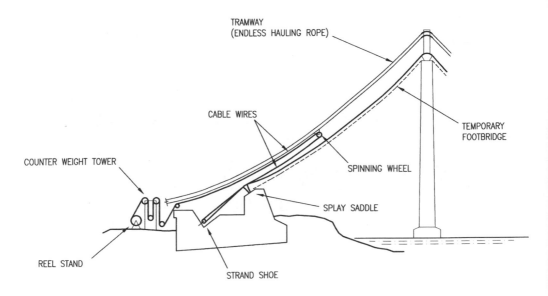

Figure 11.28.

The coils produced by the wire manufacturer contain too short a length of wire for them to be used directly in the spinning operation. Before the wire can be used it is therefore wound onto large diameter reels containing wire from a sufficient number of coils to enable the spinning wheel to make a reasonable number of trips across the bridge before the need for refilling. These reels are placed in unreeling machines able to pay out wire at a speed matching that of the spinning wheel. From the unreeling machines, the wire is led into a 'counterweight' tower, and passes around a series of sheaves to form a loop of wire supporting a free hanging counterweight. The counterweight maintains a constant back tension as the wire is pulled out across the footbridge, and if a powered unreeler is used, movement of it absorbs any small variations between the unreeler and tramway speeds. In conventional spinning, the wires are unreeled and hauled across the footbridge at a relatively low tension.

At the start of spinning each strand, the ends of the wire from two reels are led through the counterweight tower, and then through a system of deflection sheaves so that they can be led to the appropriate strand shoe location. They are then led up the anchorage chamber, and through the splay saddle to a temporary fastening point at the lower end of the footbridge. The wires are then looped into the strand shoe grooves, and a further loop formed around the spinning wheel.

The tramway is then operated to pull these loops of wire across the footbridge. As the wheel travels across the bridge (Figure 11.29), the upper (live) wires are released from the unreeler and move at twice the tramway speed, whereas the lower (dead) wires wrapped round the strand shoe are not moving. Small temporary

Figure 11.29.

sheaves positioned at intervals across the footbridge support the live wires, and enable them to be pulled out without undue friction. The 'dead' wires are allowed to fall onto the footbridge floor. As the spinning wheel passes each saddle, the dead wires are placed in their final position in the appropriate groove, with the live wires being supported by temporary sheaves. The tramway is normally operated at a speed of up to 6 m/s in the spans, but more slowly as the spinning wheel moves past the tower and splay saddles.

When the spinning wheel reaches the far anchor chamber, its loops of wire are removed and placed around the strand shoe at that end. At the same time, the empty spinning wheel, which has arrived at the spinning anchorage, has new loops of wire formed around it ready for the next trip. As each trip proceeds, the sag of the spun wires is adjusted to conform to a guide wire set slightly above the final required position of the cable. Once adjustment of all the wires of the trip has been completed, the tramway is again operated to pull a new set of wire loops across the bridge, with spinning continuing in this way until all the wires of a strand have been positioned and adjusted to a common sag. The usual practice is to spin two strands at a time, with each spinning wheel placing wires into one strand. This operation of spinning two strands is referred to as a spinning 'set-up'.

When the spinning of each pair of strands has been completed, the spun wires are freed of all temporary lashings used to control them during spinning and shaken out to allow them to hang in free catenary. Any wires falling outside a specified sag tolerance are then adjusted to come within this by cutting out a length of wire and re-splicing for any low wires, or by splicing in an extra length for high wires.

After completion of this operation a small hand press is used to compact the strand into a circular shape, which is secured by thin metal straps placed at about 3 m intervals.

The strand must finally be adjusted to the precise free cable sag required. This is done by surveying the strand to determine its actual position and then making the necessary adjustment to bring it to the correct sag for the temperature and actual saddle positions at the time of the survey. Because stable temperature conditions are essential for accurate survey, this must be carried out during the night. The strand movements necessary to produce the required sag are produced by jacking the strand shoes at each anchorage, and for the main span, moving the strand though the tower saddle with hydraulic pulling equipment.

In each cable, the first (No. 1) strand, positioned at the bottom of the cable hexagon, is spun alone and then adjusted to the required 'free cable' sag using the results of a precise survey between the towers and anchorages. All the subsequent strands are spun in pairs, and are positioned by a comparative survey relative to the bottom (No. 1) strand.

Controlled tension in situ spinning

This is an improvement on the original spinning method in which the wires are spun at a considerably higher tension, and the adjustment of individual wires is eliminated. The work on the footbridge and at the saddles is therefore greatly simplified, leading to a substantial reduction in labour requirements.

The wire is prereeled in the same way as for conventional spinning, and similar unreeling equipment is therefore required at the anchorage from which the wire is spun. However, the counterweight tower is arranged to produce a spinning tension in the range 60–80% of the wires' free hanging tension.

The tramway spinning wheels have a variable angle setting, and the outward wheel is tilted so that it places the dead wires directly into their final position in cable formers fixed to the footbridge, with the live wires being supported by temporary sheaves on these. Handling and placing the wires in the saddles is similar to that for conventional spinning. On arrival of the spinning wheel at the far anchorage, loops of wire are pulled out and placed around the strand shoe, with the wires remaining on the wheel.

The tilt of the spinning wheel is reversed for the return trip, so that as the wheel moves back across the footbridge, the live wires placed on the outward trip are lifted from their sheaves and also placed into the cable formers. As the spun wires are at between 60 and 80% of their free hanging tension, the cable formers transfer between 40 and 20% of the weight of the spun wires on to the footbridge. As the footbridge is a very flexible structure, this can cause significant deflections and hence sag variation in the spun wires, and a means of compensating for this effect is therefore required. This will only be required for the first few strands spun, as the addition of the cable strands rapidly increases the effective stiffness of the footbridge. Once spinning of a strand has been completed, it is banded and adjusted using similar equipment and methods as for conventional spinning.

This improved spinning method was originated by Japanese engineers, and was first used for the cables of the Shimotsui Seto bridge in 1985–1986. Since then it has been used for the spinning of a number of major bridge cables (Second Bosporus 1987, Storebælt East bridge 1996) and most recently (1998) for the Tagus bridge auxiliary cables.

Preformed parallel wire strands
This method represents a further step in the reduction of cable construction work at the bridge site, achieved by completely prefabricating the cable strands. Cable construction is thereby reduced to simply unreeling of the strands at one anchorage, as they are pulled across the footbridge to the far anchorage. To pull the strands across the footbridge, an overhead tramway system similar to that for spinning, but with higher pulling capacity, is preferable. A simpler hauling system can be used, but this will usually result in a lower rate of production.

To support the strands as they are pulled out across the footbridge, closely spaced rollers must be provided to minimize frictional resistance, and prevent abrasion damage. After pulling out has been completed, the strand lengths which will be placed in the saddles, must be re-shaped from their hexagonal manufactured shape into a rectangular form. Survey and adjustment of the strands to their final level is carried out using methods and equipment similar to those for spun cable strands. The advantages of this construction method are a further reduction in site labour requirements, and more importantly, reduced sensitivity to adverse weather conditions.

11.10.4 Compaction

This is commenced as soon as the erection and adjustment of all cable strands has been completed. The compaction is carried out using a machine, generally consisting of a hexagonal shaped steel frame, equipped with six hydraulic rams. Each ram pushes on a circular profile steel shoe. These shoes are overlapped and interlocked so that operation of the rams squeezes the cable wires into a nominally circular shape. The hydraulic controls are generally arranged so that the vertical and side rams can be operated either simultaneously or separately to give maximum control over the compacted shape of the cable. Figure 11.30 shows a typical compactor in operation. The compactor is operated to squeeze the cable at intervals along its entire length up to as close as practicable to the saddles. Squeezes are made at a spacing of around 750–1000 mm, with the spacing required being determined during trials at the start of the work. As soon as each squeeze has been made, and before the ram pressure is released, the compacted shape of the cable is retained by tensioning flexible steel strapping around the cable. Although the strapping requirement is only temporary,

Figure 11.30.

and is removed when the wrapping wire is applied, it is normally supplied zinc plated to provide short term protection against corrosion. Additional compaction squeezes and strapping are required at the positions where cable bands will subsequently be fitted, to ensure that the required cable shape is achieved and maintained in these areas. To control the transition from the circular compacted shape to the stepped cable profile in the saddles, a special permanent cable band without a hanger attachment is provided adjacent to each saddle.

The capacity of the hydraulic rams required to achieve the required voids ratio is a function of the size of the cable and the quality of the cable construction, in terms of the variation in individual wire sags, and the extent of wire crosses. The assessment of the required ram capacity is not amenable to theoretical analysis, and recourse must be made to previous experience.

11.10.5 Cable band and hanger erection

As soon as compaction has been completed, the cable bands and hangers can be erected. This operation is commenced at the centre of the main span, working progressively back towards each tower. In a typical sequence, the cable bands are erected first, being carried to their required position by a work car operating on the tram support strands. Once a band has been accurately located in position, its clamping bolts are tightened, preferably by direct application of tension with a hydraulically powered bolt tensioning equipment. The hangers are then erected, either by carrying out from the towers with a work car, or by direct lifting into position from a supply barge.

11.10.6 Erection of suspended deck

Prefabrication of deck

To minimize the amount of weather sensitive work at the bridge site, the deck structure is usually divided into a series of prefabricated sections, the length of which is a multiple of the hanger spacing. The only work required at site is the lifting of these sections, followed by the bolting or welding of the transverse joints between them.

As a result, the deck sections can be erected much faster than they can be fabricated and assembled, and it will be necessary to ensure that most if not all of the sections are completed and ready for lifting before deck erection is commenced. This will require the provision of a large area for temporary storage of the completed sections, with easy (usually marine) access to the bridge site.

The prefabrication and assembly of the deck sections must be very carefully controlled, so that the transverse joints are easily made after lifting, and the final deck geometry is achieved. This requires that the sections are produced to close tolerances, particularly with respect to the positions of the hanger attachments, the shape of the deck cross-section, and the transverse locations of longitudinal stiffeners at the section ends.

To ensure that the final deck geometry is achieved without the need for time-consuming corrections after erection, the ends of each abutting section must be accurately matched during assembly. This must be carried out with the sections set at their required relative vertical alignment, making due allowance for effect of any weld shrinkage on the final geometry.

During trial assembly and matching of the sections, temporary deck connectors are fitted to ensure correct alignment is maintained after erection. These temporary connections at the transverse joints maintain the deck alignment, and resist forces developed by wind loading or induced by the deviations from the final profile.

As described below, the geometry of the erected deck does not initially correspond to the final shape of the deck, with the transverse joints being initially open at the bottom of the section. Temporary connectors are therefore usually only provided on the top surface of the deck sections, and possibly also on the upper parts of the web plate of box sections.

Deck erection sequence

There are two basic possible sequences (Figure 11.31) in which the suspended deck of the bridge can be erected.

The first of these is to commence erection at the centre of the main span, working outwards until the towers are reached. The side spans may be either erected concurrently with the main span, working outwards from the anchorages, or erected after completion of the main span. Tower bending strength limitations will be an important factor in determining the choice between these options. In the early stages of deck lifting, the main cables carry only a localized load in the mid span area, with the total loading being much lower than in the final full dead load condition. As a result, the deck adopts a sagging profile, and it is not possible to close and connect

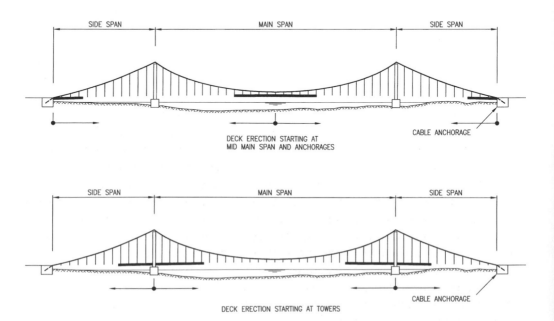

Figure 11.31.

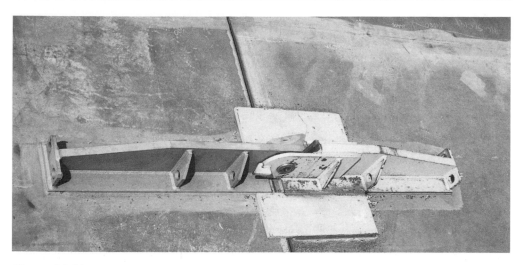

Figure 11.32.

the bottom flange joints. Temporary connections (Figure 11.32) must therefore be provided at the transverse joints to maintain the deck alignment, and to resist forces developed by wind loading or induced by the deviations from the final profile. The principal advantage of this sequence is that, as erection proceeds towards the towers, the deck profile rapidly converges towards the final profile, enabling the connection and bolting or welding of the transverse joints to be commenced soon after commencement of erection. A comparative disadvantage of this sequence is the loss of working time due to the relative inaccessibility of the part erected deck, since the main span labour force can only proceed to their working area by ascending the tower and walking down the footbridge to mid-span. A further disadvantage is that, as previously described, in the early stages of deck erection the critical wind speed for the onset of aeroelastic instability can be significantly lower than for the permanent structure.

In the second option, erection is commenced at the towers and is continued by working outwards until the centre of the main span is reached. The side spans can be erected working either from the towers or the anchorages, with their timing again being determined by considerations of tower strength. This method has the advantage that there is direct access along the already erected deck to the erection fronts. However, the effect of the part loading of the cable near the towers again causes a pronounced sagging curvature of the deck, and this persists until a fairly large proportion of the span has been erected. As a result, the joints between deck sections remain open at the bottom of the box section or truss, and the commencement of bolting or welding of the joints is therefore delayed, unless special measures are taken such as adjustment of hanger lengths. An advantage of this sequence is that there is substantially less reduction in the critical wind speed for the onset of aerodynamic instability than is the case with erection commencing at the centre of the main span.

Deck erection method

The lifting equipment used for erection is influenced by both the sequence and the type and weight of the deck structure. There are three basic methods.

1. Erection using lifting gantries supported by the main cables, and which are progressively moved along them as erection proceeds. The lifting equipment can be either rope tackles or strand jacks. In the case of rope tackles, the operating winches are normally positioned at the tower. This type of equipment is suitable for the erection of both box girder deck units, and preassembled sections of truss stiffened decks.
2. Lifting using cranes working on the already erected deck. This method is only suitable for an erection sequence commencing at the towers and for truss stiffened decks erected in relatively small sub-assembles.
3. Lifting using floating cranes. This method can be used for either basic erection sequence.

Figure 11.33.

11.10.7 Wrapping

Each length of cable between the cable bands is circumferentially wrapped using a machine to apply tightly packed turns of soft annealed galvanized wire at a tension of around 1.5 kN. Immediately ahead of the machine, sufficient paste should be applied to the cable to completely fill the interstices between the wrapping wire and the cable wires.

The machines used for cable wrapping are usually of the planetary type, a typical example of these being shown in Figure 11.33. These machines consist of a fixed structure resting on the cable, onto which a rotating 'flyer' assembly on which two, three, or four reels containing wrapping wire is mounted. The rotation of this assembly, synchronized with movement of the machine along the cable axis, simultaneously lays two, three, or four turns of wrapping wire over the paste. Spring loaded 'fingers' on the flyer press the wire turns into tight contact with the previously laid turns, and assist in controlling movement of the machine along the cable. The spring loading of the fingers enables them to accommodate slight variations in cable size and shape. The best quality wrapping, with tightly packed turns of wire is obtained by operating the machines up-slope working away from the completed wrapping. This method of wrapping is described as 'push wrapping'. However, a short length of cable adjacent to the upper cable band, and corresponding to the length of the machine, can not be wrapped in this way, and must be wrapped with the machine reversed and working over the completed wrapping. This mode of operation is referred to as 'pull' wrapping.

The cable wrapping operation is generally a critical part of the construction programme, and the earliest possible start is desirable. However, a limiting factor to this is the local completion of deck erection, as deck erection equipment can not be moved and operated over a wrapped cable.

Cable wrapping is an operation which requires the use of relatively complex mechanical equipment to apply the wrapping in an exposed environment. The application of the paste is also a difficult and dirty operation. Specifications often require that wrapping is not commenced until most of the bridge dead load is being carried by the cable. It is assumed that this requirement originated because of concern that significant increase in cable tension might result in unacceptable loosening of the wrapping, although it is doubtful whether this is really the case.

Bibliography

Bleich HH. *Die Berechnung verankerter Hängebrücken*. Julius Springer, 1935.

Crosthwaite CD. The corrected theory of the stiffened suspension bridge. *Journal ICE*, February, No. 4, 1947.

Durkee JL. Advances in suspension bridge cable construction. *Symposium on Suspension Bridges Lisbon*, Paper No. 27, 1966.

Dyrbye C and Hansen SO. *Wind Loads on Structures*. John Wiley, Chichester, 1997.

Furuya, Kitagawa, Nakamure and Suzumura. Corrosion mechanism and protection methods for suspension bridge cables. *Structural Engineering*, 3, 2000.

Gimseng NJ (Ed.) *East Bridge*. A/S Storebæltsforbindelsen, 1998.

Gimseng NJ. *Cable Supported Bridges. Concept and Design*, 2nd edn. John Wiley, Chichester, 1997.

Hardesty S and Wessman HE. Preliminary design of suspension bridges. *ASCE Proceedings*, Paper 2029, 1938.

Irvine HM. *Cable Structures*. Penerbit ITB, 1988.

Jensen G and Petersen A. Erection of suspension bridges (Figure 10). *Proceedings of an International Conference at Deauville*, 2, 351–362, 1994.

Larsen A (Ed.) *Aerodynamics of Large Bridges*. Balkema, 1992.

Larsen A and Esdahl S (Eds). *Bridge Aerodynamics*. Balkema, 1998

Melan J. *Theorie der Eisernen Bogenbrücken und der Hängebrücken*, 2nd edn. Leipzig, 1888.

Pugsley A. *The Theory of Suspension Bridges* 2nd edn. Edward Arnold, London, 1968.

Stahl FL and Gagnon CP. *Cable Corrosion in Bridges and other Structures*. ASCE Press, 1995.

Steinman DB. *A Practical Treatise on Suspension Bridges*, 2nd edn. John Wiley, 1929.

UK Highways Agency. BD49/93: Design Rules for Aerodynamic Effects on Bridges. UK Highways Agency, London, 1993.

Wyatt TA. Secondary stresses in parallel wire suspension cables. *American Society of Civil Engineers*, 128, paper 3402, 1963.

12 Movable bridges

C. BIRNSTIEL

12.1 Historical notes

Bridges with movable spans are built where an acceptable vertical profile for a fixed railway or roadway crossing of a legally navigable waterway is not feasible. An acceptable profile for a bridge provides the required underclearance for navigation, has permissible grades for vehicular and railway traffic, and minimizes adverse impact on present and future use of land contiguous to the bridge. Aesthetic and environmental considerations also influence the decision on the desirability for a movable span. Because of the higher operating maintenance and cost of movable compared to fixed bridges, and the inconvenience of bridge openings to the travelling public, bridge owners usually prefer fixed to movable bridges for new crossings and replacements. However, for crossings at which few bridge openings to permit passage of vessels would be required, overall long-term economics may favour the movable bridge. All depends on the relative amounts of vehicle and waterway traffic and the required vertical clearance in the channel.

Although movable spans were built for military bridges since ancient times, it was the advent of canal construction in Great Britain, Europe, and North America that made more movable bridges necessary. Many fixed masonry arch and timber bridges were built across those waterways, but where it was impractical to build approaches to crossings having acceptable grades, movable bridges were constructed. Because most movable bridges were manually operated, ingenious mechanisms were devised for gaining mechanical advantage. By 1800 the principal types of movable bridges had been developed, although in rudimentary form (Hovey, 1926).

The motions of all movable spans are a combination of rotation and translation; the differences between types are due to the axes selected for these displacements. In terms of displacement and axes of displacement, movable spans are commonly categorized as follows:

- rotation about a fixed horizontal axis   trunnion bascule
- rotation about a fixed vertical axis   swing
- translation along a fixed horizontal axis   retractile and transporter

- translation along a fixed vertical axis   vertical lift
- rotation about a horizontal axis that
  simultaneously translates   rolling bascule

Except for the transporter bridge, all of these movable bridge types, and most of their subtypes, had been placed into operation by 1800.

The railway boom of the nineteenth century provided another impetus for building movable bridges and introduced steam power for operating them. These railroad movable, mostly swing, bridges usually had to be replaced every few decades because of the rapid and enormous increases in locomotive and railroad car weights. In the USA, the trusses of early swing bridges had timber compression members and wrought iron tension members. Timber was replaced by cast iron as the metal industry developed and by 1874 a railroad bridge with a swing span 360 ft (110 m) long had been erected across the Missouri River at Booneville, Missouri, USA (Masterson, 1992). Swing spans were the preferred type of movable bridge until about 1900. Then movable bridge engineering practice changed, initiated by the completion of three bridges. They were:

- the double-leaf trunnion bascule of Tower Bridge across the Thames River in London, England, opened to the public in July, 1894
- the double-leaf Scherzer supporting the double-track Metropolitan Elevated Railroad across the Chicago River in Chicago, Illinois, USA, placed in operation early in 1895
- the double-leaf Scherzer rolling bascule supporting Van Buren Street across the Chicago River in Chicago, Illinois, USA, opened to traffic early in 1895. The approval enjoyed by these bascule bridges diverted the attention of bridge engineers from the swing type.

The advantages of a waterway with unlimited vertical clearance, but without a central pivot pier, were considered so important that 11 Scherzer patent bridges were built across the Chicago River between 1894 and 1907 (Wengenroth and Mix, 1975). Because of the Scherzer patents, and some maintenance problems, the Chicago City Engineer studied Tower Bridge and other European bascules and developed the Chicago type trunnion bascule. Contemporary to the adoption of the Scherzer and the Chicago type bascules was the development of the Strauss patent bascule, also in Chicago. Another movable bridge which was to have important implications for bridge design was the South Halstead Street Bridge across the South Chicago River, completed in 1895, according to a patented vertical lift design by Waddell. Then followed a 12-year hiatus in vertical lift bridge building because of Waddell's patents and other factors (Waddell, 1916). However, the events of the last decade of the nineteenth century changed the distribution of the various types of movables constructed in the USA, with a dramatic increase in the percentage of bascule and vertical lift bridges (Hardesty et al., 1975). A similar change of preferred movable bridge type occurred outside the USA. In the next section the major types of movable bridges will be briefly described together with their advantages and disadvantages.

12.2 Movable bridge types

The principal types of movable bridges may be categorized as swing, bascule, vertical lift, and retractile. Most bridges are of the first three types named. Each type has subtypes based on the manner of support for the movable part and the means of operation. Only the major subtypes will be considered herein. Another type, the transporter bridge, will not be described because it is considered obsolete. However, it should be noted that a transporter bridge with a 300 m (980 ft) span was built across the Manchester Ship Canal at Runcorn in 1905. It was in use for 50 years (Dickson, 1994). Transporter bridges were opened to traffic at Newport, South Wales in the UK and Rendsburg, Schleswig-Holstein in Germany, in 1906 and 1913, respectively, and are still in service.

12.2.1 Swing bridges

The movable span of a swing bridge, also termed the draw, rotates about a vertical axis which is often called the pivot axis. If the pivot axis is at midlength of the draw, the draw is said to have equal length arms. Sometimes, the arms are not of equal length and the draw is termed unequal-armed or bobtailed. The dead load (self-weight) of a swing span is usually balanced about the pivot. Hence, bobtailed spans require counterweights at the ends of the shorter arms for balance. However, short span swing bridges were built with only one arm and no counterweight, but special pivots were required for stability.

Figure 12.1 shows equal-arm swing bridges. The clear width of the navigation channel attainable is only about 80% of the length of an arm, or about 40% of the length of the draw, because of the space occupied by the pivot pier, the rest piers, and the fenders. In order to obtain a wider channel two swing bridges may be built in tandem. This was done at Yorktown, Virginia where a tandem swing bridge, the George P. Coleman Bridge, was opened to traffic across the York River in 1952. Each centre-bearing draw was 500 ft (153 m) long giving a 450 ft (137 m) wide clear channel (Quade, 1954). The bridge had a two-lane roadway which became inadequate with the increased traffic 40 years later and it was replaced by a wider bridge of similar design on the original piers (Green, 1996). A double-draw swing bridge with a distance between pivot axes of 167.5 m (550 ft) was built across the Suez Canal at El Ferdan, Egypt, in the early 1960s (Sedlacek, 1965). It was removed to permit widening of the canal and is being replaced by a tandem swing with bobtailed draws that will provide a 300 m (984 ft) wide unobstructed channel when fully open (Birnstiel and Tang, 1998).

12.2.2 Centre bearing swing bridges

Swing bridges are also categorized according to the manner in which the draw is supported at the pivot pier when the draw is in the open position. If the dead load is supported by a pivot bearing at the axis of rotation it is termed centre bearing. The draw weight is usually balanced on this pivot bearing, which may be mechanical or hydraulic. To prevent the draw from tipping under unbalanced loads, such as wind, balance wheels are provided that roll on a large-diameter circular track concentric

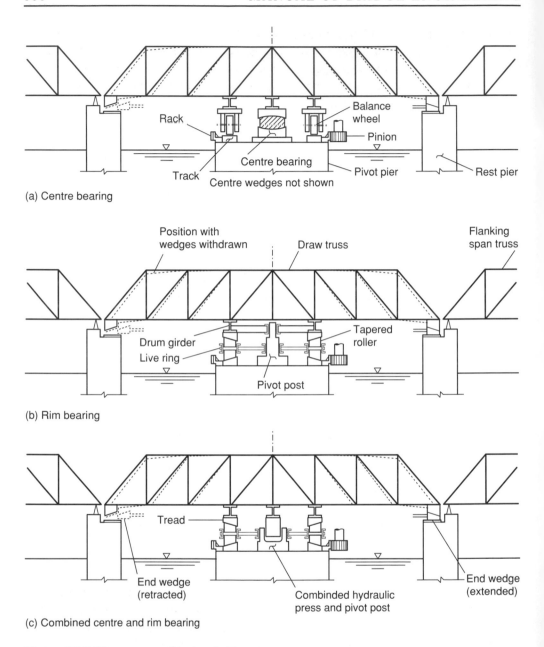

Figure 12.1 Three types of swing bridges.

with the pivot bearing. When the draw is balanced these wheels normally clear the track by about 5 mm (0.2 inch). The design intent is that the centre bearing support all the dead load when the draw is open. Figure 12.1(a) is a diagram of a centre bearing swing bridge with a plain mechanical pivot bearing, comprised of a lenticular bronze disc between hardened steel concave discs. Plain bearings are the norm. However, spherical antifriction rolling element bearings have been installed for shorter spans. The Hardesty and Hanover design for replacement of the Third Avenue Bridge across the Harlem River in New York City is a centre-bearing swing bridge with a spherical antifriction roller bearing supporting a draw weighing 5800 kips (2630 tonnes). Hydraulic bearings that support the entire dead load have been used for some centre bearing swing bridges since about 1900.

The live load on centre bearing swing bridges is usually supported by centre and end lift devices (often wedges) which are actuated when the draw is returned to the closed position. They support the free ends of the trusses and also provide a firm intermediate live load reaction for the trusses at the pivot pier.

Rotation of the draw is by means of mechanical or hydraulic machinery, or a combination thereof, except for small bridges which may be hand-powered. When the mechanical span drive is mounted on the draw one or more downward extending pinion shafts engage a rack mounted on the pivot pier and rotate the draw, as shown in Figure 12.1(a).

Alternatively, the mechanical span drive may be mounted on the pivot pier, in which case the pinion shafts extend upward to engage a rack mounted on the periphery of the drum girder. If the span drive utilizes hydraulic slewing cylinders, no rack is necessary.

12.2.3 Rim bearing swing bridges

Swing bridges in which all the dead load is supported by tapered (conical) rollers when the draw is in the open position are termed rim bearing. As shown in Figure 12.1(b), the rollers run on a circular track whose diameter is about the same as the spacing of the outer swing span trusses or girders. When the bridge is closed the rim bearing supports both dead load and live load. Rim bearings are used for wide heavily-loaded swing bridges, such as those over the Harlem River in New York City, or long spans such as at El Ferdan. Special load-equalizing framing is provided to transfer the loads from the bridge trusses (which may number 2–4) to the circular drum girder so that it is uniformly loaded along its length (at least for dead load). The load is transferred through the drum girder to a tapered tread plate supported by tapered rollers. The rollers rotate about axles that are oriented radially to the central pivot post. These axles are fixed in a live ring with provision for position adjustment of the rollers in the radial direction (toward or away from the pivot). The live ring is connected to a plain bearing (vertical axis) at the pivot post in order to maintain concentricity of the roller nest with the pivot axis. The tapered wheels transfer the load to a track plate bolted to a stool which is bolted and grouted to the pivot pier. Radical struts (spokes) connect the drum girder to a bearing at the pivot post so as to enforce the axis of rotation of the superstructure. Rotation of the draw may be by the same means as for the centre-bearing bridge.

12.2.4 Combined centre and rim bearing swing bridges

Combined centre and rim bearing swing bridges are those equipped with both a centre bearing and a rim bearing, as shown in Figure 12.1(c). The rim bearing is essentially the same as for the rim bearing bridge of Figure 12.1(b), although the rollers may be smaller and the upper set of radial struts is omitted. Usually the centre bearing is mechanical, either plain or antifriction rolling element. In this type of bridge the dead load is shared by these bearings when the span is swung open, with the rim bearing supporting most of the dead load. The distribution of live load between the centre and rim bearing is a function of the transverse rigidity of the load distribution framing. Uncertainly about load distribution is considered a disadvantage by some designers.

The centre bearing may also be a hydraulic press as shown in Figure 12.1(c). For such bridges the cylinder is pressurized during operation by water or oil so as to reduce the amount of dead load that the tapered rollers have to bear. Minimizing the load on the rollers minimizes the energy required to rotate the draw.

12.2.5 Other swing bridge types

Swing bridges have been built that are wholly or partly supported by pontoons, usually across protected waterways with slow currents. Those that were built with one end of the draw supported by a fixed pier are usually on waterways with small fluctuations of water level. Examples of pontoon-supported swing bridges were: the Galata Bridge at Istanbul, Turkey, built in 1912 and replaced in 1990 by a bascule bridge and a railroad bridge across the Suez Canal at Kantara, Egypt (Hawranek, 1936). The Galata Bridge was hinged at one end to a moored pontoon and originally swung by propellers driven by electric motors located in the pontoon at the free end of the draw. The Suez Canal bridge was supported by a fixed pier at the hinged end and was also swung by a propeller drive. Many of the shorter draws were swung open by pulling the pontoon at the free end using a chain drive.

12.3 Bascule bridges

There are two principal categories of bascule bridges; trunnion bascules and rolling bascules. The latter are often termed rolling lift bascules because the type was patented and promoted by the Scherzer Rolling Lift Bridge Company. Each of these two principal types has many subtypes, some of which will be briefly described.

12.3.1 Trunnion bascules

A leaf of a trunnion bascule rotates about a horizontal axes that is fixed in position. When the leaf is rotated to the open position the angle between the leaf and the horizontal is between 75 and 90°, making much of the distance between the bascule and the rest piers available for a navigation channel. There are many subtypes of trunnion bascules. They are categorized according to the locations of the trunnions and counterweights and articulation of the counterweight. Only one version of the simple trunnion and two subtypes of the heel trunnion will be considered herein.

Simple trunnion bascules

Figure 12.2 shows a single-leaf simple trunnion bascule with mechanical operating machinery located on the bascule pier. The trunnions are usually inserted through the webs of the bascule girders or trusses. They may rotate with the girders, or the bascule girders may rotate about fixed trunnions. The leaves of trunnion bascules only rotate, they do not translate. Counterweights are fixed to the bascule girders so that the energy required to move the leaf is minimized. The balance principal is that the trunnion axes should intersect (or nearly so) a line connecting the centres of gravity of the leaf and the counterweight.

On bridges with mechanical drives mounted on the bascule pier, curved racks are fastened to the bascule leaf, usually the bascule girders or trusses. Sometimes the racks are mounted on the leaf between the bascule girders. In either case, pinions mounted on the fixed structure engage the racks and the rotation of the pinions rotates the leaf.

Outside North American many trunnion bascules are operated by hydraulic cylinders. The cylinders are mounted on the bascule piers and the piston rods are connected to the leaves in a manner such that extending or retracting the rods rotates the leaf. An example of a large double-leaf bascule operated hydraulically is the new bridge across the Golden Horn at Istanbul, Turkey (Saul and Zellner, 1991).

For the bridge of Figure 12.2, live load on the channel side of the axis of rotation is supported by the bascule girder spanning between the trunnion and the live load

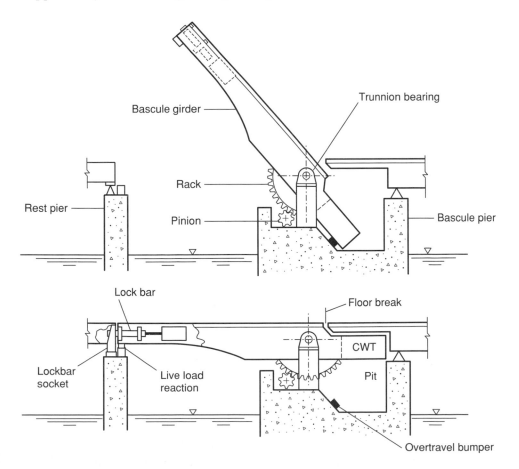

Figure 12.2 Single-leaf simple trunnion bascule bridge.

reaction on the rest pier. Live load on the shore side (rear) of the trunnion creates an uplift at the rest pier (if the live load moment exceeds the toe-heavy imbalance moment) which is resisted by wind-up in the machinery or by the span locks. It is considered desirable to minimize the extent of floor deck on the bascule leaf shoreward of the trunnion in order to avoid the tendency for uplift at the toe.

Simple trunnion bridges are often built in double-leaf form, especially for highways. Features of this type will be discussed later in connection with a description of the Tower Bridge. A complete structural design of a double-leaf trunnion bascule highway bridge is presented in Hool and Kinne (1943).

Unbalanced or partially balanced trunnion bascules, those with insufficient counterweight to balance the leaf, have been put into operation. They are built at sites where the trunnions are located at an elevation not far above high water level and it is not feasible to construct a bascule pier with a counterweight chamber which can be kept dry. Because of the large forces required to equilibrate a leaf that is in a partially or fully open position, unbalanced or partially balanced bascules are moved by hydraulic cylinders. For short spans, the larger machinery required by unbalanced leaves may be justified if it can be configured so as to be located at or above roadway level, thereby simplifying bascule pier construction and maintenance. However, if a large hollow pier is required to house the hydraulic machinery, then the rational for an unbalanced bascule with its higher first cost for machinery and higher operating cost is questionable.

An example of an unbalanced bascule is that supporting Route A1077 across New River Ancholme near South Ferriby in North Lincolnshire. The hydraulic cylinders lie outboard of the bascule girders, just below roadway level. The trunnion axis is located a few feet above the roadway.

Heel trunnion bascules

Heel trunnion bascules have the distinguishing feature that the counterweights are not fixed to the bascule girders or trusses. This offers the designer flexibility regarding location of the counterweight. They are attached to an auxiliary truss or balance beam. A parallelogram linkage connects the counterweight to the bascule girder. (The purpose of the linkage is to maintain a constant ratio of dead load leaf moment to counterweight moment about the heel trunnion axis for all angles of leaf opening.) Two subtypes of heel trunnions are the balance beam bascule (often called 'Dutch Style' bascule or lever bascule) and the Strauss patent heel trunnion.

Balance beam heel trunnion bascule

The essentials of the balance beam heel trunnion bascule are shown in Figure 12.3. The leaf rotates about a fixed axis at the heel in response to forces transmitted through the operating strut. A counterweight is fastened to the shoreward end of the balance beam or frame. The hinged bearings of the balance frame, the connecting rods, and the heel trunnions are located so as to form a parallelogram (BCDE). In this way the ratio of counterweight moment to leaf moment is maintained constant as the leaf is raised.

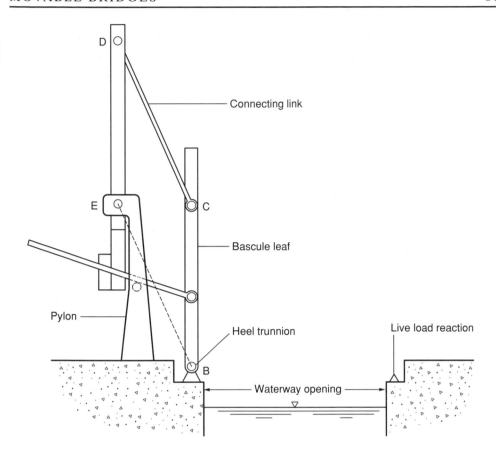

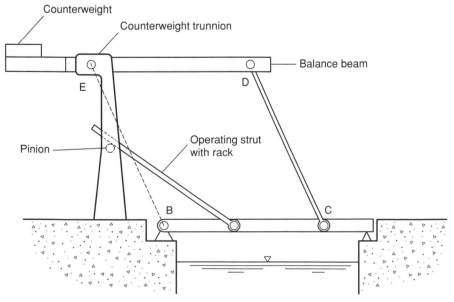

Figure 12.3 Balance beam bascule bridge.

Balance beam bascules have the advantage that most of the machinery and structure is located at or above deck level, thereby considerably simplifying foundation construction compared to trunnion bridges that require a bascule pier with a counterweight chamber. The leaf, pylon, and balance beam can be arranged such that the leaf is nearly vertical when the bridge is open with the result that almost the whole distance between the piers is available for navigation. However, the pylon and balance beam are considered unsightly by some persons. Also, some engineers are concerned by the susceptibility of the connecting rods to collision damage from vehicles. Another concern is that live load causes the leaf to deflect with resulting repetitive and cyclical rotation of the balance frame as live loads move over the span. This motion can be minimized by appropriate arrangement and sizing of members and detailing of joints. One solution is to utilize auxiliary lifting beams at deck level.

The balance beam trunnion bascule is a popular movable bridge type in the lowlands of Europe (hence the appellation 'Dutch style') and they have been powered by many means. Mechanical drives are located adjacent or inside the pylons with operating struts extending from sides of the leaf up through the forward face of the pylons as shown in Figure 12.3. Others have the machinery located atop the counterweight with the operating strut hinged to the pylon and extending upward through the balance frame. Some have had machinery mounted at the base or top of the pylon, driving curved racks fastened to the balance beams. The advantage of the latter arrangement is that operating struts are unnecessary.

Many balance beams bascules are powered by hydraulic cylinders. In one scheme the toothed operating struts of Figure 12.3 are replaced by double-acting hydraulic cylinders pivot-mounted to the pylons. An example of such a bridge is the roadway bridge at Geversdorf, near Breman, Germany. This bridge has a torsionally rigid leaf that is 11.5 m (38 ft) wide spanning 30 m (98 ft) and which is operated by one double-acting hydraulic cylinder mounted outboard of one side of the leaf (Ortmann, 1990). Alternatively, the cylinders can be mounted horizontally at a level slightly below and to the rear of the pylons with the piston rod acting on brackets fastened to the leaf girders, as at the Ennerdale Bridges across the River Hull in East Yorkshire where the rods are at a 10° angle to the horizontal when the bridge is closed.

Strauss patent heel trunnion bascule

A single-leaf Strauss heel trunnion with overhead rotating counterweight frame (rocker) is shown in Figure 12.4. The geometrical figure BCDE is a parallelogram and the centre of gravity of the counterweight at F is so located that the line EF is parallel to the line between the centre of gravity of the leaf A and the heel trunnion B. As a result the ratio between the dead load leaf moment about B and the counterweight moment remains essentially constant during rotation of the leaf. The longest single leaf bascule was the St Charles Air Line Railroad double-track Strauss heel trunnion built across the Chicago River, at Chicago, Illinois in 1919. It was later moved a short distance when the waterway was relocated and the leaf was shortened slightly.

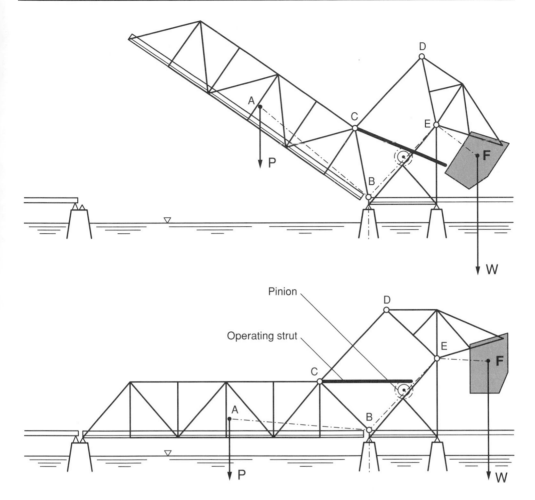

Figure 12.4 Strauss heel trunnion bascule bridge.

The Strauss leaf rotates about the heel trunnion B in response to a force transmitted to the leaf by the operating strut (a rack is fastened to the strut) which is hinged
at the top chord joint C and engages the output pinion of the span drive machinery
mounted on the counterweight frame. The trunnions at B, C, D and E are heavily
loaded during motion. As has been stated by Hovey, the reaction at the heel trunnion B when the bridge is closed may reverse, depending on the proportions of the
structure, and this effect must be considered in the heel trunnion bearing design.
Strauss heel trunnion bridges are also difficult to balance because of the motion of
the heavy operating struts.

12.3.2 Rolling bascules

Most rolling bascules have the distinguishing feature that the ends of the main spanning members (bascule girders or trusses) are cylindrical and the movable span (leaf)
rolls on these surfaces during opening and closing. Figure 12.5 shows a single-leaf

deck-type Scherzer patent bascule. As the curved ends of the girders roll shoreward the leaf tilts open to clear the channel. The leaf simultaneously rotates and translates. To close the bridge the leaf rolls toward the channel. The counterweight is fixed to the bascule girders or trusses. This bridge type was developed and promoted by the Scherzer brothers of Chicago at the end of the nineteenth century and hundreds of such bridges were built in North America and Europe. Other styles of rolling bascules, such as the Rall, were patented, but few were built.

Rolling bascules may be deck type bridges (Figure 12.5) or through trusses or half-through (Pony) girders or trusses. They may be single or double leaf. Early double-leaf bascules designed by Scherzer were constructed so as to act as three-hinged arches in resisting live load located forward of the centre of roll. This action is illustrated in Figure 12.6. The concentrated live load W is equilibrated by the pressure lines passing through the three hinges; a midspan hinge and the two hinges formed at the bascule piers by the front teeth of the track and the corresponding sockets in the treads.

Later double-leaf deck Scherzers were also equipped with supplementary live load reactions either at the rear of the leaf or at the front wall. Normally they are not active; they only became active when a leaf is lowered too far, as can happen when only one

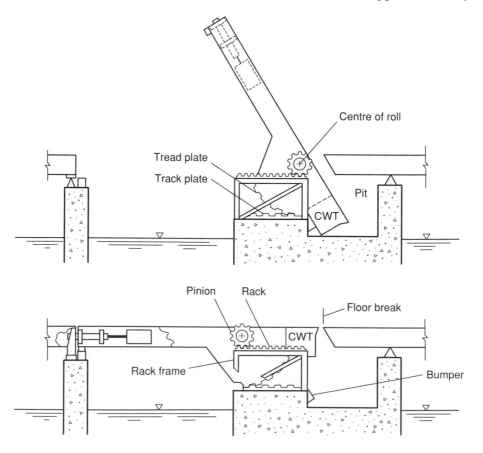

Figure 12.5 Single-leaf Scherzer rolling bascule bridge.

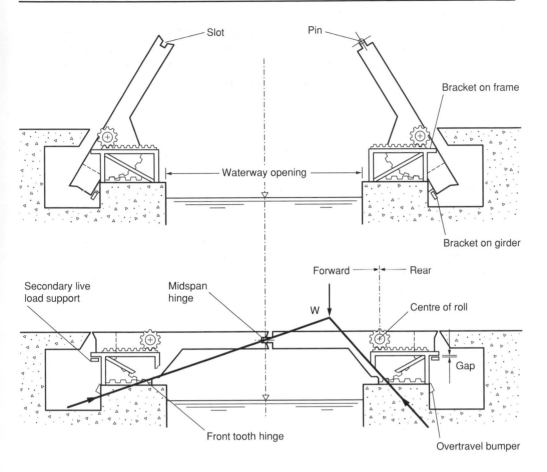

Figure 12.6 Three-hinged arch action of double-leaf Scherzer rolling bascule.

leaf is being closed while the other leaf is open. Live load rearward of the centre of roll produces moment tending to open the bridge. This action is resisted by the machinery which is 'wound up' when the midspan hinge is seated.

In Europe, rolling bascule bridges were constructed with three-hinged arch action for live load and for a portion of the dead load. In these bridges part of the counterweight is lifted by machinery at the rear wall of the bascule pit at closing thereby shifting the centre of gravity of the leaf forward toward the channel. This compresses the arch and seats the hinges prior to the addition of live load. The result is a very rigid span. Because the leaves are considerably span-heavy after the centre of gravity is shifted, live loads on the rear of the leaf do not lift the midspan hinge. Examples are the Langebro and Knippelsbro Bridges in Copenhagen, Denmark. The counterweight machinery has been described elsewhere (Rode and Nielsen, 1954).

During rehabilitation projects some engineers have altered double-leaf Scherzer bridges designed for three-hinged arch action from that behaviour to double cantilever action in order to simplify electrical control design. However, such an alteration is likely to reduce the rigidity of the closed bridge.

Many rolling bascules are operated by electro-mechanical drives mounted on the leaves as shown in Figure 12.5. A pinion on each side of the leaf engages a rack mounted on the foundation. As the pinion is turned by machinery on the span the bascule leaf rolls forward or backward, depending on the direction of rotation of the pinion. Because friction between the tread and the track may be insufficient to transmit the force needed to hold the span in the open position against wind, and to maintain the alignment of the tread and track, the track usually has upward projecting lugs which engage corresponding socket holes in the tread. In effect, the tread is a segment of a large gear and the track is a rack.

Although most rolling bascules were constructed with mechanical span drive machinery mounted on the leaf, which translates the leaf, others have machinery mounted on the flanking fixed span which is used to pull or push the leaf using toothed struts, hydraulic cylinders, or chains. Hydraulic cylinders have also been mounted horizontally below and parallel to the track and operate the leaf by moving a bracket attached to the segmental girder which extends downward below the track.

Vertical lift bridges

The movable span of a vertical lift bridge is raised in order to provide clearance for the passage of vessels. The ends of the lift span are connected to wire ropes that pass over sheaves at the tops of the towers with the far ends connected to counterweights. These counterweights balance (or nearly so) the weight of the lift span. For large lift bridges with high lifts the weight of the counterweight ropes is so large that they themselves are usually counterweighted by an auxiliary system so as to minimize power requirements for operation. Diagrams of the three most important types of vertical lift bridges are shown in Figure 12.7. The auxiliary counterweight systems are not shown in these diagrams, for clarity.

Since Waddell's design and patent of the first practical vertical lift bridge, many such bridges have been erected. Because of Waddell's patents, the relationship between his and successor firms, and firms that were founded by a former partner of Waddell, most large vertical lift bridges built in the USA prior to 1950 were designed by two consulting engineering firms.

Tower drive vertical lift

In the tower drive vertical lift bridge depicted in Figure 12.7(a) there is span drive machinery in each tower that rotates the counterweight sheaves. The forces necessary to raise the span are transmitted to the counterweight ropes by friction. The action is similar to that of a traction drive passenger elevator in a building. Both ends of the lift span should raise and lower at the same rate so that the lift span remains horizontal and does not wedge itself between the towers during motion. There are various electrical/electronic means of controlling the drives in the two towers so that skew is kept within permissible limits. It should be noted that the force necessary to raise the lift span at each end may differ due to unequal machinery friction, etc. The world's longest vertical lift span, 558 ft (170 m), is a tower drive vertical lift that

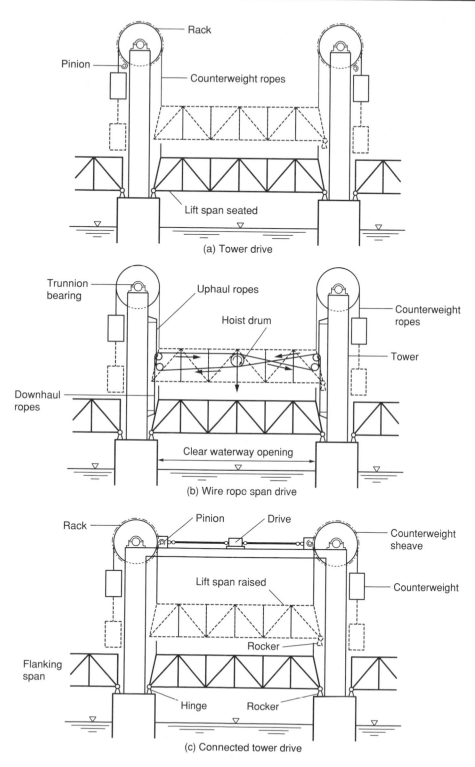

Figure 12.7 Three types of vertical lift bridges.

supports a single-track railroad across Arthur Kill between Staten Island, New York and Elizabeth, New Jersey (Hedefine and Kuesel, 1959). The longest tower drive vertical lift highway bridge is the Marine Parkway Bridge over Rockaway Inlet, in the Borough of Queens, New York City, with a lift span 540 ft (165 m) long, which was erected in 1937.

Span drive vertical lift

Figure 12.7(b) depicts a span drive vertical lift powered by wire rope hoist machinery located on the span, usually above the roadway or tracks at midspan. The wire rope system hauls the lift span up or down. The directions of rope travel shown in the figure are for the lift span being lowered. Because the four rope drums are usually geared to a common drive shaft, the span cannot skew appreciably, unless there is a problem with a haul rope. This type of span drive vertical lift bridge was perfected by the Waddell firm and many were built by the Pennsylvania Railroad.

Wire rope span drive vertical lift bridges have been built with the primary drive at midspan transmitting power via long line shafts to the secondary speed reductions and hoist drums located at the ends of the span. An example of such a bridge is the double-track Burlington Northern Santa Fe Railroad Bridge at Portland, Oregon. The lift span has a length of 516 ft (157 m) and the normal lift is 146.25 ft (45 m) giving a vertical clearance of 200.7 ft (61 m) above low water (Birnstiel and Tang, 1998).

Span drive vertical lifts that are raised by a rack and pinion drive at the towers have been built for shorter spans as, for example, those over the Illinois River in the state of Illinois, USA.

Connected tower drive vertical lift

Figure 12.7(c) is a diagram of a connected tower drive vertical lift bridge. The lift span is balanced by counterweights. However, because this type of lift bridge is most suitable for short spans of low to moderate lift, the counterweight ropes are usually not balanced by auxiliary counterweight systems. The span drive machinery is usually mounted on the structure connecting the towers. Primary machinery is located in a machine room at midspan from which line shafts extend to secondary bevel speed reducers at the towers, midway between the counterweight sheaves. From the secondary reducers shafts extend to pinions that engage the racks fastened to the counterweight sheaves. The force necessary to move or hold the lift span is transmitted between the sheaves and the counterweight ropes by friction.

Tower-hoist drive

For short and wide vertical lift bridges it is sometimes considered architecturally desirable to build separate towers at each corner of the lift span. Separate towers above the roadway level avoid the overhead clutter of the more conventional lift bridge types depicted in Figure 12.7. They are especially desirable if the towers are to be of reinforced concrete, which can be designed to contain the counterweight sheaves and the counterweights.

The span drives for tower-hoist drive bridges are essentially winches that are usually located on the tower piers below the roadway level. The span is lifted by haul ropes wrapped around a common drum and attached to each counterweight. They pull the counterweights downward to lift the span and pay out to raise it. Modern bridges of this type utilize winch drums rotated by multiple radial piston hydraulic motors. The amount of lift on opposite sides of the channel is coordinated electronically to limit skewing of the span. An example of a tower-hoist bridge is the Centenary Bridge over the Manchester Ship Canal between Eccles and Trafford Park near Manchester.

12.4 Retractile bridges

Retractile bridges translate along a fixed axis. The shorter spans are supported by wheeled bogies that roll on tracks in order to clear the waterway. Many arrangements of tracks, leaves, and auxiliary decks adopted in France have been described by Mehue (1991). Few rolling retractile bridges were constructed in the USA. Most rigid units that are mounted on wheeled trucks roll on crane or railroad rails oriented at 45° to the navigation channel. An example is the Borden Avenue Bridge across Dutch Kills in the Borough of Queens, New York City. Because about one-half of the movable span cantilevers beyond the front tracks when the bridge is rolled off the rest pier supports, retractile bridges that roll on land are not suitable for large spans.

In order to overcome the limitation of cantilevered rigid units and wheeled trucks, floating spans have been utilized that retract under the approaches (which may be fixed or also floating). They are pontoons, or are supported by pontoons, and are usually moved by wire rope hoisting systems. The longest movable span is the Ford Island Bridge at Pearl Harbor, Hawaii. The draw is a single reinforced concrete pontoon 930 ft (283 m) long that provides a 575 ft (175 m) clear waterway opening for navigation when withdrawn under the fixed trestle approach (Abrahams, 1996).

Pontoon retractile bridges are constructed when the depth to adequate foundation material is excessive or a very wide waterway channel is necessary. Such is the case for the Evergreen Point Bridge over Lake Washington at Seattle, Washington. Storms have created waves that have damaged the retractile span mechanism at least three times since it was constructed *circa* 1960 (Daniels, 1999).

12.5 Span drive machinery

Power for operating early movable bridges was mostly human and small bascule and swing bridges are still operated by hand. After steam engines became common in the early 1800s this power source was used for bridges well into the twentieth century. In Great Britain hydraulic utilities (Hydraulic Power Companies) were established in the 1800s which transmitted high pressure water in mains under the streets for industrial power. Liverpool and Kingston-upon-Hull for example, had such utilities. As the cranes and other dockland machinery were powered by the hydraulic system it was a natural extension to utilize this source for powering movable bridges. An example is the Scott Street Bridge across the River Hull at Kingston-upon-Hull in East Yorkshire. This double-leaf trunnion was operated by water hydraulic cylinders. Water from the hydraulic utility was increased in pressure by a multiplier and stored at the higher pressure in accumulator until required to move the leaves. When the hydraulic utility suspended service, electrically driven pumps were installed to supply the accumulators. At other bridges, such as Tower Bridge, the hydraulic pressure was developed by reciprocating steam pumps and stored in gravity accumulators.

In the USA it was not uncommon to power bridge machinery directly by gasoline or diesel engines, either as the primary or emergency movers. However, as electrical power became available it was adopted for operating movable bridges and very sophisticated power and control systems are now routinely installed. The topic is beyond the scope of this chapter and information may be found in FHWA (1977), Birnstiel (1990) and AASHTO (1998). The electric motor may drive gearing directly (electro-mechanical system) or drive a pump with the power being distributed hydraulically (electro-hydraulic-mechanical system). Span drives may be assembled from many combinations of equipment. In what follows three common span drive arrangements will be described.

Figure 12.8 shows three span drive arrangements; mechanical, hydraulic motor, and hydraulic cylinder, all powered by electric motors. For all types the objective is to convert the high-speed low-torque rotation of the electric motor to the low speed-high torque rotation necessary to move the massive span.

Figure 12.8(a) illustrates a basic mechanical drive that is used to move swing spans, bascules and lift spans. The electric motor is connected to the input shaft of a geared primary speed reducer. A differential is incorporated within the speed reducer in order to equalize the torque in each of the two output shafts even when their speeds differ (the function of this differential is essentially the same as that at the drive axle of an automobile). A brake, called the motor brake, is coupled between the motor and the speed reducer.

The output torques from the primary reducer are transmitted via floating shafts to the secondary speed reducer. For bascule and lift bridges these are normally parallel shaft reducers. For a swing bridge the secondary reducers would be of the bevel type so that the output shafts are oriented at 90° to the input shaft. The outputs of the secondary reducers are coupled to vertical pinion shafts. The torque in these shafts is transmitted to pinions which engage a rack. For swing spans, tower drive lift, and trunnion bascules the rack would be circular. For rolling bascules (Scherzer) and heel trunnion bascules (Strauss) the racks would be straight (radius of infinity). Another brake, called a machinery brake, or shaft brake, is shown on the input shaft to the secondary reducer. Machinery brakes should be installed in the gear train as close to the rack pinion as possible. Of course, the closer the brake is to the rack the greater the required torque rating of the brake.

A span drive in which the electric motor and the primary reducer are replaced by radial piston hydraulic motors is shown in Figure 12.7(b). The torque output of the motors will be matched to the degree that they are powered from the same pressure source and the friction losses from the pressure source to the motors are equal. The remainder of the drive is the same as in Figure 12.7(a).

Figure 12.7(c) shows a hydraulic cylinder drive suitable for slewing swing bridges and tilting bascules,. The drive has also been used for vertical lift bridges of low lift. However, the controls have to limit skewing of the lift span.

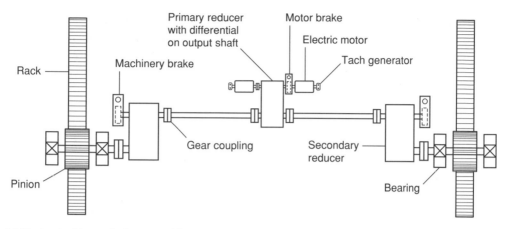

(a) Mechanical transmission span drive

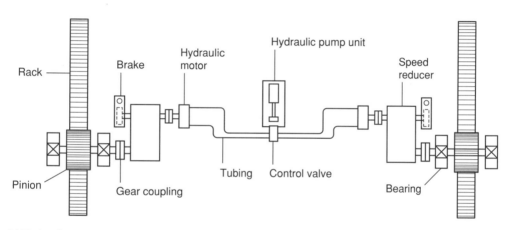

(b) Hydraulic motor span drive

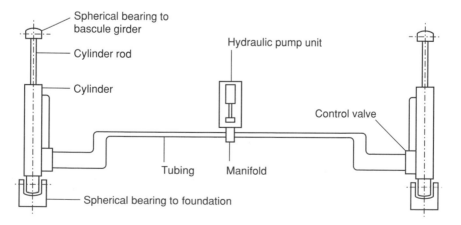

(c) Hydraulic cylinder span drive

Figure 12.8 Three types of span drives.

12.6 Stabilizing machinery

Besides the machinery that moves the movable span, machinery is required to stabilize it when at rest or during motion. Major items of stabilizing machinery will be described for swing, bascule, and vertical lift bridges.

12.6.1 Swing bridges
Centre bearings

Most mechanical pivot bearings are two-part bronze on steel bearings with a spherical interface. Some older bridges have three-part bearings comprising a steel concave disc resting on a convex bronze lens which, in turn, rests on a concave steel disc. Commercial anti-friction spherical roller bearings have also been utilized as pivot bearings.

Centre bearing swing bridges are also built with hydraulic bearings. Essentially, they are hydraulic jacks which are used to raise the draw off its central support just prior to, and during, the rotation of the draw. An air-over-oil intensifier/

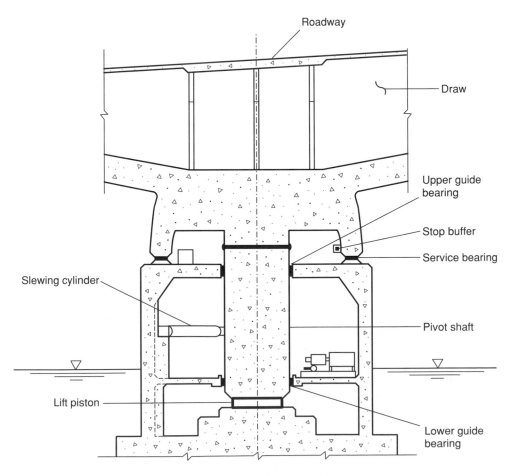

Figure 12.9 Centre bearing of Harbor Island Swing Bridge.

accumulator was introduced for a swing bridge in Hamburg about 1905. Modern hydraulic bearings rely on higher volume pumps, thereby making accumulators unnecessary. An example of a recently constructed centre bearing swing bridge with a hydraulic pivot bearing is the Harbor Island Swing Bridge across the Duwanish River in Seattle, Washington, USA, shown in Figure 12.9 (Green, 1991). It is a tandem bob-tailed swing bridge of prestressed concrete with pivots spaced 146 m (480 ft) apart.

Combined centre and rim bearing bridges may have a central pivot with a hydraulic press to reduce the load on the rim bearing during rotation. The fluid used in the early English hydraulic centre bearings was water, stored under pressure in gravity-loaded accumulators. An example of such construction is Barton Swing Aqueduct.

Rim bearing

Rim bearing swing spans rest on a circular girder called a drum girder. A tapered plate (tread plate) is fastened to the underside of the drum girder. The tapered plate really has a convex conical surface. It bears on a set of conical rollers whose axes are oriented radially, intersecting with the pivot axis. The wheels, in turn, roll on a tapered plate called a track plate that is fastened to a chair casting. The nest of tapered rollers is held concentric with the pivot axis by the live ring and radial members connected to a bearing (also called a spider) which rotates about the pivot post. Radial struts also connect the drum girder to another bearing at the pivot post in order to force the draw to rotate concentrically with the rim bearing. The pivot post does not support vertical loads from the superstructure. The design intent is that all the self-weight of the superstructure be supported by the rim bearing when the bridge is in the open position. When closed, the rim bearing supports live load and dead load.

Antifriction slewing bearings developed by the military and adopted by the mobile construction crane industry have also been used for rim bearings. Slewing bearings 5 m (16 ft) in diameter with three rows of rollers were used for the rim bearings of a tandem swing bridge in Denmark (Thomsen and Pedersen, 1998).

End lifts and centre lifts

The ends of swing bridges are lifted to accommodate the self-weight deflection of the cantilevered arms. For centre and rim bearing swing spans that have girders or trusses structurally continuous between the rest piers (continuous beams over three or four supports), the draw ends are lifted so that there will be an upward reaction at the truss ends for all combinations of live load and temperature gradient. The intent is that the trusses act as statically continuous trusses from rest pier to rest pier in equilibrating live loads. This should be so even though the truss reaction detail at the rest pier can provide only an upward reaction. Many devices have been developed for end lifts including; wedges, toggle lifts, eccentric wheels, and hydraulic jacks.

The ends of tandem swing bridges are not lifted at closing because to do so would only tend to tilt each draw. The trusses must be designed as cantilevers for the live load. However, shear locks are provided to maintain matching elevations at the floor breaks (Quade, 1954).

12.6.2 Bascule bridges
Trunnion bascules
Stabilizing machinery for single-leaf trunnion bascule bridges is shown in Figure 12.2. It comprises the trunnions, live load reactions and toe locks. The trunnion bearings are normally bronze plain bearings. There are many variations of toe locks; screw actuated, crank-type, and hydraulic.

Double-leaf bascules have, of course, no central rest piers. The live load supports are at the front wall (as for Tower Bridge) or at the rear of the counterweight. A double-leaf bascule with rear live load reactions is described elsewhere (Birnstiel, 1996). The advantage of live load reactions at the front wall is that the cantilevered length of the bascule girder is minimized. However, extra heavy live load can cause uplift on the trunnion bearing caps if the distance between the live load supports and the trunnion bearings is comparatively small. To overcome this, some bridges are equipped with primary live load reactions at the front wall and secondary live load reactions at the end of the counterweight. There is a small gap at the secondary reactions so that they will only act when the front of the leaf is heavily loaded. There is also a gap under the trunnion bearing cap.

Double-leaf bascules in the USA normally have shear locks at midspan. UK and European engineers sometimes detail the midspan joints so that they resist live load bending moment as well as shear, as for example at the New Galata Bridge (Saul and Zellner, 1991). An older, smaller, double-leaf trunnion bascule with midpsan locks arranged to develop live load bending moment resistance is the Kronprins Frederiks Bro across Roksilde Fiord, Denmark. Another, completed in 1916 and rehabilitated in 1957, is that across the Eider River near Friedrichstadt, Schleswig-Holstein.

Rolling bascules
Stabilizing machinery for rolling bascules may include tread and track, live load supports, tail locks, and midspan shear locks. There are many varieties of these devices and only a few will be described subsequently.

Tread and track are machined plates that transfer the weight of the rolling mass plus live load to the bascule pier foundation. All the force is transmitted by line loading contact between the tread and track which causes plastic deformation of the metal each time the leaf is rolled. In effect, the material is being cold-rolled during bridge operation, which lengthens and spreads the tread and track, but not by the same amount. The repeated plastic straining causes accumulated damage which limits the life of the tread and track. Tread and track should be treated as expendable items and be detailed to facilitate replacement.

Because friction between the tread and the track may be insufficient to transmit the force needed to hold the span in the open position against wind, and to maintain the alignment of the tread and track, the track usually has upward projecting lugs which engage corresponding socket holes in the tread. In effect, the tread is a segment of large gear and the track is a rack. For leaves supported by two bascule girders, both girders have tracks with lugs. For leaves with three or more bascule

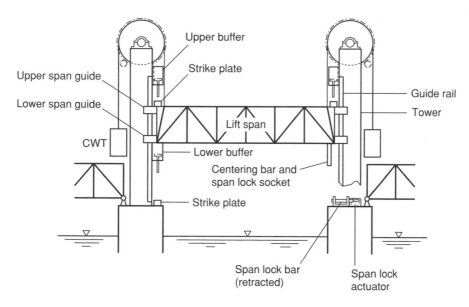

Figure 12.10 Vertical lift bridge stabilizing machinery.

girders it is usual in the USA for the tracks of all girders to have lugs. In Europe, usually only the exterior girders are equipped with lugs.

Live load supports and locks of rolling bascules are similar to corresponding equipment of trunnion bascules.

Double-leaf rolling bascules may have midspan joints similar to trunnion bascules if they are designed as double cantilevers for live load. It should be noted that many Scherzer rolling bascules (especially deck bridges) constructed between 1908 and 1917 were designed to act as three-hinged arches to resist live load on the forward part of the leaf.

12.6.3 Vertical lift bridges

Stabilizing machinery for vertical lift bridges is shown in Figure 12.10. The major stabilizing mechanical equipment are the wire rope, counterweight sheaves and trunnions, buffers and strike plates, guide rail and guides (sliding or rolling), live load reactions, centring devices, and span locks. The buffers for vertical lift bridges usually are air-operated in order to minimize maintenance. The centring device is necessary to align the lift span laterally; it is an important device for railroad bridges.

12.7 Significant movable bridges

Six movable bridges which are significant for historical or technical reasons will be briefly described subsequently. The emphasis will be on the structural and mechanical features unique to movable bridges.

Figure 12.11 Barton Swing Aqueduct – overall view.

12.7.1 Barton Swing Aqueduct

Queen Victoria formally opened the Manchester Ship Canal on 21 May 1894. The Ship Canal was in fact opened officially for business over its full length on 1 January 1894. There was a special celebratory procession on that day and one of the movable bridges opened to permit passage of the vessel carrying the official party was the Barton Swing Aqueduct. The Ship Canal was conceived as means of improving navigation along the Rivers Mersey and Irwell to permit ocean-going vessels to sail to Manchester. At Barton the Bridgewater Canal crossed the River Irwell on a stone arch aqueduct built about 1760. As the profile of the Ship Canal was to be about 8 m (26 ft) lower than that of the Bridgewater Canal, a movable aqueduct was necessary to support the Bridgewater Canal over the Ship Canal in order to permit passage of ocean-going vessels in the Ship Canal (Dickson, 1994). The alternative, two flights of locks, was not selected because of the need to conserve Bridgewater Canal water. Instead a swing bridge was built of construction that was conventional at the time (1892–1894) except that instead of supporting a roadway it supported a wrought iron trough (Williams, 1894). Gates with novel seals were provided at the ends of the trough, and at both abutments, which are closed before rotating the draw in order to keep water in the trough and in the canal when the draw is swung open.

The photograph of Barton Swing Aqueduct (Figure 12.11) was taken from the adjacent road swing bridge and shows the swing aqueduct in the closed position at the centenary. It is a combined rim and centre bearing bridge with the pivot on a long island that serves as the fender for the Barton Road Bridge and the aqueduct. The brick tower in the foreground of the photograph houses the control room on the top floor from which the bridge is operated. The lower part formerly contained the gravity accumulator which stored water under pressure for the oscillating engines

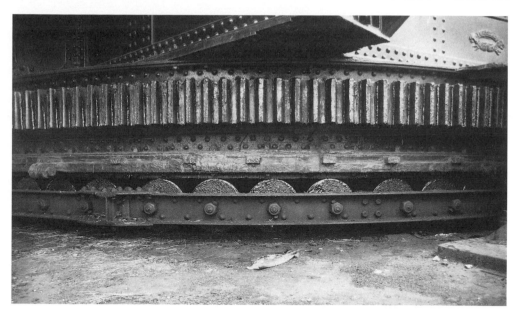

Figure 12.12 Barton Swing Aqueduct – rim bearing.

(radial piston motors) that powered both bridges. Such motors were very common at the time. The draw had four hydraulic engines, two for rotating the draw and one at each end to operate the gates. The same hydraulic engines turn the span today. However, electric pumps furnish the pressurized water instead of the gravity accumulator, which has been removed.

The draw structure comprises two Pratt-type trusses built-up of wrought iron plates and angles riveted together. They are 234.5 ft (71.5 m) long and spaced 22.17 ft (6.7 m) on centre. The 6 ft (2 m) deep trough, which takes the place of the roadway in the usual swing bridge, is built-up of 3/8 inch (9 mm) thick wrought iron plates. Because the ends of the trough are not perpendicular to the longitudinal centreline, the trusses are offset longitudinally. The draw rotates about 75° in one direction to open. The weight of the moving mass is 1400 tons (1272 tonnes), of which 57% is water.

The trusses are supported by cross-girders bearing on a circular drum box girder of 27 ft (8.2 m) mean diameter. A cross-girder is shown in Figure 12.12 bearing on the drum girder, to which the rack is fastened. The rack is mounted on the movable draw because the span drive machinery is mounted on the fixed pier. Below the drum girder, the exterior portion of the rim bearing is visible. The tapered track plate (lower path) is set in the granite blocks forming the pivot pier cap. Sixty-four tapered rollers of 14.4 inch (366 mm) mean diameter, which are held in place circumferentially by the live ring, roll on the track. They are loaded by the tread plate which receives load from the drum girder through the rust joint. The ends of the wedges (in pairs) that were used to obtain uniform bearing of the drum girder on the rollers during erection are visible.

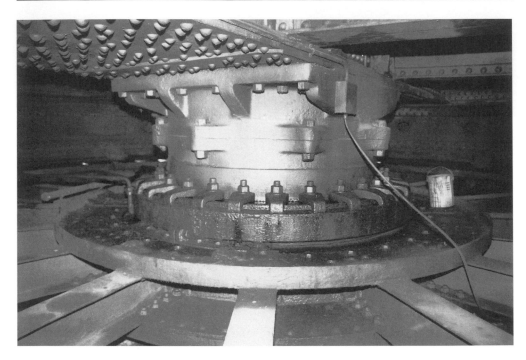

Figure 12.13 Barton Swing Aqueduct – pivot press.

Figure 12.13 depicts the centre pivot press. The radial arms (spider) connect the live ring to the pivot bearing of about 7 ft (2.1 m) diameter around the outside of the press base. The piston diameter is 4.8 ft (1.5 m). The inner ends of the dogs hold the flexible hydraulic packing in place between the base (the hydraulic cylinder) and the piston. The rollers, paths, and rust joint are not the original but are replacements *circa* 1928 (Brown, 1929). Originally the hydraulic pressure exerted an upward force approximately equal to the weight of the water in the trough.

The Barton Swing Aqueduct is an example of swing bridges built in Great Britain at the end of the nineteenth century. Distinguishing features are a combined rim and centre bearing with a hydraulic press incorporated in the pivot. Turning of the span is accomplished by geared machinery fixed to the pivot pier driven by water hydraulic engines. Swing bridges are no longer built this way but some of the Barton Swing Aqueduct features merit consideration for new heavy spans.

12.7.2 Tower Bridge

Tower Bridge was officially opened to traffic on 9 July, 1894, 6 months after the Barton Swing Aqueduct. It was the largest and most technically advanced trunnion bascule at the time. Features of this bridge were later adopted by engineers in the USA, especially the City Engineer of Chicago, and elsewhere. Figure 12.14 is a cross-section through a leaf of this double-leaf simple trunnion bascule (Cruttwell, 1894). The bascule trunnions are 226.5 ft (69 m) on centre giving a clear waterway opening of 200 ft (61 m). Each 49 ft- (15 m)-wide leaf is comprised of four bascule

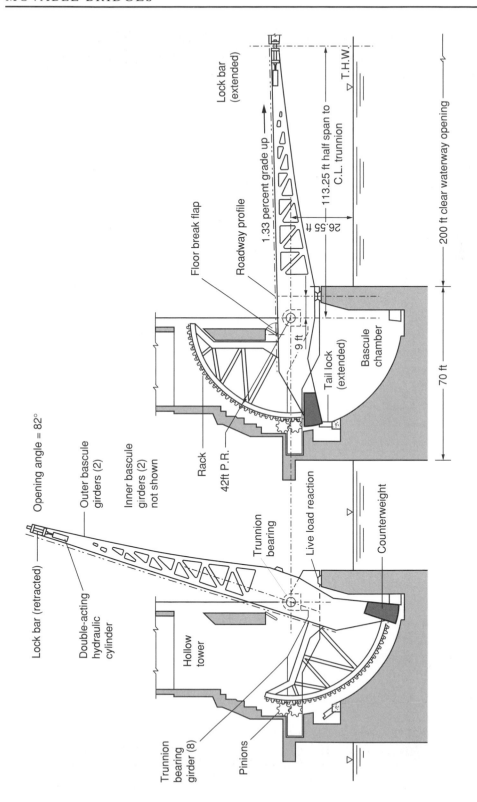

Figure 12.14 Tower Bridge – cross-sections.

girders that supported a buckle plate floor system which extended a considerable distance rearward of the trunnions. The leaf was counterweighted below deck level by cast iron and lead blocks at the rear end of girders.

The stabilizing machinery comprises the trunnion bearings, live load reactions, midspan shear locks, and tail locks. Each trunnion was 21 inches (533 mm) in diameter and 48 ft (16.8 m) long passing through the webs of the four bascule girders and keyed to each. Eight bearings supported each trunnion shaft. They were custom roller bearings with live rollers $4^7/16$ inch (113 mm) in diameter and $22^1/2$ inches (570 mm) long rolling in cast steel housings. The trunnion bearings were supported by individual girders spanning between the front and rear walls of the piers. The live load supports were placed on the front wall of the bascule pier 9 ft (3 m) forward of the trunnions which is only 8% of the bascule girder cantilever. Because the deck extended well rearward of the trunnion, rotating strut tail locks were provided. These were engaged or disengaged by double-acting hydraulic cylinders.

Span drive machinery was mechanical, powered by water hydraulic engines. Reciprocating steam pumps pressurized water to 700 psi (48 bar) which was stored in gravity-loaded accumulators (Homfray, 1896). A total of six accumulators were installed, two at the engine house at the south abutment and two in each bascule pier. The pressurized water operated the single-acting reciprocating engines which served as prime movers. There were two engines of unequal power at each side of the leaf, the smaller one for normal conditions. The intention was that under severe wind the larger engine should supplement the smaller. Through reduction gearing the engines on one side of the bridge turned one of the pinion shafts engaging racks mounted

Figure 12.15 Ennerdale Bridge.

on the exterior bascule girders. The pitch radius of the rack was 42 ft (12.8 m), about 37% of the leaf span. The machinery on the other side rotated the second pinion shaft. The span drive machinery was rehabilitated in 1976. Rotary oil hydraulic motors now power the bridge through geared speed reducers.

12.7.3 Ennerdale Bridge

The Ennerdale Bridge across the River Hull on the boundary between the City of Kingston upon Hull and the neighbouring East Riding of Yorkshire was completed in 1997. As is evident from Figure 12.15 it comprises two adjacent units of the balance beam bascule type. The bridge was constructed after a failed attempt to tunnel under the river. Each leaf is 13 m (43 ft) wide by 33 m (108 ft) long and weights 270 tonnes (300 tons). An overhead balance frame, with the connecting rods at about the 0.7 point of the leaf span, balances the deck. The machinery and controls are very sophisticated with many degrees of redundancy. There are, essentially, duplicate hydraulic and electrical power generating systems. Each bridge unit has four double acting cylinders, located shoreward and below the heel trunnions, which act on brackets extending downward from the leaf. Normally, the bridge operates utilizing all four cylinders. However, it can operate with only three acting. Each cylinder is about 400 mm (15 inches) in diameter with a stroke of 3.5 m (11.5 ft) and can exert a force of 200 tonnes (220 tons) at a maximum operating pressure of 200 bar (3000 psi). The normal leaf opening angle is 60°, which is achieved in one minute although it can be operated to a full 89° opening position.

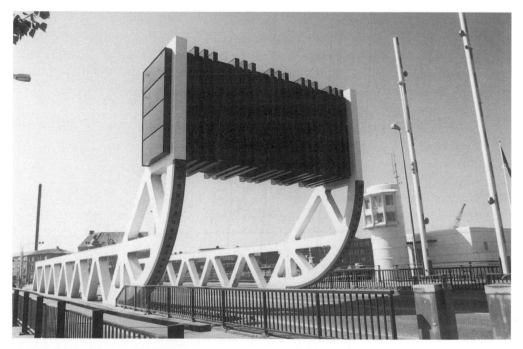

Figure 12.16 Halsskov Bridge.

The control system is a very sophisticated PLC (programmable logic controller) system with extensive use of sensors and other instrumentation. Data logging is performed for every bridge lift and is recorded on tape for archiving. A computer displays bridge parameters and messages in graphic form for use of bridge operators, but the bridges are also operable without the graphics. A program in this computer also enables engineering analysis of the hydraulic equipment and displays of the fluid power system state. It is also possible to replay past openings and evaluate past events. The use of computers for controlling and diagnosing hydraulically operated movable bridges was brought up to a new technical level during the construction of this bridge.

12.7.4 Halsskov Bridge

Rolling bascules have been popular in Europe for medium spans and much attention has been paid to the aesthetic aspects. Figure 12.16 depicts the rolling bascule across the Innerharbor between Halsskov and Korsor on the western side of Sjaelland, Denmark. It is a half-through single-leaf rolling bascule which supports a roadway and a single railroad track. The deck is of steel orthotropic plate construction and the truss members are enclosed welded boxes. There is no lateral bracing at the top chords and the bridge has a very open appearance.

The leaf is operated by hydraulic cylinders, one on each side of the bridge, alongside and below the track. The pistons act on brackets that project downward alongside the trusses near the front teeth. To open the bridge the cylinders push forward, opposite to the direction of roll. This action subjects the teeth to more force than if the bridge were opened by applying force at the centre of roll, as is confirmed by the tooth wear. However, as recommended previously, tread and track of rolling bascules should be considered expendable.

Figure 12.17 Marine Parkway Bridge.

12.7.5 Marine Parkway Bridge

The Marine Parkway Bridge depicted in Figure 12.17 is a tower drive vertical lift bridge with a lift span 540 ft (165 m) long that can be lifted 95 ft (29 m). It was completed in 1937 and is still the longest vertical lift highway bridge. The bridge supports two lanes of traffic in each direction and a sidewalk. Span drive machinery is located in each tower. The Control Room and Switchboard Room are located in one tower, above the sidewalk and roadway.

The lift span is connected to the main counterweights by 40 main counterweight wire ropes of $2^3/8$ inch (60 mm) diameter at each tower, 10 ropes passing over each of the four counterweight sheaves at that tower (Hardesty et al., 1975; Birnstiel, 1992). The span is guided up and down each tower transversely by rollers. A centring device at each end properly aligns the lift span transversely at the closed position. Auxiliary counterweights are provided to offset the shift in balance as the main counterweight ropes move from one side of the counterweight sheaves to the other. There are four buffer cylinders mounted on the lift span which are activated during closing, and two buffer cylinders mounted at the top of each tower that are activated when the lift span is opened to the maximum available lift height in the event of a control failure.

There are two, virtually identical, span drives on this bridge, one atop each tower. They are mechanically independent of each other. Electric coupling exists between the two span drives through the use of selsyn transmitters. Each span drive has two main and two emergency drive motors. They are wound rotor motors operating at 480 VAC, three phase, 60 Hz, the main motors are 200 HP (149 kW) and the emergency motors are 50 HP (37 kW), both rotate at approximately 580 RPM full load speed.

Power is transmitted from the main or emergency motors through a primary differential reduction gearset. It contains a clutch which permits the differential feature to be locked during normal operation. Each transverse output shaft from the primary reduction gearset leads to a secondary differential reduction gearset which provides equal torques to the two rack pinion shafts that it drives. The pinions engage circular racks mounted on the counterweight sheaves which transmit the power to the main counterweight ropes via friction. This is a typical arrangement of mechanical machinery in tower drive vertical lift bridges.

12.7.6 New Jersey Route 13 Bridge

New Jersey Highway Route 13 is supported over the Point Pleasant Canal, a component of the Intracoastal Waterway, by a connected-tower drive vertical lift bridge. The span drive machinery is mounted on a platform spanning between the towers and drives all four counterweight sheaves simultaneously to raise or lower the lift span. Figure 12.18 is a view of the bridge from the southwest canal bank with the lift span partially raised. The bridge was built in 1970 to replace a prior swing bridge put into operation in 1924 on opening of the canal to navigation. The towers are spaced 92 ft apart. They support a roadway wide enough for four lanes of traffic. The normal lift is 35 ft (12 m).

Figure 12.18 New Jersey Route 13 Bridge.

Each corner of the lift span is suspended from six wire ropes of 1.5 inch (38 mm) diameter that are looped over cast steel sheaves mounted atop the corresponding columns of the superstructure. The other ends of the ropes are connected to the counterweights (one counterweight for each end of the lift span). Raising and lowering of the span is accomplished by rotating the sheaves using the span drive machinery mounted atop the platform which spans between the tower legs. All the sheaves are directly connected by gearing without any differential so that, theoretically, the wire rope movements are alike at all sheaves, provided that the ropes do not slip on the sheaves.

The span drive is powered by two identical 60 HP, 460 VAC, wound rotor motors. Only one motor drives at a time, the other motor rotates, de-energized. A spring-set thrustor-released brake is mounted at each motor. Both motor shafts are coupled to the primary speed reducer input shaft by brakewheel gear couplings. The primary speed reducer has a 2:1 gear ratio and a double-extended output shaft. There is no differential in this speed reducer. Each output shaft extension is connected by a longitudinal series of floating shafts to a secondary right angle reducer with a gear ratio of 7.9:1. The secondary reducers have double extended shafts and transverse shafts coupled to them transmit power from the secondary reducers to the pinion shafts that engage the curved racks mounted on the counterweight sheaves.

In order to permit vertical adjustment of a set of counterweight ropes at one corner of the lift span with respect to the other corners, 'adjustment coupling' were installed on the output shafts of the secondary reducers. These are gear couplings with a special arrangement of flange bolts. To make adjustments, the flange bolts are removed from an adjustable coupling and the hubs rotated with respect to each other the desired amount. The flange bolts are then reinstalled. The machinery of the Route 13 bridge is typical of connected tower drive vertical lifts of the period.

12.8 Movable bridge design

Movable bridge design is multidisciplinary, involving the coordinated application of structural, mechanical, and electrical engineering. It is an art which utilizes solid and fluid engineering mechanics and electrical physics. As any art, it is based on experience. This experience is recorded in voluntarily developed design specifications and in literature. The first comprehensive design specification for movable bridges appears to be that presented by CC Schneider to a meeting of the American Society of Civil Engineers (ASCE) in 1907 (Schneider, 1908). A specification oriented to bascule and vertical lift bridges was presented by Leffler (1913). The later movable bridge specifications of the American Railway Engineering Association (AREA) and the American Association of State Highway and Transportation Officials (AASHTO) were largely based on the Schneider and Leffler specifications. The current versions of these specifications (AREA, 1997; AASHTO, 1988) are based on allowable stress design. However, an alternate AASHTO specification based on limit state concepts is in preparation. In Germany a limit state design specification is used for movable bridges (DIN, 1998). The specifications mentioned deal mainly with the mechanical, electrical, and fluid power aspects of movable bridge design. Most refer to specifications such as AASHTO (1992) for the structural design.

Movable bridge analysis and design is based on elementary mechanics except for the modern electrical and electronic controls which have become popular. Many references are available for structural analysis and design and for the design of machinery. However, a successful design requires that the engineer recognize the interrelationship between the machinery and the structure. The structural action of the movable bridge is usually different in the closed and open positions. The mechanical/structural interaction should be considered especially with regard to displacements and distortions. The machinery needs to be simple and rugged because maintenance of movable bridges is not a priority for some owners. Overpowering or overbraking of bridges should be avoided. Many machinery failures, especially on swing bridges, are due to improperly adjusted brakes with excessive braking torque capacity.

The electrical system for a movable bridge should provide for acceleration and deceleration of the span consistent with the mechanical/hydraulic machinery design. But the system involves more than speed control of span drive motors. Power and control is required for auxiliary devices, such as span locks, traffic warning and resistance gates, traffic signals, rail locks and catenary lifts. All these controls have to be properly sequenced and interlocked for safety of land and water traffic.

In some countries, such as the USA, designs for public works are prepared by governmental agencies or by engineering consulting firms engaged by them. Drawings and specifications are prepared which serve as the basis for a legal contract between the owner and the contractor who physically executes the work. The effort required to prepare the contract documents for the electrical and mechanical systems can be considerable; depending on the type of movable bridge, whether it be a new or rehabilitation project, and the construction contractor climate in the area. Engineering person-hour requirements for the rehabilitation of the machinery and controls of a medium-size dual double-leaf simple trunnion bascule highway bridge have been presented elsewhere (Birnstiel, 1996) . It is not unusual for the owner to be surprised by the high cost for design of the electrical and mechanical systems. These costs can be somewhat reduced and hidden from the owner by using design-build agreements for construction.

12.9 Construction support

In the conventional USA public works construction process, engineering services are required from the designer in order to attempt to ensure that the contractors comply with the intent of the designer as expressed in the construction contract documents. These services involve review of shop drawings and other vendor submittals, resolution of unforeseen field conditions (especially on rehabilitation projects), assistance with electrical start-up, and acceptance testing. Costs associated with construction support for the electrical and mechanical systems of the Hutchinson River Parkway Bridge rehabilitation project have been published (Birnstiel, 1996).

12.10 Inspection of machinery and controls of movable bridges

The collapse in 1967 of the Silver Bridge across the Ohio River at Point Pleasant, West Virginia, brought to public attention the state of public bridge maintenance in the USA at the time. This situation was due, in large part, to the manner in which public works were funded, with federal funds available to assist states and localities with new construction, but comparatively little for maintenance or rehabilitation of existing infrastructure. However, the failure of this eye-bar suspension bridge resulted in promulgation of legislation mandating biennial inspection of all bridges supporting public roads with the federal government bearing a large share of the cost. The original legislation was directed primarily at bridge superstructures but, in practice, inspection of machinery and controls was encouraged at regular intervals.

  As no publication addressing the inspection of machinery and controls of movable bridges was a available in English at the time , the Federal Highway Administration (FHWA) published a manual for inspectors of bridge machinery (FHWA, 1977). It has recently been revised by the American Association of State Highway and Transportation Officials (AASHTO, 1998). Parsons Brinckerhoff published a practical guide to bridge inspection which includes a chapter on movable bridges (Parsons Brinckerhoff, 1992). Various scopes of work for machinery inspection have been presented (Birnstiel, 1990).

12.11 Conclusion

Movable bridge design can be challenging, for both new and rehabilitation projects. In the USA there are some 1700 active movable bridges, and many exist elsewhere. Although changing transport patterns and economic pressures will result in the conversion of some existing movable spans to fixed spans, and the replacement of movable spans by fixed spans, there are also situations where existing movable spans are being replaced by longer movable spans in response to the needs of navigational interests. The need for movable bridge construction is likely to continue. This chapter should be considered only a cursory introduction to the subject of movable bridges.

Bibliography

AASHTO. *Standard Specifications for Movable Highway Bridges.* American Association of State Highway and Transportation Officials, Washington DC, 1988.

AASHTO. *Standard Specifications for Highway Bridges.* American Associations of State Highway and Transportation Officials, Washington DC, 1992.

AASHTO. *Movable Bridge Inspection, Evaluation and Maintenance Manual.* American Association of State Highway and Transportation Officials, Washington DC, 1998.

Abrahams MJ. Ford Island Bridge, Pearl Harbor, Hawaii. *Proceedings, Sixth Biennial Symposium, Heavy Movable Structures*, 30 Oct.–1 Nov., 1996, at Clearwater Beach, FL.

AREA. Chapter 15, Part 6 – Steel structures, *Manual for Railway Engineering.* American Railway Engineering and Maintenance-of-Way Assoc., Landover, MD, 1997.

Birnstiel C. Movable bridge machinery inspection and rehabilitation. *Bridge Management* (Ed. JE Harding, GAR Parke and MJ Ryall). Elsevier Applied Science, London, 1990.

Birnstiel C. *Marine Parkway Bridge: Inspection of Movable Span Machinery and Electrical System.* Charles Birnstiel, Consulting Engineer, PC, 1992.

Birnstiel C. Bascule Bridge machinery rehabilitation at Hutchinson River Parkway Bridge. *Bridge Management 3* (Ed. JE Harding, GAR Parke and MJ Ryall). E & FN Spon, London, 1996.

Birnstiel C and Tang MC. Long span movable bridges. Paper T154-6, *Structural Engineering World Wide 1998,* Elsevier, Amsterdam, 1998.

Brown RD. The raising of Barton Swing Aqueduct and the renewal of paths and rollers. *Selected Engineering Paper No. 67*, The Institution of Civil Engineering, 1929.

Cruttwell GEW. The Tower Bridge: superstructure. *Minutes of Proceedings Institution of Civil Engineers*, 127, 35–53, 1896.

Daniels SH. Winds wreak havoc on pontoon span with troubled history. *Engineering News – Record*, New York, **242**, No. 11, 15 March, 13, 1999.

Dickson H. Bridges of the Manchester Ship Canal – past, present and future. *The Structural Engineer*, **72**, No. 21, November, 1994.

DIN. *Hydraulic steel structures: criteria for design and calculation*, DIN 19074. Edition 1998-05, Deutsches Institute für Normung, Berlin.

FHWA. *Bridge Inspector's Manual for Movable Bridges.* Federal Highway Administration, US Department of Transportation, Washington DC, 1977.

Green P. Seattle slews long heavy span. *Engineering News – Record*, New York, 28–30, 1991.

Green P. Float out the old, float in the new. *Engineering News –Record*, New York, 339, No. 55, July, 8, 1996.

Hardesty ER, Fischer HW and Christie RW. Fifty-year history of movable bridge construction – Part I. *Journal of the Construction Division*, ASCE, 101, No. CO3, September, 1975.

Hawranek A. *Begwegliche Brücken*, Julius Springer, Berlin, 1936.

Hedafine A and Kuesel TR. How the world's longest vertical lift bridge will work. *Engineering News – Record*, New York, 11 July, 38–45, 1959.

Hool GA and Kinne WS. *Movable and Long-Span Bridges,* 2nd edn. McGraw-Hill, New York, 1943.

Homfray SG. The machinery of the Tower Bridge. *Minutes of the Proceedings Institution of Civil Engineers*, 127, 54–59, 1896.

Hovey OE. *Movable Bridges*, Vols I & II. Wiley, New York, 1926.

Leffler BR. Specifications for railroad bridges movable in a vertical plane. *Transactions, ASCE*, LXXVI, Paper No. 1251, 370–454, 1913.

Masterson VV. *The Katy Railroad and the Last Frontier.* University of Missouri Press, Columbia and London, 1992.

Mehue P. Movable bridges. *Selected Papers from IABSE Symposium Leningrad 11–14 September 1991.* Association Francaise pour la Construction, Paris, 1991.

Ortmann R. Bascule bridge over the Oste at Geversdorf. *Mannesmann Rexroth Report RE 09719/10.90,* Lohr a. Main, Germany, 1990.

Parsons Brinckerhoff. *Bridge Inspection and Rehabilitation* (Ed. LG Silano), 143–166. Wiley-Interscience, New York, 1992.

Quade MN. Special design features of the Yorktown Bridge. *Transactions, ASCE*, 119, 109–123, 1954.

Rode JG and Nielsen SV. Gadebroens Klapfag. *Ingenioren,* Copenhagen, 9–14 November, 1954.

Saul R and Zellner W. The Galata bascule. *Report of the IABSE Symposium Leningrad*, IABSE, Zurich, 64, 557–562, 1991.

Schneider CC. Movable bridges. *Transactions, ASCE*, LX, Paper No. 1071, 258–336, 1908.

Sedlacek H. Die neue Drehbrücke über den Suez-Kanal bei El Ferdan/Agypten. *Der Stahlbau*, Berlin, **34**, No. 10, October, 1965.

Thomsen K and Pedersen KE. Swing bridge across a navigation channel, Denmark. *Structural Engineering International. IABSE*, 8, Zurich, 1998.

Waddell JAL. *Bridge Engineering*, Vols I & II. Wiley, New York, 1916.

Wengenroth RH and Mix HA. Fifty-year history of movable bridge construction – Part III. *Journal of the Construction Division, ASCE*, 101, No. CO3, September, 1975.

Williams L. The Manchester Ship Canal. *Engineering*, 1894.

13 Modern developments

L. HOLLAWAY AND H. SPENCER

13.1 Aluminium

13.1.1 Introduction

The first recorded application of aluminium alloys in bridge construction was the re-decking of the Smithsfield bridge in Pittsburg in 1934. This was a heavily used urban bridge which required an upgraded load; it carried both cars and tramway. The use of aluminium for both the decking and the deck box girders enabled this to be achieved without strengthening the bridge structure or foundations. Despite the advantages confirmed by this installation, progress in widening the market for aluminium bridges has been slow.

Between 1946 and 1962 a number of lift and swing bridges were erected utilizing the light weight of aluminium but after that there was only spasmodic interest. Recently, however, several aluminium bridges have been built or re-decked in Scandinavia and in the USA. More importantly, the designs are applicable to wider use and not only exploit the weight saving but also the extrudability of special aluminium sections. These properties, plus good corrosion resistance, have established aluminium alloys as the preferred material for equipment, particularly for heavy duty gangways on oil platforms and also on lighter marine walkways.

Even so, general acceptance of the material in bridge construction has not developed. It is not possible to be precise about the reluctance to accept aluminium in this potential field of engineering. Cost is frequently mentioned but although aluminium is more expensive than steel on a tonne to tonne basis it can be shown that by careful design and by considering convenience factors such as ease of erection and reduced decommissioning times, cost parity can be attained. Full life cycle costing taking into account low maintenance requirements would make the comparison even more attractive.

13.1.2 Strength and stiffness

Structural design codes have existed since 1950, first with the Institution of Structural Engineers' report on aluminium followed in 1969 by the British Standard (BS) CP118 and finally in 1991 BS 8118 (*Structural Use of Aluminium*) which is based on limit state philosophy. Future code changes are likely with the harmonization of BS 8118 into a Euro-Norm specification.

The principal reasons for using aluminium alloys in structural engineering are their high strength-to-weight ratios and their excellent durability. Pure aluminium is a relatively low strength material but its strength can be significantly increased by lightly alloying. All of the structural aluminium alloys contain over 92% pure aluminium, the balance varying between copper, zinc, magnesium and silicon. As well as increasing strength, these other alloys can influence other properties of pure aluminium to varying degrees.

If overall tensile stress is the main requirement, then aluminium structural alloys can give equal performance to steel on the basis of their 0.2% proof stresses against the yield stresses of steel. The transition between elastic and plastic behaviour is not so pronounced for aluminium alloys as it is for structural steels. The arbitrary standard for aluminium specified in all international structural codes is 0.2% proof stress which is the stress level at the onset of 0.2% permanent strain.

The moduli of elasticity and rigidity for aluminium alloys are relatively low, being only one third of those for mild steel. There is very little spread in the range of values (1.4% difference) so that for general purposes the elasticity and rigidity moduli can be taken at 70 000 and 26 500 MPa, respectively. There are advantages in this flexibility, the aluminium alloys are used in applications where energy absorption is required and where adequate deflection is required within the elastic limit of the material. The latter is usefully applied in the design and application of mating and clip fit extruded sections.

In bridge structures where lower deflection levels are specified aluminium sections can and must be designed to meet these values without significant loss in weight advantage over steel. Where the bending conditions of end fixity, span and loading are equal, deflection is dependent upon EI and parity with steel can be obtained where $E_{al}I_{al} = E_{st}I_{st}$. This means that the aluminium section inertia should be three times that of the steel section. This can be achieved in several ways as shown in Figure 13.1 Increasing thickness only is not economical as the weight is also increased by the same factor. Increasing depth and/or width, depending upon the significant axis, will give better results.

The deepening of an aluminium section to obtain deflection parity with steel will also influence the bending modulus Z which will also be higher than that of the steel section.

The elastic ordinate on a stress–strain curve is at a more acute angle than that of steel so that the stress–strain relationships are different and the imposed stress for a given strain condition is much less for aluminium than it is for steel as illustrated in Figure 13.2.

The behaviour of aluminium alloys under axial compression loading is governed by buckling stresses. Buckling curves are available for each alloy and temper and are set out on the basis of stress against slenderness ratio, $\lambda = Kl/r$, where K is the end fixity factor, l is the actual length of the column and r is the radius of gyration of the section. It follows, therefore, that the ratio is influenced by the geometric properties of the section for any given set of conditions. The radius of gyration is not greatly influenced by thickness and it is therefore more efficient to use a larger section than to increase thickness.

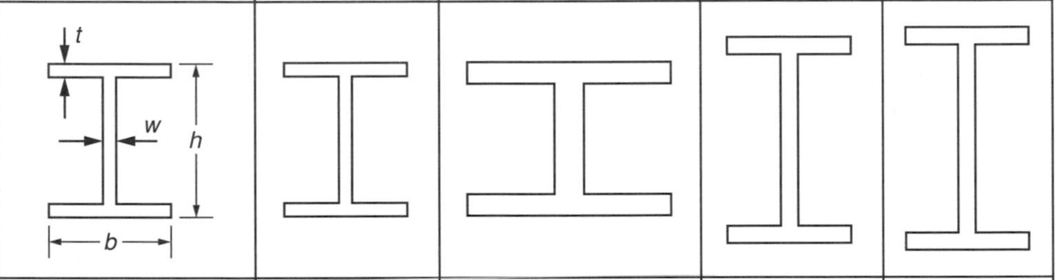

	Steel	Aluminium alloy	Aluminium alloy	Aluminium alloy
Moment of inertia in mm^4	38.9 *E* 6	116.6 *E* 6	116.7 *E* 6	117.3 *E* 6
EI (N/mm^2)	8.17 *E* 12	8.16 *E* 12	8.17 *E* 12	8.21 *E* 12
h (mm)	240	240	300	330
b (mm)	120	240	200	200
t (mm)	9.8	18.3	12.9	10
w (mm)	6.2	12	6	6
g (kg/m)	30.7	30.3	18.4	15.8

Figure 13.1 Comparison between four beams which give the same deflection (by kind permission of TALAT).

Structure in bending	AA 6082 T6	St 52
Factor against yielding	4.0	2.1
Factor against fracture	4.6	3.2–3.9

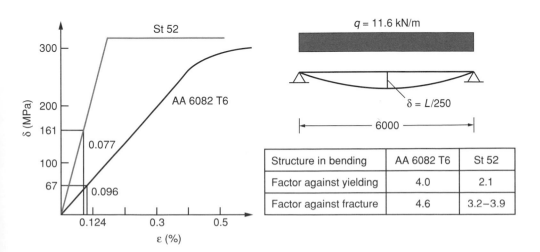

Figure 13.2 Stress and strain for beams made of St52 or AA6082-T6 (by kind permission of TALAT).

Local buckling, which is a type of local instability, can occur in very wide thin elements under compressive load and at a stress below the elastic limit. It takes the form of waves or wrinkles and is best avoided. The onset of buckling does not necessarily lead to overall failure because some post-buckling strength is retained. The extrusion limits on wide, thin elements prevent the production of the more inherently unstable sections but it is still advisable to check sections against this form of buckling. There are several methods of calculating section resistance which is always based on the performance of the worst element. Resistance to local buckling can be very simply improved by thickening the element, incorporating mid-width support bosses or surface swaging to break up the unsupported width. In the case of the extruded sections, both swaging and boss supports can be simply incorporated.

The critical ratio is between element width and thickness but is restricted to the thinnest part of the element and does not include bend radii or stiffening ribs or lips. The relevant widths for the flange and web of an I beam under bending load conditions are shown as B_f and B_w in Figure 13.3(a). These ratios are further factored by the position of the neutral axis (for the web only) and also for the material proof stress. The final values are entered on a series of curves from which the percentage thickness that will resist buckling can be found. This is applied to the section as in Figure 13.3(b) and the section resistance is then calculated on the modified shape.

There are three categories of classification:

- compact, in which local buckling will not occur
- semi-compact, in which buckling will only occur in the plastic zone (this class is not applicable to axially loaded sections)
- slender, in which buckling can occur below proof stress level.

Hollow symmetrical sections have the greatest resistance to buckling. The aluminium alloy sections and plate used on bridge construction are generally robust and are

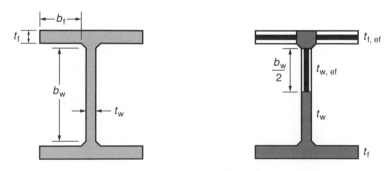

(a) Gross cross section, notations   (b) Effective cross section for bending moment

Figure 13.3 Effective cross-section – I-Beam with equal flanges (by kind permission of TALAT).

Table 13.1 Properties of aluminium vs steel

Properties	Units	Aluminium	Steel
Density	kg/m^3	2 700	8 100
Elastic modulus	MPa	70 000	203 000
Shear modulus	MPa	26 500	79 000
Coefficient of expansion	10^{-6}/°C	23	12
Thermal conductivity	W/M°C	244	47
Electrical conductivity	%IACS*	61	11
Melting point	°C	660	1370
Fire	–	Non-combustible	Non-combustible

*IACS – International Annealed Copper Standard.

unlikely to be the subject of local buckling constraints. This, together with the relatively lower working stress levels associated with designing to stiffness rather than strength, should again make local buckling very unlikely. All the sections contained in BS 1161 *Aluminium Alloy Sections for Structural Purposes* are of the compact class. For the purposes of checking, the approved method is set out in BS 8118 *Structural Use of Aluminium*.

Table 13.1 provides aluminium and steel mechanical properties.

13.1.3 Alloys

The choice of aluminium alloy will normally depend upon the service requirements of strength, durability and the fabrication required. These are set out in Table 13.2 for those aluminium alloys suitable for bridge construction. The properties for alloy 5083 in plate or sheet form will be higher than those quoted for extrusions but will depend upon the chosen temper.

13.1.4 Temperature

The strength of aluminium alloys varies with temperature so that below freezing and down to −200°C the structural strength and elastic modulus are increased, whilst at temperatures higher than 100°C they are reduced. A further important characteristic, is that, unlike carbon steels they are not brittle at low temperatures.

The linear coefficient of expansion can be standardized at $23.5 \times 10^{-6} \times \Delta°C$ for all aluminium alloys, where Δ is the temperature variation. Although the coefficient of expansion is relatively high, the lower modulus of elasticity enables temperature induced stresses to remain at a low level, approximately two thirds those of similar steel components.

The thermal characteristics of aluminium alloys are such that where temperature variations are high and the aluminium painted black or with a dark pigmented colour the surface temperature can be much higher than ambient and longitudinal expansion should be checked. Natural mill-finished aluminium or light coloured surfaces reflect the heat away from the material and it is, therefore, not necessary to check for expansion.

Table 13.2 Service requirements for aluminium alloys

Alloy	Major alloying elements	Type of alloy	Temper	Mechanical properties			
				0.2% Proof stress (MPa)	ULT stress (MPa)	Elongn (%)	Shear stress (MPa)
6063	Mg/Si	H	T4	65	130	12	78
			T6	170	215	6	129
6005A	Mg/Si	H	T4	90	180	13	108
			T6	225	270	6	162
6082	Mg/Si	H	T4	110	205	12	123
			T6	260	310	8	186
7020	Zn/Mg	H	T4	190	300	12	180
			T6	290	350	8	210
5083	Mg	N	O	125	270	10	162
			F	110	270	10	162
			H112	125	270	10	162

Values quoted are for ultimate shear stress.
H, heat treatable, N, non-heat treatable.
Proof shear values can be taken as 60% of 0.2% proof stress value.

Alloy		Weldability	Durability	Formability	Machinability	Brinell hardness	Previous designation
6063	(T4)	V	V	V	G	50	H9
	(T6)	V	G	G	G	75	
6005A	(T4)	V	V	V	G	70	H10
	(T6)	V	G	G	G	80	
6082	(T4)	V	V	V	V	70	H30
	(T6)	V	V	G	E	95	
7020	(T4)	V	G	G	F	110	H17
	(T6)	V	G	P	G	120	
5083	(O)	E	G	G	G	70	N8
	(F)	E	G	G	G	70	
	(H112)	E	G	G	G	75	

E= excellent, V= very good, G = good, F = fair, P = poor.

13.1.5 Fatigue

Fatigue performance can be difficult to quantify at the design stage particularly
if there is no accurate loading pattern available for the type of structure under

consideration. There has been a great deal of research carried out over the last 20 years to collate data from actual installations and to improve structural codes. The University of Munich has set up a data bank to specifically monitor aluminium bridges. Another result of this research work has been to revise the fatigue chapter in BS 8118 *Structural Use of Aluminium*. This covers the behaviour of parent metal, mechanical joints and welded joints, in all 29 classes of joint are covered. The areas that should be considered are those where there are likely to be high stress concentrations, discontinuities or notches. The first two can generally be covered by design and the latter by careful control of fabrication. The common forms of endurance curves for most aluminium alloys arc similar to those for non-ferrous metals. They differ from typical curves for mild steel in that in most cases they do not become asymptotic to the N axis even after the tests have been continued well beyond the number of cycles commonly applied.

13.1.6 Durability

Aluminium and its alloys have, in general, excellent durability and corrosion resistance. There are, however, basic variations in the behaviour of the different alloys which can be further influenced by the way in which they are used.

Aluminium's natural affinity with oxygen results in the formation of an oxide layer when exposed to air. The resulting film is generally 500 nm thick, extremely hard, chemically stable, corrosion resistant and adheres strongly to the parent metal surface, producing an integrated material. Once formed, it prevents further oxidation and if damaged in any way, will reform always providing oxygen is available. The only practical reason for removing this film is to facilitate anodizing or welding.

The behaviour under atmospheric exposure of the 5000 and 6000 series alloys can be described as self-stifling. If the surface layer is pitted by any of the air-borne pollutants usually found in industrial or marine atmospheres, such as sulphuric acid and sodium chloride, the resulting chemical reaction produces a larger volume of powdered corrosion product than the volume of the original pit, thereby sealing off the surface of the aluminium and inhibiting any further corrosive reaction. In general, the ratio of corrosion product to pit volume is 240:1, this is illustrated in Figure 13.4.

With time, existing pits, which are usually of a shallow hemispherical shape, are sealed and the rate of formation of new pits is reduced so that eventually all reaction can be assumed to have ceased. This process can be described as weathering, for the depth of pitting is extremely small. The level of pollution, of course, will determine the general appearance which will appear to be a soft blueish-grey colour in rural areas and a dark grey to black in industrial areas. Regular maintenance and washing down should prevent the permanent discoloration from industrial pollutants.

The alloys in the 2000 and 7000 series do not suffer from pitting but exhibit a layered form of corrosion and are generally less resistant than the 6000 series. Some form of surface protection is therefore required in aggressive environments, paint being the usual choice.

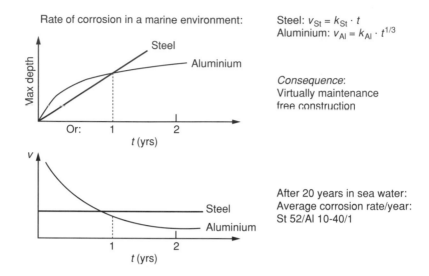

Figure 13.4 General corrosion behaviour of Al and steel – a factor of maintenance costs (by kind permission of TALAT).

   All alloys are assessed for their suitability in the various types of environmental conditions which for simplicity are divided into three categories. These are rural, marine and industrial and are shown in Table 13.3.

   When dissimilar metals are coupled together in the presence of moisture, there is a likelihood of a galvanic reaction in which one metal will be corroded. In this situation, an electrolytic couple is formed in which a current flows from the less noble metal to the more noble one with corrosion concentrated on the less noble metal. This behaviour is usually consistent with the relative placing in the electro-chemical series as is observed in Table 13.4. As can be seen aluminium will be corroded by those metals listed below it, with the exception of type 304 stainless steel. The severity of the galvanic action also depends upon the degree of separation, electrical resistance of the metal path, conductivity of the solution and the area ratio between the

Table 13.3 The requirements of BS 8118 Structural Use of Aluminium

Alloy series	Rating	Comments
5000	A	Suitable for all site conditions including immersion in seawater
6000	B	Suitable for most site conditions excluding immersion in seawater Material below 3 mm thick in aggressive environments should be painted
7000	C	Should be painted in all but mild atmospheric conditions

Table 13.4 Electro-chemical series

Base	Magnesium
	Zinc
	Aluminium
	Cadmium
	Mild steel
	Cast iron
	Lead
	Tin
	Nickel
	Brasses
	Copper
	Bronze
	Monel
	Silver solders (70% Ag, 30% Cu)
	Nickel
	Stainless steel (Type 304, passive to aluminium)
	Silver
	Titanium
	Graphite
	Gold
Noble	Platinum

two dissimilar metals. In practice, however, reaction between the metals can be avoided by insulating them from each other with an electrically inert non-absorbent barrier. An excellent example of this kind of connection in a working joint between aluminium and steel is the deck–superstructure connection on large liners.

In reviewing the bridges built in the 1950s, the area which requires remedial work on several bridges is the bi-metallic connections through either poor design or bad installation.

13.1.7 Fabrication

Apart from one or two modifications, fabrication practice for structural aluminium does not vary greatly from other materials. Cutting by high speed band saw or circular saw is possible with tungsten carbide tipped blades. Wide-spaced teeth are necessary to ensure metal clearance and prevent clogging. Machining is preferably carried out on high tempered strong material with routing generally used for hole shaping in plate and extrusion. Edge planing or plasma cutting can be used for weld edge preparation. Aluminium responds well to brake press forming and section bending although spring back can occur on the stronger alloys and should be allowed for in batch forming. Whenever possible lubrication should be applied and on high speed operations coolants should be provided to avoid unnecessary expansion.

Riveting is still an acceptable method of construction, using aluminium rivets. Steel rivets are acceptable in mild environments but would require over painting to minimize bi-metallic effects. Clench-type rivets with their single-sided closing are available in aluminium and have a proven record in light to medium structures.

Most modern aluminium structures would be of welded construction, using an inert gas shielded system. It is necessary to remove the natural surface film to avoid oxide contamination of the weld. The metal inert gas MIG (argon or argon–helium mixture) is ideal for automated welding whilst tungsten inert gas TIG (argon or argon–helium mixture) method although slower is better suited to manual control. Figure 13.5 shows, diagrammatically, the welding methods. All high tempered aluminium alloys have reduced post-welded strength in the heat-affected zone (HAZ) reducing properties to the T4 solution heat treated level for heat treated alloys and the O annealed condition for non-heat treatable alloys. The effect of this reduced strength can be offset in extruded joints by increasing the material thickness of the welding flange in way of the HAZ, shown in Figure 13.6.

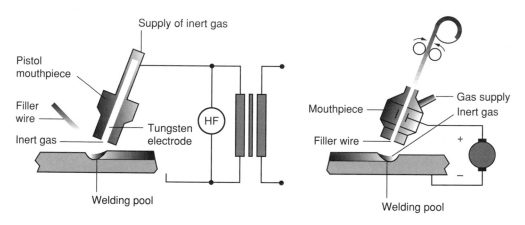

Figure 13.5 Welding methods (by kind permission of TALAT).

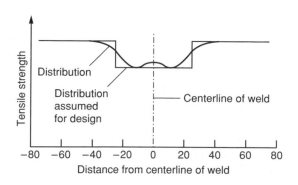

Figure 13.6 Tension member – influence of welds (by kind permission of TALAT).

The present development of friction stir welding procedures with reduced heat input should allow higher post-welding properties to be used.

There is a standard range of structural sections similar to those produced in steel. A series of lipped sections are available, however, which are specially designed for economic applications. Both series are listed in BS 1161 *Specification for Aluminium Alloy Sections for Structural Purposes*. Where lipped sections are used, weight savings of up to 15% over standard sections can be achieved.

13.1.8 Early aluminium bridges

The Arvida bridge over the Saguenay river in Canada is of aluminium construction, and was opened in 1950, it is shown in Figure 13.7. It is a single 90 metre parabolic arch design. The rise is 14.4 m and the total bridge length including approach spans is 155 m with a width of 8 m. It is of riveted construction with 19 mm diameter cold driven annular points in alloy 2117. Sheet and plate conform to HC15 which is an aluminium copper alloy sheathed on both sides with commercial pure aluminium. The extrusions were not sheathed and conformed to HE15 material. The current equivalent to this alloy is 2014A. The two arched ribs are of box girder construction and were prefabricated into units of 6.5 tonnes for easy transportation and handling on site. A 200 mm deep concrete slab was laid on top of the structure with a 62 mm bituminous wearing surface. The total aluminium weight of 200 tonnes was only 43.5% of an equivalent steel construction.

Figure 13.7 The Arvida bridge over the Saguenay (courtesy of Alcan Aluminium Co., Canada).

The bridge is still in full service and although maintenance requirements have been low some remedial work was recently required on one of the approach spans.

Although this is an example of a very successful bridge, the choice of alloy and fabrication methods were based upon knowledge existing in 1947 and a similar bridge built today would be of welded construction and the material would be one of the 6000 series alloys.

In the USA during the late 1950s a federal highway building programme encouraged the aluminium industry to become actively involved in the design of aluminium road bridges of which seven were built. They were all of aluminium–concrete construction, designed to be fully composite. The bridges varied in the type of aluminium support structures but the concrete decks usually consisted of 200 mm thick lightweight concrete with a 60 mm concrete–bitumen wearing surface. Two bridges are typical of the structural design options.

An overpass bridge at Des Moines was built in 1958 with a traditional girder system. It was a 70 m four span design with five support girders in aluminium alloy 5083. It was of welded construction each girder being 900 mm deep and 400 mm wide and of I beam construction. The flange and web thicknesses were 25 mm and 12 mm, respectively. Transverses were also of I construction and were intercostal to the main girders. Shear connectors were welded to the longitudinal girders. The bridge was regularly inspected during its life and reports indicate some deterioration in the concrete deck and a few minor fatigue cracks in the aluminium girder–diaphragm connections. The cracks were monitored over several inspections and showed no sign of any significant propagation but were then stabilized by drilling holes at the tail of the cracks. The only major damage to the bridge occurred when the bottom flange of one of the exterior beams was hit by an over-height lorry. The beam was partially straightened and some of the flange material at the point of impact was removed to eliminate the gouge. Tests to establish the neutral axis of the composite structure located it in the lower regions of the concrete deck. This was higher than expected for in a similar steel structure the neutral axis would have been in the web section of the girder.

A highway re-alignment in 1993 would have necessitated the widening and lengthening of the existing bridge. This was not economically viable so the bridge was dismantled and a new bridge was constructed. This presented a unique opportunity to fatigue test a structure that had been in service for 35 years. The girders were sent for test to the Center for Advanced Technology for Large Systems at Lehigh University in Bethlehem, Pennsylvania. A detailed inspection after dismantling confirmed that the original treatment of this aluminium structure by way of concrete was still in prime condition and the zinc chromate wash followed by a coat of alkaline resisting bituminous paint had prevented any corrosive reaction between the concrete and the aluminium.

The 90 m long three span Sykesville bridge has an aluminium superstructure and a concrete deck. The structure is made up of five interconnected inverted triangular girders, fabricated from 2.5 mm aluminium sheet. They are of riveted construction with both longitudinal and vertical stiffening. All aluminium material is of 6061 designation. The girders are 1.47 m high and 2 m wide and are connected by a transverse section at the bottom flanges and by a corrugated sheet at the top. This decking provides shear connections for the concrete deck although further steel shear connectors are fitted at the girder ends to absorb any thermally induced stress.

Built in 1961, the bridge is scheduled for replacement giving a life of over 40 years. The bridge is very lightweight but fabrication costs are high for this type of riveted construction which would not be considered today.

The behaviour of aluminium–concrete composite structures has been the subject of several research programmes by Naples University.

13.1.9 Military bridges
Speed of construction and reliability are two of the main requirements for military bridges. These can translate into weight and strength. Early aluminium bridges were therefore built in the high strength alloy 2014A but as this alloy could not be welded the bridges were of bolted or riveted construction. The reliability of new welding procedures and the introduction of improved self-ageing alloy 7020 type, made it possible to have even lighter bridges as post-welded strengths were increased. This alloy may, under certain conditions, suffer from stress corrosion but this has not stopped its use for military bridges for the past 30 years. An operational comparison between steel and aluminium bridges confirms the high performance of aluminium. The results were that aluminium was 50% lighter required 50% less transport vehicles and only a quarter of the manpower to complete the installation in only 20% of the time.

13.1.10 Footbridges
Aluminium alloy footbridges have mainly been used for difficult or exposed sites where installation necessitated low weight and where good corrosion resistance met the need for low maintenance. Erection methods varied but where access was extremely difficult helicopters have been used to position the bridge. Designs have been very varied and cover most forms of structure, for example, arches, box girders, Vierendeel girders, suspension bridges and pontoons.

There are, however, very few system bridges that use standard units suitable for variable spans. The usual design for this type of bridge is box construction with the walkway internal to the box girder.

The availability of special sections has always been of interest to architects who are commissioned when a bridge has not only to be functional but also has to fit into a particular environment. One example is the recently completed bridge linking West India Quay in Docklands. It is a pontoon construction, spanning 90 m and weighing 13 tonnes. It was designed by the architectural practice Future Systems and incorporates special aluminium sections

The weight of aluminium pedestrian bridges depends upon their span and load factor but values taken from typical bridges would indicate a weight of between 52–160 kg/metre.

13.1.11 Recent aluminium bridges

The most recently completed aluminium alloy road bridge is over the Forsmo river just below the Arctic Circle in Northern Norway; Figure 13.8 shows this bridge. The two-span bridge has a length of 39 m, a road width of 7.4 m and weighs 28 tonnes. It is of all-welded construction and material specifications are 6082 and 6005 for extrusions and 5083 for plate. The design is based on two 1.5 m deep longitudinal box girders spaced apart but transversely connected at 3 m intervals. The girders are constructed from flat extrusions with integral stiffeners running longitudinally. The orthotropic bridge deck was constructed of 250 mm wide by 123 mm deep double skin sections with integral internal stiffeners. These extrusions run transversely across the deck and are welded to the girders to act as the top flange. The integral stiffening of the bridge deck sections is designed to help carry the wheel loads and to resist any local deformation. The top flange of the section is thickened beyond the requirements of the wheel loads to provide overall strength to the deck and to the longitudinal girders. The bridge was designed to spread loads through the structure and as far as possible avoid any high stress concentrations.

The previous bridge had been dismantled and the original piers replaced so that the site was ready for installation of the new bridge. This was carried out by two mobile cranes, one taking the full load of the bridge and the other guiding the bridge into the final position. This took less than one hour, following which the bridge railings were fitted and an asphalt surface laid down. The bridge was opened to traffic within 24 hours of installation. An ongoing test programme covers static loading, dynamic loading and the recording of temperature induced stresses.

Figure 13.8 The Forsmo river aluminium road bridge (courtesy of Hydro Aluminium Structures).

13.1.12 Aluminium bridge decking

The use of aluminium for a replacement road deck on an existing bridge is well established in Sweden where over 30 bridges have been re-decked using the Svensson system over the past 10 years. It is a patented design based on an extruded aluminium orthotropic section and approved by the Swedish road authority after analysis by the Royal Institute of Technology in Stockholm. This type of deck has potential value in those cases where the load capacity of a bridge is to be increased and where the support structure and foundations are in good condition.

One recent application is the Tottnaes bridge near Stockholm. It provides the only fixed link between the Swedish mainland and the island of Toroe. In the 1980s it became evident that due to the deterioration of the concrete bridge deck, the bridge was in need of immediate repair. A new deck consisting of aluminium extrusions was constructed and mounted onto a secondary structure of steel girders. The replacement system was developed in Sweden and used an orthotropic plate that consisted of hollow aluminium extrusions. These were fitted together using a tongue and groove system in the upper flange. This type of construction transferred shear force from one extrusion to another, whilst at the same time allowing each extrusion to rotate independently of the neighbouring ones. The hollow section created an extrusion with a low unit weight and a high degree of torsional stiffness and was designed to suit two support girder widths of 1.2 m and 3 m, with two sizes of extrusion; this is illustrated in Figure 13.9. This enables the distribution of concentrated loads to be carried by more than one extrusion at the same time. High point loads are supported by the internal stiffeners of the deck section, thereby avoiding local distortions. The deck system weighs between 50 and 70 kg/m in contrast to 600–700 kg/m for standard concrete decks.

The system is designed to meet the requirements of the Swedish National Road Authority which specifies both static and dynamic loads. The initial design was checked by the utilization of a finite element program where the deck extrusions were

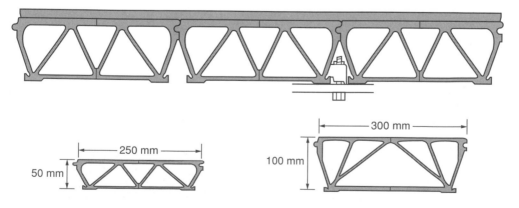

Figure 13.9 Section through an Aluminium bridge deck extrusion (by kind permission of TALAT).

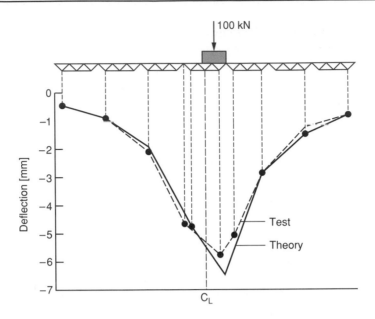

Figure 13.10 Comparison of deflections obtained from the FEM analysis and full-scale tests (by kind permission of TALAT).

simulated by beam elements having the same flexural and torsional stiffness as the proposed section and the connecting elements with a high degree of flexural stiffness. A 100 kN load was placed in the most unfavourable part of the deck and the largest deformation reached a maximum of 6.5 mm. The actual static load tests demonstrated good agreement with the theoretical investigation; this is shown in Figure 13.10. The load was finally increased up to a maximum value of 320 kN when a deflection of 27 mm was recorded in the deck element. This load, of course, was well in excess of the 100 kN design load.

Dynamic load tests were carried out to establish fatigue performance. The load amplitude was 96–100 kN at a frequency of 1–2 Hz for 2 million cycles. No cracks or signs of fatigue were evident and the maximum residual deflection after the test was 0.4 mm.

The sections were extruded from alloy 6063 in the T6 temper. This is a magnesium silicon alloy of medium strength, excellent corrosion resistance and with very good extrudabiity. The release mechanical properties of this alloy are 0.2% proof stress 170 MPa, 215 MPa ultimate stress and an 8% elongation.

The decking was supplied in prefabricated units with the special Acrydur wearing surface already applied. This resulted in a speedy installation which was essential as the bridge is the only access to the island to Toroe.

13.1.13 Review
Medium and high strength aluminium alloys are recognized as structural materials and have been well accepted for civil and structural applications. The life span of

most of the early aluminium bridges built between 1947–1961 has been good and maintenance requirements have been low.

This review has, however, concentrated upon the very few early bridges that required remedial work. Where problems have existed they can be attributed to the state of knowledge at that time and can be listed as:

- wrong choice of aluminium alloy for a particular application
- use of steel rivets for applications in aggressive environments
- poorly designed or installed bi-metallic joints.

The information obtained from the performance of the early bridges has been very valuable and the improved knowledge now available on design, fabrication and material performance would make the future re-occurrence of the above faults most unlikely.

Actual bridge fabrication and installation costs are difficult to standardize, with differences in types of construction, size, loading requirements and location. The data in Figure 13.11 is obtained from an oil platform connecting bridge and is presented as comparisons of weight savings against cost ratios. Although a saving of 63% for cost parity with steel seems rather high it is by no means impossible. Some existing bridges have attained this level and weight savings of 55–60% are quite normal for some other types of structures.

There is now a much wider interpretation of cost where not only metal price and fabrication are considered but also all the costs involved in installation and maintenance in service. The wider implications of public disruption costs have yet

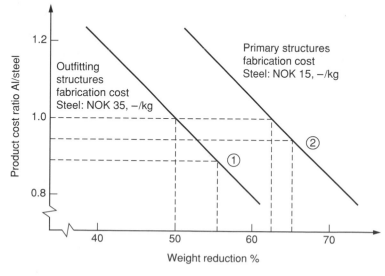

Examples: ① Sture oil terminal: Pipe supports and access systems
② Bridge structure: Connection bridge (105 m long) between 2 platforms

Figure 13.11 Comparisons of weight savings against costs ratio.

to be fully considered but they could be very high on any lengthy de-commissioning. The generally fast replacement times in evidence for aluminium prefabricated bridges and deck structures could reduce this hidden but significant cost.

13 2 Advanced polymer–fibre composites

13.2.1 Introduction

There has been considerable activity in the utilization of advanced polymer composites in the construction industry within the past 5–10 years. The developments that have taken place during this period have been considerable and the requirements that have initiated this state have revolutionized some manufacturing techniques

This section will demonstrate that the material has many advantages over the more conventional civil engineering materials. Furthermore, their properties are such that combinations with either reinforced concrete (RC), steel or cast iron produces, from a retrofitting or a new constructional unit point of view, an enhanced product which can be engineered to reflect the most advantageous properties of either of the component materials.

In 1987 an informal study group was established by the United Kingdom's Institution of Structural Engineers as it was believed that professional bodies such as the Institution should make an important contribution to the development, and to provide an understanding of composites and the associated development of standards in Europe. As a result of the work of the group, a report was produced in March 1989 (The Institution of Structural Engineers, 1989) describing and defining advanced polymer composites as:

> Composite materials consist normally of two discrete phases, a continuous matrix which is often a resin, surrounding a fibrous reinforcing structure. The reinforcement has high strength and stiffness whilst the matrix binds the fibres together, allowing stress to be transferred from one fibre to another producing a consolidated structure.
>
> In advanced or high performance composites, high strength and stiffness fibres are used in relatively high volume fractions whilst the orientation of the fibres is controlled to enable high mechanical stresses to be carried safely. In the anisotropic nature of these materials lies their major advantage. The reinforcement can be tailored and orientated to follow the stress patterns in the component leading to much greater design economy than can be achieved with traditional isotropic materials.
>
> The reinforcements are typically glass, carbon or aramid fibres in the form of continuous filament, tow or woven fabrics. The resins which confer distinctive properties such as heat, fibre or chemical resistance may be chosen from a wide spectrum of thermosetting or thermoplastic synthetic materials, and those commonly used are polyester, epoxy and phenolic resins. More advanced heat resisting types such as vinylester and bismaleimides are gaining usage in high-performance applications and advanced carbon fibre–thermoplastic composites are well into a market development phase.

It is important to understand that the term 'Polymer Composite Material' encompasses a wide range of fibre–matrix materials each with their own unique characteristics.

It is not the function of this chapter to discuss these and reference should be made to Hollaway (1993) for further information. Only the relevant polymer composites used in bridge management will be discussed here.

13.2.2 Mechanical properties of the composite

The matrix material which has and can be used in bridge construction is invariably the thermosetting polymer and would be either the epoxies, the vinylesters or the polyesters. The fibre would be either carbon, aramid or glass fibre depending upon the required function of the composite. The polymer composite must be designed in conjunction with the structural unit to provide an efficient and economic solution.

The strength and stiffness of composites depend upon a number of parameters, these are:

- the stiffness and strength of the component parts
- the fibre orientation
- the fibre–polymer composite fraction
- the method of manufacture.

The stiffness and strength characteristics of the component parts of the composite, namely the carbon and glass, aramid and polyester fibres and the epoxy and vinylester resins are given in Tables 13.5(a) and (b), respectively. The greater the stiffness

Table 13.5(a) Typical average mechanical properties for polyester, glass, aramid, and carbon fibres

Material	Fibre	Elastic tensile modulus (GPa)	Tensile strength (MPa)	Density (g/cm^3)	Max. temp. (°C)
Glass	E	72.4	2400	2.55	250
	A	72.4	3030	2.50	250
	S-2	88.0	4600	2.47	250
Polyester		8.0	700	1.00	200
Aramid	29	83.0	2500	1.44	180
	49	126.5	2500	1.44	180
Carbon	XAS HS	235.0	3800	1.79	600
Toray	T-300	230.0	3530	1.77	600

Table 13.5(b) Typical mechanical properties of unreinforced polymers

Materials	Specific weight	Ultimate tensile strength (MPa)	Modulus of elasticity in tension (GPa)	Coefficient of linear expansion (10^{-6}/°C)
Thermosetting				
Polyester	1.28	45–90	2.5–4.0	100–110
Vinylester	1.28	60–90	3.5–4.0	–
Epoxy	1.30	90–110	3.0–7.0	45–65

and strength values of the fibre, the higher will be the corresponding value of the composite. However, as the value of the modulus of elasticity of the fibre increases, the fibre becomes more brittle and consequently more difficult to handle due to its breakdown during the fabrication process. Furthermore, for civil engineering purposes, the strength of the fibre and therefore the composite, exceeds the requirements of that industry and consequently, the lower stiffness carbon fibres are more than adequate for most purposes in construction.

The fibre orientation in the matrix will greatly influence the strength and stiffness of the composite. If the fibres are randomly orientated the maximum proportion of glass or carbon fibres to polymer by weight is about 50%. When the fibres are orthogonally arranged the equivalent percentage would be about 60%. Finally, the condition in which the fibres are laid in one direction and because they are then packed more closely, the percentage can be up to 85%. However, the practical value is generally nearer to 65–70%. Tables 13.6(a) and (b) gives some typical mechanical properties of glass fibre and carbon fibre reinforced polymer composites respectively. The large difference in properties in the two directions of a unidirectionally aligned fibre composite, specifically in this case a glass fibre composite, should be noted. The advantages of composite materials become apparent when the modulus and strength are considered in terms of specific modulus and specific strength respectively, both of which are high. The value to the engineer of high specific strength and modulus carbon fibre composites with a fibre–matrix ratio by weight of 55% or more becomes apparent when designing for retrofitting to upgrade structures or when designing

Table 13.6(a) Typical mechanical properties for glass fibre reinforced polymer composites

Material	Glass content (% by weight)	Specific weight	Tensile modulus (GPa)	Tensile strength (MPa)
Unidirectional rovings (parallel to fibre)	50–80	1.6–2.0	20–50	400–1250
(perpendicular to fibre)	50–80	1.6–2.0	2.5-4.0	35–45
Randomly orientated fibres	25–50	1.4–1.6	6–11	60–180
Bi-directional fibres	45–62	1.5–1.8	12–24	200–350

Table 13.6(b) Typical mechanical properties for unidirectional carbon fibre reinforced polymer composites

	Carbon content (% by weight)	Specific weight	Tensile modulus (GPa)	Tensile strength (MPa)
Unidirectional carbon fibre composite	60–68	1.5–1.6	120–135	1200–1500

wholly polymer composite structures to erect them in difficult terrain; examples of these systems are given later in this chapter and in Chapter 19.

The method of manufacture, as stated earlier, is important from the point of view of strength and stiffness of the composite. Generally, the fully automated procedure for the manufacture of composites gives the highest mechanical property values in the direction of the fibres followed by the semi-automated and then the hand lay-up method the lowest value; the reason for this situation is mainly because of the compaction of the fibres in the polymer. It should also be realized that the randomly orientated fibres provide equal mechanical properties in all directions, whereas the cross-ply fibres provide strength and stiffness in those two orthogonal directions but these properties decrease in all other Cartesian coordinate directions relative to these. The unidirectionally aligned fibre has maximum strength and stiffness in this direction but at right-angles to this, these values are those of the matrix or less.

Creep of composites

The most important components which affect the creep characteristics of composites are: (a) their load and environmental history, (b) the nature of the applied load, (c) the dependence of the rate of loading on the system, and (d) the temperature and moisture environment in which the system is situated. In addition, the creep characteristics of the polymer composite are dependent upon the direction of alignment, the type and the volume fraction of the fibres.

Polymeric materials show an approximate linear relationship between the applied stress and the creep strain. A typical value for creep strain in air and under an initial stress level of 50% of ultimate and over a time period of 10^6 hours would be of the order of 0.2%

13.2.3 Inservice properties of the composite
Quality control

The quality of the finished product will be dependent upon the quality assurance of that product. To ensure that the components met all the necessary requirements, it is essential to write a specification covering all aspects of performance and to ensure that the finished products meet this specification. From a polymer composite point of view the specification should include:

(i)   raw material quality
(ii)   adequate design of all components
(iii) sufficient detail consistent with design requirements for the manufacture of the finished components.

As there are different manufacturing processes for polymer composite materials, the quality control problems for the various processes are different.

The quality control for reinforced polymer units is covered by BS 4549-1 1997 and is based upon performance criteria. This control must be assessed by a routine programme of testing.

Durability
The term durability is used to denote the period of time over which a material will
perform its allotted task in its given environment. The specification of the durability should state that the components must conform to the performance specification
for the expected life of the structure.

Durability is often difficult to assess and requires a keen judgement of what constitutes sufficient duration and adequate performance.

The durability of each composite used in a particular situation is given in the relevant section under that specific material item.

Fire behaviour of polymers
A major concern of the engineer using polymers in the construction industry is the
problem associated with fire. Because the organic materials are composed of carbon,
hydrogen and nitrogen atoms they are flammable to varying degrees. It is possible,
however, to incorporate additives into the resin formulations or to alter their
structure, thereby modifying the burning behaviour and producing a composite with
much enhanced properties. In addition because many composites used in bridge
engineering have high fibre volume fractions it is not possible for the fire to
penetrate the interior of the composite through the burning of the polymer and
consequently the composite will not burn easily.

13.2.4 Polymer composite manufacturing processes associated with bridge engineering

There are various techniques for the manufacture of advanced polymer composites
all of which will have an influence on the mechanical properties of the final composite product. The methods may be considered under three broad headings:

- the manual process
- the semi-automated process
- the automated process.

The manual process covers methods such as hand lay-up, spray-up (both of which
are known as contact moulding), pressure bag and autoclave mouldings. The semi-
automatic process includes compression moulding and resin injection. The automated
processes are those such as pultrusion, filament winding and injection moulding.

This chapter is not primarily concerned with the manufacturing aspects of
composite materials, these have been discussed in Hollaway (1993). It is, however,
important to recognize the profound effect that the manufacturing processes have
on the quality of the final products and on their properties due to the varying degree
of compaction provided by each of the manufacturing methods, the degree of
pre- and post-curing of the resin and the overall effect on the micro-structure and
internal stresses developed during the fabrication procedure.

The techniques within the three broad fabrication processes and which will be
discussed in more detail here, are methods currently used in bridge engineering to
manufacture composites for specific applications, for instance, in flexural and shear

retrofitting to the bridge beams, in column wrapping to improve their local buckling and overall compressive strength and for composite rebars for the reinforcement of concrete structures. Examples are:

- The wet lay-up process – some of the commercial methods available to manufacture the composite and thence to undertake the retrofitting of these materials to structures are:
  (a) the 'Replark' method
  (b) the Dupont method
  (c) the Tonen Forca method.
- The semi-automated processes – these methods are used for the manufacture of the composite material and thence to retrofit them to concrete, steel and cast iron:
  (a) the resin infusion under flexible tooling (RIFT) method
  (b) the XXsys method.
- The automated process – these methods are used to produce polymer composite materials to be fabricated into structural units to be used in the construction industry:
  (a) the pultrusion technique
  (b) the filament winding technique
  (c) the resin transfer moulding (RTM) and the Seemann composites resin infusion manufacturing process (SCRIMP) method.

The wet lay-up method

(a) Replark method

The Replark fabrication technique is marketed and distributed in this country by Sumitomo Corporation; the process uses Mitsubishi manufactured fibres. The Replark material is manufactured from unidirectional carbon fibre impregnated with a small amount of polymer. Replark can be readily applied to concrete with epoxy resin (Epotherm). The fabrication technique to form this material is basically a wet lay-up method in which the mould is usually the structural unit to be retrofitted with polymer composite. However, planar and non-planar composites can be manufactured independently and used as structural units. The application procedure is:

- to grind the surface of the concrete on which the composite is to be laid-up thus providing a clean rough surface for bonding
- to prime the surface of the concrete with a compatible resin to that of the composite
- to apply an epoxy putty filler if necessary
- to apply the first resin coat (undercoat)
- to apply the carbon fibre sheet (Replark) to the resin coat
- to apply the second resin application (overcoat)
- to apply a protective coating if required.

(b) The Dupont method

Dupont produce a system using Kevlar fibres which is marketed as a repair system for concrete structures. The application of the material to the surface to be retrofitted is similar to the above.

(c) The Tonen Forca (Towsheet) method

The Tonen Forca is an unidirectional carbon fibre sheet in an epoxy laminated system marketed in this country by Kyokuto Boeki Kaisha Ltd.. The system was originally developed by Mitsubishi Chemical Corporation and is therefore similar to the Replark system.

The semi-automated process

(a) The resin infusion under flexible tooling (RIFT) process

To retrofit carbon fibre composites to steel, cast iron and concrete bridges built from these materials, the semi-automated RIFT process has been developed to allow high quality composites to be formed *in situ* and bonded to the structure. In this process dry fibres are preformed in a mould in the fabrication shop and the required materials are attached to the preform before packaging and sending to site. The preform is attached to the structure and a resin supply is channelled to the prepreg. The prepreg and resin supply is then enveloped in a vacuum bagging system. As the resin flows into the dry fibre preform it develops both the composite material and the adhesive bond between the CFRP and the structure. The process provides high fibre volume fraction composites of the order of 55% which have high strength and stiffness values.

(b) The XXsys method – continuous carbon fibre jackets

The XXsys method was developed in the USA by XXsys Technologies, Inc, San Diego, California, as a semi automated process for seismic retrofitting and strength restoration of concrete columns using continuous carbon fibre. The XXsys carbon fibre composite jackets are installed with a fully automated machine called Robo-Wrapper^{TM} and portable oven for curing. The technology associated with the technique is based upon the filament winding of prepreg carbon fibre tows around the structural unit thus forming a carbon fibre jacket; currently, the structural unit to be upgraded would be a column. The polymer is then cured by a controlled elevated temperature oven and can, if desired, be coated with a resin to match the existing structure. An advantage of this automated process is that the carbon fibre prepreg is impregnated with the polymer under factory controlled conditions, providing good quality control and as a consequence a high strength to weight ratio. The equipment is erected on site with minimum disturbance to traffic and the whole operation is undertaken in minimum time; the latter will, however, depend upon the size of the job. The carbon fibre jacket which is eventually formed around the column will increase the shear capacity of the column and will confine the concrete and greatly enhance its ductility in the flexural plastic hinge region. Furthermore, it will provide lap splice clamping and will prevent local buckling of the vertical reinforcement. For corrosion-damaged columns, the jacket restores shear capacity and will prevent spalling of the cover concrete.

The fully-automated process

(a) The pultrusion process

The pultrusion technique is one of the fully automated continuous processes used in the reinforced plastics industry. Constant section shapes are produced by placing strands of fibre interleaved with any additional required layers of fabric, through a heated die, the strands having been previously impregnated with resin. Alternatively the fibres can be impregnated with the resin by injecting the latter into the heated die as the fibres pass through it. The process has been discussed by Hollaway (1986).

The finished pultrusion products are generally straight and can have most geometrical cross-section shapes. Recently some products have been manufactured which are curved in the longitudinal direction.

The process requires considerable experience to produce complicated geometrical sections particularly those which incorporate right-angled bends. It is necessary to ensure that the fibres can flow well in all cross-sections of the manufactured unit and that they are well compacted around any bends in that section to prevent voids forming in the unit; thus avoiding any weaknesses in the section.

The technique generally refers to the utilization of thermosetting polymers such as polyester, vinylester and epoxy resins in conjunction with fibres. A technique has been developed for the production of thermoplastic polymer and fibre composites by a similar process and generally also referred to as pultrusion. Composites made by this method have been used in composite rebars (see Section 13.2.10).

(b) The filament winding

In filament winding, continuous strands or rovings of reinforcement are passed through a bath of activated resin and then wound onto a rotating mandrel. The angle of helix is determined by the relative speeds of the traversing bath and the mandrel. If resin pre-impregnated reinforcement is used, it is passed over a hot roller until tacky and is then wound on to the rotating mandrel. After completion of the initial polymerization, the composite is removed from the mandrel and cured; the composite unit is then placed into an enclosure at 60°C for 8 hours.

(c) The resin injection moulding (RTM) and the Seemann composites resin infusion manufacturing process (SCRIMP)

A fibre preform is placed within a mould cavity or on a tool covered with a vacuum bag. Thermosetting resin is then injected into the mould where it saturates the preform and fills the mould. Once resin injection is complete, the resin system is designed to undergo a curing reaction and produce the finished part. The mould filling process, which can take a number of minutes is critical to obtaining a good quality product. The resin must fully wet out the preform so that the part contains no voids or dry spots. Any voids present can result in defects that can diminish the strength and quality of the cured part.

The SCRIMP process is a vacuum-assisted RTM process.

13.2.5 End use of advanced polymer composites
The rehabilitation of the deterioration of the infrastructure

Reinforced concrete structures may, for a variety of reasons, be found to be unsatisfactory. In the design and construction phase causes of deficiency include marginal design/design errors causing inadequate factors of safety, the use of inferior materials, or poor construction workmanship/management, causing the design strengths not to be achieved. In service, increased safety requirements, a change in use or modernization causing redistribution of stresses, an increase in the management or intensity of the applied loads require to be supported, or an upgrading of design standards may render all or part of a structure inadequate. Increased loading may also result from a less favourable configuration of existing loads. In addition, the load carrying capacity of a member may be compromised by material deterioration, such as corrosion of the internal reinforcement particularly in marine or industrial environments, carbonation of the concrete or alkali-silica reaction, or structural damage, caused by fire, impact, explosion, earthquake and overloading. On highway structures, corrosion of the internal reinforcement is exacerbated by the application of de-icing salts. For prestressed concrete beams, strengthening measures may be required to prevent further loss of prestress.

These inadequacies may manifest themselves by poor performance under service loading in the form of excessive deflections and material failure, or through inadequate fatigue or ultimate strength. When maintenance or local repair will not restore a deficient structure to the required standards, there are two possible alternatives; complete or partial demolition and rebuild, or commencement of a programme of strengthening. In this context, strengthening is defined as rehabilitation to restore the original structural performance, or upgrading to attain higher strengths or stiffness requirements. The choice between strengthening or demolition depends on many factors, such as material and labour costs, time during which the structure is out of commission and distribution of other facilities. However, the financial benefits of strengthening as opposed to demolition can often be considerable, particularly if a simple, quick strengthening technique is available. In addition, if the structure in question has historical importance, the possibility of demolition may be precluded.

Strengthening can be carried out by several techniques to achieve the desirable improvement. These include increasing the size of the deficient member through the provision of additional reinforced or prestressed concrete layers using stapling and pressure grouting, the introduction of additional supports, beams or stringers, the replacement of non-structural toppings with structural toppings or lighter materials, overslabbing or polymer impregnation. For bridge structures, traffic management measures may be imposed to relieve loading on weak members. Costs of the methods of strengthening vary considerably depending on the size of the structure, the extent of the strengthening work required and in the case of bridges, the volume of traffic carried over and under it. In techniques where additional material is applied to the original member, the main problem is that of ensuring adequate connection and composite action between the reinforcing element and the existing structure. External post-tensioning by means of high strength strands or bars has been successfully used to increase the strength of beams in existing bridges and buildings. However, this method does present some difficulties in providing anchorage for the post-tensioning strands, maintaining the lateral stability of the girders during post-tensioning and protecting the strands against corrosion.

The development of structural adhesives has lead to the evolution of a further method of structural repair in which steel plates are externally bonded to the structure *in situ*, effectively increasing the area of reinforcement provided. The plates then act compositely with the original member, producing a section with improved flexural strength and stiffness. If propping is used during the bonding operation, the plates can help to support dead as well as live loading; otherwise, additional live loading capacity only is provided. The success of strengthening methods depends critically on the performance of the adhesive used. When bonded to the tensile faces of the concrete members, the plate is in a position where it can have the maximum effect on the ultimate strength, stiffness, and hence deflections and also on the initiation of cracks.

Although plate bonding was pioneered using steel plates, the technique has several disadvantages inherent in the use of steel; these are discussed in Chapter 19. The replacement of steel with fibre-reinforced polymer (FRP) materials has therefore been proposed in an attempt to overcome some of the shortcomings of the strengthening technique.

Carbon fibre vinylester composite plates bonded to the soffit of a reinforced concrete beam will not suffer any ultraviolet light degradation as the material is not exposed directly to the sunlight.

The following sections will discuss the various polymer composites methods, based upon their manufacturing techniques, available to the engineer to rehabilitate or strengthen beams to their former design values or to provide an increase in flexural, shear and compressive–buckling capabilities.

Plate bonding to improve the flexural characteristics of the beam.

The potential of external plating and its application as a strengthening technique has only been made possible by the development of suitable adhesives; these have been discussed in Chapter 19. In this chapter it will be assumed that the method of application of the adhesive to the two adherents has been determined.

In the mid-1980s it was proposed that fibre-reinforced polymer plates could prove advantageous over steel in strengthening applications. Unlike steel, FRPs are unaffected by electrochemical deterioration and can resist the most corrosive effects of acids, alkalis, salts and similar aggressive materials under a wide range of temperatures (Hollaway, 1993). Consequently, additional corrosive resistant applications are not required in the case of composites, making preparation prior to bonding and maintenance after installation less arduous than for steel.

The reinforcing fibres in the plates can be introduced into a certain position, volume fraction and direction in the matrix to obtain maximum efficiency, allowing the composites to be tailor made to suit the required shape and specification. The resulting materials are non-magnetic, non conductive and have high strength and stiffness in the fibre direction at a fraction of the weight of steel. They are consequently easier to transport and handle, require less falsework, can be used in areas of limited access and do not add significant loads to the structure after installation. Continuous lengths of FRP manufactured by the pultrusion process (described in

Section 13.2.4) can readily be produced which, because of its geometric dimensions with a small thickness usually of the order of 1.0–1.5 mm, can be delivered to site in rolls. The inclusions of joints during installation is thus avoided. With the exception of glass composites, FRP (carbon fibre and aramid fibre composites) generally exhibit excellent fatigue and creep properties and require less energy per kg to produce and transport than metals. As a result of the above there is less disruption in the process of installation than steel allowing faster, more economical strengthening. In the UK between 1993 and 1997 considerable R&D effort was put into the investigation of polymer composite plate bonding for upgrading reinforced and prestressed concrete beams. ROBUST, a UK Government/Industry collaborative research funded through the DTI/EPSRC Structural Composites LINK scheme was set up set up to undertake investigative work in this area. A comprehensive review of composite plate bonding, based upon the investigative work undertaken during the ROBUST project, has been given by (Hollaway and Leeming, 1999).

For flexural strengthening, in which a single composite plate would be bonded to the soffit of the beam, the manufacturing technique for the plate would most likely be the pultrusion method. The fibres in the composite would be unidirectionally aligned thus achieving the maximum possible strength and stiffness in that direction. It should be mentioned that, at right angles to this direction, the composite would be weak in strength and stiffness, having values of less than that of the polymer component. Therefore, care should be taken in the handling of the composite particularly at 90° to the longitudinal direction. The composite would have a removable peel-ply membrane on both surfaces which is attached during its manufacturing process The peel ply, on one side of the composite, would be removed immediately before bonding, thus providing a clean and roughened surface on to which the adhesive is applied. The peel-ply on the other (external) side of the plate could remain during the remaining life of the structure or it could be removed after the adhesive has polymerized. An added advantage of the peel-ply is that it gives an added stiffness to the plate in the 90° direction. If the composite is not supplied with a peel-ply, surface treatment is necessary and this would be performed by a very light gritblasting followed by degreasing with acetone to remove surface contamination and to produce a surface more receptive to bonding. It has been concluded by many authors that the subsequent strength of joints fabricated from treated FRP is dependent upon the extent to which contaminants, in particular release agents, are removed prior to bonding and not the severity of abrasion.

The flexural strengthening generally takes place on an uneven concrete surface even after the concrete surface preparation of grit blasting and removal of any contaminants. Consequently, the thickness of the adhesive would be between 1 and 2 mm to allow for this unevenness. Furthermore, a 'propping' system, under the entire length of the plate, would be required until the adhesive has polymerized, this small pressure would enable a more satisfactory adhesive joint to be formed. The 'propping' system could be effected by vacuum bagging or more simply by a physical prop, always providing access in permitted under the beam.

If it were necessary for the composite to navigate a corner or non-planar section of the concrete system, the pultrusion manufacturing technique could not normally be used in this situation. Although pultruded angles or other more complicated sections are able to be formed (provided the dies are available), the engineered angles would not necessarily fit the corner of the concrete. In this case a prepreg could be employed and the Replark or the RIFT procedures would be used to apply the composite to the surface of the concrete. Figure 13.12 gives an example of flexural plate bonding using a pultruded carbon fibre in an epoxy resin. The system uses Sika Ltd. CarboDur plate and Sikadur 31 PBA adhesive.

Plate bonding to improve the shear capacity of the beam

Only a small amount of site work has been undertaken to date to upgrade beams in shear using composite materials, research into fabrication techniques for retrofitting the material are currently being conducted at a number of research institutes and in academia.

The techniques discussed in Section 13.2.4, namely the pultrusion, the RIFT and the REPLARK methods, could all be used for retrofitting shear plates. In the pultrusion system the fibre orientations in the composite would be at an angle (usually ±45°) to the longitudinal direction. To form angle ply pultruded units it would be necessary to provide some unidirectionally aligned fibres in the composite to enable the tensile force from the pultrusion puller to be taken by the uncured composite.

Figure 13.12 Composite plate bonding for flexural strengthening (by kind permission of Sika Ltd., Welwyn Garden City, Herts, UK).

Peel-ply membrane would also be placed on the composite at the time of manufacture. One of the problems of using pultrusion in this situation is the difficulty of obtaining sufficient bond length for the composite and it would probably be necessary to combine this technique with the REPLARK or RIFT methods allowing overlay of the pultrusion with composites made by these latter processes. However, both the RIFT and the REPLARK methods could be used independently and both have been used in the field. Figure 13.13 shows the REPLARK material being used as external vertical strips to an RC beam to upgrade a beam in shear.

Hutchinson *et al.* (1997) has described tests that were undertaken at the University of Manitoba to investigate the shear strengthening of scaled models of the Maryland bridge which required shear capacity upgrading in order to carry increased truck loads. The bridge had an arrangement of stirrups which caused spalling of the concrete cover followed by straightening of the stirrups and sudden failure. CFRP sheets were effective in reducing the tensile force in the stirrups under the same applied shear load. The CFRP plates were clamped to the web of the Tee beams in

Figure 13.13 The REPLARK method being used as external strips to an RC beam for upgrading a beam in shear (by kind permission of Sumitomo Corporation (UK), London, UK).

order to control the outward force in the stirrups within the shear span. This allowed the stirrups to yield and to contribute to a 27% increase in the ultimate shear capacity. Hutchinson showed that CFRP sheets are more efficient than the horizontal and vertical CFRP sheet combination in reducing the tensile force in the stirrups at the same level of applied shear load.

Taljsten (1997) studied the shear force capacity of beams when these had been strengthened by CFRP composites applied to the beams by three different techniques; these were:

* wet lay-up system by two different approaches
* prepreg in conjunction with vacuum and heat
* vacuum injection.

The results of the four-point loaded tests showed, in all cases, a very good strengthening effect in shear when the CFRP-composites were bonded to the vertical faces of the concrete beams. The strengthening effect of almost 300% was achieved, although it must be stated that this value is dependent upon the degree to which the beam was reinforced in shear initially. It was possible to reach a value of 100% with an initially completely fractured beam. Generally it was easier to apply the wet lay-up system and Taljsten suggested that although the prepreg and vacuum injection methods gave higher material properties than those of the wet lay-up method, the site application technique seemed to be more controllable for the wet lay-up process.

Composite material wrapping degraded bridge columns

It is already known that confinement of concrete enhances its durability and strength. Currently, to enhance reinforced concrete columns, additional longitudinal steel bars and concrete are added around existing columns. Another method consists in placing a steel jacket around a column. The two methods are difficult to achieve, the first one requires the construction of new formwork around the existing column and produces modifications in its external dimensions. In the second method, the installation of a steel jacket requires difficult welding work and, in the long term, the potential problem of corrosion remains unsolved.

Confinement of concrete columns with polymer composite strands or sheets of composite prepreg shows many advantages in comparison with other confinement methods. These include the advantageous properties of the composite materials such as high specific strength and stiffness and the relative ease of applying the composite material to construction site situations.

There are a number of bridge pier wrapping composite systems that have been used throughout the world. The majority of the work on confinement has been for seismic loading, predominately in Japan and the USA. The available composite systems include epoxy with either glass fibre, aramid fibre or carbon fibre fabric materials. A column consisting of a hybrid of fibre–polymer composite and concrete can deform much more under an extreme stress state than a conventional material system before failure. Furthermore, confining the concrete laterally provides an order of magnitude in the improvement of the ultimate compressive strain.

Figure 13.14 Column wrapping using REPLARK technique – siesmic strengthening of railway viaduct column, Sakurai, Nara, Japan (by kind permission of Sumitomo Corporation (UK), London, UK).

The two fabrication systems that are available for site work are the REPLARK prepreg described earlier (Figure 13.14 shows a column being wrapped using this method) and the XXsys methods also described earlier (Figure 13.15 shows a column being wrapped using this method).

A Sakawa River Bridge project which has recently been completed in Japan is believed to be one of the largest seismic retrofitting projects to date using carbon fibre composites. The bridge consists of 7 m diameter hollow concrete columns varying from 30 to 60 m high. Some parts of the columns where the concrete–steel splice joints are located were jacketed by the utilization of the REPLARK process in a carbon fibre wrap; 50 000 m^2 of sheet material were used to upgrade eight columns.

In a demonstration project in 1997 six columns, supporting a bridge, under a Seattle freeway were wrapped with multi-layers of carbon composites using the XXsys technique; recently the method has also been used to strengthen an historic arch bridge at the Arroyo Seco near the Rose Bowl stadium, Pasadena.

At Sherbrooke University, Canada, a research project demonstrated that E-glass–epoxy wrap can enhance the tolerance of reinforced concrete columns to axial loads by 45% and this lead to glass fibre–epoxy wraps being used by the city council to undertake rehabilitation work on columns, beams and slabs in the car parks of that city.

Figure 13.15 Column wrapping using the XXsys technique (by kind permission of XXsys, Technologies, Inc., San Diego, CA, USA).

In the UK the REPLARK method has been used to determine whether it is practicable, with currently available technology, to strengthen bridge piers against vehicle collision; the demonstration structure is the bridge at Bible Christian over-bridge in Cornwall. The three composite systems used, on three separate piers, by the Highway Agency were the glass fibre–epoxy, the aramid fibre–epoxy and the carbon fibre–epoxy composites. The major direction for the unidirectionally aligned fibre composite prepreg was along the length of the column (column height) but some prepreg composites were wrapped around the columns to provide the hoop strength. It should be mentioned that it is necessary to remove, by grinding, any sharp corners or protruding areas as these sharp corners or edges will damage the fibre during the lay-up period.

13.2.6 Manufacture of structural components

The construction industry requires durable structures which can be rapidly installed with minimum disruption and formability. This is particularly true with bridge structures where any long maintenance period causes disruption to the flow of traffic and is extremely expensive. Polymer composites fulfil these requirements provided the initial design of the basic building modular system is properly designed. The system manufactured from this material is highly durable and is lightweight. The composite material is strong but its low modulus of elasticity must be seriously considered in the design and if possible it should be increased by shaping the modular system to give an enhanced EI. value. Ideally the modular system should be able to be used on its own or in conjunction with construction materials and techniques.

The first major UK, all composite, versatile modular system was produced by Maunsell Structural Plastics, Croydon, UK and was introduced into the construction industry as the Advanced Composite Construction System (ACCS). It was first used in modular form as the bridge enclosure (Section 13.2.7) to the A19 Tees Viaduct at Middlesborough and was also used in the form of an all composite box beam road bridge (described later). The recently developed and highly optimized SPACES system (described later) was born out of the techniques developed during the development and erection of the first bridge enclosure at Middlesborough. The ACCS modular system is extremely versatile in its use, extending from bridges (including walkways) to building structures and to Modispine cable support system which was developed specifically for use in tunnels.

The Advanced Composite System (ACCS)

The ACCS module consists of a number of interlocking fibre reinforced polymer composite units which can be assembled into a large range of different high performance structures for use in the construction industry (Section 13.2.7 discuss-es one such use as load bearing beams in bridge engineering). Figure 13.16 shows the Maunsell plank and ten planks fabricated into a box beam. The system is manufactured by the pultrusion technique using isophthalic polyester resin and unidirectional, bidirectional and chopped strand mats glass fibre reinforcement for

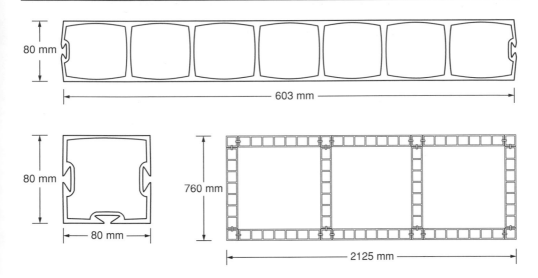

Figure 13.16 Maunsell plank and a box beam fabricated from 10 Maunsell planks.

the main structural members. For bridge enclosures, the connectors and ancillary members, which support the ACCS planks from the bridge soffit, are manufactured using either the sheet moulding compound (SMC) or the dough moulding compound (DMC) materials, and the techniques, from which the structural members are formed using these materials are either, the compression moulding, extrusion or injection moulding methods (Hollaway 1993). The main advantages of utilizing advanced composites in bridge enclosures are their high strength to weight ratio and their corrosive resistance.

The production and material content of the ACCS plank are optimized to provide highly durable and versatile components and, in addition, structures can be formed quickly from a small number of standard components. As the material is lightweight, transportation and erection on site is efficient. The design methods for the ACCS modules are based upon the limit state design principles. This principle provides a logical design procedure which identifies the limit state at which a structure ceases to fulfil its design functions. The aim of limit state design is to achieve acceptable probabilities that the relevant limit states will not be reached during the intended life of the structure. The assessment of the probabilities sets up a framework within which the uncertainties of that data, loading, stress analysis, etc. can be quantified and understood.

With any system that is being developed it is vitally necessary to draw up a manufacturing specification, testing, assembly and erection procedures and this was undertaken during the development of the ACCS.

Due to the unique characteristics of GFRP material it is recommended (Hollaway, 1990) that design standards are produced for specific applications thus ensuring a consistent approach to design.

13.2.7 Advanced composite construction bridge beams
Bridge enclosures and fairings

The need for regular inspection and maintenance of bridge structures is causing increasing concern because of the disruption caused to travellers if closure is required to maintain a bridge. Furthermore, the cost of closure will be extremely large. Moreover, stringent standards are increasing costs of maintenance work over or beside busy roads and railways. Most bridges designed and built over the last 30 years do not have good access for inspection and in Northern Europe and North America deterioration caused by de-icing salts is creating an increasing maintenance workload.

The concept of 'bridge enclosure' was developed jointly by Transport Research Laboratory (TRL, formerly TRRL) and Maunsell, Croydon UK, in 1982 to provide a solution to the problems. The function of these enclosures is to erect a 'floor' underneath the girder of a steel composite bridge to provide inspection and maintenance access. In addition to providing these structural requirements, enclosures allow greater freedom of aesthetic expression independent of the strength requirements. The floor is sealed on to the underside of the edge girders to enclose the steelwork and to protect it from further corrosion. Research work undertaken at the TRL (McKenzie, 1991, 1993) has shown that once the enclosures are erected the rate of corrosion of uncoated steel in the protected environment within the enclosure is 2–10% of that of painted steel in the open. It should be emphasized that no de-humidifying equipment is needed to prevent corrosion. Although this enclosure space has a high humidity, chloride and sulphur pollutants are excluded by seals so that when condensation does occur (as in steel girders) the water drops onto the enclosure floor which is set below the steel girders and there it escapes through small drainage holes. The floor and fixings are non-corrosive and no water is able to pond against the steel and hence corrosion is prevented.

Enclosures will undoubtedly have even more important implications for future design of long span bridges. Currently steel box girders are often used for the deck girders of such bridges in order to provide an aerodynamic shape to minimize exposed steel areas and to give adequate torsional stiffness. However, the development of cable-stayed bridges and the reduction in the fabrication costs of steel girders compared with the labour intensive steel boxes has resulted in a recent increase in the use of plate girders for long span bridges. The addition of fibre polymer composite enclosures around such structures not only enable maintenance costs to be greatly reduced, but also enables the shape of the cross-section to be optimized by extending the enclosure into a fairing to give minimum drag consistent with aerodynamic stability. The nine structures on the approach roads to the second Severn crossing Figure 13.17 is one recent example where the GFRP enclosure is extended into a fairing.

Polymer composites are ideal materials from which to manufacture enclosure floors because they add little weight to the bridge and are highly durable particularly as the polymer composite, being under the bridge soffit, is protected from direct ultraviolet light. This form of degradation, however, is no longer the problem that it was

Figure 13.17 Approach Road to second Severn crossing showing the Maunsell composite enclosure and fairings (by kind permission of Maunsell Structural Plastics, Beckenham, Kent, UK).

formerly due to improved resin formulations and the possibility of incorporating ultra-violet additives to the resins. Most bridge enclosures which have been erected in the UK have utilized polymer composites. The first major example of this technique was in 1988–1989, when the A19 Tees Viaduct at Middlesborough, Figure 13.18, was fitted with the Maunsell 'caretaker' system. This was followed by further retrofit projects, one at Botley, Oxford (1990) where the hand lay-up GFRP method was used, and Nevilles Cross (1990) near Durham where the pultruded GFRP system was fitted to an existing bridge over the main east coast railway line. Two new bridges were then built with enclosures, one at Bromley in South London (1992), which is shown in Figure 13.19, and utilized the Maunsell 'caretaker' system, the other was in 1993, at Winterbrook, Figure 13.20, and was manufactured by the hand lay-up GFRP method. The bridge carries the A4130 Wallingford by-pass over the river Thames. The design of the bridge structure was undertaken by the Bridge Department of Oxford County Council and the enclosure was designed by Mouchel Consulting. The structural steelwork of the bridge was enclosed by a number of GFRP panels and as a curved profile of the bridge was required, the panel offsets from the plate girders were varied. Each panel was manufactures from single laminates, major ribs on the panel perimeter, minor ribs elsewhere and stainless steel inserts.

Figure 13.18 Maunsell caretaker system used on the A19 Tees Viaduct at Middlesborough (by kind permission of Maunsell Structural Plastics, Beckenham, Kent, UK).

Figure 13.19 Enclosure at Bromley South Railway Station (by kind permission of Maunsell Structural Plastics, Beckenham, Kent, UK).

Figure 13.20 The A4130 Wallingford by-pass-bridge over the River Thames (by kind permission of Oxfordshire County Council, Oxford, Oxfordshire, UK).

Since 1994 health and safety legislation and design guidance published by the Highway Agency has had an impact on the design of bridge enclosures. Currently their designs are covered by the Highway Agency Standard BD67/96 (Highway Agency 1996). The requirements for wind loading would be covered by BD 37 or 38/88, Loads for Highway Bridges In addition, when enclosures are placed under railway bridges, aerodynamic pressure caused by the displacement of air due to the passage of a train is significant and, therefore, the allowable deflection of, and the design of the fixings for, the enclosure must be carefully considered.

The construction (Design and Management) Regulations 1994 and the Confined Spaces Regulations 1997, relate to Health and Safety Legislation, are particularly relevant to the design of enclosure systems. The latter regulation specifies the risks of working in confined spaces from the point of view of injury from fire or explosion, loss of consciousness due to increased body temperature and asphyxiation.

The structural form shown in Figure 13.21 is known as the SPACES system and was conceived by Maunsell Group, London, UK. It is being developed and promoted by a partnership of material suppliers, manufacturers, fabricators, bridge designers and constructors. The system has been developed as a total structural component combining key technologies and involving many of those developed for the offshore industry.

The SPACES system is a predominately factory manufactured product where quality and reliability can be achieved more readily than they can be on site. The system consists of a steel space frame acting compositely with a concrete deck slab. The space frame is enclosed by an aesthetic, aerodynamically profiled shell manufactured from polymer composite material. The enclosure, therefore, provides

Figure 13.21 The SPACES system (by kind permission of Maunsell Structural Plastics, Beckenham Kent, UK).

permanent protection for the steelwork and safe access for inspection and maintenance of the superstructure and bearings. The key to the economic success of the
SPACES concept is the excellent long-term performance provided by the advanced
composite enclosure skin as well as the development of a robotics welder for the joints
between the steel tubes. The system application is wide ranging from short 60 m spans
to long span cable support decks. The structure enables designers to adopt a 'Systems Approach' to bridge engineering for a complete range of spans, thus providing
inherent reliability and geometric flexibility whilst allowing a much greater aesthetic freedom than is normally available to bridge designers.

The enclosure skin, which is provided by the ACCS, has excellent long-term performance properties and the structural form, materials and technology of the SPACES
concept provide excellent long-term durability. In addition, and as with all enclosure
systems, the bridge owners will enjoy the benefits of permanent inspection and maintenance access. Furthermore, because of the superior resistance to corrosion and
fatigue of the enclosure skin, a minimal maintenance programme will result in
significant cost savings over the life span of the structure. It will also avoid road
closures thus reducing traffic disruption costs.

A typical specification for composites used in bridge construction would be formulated under the following headings and appended to the general specification for
the whole work.

(1) Specification for the materials used
(2) Production type approval tests on small test coupons taken from an initial
    production batch, such tests may include:
    (i)   short-term flexural strength and stiffness tests
    (ii)   long-term flexural stiffness test
    (iii) long-term flexural strength test
    (iv)   weathering test.
    (v)   spread of flame test
    (vi)   lap joint test.
    (vii) paint pull-off test.
    (viii) testing of resins, additives and glass fibre
(3) lapping of reinforcement
(4) cutting of sections
(5) dimensional check – tolerances
(6) non-destructive testing
    visual inspection
    ultrasonic examination
(7) destructive testing
    flexural test
    interlamina shear test
    glass content
    flammability tests.
(8) Handling and storage of materials
(9) Packaging of component.

Bridges – all composite construction

The development of bridges constructed entirely out of fibre reinforced polymers commenced with the prototype footbridges in Europe and North America in the late 1970s. The first GFRP highway bridge is believed to be the 10 metre span bridge constructed in Bulgaria in 1981–1982, using hand lay-up techniques. The bridge was built, rather like Ironbridge 200 years earlier, to show what could be achieved with the material.

The second all GFRP bridge is the Miyun Bridge in Beijing, China. It is a prototype structure completed in October 1982 (Shu, 1983). This bridge is the culmination of 25 years of Chinese research into the structural use of polymers. The Shanghai GFRP Research Institute has carried out a great deal of this research which has included ageing tests on the polyester matrix and glass fibre materials.

Although the materials for GFRP bridges are always likely to be more expensive than steel or concrete, the savings in fabrication costs may be considerable if highly automated production of large advanced composite members is developed. It is possible that complete box girder structures may be pultruded in the future, with a manufacturing facility being setup on site for large projects. Speed of construction, savings in erection costs and savings in foundations will also contribute to economy. However, the biggest attraction is likely to be the low maintenance costs of such structures.

In long-span bridges the deck weight is an important part of the total design load. In addition, the form and stiffness of the deck are important with respect to aerodynamic stability. As the trend to increase spans of bridges beyond previous limits continues into the twenty-first century, it is clear that either existing materials will require to have their strengths enhanced or it will be necessary to use materials with high specific strength and stiffness properties. These latter materials are in the form of fibre–matrix composites; the highest strength property that any material can achieve is when that material is in a fibrous form and the strength direction is measured along its longitudinal axis.

A single bascule lift bridge at Bonds Mill in Gloucestershire is shown in Figure 13.22. It was developed and manufactured by Maunsell Structural Plastics from fully epoxy bonded ACCS multicell box beam and 90 kg/m^3 epoxy foam infill in the compressive flange and webs cells of the ACCS modules. The weight of the bridge is 4.5 tonnes. The running surface of the polymer composite bridge was made from ACME panels (a proprietary system of epoxy coated panels with grit embedded into it) which were bolted onto the top flange of the GFRP box beam.

A requirement of this road bridge was that it should carry concentrated wheel loads and resist the large number of load cycles without fatigue damage. The key to solving this problem was the development by CIBA Polymers of a slow foaming epoxy which could be used to fill the 80×80 mm $\times 9$ m long cells of the ACCS modules. This material provided uniform support to the thin walls of the ACCS units allowing load transfer without high local bending stresses. The project included research at the University of Surrey undertaken through a Highways Agency/EPSRC LINK programme, into the effects of local wheel loads on the fibre reinforced plastics structures. Figure 13.23 shows the experimental set-up for the wheel load test.

Figure 13.22 Single bascule lift bridge at Bonds Mill – Gloucester – Manufactured from 10 Maunsell planks per box beam (by kind permission of Maunsell Structural Plastics, Beckenham Kent, UK).

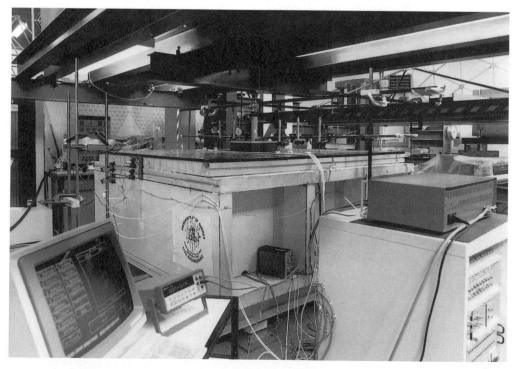

Figure 13.23 The experimental set-up for the wheel load test at the University of Surrey.

Figure 13.24 Composite bridge situated near St Moritz Switzerland (by kind permission of Fiberline Composites A/S, Kolding, Denmark).

An example of the advantages of using high-specific strength and stiffness materials for bridge building is found in a pedestrian bridge erected across a river at a ski resort near St Moritz, Switzerland. The bridge was designed by a consortium of the Municipality of Pontresina, ETH Technical University of Zurich and the pultruder, Fiberline Composites (Fiberline Design Manual, 1995). The bridge is removed each spring before glacier melt water washes tonnes of stone and gravel down stream and with a total weight of only 2.5 tonnes the composite bridge is transported and installed by helicopter in this remote area. The bridge is constructed of two sections spanning a total length of 25 metres. The load carrying capacity of the bridge is 500 kg/m^2 in conjunction with a rolling load of 1 tonne passing over it. The bridge is shown in Figure 13.24

One section of the bridge is fabricated using a combined connecting system of adhesive and bolts; if the adhesive performs as designed, it is proposed to remove the bolts during subsequent studies.

Another example of a light-weight footbridge, being installed by helicopter due to its positioning requirement in a remote area in Wales, is the Maunsell bridge using the Maunsel plank system. The area was inaccessible to heavy vehicles and the only real option was to air-lift a light weight bridge; Figure 13.25 illustrates the difficult terrain.

The Clear Creek FRP footbridge was built in the Daniel Boone Forest, Kentucky, to provide access from Clear Creek Furnace Picnic Area to the Sheltowee Trace National Recreation Trial. The bridge had a span of 18.3 metres and was constructed from standard Strongwell (formerly Morrison Moulded Fibre Glass) hybrid sections. The 600 mm deep GFRP I sections produced by Strongwell, were modified to incorporate carbon fibres into the flanges of the beam in order to increase its stiffness. In addition, in order to attain the design deflection of 175 mm, GFRP sucker rods, which are used in the oil industry, were anchored to the abutments and supported the bridge girders 3 metres from each abutment. The design load was 128 kg/m^2 which gave a mid-span deflection of 100 mm. Figure 13.26 is a photograph of the bridge.

Bridge decks

The largest single composite moulding produced in the Western world to date has been constructed by Hardcore DuPont Composites; it was fabricated using Seeman composite resin infusion moulding process (SCRIMP) method. The 'cellcore' box structure is a hollow composite unit formed over a foam core into girders and beams in the same plane. A composite skin is then added to the top and bottom faces. Finally the surface is filled with Transpo basaltic filler. It was designed by Hardcore and was provided with an epoxy overlay surface to provide a self supporting skin. The structural deck system is a prototype and has been used to construct a complete vehicular interstate highway bridge in New Castle Delaware, USA. The deck replaced an old rotting wooden bridge which was approximately 6.1 m long × 7.6 m wide with a 400 mm thick composite deck weighing 5 tonnes. The abutments of the former timber bridge were re-used and a neoprene elastomeric bearing pad separated the concrete abutment from the surface of the composite structure. The composite deck is rated to an AASHTO HS-27 load rating.

Figure 13.25 Footbridge in Wales (by kind permission of Maunsell Structural Plastics, Beckenham, Kent, UK).

Figure 13.26 Composite bridge at Clear Creek, Daniel Boone Forest, Kentucky (by kind permission of University of Kentucky, USA).

Hardcore Dupont Composites have also completed a deck component for the Delaware River and bay Authority's (DRDA) Magazine Ditch Bridge demonstration project; the deck weighed 15.4 tonnes. The bridge deck is mated with post-tensioned concrete longitudinal beams. The edge beams and deck were clamped together to create the AASHTO-22.9 metre long beam.

Figure 13.27 Aberfeldy footbridge over the River Tay Scotland (by kind permission of Maunsell Structural Plastics, Beckenham, Kent, UK).

Cables for supporting long-span bridges and cable-stayed bridges
The advantages of the strength-to-weight ratio of fibre-reinforced polymers are most important in very long span structures and these characteristics are best illustrated by their potential for forming the main supporting cables of suspension bridges. The theoretical limit of suspension bridge spans constructed from currently available high strength steel wire is of the order of 5000 m; the cables can only just support their own weight at this span. If, however, aramid or carbon fibre–polymer composites were utilized for the construction of the cables this value would increase to over 10 000 metres (Richmond and Head, 1988). Meier (1987) proposed to use carbon fibre reinforced polymer cables for a 10 000 m span bridge across the straight of Gibraltar. Assuming that such structures were technically feasible, investigations suggest that the economic span of advanced composite cables in suspension bridges would be around 4000 m.

At the end of the 1980s British Ropes undertook research investigations into the possible use of aramid and carbon fibres in bridge cables; tests were carried out using both static and dynamic loading. The latter test loads were applied to assess the potential fatigue performance of the fibre reinforced polymer relative to steel. The results showed that the two types of composites were much more durable in fatigue than steel strands; typical values of five to ten times that of steel strand were achieved.

The cable-stayed advanced composite Aberfeldy bridge over the river Tay in Scotland, for the Aberfeldy Golf Course was designed in 1992, and is shown in Figure 13.27. The tower and footbridge, which has a main span of 63 metres and an overall length of 113 m, were manufactured by bonding together composites of the ACCS using the DTp/EPSRC LINK project results mentioned below. The cable stays are made from aramid fibres. It was a fundamental advance not only in the main span length but in the technology of the ACCS which now makes possible the assembly on and off site, of composites made by automated plant, into three-dimensional structural array with the safety, reliability and cost-effectiveness demanded of all modern bridge structures

The conception and design of the Aberfeldy bridge was by Maunsell Structural Plastics and it was erected by final year project students of the University of Dundee. It required a significant R & D input during the design stage to enable the incorporation of the new construction techniques made possible only by the characteristics of the materials. A major research programme was undertaken within a LINK programme, sponsored by DTp, EPSRC and Industry, in which an extensive test programme was undertaken at the University of Surrey, of a full size prototype highway bridge; Figure 13.28 shows the box beam, fabricated from ten ACCS modules, under a continuous static load test of 20 tonnes for 9 months.

In 1997 the architect to the City of Kolding, Denmark, conceived and designed, in conjunction with Fibreline Components, Denmark, the Fiberline cable-stay bridge of 40 metre span to cross an overhead electrification main railway line which runs in a narrow cutting bordered on one side by a salt-water fjord. The bridge is shown in Figure 13.29.

Figure 13.28 An 18-metre long box beams manufactured from 10 ACCS modules being monitored under continuous 20 tonnes load at University of Surrey.

The individual bridge components were manufactured, machined and assembled in the factory and transported to site in three sections; the erection time on site was 18 hours.

The materials for the decking and parapets were manufactured from pultruded GFRP and the support columns were also manufactured from GFRP pultruded profile sections using standard geometric shapes. The only potentially corrodible materials utilized in the bridge construction were the holding down bolts at foundation level. The bridge has a load carrying capacity of 500 kg/m^2 and also allows vehicles of 5 tonnes to pass at the same time for snow clearing.

The main reasons for choosing composites in this case were:

- the material is non-conductive and therefore will pose no danger regarding any accidental contact with the overhead lines
- the corrosive resistance properties of the material will prevent any deterioration of the structure when exposed to the salt environment; the bridge should require only cosmetic maintenance over a 50-year period
- the light-weight bridge reduced the time and cost of foundation construction, of assembly of bridge and of transportation of components.

The Konaji Bridge is 100 km north of Tokyo and is a cable stay footbridge built in 1992 over an existing road leading to the Iwafune golf club. The length and width of the bridge are 28.2 m and 3.0 m, respectively, and because of the short span, the stay cables were placed on one side of the tower which is inclined away from them. Each stay cable consisted of seven 8 mm diameter rods, four of which were manufactured from GFRP and three from CFRP. The cables were encapsulated in a GFRP tube to protect the fibres from the external environment.

Figure 13.29 Cable stay all composite footbridge at Kolding, Denmark (by kind permission of Fiberline Composites A/S, Kolding, Denmark, Denmark).

A pultruded carbon fibre CFRP rope has been developed and has been used as one pair of cables, (the other 11 pairs are steel cables) in the Stork cable stayed bridge over railway lines in Winterthier, Switzerland (Sennhauser *et al.*, 1997). The carbon fibres are aligned parallel and are continuous, forming ropes which have diameters of 5 mm, the fibre volume fraction is in the range of 65–70%. The ropes, which are cut to lengths of 35 metres, are assembled into cables as parallel rope bundles in a specific pattern to avoid the strengthening loss found in single wires.

The low density of the CFRP material is an advantage, since a stay cable is uniformly loaded by its own weight, thus causing it to sag, and this sag will be greater for higher density materials, causing the cable to become 'soft' under load. This characteristic reduces the usefulness of the cable as a supporting member of the bridge structure. Lightweight CFRP cables, with smaller sag, straighten immediately under vehicle load enabling them to have a 'stiffer' characteristic. With increasing horizontal span the cables have a higher relative equivalent modulus (defined as applied stress–cable strain) compared with steel.

The key problem, impeding the widespread use of CFRP cables is the anchorage system. EMPA have developed a gradient casting material (load transfer media, LTM) to fill the space between the metallic cone of the termination and CFRP ropes.

In addition, to its anchorage potential the LTM must possess electrical insulation properties to inhibit the onset of galvanic corrosion, the major cause of material failure, in the metallic termination point. The future of the CFRP cable technology depends upon cost-benefit approach of potential user. Replacement and maintenance of steel suspenders and stay-cable have resulted in very high costs in the past 20 years. However, as initial cost is the major factor by bridge clients in decision making, it is difficult for the relatively high cost of CFRP to compete against steel unless the whole life cost of the bridge is considered.

13.2.8 Advanced composite–concrete construction for beams

Currently, work is being undertaken at the University of Surrey to develop a duplex beam construction consisting of concrete and composite units. The component parts of the beam are positioned to enable them to develop to the full, their unique mechanical and physical characteristics; Figure 13.30 shows a cross-section of this type of beam; this latter is discussed below. Such beams could also ideally be constructed as a beam and slab system and would be suitable for bridge slab construction.

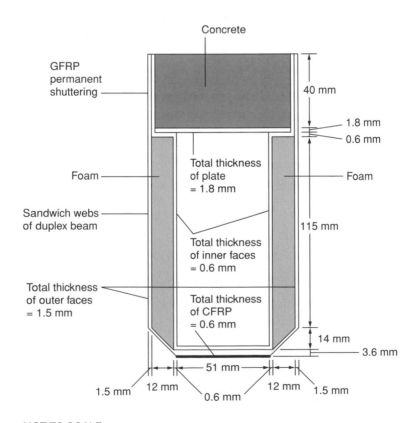

Figure 13.30 Cross-section of the University of Surrey duplex beam.

The reinforced concrete beam is one of the main structural forms currently used in the construction industry. However, a disadvantage of this form of construction is the low tensile strength capacity of the concrete with the consequence that over half of the construction material has no function in supporting external load but adds considerably to the dead weight of the system. Furthermore, the concrete acts as a cover to the steel but because of the low tensile strength of the former, it cracks and can expose the steel at the crack location to the environmental influences. Therefore, a more efficient construction is the development of a composite unit in which the materials forming the beam are used to their best advantage.

From an economic point of view structural beams, which are manufactured from a polymer composite, would utilize thin-walled box sections which are the most efficient form (Ashby, 1991) but they do have some disadvantages which include the weakness of the compressive flange to buckling and a catastrophic collapse due to the linear elastic characteristic of the material at failure.

The University of Surrey's duplex beam system, (Figure 13.31) uses to their best structural advantage the above two material components (namely the composite and the concrete). The composite forms the tensile component and is extended into the compressive zone to form a permanent shuttering for the concrete. The design for this type of beam is governed by stiffness and requires special design considerations to satisfy displacement requirements. The duplex beam system is manufactured from low temperature cure carbon and glass polymer prepreg, polymer foam and *in situ* plain concrete.

The first known work utilizing this composite–concrete combination was at EMPA, Switzerland in the mid-1980s, where a filament wound rectangular cross-section unit

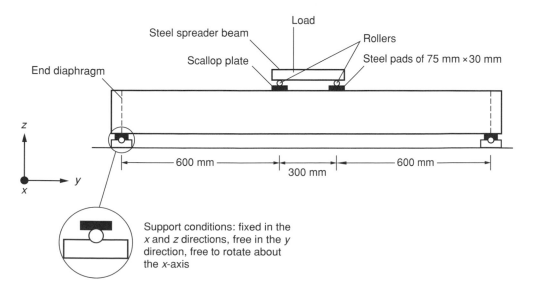

Figure 13.31 University of Surrey – 1.5 metre span experimental duplex beam.

with concrete positioned in the top surface of the composite was used as a beam. Meier *et al.* (1983), undertook fatigue tests on the filament wound GFRP composite only. Trantafillou and Meier (1992), introduced the system of a polymer composite single skin rectangular section on the top of which was placed *in situ* concrete retained by the permanent polymer composite shuttering. A theoretical analysis was introduced detailing the depth of the neutral axis, flexural rigidity, and relevant failure modes such as tensile failure of the lower flange web shear fracture and web shear buckling.

Beam manufacture

Figure 13.30 shows the cross-section of one of many 1.5 metre long laboratory rectangular beams which are being used in the current investigations. The webs below the neutral axis are formed as sandwich constructions, the inner and outer faces of the sandwich system are manufactured from Advanced Composites Group Ltd, Derbyshire, material as bi-directional glass prepreg and are situated in the webs at ±45° to the vertical; they have three and seven composite layers, respectively. The foam cores are made from Airtex R63.80 PVC Rigid Foam and are 12 mm thick. The GFRP face materials of the sandwich construction extend into the compressive region to form the permanent shuttering for the concrete. The soffit of the beams are fabricated by wrapping the ±45° GFRP prepregs around the base of the beam; laminates of unidirectional carbon fibre prepregs are interspersed in four layers between the GFRP laminates. The FRP is laid up onto a timber mould which is then enveloped in a heated vacuum bagging set at a maximum temperature 60°C and cured at this temperature for 16 hours. The manufacturing procedure, although undertaken in the laboratory in this instance, is not envisaged to cause problems when undertaken as a site production procedure.

The male timber tool was manufactured from medium density fibreboard and a glass reinforced self-adhesive PTFE film was applied to the surface of the tool for ease of de-molding the composite. To enable a bond to be developed between the permanent shuttering and the concrete a number of methods were investigated; these involved:

- placing indents at 200 mm centres along the length of the beam, in addition, a peel ply was applied to the internal surface of the shuttering
- injecting a resin adhesive into the gap between the vertical sides of the concrete and permanent shuttering
- placing bolts into the permanent shuttering at uniform distances along the longitudinal length of the beam.

It has been shown (Lee *et al.*, 1999) that with the above beam dimensions and with composite action being maintained throughout the loading to failure of the duplex beam, the failure criteria would be by crushing of the concrete.

13.2.9 Fibre composite tendons for prestressing concrete

Currently, one important use of fibre-reinforced polymeric composites, in concrete structures, is to provide the prestressing system; the most suitable fibres for pre-stressing are glass, aramid and carbon. Their ultimate strengths are high and their linear elastic response to ultimate load contains no significant yielding. By taking out some of the elongation of the tendons during prestressing operation, the use of the high strength but lower modulus materials become feasible. Prestressing tendons are subjected to very high permanent stresses, so making efficient use of an advanced technological material in which the strength is concentrated in one direction. The resistance to creep, however, becomes of paramount importance and an ability to apply the prestressing force and to anchor the tendon must be provided.

There are three systems which are commercially available; these are Polystal, Parafil and Arapree.

Parafil ropes

Parafil ropes consist of a closely packed parallel laid high strength synthetic core pro-tected by an abrasion resistant polymeric sheath. There are three types of Parafil dependent upon the type of fibre used, these are polyester, standard modulus aramid (Kevlar 29) and high modulus (Kevlar 49). There are four types of sheath available, these are polyethylene, Polyethylene-EVA copolymer, Polyester elastomer and flame retardant. The Parafil, which is the registered trademark product range, is shown in Table 13.7.

The most suitable rope for use as prestressing tendons is the Type G rope, which contains Kevlar 49 as its core yarn. The elastic modulus is about 120 GPa, with a short-term ultimate tensile strength of 1930 MPa. The rope will creep to failure at high loads. Extrapolation from short-term tests combined with predictions based upon the reaction rate theories of chemical processes, predict that a Parafil rope will sustain a load of 50% of the short-term strength for 100 years (Chambers, 1988). Measure-ments made on Parafil Type G have produced the following creep coefficient equation (Burgoyne and Guimaraes, 1992):

$$\varepsilon_t = (0.012 \pm 0.003) \log_{10} t$$

where $\varepsilon_t = \{\varepsilon_0 (t)\}/\{\varepsilon_0\}$ = creep coefficient, $\varepsilon_0(t)$ = creep strain at time t and ε_0 = initial strain.

Table 13.7 The three standard types of Parafil ropes (by kind permission of Linear Composites Ltd)

| Yarn | Sheath | | | |
	Polyethylene	Polyethylene–EVA copolymer	Polyester elastomer	Flame retardant
Polyester	Type A	Type A/C	Type A/H	Type A/X
Standard modulus aramid	Type F	Type F/C	Type F/H	Type F/X
High modulus aramid	Type G	Type G/C	Type G/H	Type G/X

Observations from strain rupture work and creep data analysis show that type G has a limiting creep strain, irrespective of the initial stress, between 0.10 and 0.12% (Technical note PF2, Linear Composites Ltd).

The stress relaxation of type G Parafil can be given in the following relationship (Chambers, 1986).

$$r = 1.82 + 0.0403f + 0.67 \log_{10}(t - 100)$$

where r = stress relaxation expressed as % (normal break load – NBL), f = initial stress expressed as % (NBL) and t = time in hours.

At say 60% NBL the relaxation over 100 years is 8.2% NBL. This equates to a relaxation of 13.6% initial stress.

Tendons used in cable stay bridges are expected to have high durability in normal environments. Kevlar is degraded in ultraviolet light, but the fibre is shielded by the sheath and therefore, this type of degradation is not a problem. In addition, Kevlar fibres suffer from hydrolytic attack by strong acids and alkalis, but the tendons would not be bonded directly to the concrete and are shielded and, therefore, they will again not suffer an attack from this cause.

Corrosion resistance. Parafil ropes are manufactured from materials which possess a high degree of mechanical toughness and are inert chemically including resistance to the corrosion action of salt water, most inorganic salts and acids. Furthermore, there is poor adhesion between ice and the smooth water repellent surface of the Parafil ropes.

Anchorages. The ability of an element to carry significant tensile force is only as good as the mechanism for getting the force into, or out of, the tensile member. A considerable amount of research has been undertaken to address this problem. The shape of the wedge must be carefully chosen to provide a uniform load over a considerable joining length of the bar. Chabert (1988) has considered complicated technologies for anchoring using external wedges and the Technical Note PF3 issued by Linear Composites Ltd. provide information on their method.

These ropes are not strictly reinforced polymer composites because the reinforcing fibres are not embedded in a polymer matrix. The sheath is used merely to protect the fibres from abrasion and weathering.

Polystal

Polystal tendons consist of bundles of bars or rods, each containing E-type glass fibre filaments in an unsaturated polymer resin. The diameter of a typical bar would be 7.5 mm and would have a fibre volume fraction of 68%. Nineteen of these bars would be grouped together to give a tendon working load capacity of 600 kN. Polystal is produced by Bayer AG in association with Strabag AG, Germany.

It should be noted that glass fibres, under long-term loading of magnitude greater than 20% of their ultimate value would suffer from stress corrosion (or stress

Table 13.8 Typical average values of residual strengths at elevated temperatures

Materials	Residual strengths (%)			
	150°C	200°C	300°C	400°C
Prestressing strand	100	90	70	50
Reinforcement	100	100	100	80
Arapree*	100	95	85	55
Polystal	95	90	80	55
FRP rebar	90			

*Heated for 30 min.

ageing) in which cracks develop at the surfaces of flaws and propagate through the thickness.

Polystal tendons would have to be protected against overheating particularly in the anchorage zone. This would be undertaken by structural mean such as increasing the concrete cover. Table 13.8 gives typical average values of residual strengths of various prestressing tendons.

Arapree

Arapree consists of aramid filaments embedded in an epoxy resin. Although aramid is very strong and can resist hard treatment, it has been shown that effective use could be improved by impregnating the bare fibre bundle in resin in order to facilitate handling, to improve alkali resistance and to activate the real tension strength of the material. Arapree, therefore, is manufactured from a pultrusion of the aramid fibre Twaron in an epoxy resin. It was developed by AKZO in association with Hollandsche Beton Group (HBG) in the Netherlands and is now produced by Sireg S.p.A. in Italy. The tendons rely upon the bond between the concrete and the pultrusion resin and this is provided by silica particles bonded to the surface of the composite. The properties of Twaron are very similar to those of Kevlar including the relaxation figures but it gives a higher overall strength compared to Parafil; the local failure of some fibres would not cause its strength to be lost over the whole length of the tendon.

The standard types of Arapree elements are circular and rectangular in cross-section. Both consists of up to 400 000 filaments of aramid. The ultimate tensile strength and modulus of elasticity of 200 000 filaments is 67 kN and 130 GPa, respectively; Table 13.9 shows the mechanical properties of the material.

Arapree tendons exhibit excellent resistance to chlorides and many other environments aggressive to steel. Specifically, the insensitivity to chlorides, such as de-icing salts, offers opportunities to overcome a range of existing deterioration problems in concrete structures.

As Arapree is an organic material, for service temperatures higher than 100°C the strength will start to decrease and at 150°C when loaded continuously for 10^3 hours it will have decreased to 90% of its initial value.

Table 13.9 Mechanical properties of Arapree rectangular strips at 20°C (derived from the manufacturer's data sheets)

Property		Value
Uniaxial tensile strength	MPa	3000
Modulus of elasticity	GPa	125–130
Failure strain	%	2.4
Density	kg/m^3	1250
Transverse compressive strength	MPa	150
Interlaminar shear strength	MPa	45
Poisson's ratio		0.38

The first prestressed concrete bridge to be built using glass fibre reinforced prestressing strand was a small footbridge in Dusseldorf which was completed in 1980, Weiser (1983). This bridge was essentially designed as a reinforced concrete bridge allowing some of the tendons to be removed for testing.

A number of bridges have been built world wide utilizing prestressed FRP cables but generally using FRP rebars for the unstressed concrete slabs. A total of five road bridges and footbridges have been built in Germany and Austria utilizing glass fibre composite tendons, Polystal (Wolff and Miesseler, 1993). The first highway bridge which was opened to traffic in 1986 was the Ulenbergstrasse Bridge in Dusseldorf. The bridge is 15 metres wide and has spans of 21.3 and 25.6 metres. The slab was first post-tensioned with 59 Polystal prestressing tendons, each made up from 19 glass-reinforced polymer rods of nominal diameter 7.5 mm. These tendons were anchored to a designed block and each tensioned to a working load of 60 kN; four tonnes of glass reinforcement polymer prestressing tendons were used. This bridge has been monitored and test loaded periodically since it was opened. In Japan the emphasis has been on the development of carbon and aramid fibre tendons where a total of ten bridges have been built since 1988 (Tsuji *et al.*, 1993) and (Nortake, 1993). Carbon fibre has also been used on one bridge in Germany and aramid fibre tendons for a cantilevered road way in Spain (Casas and Aparicio, 1990).

One bridge in North America, at South Dakota, has been stressed using glass and carbon fibre composite tendons (Iyer, 1993), and a bridge in Calgary, Canada has been built using carbon fibre composite strands (Anon, 1993).

13.2.10 Composite rebars for concrete in bridge construction

Concrete structures have, traditionally, been reinforced with steel rebars which are very durable with the concrete providing a benevolent alkaline environment. However, in situations when the structure is exposed to highly aggressive environments, the concrete is unable to provide sufficient protection to ensure the required service life of the structure. The situation is exacerbated when the reinforced concrete, cracks under load in the tension region, and the steel in the cracked zones is wholly exposed to the hostile environment.

As has been stated already, in Section 13.2.4, the corrosion of the steel reinforcement in concrete is a major concern in some civil engineering construction environments such as coastal and marine environments, chemical plants and bridges. In the case of bridges the degradation of the steel would most likely be from de-icing salts.

Several methods have been employed to alleviate the corrosion of steel, all with varying degrees of success, these have included:

- the increase to the cover concrete
- the reduction in the permeability of concrete
- the application of cathodic protection
- an epoxy resin coating of the surface of the reinforcement.

A further option would be to replace the steel rebars with polymer composite rebars.

Eurocrete investigated the utilization of glass fibre–polymer composites as a replacement material for the steel rebars. (Eurocrete was a multimillion pan-European project funded through the DTI/EPSRC LINK scheme, which also had EUREKA status. The project undertook R&D work into the embedment of non-ferrous reinforcement for concrete. The development of design guidance, extensive durability testing as well as the construction of typical demonstration structures.)

The most suitable fibres currently, for the reinforcing or prestressing of concrete are glass, carbon and aramid. The reinforced polymer matrix rebars would be manufactured by the pultrusion process using thermosetting polymers such as polyester, vinylester or epoxy but the final choice will depend upon their durability and cost. Theoretically there are no limits to the sizes and shapes which could be made. The bars and rods are manufactured in straight lengths and, unlike steel rebars, cannot be bent into the standard 90° and 180° hooks. There are, however, various ways to overcome this shortcoming. For example, the 90° hooks with the same dimensions as the main reinforcement can be lap spliced in the field on to the main reinforcement. To form stirrups, filament wound glass or carbon fibre–polyester or vinylester thermosetting polymer composite sections can be manufactured using the relevant size mandrel and when cured the formed section member can be cut to the required size to form the stirrups; care must be taken however, to seal the cut ends with a resin to prevent moisture entering through the interface between the two components and causing leaching, Hollaway, (1993).

The composite rebars can readily be cut on site by the use of a portable saw using masonry blades or by hacksaw. However, the cut surfaces must always be sealed to prevent leaching.

As the mechanical properties of the composite rebars are a function of the amount and type of fibre used their values will vary but the strength of FRP reinforcement will generally tend to be between that of high yield reinforcing steel and prestressed strand and will lie between 550 and 1500 MPa and the stiffness will lie between 41–55 GPa (Table 13.10).

The single most important aspect of polymer composite rebars embedded in concrete is the durability of the rebars. It is most likely that the resin in this situation would be the polyester or vinylester, the fibre would be the glass or carbon fibre,

Table 13.10 Comparison of typical properties of steel and GFRP rebars

	Steel	GFRP
Tensile modulus of elasticity (GPa)	200	41–55
Tensile strength (MPa)	500–700	550–1500
Yield strength (MPa)	280–420	
Compressive modulus of elasticity (GPa)	200	34–47
Compressive strength (MPa)	500–700	320–470
Specific gravity	7.9	1.5–2.0

however, the vinylester resin is more stable than the polyester and the carbon fibres are not attacked by the alkaline medium. It is well known that some resins and fibres can degrade in the high alkaline concrete environment, consequently, it is necessary to choose the correct fibre and resin systems.

The GFRP rebars do not suffer from ultraviolet light degradation when they are used as internal reinforcement for concrete structures.

Glass fibres are relatively insensitive to high or low temperatures. The effect of the high temperature is more severe on the resin matrix than the fibre. Tests conducted in Germany have shown that when E-glass fibre–polymer composite rebars were stressed to 50% of their tensile strength the bars, after half an hour exposure to a temperature of 300°C, maintained 85% of their room temperature strength. This performance is better than prestressing steel. However, at 500°C the two materials approach the same value of 50% of their ambient temperature value. The problem of fire for concrete members reinforced with GFRP rebars is different from that of composite materials subject to direct fire. In the former case the concrete will serve as a barrier to protect the rebars from direct contact with flames. However, as the temperature in the interior of the beam changes the mechanical properties of the GFP rebars may change, depending upon the rise of temperature.

Thermoplastic polymers would be a possible group of materials from which to manufacture FRP rebars, the choice of this material would enable them to be formed on site as thermoplastics can be shaped by heat. Pultruded fibre–thermoplastic matrices have been manufactured for this purpose.

Simple design methods for the use of non-ferrous reinforcement are being developed in a number of countries; these methods will then be incorporated into design codes for the steel-reinforced structures The Japanese Ministry of Construction has published draft guidelines for design, the Canadian Bridge Code will shortly have a section dealing with FRP rebars, and guidance is being prepared by the American Concrete Institute. Modifications have been developed during the EUROCRETE project which will be incorporated into BS 1880. Some of the most significant points developed by the Eurocrete consortium and other investigators are given below:

Flexure
When considering flexural situations the basic design principles are independent of the type of reinforcement used. The stress–strain relationship for FRP is essentially

linear to failure and therefore it is likely that with the high strength to failure but low elastic modulus of the material, the reinforced concrete beam would fail by compression of the concrete. Generally the conventional design assumption that 'plane sections remain plane' is valid and hence the current conventional design is adequate. With low percentage ratios of reinforcement, FRP RC sections would be expected to be stronger than steel RC sections. However, the FRP RC structural elements would be expected to deflect much more than the steel RC elements after concrete cracking. Consequently, if deflections are to be limited the cross-sectional area of FRP bars should be increased in order to increase axial stiffness.

Shear

Due to the low stiffness of FRP the contribution of the tensile reinforcement to the shear capacity of the cross section is reduced. Consequently, the design equation for the permissible shear is required to be modified to include an effective area of reinforcement based on the modular ratio.

A stress concentration problem could result in the shear links due to the change in direction of force at the corners of the links. This can, however, be overcome by increasing the radius of curvature at the corners or by decreasing the link thickness.

Deflections

The deflections of beams calculated by the approach given in BS 1880, show that they can be estimated with reasonable accuracy, albeit with larger values than those for high tensile steel. The system may be considered to be an under reinforced steel RC beam in which the large deflection and cracks provide considerable warning prior to the flexural failure.

Bond

Providing the surfaces of the FRP rebars are roughened the performance will be as good as can be expected and will improve the greater the degree of roughness. Pultrusion sections are produced with a highly smooth surface but these can be roughened; one method is to apply a peel ply to the surface of the section during manufacture. If the concrete strength is generally below 25 MPa the pull-out bond failure will be in the concrete. The CFRP rebars will show a slightly superior pull-out bond strength to that of GFRP rebars, with values of the order of 13 MPa, but this will depend upon a satisfactory surface roughness being achieved. The bond strength in flexural situations will be of the order of one third of this value. Another option would be to use Arapree. This material has been discussed earlier in the chapter.

Columns

As FRP rods have low compressive strengths in comparison with those for steel, the guide lines recommend that the contribution of compressive reinforcement in columns should be ignored.

A significant design consideration for reinforced concrete will be the influence that fire will have on the reinforced system. The fire testing of composites generally takes the form of surface spread of flame and the integrity of the structure over half an hour. When composites are heated to above their glass transition T_g temperature they become soft and if this occurred in a reinforced concrete situation the bond between the fibre and polymer will weaken and the system will not then act compositely.

Applications

The first footbridge using glass-fibre composite reinforcement in Britain was built in 1995 at Chalgrove in Oxfordshire under the auspices of Eurocrete. The bridge was precast by Tarmac Precast and has a span of 5 m, a cross-section of 1.5 × 0.3 m; the concrete is a grade 40. The bridge was transported to site by Laing Civil Engineering. The rebars were manufactured by GEC (now Fibreforce Ltd.) and were 13.5 mm in diameter, and were used as mesh reinforcement with the rods placed orthogonally at 150 mm spacing at the top and bottom of the beam. The bridge was test-loaded to 125% of the design load in accordance with BS 1880, using steel dead weights, and was monitored using vibrating wire gauges and fibre-optic cables.

13.2.11 Intelligent structures

Smart structures and materials have emerged during the past few years as one of the important technologies for the twenty-first century. The ideas are simple although the technologies for obtaining the intelligent structure can be complicated. At the structural level an integrated sensor system provides data on the structural loading and on the environment, in which the structure is situated, to a processing and control system which incorporates signal integrated actuators to modify the properties of the structure in an appropriate way. Such systems can offer immense benefits to bridge engineering. The sensing and response functions are built into the material itself possibly using a chemical or morphological structure to provide the response. There are a number of different disciplines involved in achieving a high level of sophistication in the art of intelligent structures before any meaningful activities in smart structures and materials can take place. Included in these disciplines are three of particular importance, these are: material systems, adaptive control systems and artificial intelligence systems. The use and creation of materials has been an important human activity throughout history, whilst the use of adaptive control systems has only become of significance since the beginning of the industrial revolution and the artificial intelligence depends upon the development and availability of a computer.

A smart material can 'sense' changes in the environment and make a response by either changing its material properties, geometry, mechanical or electromagnetic response. Both the sensor and actuator functions, which comprise the 'brain' of the material, must be integrated with the appropriate feedback. Piezoelectric ceramics have proved to be effective both as sensors and actuators for a wide variety of applications. Such materials can respond by either changing the stress–strain fields to a desirable value (active noise and vibration control for example) or changing its surface stress–strain distribution such as the external field which itself could be a sensing signal.

The development of materials with built in optical sensing systems constitutes a necessary phase in the evolution of smart structure technology. Structures from such materials could continuously monitor their internal strains, vibration temperature and structural integrity. In the case of advanced composite materials this intrinsic sensing system might also be capable of improving quality control during fabrication. This clearly has both safety and economic aspects for it and could lead to greater confidence in the use of advanced composite materials and material savings through avoidance of over design.

Optical fibre methods which have been directed towards the development of smart aerospace and hydrospace vehicle material evaluation and control during the past 15 years, could be applied, after modification, to the evaluation of some civil engineering structures. The advantages of optical fibre technique for civil applications include the general robustness of the optical fibre and cable material under harsh environmental conditions and the general geometry of the fibre-sensing systems which allows multiple sensor locations to be placed along a single linear fibre of extended length. These advantages are particularly attractive for the instrumentation of civil engineering structures which are exposed to external environmental effects over practical lifetimes of 50–100 years. A method of monitoring strains is to use a fibre optic differential interferomentric measuring system. Single mode fibres embedded in a material can be used to detect both strain and temperature fluctuations although fluctuations in non-laboratory environments could mask the resulting temperature induced strain. A range of measurement systems based upon optical fibres are available. A review of these was presented by Hofer (1987).

Fibre-optic sensors make ideal sensing systems for composite materials as they are compatible with them, and are extremely small and lightweight, resistant to corrosion and fatigue, immune to electrical interference. With increasing use of composites in bridge engineering the development of smart composites will accelerate this trend as it is extremely difficult, if not impossible, to incorporate the same capabilities into competitive materials. Although major advances have been made in the last decade in all enabling technologies associated with smart structures, the technical challenges remain formidable. In civil engineering, the problem associated with manufacturing, where the sensors are embedded into the material must be able to resist the pressures of manufacture. Particular attention must be paid to the sensor choice, fibre coating and movement of damage of the device during manufacturing. In general, the choice of the smart structure 'system' is extremely critical and work is required to help the designer and fibre optic engineer to select the most appropriate materials for a particular fabrication route and application.

A strain-measuring device, developed for polystal, uses optical fibres which are incorporated in the tendons to enable monitoring of their performance to be made. If the optical fibres break, or neck, a comparison between the reflected and transmitted light signals would allow the position of the break to be ascertained. The inclusion of copper wire sensors could also measure fractures in the tendons. Pairs of copper wire would act as capacitors, with the composite tendon acting as the insulator. Stress changes would not be expected to produce measurable changes in capacitance but a break in one or more wires should be measurable.

The electrical resistance (er) strain gauge is a possible candidate for the long-term monitoring of strains in civil engineering structures and bridges. The use of the er gauges to measure strains on or around the reinforcement of RC beams is obviously attractive, but care must be taken to ensure that the presence of the gauges and their wiring does not disturb the bond characteristics of the surface of the bar; bond between the reinforcement and the surrounding concrete is a key parameter governing the behaviour of a reinforced concrete member. This generally will preclude mounting strain gauges on the surface of a bar and it would suggest that if rebar strains are to be monitored the strain gauge should be mounted in a duct running longitudinally through the centre of the reinforcement. This technique was pioneered by Mains (1951) and used by Scott and Gill (1992).

The application of er strain gauges to the prestressing steel tendons or the measurement of the prestressing forces with the aid of load cells is not possible in the case of prestressing with post-bond. Furthermore, it is not a durable solution in the case of prestressing without bond. However, if the prestressing bars are manufactured from fibre reinforced polymer material (e.g. polystal) a permanent control of the prestressing element over its entire length using optical fibre or copper wire sensors is feasible. Indeed, it is possible to monitor individual elements as the sensors would be integrated into the tendons during their fabrication.

Bibliography

Anon. Carbon fibre strands prestress Calgary span. *Engineering News Record*, 18 October, 21, 1993.

Ashby J. Materials and shape. *Acta Metall. Mater.* 39, No. 6, 1025–1039, 1991

British Standards Institution. *Structural Use of Aluminium*, BS 8118 Parts 1 & 2. British Standards Institution, London, 1991.

Burgoyne CJ and Guimaraes GB. Creep behaviour of a parallel-lay aramid rope. *J. Mater. Sci.* 27, 2473–2489, 1992.

Casas JR and Aparicio AC. A full-scale experiment on a prestressed concrete structure with high strength fibres: the North ring road in Barcelona, paper T15. *FIP-XI International Congress*, Hamburg, Jume 1990.

Chabert A. Technologie et proprietes d'emploi des ancrages. *Symposium on les materiaux nouveau pour la precontrainte et la renforcement d'ouvrages d'art*, ENPC, Paris, October, 1988.

Chambers JJ. Parafil-lay aramid ropes for use as tendons in prestressed concrete. Doctoral thesis, University of London, 1986.

Chambers JJ. Long term properties of Parafil. *Proc. Symp. Engineering Applications of Parafil ropes*, 21-28, 1988.

Culshaw B. *NDT International*, 18, 5, 265–268, 1985.

Design Manual for Roads and Bridges: BD 67.96 Bridge Enclosure. HMSO, London.

Fiberline® Design Manual for Structural Profiles in Composite Materials, Fiberline Composites, A/S Kolding, Denmark, 1995.

Highways Agency. Design Manual for Roads and Bridges, BD 37 or 38/88 – Loads for Highway Bridges. HMSO, London, 1988.

Hofer, B. *Composites*, 8., 4, 309–316, 1987.

Hollaway L. Pultrusion In *Developments in Plastics Technology - 3* (Ed. A Whelan and JL Croft), Chapter 1. Elsevier Applied Science, Oxford, 1986.

Hollaway L. *Polymer Composites for Civil and Structural Engineering*. Blackie Academic and Professional, London, 1993.

Hollaway L and Leeming MB (Eds) Strengthening of Reinforced Concrete Structures using Externally Bonded FRP Composites in Structural and Civil Engineering. Woodhead Publishing, Cambridge, 1999.

Hutchinson R, Abdeirahman A and Rizkalla S. Shear strengthening using CFRP sheets for a prestressed concrete highway bridge in Manitoba, Canada. In *Recent Advances in Bridge Engineering – Advanced Rehabilitation, Durable Materials, Non-destructive Evaluation and Management* (Ed. U Meier and RBetti). Proc. of Workshop held at EMPA Switzerland, 97–104, 1997.

Institution of Structural Engineers. Uses of advanced composites in structural engineering. First report by Informal Study Group on Advanced Composite Materials and Structures, London, UK, 1989.

Iyer SL Advanced composite demonstration bridge deck. In *Fibre-reinforced Plastic Reinforcement for Concrete Structures* (Ed. A Nanni and CW Dolan), SP 138, 83. American Concrete Institute, 1993.

Lee C, Hollaway LC and Thorne AM. The manufacture, testing and numerical analysis of an innovative polymer composite/concrete structural unit. *Proc. Instn Civ. Engrs Structures and Buildings*, 134, Aug., 231–241, 1999.

Mains RM. *Jnl. Am. Conc. Inst*. 3, 225–252, 1951.

Mazzolani FM. *Aluminium Alloy Structures*. HE & FN Spon, London, 1995.

McKenzie M. Corrosion protection: The environment created by bridge enclosure. Research Report 293, TRRL, 1991.

McKenzie M. The corrosivity of the environment inside the Tees Bridge Enclosure: Final year results. Project Report PR/BR/10/93, TRRL, 1993.

Meier U. Proposal for a CFRP bridge crossing the Straits of Gibraltar at its narrowest point. *I Mech E*, 2.1, B2, 7378, 1987.

Meier U, Müller R and Puck A. FRP-Box beams under static and fatigue loading. *Proc. of Int. Conf. (TEQC 83) in Testing Evaluation and Quality Control of Composites* (Ed. T Feest) held at University of Surrey, Butterworth Scientific Ltd, Sevenoaks, 324, 1983.

Nortake K. Practical applications of aramid FRP rods to prestressed concrete structures. In *Fibre-reinforced Plastic Reinforcement for Concrete Structures* (Ed. A Nanni and CW Dolan), SP 138, 83. American Concrete Institute, 853, 1993.

Richmond B. and Head PR. Alternative materials in long span bridge structures. *Kerensky Memorial Conference*, London. June, 1988.

Saunders WS and Abendroth RE. Construction and evaluation of a continuous aluminium girder highway bridge. *6th International Converence on Aluminium*, Weldments, 1995.

Scott RH and Gill PAT Possibilities for the use of strain gauged reinforcement in smart structures. *First European Conf. on Smart Structures and Materials* (Ed. B Culshaw, P.T.Gardener and A. McDonach). Pub. Inst. of Physics Publishing and EOS/SPIE Orsay Cedex and Billingham, 1992.

Sennhauser U, Anderegg P, Bronnimann R and Nellen PhM. Monitoring of Storck Bridge with optical and electrical resistance sensors. *Proc. of US–Canada–Europe Workshop on Recent Advances in Bridge Engineering – Advanced Rehabilitation, Durable Materials, Non-destructive Evaluation and Management* (Eds U. Meier and R Betti). EMPA, Dubendorf, Switzerland, 368–375, 1997.

Shu Y. Chinese crossing first for Plastics Pioneers. *New Civil Engineer*, 14 April, 1983.

Taljsten B. Strengthening of concrete structures for shear with bonded CFRP fabrics. *Proc. of US–Canada–Europe Workshop on Recent Advances in Bridge Engineering – Advanced Rehabilitation, Durable Materials, Non-destructive Evaluation and Management* (Eds U. Meier and R Betti). EMPA, Dubendorf, Switzerland, 1997.

Triantafillou TC and Meier U Innovative design of FRP combined with concrete. In *Advanced Composite Materials in Bridges and Structures* (Eds KW Neale and P Labossiere). The Canadian Society of Civil Engineering, Montreal and Quebec, 491–500, 1992.

Tsuji Y, Kanda M and Tamura T. Applications of FRP materials to pretressed concrete bridges and other structures in Japan. *PCI Jnl*, July–Aug., 50, 1993.

Weiser M. *Erste mit Glasfaser – Spanngliedern vorgespannte Betonbrucke*. Beton-und Stahlbeton-bau, 1983.

Wolff R and Miesseler HJ Glass fibre prestressing system. In *Alternative Materials for the Reinforcement and Prestressing of Concrete* (Ed. JL Clarke), 127–152. Blackie Academic and Professional, London, 1993

14 Substructures

P. LINDSELL

14.1 Introduction

Bridges are usually constructed as part of a roadworks contract, so that the cost of the bridges may represent only a small part of the total contract value. The construction of the substructures has a major disruptive influence on the overall contract programme, since it is normally concurrent with the major earth-moving and drainage operations.

The cost of providing the substructure to a bridge deck often represents more than half of the total bridge price. In spite of this, present design practice and rules require that as much as 90% of the total design time is spent on the analysis and refinement of the superstructure. One reason for the emphasis on bridge deck analysis is that applied design loads on bridge decks were originally specified by the Code of Practice BS153: Part 3A (British Standards Institution, 1972) and subsequent design and assessment loadings have been continually updated by the Department of Transport over the last 25 years in recognition of the increases in traffic intensity and vehicle weights (Department of Transport, 1977, 1997; British Standards Institution, 1978).

Poorly designed substructures that cause unnecessary direct costs and consequential indirect costs will undoubtedly be penalized by an overall increase in the tender price. A simplification in design or detail that leads to a speeding up of the construction process should be welcomed, as the real saving may not necessarily be in the individual price of a bridge. It may, for example, be in a reduction of the operating costs involved in the hire of an earthworks fleet.

Substructures for bridges fall into two distinct categories, end supports and intermediate supports. The end supports are normally described as the 'abutments', whilst the general term for the intermediate supports of a multi-span bridge is the 'piers'. The abutments and piers are usually constructed from *in situ* concrete, but precast sections can be employed to speed up the construction process.

14.2 Abutments

The selection of the appropriate abutments for a bridge should be made at the same stage as the choice of the deck superstructure. There are many types of abutment in use in this country and a comprehensive survey by the Building Research Establishment (Department of the Environment, 1979) revealed a wide variation in the basic assumptions made in the design of these structures.

Mass concrete bankseats and skeletal abutments are commonly employed where open side-spans are required. The bankseat is more suited to the top of a cutting slope where simple footings can be used just below existing ground level. However, it can also be applied to the embankment situation where pile supports driven through the fill may be necessary. The cost of piling and the effects of downdrag on the piles will influence the relative economy of this choice of end support.

In the embankment case, a skeletal abutment founded close to the previous existing ground level can be more economic, but there are complications in the construction of this type of structure. The contractor is faced with a severe restriction on backfilling operations and casting of the transverse capping beam causes further delay to the construction sequence. The magnitude and distribution of the earth pressures acting on the buried columns are particularly in doubt and there is limited experimental evidence available to form the basis of a realistic theoretical analysis.

A solid wall abutment design is often favoured in practice because it produces a minimum span length for the superstructure. However, it does restrict the aesthetic appeal of a bridge and it can produce a tunnel-like appearance on wide structures. An open side span solution produces a more attractive appearance and can be used to assist with farm access or provide additional areas for flood relief in a river crossing. In the case of bankseat and skeletal abutment forms of construction, the savings arising from lower material content in the abutments and the smaller amounts of granular backfill have to be off-set against the cost of the additional deck area required.

14.2.1 Cantilever abutments

The survey by the Building Research Establishment (Department of the Environment, 1979) confirmed that the T-section reinforced concrete cantilevered wall is the most common form of construction for the solid wall type of abutment. There are several variations on the basic theme to cater for different requirements. Propped cantilever walls are often used for right bridge decks with spans below 12 m, sloping abutments for aesthetic or clearance reasons and counterfort walls for heights of 10 m and above.

The minimum headroom for new highway bridges is typically 5.1 m, so the overall height of a solid wall abutment is automatically in the region of 7–9 m. This height is beyond the economic range of mass concrete walls and has encouraged the use of reinforced cantilever abutment walls to be widespread throughout Britain. The simplicity of this form of construction and the similarity with cantilever retaining walls also accounts for its economic success and popularity.

14.2.2 Free cantilever

This type of cantilever abutment is the most common form of construction for heights of 6–9 m and, in spite of the size, the main concrete wall is often poured in one lift. The wall stem generally ranges from 0.9 to 1.2 m in width, so that it is wide enough to allow a person to climb into the reinforcement cage during construction. The overall width of the base will generally be 0.4–0.6 times the height and the toe may project 1.0–2.0 m in front of the wall. However, the physical dimensions and proportions of the base will depend upon the soil foundation conditions and the resistance to sliding. A typical example of a cantilever abutment wall with horizontally cantilevered wing walls is illustrated in Figure 14.1.

Where an abutment wall can physically yield, active earth pressure conditions are assumed for overturning, sliding and bearing pressure calculations. For walls that are rigidly supported, for example, on a combination of vertical and raking piles, then

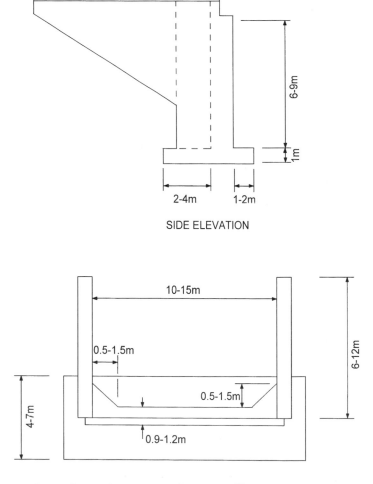

Figure 14.1 Typical cantilever abutment with wing walls.

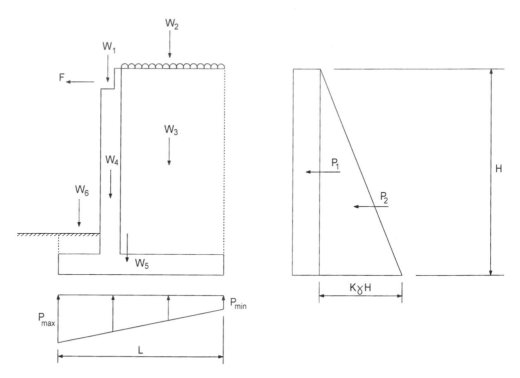

Figure 14.2 Idealized forces on an abutment wall.

the lateral earth pressure behind the abutment wall is usually assumed to be in the at-rest condition. In all cases, the wall stem design is normally based on at-rest conditions to allow for high pressures during compaction of the backfill material.

The design forces are often calculated on the basis of a metre wide strip, assuming the abutment wall acts only as a vertical cantilever. If wing walls are attached to the rear of the abutment as in Figure 14.1, then there is a strong case for considering the three-dimensional behaviour of the structure. The combined effects of the wing wall weight and substantial corner splays can reduce the requirement for vertical reinforcement in the main abutment wall to a nominal percentage of the cross-sectional area.

An idealized system of vertical and horizontal forces acting on a simple cantilever wall is illustrated in Figure 14.2. Passive pressure at the front of the wall is generally unreliable and may be completely removed by the introduction of highway services along the toe of the foundations.

14.2.3 Counterfort wall

This type of cantilever abutment becomes economic for heights greater than 10 m, where the percentage reinforcement in a free cantilever becomes very large. Triangular-shaped counterforts are added to the rear of the abutment wall slab to provide further flexural rigidity and resist the lateral earth pressures developed by the depth of backfill material. The construction is complicated by the steel and formwork around the counterforts and physical compaction of the backfill is more difficult.

The counterforts are spaced at about one half the height of the wall and they are designed as vertical cantilevers. The wall slab may be treated as a slab clamped on three sides, although it naturally spans the shorter horizontal distance between the counterforts and the wall thickness can be reduced accordingly. The heel of the base slab also spans between the counterforts. However, there is little scope for reducing the thickness because the anchorage length for the main tensile reinforcement at the rear of the counterforts is a limiting factor.

14.2.4 Propped cantilever

There is little longitudinal movement in bridge decks up to 12 m in span, and it is possible to use the deck as a strut for square bridges or bridges with small angles of skew. The abutments can be designed as a propped cantilever but, due to the rigid nature of the structure, at-rest earth pressures are normally assumed for the design of reinforcement in the rear of the wall and footing, and both stability and bearing pressure calculations. Complete fixity of the base is unlikely, so that the front face reinforcement in the abutment is estimated by assuming the wall is pinned at the deck and base levels.

A flexible packing is normally used to separate the deck from the top of the abutments and the curtain walls are designed to withstand the propping force. It is often necessary to specify that initial backfilling should be limited to 50% of the abutment height to avoid rotation and horizontal deflection of the abutments. Completion of the backfill behind the abutment walls is then delayed until the deck has been constructed.

14.2.5 Open abutments

The term 'open abutments' is often used to denote the type of end supports required to extend the central span of a bridge to create adjacent 'open side spans'. Two basic types of abutment are used in this situation. A mass concrete bankseat situated at the top of the slope containing a side span or a buried reinforced concrete skeletal or 'spill-through' abutment founded at previous existing ground level beneath an embankment slope.

In general, it is possible to choose between a single span deck with solid cantilever abutments, or a three-span deck with intermediate piers and end abutment supports. The relative costs of two large cantilever abutments, associated wing walls and selected granular backfill may therefore be compared with the price of two intermediate piers, two end abutments and two additional deck spans. However, there are other considerations involved in choosing a three-span open structure, such as aesthetics, sight-lines, flood relief and pedestrian safety.

14.2.6 Bankseats

Simple mass concrete or lightly reinforced sections may be used for abutment supports at the top of cuttings where the foundation level is close to existing ground level. This type of structure is relatively small, and is usually 'stepped out' in section to reduce the foundation pressure and confine the resultant reaction on the base within the middle third. Small wing walls may be conveniently hung from the back of the bankseat to contain the immediate area of backfill behind the wall.

Bankseats may also be used on embankments, in which case they can either be allowed to settle with the fill or supported directly on pile foundations. In the latter case, settlement of the embankment can cause downdrag on the piles and this reduces the payload of the pile group. Driving raking piles for a bankseat can be particularly awkward at the edges of an embankment.

The cost of using a bankseat, intermediate pier and extra deck for the side span is often less than a solid abutment wall with large wing walls. This is particularly true for narrow bridges, but the closed abutment is generally more economic for wide structures since the cost of the wing walls is constant and represents a decreasing proportion of the solid abutment as the width increases.

14.2.7 Spill-through abutments
This form of abutment is illustrated in Figure 14.3 and consists of two or more buried columns supported on a common base slab and capped by a cill beam to carry the deck construction. The backfill spills between the legs and needs careful compaction around the columns to minimize long-term settlement. It is often used in the embankment situation where it is possible to obtain a suitable foundation at original ground level. In this case, it can form an economic alternative to a bankseat supported on piles driven through the embankment fill.

Design assumptions for this type of abutment vary widely, since very few field investigations have been undertaken to determine the long-term movements and earth pressures on the buried structure (Lindsell and Buchner, 1987, 1994). One conservative simple design approach is to assume full active earth pressure across the entire width of the abutment, regardless of the soil that spills between the columns. Normally, the columns and cill beams are considered to be loaded by active earth pressure, but some arbitrary allowance is made for 'drag effects' or arching of the fill between the columns.

No additional calculations are really necessary for traction and braking forces acting towards the backfill. The total reaction from 'at rest' earth pressure acting on the rear face of the capping beam and curtain wall is often sufficient to balance the longitudinal forces. If not, monitoring studies have demonstrated that the columns will rotate about a point above the base foundation and derive a very effective restoring moment from the lateral soil pressures with which to counteract horizontal loads at bearing level (Lindsell and Buchner, 1987).

14.2.8 Wing walls
The primary function of the wing walls on an abutment is to contain the backfill material at the rear of the abutment wall and minimize settlement of the carriageway. The combination of soil containment and compaction of the backfill material may lead to high lateral earth pressures. Consequently, the horizontal forces acting on both the abutment wall and the adjacent wing walls can be similar and a very

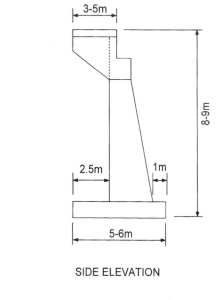

SIDE ELEVATION

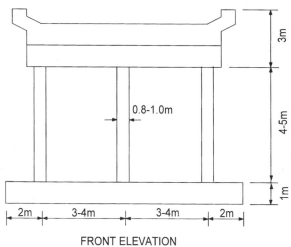

FRONT ELEVATION

Figure 14.3 Typical spill-through abutment.

significant factor in selecting an appropriate design. The wing walls may be constructed as free-standing, independent structures or they can be designed as an integral part of the abutment wall construction.

14.2.9 Free-standing walls

These walls are designed as a nominal cantilever retaining wall with a separate foundation from the main abutment. Differential settlement and tilting between the abutment and wings may occur. Hence, construction joints between the two structures require careful design to both permit and conceal the relative movements.

To suit the local terrain, the wing walls can be arranged parallel to the abutment wall and this allows simple compaction of the backfill with no complications in the design, regardless of the skew angle of the deck. Alternatively, the wings can be designed to follow the direction of the over-road, and have the dual function of supporting the parapet fencing and the backfill. It is more difficult to place the backfill material with this configuration and higher earth pressures will result because of the restraint against sideways movement. Consequently, this form of design may be more expensive to construct and a cheaper arrangement can be to use wings splayed at 45° to the abutment and tapering in height.

14.2.10 Cantilever walls

A second approach to the design of wing walls parallel to the over-road is to use horizontally cantilevered wings. This form of construction is practical for lengths up to 12 m from the abutment, but care must be taken in designing the junction between the wing and abutment wall. The structure has the advantage of being founded on a common base so that it settles as one unit, but compaction of the backfill may be difficult around the wings. The rigid nature of this type of design encourages high earth pressures and at least 'at-rest' earth pressures should be considered in a design (Ingold, 1979; Jones and Sims, 1975).

This type of abutment and wing wall system forms a three-dimensional structure. A traditional metre-strip assumption is widely used but it may not be not an appropriate basis for a design (Lindsell and Buchner, 1994). The vertical and horizontal bending actions in the abutment are significantly modified by the presence of the wing walls and an overall reduction in the steel requirements is possible if the wings are used to their full advantage.

The self-weight of the wing walls should be considered since they have a major influence on the stability and bending moments in the abutment wall. Horizontal forces on the wings are transmitted into the abutment corners and are distributed across the abutment wall. Therefore, the corner splays between the abutments and wing walls can be designed as vertical torsion blocks to carry the high torsional moments generated by the wing wall loading.

14.2.11 Design calculations

The primary function of an abutment wall is to transmit all vertical and horizontal forces from a bridge deck to the ground, without causing overstress or displacements in the surrounding soil mass. The abutment wall also serves as an interface between the approach embankments and the bridge structure, so it must also function as a retaining wall.

The degree of interaction between a bridge deck and an abutment wall depends largely on the nature of the bearing supports. The effect of bearing type, end fixity or free supports can be readily idealized in the design process. However, the influence of ground movements due to settlement, mining subsidence or earth tremors is more difficult to anticipate and such effects require individual consideration for a particular structure.

14.2.12 Applied loadings

The 'equivalent fluid' concept is normally used for calculating the earth pressures on an abutment, but selection of the appropriate intensity depends on the degree of restraint offered by the wall and the particular calculation being considered. Traditional practice has followed retaining wall design. Therefore, in a situation where a wall can move by tilting or sliding and the backfill is a free draining granular material, active pressures are assumed. A common design approach is to use an equivalent fluid pressure of $5H$ kN/m^2, where the coefficient of active earth pressure, K_a, is nominally 0.25.

Modern compaction techniques for placing the backfill material and the use of more rigid types of construction have caused many designers to estimate design pressures for the at-rest condition. The value of the earth pressure coefficient at-rest, K_o, is often taken to be 1.5–2.0 times the active coefficient K_a.

The primary vertical loading acting on an abutment is due to the dead load and live load reactions from the bridge deck. Additional loading arises from the self weight of the abutment, backfill and live loading immediately behind the abutment. The effect of *HA* and *HB* loadings on the carriageway behind the abutment is arbitrarily treated as an additional surcharge loading.

Longitudinal movements in the bridge deck due to creep, shrinkage and temperature changes cause forces at bearing level on the abutments. The magnitude of these forces depends upon the shear characteristics or frictional resistance of the bearings. The coefficient of friction of most bearings lies in the range $\mu = 0.03$–0.06. The frictional force is derived from the nominal dead load and the superimposed dead loads on the deck.

Traction and braking forces due to live loads on the deck are carried at the fixed bearings and may represent a substantial overturning moment on a tall abutment. Although these forces are applied to localized areas of the deck, they can usually be treated as a uniform load across the width of the abutment.

14.2.13 Bearing pressures

The size of an abutment base is largely controlled by the allowable bearing capacity of the ground. The variation in ground pressure across the base illustrated in Figure 14.2 is assumed to be linear and the width of the base selected to ensure there is no tension at the rear of the heel. The summation of the horizontal and vertical forces acting on an abutment are obtained from Figure 14.2 and the base width and proportions are selected so that the resultant reaction falls within the middle third of the base.

14.2.14 Base design

The toe of a base slab is designed to resist the peak ground pressures acting on the base, although some relief can be obtained from self weight of the toe and any superimposed fill. The heel must also be designed to resist upward ground pressure, but in this case the resultant moments may be reversed by extreme loading conditions caused by fill, live load surcharge and self weight. The base slab may be supported on piles and in this situation the bearing pressures would be replaced by calculated loads in each pile.

14.2.15 Wall design

The stem of an abutment wall is designed to resist the bending moments and shears produced by horizontal forces. Direct stresses due to vertical loads are normally very low and may be neglected for wall design. Significant in-plane stresses can develop at the root of torsion blocks on horizontally cantilevered wing walls. In the case of simple vertical cantilever walls, the critical section for moments and shear forces occurs at the root of the wall.

Concentrated horizontal loads can occur at bearing level and may also be present at the rear of the curtain wall due to traction and braking effects. A standard method is to distribute these loads vertically at 45° when calculating bending moments in the wall.

14.2.16 Stability

The stability of an abutment should be checked for three basic modes of failure.

Sliding

When passive resistance in front of the toe can be relied upon, the minimum factor of safety taken in design is normally 2.0. If the passive pressure contribution is neglected, then a minimum factor of safety against sliding is usually 1.5. A shear key is sometimes provided in the base slab when the resistance to sliding is inadequate.

Typical design values for the coefficient of friction between the base slab and the soil are:

$$\text{Clay } 0.2; \quad \text{Sand } 0.4; \quad \text{Gravel } 0.4.$$

Overturning

Overturning is checked by taking moments about the toe when the most adverse load combination is acting on the structure. A minimum factor of safety of 2.0 is normally adopted providing the resultant reaction lies within the middle third. If there is 'tension' in the bearing pressure at the heel, then a higher factor of safety may be used as a further precaution against failure.

Slip failure

A slip circle analysis is essential for a bankseat form of construction and may be necessary for other types of abutment when the soil strata well below the structure is weaker than the soil layers at foundation level. Where soil strengths are based on tests, then a minimum factor of safety would be 1.5. Particular care is needed during construction if an intermediate pier foundation is being excavated at the toe of a cutting slope, when there is a bankseat positioned at the top.

14.2.17 Construction

The construction sequence and concreting procedures are often left to the contractor's discretion, although the designer will normally specify the location of construction joints and design the reinforcement cages to suit. The timing of the granular backfilling behind an abutment should also be considered in the overall design of the bridge, particularly if 'rigid frame' or 'propped abutment' designs are used.

The construction of the bridge deck may represent the most severe load case in the service life of an abutment wall (Lindsell and Buchner, 1987). Potential interaction between an *in situ* concrete deck, the falsework and an abutment can be critical due to thermal expansion and early shrinkage of the deck. These factors should be anticipated at the design stage, but further consideration by resident engineering staff is necessary during the planning of the construction stages on site.

14.2.18 Temporary works

The excavation costs associated with the construction of a deep foundation base for an abutment and any independent wing walls can represent a substantial proportion of the substructure price. Temporary stabilization of the adjacent soil may be necessary in the form of sheet piling or de-watering operations. Such matters should be considered at the preliminary design stage, before selecting the most appropriate form of deck and substructure.

The principal cost of constructing the main abutment stem and wing walls relates to the formwork and supporting falsework. The effective concreting pressures on the formwork will depend upon the wall height, rate of pouring, the width of the section and the compaction procedures. Hence, the physical construction and the overall costs of the formwork and falsework will depend upon these parameters. Any simplifications in the abutment wall geometry and wing wall details should reduce the reinforcement detailing and formwork costs.

14.2.19 Construction joints

A series of horizontal and vertical construction joints are normally required to build an abutment wall in two to three stages. A horizontal joint at 100–150 mm above the base slab is essential to permit construction of the main wall of a cantilever abutment or the individual columns in a spill-through abutment. A further horizontal joint is usually introduced at 100–150 mm above the bearing shelf to enable the curtain wall at the rear of the abutment to be constructed as a third stage. However, a spill-through abutment design may also require an additional horizontal joint at the top of each column support to form the transverse cross-beam, which constitutes the bearing shelf.

Wide abutments and integral wing wall designs may require the introduction of several vertical construction joints. Longitudinal vertical joints may also be necessary through the base slab or bearing shelf sections of an abutment.

The upper 0.5 m of a wing wall often contains a string course section, which carries the bridge deck parapet fencing to the ends of the structure. The stringcourse is frequently constructed separately to the main wing wall stem, requiring a further horizontal joint at the top of the wing wall section.

The location of such construction joints and the timing of the construction sequence may have significant structural effects and influence the durability of an abutment.

Substantial stresses may be induced by the thermal expansion and contraction of new concrete cast against previous sections. Additional stresses are likely to be induced by differential shrinkage between the fresh concrete and the older sections.

14.2.20 Durability aspects

Horizontal construction joints in abutment walls will inevitably lead to vertical cracking in the concrete section immediately above the joint line. A pattern of vertical cracks will emanate from the construction joint just above the base slab and may extend as far as 4–5 m up the abutment stem. In a typical highway location, the long-term effect of salt spray from passing traffic may lead to early corrosion of the front face reinforcement.

The introduction of a horizontal joint immediately below the stringcourse on a wing wall can have serious consequences. Differential shrinkage stresses will probably initiate a regular pattern of vertical cracks running completely through the string course section. Traffic passing over the structure will splash road salts directly onto the string course and there will be a high risk of corrosion in the parapet reinforcement. Such effects are easy to predict and steps should be taken at the design stage to minimize the consequences. The concrete mix should be designed to resist the penetration of chlorides. The effects of differential shrinkage may be reduced by careful selection of materials and timing of construction events. Crack widths and spacings can be controlled by judicious spacing and location of the internal reinforcement in the area immediately above a horizontal joint line.

Vertical construction joints can also lead to durability problems in abutment walls. Similar consideration should be given to the control of crack widths and penetration of road salts into the joints.

One of the primary areas of concern in all types of abutment walls is the top of the bearing shelf. Water running from the carriageway surfaces above will normally flow onto the bearing shelf and every effort should be made at the design stage to drain this water away from the abutment wall. The majority of existing abutment walls suffer from water staining and chloride induced corrosion, where inadequate provision has been made for effective drainage of the surface water.

The rear faces of abutment walls are normally protected with two coats of bitumen paint, prior to backfilling with granular material. Consideration should also be given to surface protection of the ballast wall and bearing shelf concrete, which may be subject to continuous water leakage from the carriageway surfaces.

14.3 Bridge columns and piers

Intermediate supports for bridge decks may be grouped into columns or leaf piers. A leaf pier is the term used to describe a reinforced wall with the largest lateral dimension more than four times the least lateral dimension. Individual columns may be used with separate bases and direct contact with the bridge deck. Alternatively, columns may be grouped together to form transverse portal frames with a capping beam and a common footing.

14.3.1 Leaf piers

This type of intermediate support is common in modern bridge construction, since it is economic to construct. It is usually designed as a solid reinforced concrete wall, with the overall length of the pier equal to the transverse width of the bridge deck. Hence, typical leaf piers for an overbridge can be 4–10 m in length, when supporting a single carriageway. The least lateral dimension may be 0.5–1.0 m in width and it is common practice to taper the pier dimensions from the base up to the soffit of the deck. The greatest transverse dimension is required at the top in order to provide sufficient bearing support for the superstructure.

Vertical and horizontal loads transmitted from the superstructure disperse rapidly from the top of a pier. Hence, the overall design of a leaf pier is normally conducted on a metre strip basis, assuming a uniform distribution of axial and bending effects. The magnitude of the axial compressive stresses in a concrete pier is normally between 0.5–1 N/mm^2 under dead loading and it is unlikely to be more than 2 N/mm^2 under the most severe live load conditions. The degree of bending will depend upon the articulation of the deck and the length of the superstructure.

14.3.2 Columns

Individual concrete columns are often used to support footbridges and bridge decks with high skew or greater height than the minimum headroom clearance. Columns may be vertical, inclined or even curved in shape to produce greater aesthetic appeal.

A column section is normally required to resist bi-axial bending and significant axial loading. Concrete columns are therefore often circular or square, but hexagonal and octagonal sections are also common. Typical dimensions for a square section may range from 0.4×0.4 m to 0.8×0.8 m, depending upon the loading and height of the column. Axial stresses may amount to 3–5 N/mm^2 under full service loading, so that section dimensions are usually selected to avoid the need to consider axial stability. A typical bridge column with a height of 6–8 m is normally designed as a 'short column'. This situation can be achieved by an appropriate choice of cross-section and articulation conditions to control the effective length and eliminate slenderness effects.

14.3.3 Portal frames

Concrete columns are often grouped into pairs or sets of three, by placing them on a common foundation and capping them with a transverse cill beam. The cill beam provides continuous intermediate support for the superstructure, in a similar manner to a leaf pier. The transverse frame action developed by the capping beam, the columns and the foundation slab produces much greater transverse flexural rigidity than a group of individual columns, but has no significant effect upon the longitudinal flexibility. These characteristics are important where wind loading and vehicle impact effects have to be considered, but the superstructure design requires flexible intermediate support to cater for longitudinal movements.

14.3.4 Articulation

The connection between an intermediate support and the deck largely determines the type of loading to be carried by the columns or piers. Similarly, the structural connection between a column or pier and the footing controls the degree of axial loading and the bending effects at the critical section immediately above the base.

The common forms of articulation for longitudinal movements in a bridge deck are illustrated in Figure 14.4. Columns or piers of type 1, 2 and 3 are required where large movements have to be accommodated. The loading in any columns of this type will be predominantly axial, with relatively small bending moments. The bases of columns with type 2 or 3 end conditions will carry bending moments but a type 1 column will transmit axial load only.

Columns or piers with type 4 and 5 end conditions will provide restraint to longitudinal movements of a bridge deck. Therefore longitudinal braking forces and temperature effects will cause significant bending moments at the root.

Biaxial bending effects on columns may be particularly severe in a portal frame design or wide bridge decks with rigid connections into the tops of the columns. Therefore, it is very important that the selection of bearings and end conditions should take all loading combinations and ranges of movement into account at the design and construction stages.

14.3.5 Slenderness

The influence of buckling effects may become important if the unrestrained length of a column is large with respect to the column cross-section. Where the ratio of effective length (l_e) to thickness (h) is limited to 12–15, then the effects of slenderness are relatively small. This slenderness ratio will normally be satisfied for standard two-level structures, but may be exceeded at three-level interchanges and bridges crossing a deep-sided valley. When tall columns are necessary, the sections may be designed to be flexible so that movements at deck level can be accommodated with articulation conditions types 4 or 5. This should not present a stability problem if the effective length is less than the actual length (l_o).

The assessment of the effective length of a column or pier is fundamental to the design of a section (Cranston, 1972; Jackson, 1988). Figure 14.5 illustrates the potential buckling modes and conservative estimates of the effective heights for the five articulation arrangements shown in Figure 14.4. The slenderness ratio is calculated by dividing the effective height by the thickness of the pier or column. In the case of tapered piers it is conservative to take the average thickness.

Longitudinal movements in a deck can have a significant effect upon columns designed with articulation types 3, 4 and 5. A bridge deck may provide full restraint in the transverse direction, so the effective length is less than the actual length. In the longitudinal direction, thermal movements may create sidesway and the effective length of a type 3 or 4 column could then be twice the actual length. In addition, the effective length will increase due to flexibility of the base slab and soil foundation which are unlikely to meet the fully fixed assumption illustrated in Figures 14.4 and 14.5.

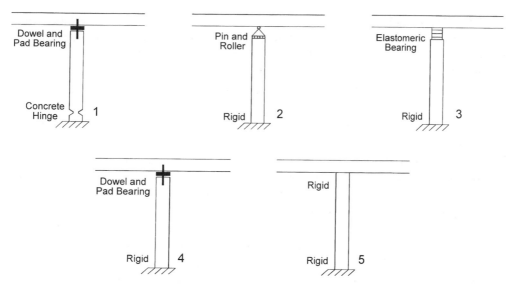

Figure 14.4 Articulation of columns and piers.

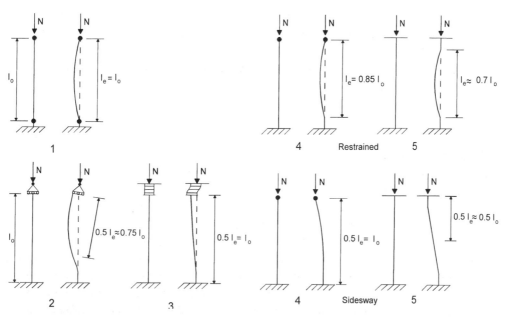

Figure 14.5 Buckling modes.

The overall effective length of a pin ended column with a flexible foundation may then amount to $2.5l_o$, when sidesway occurs during seasonal movements of the deck.

14.3.6 Design considerations

Many columns and piers in the UK were designed for bridges on the basis of working stresses and the principles contained in CP114 (British Standards Institution, 1969) during the period 1955–1980. Permissible stresses for steel and concrete were given in BE1/73 (Department of Transport, 1973) and reference made to CP114 for the permissible loads on short and long columns. Where the slenderness ratio was less than 15, then the column was treated as 'short'.

As the slenderness ratio increases above 15, the deflection effects become increasingly significant in practical reinforced concrete construction. The permissible axial loads were therefore reduced by a set of coefficients to allow for slenderness effects (British Standards Institution, 1969).

The introduction of limit state design for reinforced concrete columns commenced with the first edition of BS5400: Part 4 in 1978 (British Standards Institution, 1978). This fundamental change in design philosophy affected the analysis of slender piers and columns at ultimate conditions. However, little guidance was given in BS5400 for the serviceability situation. The uniaxial compressive stress in concrete is limited to $0.5f_{cu}$ and this is likely to cause the serviceability condition to be critical. However, this depends upon the value of the modular ratio (m) adopted in the calculations and little specific guidance is given in the code. A value of $m = 15$ is recommended unless the dead load is less than 30% of the live load.

An initial analysis would normally be carried out on an elastic basis calculating stiffnesses from the gross concrete cross-section of the trial column. Other options permitted in the BS5400 allow the use of the transformed area or a cracked section if a good estimate can be made of the steel percentage in the final design. It is suggested that providing the slenderness ratio is less than 25, then no allowance for instability effects is necessary at the service load.

Columns and piers subjected to high bending moments are liable to suffer from tensile stresses at the extreme fibres. Crack widths for columns may therefore be checked using the simple beam formula. Where piers are used in a similar manner to wall construction, it seems appropriate to treat them as vertical slabs for the purposes of crack control, since axial compressive stresses are generally very low.

The modern design of concrete columns is based upon ultimate strength criteria with checks for crack widths and stresses under service conditions. The principles were originally developed by Cranston (1972) during the drafting period for CP110 (British Standards Institution, 1972), which was intended for framed building structures. Consequently, the effective column heights did not deal specifically with bridge columns of types 1, 2 and 3 illustrated in Figure 14.4. Additional information was introduced in BS5400: Part 4: 1984 (British Standards Institution, 1984) for bridge column analysis at the ultimate limit state. Specific consideration was given to defining effective lengths for various bearing types and support conditions.

Buckling is dealt with by the 'additional moment' concept, which has been developed from a greater knowledge of slender column behaviour in framed buildings. The basic principle is to estimate the deflections of the columns at collapse and calculate the additional moments arising from the deflections.

BS5400 requires a check on a section for the sum of the initial moments, calculated ignoring displacement effects, combined with additional moments created by deck movements. The magnitude of the initial moments depends upon the type of bearings and the degree of end restraint to the columns (Jackson, 1988).

14.3.7 Applied loadings

Longitudinal and transverse forces due to wind, impact, braking, shrinkage and thermal effects all require detailed consideration. On wide bridges and portal frame designs, the effects of shrinkage and temperature will play an important part in the design of bearings and columns. Biaxial bending will be produced in such situations and the design must cater for the transverse behaviour. Skew bridges are frequently provided with circular or hexagonal columns to allow the biaxial bending effects to be simply treated.

14.3.8 Construction effects

The construction of individual concrete columns is usually undertaken in two stages. A foundation slab with a construction joint formed at 100–200 mm above the root of the column. For column heights of 5–10 m, the remaining height is then poured in one lift. Although the critical section for bending effects is theoretically at the root of a column, in reality the construction joints just above the base slab will constitute a plane of weakness. Flexural cracks are likely to develop at this construction joint and ultimate collapse would probably commence at the joint section.

Concrete leaf piers are constructed in a similar manner, but an added complication may arise on wider structures. Where pier widths exceed 4 m, vertical construction joints may be introduced during casting of the vertical wall sections. Differential shrinkage effects may become significant, since the temperature and age differences between the foundation concrete and the wall sections at the time of casting can be substantial. Consequently, vertical cracks are commonly found in the wall sections, immediately above the horizontal construction joint. The crack widths can be serious, unless adequate quantities of horizontal reinforcement are provided in the first few metres above the base.

The construction of portal frame supports may require a third stage in the casting process in order to form the capping beam. Additional secondary effects in the transverse direction can be particularly severe, due to the temperature rise and fall in the capping beam immediately after the initial 'set' of the concrete. Subsequent shrinkage of the capping beam and temperature differences between a backfilled foundation slab can also create significant transverse bending of the edge columns in a transverse frame.

The construction effects produced by the addition of the superstructure may also represent substantial temporary loading conditions on all types of intermediate support. Where early cracking occurs at construction joints and exposed surfaces, there is a long-term risk of water ingress and the development of a local plane of weakness under ultimate load conditions. The design of concrete columns and piers should therefore take into account the likely sequence of construction and appropriate reinforcement quantities and detailing provided to cater for the anticipated secondary effects.

Bibliography

British Standards Institution. *Specification for Steel Girder Bridges*. *BS153: Part 3A: Loads*, London, 1972.

Department of Transport. Technical Memorandum (Bridges) *BE1/77: Standard Highway Loadings*. February 1977.

British Standards Institution. *Steel, Concrete and Composite Bridges*. *BS5400: Part 2: Specification for Loads*, London, 1978.

Department of Transport. *The assessment of highway bridges and structures*. *BD21/97*, 1997.

Department of the Environment. *Bridge foundations and substructures*. Building Research Establishment, HMSO, London, 1979.

Lindsell P and Buchner SH. Long-term monitoring of spill-through bridge abutments. Structural assessment – the use of full and large scale testing. *Proceedings of the Building Research Establishment I.Struct.E.*, Butterworths, London, 228–234, 1987.

Lindsell P and Buchner SH. Structural behaviour of bridge abutments. Bridge Assessment, Management and Design. *Proceedings of Centenary Bridge Conference,* Cardiff. Elsevier, 411–416, 1994.

Ingold TS. Lateral earth pressures on rigid bridge abutments. *The Highway Engineer*, 26, No. 12, December, 2–7, 1979.

Jones CJFP and Sims FA. Earth pressures against the abutments and wing walls of standard motorway bridges. *Geotechnique*, **25**, No. 4, 731–742, 1975.

Cranston WB. Analysis and design of reinforced concrete columns. Wexham Springs, Research Report 20, Cement and Concrete Association, 1972.

Jackson P.A. Slender concrete bridge piers to BS5400. *The Journal of the Institution of Highways and Transportation*, January, 1988.

British Standards Institution. *CP114: Part 2: 1969, The structural use of reinforced concrete in buildings*. London, 1969.

Department of Transport. Departmental Standard BE1/73, *Reinforced concrete for highway structures*. London, 1973.

British Standards Institution. *Steel, Concrete and Composite Bridges*. *BS5400: Part 4: 1978, Code of Practice for the Design of Concrete Bridges*. London, 1978.

British Standards Institution. *CP110, The structural use of concrete*. London, 1972.

British Standards Institution. *Steel, Concrete and Composite Bridges*. *BS5400: Part 4: 1984, Code of Practice for the Design of Concrete Bridges*. London, 1984.

15 Bridge accessories

P.A. THAYRE, D.E. JENKINS, R.A. BROOME AND D.J. GROUT

15.1 Introduction

This chapter looks the various additional elements that are so important in bridge design and construction. Each of the following is examined in turn:

- parapets
- expansion joints
- drainage
- waterproofing.

15.2 Parapets

Parapets are provided on highway bridges and similar structures such as retaining walls for the safety of errant vehicles and their occupants, pedestrians and other road users. They also serve to provide protection to areas beneath and adjacent to structures, for example other roads, railways, buildings, etc.

The variation in types of vehicles using highways is considerable and constantly changing. Vehicles vary in size, both length and height, overall weight, axle spacing and weight distribution. A vehicle's response to an impact can also vary considerably from model to model due to such features as engine position, provision of crumple zones, suspension type, etc.

Parapets should be designed to have a reasonable appearance and should not detract from the overall aesthetics of the structure to which they are attached. They should further not cause forces to be applied to the supporting structure that would significantly affect its structural requirements.

To satisfy the vast array of requirements for parapets in a single design would be almost impossible and it is therefore normal to specify a parapet design for a specific containment requirement. In the UK, three levels of containment are considered: low, medium and high. European standards provide for further sub-divisions of these general containment standards.

The specific requirements for the UK containment levels are that a parapet shall be required to resist penetration from the following vehicle impact characteristics:

(a) Low level of containment
  Vehicle   Saloon car
  Mass   1500 kg
  Height of centre of gravity   480–580 mm
  Angle of impact   20°
  Impact speed   80 km/h (50 miles/h)

(b) Normal level of containment
  Vehicle   Saloon car
  Mass   1500 kg
  Height of centre of gravity   480–580 mm
  Angle of impact   20°
  Impact speed   113 km/h (70 miles/h)

(c) High level of containment
  Vehicle   Four axle rigid tanker or equivalent
  Mass   30 000 kg
  Height of centre of gravity   1.65 mm
  Angle of impact   20°
  Impact speed   64 km/h (40 miles/h)

In addition to the containment requirements parapets must remain intact and redirect the impacting vehicle in a safe manner. Damage to the vehicle must also be of an acceptable nature and forces imparted to the vehicle's occupants must be kept to a minimum. Although parapet designs are completed to satisfy the specific requirements of the particular level of containment, consideration should be given to their response to other vehicles on the highway. With this in mind European standards require parapets to be tested with a small car (900 kg) in addition to the prescribed containment requirements.

The low containment standard of parapet should only be used in situations where vehicle speed is controlled by a mandatory speed limit not exceeding 80 km/h (50 miles/h). The normal containment standard of parapet should be satisfactory for most other situations with the exception of high risk sites where the high containment may be deemed appropriate. High risk sites must be assessed individually for the need for a high containment parapet but may include major rail crossings, major highway junctions or hazardous material storage areas.

Parapets can be constructed of a variety of materials but the majority of modern parapets are of metal (generally steel or aluminium), reinforced concrete, a combination of reinforced concrete and metal or to a lesser extent masonry.

Metal parapets are generally of post and rail construction, with the posts spaced at approximately 3.0–3.8 m centres. The space between the horizontal rails is normally limited to a maximum of 300 mm to prevent penetration of the front of an impacting vehicle. Metal parapets have been designed with vertical infill bars between upper and lower horizontal rails but these have been found to only be suitable for low containment standards since they tend to arrest the vehicle rather than redirect it. Combined reinforced concrete and metal parapets normally consist of a reinforced concrete plinth of between 400 mm and 900 mm in height surmounted by one or two horizontal metal rails.

Plain masonry parapets are inclined to fracture into small blocks or even individual bricks when impacted. The detached sections present a serious risk of causing a secondary accident and such parapet should generally only be considered suitable for low containment situations and in areas where detached masonry would not be hazardous. If a masonry finish is required for aesthetic reasons this can be achieved by the use of cladding on a reinforced concrete parapet. In such situations the cladding, which should not generally be placed on the traffic face of the parapet, must be securely anchored to the concrete core of the parapet. The cladding should not be considered structural and should be fixed in a manner to avoid increasing the ultimate capacity of the concrete parapet at its designed failure position.

Where parapets are used adjacent to areas accessible to pedestrians, post and rail metal parapets should be provided with cladding to deter climbing and to prevent children climbing through between the rails. This cladding which can be either solid sheeting or mesh will normally extend for the full height of the parapet and the mesh should have a small aperture size to prevent objects passing through and causing a danger to users of the area beneath the structure. Where solid sheeting is used it should not have a reflective surface which may create a hazard for any road or rail user. Railway authorities have particularly stringent requirements with regards to cladding requirements. Where parapets are required on structures frequently used by horses consideration should be given to providing increased height parapets and a solid section should be provided at the base of the parapet to prevent the horse from seeing traffic passing beneath and possibly being spooked by it.

The reaction of a parapet to impact loads is very complex and difficult to analyse. The design loads for parapets in the UK have been developed from a series of full scale dynamic tests on prototype and production parapets. The response of a relatively rigid concrete or combined concrete and metal parapet can be predicted reasonably accurately and therefore designs based on loads derived from previous tests may be acceptable without further testing. However, the response of flexible metal frame parapets can vary considerably for relatively small changes in the design and therefore all such designs should be verified by full scale testing.

The dynamic test which is conducted on a representative panel of parapet is designed to check the parapets response to the standard vehicle impact for the appropriate containment category. Standard production vehicles in roadworthy condition should be used for the test. To satisfy the requirements of the containment capacity the vehicle must be retained and redirected in the prescribed manner. Deflection of the parapet must be less than the appropriate limit and damage to the parapet and the vehicle must not be severe. No significant parts of the parapet or the vehicle should become detached due to the impact. European standards require an assessment of the likely affect on occupants within the vehicle. This is by the calculation of the Acceleration Severity Index (ASI), Theoretical Head Impact Velocity (THIV) and Post Impact Head Deceleration (PHD). They also require the measurement of a Vehicle Cockpit Deformation Index to assess the severity of the vehicles interior deformations.

The designs of bridge parapets to UK standards are conducted using limit state principles and are intended to provide a parapet of minimum strength for the containment required thus imposing the minimum loads on the supporting structure. The design of the attachment systems, anchorages and the support structure are completed using the actual failure loads of the parapet sections as their design loads. Progressively larger load factors are applied to each component to ensure a progressive failure with the first part to fail in an impact being the design failure section of the parapet which is usually the parapet post for a metal parapet or the wall stem for a concrete parapet. This allows replacement or repair to be achieved simply without affecting the supporting structure. Attention should be given to the detailing of reinforcement in *in situ* reinforced concrete parapets to allow the replacement of damaged sections of parapet without the need to breakout large sections of the supporting structure.

Since extensive damage can occur to a parapet in an impact, structural members of a bridge should not be used as parapets. Parapets should be designed such that composite action with the supporting structure does not occur. This is particularly relevant to those of reinforced concrete which should normally be constructed in short panels with movement joints between. Transfer of longitudinal forces between panels should be avoided but shear transfer may be allowed if required.

Joints should be provided in parapets to allow for movements in the supporting structure. In metal parapets, joints in the rails should be capable of transmitting the full maximum design requirements in bending and a proportion (approximately 60%) of the tensile strength of the rail section. Where the joint movement is large (greater than 50 mm) it may not be practical to provide tensile load transfer. In this case, end posts should be provided close to each side of the joint. Flexural continuity should still be maintained across the joint wherever possible.

The ends of parapets can pose a considerable risk to vehicles which may collide with them. It is normal to provide protection to exposed ends of parapets by installing a length of safety fence in advance of the parapet. However, the safety fence is significantly more flexible than the bridge parapet and therefore a transition section with progressively increasing stiffness should be provided to redirect the vehicles smoothly. Similar transition sections should also be provided between parapets of differing containment standards. Where the height of a parapet is reduced by the termination of the upper rail and at the connection to a safety fence the terminating rail should be flared back away from the approaching traffic.

Parapets which use baseplates such as metal post and rail systems, and precast concrete units should be fixed to the main structure using stainless steel bolts engaging with an anchorage. The anchorage within a concrete supporting structure should be either a cast in cradle anchorage or cast in or drilled individual bolt anchorages. The design of the anchorages should take into consideration the effects of possible overlap of stress cones from individual bolts and adequate reinforcement should be provided in the supporting structure. Due to the corrosive environment within which the anchorage is placed, components of the anchorage adjacent to the surface should be constructed of corrosion resistant materials. All anchorages should be provided with an internally threaded socket to receive the holding down bolt. Studs should not be used since they are difficult to replace should they be damaged in a parapet impact.

Metal parapets should be constructed of corrosion resistant materials, e.g. aluminium, or provided with a suitable protective system. Hollow sections may be subject to internal corrosion and the protective treatment selected should take this into consideration. Hollow sections should be either drained to resist corrosion and prevent damage due to the freezing of water which may accumulate within them or be sealed with all joints being made with continuous welds of structural quality.

Footbridges and cycleway bridges from which vehicles are excluded should be provided with pedestrian parapets. Pedestrian parapets should also be provided on retaining walls which support footpaths or cycleways remote from highways. Similar to vehicle parapets these may be constructed in a variety of materials. They may also be of framed construction with suitable infilling, solid or a combination of both. Infilling may be by solid sheeting, mesh or vertical bars, and should be installed such as to deter climbing of the parapet. Where vertical infill bars are used the space between the bars should not exceed 100 mm. Pedestrian parapets on structures over or adjacent to railways should generally be of increased height and be solid or provided with solid sheet infilling. As with vehicle bridges, parapets on footbridges should be increased in height if frequently used by equestrians.

15.3 Expansion joints

15.3.1 Introduction

Wherever two elements of a structure are moving relative to each other, it is often necessary to provide an expansion joint which seals the gap between the two elements whilst accommodating the relative movements. For bridges the gap is usually that between the deck end and the abutment ballast wall. However, on long viaducts additional joints will often be needed between sections of deck in order to limit the movement at any one point. Expansion joints are by virtue of their function a point of weakness within a bridge and history has highlighted many examples of joints leaking. As water, laden with de-icing salts, has leaked onto bearing shelves or pier supports, corrosion of the reinforcement has frequently resulted. The repairs which are required cost significantly more than the capital cost of the joint alone, especially when the cost of traffic delays is taken into account. It is therefore important to give full consideration to the design, detailing and installation of bridge expansion joints in order to minimize the risk of the bridge owner being faced with high repair bills in the future.

The vulnerability of expansion joints is one of the driving forces behind the increased use of integral bridge construction, in which the need for expansion joints is eliminated by connecting the deck directly into the abutments. In the UK, the use of integral bridges is now mandatory on Highways Agency schemes wherever the bridge is less than 60 m long and has a skew angle of up to 30° (Highways Agency, 1995). The elimination of expansion joints is generally to be encouraged. However, the same load effects and causes of movement for which an expansion joint is designed will be present in an integral bridge, and will need to be considered in its design. Even so, for many bridges, especially those already built, integral construction will not be an option, and so there will always be a need for expansion joints.

An expansion joint must exhibit a number of characteristics for it to perform satisfactorily. It must sustain loads and movements without being damaged itself or causing damage to other parts of the structure. It should be watertight, give a good ride quality and not be hazardous to road users, who could include cyclists, pedestrians and equestrians. The skid resistance of the joint should match that of adjacent surfacing, and noise emission from the joint should be limited, especially if it is to be used in residential areas. Finally, any joint should be easily inspected and maintained.

15.3.2 Types of joint

Different types of expansion joint are presented in Figure 15.1, together with an indication of typical movement ranges for each type of joint. However, the exact details will vary between manufacturers, who should be contacted to discuss particular details of their systems. The variations for each joint type are highlighted in the figure.

The list of joint types presented is not meant to be exhaustive. However, it does identify all the generic joint types for those systems which currently hold Departmental Type Approval for use on Highways Agency schemes in the UK. In order to obtain this approval joints are required to undergo testing to verify their characteristics and so give some assurance as to their performance. Joints not on the list of approved/registered products (Highways Agency, 1998a), which is updated annually, cannot be used on Highways Agency schemes without obtaining a Departure from Standard.

Other types of joint which have been used, but of which there is limited practical experience to date, include the bearing level joint and the Clwyd Buried Plug joint. The former is located at the level of the bearing shelf and has a small capacity in the range 5 mm to 10 mm. It comprises a vertically installed inert board, located in a recess in the deck end and back of the abutment, both of which are aligned. The second type has been designed and used by Clwyd County Council. In effect it is a buried or sub-surface 'asphaltic plug' joint with a movement range of up to 20 mm. It should be particularly suitable for use with porous asphalt surfacing.

The final choice of joint type will depend on a number of factors. Joint types should not be mixed on an individual joint, and this will often dictate the form of maintenance works on an existing joint. For new applications the joint must clearly be able to accommodate the predicted movements, but there are other factors related to the joint's performance which should be considered. These include the treatment of the verges and footways, which may contain many services, the road alignment (gradient, cross-fall and curvature), the proximity of junctions (where longitudinal loads will be more frequent) and how heavily trafficked the joint will be. All of these factors can affect the performance and hence life of an expansion joint, and need to

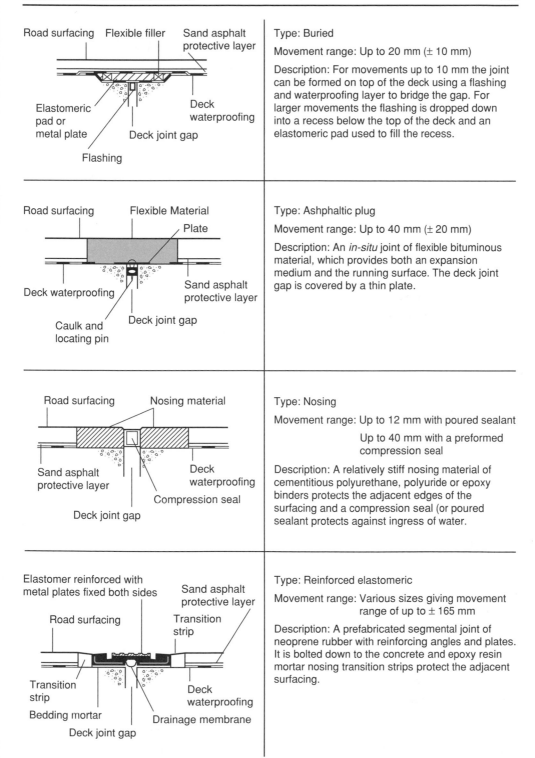

Road surfacing   Flexible filler   Sand asphalt protective layer

Elastomeric pad or metal plate

Deck waterproofing

Deck joint gap

Flashing

Type: Buried

Movement range: Up to 20 mm (± 10 mm)

Description: For movements up to 10 mm the joint can be formed on top of the deck using a flashing and waterproofing layer to bridge the gap. For larger movements the flashing is dropped down into a recess below the top of the deck and an elastomeric pad used to fill the recess.

Road surfacing   Flexible Material   Plate

Deck waterproofing

Sand asphalt protective layer

Caulk and locating pin

Deck joint gap

Type: Ashphaltic plug

Movement range: Up to 40 mm (± 20 mm)

Description: An *in-situ* joint of flexible bituminous material, which provides both an expansion medium and the running surface. The deck joint gap is covered by a thin plate.

Road surfacing   Nosing material

Sand asphalt protective layer

Deck waterproofing

Compression seal

Deck joint gap

Type: Nosing

Movement range: Up to 12 mm with poured sealant

Up to 40 mm with a preformed compression seal

Description: A relatively stiff nosing material of cementitious polyurethane, polyuride or epoxy binders protects the adjacent edges of the surfacing and a compression seal (or poured sealant protects against ingress of water.

Elastomer reinforced with metal plates fixed both sides   Sand asphalt protective layer

Road surfacing   Transition strip

Transition strip

Deck waterproofing

Bedding mortar

Drainage membrane

Deck joint gap

Type: Reinforced elastomeric

Movement range: Various sizes giving movement range of up to ± 165 mm

Description: A prefabricated segmental joint of neoprene rubber with reinforcing angles and plates. It is bolted down to the concrete and epoxy resin mortar nosing transition strips protect the adjacent surfacing.

Figure 15.1 Types of expansion joint (continued overleaf).

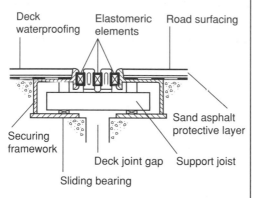

Type: Elastomeric in metal runners

Movement range: Single element up to 80 mm
(± 40 mm)
Multi-element up to 960 mm
(± 480 mm)
Embedded up to 150 mm
(± 75 mm)

Description: An elestomeric seal is fixed between two metal runners cast into recesses in the abutment and deck concrete. By introducing intermediate runners, multi element joints can be provided (as illustrated) with greater capacity. As an alternative the rails can be embedded in a resin bonded to the concrete or a rubber element bolted to the concrete.

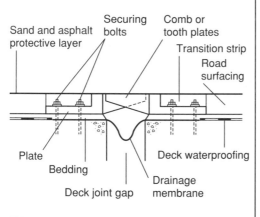

Type: Cantilever comb or tooth

Movement range: Typically up to 600 mm
(± 300 mm)

Description: A prefabricated joint in which metal comb or tooth plates slide back and forth between each other across the gap. They are bolted down to the concrete and a drainage membrane is provided underneath to collect water.

Figure 15.1 (Continued) Types of expansion joint.

be taken into account when considering the whole life cost of the joint. While a joint chosen purely on the basis of its initial capital cost and speed of installation may perform perfectly well, some joints have been shown to fail prematurely in adverse conditions. Replacing joints can prove to be very expensive once all the associated traffic management and delay costs are taken into account. These can easily outweigh the initial capital expenditure in whole life costing terms.

15.3.3 Design
Design loads and movements for expansion joints are laid down in BD 33 (Highways Agency, 1994a). Joints should be designed for both serviceability and ultimate limit states, so ensuring that the joints will function correctly without requiring excessive maintenance and will sustain ultimate design loads and movements.

The design loads specified in BD 33 are a 100 kN single wheel load or a 200 kN single axle load, with a 1.8 m track, for vertical effects, and a 80 kN/m horizontal load. Partial load factors, γ_{fL} are given as 1.50 and 1.20 for vertical loads at ULS and SLS, and as 1.25 and 1.00, respectively, for the horizontal loads. There are two important points to note. Firstly, the supporting structure should be designed to sustain the above loads. Secondly, to the above loads should be added those resulting from strains developed in the joint fillers over their design range of movement. This can be quite sizeable for the more rigid types of joint and can, for example, influence the ballast wall design. Some elements of expansion joints could be subject to fluctuating traffic loads, and in these cases fatigue lives should be evaluated.

Movement calculations are based on partial load factors of unity at both ULS and SLS. Movements of the joint can arise from a number of sources, all of which should be summed to give a total movement range, based on which a choice of joint type can be made. As not all movements are reversible, it is preferable to evaluate and specify limits to both the closure and opening of the joint, since by quoting only a total range of movement it is not apparent whether movements in each direction balance. For certain joints, it is also common practice to vary the expansion joint gap to suit the prevailing effective bridge temperature. For example, if a joint is installed in summer, when the deck has expanded, a smaller gap would be specified than for colder conditions in winter. In this way it is possible to enhance the movement capacity of a joint that is simply installed at some mid position of its range regardless of the prevailing conditions. However, in specifying the joint it should be made clear exactly what movement is being quoted.

Temperature movements as discussed above depend on the effective bridge temperatures experienced by the deck and should be evaluated in accordance with the relevant bridge design standard. Irreversible movements result from creep and shrinkage of the concrete and these must be evaluated using design standards for concrete or composite bridges. Lateral movement of the joint on curved or skew bridges should also be considered, as this can affect the joint design. Settlement of supports can also cause movements, as can sway of the bridge under longitudinal braking or traction loads depending on the bridge's articulation. On bridges with flexible or deep decks, rotation of the deck ends under live load can also cause significant movement at the joint level. This explains why even a joint above a fixed bearing will experience some movement. A similar effect for permanent loads is generally avoided by installing the expansion joint as late as possible, when the majority of permanent movements have already taken place.

15.3.4 Drainage
Expansion joints do not generally fail because their movement capacity is exceeded. Safety factors built into their design ensure this. Occasionally elements of joints may deteriorate more rapidly because they are locally subject to higher than expected loads, perhaps due to increased dynamic factors on wheel loads caused by uneven surfacing. The main cause of failure is that water starts to leak through the joints. This can arise from poor detailing of the joint, poor workmanship during its installation

and simply the inherent difficulty of completely sealing any joint between two elements moving relative to each other. The management of water on the bridge deck is key to the successful functioning of an expansion joint, and should be addressed early in any bridge's design and not as an afterthought.

At the road surfacing level, road gullies located on the uphill side of a joint can help limit the flow of water over the joint. However, the benefit of providing gullies on the deck can be offset by the difficulty of discharging the water, since the carrier pipes would need to be sleeved through the expansion joint within the verge, so creating another potential source of leakage. For joints with a seal at road level, the seal should be watertight and continuous. For reinforced elastomeric and cantilever comb or tooth joints a secondary drainage system, in the form of a continuous membrane, should be installed immediately below the expansion joint. Buried joints can crack and in some situations reinforcement is placed in the base course to control this. Asphaltic plug joints can also crack or debond. As a consequence, it is recognized that whilst every effort should be made to make expansion joints watertight, there is a risk of water from the surface leaking through the joint over the course of time.

In order to collect water that does leak through the expansion joint, it is common practice to provide a drainage system beneath the deck joint gap. This system should have adequate access for inspection and maintenance and discharge from the structure into a suitable road drainage system or soakaway. Water passing through the surfacing and running along the waterproofing should ideally be collected and discharged through a sub-surface drainage system before it reaches the expansion joint. The interface between the expansion joint and the deck waterproofing should be watertight. Some joints, such as buried joints, do not impede sub-surface water. However, other types of joints, such as Elastomeric in Metal Runners, actually require upstands to be build on the deck onto which the joint is attached. In this situation the upstand acts as a dam to the sub-surface water, which must therefore be collected and removed. If this water was allowed to collect, hydraulic pressures would develop from wheel loading and these pressures would cause failure of the joint's bond or seal between it and the waterproofing system. This would result in leakage into the deck end gap. The means of collecting the water is discussed elsewhere in this chapter.

In passing, it is worth commenting that where porous asphalt is widely used as a surfacing material and is incorporated on bridge decks, the means of collecting and discharging subsurface water becomes even more critical.

15.3.5 Detailing

Good detailing of expansion joints will help go a long way to providing a low maintenance and effective joint. BD 33 incorporates a number of requirements relating to the safety of bridge users. For example, the maximum width of any open gap not bridged by a load bearing element should be 65 mm, and no gaps are permitted where pedestrians have access to the bridge. This is overcome either by installing a load bearing seal or a cover plate. Cover plates should be placed in shallow recesses in the footway and be profiled to ensure they are not slippery. They are bolted on one side of the joint, but free to slide on the other side. As these plates have to sustain accidental wheel loads they are typically 12 mm thick. Thinner cover plates are often provided over the parapet string course to mask what would otherwise be an open expansion joint. Kerb cover plates should be provided to protect the expansion joint at these locations, which could be prone to impact. Special attention needs to be paid to tooth and comb joints where the gaps are orientated generally in the line of traffic in order to ensure that cyclists can ride across them.

The detailing of the joint in the verges and footways must be considered, as the footway level is often a few hundred millimetres above the level of the top of the concrete deck. For buried joints the joint is installed at top of deck level and verge construction continues across it. For asphaltic plug joints, the joint material is brought up to the top of footway level, so giving a much deeper joint than over the carriageway. Nosing joints generally follow the footway profile as well. Reinforced Elastomeric Joints can either be profiled to suit the footway levels or can be installed at a lower level as a continuation of the carriageway joint. In the latter arrangement a sliding plate must be provided to span the gap. Elastomeric Metal in Runner joints can similarly be installed with a prefabricated profile to match the surface level or can be kept at the lower level with a surface sliding plate bridging the gap.

The detailing of any services as they pass through expansion joints is also important, and the need to accommodate services may well dictate which of the above options for detailing joints in verges is adopted. Some joints require certain clearances to any service ducts to be provided and this should be checked. Service ducts should be adequately spaced to allow the flow of joint material around them or the positioning of fixings between ducts. It would also be prudent to install empty ducts, as the retrospective installation of ducts is not straightforward. The detailing of the ducts across the joint must allow for movement to take place. This can be done, for example, by stopping off each duct either side of the joint within a sleeve which spans across the joint. Continuous ducts can be used, but it must be ensured that they are sleeved through any abutment or deck upstands. It is also important that seals are provided between the ducts and any sleeves, otherwise any water within the verge will pass between the duct and the sleeve and down the joint. The seals will need to allow relative movement between the duct and the sleeve.

15.3.6 Installation and maintenance
Two of the observed causes of expansion joint failure are faulty installation and poor materials. Care should be taken in installing expansion joints and manufacturers' recommendations followed. Trained operatives should be employed and particular attention paid to known points of weakness such as the interface with the bridge deck waterproofing.

Expansion joints should be installed as late as possible in a bridge's construction to allow for shrinkage, creep and settlement movements to have taken place, at least in part, before the expansion joint gap is fixed. For joints fixed to the deck with cast-in anchors, the joint system is installed before the surfacing is laid. The anchors are cast into boxed out recesses left in the deck or abutment, having tied the anchors into the main reinforcement with suitable additional bars. Other joint types, which are either bonded or bolted to the concrete, are generally installed after the deck surfacing is in place, having previously covered the deck joint gap and waterproofing with a thin layer of plywood of width equal to that of the joint. The surfacing is then saw cut and removed, so enabling the joint to be installed.

Where it is necessary to set expansion joint gaps according to the effective bridge temperature as discussed earlier, reference can be made to the Transport Research Laboratory's Report No SR 479 (Emerson, 1979).

Expansion joints should be designed to ensure that all elements subject to wear can be easily replaced or reset, preferably in off-peak hours. Joints should be regularly inspected to ensure that they are continuing to function properly, and have not closed up or are leaking. Any blocked drainage should be cleared expeditiously due to the consequences of water being allowed to leak onto other bridge elements. Any silting up of joints also needs to be cleared to prevent the transmission of high forces across the joint.

15.4 Drainage

In the initial concept stage of the design of a bridge structure, consideration should be given to each element of the structure to ensure that the effects of water, particularly when contaminated with road salts, do not seriously affect the life of the structure. This can be achieved by effective drainage of these elements.

15.4.1 Abutment and retaining wall drainage
Back of wall

Back of wall drainage is provided to prevent the build up of hydrostatic pressure on the rear face of the walls. It usually takes the form of a 150 mm diameter porous drain pipe located on top of the wall base which can be enclosed in no fines concrete. Above this, for the full height of the wall, a drainage medium is provided. This can take the form of hollow concrete blocks or a free draining granular material and, in the UK, their specification requirements are given in BD 30 (Figure 15.2).

Weep holes should be provided in all high walls in case the porous drain pipes become inoperative for any reason. They should outfall just above the external paved surface and ideally, have a reverse fall to prevent continual dripping. In urban areas, small diameter pipes, say 50 mm of 3 m centres should be used, larger pipes often attract vermin and are also frequently blocked by litter, such as cans and bottles

The maintenance requirements for the drain must be considered in the design. The pipes should be straight, albeit on a gradient, with catch pits or manholes at each end to enable the pipe to be cleaned by rodding or water jetting. Where the pipe is longer than approximately 80 m, an intermediate manhole should be provided as, above this length, rodding and waterjetting becomes difficult in a rough pipe (Figure 15.3).

Depending upon their length, wing walls should be treated in the same way as abutment walls. However, for short walls, it is common practice to omit the pipe and allow water to flow to the abutment drainage media through its own drainage medium which is then collected by the abutment drain.

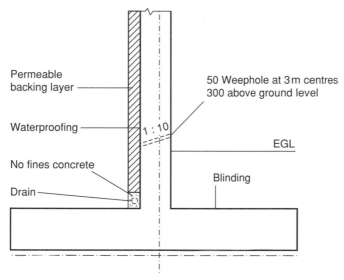

Permeable backing layer

Waterproofing

No fines concrete

Drain

50 Weephole at 3 m centres
300 above ground level

1 : 10

EGL

Blinding

Typical Section Through Abutment

Figure 15.2.

Where continuous bored piles, or similar are used as the structural abutment, water penetrating the piled wall can be collected in a channel between the piles and the cladding and discharged as described above (Figure 15.4).

15.4.2 Carriageway drainage

Carriageway drainage on bridge decks can cause problems where gullies or down pipes penetrate the deck and its associated waterproofing. Where possible this should be eliminated by the use of gradients, hard strips acting as reservoirs or drainage channels.

Where drainage is to be provided, kerb gulley inlet type, although more expensive, are favoured because they can be installed on top of and without penetrating the waterproofing. This system is also suitable where porous asphalt surfacing is specified (Figure 15.5). Where this system is not suitable and a gulley and pipe system is required care should be taken to ensure that all penetrations are well detailed to prevent water from escaping around rather than through the system.

Subsurfacing drainage

Water can collect between the bridge deck surfacing and the waterproofing, and cause the surfacing to bubble and delaminate in freeze/thaw conditions. This build up can be released by the use of conical drainage units though the deck at low points on the structure. These units are particularly useful adjacent to deck expansion or rotational joints. Ideally, the outlet tubes should be cast into the deck and not drilled though as an after thought. The outlets should be piped to a positive drainage system and not allowed to drip onto a carriageway or side slope paving area where freezing can occur.

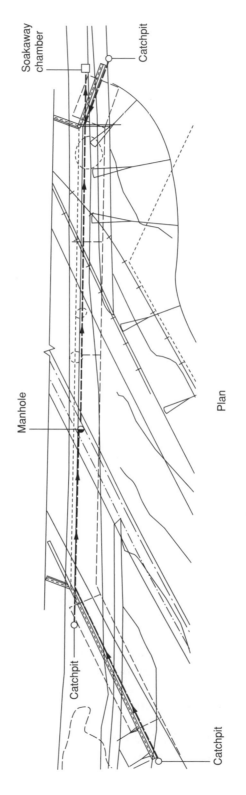

Figure 15.3.

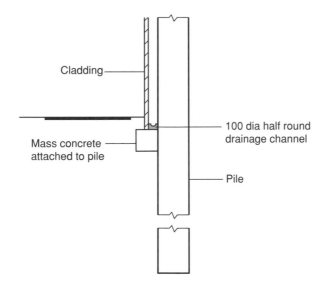

Cladding

Mass concrete
attached to pile

100 dia half round
drainage channel

Pile

Figure 15.4.

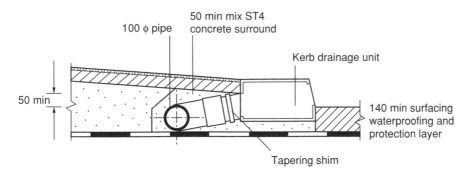

50 min mix ST4
concrete surround

100 φ pipe

Kerb drainage unit

50 min

140 min surfacing
waterproofing and
protection layer

Tapering shim

Figure 15.5.

15.4.3 Abutment shelf drainage

In the UK, BD 57 requires the bearing shelf area to be waterproofed and drained to prevent deterioration of the end of the deck and bearings from contaminated water (Figure 15.6).

This area can be used as a collecting area for drainage from the deck joints and sub surfacing drainage units where they can be connected to preferably sealed drainage piping or by guttering. The disadvantage with guttering is that, when it becomes blocked, water can drip onto susceptible areas. The outlet for this system can discharge onto the floor of the shelf, provided the resultant splash is contained.

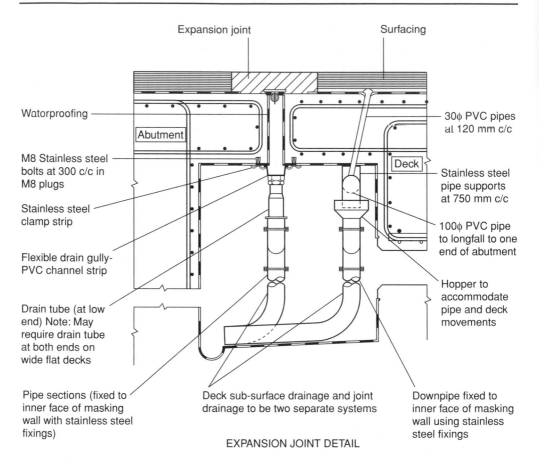

Figure 15.6.

The floor of the shelf must be provided with a fall and with a drainage channel to collect the water. This channel is to be connected to a positive drainage system which should be connected to the highway drainage system or an adjoining soakaway. Provision should be made for the future maintenance of this system by the inclusion of catch pits or manholes.

15.5 Waterproofing

15.5.1 Introduction

The use of de-icing salts for winter maintenance on highway bridges is causing considerable damage to the concrete bridge stock. The chloride salts penetrate the concrete and cause corrosion of the reinforcement. In particular, elements which are most at risk are reinforced concrete bridge decks, tops of piers, columns, crossbeams and abutments which carry simply supported decks, the splash zone of piers, columns, abutments and parapets, and surfaces below ground level.

Since 1965 waterproofing of bridge decks has been mandatory for UK motorway and trunk road bridges, and has been used on most other bridges. It has been recognized that the provision of an effective waterproofing system on a bridge deck is of crucial importance, and this may help explain why salt attack on UK bridge decks has generally not been as severe as in other countries.

BA 57 (Highways Agency, 1995) recommends that in addition to the bridge deck the following concrete surfaces should be waterproofed:

- vertical faces at deck ends and abutment curtain walls
- top faces of piers and abutment bearing shelves
- inaccessible areas which may be subject to leakage, for example beam ends.

Splashing and spraying of salt water can also cause deterioration and damage to the bridge structure. Particularly prone areas are bridge abutments, piers, parapet edge beams, deck soffits and the splash zones of river piers and abutments. BA 57 (Highways Agency, 1995) recommends that special precautions should be taken in these areas by the application of a protective coating. Currently the only protective coating endorsed by the Highways Agency to provide protection against the ingress of chlorides is silane, which is chemically impregnated into the concrete. Impregnation procedures are given in BD 43 (Highways Agency, 1990a) and BA 33 (Highways Agency, 1990b). It should be noted that in some countries the protective coating comprises a proper waterproofing system applied to all concrete surfaces.

Since 1975 waterproofing systems for bridge decks are required to pass a series of tests in order to obtain a British Board of Agrément (BBA) Roads and Bridges Certificate. However, despite these requirements there was evidence that some systems were not adequately performing in service. Various problems included leakage, poor bond, embrittlement and disintegration. A major research programme at TRRL was undertaken in the late 1980s including field and laboratory trials (Price, 1989, 1990, 1991), and some of the waterproofing systems previously considered suitable were found to be unsuitable. In particular, some systems were found to be permeable in the long term, and many systems were damaged by the direct application of surfacing, although the use of a sand asphalt carpet prevented most of the damage. This additional red tinted bituminous protection is generally required to prevent the penetration of base course aggregates into the membrane. However, in specific instances where there are limitations on the total surfacing thickness, the additional protective layer can be omitted provided the waterproofing system passes an aggregate indentation test at 125°C, and gradients and/or sub surface drainage is provided.

Debonding of waterproofing due to poor adhesion at the concrete/membrane and membrane/asphalt interface was also reported as a problem and lead to a further TRRL study (Stevenson and Evans, 1992).

Nowadays, waterproofing systems for highway bridge decks must have Registration (Highways Agency, 1998b) granted by the Highways Agency as the Overseeing Organization. The systems have to pass various tests to be included on a list of approved/registered products (Highways Agency, 1998a). Procedures and certification test requirements are specified in BD47 (Highways Agency, 1999a) and BA47 (Highways Agency, 1999b), and as from 1 June 1998 all waterproofing systems used on Highways Agency schemes must satisfy these requirements. In addition, the Specification for Highway Works (Highways Agency, 1998a) and associated Notes for Guidance (Highways Agency, 1998b) gives further requirements including the provision of PWS (Proprietary Waterproofing System) Data Sheets by the supplier.

Buried concrete surfaces below ground should also be protected from attack by salt and other chemicals in the soil. It is normal practice to apply waterproofing to all buried concrete surfaces greater than 300 mm below finished ground level, such as the rear faces of abutments and wing walls. Waterproofing in these situations prevents water seeping through cracks and other defects to the exposed faces. As in the case of bridge deck waterproofing, the Specification for Highway Works (Highways Agency, 1988a) and associated Notes for Guidance (Highways Agency, 1988b) gives requirements for these waterproofing materials.

15.5.2 Types of waterproofing

In the case of bridge decks, waterproofing membranes normally comprise sheet systems or liquid systems which are bonded to the concrete surface. In addition, mastic asphalt has been used in the past but is now rarely used and is not recommended.

Sheet systems are mainly bituminized fabrics, polymer or elastomer based membranes. They can consist of sheets or boards, either incorporated into the membranes, or used to protect them. Sheet membranes are pre-formed factory manufactured and are generally hand applied by the pour and roll method or by torch or are self adhesive. They are bonded to the surface in overlapping strips and must not tear, puncture, rupture, or come apart at the seams, corners and edges of the structure.

Liquid systems are mainly acrylic, epoxy, polyurethane or bitumen based. They are one or two part moisture or chemically curing solutions and can be applied by brush, squeegee or spray to form a seamless coating on the concrete surface. In particular, the sprayed applied membranes can be rapidly applied and have good adhesion to the concrete.

Sheet systems generally cost less than liquid systems but spray applied membranes are quicker to lay which may well justify the extra cost. Whilst sheet systems are of uniform thickness, liquid systems rely on strict quality control to ensure minimum thicknesses are achieved. Liquid systems are also more suitable for irregular areas and at drains as sheet systems require special cutting and trimming.

In the case of bridge elements below ground, waterproofing systems normally comprise cut back bitumen or proprietary materials such as rubberized bitumen emulsions, pitch epoxies and sheet membranes or coal tar, with compatible primers where required. The current trend is for proprietary materials with bitumen emulsions being the most popular. The rubber content of rubberized bitumen emulsions can be varied to suit the desired viscosity and thickness of waterproofing. To ensure the effectiveness of the waterproofing system a minimum dry film thickness of 0.6 mm is recommended. Pitch epoxies are sometimes used in aggressive soil conditions, and self-adhesive sheet membranes where there is a high hydrostatic head.

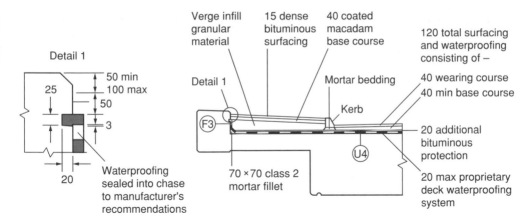

Figure 15.7.

15.5.3 Detailing

The detailing of deck waterproofing is important to ensure the effectiveness of the waterproofing. All detailing should allow for the different types of waterproofing, as the type is generally unknown at design stage. The waterproofing system should be continuous over the deck between parapet upstands, and should include all footways, verges and the like as shown in Figure 15.7.

Careful consideration should be given to the detailing of waterproofing at deck movement joints and drains to ensure that water does not penetrate beneath the waterproofing system. Waterproofing should also be detailed along the sides and floors of all service bays within bridge decks.

15.5.4 Installation and maintenance

The effectiveness of the waterproofing system depends on good installation. Care should be taken using trained operatives familiar with the manufacturer's instructions.

Waterproofing systems should perform satisfactorily if installed as specified by the supplier in the PWS Data Sheet. This sheet should specify a Class U4 concrete finish, and the bridge should have fall and gradients to prevent water accumulating in the surfacing above the waterproofing system. The PWS Data Sheet should also specify that the minimum installation temperature of the waterproofing system should be 4°C and rising, because curing time is increased and some materials stiffen at low temperatures. In particular, some liquid applied systems are more tolerant at low temperatures than others. To ensure the durability of the waterproofing, it is important that uniform adhesion at all interfaces is achieved. If it is deemed necessary to verify the integrity of the waterproofing system various test methods should be considered. These include high voltage holiday detection and low voltage breach detection to check for leakages, and hammer tapping and transient thermography to check for debonded areas.

It is advisable to consider re-waterproofing a bridge deck whenever re-surfacing work is carried out otherwise the waterproofing layers could be damaged causing chloride attack of the underlying structure. It is essential that any repairs to joints and drains is carried out carefully to maintain the integrity of the waterproofing system. Where existing membranes are removed, the concrete surface may be uneven and not have a U4 finish, and be contaminated with primer and bonding agent. Grit blasting is often preferred to remove these existing products as they may not be compatible with the new materials. The grit blasting also prepares the concrete surface so that it is suitable for the chosen waterproofing system.

It is intended that the red sand asphalt layer above the waterproofing on modern bridges acts not only as a protection layer but as an indicator layer. This indicator layer enables the waterproofing to remain in place when planing for re-surfacing work. On some bridges a red indicator mesh has been laid over a layer of black sand asphalt in lieu of the red sand asphalt layer.

Generally waterproofing systems perform satisfactorily for at least 20 years, but longer durability periods are required for decks to be re-surfaced without removing the waterproofing.

Bibliography

For further information on the details and performance of different types of expansion joint the reader is referred to the Highways Agency's Standard and Advice Note on the subject, BD 33 (Highways Agency, 1994a) and BA 26 (Highways Agency, 1994b), and also to the comprehensive 'Practical Guide to the Use of Bridge Expansion Joints', published as TRL Application Guide 29 (Barnard and Cunninghame, 1997a). This in turn is based on the findings of a detailed study carried out by a Working Group under the auspices of the County Surveyors' Society and Transport Research Laboratory. The full details of the study are given in the interim report *Bridge Deck Expansion Joints* (Cunninghame, 1994) and final report *Improving the Performance of Bridge Expansion Joints* (Barnard and Cunninghame, 1997b).

For further information on bridge deck waterproofing and waterproofing/ protection of below ground concrete reference should be made to the comprehensive guide *Water Management for Durable Bridges* published as TRL Application Guide 33 (Pearson and Cunninghame, 1998).

Practical advice and guidance on problems of bridge deck waterproofing which have been experienced is given in the notes produced by the Structures Working Group of the Highways and Infrastructure Board of the Institution of Highways and Transportation and IHIE (1996).

Barnard CP and Cunninghame JR. *Practical Guide to the Use of Bridge Expansion Joints.* TRL Application Guide 29. Transport Research Laboratory, Crowthorne, 1997a.

Barnard CP and Cunninghame JR. *Improving the Performance of Bridge Expansion Joints:* Final Report of the Bridge Expansion Joint Working Group. TRL Project Report TRL 236. Transport Research Laboratory, 1997b.

Cunninghame JR (Ed). *Bridge Deck Expansion Joints Interim Report.* TRL Project Report PR/BR/4/94. Transport Research Laboratory, Crowthorne, 1994.

Emerson M. Bridge temperatures for setting bearings and expansion joints. TRL Report SR479. Transport Research Laboratory, Crowthorne, 1979.

Highways Agency. *Design Manual for Roads and Bridges (DMRB)*, Volume 2: Section 1 BD 30 Backfilled Retaining Walls and Bridge Abutments. Department of Transport, London, 1987.

Highways Agency. *Manual of Contract Documents for Highway Works (MCHW)*, Volume 1, Specification for Highway Works, Series 2000. HMSO, London, March 1988a.

Highways Agency. *Manual of Contract Documents for Highway Works (MCHW)*, Volume 2, Notes for Guidance on the Specification for Highway Works, Series NG2000. HMSO, London, March 1988b.

Highways Agency. *Design Manual for Roads and Bridges (DMRB)*, Volume 2: Section 4: Paints and Other Protective Coatings, BD 43 Criteria and Material for the Impregnation of Concrete Highway Structures. Department of Transport, London, 1990a.

Highways Agency. *Design Manual for Roads and Bridges (DMRB)*, Volume 2: Section 4: Paints and Other Protective Coatings, BA 33 Impregnation of Concrete Highway Structures. Department of Transport, London, 1990b.

Highways Agency. *Design Manual for Roads and Bridges (DMRB)*, Volume 1: Section 3: Materials and Components. BD 33 Expansion Joints for Use in Highway Bridge Decks. HMSO, London, 1994a.

Highways Agency. *Design Manual for Roads and Bridges (DMRB)*, Volume 2: Section 3: Materials and Components. BA 26 Expansion Joints for Use in Highway Bridge Decks. HMSO, London, 1994b.

Highways Agency. *Design Manual for Roads and Bridges (DMRB)*, Volume 1: Section 3: General Design, Part 8, BA 57 Design for Durability. HMSO, London, 1995.

Highways Agency. *Manual of Contract Documents for Highway Works (MCHW)*, Volume 0: Section 3: Advice Notes, Part 1, SA 1 Lists of approved/registered products. HMSO, London, 1998a.

Highways Agency. *Manual of Contract Documents for Highway Works (MCHW)*, Volume 0: Section 2: Implementing Standards, Part 1, SD 1 Implementation of Specification for Highway Works and Notes for Guidance. HMSO, London, 1998b.

Highways Agency. *Design Manual for Roads and Bridges (DMRB)*, Volume 2: Section 3: Materials and Components, Part 4, BD 47 Waterproofing and Surfacing of Concrete Bridge Decks. HMSO, London, 1999a.

Highways Agency. *Design Manual for Roads and Bridges (DMRB)*, Volume 2: Section 3: Materials and Components, Part 5, BA 47 Waterproofing and Surfacing of Concrete Bridge Decks. HMSO, London, 1999b.

Highways and Infrastructure Board Structures Working Group. Bridge deck waterproofing, notes for bridgeworks. *Highways and Transportation, the Journal of the Institution of Highways and Transportation and IHIE*, 1996, 43, No. 04, April, 29–32.

Pearson S and Cunninghame JR. *Water Management for Durable Bridges*. TRL Application Guide 33. Transport Research Laboratory, Crowthorne, 1998.

Price A R. *A Field Trial of Waterproofing Systems for Concrete Bridge Decks.* TRRL Research Report 185. Transport and Road Research Laboratory, Crowthorne, 1989.

Price AR. *Laboratory Tests on Waterproofing Systems for Concrete Bridge Decks*. TRRL Research Report 248. Transport and Road Research Laboratory, Crowthorne, 1990.

Price AR. *Waterproofing of Concrete Bridge Decks: Site Practice and Failures*. TRRL Research Report 317. Transport and Road Research Laboratory, Crowthorne, 1991.

Stevenson A and Evans W. *The Adhesion of Bridge Deck Waterproofing Materials, TRL Contractor Report 325.* Transport Research Laboratory, Crowthorne, 1992.

16 Protection

M. MULHERON

16.1 Introduction

In recent years the repair and maintenance of bridge structures showing signs of deterioration has become a major concern. This deterioration represents not only an enormous technical challenge to those responsible for the maintenance of an ageing infrastructure but also a huge cost to the nations economy (Anon, 1999). Clearly the protection of bridge structures is of prime importance to both reduce the likelihood of deterioration and, where it does occur, reduce its extent. However no client is prepared to pay for the measures necessary to produce a structure which has infinite life, nor would such a structure necessarily be desirable. Instead most operators seek protection methods, and associated maintenance strategies, that represent an efficient use of the available funds both in terms of cost per structure and serviceability over the intended design life.

While the deterioration of both old and new bridge structures remains a pressing problem it is important to remember that there are many examples of bridge structures showing excellent durability. Such structures demonstrate that the application of appropriate design codes coupled with good construction practice and an appropriate maintenance programme can yield durable and cost-effective structures.

16.1.1 Causes and consequences of deterioration

An important outcome of any analysis of the stability of the materials used in the construction of bridge structures is that they are thermodynamically unstable with respect to their environment. For example metallic iron is unstable with respect to the Fe(II) and Fe(III) ion species and will readily oxidize to FeO or Fe_2O_3, or their hydrates, in the presence of liquid water and oxygen. Similarly the calcium hydroxide formed within hardened Portland cement paste readily reacts with a range of materials, such as water, carbon dioxide, sulphates, etc., to form more stable, but less useful, products. In addition many of the engineering polymers used as protective coatings undergo long-term swelling and degradation when in contact with water. While such deterioration is inevitable investigations of deteriorated structures suggest that the majority of problems arise from design errors, errors in the specification or construction method employed, and poor workmanship or quality control (Shaw, 1987).

Regardless of the underlying cause the deterioration of bridge structures occurs primarily by either physical processes or as a result of chemical reactions. In most cases the deterioration seen by the bridge engineer is the result of physical effects induced by some underlying chemical, or electro-chemical, reaction. For example the deterioration of reinforced concrete structures due to corrosion of the embedded steel manifests itself as physical cracking and spalling of the cover concrete. This results from the formation of voluminous oxide products which have the effect of placing the concrete surrounding the steel into tension causing the formation of cracks in the concrete along the line of the reinforcement. It should be noted that the onset of aqueous corrosion of the steel can occur many years before any outward signs of physical deterioration can be seen (Figure 16.1). The time between initial construction, t_0, and the initiation of aqueous corrosion, t_1, represents a period during which a reinforced concrete structure is often considered as being 'maintenance-free'. In practice the period $t_0 \rightarrow t_1$ can be significantly increased by a combination of good design, quality construction and careful upkeep of the water management system, see Section 16.2. Once the initiation of corrosion has occurred some unplanned crack-

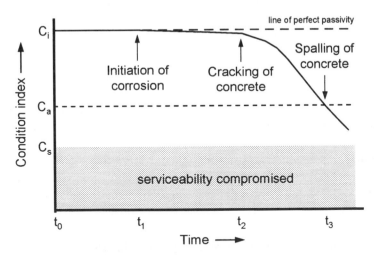

Figure 16.1 Idealized change in condition of a reinforced concrete structure as a result of corrosion of the steel reinforcement. During the early stages of the structures life, $t_0 \rightarrow t_1$, the steel remains passivated by the high pH of the surrounding concrete. Initiation of corrosion, whether by carbonation of the cover concrete or the ingress of chloride ions, occurs at time t_1 and the condition index begins to deviate from the line of perfect passivity. Subsequent corrosion results in the formation of an expansive oxide film at the steel/concrete interface inducing tensile stresses in the cover concrete which eventually cracks, t_2. Once cracked the rate of corrosion generally increases and eventually spalling of the cover concrete occurs, t_3. (Note: C_i = initial condition, C_a = limit of acceptable condition, C_s = limit below which serviceability is compromised.)

ing of the cover concrete becomes almost inevitable (Woodward and Williams, 1988). However if the bridge engineer takes steps to control the rate of underlying aqueous corrosion, e.g. by keeping the structure as dry as possible, then the time before cracking of the cover concrete occurs, t_2, can be significantly increased. Thus by using appropriate maintenance works a bridge can be successfully operated for many years in a state of controlled deterioration.

The use of controlled deterioration represents a valuable tool in the protection of bridge structures and finds its logical outcome in the process of corrosion allowance. This is the fundamental basis upon which steel bridges manufactured from weathering steel are designed, see Section 16.6.1. It is arguable that the changes in modern design codes for reinforced concrete, to higher cover depths and concrete grades for a given exposure class, are intended to allow for the deterioration which is expected to occur. Ideally such allowance will seek to delay spalling of the cover concrete or major loss of steel section to the point where the structure is no longer required. This approach does not seek to stop the initiation of deterioration but aims to manage its impact on the structural reliability. However problems can arise when the local rate of deterioration within part of the structure exceeds that shown generally raising the possibility of unexpected failure. The collapse of the Ynys-y-Gwas bridge greatly increased concern regarding the durability of conventional post-tensioned construction and highlighted the fact that highly localized deterioration can result in structural instability (Concrete Society, 1995).

As a consequence of the fundamental instability of construction materials, coupled with the various physical effects which can cause degradation, engineers adopt a range of protection measures with the aim of holding deterioration at economically tolerable levels. A logical outcome of this is the need to plan for the maintenance that will inevitably be required. Thus increases in initial production costs must be balanced against the lower operating costs associated with the use of more corrosion resistant materials and designs. This approach finds its most coherent form in life-cycle analysis in which a structure is considered to be an asset whose whole-life cost includes both the initial and recurrent costs. At present this approach is in its infancy in the UK but can still provide a useful framework within which various designs and associated protection options can be compared.

16.1.2 Protection strategies for combating deterioration

While accepting there is no universal method for stopping bridges from deteriorating with age it is possible to identify a number of strategies for protecting bridge structures.

Remove the environment

Most of the processes which lead to the degradation of bridge structures require the presence of water. As a consequence if a structure is either never exposed to a wet environment, or that when wetting does occur rapid drying can take place, then deterioration will be kept to a minimum. This approach is fundamental to the concept of all water management systems (Section 16.2).

Alter the environment

The extent and rate of deterioration of any structure can often be controlled by influencing the environmental conditions. Removing oxygen and oxidizers finds application in controlling metallic corrosion in closed systems where once the initial supply of oxygen is consumed then the corrosion process effectively comes to a stop. In semi-closed systems, where the supply of oxygen and water is restricted by physical factors, it is possible to control aqueous corrosion by the application of corrosion inhibitors, see Section 16.4.4. Where the supply of oxygen and water is effectively infinite it is still sometimes possible to restrict the rate of deterioration. A good example of this are metals protected by passive film formation such as stainless steels and weathering steels, see Section 16.6.

Protect the structure from the environment

For many bridge engineers the most obvious, and direct, method of protecting structures from deterioration is the use of coatings. Inert barriers rely on the exclusion of the environment but suffer from the inherent problem that where the coating becomes cracked, or broken, then any protection is compromised (Hartley, 1994). One solution to this problem is the use of sacrificial layers which actively deteriorate in preference to the underlying structure, e.g. galvanized steel. Such a coating has a major advantage over inert barrier coatings in that where the coating is broken, or scratched, the exposed steel is cathodically protected by the preferential corrosion of the more anodic zinc. Hydrophobic surface treatments do not attempt to form an impermeable layer but rather alter the wetting characteristics of the surface. The use of silane and siloxane surface treatments to protect concrete and masonry surfaces is now wide-spread, see Section 16.3.1. These have the advantage of allowing water vapour to pass unhindered hence promoting drying of the underlying substrate (Nwaubani and Dumbelton, 1997).

Alter the material or structure

Arguably the most cost-effective method of protecting bridge structures from deterioration is the selection of appropriate materials for the particular environment to be encountered. For example blended cement concrete incorporating blast furnace slag has been used in the construction of several prestige bridge projects (Deason *et al.*, 1993) due to its potential to resist the ingress of water borne chloride ions. However, such blends are inherently prone to poor curing (Parrot, 1991) which can lead to unexpectedly permeable cover concrete. Thus simply specifying a potentially durable material does not itself guarantee a durable structure.

It is an obvious, but often forgotten, point that if a component is difficult to construct it will almost certainly be constructed badly. However perfect the underlying philosophy if a design incorporates details that are difficult to construct then this can introduce defects which compromise the overall durability of the final structure. As a consequence a design which complies with the appropriate design codes but which is unnecessarily complicated to build under site conditions should be considered a bad design. This requires the engineer to think beyond the normal

Figure 16.2 Provision of safe and simple access to bridge bearings and half-joints.

limits of the available design codes and take control of the ease, and reliability, with which the structure can be constructed.

Finally it is important to remember that all bridge structures will deteriorate to some extent and will inevitably require inspection and maintenance. As a consequence good design should ensure safe and simple access to critical parts of the structure (Figure 16.2). This means that not only can component parts of a structure be inspected but regular maintenance and repairs can be carried out easily and under good conditions. Indeed where deterioration of a component is likely then provision should be made for its replacement. A simple example of this is in the area of bridge bearings which have a life expectancy considerably less than the bridges they support. Despite this it is surprising how many engineers do not design the bearings to be replaced without impeding the flow of traffic. The simple inclusion of a raised plinth, to ensure good drainage, coupled with an integral platform within the bearing support structure to allow the placement of jacks can significantly reduce long-term maintenance costs with little impact on the initial cost of construction.

16.2 Water management

The central role of the uncontrolled flow of water in the deterioration of bridge structures has highlighted the need for designers and engineers to carefully manage the flow of water as it passes onto, over and eventually away from a structure. An important outcome of water management concepts is the recognition that there are essentially three philosophies that may be adopted by those seeking to protect structures from deterioration:

- Prevent contact between the structure and water. This approach seeks to prevent water from entering those parts of the structure which are known to be prone to deterioration. Such exclusions are often based on providing a number of barrier layers or associated protective measures. The aim being that if water breaches one layer then it will be kept back, for an acceptable period, by the other layers of protection. An inherent danger of such 'exclusion' systems is that should water enter from unexpected areas, or through failures in continuity of the protective system, then it can remained trapped within the structure providing ideal conditions for long-term deterioration. The problem of waterproofing actively moving expansion joints has produced a number of commercial solutions but it is no coincidence that over the past 10 years there has been a considerable interest in integral bridge forms which eliminate the need to seal, and maintain, moving joints.
- Control the flow of water and promote rapid drying after wetting has occurred. This approach acknowledges that, however well designed the water management system, water will eventually find its way onto critical parts of the structure. This is especially true if the effects of water vapour and sprays are taken into account. Having accepted that is fundamentally impossible to totally exclude water steps can then be implemented to ensure that any water that does enter a structure is removed as quickly as possible. This both keeps the contact time with water to a minimum, and prevents the formation of high salt concentrations due to evaporation from stagnant pools of water. This approach has two main advantages. Firstly it encourages the designer to explicitly consider how water falling onto a structure moves over, through and off it. Secondly it tends to produce structures which are more easily inspected since provision has to be made for maintenance of the drainage system. A simple criteria for the success of a design based on this approach can had by inspecting the structure 2 hours after significant rainfall. Where water remains ponding on, or adjacent to, critical areas of the structure the design can be said to have failed.
- Select materials that can endure in the environment that exists. This third approach is arguably not true water management since it accepts that the structure exists in a wet environment and relies on the selection of materials that are inherently stable in that environment. Unfortunately this approach is subject to a number of problems. Firstly it is no simple matter to specify a construction material which will endure a wet environment without some additional protection measures. Secondly in selecting a material that should be perfectly durable for a particular environment there is the risk that the environment will change over the life of the

structure. Thirdly some forms of deterioration are conjoint phenomena requiring a combination of events to initiate failure. Thus the stress corrosion of stainless steel (Page and Anchor, 1988) requires the presence of both a suitably aggressive environment and some minimum applied stress. By failing to remove the water from a structure this form of deterioration becomes possible.

16.2.1 Drainage

The primary drainage system has two main functions. First it should ensure the removal of standing water from the main deck. Second it should channel water away from the main structural elements so that they remain dry for as long as possible. To achieve these aims a well designed draining system must fulfil the following criteria:

- Have the capacity to collect any water that arrives on the surfaces of a bridge, including direct rainfall, condensation and spray from moving vehicles.
- Control the flow of this water off the running surface and divert it through, or around, the structure without it coming into contact with the main structural elements.
- Dispose of the water away from the structure so as to both protect its own substructure and foundations and ensure that it does not cause problems for adjacent roadways or other structures.

The capacity of a drainage system must reflect both the area of the structure likely to collect water and the rate at which it arrives. Whilst it is possible to calculate the capacity based on annual rainfall data it is more usual to design for the more extreme values likely to be encountered over the design-life of the structure, e.g. the once in a 100-year storm. Inadequate capacity of the drainage system can seriously influence the safety of vehicles using the running surface as standing water can lead to skidding and lack of visibility. In contrast inadequate capacity only indirectly affects the long-term durability of a structure since over the life of a structure it will remain flooded for only a few hours at a time. Assuming that the structure subsequently dries quickly then this extra period of wetting is unlikely to significantly effect the overall extent of deterioration. Ultimately such problems of deck drainage can be avoided by the use of open, metal-grill style, running surfaces as are encountered in parts of North America. Such decks are, however, subject to significant speed restrictions due to the inherently low skid resistance of such a surface and are unsuitable for many applications.

Once on a structure the flow of water can be controlled by a number of physical methods that can be incorporated into a bridge structure. First and foremost of these are the natural gradients that exist both in the vertical and horizontal alignment of the bridge deck and the cambers of the running surface. Where the deck is inclined at an angle then this must be taken into account when designing both the position and capacity of the main drainage points on the deck. Since it is not feasible to rely solely on these natural gradients to control the movement of the water most successful drainage systems rely on water being directed off of the deck by the presence of suitable kerbs and gullies. In utilizing such methods it is important to remember the role of joints which, if not properly sealed, will allow water to leak down onto bearing shelves.

Having designed a drainage system that carefully collects and diverts water from a structure it is important to ensure that the water is channelled away from the structure and is not simply allowed to drip over retaining walls or other parts of the substructure. Where this does happen staining and long-term frost damage can occur. As a consequence weep-holes and drainage pipes should be designed such that their outfall lands away from the structure. This can be achieved by ensuring that such outlets protrude sufficiently to ensure any discharged water does not flow over the lower part of a wall. Alternatively the drainage pipes should be extended to direct the water into some suitable gully or drainage channel. Water that leaves the drainage system can either be deposited into a main drain, incorporating suitable storm storage capacity or into local soak-aways.

In attempting to control the flow of water it is important to distinguish between the behaviour of bulk water in drains and gullies from that of thin films of water on the surfaces of a structure. Bulk water can generally be relied on to flow in the direction of gravity following the line of least resistance and hence can be directed with some degree of certainty. Although this is subject to limits imposed by the capacity of the drainage system and the vagaries of wind action coupled with the influence of vehicle movements which can simultaneously transport water onto a structure and create sprays and fine mists of water.

In contrast the behaviour of water in thin surface films is dominated by the influence of surface tension and consequently will 'hang' up-side down on, and flow over, inclined surfaces rather than drip or fall from the surface. On concrete surfaces this effect is magnified by the capillary absorption exerted by the structure of the cement paste which both encourages the water to spread onto the surface of the structure and helps hold it in place against the pull of gravity. There are a number of important outcomes of this behaviour. First is the need to include grooves into the soffit of edge beams to encourage water running down the sides of beams to run-off the edge of the beam as drips rather than run onto and flow along the bottom of the beam (Figure 16.3). It is important to ensure that any water that is forced to drop from the beam does not subsequently fall onto other, more vulnerable, parts of the structure and promote deterioration. Secondly it means that water can unexpectedly find its way into parts of a structure that would normally be considered safe from water ingress, e.g. bearing shelves. Once water has entered such areas it is important that it can be quickly removed. As a consequence it is important to design bearing shelves to be self-draining. Typically this can be achieved by placing the bearing on a suitable raised plinth that has adequate air-space around it to help promote drying (Figure 16.4).

Having made adequate provision for drainage it is important that consideration is also given to how the system can be maintained. This is essential since even after only a few years in operation road debris, dead animals, litter and dust can collect in the drainage system and either reduce the maximum capacity or block the drain

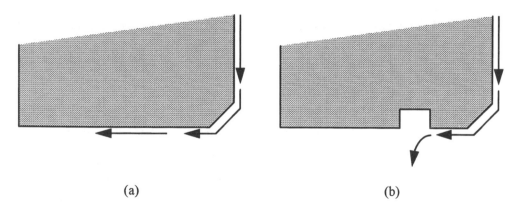

(a)   (b)

Figure 16.3 Idealized flow of water over the edge of a concrete beam (a) on to a flat soffit, and (b) on to a flat soffit with integral drip feature.

entirely (Figure 16.4). Thus the drains must be sized and angled not just to control the flow of water but also to allow the system to be easily rodded, water jetted or otherwise cleaned. A drainage system that cannot be maintained using easily available equipment and standard methods will inevitably contribute to poor long-term durability.

Figure 16.4 Raised plinth for protection of bridge bearing. The water ponding around the raised plinth is the result of either inadequate design, or maintenance, of the drainage outlet which has become blocked with debris. This photograph was taken 6 years after the bridge was first constructed.

16.2.2 Waterproofing membranes

It is the role of the waterproofing membrane to augment the drainage system by forming an impermeable layer over the upper surfaces of a bridge structure. This prevents the water from running into the underlying materials and instead pass into the main drains. To achieve this the waterproofing membrane should be continuous over the part of the structure to be protected and seamlessly fit into the main drainage channels. The effectiveness of the concept of waterproofing layers can be illustrated by comparing concrete bridges in the UK with similar structures in the USA which, until recently, did not employ such layers and have deteriorated more rapidly than originally envisaged.

It is important to distinguish true waterproofing membranes from the layers of asphaltic or bituminous bound material which make up the running surface of many bridge decks. While such materials are inherently water resistant they are rarely designed to be complete barriers against water ingress. Indeed porous asphalt surfacing deliberately holds water within the surface of the road to prevent water ponding and reduce dangerous road spray.

Typical materials used for waterproofing are organic in origin and range from simple bituminous layers through to complex binary polymers. All such materials must meet a number of technical requirements:

- They must be capable of forming a continuous film on the surfaces to which they are applied.
- They should have good stability to long-term contact with saline water and common road-side pollutants such as petrol, diesel and other organic solvents.
- They should be able to resist the effects of light construction traffic without failure.
- They should be capable of resisting the application of bituminous and asphaltic layers used to create the final running surface.
- They should be capable of resisting the normal thermal and load induced movement of the underlying structure without the formation of holes or other breaks which might allow the passage of water.

There are currently a range of waterproofing products in use around the world and these can be classified into two distinct types.

- Preformed membrane layers are heavy-duty films, typically 1–3 mm thick, manufactured from a variety of polymers which have been reinforced with fibres and other products to improve their mechanical properties and handling characteristics. They are supplied on rolls several metres wide and tens of metres long. These membranes are effectively glued onto the surface of the bridge deck using a compatible primer and adhesive. Great care must be taken to ensure that no sharp objects remain on the deck surface which could puncture the membrane under traffic. Once in place such membranes are normally sufficiently tough to withstand pedestrian movements and limited traffic from rubber wheeled vehicles. In practice the asphaltic or bituminous materials that are to be used in the road surface are applied a soon as possible. A good bond between the membrane and the road material can be generated where the heat from the bitumen melts the topmost surface of the membrane.

Given the finite width of the rolls on which the material is supplied relative to the width of a typical bridge deck there is always a need to lap adjacent layers to ensure continuity of the completed layer. Alternatively special lapping rolls may be glued into place to seal over the joint. Great care and attention is also needed to fit the membrane system so that it fits directly into the main drains. Whilst careful adherence to cleanliness and the manufacturers instruction can produce joints which are no more prone to water penetration than the membrane itself their production is by no means certain and inspection, and testing of the completed membrane is mandatory.

• Liquid applied layers are, typically, two part liquid resins which are sprayed using specialized machinery onto the cleaned deck where they react to form a thin, continuous polymer membrane that is impervious to the movement of liquid water. Such coatings have the advantage over reinforced membranes in that they can be sprayed to form seamless layers that are directly lapped into the drainage system. Once in place however these relatively thin layers are prone to damage from both foot and vehicular traffic, from which they should be protected, until covered by the layers which will make up the road above it. Since 1986 a red sand asphalt carpet has been used in the UK as a protection for such layers when they are overlaid with basecourse mixtures containing large aggregate particles. To be effective it is important that the red sand asphalt carpet is laid and compacted at temperatures high enough to form a dense layer that bonds firmly to the waterproofing.

Regardless of the type of waterproofing membrane employed the need for continuity of the water proofing layer is compromised by the presence of joints provided either to simplify the construction process or to allow movement of the structure due to thermal and moisture gradients. Such expansion joints are always a potential problem for the efficient management of water flows since any sealing system must both fill the joint, preventing water ingress, and also accommodate subsequent expansion and contraction.

16.2.3 Expansion joints

The use of half-joints was common practice in the 1960s through to the 1980s allowing for both ease of construction, e.g. drop-in prestressed beams on RC cantilever supports, and simple articulation to allow for thermal movements. A particular problem affecting such structures is the leakage of water contaminated with de-icing salts through the joints at the end of spans. In principle this should not be a problem since all such joints are designed with either static or moving seals. Unfortunately the development of effective methods of sealing such joints remains problematic with many joints failing after only 5–10 years in service (Figure 16.5).

There are currently three main types of expansion joint in use in the UK (Figure 16.6). In a survey of the condition of such expansion joints in 250 road bridge struc-

Figure 16.5. Failure of expansion joint in road bridge (a), and the associated leakage of water through the unprotected half-joint (b). These photograph were taken less than 6 years after the bridge was first opened to traffic.

tures (Johnson and McAndrew, 1993) only 25% of the joints examined were free of defects and many instances of poor detailing were found. About half the asphaltic plug joints surveyed were found to leaking and tracking was common in heavily trafficked lanes while cracking and debonding was observed more on lightly trafficked areas. Nearly 65% of the reinforced elastomeric joints surveyed were leaking and half had cracked transition strips and many were not flush with the road surface so that ride quality was affected and impact damage occurred on the edges of the joint. The elastomeric joints in metal runners (Figure 16.6b) were the most waterproof with only 40% leaking. It is clear that these findings have significant implications both for the maintenance costs of existing bridges and the designer of new structures. One solution to this problem is the avoidance of joints best summarized as 'the best expansion joint is no expansion joint at all' (Iles, 1994). This has lead to significant developments in both the USA and UK in the design and construction of integral bridges. Naturally, the elimination of expansion joints means that the thermal changes have to be resisted by thrust into the ground at the ends of the bridge.

16.3 Coatings for concrete structures

In the late 1950s designers took to the use of reinforced and prestressed concrete with enthusiasm since it was believed they were essentially maintenance-free materials due to the passivity induced by the high pH of the cementitious materials surrounding the steel. Experience has shown that this belief was incorrect and many

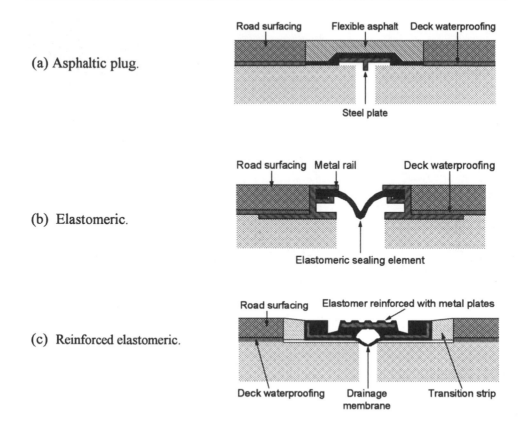

Figure 16.6 General form of the main types of expansion joints found in UK road bridges.

of the concrete structures built in the past 40 years are now showing signs of distress. The causes of this failure of durability are well documented and it is becoming widely accepted that concrete structures are as much in need of coatings as their steel equivalents. Indeed all new concrete structures built in the UK to the Specification for Highway works must be given two coatings of a silane waterproofing treatment. This builds on German experience where concrete road bridges are coated on all exposed surfaces with a variety of surface treatments designed to reduce the ingress of water, chloride ions and carbon dioxide (Anon, 1985a). Any such coating can be expected to have a life of 5–20 years depending on its type and application which clearly implies a continuing maintenance commitment over the life of a typical road bridge.

16.3.1 External protection coatings

External coatings are applied to the surface of concrete structures for a variety of reasons. In considering only those that are concerned with improving durability it is possible to distinguish three main types of treatment based on the method by which they protect the underlying substrate (Leeming and O'Brien, 1987) (Figure 16.7). Thus barrier layers rely on the formation of a substantially pin-hole free surface layer

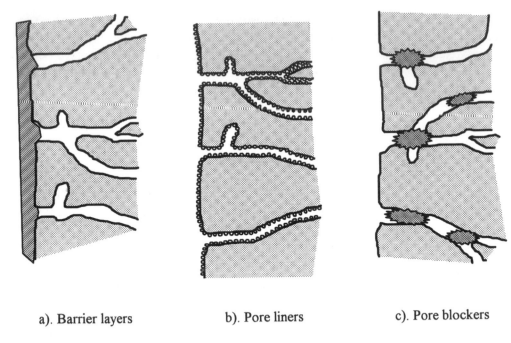

a). Barrier layers   b). Pore liners   c). Pore blockers

Figure 16.7 The three main types of treatment used to protect concrete structures.

which acts to physically exclude liquids and gases from entering the concrete. In contrast pore liners penetrate into the capillary pores of the concrete where they react to produce a water repellent layer that delays the ingress of water into the pore structure under all but driving rain, or standing water, conditions. Pore blockers also penetrate the capillary pores of the concrete but operate by physically blocking the capillary pore system.

Barrier layers
Barrier layers all form of an impervious film over the surface of the concrete and can be divided into three main types; coatings, surface sealers and renders. Coatings are similar to the paint layers used to protect steel surfaces in that they are applied as viscous liquids by spray, brush or roller application and subsequently cure, by solvent loss or chemical reaction, to produce layers with a dry film thickness of between 150 and 500 μm. Typical materials used for coatings include bitumen and man-made polymers such as epoxy resins, polyurethane's, and alkyds. Sealers are similar to coatings in their physical form and method of application but are distinguished by their claim to penetrate into the surface of the concrete. As a consequence sealers are characterized by their good adhesion to the substrate and are often used as primer layers in multi-coat systems. Sealers are typically based on materials such as low viscosity epoxy resins, acrylics and linseed oil. Renders are distinguished by their thickness, which can be many millimetres, and by the fact that they are applied by trowel. They are usually extended by the addition of inert fillers which reduce the

cost per unit volume and increase the resistance to weathering. Due both to the viscosity of such materials and their method of application their adhesion to the substrate is usually inferior to coatings and sealers. Most modern render are based on polymer-modified cement mortars.

The most common form of barrier layer used for protecting concrete bridges are coatings. For such surface coatings to be effective good workmanship is essential and the concrete surface must be carefully prepared, by filling and levelling, before the coating is applied. A minimum of two coats will normally be required to ensure uniform coverage and the correct film thickness. The use of multi-layer coating systems is generally preferred over one-coat, high-build coatings which are prone to defects and pin-holes when poorly applied. Multi-layer systems are inherently resistant to such problems, since the chances that the defects in the various layers match-up and expose the underlying concrete are small. Regardless of the type, or quality, of coating specified it is important to remember the long-term exposure to weathering produces a stiffening of the material and reduction in elongation at break (Le Page, 1996). This implies a need to maintain and, where necessary, replace such coatings over the life of a structure.

Coatings are commonly used for their ability not only to prevent the ingress of liquid water but also to restrict the passage of carbon dioxide. Such anti-carbonation coatings are carefully formulated to allow some permeability to water vapour to avoid a build-up of vapour pressure behind the coating and allow long-term drying of the structure. This is assisted by the fact that the molecular size of carbon dioxide is larger than that of water. Such coatings can be characterized in terms of their diffusion resistance, u, which is a dimensionless parameter indicating how many more times impermeable a coating is than air under equal conditions.

Preventing carbonation and water/chloride ingress is clearly beneficial in reducing the risk of reinforcement corrosion (Robery, 1988). Keeping the concrete dry is also beneficial in restricting damage due to both freeze-thaw cycling and alkali silica reaction. However such protection relies on the coating remaining unbroken. As a result the ability of a coating to bridge over a crack as it forms in the underlying substrate is one of several parameters influencing coating selection (Harwood, 1990). Some multi-layer systems are formulated to ensure that the primer coat is not too adherent since this helps promote the debonding necessary on either side of a newly formed crack to form a sufficient gauge length of material to allow crack bridging to occur. Having bridged a newly formed crack the coating must also be capable of withstanding the repeated cyclic opening and closing of an active crack. The ability of a surface coating to both bridge a crack which forms in the substrate and accommodate subsequent movement of the crack is a complex function of coating thickness, material type, bond strength and temperature (Le Page, 1996). Figure 16.8 shows the influence of temperature on the behaviour of a 300 μm thick coating subject to an initial crack opening of 0.3 mm which subsequently varies in width between 0.3 and 0.6 mm. It is clear that there is a critical temperature above which the coating accommodates many thousands of crack opening cycles prior to failure but below which it fails to either accommodate or bridge the crack. It is note

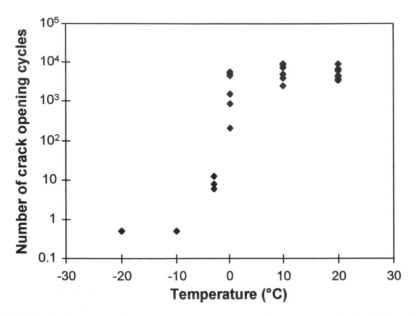

Figure 16.8 The influence of temperature on the crack accommodation behaviour of a 300 μm thick coating subject to an initial crack opening of 0.3 mm and which subsequently varies in width between 0.3 mm and 0.6 mm (Le Page, 1996).

that coatings made-up of several individual layers are more tolerant to cracks in the underlying concrete since a defect in one layer will not necessarily initiate failure of the adjacent layers (Le Page, 1996).

Pore liners

Pore liners are low viscosity fluids that are able to penetrate into the concrete where they are adsorbed onto the surfaces of the capillary pores and react to form a water repellent layer. Typical penetration depths on normal concrete under site conditions are in the range 2–10 mm although penetration depths exceeding 50 mm are attainable under laboratory conditions. The materials used as pore liners are generally based on monomeric alkylalkoxysilane and oligomeric alkylalkoxysiloxane, commonly known as silane and siloxane, respectively. Certain silicone compounds can also be used to produce the same effect. However silicones are generally less effective than silane treatments which, because of their small molecular size, can penetrate the concrete more easily. Silanes become reactive in the presence of moisture, the speed of reaction being governed by the surrounding pH. In normal alkaline concrete silanes react with the pore lining quite rapidly (McGill and Humpage, 1990). However the silane molecule is very volatile and can evaporate before it has time to react with the surface. This is less of a problem with the oligomeric siloxanes which are less volatile and so have longer time to react with the surface. Work on stone faces has demonstrated that the efficiency of a hydrophobic surface treatment is greatly increased by curing at elevated temperature and humidity (Bohris, 1996).

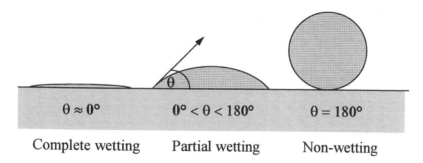

Figure 16.9 Wetting characteristics of a surface as a function of the contact angle, θ, between the liquid and the solid.

All pore liners work by altering the contact angle, θ, between the concrete surface and liquid water (Figure 16.9). Under normal conditions water readily wets the concrete surface, $θ ≈ 0°$, and capillary forces promote the ingress of water into the pore system within the cement paste. By treating the surface produce a hydrophobic lining on the surface the concrete effectively becomes unwettable, $θ ≈ 90–180°$. As a consequence any water falling on vertical and inclined surfaces simply runs off. This significantly delays the ingress of water into a structure under all but standing water conditions. Unlike pore blockers these treatments allow the passage of water vapour promoting long-term drying of the concrete reducing the likelihood of corrosion of any embedded steel (Darby *et al.*, 1996). Another advantage of these materials is that they are clear and colourless and so do not prevent subsequent visual inspection of the treated structure for cracks and other defects. Against this must be set the disadvantage that once applied and cured these treatments are effectively invisible and can only be detected by their impact on the wetting behaviour of the surface. For this reason ISAT testing of the surface before and after treatment is often used to ensure that the application process has been adequate. Since the ISAT method is effectively non-destructive it can be repeated at intervals to determine the effectiveness of the treatment over time and identify any need for reapplication.

Pore blockers
Pore blockers are a family of products, based originally on calcium and sodium silicates, which are claimed to penetrate 1–3 mm into the surface of the concrete and form reaction products which block pores within the concrete and densify the matrix. Due to the partly crystalline nature of the products that form these have become known as 'crystal growth' materials. There is little evidence that such treatments are effective in reducing either carbonation or chloride ion ingress into the bulk of the concrete although more recently introduced products based on lithium polysilicates appear to offer improved performance. Other pore blocker treatments make use of low viscosity resinous materials such as epoxy resins and acrylics which penetrate into the pores and harden *in situ*.

16.3.2 Internal protection coatings

Recognition of the fact that concrete cover sometimes fails to provide the expected protection to embedded reinforcement has given rise to the development of coated rebars (Walker, 1989). The object is to ensure that should chloride ions reach the depth of the steel then there is another layer of defence stopping the onset of corrosion and the subsequent development of cracking and spalling of the concrete cover. However a designer considering the use of coated reinforcement must weigh the perceived benefits against the extra cost of coated steel products which are typically twice that of equivalent uncoated steel bar.

In the USA many concrete road bridges have been manufactured using fusion-bonded epoxy coated reinforcement with a coating thickness of between 1–2 mm (Hartley, 1994). Such an impervious coating relies on being a pin-hole free barrier if it is to provide adequate protection. There has been some concern that during the construction process such coatings can become damaged exposing small areas of steel. This is potentially disastrous since such defects could become the sites of active anodes with the remainder of the coated bar becoming cathodic. This unfavourable anode/cathode area can lead to significant localized pitting. An assessment of structures containing coated reinforcement in the Florida Keys area – characterized by a marine environment – has revealed considerable damage to the coated bars after periods of only 10–15 years (Hartley, 1994). It is not yet clear whether this reflects a fundamental flaw in this type of corrosion protection system or that early coating methods did not provide sufficient adhesion of the coating to the steel.

In Japan there has been a considerable interest in the use of galvanized steel reinforcement with a 60–100 μm thick layer of zinc. Such coatings are equally likely to be subject to damage during the construction process but because the zinc is anodic relative to the steel any steel exposed by cracks in the coating will be cathodically protected by preferential dissolution of the zinc. Again long-term evidence regarding the effectiveness of such systems is limited although there is evidence that galvanized reinforcement can increase the time to the initiation of corrosion of rebars in chloride infested concrete (Treadaway, 1980). However as the concentration of chloride ion increases the rate of loss of the zinc layer increases appreciably reducing its effective lifetime (Treadaway and Davies, 1989). In addition there remains the possibility of an interaction between the zinc and the steel to form brittle compounds, which reduce the ability of the reinforcement to deform in a ductile manner (Comite Euro-International Du Beton, 1995).

16.4 Protection systems for reinforced concrete structures

While coating systems represent a useful method for protecting new and existing concrete bridge structures they should not be viewed in isolation. It must be remembered that from a theoretical viewpoint reinforced concrete is an almost ideal combination of materials. The steel carries tensile forces while the concrete both resists compressive loads and protects the embedded reinforcement from corrosion. Thus reinforced concrete bridges should be designed, and constructed, to make the best use of this inherent durability and augment it with other methods that, taken together, form a coherent protection system. This includes proper specification of the concrete, adequate cover to the reinforcement, the incorporation of electrochemical techniques and the use of chemical treatments such as corrosion inhibitors to protect the steel.

16.4.1 Concrete quality

The provision of good quality concrete starts with the appropriate choice of constituent materials including the cement type, aggregate source (and grading) and use of admixtures. These must be combined in an appropriate mix design that ensures a workable concrete with a cement content typically in excess of 300–350 kg/m^3 and a free water/cement ratio less than 0.45. Suitable limit values can be found in appropriate design codes and should reflect the likely exposure conditions of the structure being constructed (European Standard DD ENV 206:1992). Once trial mixes have confirmed the suitability of the mix design it is important to enforce high standards of batching and mixing. All of this effort will be in vain if the mix is not properly compacted and subsequently cured for an adequate time at the correct relative humidity and temperature.

In trying to optimize the physical and mechanical properties of the fresh and hardened concrete it is possible to use a wide range of chemical admixtures such as air-entraining agents (to improve frost resistance), set retarders and accelerators (to control the onset of hardening), and water reducing agents (to aid workability). Under most circumstances such additives can help to ensure that the fresh concrete delivered to the formwork is of a consistent workability and stability under the prevailing site conditions. However they should not be used to compensate for fundamental inadequacies in the constituent materials or mix proportions. It is also important to assess the impact of temperature, and time, on the properties of the admixed concrete. This helps to avoid creating a mix whose fresh properties change rapidly over the time frame in which construction crew are trying to place the concrete.

In designing reinforced, and prestressed, concrete bridge structures it remains common to specify the concrete in terms of its 28-day cube strength. However in many instances this is not the correct design parameter since it is the permeability and composition of the hardened concrete which tends to control the long-term durability. About 80% of all concrete used in construction in the UK is made using Ordinary Portland Cement (BS12 42.5) although rapid-hardening, low-heat and sulphate-resisting cements are used for particular situations and environments. In addition a number of blended cements are available that can help to improve the long-term durability of the hardened concrete. It may be noted that any gains in durability are often at the expense of short to medium-term strength gain, due to the lower reactivity of the components blended with the Portland cement (Dewar, 1988). This has implications both for the speed of construction, by delaying the removal of shuttering, and the type, and length, of curing. As a consequence it is vital that the designer of any concrete bridge structure appreciates the specific properties and performance of the available cement blends (Jones, 1994).

In general blends of Portland cement with ground, granulated, blastfurnace slag, GGBS, show excellent resistance to the ingress of chloride ions but have extremely poor resistance to carbonation. Such blended cements are commonly used for bridges in marine environments, e.g. the Humber bridge and the more recent Queen Elizabeth II bridge at Dartford, where their resistance to sulphate attack is an advantage (Deason *et al.*, 1993). However they are considered unsuitable where the primary cause of deterioration is likely to be carbonation induced reinforcement corrosion. Properly cured blends of Portland cement with pulverized fuel ash, PFA, resist chloride ion ingress quite well whilst retaining reasonable resistance to carbonation. Portland cement concretes on the other hand have the best overall resistance to carbonation attack but show poor resistance to chloride and sulphate bearing environments.

In the design of durable reinforced concrete it is important not to forget that there are a number of possible choices regarding the type of reinforcement. In recent years there has been considerable interest in the development of stainless steel reinforcement and tests have shown that this provides excellent resistance to chloride ion induced corrosion (Treadaway and Davies, 1988; Treadaway *et al.*, 1989). However this good long-term performance comes at a high initial cost since stainless steel reinforcement is typically an order of magnitude more expensive than mild steel ribbed bar of equivalent diameter. The use of galvanized steel reinforcement has been found to produce good resistance to corrosion in marine exposure conditions (Treadaway *et al.*, 1980). An alternative to the use of galvanized steel is epoxy coated steel reinforcement (Hartley, 1994). However there are concerns about the long term durability of such coated bar especially where chloride ions are present in the concrete and can reach the steel through breaks in the epoxy layer. More recently both non-ferrous and non-metallic reinforcement materials have become available but these are still the subject of much research and currently have little proven advantage over the use of steel. Indeed many of the composite materials being promoted for study are themselves inherently unstable with respect to their interactions with wet, alkaline environments such as are found in, and around, concrete bridge structures (Clarke, 1993).

16.4.2 Role of cover concrete

A major cause of the durability problems encountered in concrete structures has been inadequate cover to the reinforcement which reduces resistance to both carbonation and chloride ion induced corrosion of the reinforcement. Inadequate cover detracts from the effectiveness of specifying low water/cement ratio mixes with pozzolanic, or other additives, and incorporating careful long-term curing. A survey in Australia of a number of RC structures (Griffiths, 1987) found that at a large proportion of the sites where corrosion induced damage had occurred the maximum cover was less than 10 mm, the average cover being around 5 mm. An analysis of similar data obtained from both bridges and buildings (Marosszeky and Chew, 1990) suggests that average depths of cover are much higher in bridge structures (Table 16.1). This has been attributed to the better control of the construction process usually exercised on bridge projects. In considering the consequences of inadequate cover Beeby (1993) has argued that it is fundamentally incorrect to deal with the problem by simply specifying larger covers, or higher quality concrete, as has occurred in recent revisions

Table 16.1 Cumulative distributions of cover in buildings and bridges (Beeby, 1993)

Measured cover	Buildings	Bridges
Cover > nominal	38%	51%
Cover > 0.9 nominal	52%	87%
Cover > 0.8 × nominal	62%	93%
Cover > 0.7 × nominal	74%	95%
Cover > 0.6 × nominal	82%	96%

of design codes. Instead it is better to ensure that all those involved in the placing of concrete are aware of the problems of low cover and that they check covers conscientiously during the construction phase. Thus while design codes may be revised to provide structures which are potentially more durable such gains can only be achieved if the standard of workmanship is high and good quality control procedures are in place to ensure the reliability of the construction process.

16.4.3 Electrochemical methods – cathodic protection

The use of cathodic protection for both new and old reinforced concrete structures is becoming increasingly common since it offers a secure means of stopping corrosion of the embedded steel (Hayward, 1997). Cathodic protection is based on the principle of lowering the normal rest potential of a metal (Figure 16.10 – point A), to bring into the immune region of the Pourbaix diagram (Figure 16.10 – point B), where the metallic state is the stable one. In this region the metal is effectively held in a state where anodic dissolution is impossible since only cathodic reactions are allowed to occur on the surface of the protected metal. There is an interesting side effect of doing this since the cathodic reaction results in the production of hydroxyl, OH^-, ions at the metal surface. This has the effect of raising the local pH (Figure 16.10 – point C), and as a consequence the surface of the metal becomes protected by the build-up of basic salts. In the event of failure of the cathodic protection system this provides additional protection by ensuring the steel surface remains in the passive condition (Figure 16.10 – point D).

There are two ways to cathodically protect steel reinforcement within a concrete structure.

Galvanic coupling (sacrificial protection)

By electrically connecting the reinforcement to a metal higher in the electro-motive force series the steel will be protected by preferential corrosion of the sacrificial anode. Galvanized steel reinforcement works in this way and typically has 50–150 μm of zinc coating on the surface of the steel (Treadaway *et al.*, 1980). The sacrificial corrosion of the layer of zinc lowers the potential of the steel so that any exposed parts of steel do not corrode. The system is not as wasteful as it first appears as the zinc soon acquires a coating of basic salts. In addition the cover concrete acts as a buffer zone reducing the supply of oxygen, water and chloride ions to a level where the thin zinc

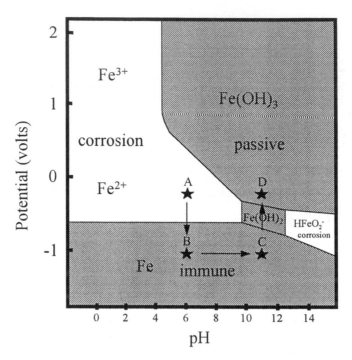

Figure 16.10 Pourbaix diagram for iron in oxygenated water showing the process of cathodic protection.

coating can last for many years (Treadaway *et al.*, 1980, 1989; Treadaway and Davies, 1988). The problem with this approach lies in that once the galvanized layer has been consumed then normal corrosion of the steel can commence albeit the onset will be much delayed. Some concern has been expressed that the hot-dipping process used to cover the steel bar with the zinc can lead to the formation of brittle inter-metallic compounds within the interfacial layer between the steel and zinc leading to embrittlement (Comite Euro-International Du Beton, 1995).

Impressed current

The cathodic protection of steel reinforcement by means of an impressed current supplied from an external power supply involves connecting the negative terminal of the power supply to the reinforcement cage tank, and the positive to an inert, titanium anode mounted on the surface of the concrete (Figure 16.11). The electric leads connecting the steel reinforcement and titanium anode are carefully insulated to prevent current leakage. The current passes to the steel which becomes the site of cathodic reduction reactions with the simultaneous formation of hydroxyl ions and a rapid rise in the local pH.

16.4.4 Corrosion inhibitors

One method of increasing the tolerance of reinforced concrete to carbonation and the presence of chloride ions is the application of corrosion inhibitors. These are sub-

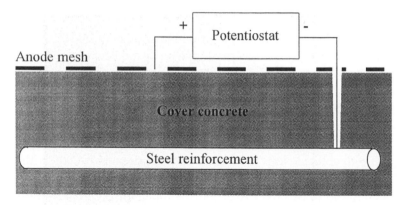

Figure 16.11 Cathodic protection of steel reinforcement in concrete.

stances that when present on the surface of the steel decrease the rate of corrosion without significantly changing the concentration of other corrosion agents (International Standards Organization, 1999). This differentiates them from coatings and other materials that act to restrict the supply of water; oxygen and chloride ions around the steel bars. There are a number of commercially available corrosion inhibitors and these can be categorized into three main groups:

- inorganic inhibitors, e.g. nitrites
- organic inhibitors, e.g. amines, and
- vapour phase inhibitors, e.g. amino alcohol.

All these materials act to interfere with one, or more, stages of the corrosion process and can, under ideal conditions reduce the corrosion rate by at least one order of magnitude. Of course a reduction of only 50% in the corrosion rate might provide a cost effective increase in the life and reliability of a particular structure.

In attempting to slow the corrosion rate inhibitors can either suppress anodic dissolution of the steel, anodic inhibitors, or suppress the various electron consuming reactions which can occur, cathodic inhibitors. One problem inherent in the use of anodic inhibitors is that it is essential they are present in sufficient concentration to close down all the anode sites. Failure to achieve this can result in a few anodes remaining active, the subsequent anode/cathode ratio being such as to promote aggressive, and highly localized pitting of the steel. In contrast cathodic inhibitors are essentially safe at all concentrations since they always act to increase the anode/cathode ratio reducing the total available current for anodic dissolution. Some 'mixed' mode, ambiodic, inhibitors are capable of suppressing the reactions at both anode and cathode sites simultaneously.

When used to protect steel in concrete corrosion inhibitors can be either added as admixtures to the fresh concrete, or applied to the surface of the hardened concrete (Treadaway and Russell, 1968).

Admixed inhibitors

Admixing allows a uniform distribution of inhibitor throughout the concrete and in high performance concretes the low permeability of the cover concrete prevents the inhibitor from being lost. The main problem to overcome is preventing any adverse effect on the fresh and hardened properties of the concrete. There are several types of inhibitor specifically used as admixtures for new concrete the most common being based on calcium nitrite. It is now well established that nitrites are anodic inhibitors which interfere with reactions at the steel/concrete interface and are consumed as they compete with chloride ions to block the anodic sites. However, with a sufficient dosage, they should be able to maintain a constant availability at the steel surface and have been found to be effective as long as the chloride/nitrite ratio stays below about 1.8 (Virmani and Clemena, 1998). A dosage rate of 10–30 litres/m^3 is generally specified, depending on the expected maximum chloride level at the rebar. This is usually the benchmark against which other inhibitors are tested.

Figure 16.12(a) illustrates the corrosion behaviour of steel bars embedded at 10, 25 and 40 mm in OPC concrete (w/c= 0.60) and subject to weekly wet/dry ponding with a 5% solution of sodium chloride solution. The bars at 10 mm start to corrode after only 4–6 cycles, the bars at 25 and 40 mm started to corrode after 12–16 cycles and 54–58 cycles, respectively. This clearly demonstrates the role of cover depth as a means of protecting steel from the aggressive action of ingressing chloride ions. This behaviour can be compared with that of similar bars in an equivalent concrete containing 4% (by weight of cement) admixed calcium nitrite (Figure 16.12b). It can be seen that the presence of the inhibitor increases the time to initiation of corrosion for the bars at all the depths tested. It is of note that linear polarization measurements showed that once initiated the subsequent rates of corrosion of the bars in the concrete containing the inhibitor were less than those in plain OPC concrete (Mulheron and Nwaubani, 1999).

Surface applied inhibitors

The main advantage of surface applied corrosion inhibitors is that they can be added after construction is complete. Thus if a structure is showing signs of reinforcement corrosion then the inhibitor can be applied as a surface treatment which subsequently penetrates the concrete, repassivating the reinforcement and stopping further corrosion. This requires the inhibitor molecule to be sufficiently mobile to move through the pore structure of the cover concrete and for this reason they are sometimes referred to as penetrating corrosion inhibitors. It is commonly assumed that such materials rely on capillary action. However vapour phase inhibitors, which are characterized by their high vapour pressure, can also migrate as vapours provided the pores are sufficiently empty to allow vapour transport mechanisms to operate.

Amino alcohols are ambiodic, forming a film on the steel surface that blocks both anodic and cathodic reactions. They can be applied as a coating on the surface of the concrete, as a 'plug' of material in a hole, or mixed into repairs. Under laboratory conditions this inhibitor is able to migrate through the concrete cover of a relatively porous, poor quality concrete (Mulheron and Nwaubani, 1999). However it

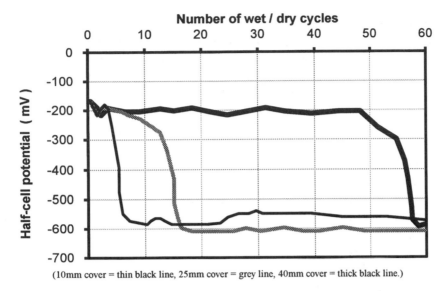

Figure 16.12(a) Change in half-cell potential with number of wet/dry cycles for steel bars embedded in OPC concrete (free w/c = 0.60).

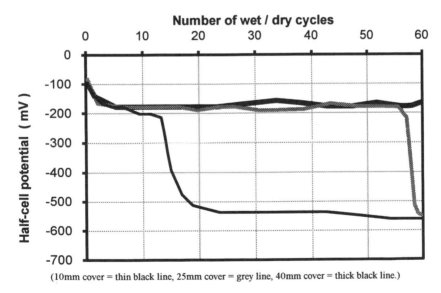

Figure 16.12(b) Change in half-cell potential with number of wet/dry cycles for steel bars embedded in OPC concrete containing 4% admixed calcium nitrite (free w/c = 0.60).

is uncertain whether migratory inhibitors can penetrate into concrete structures with low permeability. In addition there is concern that if such molecules are able to move into the cover concrete that over long time periods they may subsequently move out and cease to protect the reinforcement.

One issue with corrosion inhibitors is to determine how effective they are. Whilst the results of laboratory tests suggest that both admixed and surface applied inhibitors can significantly reduce corrosion rates in the presence of chloride ions (Treadaway and Russell, 1968; Virmani and Clemena, 1998; Berke, 1999; Mulheron and Nwaubani, 1999). The results of field trials are more variable with often little, or no, reduction in the long term rate of corrosion being reported (Prowell *et al.*, 1993; Broomfield, 1997; Sohanghpurwalla *et al.*; Sprinkel and Ozyildirim, 1998). In part this reflects the practical problems of ensuring that the inhibitor is able to reach the surface of the steel. However in some trials (Broomfield, 1997) the level of chloride ion present was greater than 1% by weight of cement which is beyond the limit at which many inhibitors can reasonably be expected to suppress corrosion (Mulheron and Nwaubani, 1999). In addition some of the structures which have been treated were already showing signs of reinforcement corrosion with associated cracking and spalling of the cover. Such structures are beyond the point at which a non-invasive repair using a migratory corrosion inhibitor would be considered practical by their manufacturers. Indeed the balance of evidence suggests that surface applied corrosion inhibitors are most effective when applied to a reasonably permeable cover concrete which is either carbonated or has a chloride ion content below 1% by weight of cement. Where reinforcement corrosion is well established and spalling of the concrete cover has taken place then the application of surface applied corrosion inhibitors is of little value and invasive repair methods must be employed. However in such situations corrosion inhibitors, such as calcium nitrite, can be admixed with the repair mortar or overlays with the intention that the inhibitor will protect the steel in adjacent areas.

16.5 Coatings for steel structures

Each year corrosion destroys the equivalent of 20% of the annual world production of ferrous metals (Porter, 1979). This stems from the fundamental instability of iron and its alloys with respect to the iron oxides from which they are extracted. Under dry conditions the direct oxidation of iron by gaseous oxygen is a slow process at room temperature as the rate becomes limited by the oxide film that forms on the surface of the metal. In the presence of water, however, a new mechanism becomes available in which the anodic dissolution of iron is driven by the cathodic reduction of oxygen and water. That the hydrated iron oxide that forms is highly expansive, porous and poorly adherent ensures continued corrosion of the metal surface. The rate of such aqueous corrosion is many times that of direct oxidation and occurs in the temperature range commonly encountered by the majority of steel structures. As a consequence steel structures have an inherent, and absolute, need to be protected from aqueous corrosion.

While changes in the composition of a steel can increase its resistance to aqueous corrosion, see Section 16.6, the most common form of protection is the application, and maintenance, of a suitable coating. This approach is not new. Nearly 2000 years ago the problem of 'spoiled iron' was described by Pliny (Plinius, 1968) who noted that the rusting of iron could be prevented by the application of a coating of vegetable pitch, mixed with gypsum and white lead. This recipe works as well today as it did then but has been superseded by a range of coating systems that are both easier to apply and more effective.

Coatings for steel structures can be clearly split into two main types.

- Barrier layers act to exclude water and oxygen from the surface of the steel, e.g. bituminous membranes, greases, varnishes and paints. Such layers depend on their continuity for effective protection and where the coating becomes cracked, or broken, then any protection is compromised. Even in the absence of cracks many barrier layers provide only imperfect exclusion of oxygen and water over long exposure periods. Thus, at best, they may be regarded as introducing a large ionic resistance into the corrosion cell. For example, the resistivity of paint films is some 10 000 times that of water and so the corrosion rate is reduced by this amount (Fontana, 1986).
- Sacrificial layers act to exclude water and oxygen but also provide electrochemical protection to the underlying substrate. The classic example of this type of coating is galvanized steel in which a thin coating of zinc is applied to a steel substrate. Such coatings have a major advantage over inert barrier layers in that where the coating is broken, or scratched, the exposed steel is cathodically protected by the preferential corrosion of the more anodic zinc.

In reality the protection of steel bridges employs multi-layer coating systems that incorporate aspects of both barrier and sacrificial methods. Such duplex systems are preferred because the combination can produce a synergistic effect where the effective life of the coating system, measured in terms of the time to first maintenance, is longer than might be achieved with either a wholly barrier system or sacrificial layer.

16.5.1 Selection of coatings

From the end of World War II through to the mid 1970s the majority of steel bridges were protected by multi-layer alkyd paint systems applied directly over millscale adherent to the steel. These paints contained lead and chromate additives in the primer layers to help prolong the life of the coating. The use of metallic zinc and aluminium coatings was also employed in this period due to their superior time to first maintenance. However such metal coatings found limited application due to the relatively slow rate of coating application achievable with the existing technology, higher initial costs and difficulties associated with the *in situ* maintenance of factory applied metal layers.

Today steel bridges are protected by a wide range of commercial coating systems incorporating multiple layers of polymer-based paints, metal layers and powders, and inhibitive admixtures. Typically as much of the coating system as possible is applied under factory controlled conditions with one, or two, top-coats being applied on site to seal any defects created during handling and provide a uniform, and aesthetically pleasing, appearance. The decision to employ factory controlled conditions for the application of a coating carries a number of potential advantages over site application methods. Firstly it enables good control of the initial surface preparation, which is a crucial factor in ensuring adequate bond between the steel and the primer layers. Secondly it ensures that the selected coating can be applied under controlled conditions, in terms of temperature and humidity, and to high tolerances of workmanship and thickness. Despite these advantages it must be remembered that a major requirement of any factory applied coating system is that it must not be susceptible to damage during transit and handling. Where damage does occur the coating system must be both tolerant of such defects and be repairable, under site conditions, to some close approximation of its original condition.

Given the diversity of available coating systems, the conditions under which they applied, environmental issues, and the need to minimize maintenance costs, the choice of a suitable coating is never straight forward. As a consequence the proper selection of a surface coating requires an appreciation of the many factors affecting a coatings performance and the methods by which it can be maintained throughout the life of the structure. Among the purely technological factors affecting coating selection the prevailing environment is usually the most important one. Thus when the corrosion rate is uniform, and below a certain threshold level, no coating is required. Conversely where the environment is too severe alternate materials should be considered instead of relying on a surface coating.

Environmental concerns and health and safety requirements are becoming increasingly important in the selection process. As a consequence the use of lead-based additives in coatings is now banned in many countries. Similarly limits on the allowable levels of volatile organic compounds (VOCs) are steadily reducing with many of the traditional solvent-based paint products being replaced by water-based equivalents. Thus coating selection criteria are not static but are evolving to meet the more stringent regulatory controls whilst at the same time attempting to ensure adequate, and cost effective, long-term performance. This presents problems for those concerned with the specification of coatings for steel bridges since although water-based paints may meet the environmental requirements their long-term durability remains unproven.

In selecting a suitable coating system for the particular environment it must be remembered that in most cases more money will be spent on maintenance of the coating than on its original application (Chandler, 1979). To this end it can be useful to employ layers that breakdown in such a way that it is obvious before it becomes complete, e.g. coloured primer coats on sprayed metal coatings. The early detection of deterioration is only a first step since the outer layer must also be capable of being repaired to a satisfactory standard under the prevailing site conditions. Of course any maintenance to the coating system will be ineffective if the fundamental cause of deterioration, such as uncontrolled water leakage, is not identified and corrected. However maintenance is subject to the availability of resources and there can be no guarantee that maintenance will be carried out when the need first becomes apparent. If maintenance is delayed for too long, and corrosion takes a firm hold, then any subsequent repair may prove ineffective in stopping subsequent corrosion.

Clearly any maintenance operations that are carried out must not carry any undue risks to either the maintenance operators or the local environment. This is particularly important when dealing with coatings that contain heavy metals, such as lead and chromate. These can be hazardous to human health in even small quantities if ingested or inhaled. As a consequence it is vital to take measures to protect all workers exposed to such hazards (Hoffner, 1995). The risk to personnel can be reduced by the use of over-coating techniques (Hopwood, 1995). These painting operations require only the partial removal of the existing coating and subsequent application of two or three coats of new material over a mixture of clean steel, corroded surfaces and existing coating. However such coatings are prone to early disbondment due to incompatibility with the original layers and are not a viable alternative to the full removal and replacement of a coating system that has undergone significant deterioration.

16.5.2 Paints

Paints are the most complex of all anti-corrosion coatings and consist of an organic polymer film, or binder, containing a filler or pigment dispersed throughout its thickness. Most polymer binders are initially dissolved in an organic solvent which prevents polymerization, aids spreading of the paint and then evaporates as the paint 'dries'. Due to concern about the environmental and health risks associated with the uncontrolled release of VOCs manufacturers have introduced a range of water-based paint systems. These are still relatively unproven in terms of their long-term durability under bridge exposure conditions. An alternative to these new systems are the more established solvent-free paints that are catalysed just prior to application and polymerize on the surface of the steel at a rate depending on temperature and catalyst concentration.

The pigments used in commercial paint systems for bridge structures are essentially of three kinds; inert, inhibitive and sacrificial. Inert fillers simply thicken the paint and serve only to lengthen the diffusion path of water to the metal surface. Inhibitive pigments act as a controlled source of corrosion inhibitor, see Section 16.4 and are used in primer coats in contact with the metal. Sacrificial fillers such as metallic zinc corrode in preference to the steel substrate and so provide protection from corrosion.

Experience suggests that poor paint performance usually stems from either poor preparation of the substrate surface or poor application of the paint rather than from the type of paint used. However when both the preparation and application of a paint coating are satisfactory the type and quality of the paint itself strongly influences the long-term rate of deterioration.

Surface preparation

Surface preparation is a vital factor in coating durability and involves the production of a clean, rough surface. The best method involves grit or sand-blasting the steel substrate to remove any residual millscale and leave a bright metal surface. Pickling and other chemical processes can also be effective when used under factory conditions. Prior to coating the surface should be firm and physically free from any layers that will prevent the paint film wetting the surface and adhering. The surface must also be chemically clean and free from corrosion producing matter such as ferrous salts and weld deposits. Whilst visual inspection can ensure the absence of major contamination, the presence of residual grease layers, and salt contamination, are not easily detected. Fortunately there are a range of methods available for detecting the presence of undesirable contamination (Bayliss, 1979).

Assuming that the surface is physically and chemically clean the most important parameter is that the surface must be of the correct roughness. Obviously the surface must not be too rough in relation to the thickness of the coating or it will encourage wide variations in the thickness of the final coating and help encourage the formation of holes. Less obviously the surface needs to be of a sufficient roughness to prevent loss of adhesion by peeling. Modern paint films adhere remarkably well to smooth surfaces but their cohesive strength is such that leverage under the film can easily strip a coating if there is no mechanical key to resist disbondment.

Coating application

A major problem with all paint systems is their susceptibility to poor application technique. This rarely produces an obvious failure of the coating but rather reduces the long-term durability of the coating system. To help overcome such problems and ensure complete coverage of the substrate and freedom from pores, or 'holidays', in the coating it is best to apply several layers of paint, by either brush, roller or spray, taking care that successive layers bond to one another. The thinning of high build paints, to make them easier to apply is a common mistake as is the inadequate stirring of heavily pigmented materials and the incorrect mixing of two component systems. An important outcome of correct application is the achievement of the specified coating thickness. This is an important factor in determining the life of a coating. For any given paint and environment the time to first maintenance generally increases with coating thickness and decreases with increasing severity of exposure. From a practical perspective measurement of the wet film thickness can enable the operator to predict the final dry film thickness at a stage when it can still be corrected. This requires a well established relationship between wet and dry film thickness.

Coating types

A wide range of coating types have been used to protect steel bridges over the past 50 years. Early alkyd paint systems have achieved times to first maintenance ranging from 2 to 15 years. This variable performance reflects the wide range of exposure conditions to which they have been subjected and the inherent limitation of such paints under damp conditions. In terms of overall performance alkyd paint systems have given similar performance to systems incorporating micaceous iron oxide. Traditionally these consisted of a lead based primer with two coats of micaceous iron oxide applied over a grit-blasted surface. More recently the lead has been replaced by a zinc phosphate primer. Chlorinated rubber paints have a good reputation for use in severe environments where they perform better, thickness for thickness, than alkyd based coatings with times to first maintenance of 5–10 years. Despite this, as a class, chlorinated rubber paints do not perform substantially better than alkyd paints under moderate exposure conditions and rarely exceed 10–15 years before requiring maintenance.

Two part epoxy resin paint systems have shown variable performance over the past 25 years with early formulations being sensitive to application conditions and overcoating time. Poor performance also resulted from the tendency of the epoxy resin top-coat to degrade with exposure to ultraviolet radiation. In many modern systems epoxy layers are protected by a polyurethane top-coat. More recently epoxy mastic bridge coatings have gained popularity in the USA (Federal Highway Administration, 1995a). These two component coatings are based on one of several epoxy resins cured using a polyamide or amine curing agent and are often heavily pigmented with aluminium flakes or powder. As a class they offer low solvent content, short recoat times, and low cost. They are also claimed to provide high build and be tolerant of surface defects and so require less surface preparation prior to application however the high viscosity of some of these products can lead to inconsistent paint thickness. A major problem with such coatings is their poor resistance to under-film corrosion at defects. Under marine exposures epoxy mastic coatings have shown significant failure after only 18 months in the presence of defects (Federal Highway Administration, 1991a). Even after careful application to near-white blast-cleaned steel such coatings showed significant failures after only 3–5 years (Federal Highway Administration, 1991b).

Zinc-rich paints for steel bridges are characterized by the high concentration, typically greater than 80% by weight, of metallic zinc pigment they contain mixed into either an organic polymer binder or inorganic silicate binder. Both types are used as primer coats applied directly onto the surface of the cleaned steel to provide sacrificial protection at any breaks in the upper layers. A typical three-coat system consists of a zinc primer covered with an epoxy intermediate layer and a polyurethane top-coat. Such coatings provide excellent long-term corrosion control even in salt-rich environments. The main difference between such zinc-rich coating systems and systems based on barrier protection is their inherent resistance to disbondment and under-film corrosion. The use of water-based inorganic zinc coatings has gained popularity in recent years as they contain no VOCs. However experience with these coatings has been mixed and they should only be used in areas where low air flow and high humidity during application and curing can be avoided (van Eijnsbergen, 1976).

16.5.3 Metal coatings
Metallic coatings for steel can provide either direct exclusion of oxygen and water or sacrificial protection. Direct protection of steel by layers of chromium and nickel rely on a continuous layer covering the substrate. However where the coating is damaged corrosion of the steel occurs and spreads outwards lifting and destroying the film as it goes. As a consequence such metals are never used in the protection of engineering structures. In contrast coatings of zinc and aluminium have been used for the protection of a large number of bridge structures (Porter, 1979). These sacrificial coatings do not have to be continuous since at breaks in the layer the zinc, or

aluminium, will corrode preferentially. There is little doubt that properly applied metal coatings provide longer lives to first maintenance than conventional paint systems, especially in aggressive environments. In practice the sacrificial layers are themselves protected, or sealed, with a paint layer which helps to extend the effective life of the coating (van Eijnsbergen, 1976). This in turn increases the time to first maintenance, which can exceed 20–25 years, and so enables costs to be reduced over the life of the structure. Given the increasing costs of labour for painting operations, and the high indirect costs associated with interruption of service, the long-life of sacrificial metal coatings has made them increasingly attractive to bridge engineers. One important feature of sacrificial metal coatings is their inherent resistance to damage introduced during transport, handling and in-service. This is in marked contrast to many paint systems where such damage can initiate the formation of corrosion cells.

It is generally accepted that for adequate corrosion protection the thickness of zinc or aluminium applied should be not less than $100\,\mu m$ (HMSO, 1998) and thicker coatings are often desirable and more economical. The choice of coating thickness will depend on both the corrosion resistance required and the environment. For example in rural environments zinc corrodes at 1–$3\,\mu m$/year rising to 3–$6\,\mu m$/year in urban environments depending on the acidity level and around $5\,\mu m$/year in marine exposures. These values are one-tenth of those of bare mild steel (Porter, 1979) and one-fifth of that of weathering steels (Kilcullen and McKenzie, 1979) under the same conditions. Based on a notional $100\,\mu m$ thick coating of zinc this equates to a minimum coating life of 20 years even under severe exposure conditions (British Standards Institute, 1977).

Application

Zinc coatings can be applied by a range of techniques but for steel bridges are usually applied by either hot-dip galvanizing or metal spraying. The choice of coating method depends on the size and shape of the article to be protected, the severity of the exposure, and the requirements for joining the treated components.

The galvanizing of bridge components is usually carried out after fabrication and requires careful cleaning of the steel surface by acid pickling, followed by fluxing and dipping into a bath of molten zinc at around 450°C. The zinc coating which forms is bonded to the underlying steel by a layer of iron-zinc alloy which forms during the coating operation and which is as protective as the pure zinc outer layer. The thickness of the alloy layer increases with dipping time and its presence significantly increases the time to first maintenance of the overall coating system. Galvanized coatings have the less obvious advantage of having suitable characteristics for high strength friction-grip joints (Moore, 1970). Despite its many advantages there are physical limits on the size of component that can be successfully, and economically, galvanized. In addition great care must be taken to avoid warping, or distortion, of the component as a result of thermal strains induced by the relatively long contact times with the molten zinc.

Full-size bridge components, and complete bridges, can be coated with both aluminium and zinc coatings utilizing metal spraying techniques. Such 'metallized' coatings may be applied under factory conditions or in special site-spraying shops in the field. The basic technique uses a controlled heat source, such as a gas flame or electric-arc, to melt the coating metal in an air-stream which propels the droplets of semi-molten metal onto the surface being coated. Using modern equipment it is possible to achieve application rates similar to conventional air-spray paint methods. In contrast with galvanizing there is practically no significant heat input to the material being sprayed. This is advantageous in that it reduces the risk of thermally induced distortions of the component but means that the metal coating does not react with the underlying steel surface as occurs during galvanizing. As a consequence the bond between the metal coating and the steel is purely mechanical in nature and depends crucially on good surface preparation. The steel must be prepared by abrasive blasting to achieve a near-white surface with a roughened texture that provides a good mechanical key to the sprayed metal. It should be noted that conventional peening with rounded shot produces a surface which is too smooth for the sprayed metal to bond to satisfactorily. Because of the nature of the spraying process metal coatings are inherently porous and the final coating must be sealed with a suitable sealer and top-coat paint combination. The top-coat is necessary partly because of the desire for uniform appearance but principally because the need for maintenance can be seen when the underlying sealer coat becomes exposed. If the top-coat is well maintained the coating system will last almost indefinitely but even if it is not recoated the metallic layer still provides many years of protection.

Coating types

Metal coated steelwork has been widely used in the construction of bridge structures and can be divided into four main types:

- hot dip galvanizing without paint
- hot dip galvanizing with paint
- zinc spray with paint; and
- aluminium spray with paint.

Experience with unpainted galvanized steel structures suggest they have lives in excess of 20 years in rural areas but much less in industrial areas and those subject to local pollution. Painted galvanized steel structures have shown similar performance in mildly polluted environments dropping to 8–12 years in severe exposures where failure of the top-coat was not adequately addressed. British (Deacon *et al.*, 1998) and American (Lieberman *et al.*, 1984) experience suggests that sprayed metal coatings that are suitably sealed and over-coated are some of the most corrosion resistant of all coatings for the protection of steel structures. Recent studies (American Welding Society) suggest that properly applied metal coatings of zinc, 85% zinc/15% aluminum, and aluminium provide at least 20 years of maintenance free corrosion protection in wet, salt-rich environments and 30 years of protection in less severe environments.

16.6 Special steels

Steels are a range of alloys of iron and carbon with small additions of other elements to improve their ease of fabrication and ultimate properties. The carbon content of a steel exerts a significant influence on its physical and mechanical properties. For example an annealed high carbon steel, with 0.4% carbon, has twice the tensile strength of an equivalent low carbon steel with 0.2% carbon but a much reduced elongation at break. As a class steels are characterized by their excellent combination of ductility and strength, ease of fabrication, ready availability and low cost. However they have only limited resistance to corrosion under normal atmospheric exposure. This is because the rust which normally forms on the steel is both porous and non-adherent to the underlying surface so that, in the presence of oxygen and water, corrosion continues unabated. Polluted, industrial environments and marine atmospheres are particularly aggressive and plain carbon steel will deteriorate rapidly if not actively protected. It is of note that the carbon content of a steel has little direct effect on the corrosion resistance. However increasing carbon content results in a two phase microstructure which can introduce the possibility of highly localized galvanic corrosion within the microstructure.

Temporary protection of steel can be achieved by the application of a variety of oils and greases which limit the availability of water and present a diffusion barrier to oxygen. More permanent uses of greases include commercial systems such as Densopaste, a mixture of a petrolatum base, inert fillers and corrosion inhibitors, and Densotape, a reinforced fabric impregnated with Densopaste. These are used for the protection of steel cables and other structural elements, such as external prestressing elements. They have the advantage that they can be applied to complex shapes, offer excellent long-term protection against corrosion and, when required, can be easily removed to allow inspection of the protected components and subsequently replaced.

While petroleum and bituminous based coatings are used to protect many steel components they find only limited application in most bridge structures where other coating systems are preferred (Section 16.5). However all coating systems imply the need for a continuing maintenance commitment over the life of a structure and this in turn has implications for future financial resourcing. Thus in cases where a structure is difficult to get to, due to its remote or dangerous location , or where on-going maintenance of a coating system would unacceptably restrict the use of the structure other approaches can be employed. For example the very low cost of mild steel means that it sometimes more economic to use unprotected steel components with corrosion rates up to 0.2 mm/year being allowed for by specifying an increased initial wall thickness. Whilst such high section losses would not be appropriate for steel bridges such 'corrosion allowance' is the basis of the use of weathering steel.

16.6.1 Weathering steel

By careful control of their composition, fabrication and heat treatment the mechanical properties of plain carbon steel can be optimized for a range of applications. An extended range of properties can be obtained by the use of low alloy steels. These were developed primarily to improve mechanical properties and the ease, and effectiveness, of heat treatment, e.g. high strength low alloy steels with a minimum yield strengths of 275 MPa in the as-rolled condition for construction purposes. In general the corrosion resistance of low alloy steels is not very different from that of plain carbon steels. However their resistance to atmospheric corrosion can be improved by small additions of elements such as chromium, nickel, copper and phosphorus. A typical example is 'Corten' in which 2% of copper and small amounts of phosphorus are added to a plain carbon steel. This alloy was originally developed by the United States Steel Corporation in the late 1920s and was used initially for its increased strength but its resistance to atmospheric corrosion was also found to be superior to that of mild steel. This is because the presence of the copper and phosphorous modifies the rust that forms on the surface of the steel to one which grows slowly to produce a compact, dark brown protective film. For this reason such alloys are commonly referred to as weathering steels. The change in the normally porous, non-adherent and non-protective oxide to one that is both adherent and protective is an example of a conversion coating and is distinct from the passive film that forms on the surface of stainless steels.

The protective properties of the rust film that forms on weathering steels depends crucially on the exposure conditions and typically takes 2–3 years to develop fully. During this period the initial rate of corrosion is similar to that of an unprotected plain carbon steel but steadily decreases with time. Consequently it can be very difficult to predict the long term performance of such steels on the basis of tests carried out over only a few of years. However long-term site exposure tests in the USA on architectural grade weathering steel, Corten A, produced a loss of only 40 μm after 15.5 years exposure as compared with 720 μm loss by mild steel (Larrabee and Coburn, 1961). Work in the UK (Chandler and Kilcullen, 1968) has confirmed that under suitable exposure conditions the rate of deterioration of weathering steels is much lower than corresponding rates for mild steel. However the rates observed in the UK were considerably higher than those reported in the USA.

Weathering steels have been used extensively in the USA where over 2300 bridges have been built with this material over the last 30 years (Nickerson, 1995). It has been found to be particularly suitable for remote rural and agricultural environments which are free from the effects of acid rain and other pollutants. Care must be taken in selecting this material; for use in industrial and marine environments where the passive film that should protect the steel can sometimes fail to form and consequently the steel corrodes at rates similar to unprotected mild steel. Results from 5 years exposure in a variety of industrial, marine and rural sites in the UK (Kilcullen and McKenzie, 1979) found the rates of corrosion of mild steel to be significantly higher (9–63 μm/year) than that of structural grade weathering steel, Corten B (7–36 μm/year) which was in turn higher than Corten A (5–33 μm/year). These results suggest that only in the severest environment does the corrosion rate of a weathering steel exceed 25 μm/year after 5 years exposure and is often considerably less.

However the case for using weathering steel must be based on economic considerations and not on whether it performs better than an uncoated plain carbon steel which would never be used in practice for bridge construction. Since weathering steels are known to corrode at a finite rate under all exposure conditions then a judgement must be made on whether or not the extra cost of specifying such materials can be offset over the life of the structure against the cost of protecting a plain carbon steel by some means or other. For many projects the use of weathering steel, to postpone painting for 30 or more years can represent a considerable saving bearing in mind the cost of both the initial painting and maintenance in that period. It is probable that with suitable monitoring of the structure the postponement of painting can be for periods considerably greater than 30 years or even for the whole life of the structure.

Practical experience of bridge structures in the USA and other countries over the past 30 years has generally been favourable (American Iron and Steel Institute, 1982) with considerable experience now being accumulated on maintenance coatings which can be applied to salt-contaminated weathering steel (Federal Highway Administration, 1995b). Indeed a recent study of 63 steel bridges (Nickerson, 1995) that have been in service for 18–30 years suggests that uncoated weathering steel bridges designed and detailed in accordance with standard recommendations (Federal Highway Administration, 1989) can perform well. The positive performance of these bridges indicates that, at a minimum, the original selection of uncoated weathering steel eliminated the need for initial painting and at least one additional maintenance painting.

It is important to appreciate that weathering steel structures must be protected from the ingress, and ponding, of liquid water as the presence of a film of surface water completely alters the corrosion process and can lead to rapid loss of metal section. Indeed the inspection of a number of foot and road bridges manufactured from weathering steel has confirmed that where water leakage and restricted ventilation occur the steel can develop normal, unrestrained corrosion (Goodman, 1979). The potential of weathering steel to fail to form a passive protective layer has significant implications for structures designed to use this material as crevices, such as riveted joints, must be avoided since they allow the ingress of water with highly accelerated crevice corrosion resulting. Similar problems can be experienced in areas subject to the build-up of debris which acts to shield the surface of the steel. This is a particular problem in marine environments where the presence of chloride ions in sheltered sections of steel it can lead to some localized attack such as a mild form of pitting. This has been observed both in marine environments and on specimens on the underside of motorway bridges where de-icing salts are used in winter months (McKenzie). Based on the above it is clear that where a bridge is to be exposed to constant damp, or aggressive, conditions then inherently more stable steel alloys are required than can be provided for by any grade of weathering steel.

16.6.2 Stainless steel

In some cases the film of oxide that forms on the surface of a metal is very dense and closely adherent and so provides a barrier to the diffusion of oxygen and further corrosion. Such passive film formation represents an example of kinetic control of the corrosion process. Whilst generally resistant to chemical forms of attack the oxide films which characterize this class of materials are, by their ceramic nature, inherently prone to fretting corrosion, as a result of rubbing contact, and erosion by rapidly moving liquids containing suspended solids. Typical examples of metals protected by passive oxide films include aluminium, steel in alkaline concrete and stainless steel.

Stainless steel is a generic name for a series of alloys containing from 11.5–30% chromium, 0–22% nickel and various other alloying additions. Stainless steels do not resist all environments and some grades are susceptible to localized corrosion such as intergranular corrosion, stress-corrosion cracking and chloride induced pitting. This is an important point since some failures can be attributed to the indiscriminate use of stainless steel on the basis that it is a 'corrosion free' material when experience suggests it is not (Anon, 1985b). Therefore whilst stainless steels represent a class of highly corrosion resistant materials they should be used carefully with due regard for their limitations.

There are basically four types of stainless steel:

- martensitic stainless steels can be quench hardened in a similar way to conventional carbon steels. The strength of these alloys increases and ductility decreases with increasing hardness and they tend to be less corrosion resistant than ferritic and austenitic alloys. Such steels are magnetic and are used in applications requiring moderate corrosion resistance plus high strength or hardness. A typical composition is 13% chromium and 0.3% carbon
- Ferritic, non-hardenable, stainless steels cannot be hardened by heat treatment and so must be hardened by cold-working. Such steels are readily formed and have good resistance to atmospheric corrosion and so have wide use for trimmings and claddings. They tend to show better resistance to stress corrosion than austenitic steels especially in chloride containing waters. Unfortunately their structural stability is often poor and they should be selected with care
- austenitic stainless steels, often referred to as simply '18-8' because their basic composition is 18% chromium and 8% nickel, are essentially non-magnetic and cannot be hardened by heat treatment and so must be hardened by cold-working. Such steels possess the best corrosion resistance of all types of stainless steel and for this reason are used for the more severe corrosion conditions found in the process industry. Corrosion resistance generally increases with nickel and chromium contents and these can rise to as high as 22% and 30%, respectively in the most corrosion resistant alloy compositions
- age-hardened stainless steels can be hardened by solution-quenching followed by suitable heat treatment to produce high tensile strengths. With a few exceptions the corrosion resistance of this type of steel is less than that of the austenitic varieties especially in the more severe environments.

Currently stainless steel is used for only a limited number of bridge components, such as handrails, impact barriers and structural elements such as prestressing tendons and cables. However the inherent resistance of some stainless steels to chloride induced corrosion makes them attractive despite their relatively high initial cost. One problem with incorporating stainless steel components into new or existing steel structures is the risk of bimetal or galvanic corrosion. This occurs where the passive stainless steel is in contact with a plain, or low alloy, steel in the presence of a common electrolyte. This can result in rapid localized dissolution of the carbon steel which becomes anodic relative to the stainless steel. This problem can be avoided either by insulating the two metals to prevent the metals from coming into contact or building the entire structure from stainless steel.

16.7 Protection from scour and erosion

Many bridges are required to span over rivers, waterways and flood valleys. As a consequence such bridges, and their supporting piers, columns and abutments, must resist the action of flowing water. This resistance must encompass both the normal variations in seasonal flows and also those which occur under flood conditions. The maximum flow depending on geographical location, the topography of the local catchment area and the prevailing weather system. However in designing a structure to resist the forces exerted by the most severe 100 year flood account must also be taken of the cumulative effect of the much smaller forces exerted by the normal flow over the life of the structure.

Where a bridge and its approaches exert no significant constriction of natural flows, then the risk of long-term damage is relatively small. In contrast where a bridge, and its associated training works, restrict the flow of water or where a bridge is situated at a local narrow point in the watercourse there is a risk of severe hydrodynamic forces on the structure and scouring of the supports. This risk is greatly magnified under flood conditions which can, in the extreme, lead to failure of the structure and loss of life. For example the collapse of the I90 highway bridge crossing over Schoharie Creek on 5 April, 1987 resulted in the loss of 10 lives when two spans of the bridge fell into flood waters after a pier was undermined by scour (Boehmler, 1998). A further seven lives were lost on March 10, 1995, when the two I5 bridges over Los Gatos Creek near Coalinga, California failed because of scour from a large flood (Richardson, 1997).

16.7.1 What is scour?

Scour may be defined as the removal of stream bed material, such as sand and rocks by the action of river or tidal currents. Scour occurs on a continuous basis over time but is especially strong during flood conditions since the swiftly flowing water has more energy to lift and carry material downstream. Scour is a particular problem around bridge structures as it has a tendency to expose, or undermine, foundations that would otherwise remain buried (Figure 16.13).

Scour is a general term used to identify the process of stream bed erosion under and near bridges and may be divided into a number of distinct types (Noble and Boles, 1989). Degradational scour is the removal of sediment from the river bed by the flow of water. This removal of material and resultant lowering of the river bottom is a gradual process, but may remove large amounts of sediment over time. In addition to this general scour, there are two other types that can occur at bridge sites, these are contraction scour and local scour.

Contraction scour is the removal of material from the river bed caused by the acceleration of water approaching and flowing under a bridge. This process occurs when

Figure 16.13 Example of flood induced scour under bridge foundation (Alhinewa bridge, Libya – photographs courtesy of F. Gergab).

the width of the bridge opening is narrower than the natural river channel. The faster flows enables the water to carry a greater volume of material from under the bridge than would occur if the flow was uncontrolled.

Local scour is the removal of sediment associated with vortex systems induced by obstruction to the main flow such as from around bridge piers, abutments and embankments. This results in highly localized removal of material adjacent to the structure to form so called scour holes. The vortex which contributes to the formation of these holes originates at the upstream nose of the obstruction where the flow acquires a downward component which reverses direction at the stream bed. The resultant spiralling eddy is frequently referred to as a horseshoe vortex and rapidly removes material from around the foundation. The depth of scour is affected by the geometry of the pier and its foundation and depends on the pier width, length, shape and alignment. Other factors include the velocity and depth of flow, the type and size of bed material and the rate of bed transport. Pile foundations are often assumed to give more security against scour than spread footings (Lagasse, 1995).

In most cases the scour observed at any given bridge is likely to reflect a combination of these three types of scour, as well as other, less common, effects.

16.7.2 Avoidance of problems

In assessing the failure of the Schoharie bridge crossing it was discovered that the footings were particularly vulnerable to scour because of inadequate rip-rap around the base of the piers and a relatively shallow foundation. Thus problems of failure can be addressed by either changing the structure or replacing any material that is washed away. The first solution typically involves altering the foundation. This can be accomplished by enlarging the footing, strengthening or adding piles, or providing a sheet-piling barrier around the pier foundation (Fotherby, 1993).

Replacement of material generally involves the placement of erosion resistant material such as rip-rap or broken concrete around the pier or abutment to offer a barrier to scour. Where possible rip-rap should not extend above the original stream bed because it could act as an obstruction to stream flow. The stone used for the rip-rap around abutments and piers can vary in size but optimum values can be estimated using the equation:

$$da = 2.5D \left(\frac{V}{\sqrt{gD}} \right)^3$$

where, da is the average stone diameter, D is the estimated depth of flow, V is the average velocity of flow and g is the acceleration due to gravity. The above equation may be used to determine the maximum sized stone at the location of most severe attack, and smaller sized stone can be used where the conditions are less severe.

In addition to alterations to the existing foundations many bridge sites require the use of some type of training works to protect the bridge and its approaches from damage by floodwater. They are used both to stabilize eroding river banks and channel location in the case of shifting streams, and also direct flow parallel to abutments and piers and thereby minimize local scour.

16.8 Summary

The designer of any bridge structure is rarely faced with a straight choice between steel or concrete. In reality a range of solutions usually suggest themselves, although many will be rejected on the grounds of aesthetics, cost or the problems associated with construction. This will leave a number of structurally viable alternatives which after detailed consideration can be distinguished between on the basis of initial capital cost and the recurrent costs incurred in routine inspection and maintenance.

This chapter has reviewed some of the methods available to reduce the deterioration of steel and concrete structures. Unfortunately it is not possible to state any general rule governing when steel or concrete should be the preferred design since there are as many exceptions as there are structures! Despite this it can be stated unequivocally that no material is maintenance free and so the designer remains responsible for ensuring that the final structure will be durable. Good design, irrespective of the choice of material, involves not just a knowledge of structural behaviour but also an appreciation of the effect of small design details on the likely long-term performance. This should allow the designer to incorporate modifications that will reduce the risk of unexpected deterioration and at the same time make the routine inspection and maintenance that will inevitably be required as simple and cost-effective as possible.

In conjunction with good design there exists a range of methods for protecting bridge structures from the extremes of any natural exposure. Perhaps the most common, and universally applicable, of all protection techniques is the use of surface coatings. These all attempt to reduce, or eliminate, contact between the structure and the water in the environment without which many forms of deterioration cannot take place. The use of coatings, whether as barriers or sacrificial layers, implies an on going need for maintenance and this must be acknowledged in the original selection process if a reliable life-cycle costing is to be obtained.

The selection of inherently durable materials is always an important part of the design process but must acknowledge that all known construction materials suffer from some form of deterioration over time. As a consequence it is necessary to select materials that will provide adequate times to first maintenance based on the expected exposure conditions and the available funds for both initial construction and ongoing maintenance. Thus the decision to use blended cement concretes, weathering steel or even stainless steels requires a good understanding not only of the potential benefits of such materials but also the situations under which their use might lead to unexpectedly high rates of deterioration. In this respect a sound knowledge of the causes of deterioration of structural materials and the methods which can be used to limit that deterioration is an essential first step.

Finally it is true to say that bridge engineers have never been faced with a greater array of materials and methods for protecting bridge structures from deterioration with new options appearing almost daily. As a consequence it is important not to forget that concrete and steel bridges designed and constructed in accordance with normal standards and recommendations should be capable of providing adequate, cost effective lives.

Bibliography

American Iron and Steel Institute. Performance of weathering steel bridges – a first phase report, August, 1982.

American Welding Society. Corrosion tests of flame-sprayed coated steel – 19-year report. American Welding Society Technical Report C2.14-74, 1974.

Anon. Berlin bridges take cover. *New Civil Engineer*, 9 May, 15, 1985a.

Anon. Condensation suspect in Zurich pool collapse. *New Civil Engineer*, 16 May, 4–5, 1985b.

Anon. Concrete corrosion – a $550m-a-year problem. *Research Focus*, 37, May, 4, 1999.

Bayliss DA. Quality control of protective coatings. *Corrosion in Civil Engineering*, Institution of Civil Engineers, London, February, 121–130, 1979.

Beeby AW. Design for life. In *Concrete 2000* (Ed. RK Dhir and MR Jones), 37–49, 1993.

Berke N. Calcium nitrite inhibitors for the prevention of chloride induced reinforcement corrosion. *Inhibitors for the Prevention & Cure of Reinforcement Corrosion in Concrete*, Society of Chemical Industry Seminar, London, 30 September, 1999.

Boehmler EM. Evaluation of scour potential at susceptible bridges in Vermont. US Geological Survey, http://nh.water.usgs.gov/CurrentProjects/vtscour.htm, 1998.

Bohris AJ, McDonald PJ and Mulheron, M. The visualisation of water transport through hydrophobic polymer coatings applied to building sandstones by broad-line magnetic resonance imaging. *Journal of Materials Science*, 31, 22, 5859–5864, 1996.

British Standards Institute. BS 5493: 1977, *Code of Practice for Protective Coating of Iron and Steel Structures Against Corrosion*, 1977.

Broomfield JP. The pros and cons of corrosion inhibitors. *Construction Repair Journal*. July/August, 1618, 1997.

Chandler KA. BS 5493: Code of Practice for Protective Coating of Iron and Steel Structures. *Corrosion in Civil Engineering*, Institution of Civil Engineers, February, 23–30, 1979.

Chandler KA and Kilcullen MB. Survey of corrosion and atmospheric pollution in and around Sheffield. *British Corrosion Journal*, 3, 80–84, 1968.

Clarke JL. Non-ferrous reinforcement for structural concrete. In *Concrete 2000* (Eds RK Dhir and MR Jones). E & FN Spon, London, 1993.

Comite Euro-International Du Beton. *Coating Protection of Reinforcement – State of the Art Report*, Thomas Telford Publications, London, 1995.

Concrete Society. The relevance of cracking in concrete to corrosion of reinforcement. Technical Report No. 44, 1995.

Darby JJ, Hammersley GP and Dill MJ. The effectiveness of silane for extending the life of chloride contaminated reinforced concrete. *Bridge Management 3* (Eds JE Harding, GAR Parke and MJE Ryall). E & FN Spon, London, 838 848, 1996.

Deacon DH, Iles DC and Taylor AJ. Durabiliy of steel bridges: a survey of the performance of protective coatings. The Steel Construction Institute, Technical Report SCI Publication 241, 1998.

Deason PM, Miller M. and Nicklinson A. The Queen Elizabeth II Bridge. In *Concrete 2000* (Eds RK Dhir and MR Jones), 929–941. E & FN Spon, 1993.

Dewar JD. Composite cements, ground granulated blastfurnace slag and pulverised fuel ash in ready-mixed concrete. *Municipal Engineer*, 5, 207–216, August, 1988.

European Standard DD ENV 206:1992, Concrete – Performance, production and conformity, 1992.

Federal Highway Administration. Performance of alternative coatings in the environment, Volume II, five-year field and bridge data of improved formulations, FHWA-RD-89-235, 1989.

Federal Highway Administration. Evaluation of volatile organic compound (VOC)-compatible high solids coating systems for steel bridges. FHWA-RD-91-054, 1991a.

Federal Highway Administration. Environmentally acceptable materials for the corrosion protection of steel bridges – Task C, Laboratory testing. FHWA-RD-91-060, 1991b.

Federal Highway Administration. Epoxy mastic bridge coatings. FHWA Bridge Coatings Technical Note, November, 1995a (http://www.tfhrc.gov/hnr20/bridge/mastic.htm).

Federal Highway Administration. Maintenance coating of weathering steel: field evaluation and guidelines. FHWA Report RD-92-055, March, 1995b.

Federal Highway Administration. Uncoated weathering steel in structures. FHWA Technical Advisory Note, T5140.22, October, 1989.

Fontana MG. *Corrosion Engineering*, 3rd edn. McGraw-Hill Series in Materials Science & Engineering, 1986.

Fotherby LM. Alternatives to rip-rap for protection against local scour at bridge piers. *Transportation Research Record*, 1420, 32–39, 1993.

Goodman DF. Corrosion protection as seen by an engineer in a large organisation. *Corrosion in Civil Engineering*, Institution of Civil Engineers, February, 11–21, 1979.

Griffiths D, Marosszeky M and Sade D. Site study of factors leading to a reduction in durability of reinforced concrete. In American Concrete Institute Special Publication SP100, Detroit, 1987.

Hartley J. Improving the performance of fusion-bonded epoxy-coated reinforcement. *Concrete*, Jan./Feb., 12–15, 1994.

Harwood PC. Surface coatings – specification criteria. *Protection of Concrete* (Eds by RK Dhir and JW Green), 201–210, 1990.

Hayward D. Corrosion control. *New Civil Engineer*, 29 May, 37–38, 1997.

HMSO. *Manual of Contract Documents for Highway Works*, Vol. 1, *Specification for Highway Works*. Stationery Office, London, 1998.

Hoffner K. Safety and health on bridge repair, renovation and demolition projects. US Department of Transportation, Federal Highways Authority Final Report DTFH-95-X-00004, 1995 (http://www.tfhrc.gov/hnr20/bridge/repair/intro/intro.htm)

Hopwood T. Overcoating research for steel bridges in Kentucky. *Fourth World Congress on Coating Systems for Bridges and Steel Structures*, Singapore, February, 1995.

Iles D. Durability and integral bridges. *New Steel Construction*, 29, Feb., 1994.

International Standards Organization, ISO 8044:1999: *Corrosion of Metals and Alloys – Basic Terms and Definitions*, 3rd edn, 1999.

Johnson ID and McAndrew SP. The condition and performance of bridge expansion joints. Transport Research Laboratory Project Report No. 9, 1993.

Jones MR. Performance in carbonating and chloride-bearing exposures, *Euro-Cements: Impact of ENV 197 on Concrete Construction* (Eds RK Dhir and MR Jones). E & FN Spon, London, 149–167, 1994.

Kilcullen MB and McKenzie M. Weathering steels. *Corrosion in Civil Engineering*, Institution of Civil Engineers, February, 95–105, 1979.

Lagasse PF, Thompson PL and Sabol SA. Guarding against scour, *Civil Engineering*, 65, No. 6, 56–59, 1995.

Larrabee CP and Coburn SK. The atmospheric corrosion of steels as influenced by changes in chemical composition. *Proceedings of the First International Congress on Metallic Corrosion*, Butterworth Press, London, 276–285, 1961.

Le Page BH. The assessment and behaviour of crack bridging and crack accommodating protective coatings on reinforced concrete. PhD thesis, University of Surrey, Guildford, 1996.

Leeming MB and O'Brien TP. Protection of reinforced concrete by surface treatments. CIRIA Technical Note 130, 1987.

Lieberman ES, Clayton CR and Herman H. Thermally-sprayed active metal coatings for corrosion protection in marine environments. Navel Sea Systems Command, final report, Contract No. 0040682C3258, January, 1984.

Marosszeky M and Chew M. Site investigation of reinforcement placement in buildings and bridges. *Concrete International*, April, 12, 4, 1990.

McGill LP and Humpage M. Prolonging the life of reinforced concrete structures by surface treatment. *Protection of Concrete* (Eds by RK Dhir and JW Green), 191–200, 1990.

McKenzie M. The corrosion of weathering steel under real and simulated bridge decks. Transport Research Laboratory Research Report RR233/BR, TRL, Crowthorne, 1990.

Moore R. Galvanized steel in friction grip connections. *Construction Steelwork Metals and Materials*, 1970.

Mulheron MJ and Nwaubani SO. Corrosion inhibitors for high performance reinforced concrete structures. RILEM TC-AHC:158, The role of admixtures in high performance concrete, Monterrey, Mexico, March, 1999.

Nickerson RL. Performance of weathering steel bridges – a third phase report. American Iron and Steel Institute TSC-95, 1995.

Noble DF and Boles CF. Major factors affecting the performance of bridges during floods. Virginia Transportation Research Council, Charlottesville, VA, 1989.

Nwaubani SO and Dumbelton J. Influence of polymeric surface treatments on the permeability and microstructure of high strength concrete. *Polymer in Concrete, 3rd Southern African Conference and ICPIC Workshop*, Johannesburg, 388–402, July, 1997.

Page CL and Anchor RD. Stress corrosion cracking of stainless steels in swimming pools. *The Structural Engineer*, 66, 24, December, 416 1988.

Parrot LJ. Factors influencing relative humidity in concrete. *Magazine of Concrete Research*, 43, 154, March, 45-52, 1991.

Plinius G. *Natural History* – Book 34, Loeb Classical Library, No. 394 (translated by H. Rackman). Harvard University Press, 1968.

Porter FC. Protection of steel by metal coatings. *Corrosion in Civil Engineering*, Institution of Civil Engineers, February, 107–120, 1979.

Prowell EA. Concrete bridge protection and rehabilitation: chemical and physical techniques – field validation. SHRP-S-658. Strategic Highway Research Program, National Research Council, Washington DC, 1993.

Richardson EV, Jones JS and Blodgett JC. The findings of the I-5 bridge failure. *Proceedings of the Congress of the International Association of Hydraulic Research*, A, 117–123, 1997.

Robery PC. Requirements of coatings. *Journal of the Oil and Colour Chemists Association*, 12, 403–406, 1988.

Shaw JDN. Concrete decay: causes and remedies. In *The Durability, Maintenance and Repair of Concrete Structures*. University of Surrey, Guildford, 1987.

Sohanghpurwalla AA, Islam M. and Scannell W. Performance and long term monitoring of various corrosion protection systems used in reinforced concrete bridge structures. *Proceedings of International Conference Repair of Concrete Structures, from Theory to Practice in a Marine Environment*, Washington, October, 1996.

Sprinkel M and Ozyildirim C. Evaluation of exposure slabs repaired with corrosion inhibitors. *Proceedings of the International Conference on Corrosion and Rehabilitation of Reinforced Concrete Structures*, Orlando, FL, December, 1998.

Treadaway KWJ and Davies H. Corrosion-protected and corrosion-resistant reinforcement in concrete. BRE Information Paper IP 14/88, Building Research Establishment, Garston, Washington, November, 1988.

Treadaway KWJ and Russell AD. Inhibition of the corrosion of steel in concrete. *Highways and Public Works*, 36, August, 19–21, 1968.

Treadaway KWJ, Cox RN and Brown BL. Durability of corrosion resisting steels in concrete. *Proceedings of the Institution of Civil Engineers*, Part 1, 86, April, 305–331, 1989.

Treadaway KWJ, Cox RN and Brown BL. Durability of galvanised steel in concrete. American Society for Testing and Materials Special Publication 713, Philadelphia, PA, 102-131, ASTM, 1980.

van Eijnsbergen JFH. Twenty year of duplex systems – galvanising and painting. *Eleventh International Galvanizing Conference*, Madrid, 1976.

Virmani YP and Clemena, GG. Corrosion protection – concrete bridges, 30. Federal Highways Administration Report FHWARD-98-088, Washington DC, 1998.

Walker M. Reinforcement protection. *Concrete Forum*, Jan., 21–24, 1989.

Woodward R and Williams F. Collapse of Ynys-Y-Gwas bridge, West Glamorgan. *Proceedings of the Institution of Civil Engineers*, 84, 1, August, 635–669, 1988.

17 Bridge management

P. VASSIE

17.1 Introduction

Bridge management (Das, 1996, 1998) is concerned primarily with existing bridges and the objectives are to ensure they achieve their design life, remain open to traffic continuously throughout their life, and that their risk of failure is always very low. These objectives are to be achieved at a minimum life-time cost.

The term 'Bridge management' encompasses a wide range of activities that are commonly encountered in the day to day management of bridges such as inspection, assessment of load carrying capacity and various types of testing. The results from these activities are used to prioritize the maintenance requirements. These aspects are covered comprehensively in other chapters of this book so only those features that are pertinent to bridge management will be discussed in this chapter. Other important subjects associated with the management of individual bridges are the evaluation of their current condition and their rate of deterioration. This information is needed to help decide the most appropriate time to carry out maintenance work.

Many aspects of bridge management relate more specifically to the management of a stock of bridges. Examples of topics in this area include maintenance planning, prioritizing and budgeting. Techniques that have been developed to aid the management of bridge stocks include whole-life costing, cost-benefit analysis and risk analysis. These techniques often make use of economic or probabilistic models.

Bridge management involves making decisions (Frangopol and Estes, 1997) such as when a bridge should be maintained and what type of maintenance should be carried out. These are complex decisions which can have a major influence on life time costs and serviceability.

In order to make the best decision the consequences of all possible decisions must be evaluated against criteria based on the objectives. Decisions are sometimes made from limited or inappropriate information in which case they will tend to be conservative with the result that more maintenance will be recommended than is actually required. Studies of bridge management decisions have established the type of information and algorithms that are needed to make sensible decisions. These data and algorithms form a vital part of computer based bridge management systems that have been developed during the last decade. How to make decisions is the key theme of this chapter.

17.2 Project and network level bridge management

Some aspects of bridge management are concerned primarily with the management of individual bridges (project level bridge management) whereas other aspects are concerned with the management of a stock of bridges (network level bridge management).

Project level aspects such as the condition, load carrying capacity and non-destructive test results clearly have a major influence on the timing and type of maintenance because they relate to a particular bridge. These decisions do not, however, depend entirely on factors associated with the particular bridge; they also depend on factors associated with other bridges in the stock. There are two ways in which conditions within the stock of bridges often affect the timing of maintenance for a particular bridge:

- When the maintenance budget is insufficient to undertake all the identified maintenance requirements the work is prioritized; in a particular year the maintenance on one bridge may be deferred in favour of another bridge with a higher priority.
- The flow of traffic around the road network can be influenced by construction work and maintenance work to the pavement, bridges and street furniture hence these factors can influence the timing of maintenance work to a particular bridge.

The type of maintenance depends more closely on factors related to the particular bridge, but when there is more than one feasible maintenance type the decision can depend on the maintenance management policy for the stock.

Network level bridge management is more closely associated with the overall condition and serviceability of the stock and somewhat less concerned with the maintenance of individual bridges, although it is important to note that most of the input information for network level algorithms is derived from project level inspections, assessments and test results.

In this chapter project level aspects of bridge management will be considered next followed by the network level features, but remember that both are vital parts of bridge management and are closely inter-related.

17.3 Project level bridge management

17.3.1 Inspection

The main purpose of bridge inspections (Department of Transport, 1983; Highways Agency, 1994) is to note any defects that could affect the safety or serviceability of the bridge or lead to a reduction in the life by accelerated deterioration. In the UK three types of inspection are normally carried out. These are called general, principal and special inspections. General inspections are based entirely on visual observations made from any readily accessible position using aids such as binoculars and lamps to observe distant and dark locations. General inspections are made every 2 years.

Principal inspections are carried out every 6 years and are broadly similar to general inspections except that observations must be made at a distance of less than a metre from every part of the bridge. This requirement means that it is often necessary to use a hoist. The close proximity of inspector and bridge means that more detailed observations can be made so the principal inspection report is more comprehensive than the report for a general inspection. Sometimes a limited amount of testing is carried out during the principal inspection of concrete bridges to test for reinforcement corrosion. These tests measure the cover depth and half cell potential of the reinforcement, the chloride content of the concrete and the depth of carbonation.

An important aspect of the inspection procedure, particularly for bridge management, is the assessment of the condition (US Department of Transportation, 1988) of each element and component of the bridge, e.g. deck, piers, abutments, bearings, expansion joints, protective systems such as paint. Numerical values are given to represent the extent and severity of defects to each part of the bridge. This assessment is made by the inspector and involves matching their observations with pre-defined descriptions of defects of different extent and severity.

A more useful evaluation of condition for bridge management purposes is to associate the condition with one of the major phases in the deterioration of a bridge element or component. Each phase is associated with a particular type of maintenance. The main phases of deterioration are:

(a) progressive breakdown of protective systems such as paint, waterproofing membranes, expansion joints
(b) physical deterioration of bridge elements or components leading to a reduction in life which commences after protection is lost
(c) significant damage has occurred with possible hazards to users from falling lumps of concrete, for example
(d) substantial damage has occurred, which may have affected the strength of the bridge, producing a request for a special inspection and assessment of load-carrying capacity.

For a bridge in state (a) preventative maintenance techniques should be effective and these are normally quick and easy to carry out with the result that costs and disruption to bridge users are low. For a bridge in states (b), (c) or (d) preventative maintenance will not be effective and the objective of maintenance is to stop or drastically reduce the rate of deterioration, reinstate the protective system and make the bridge safe for users. If the load carrying capacity is assessed at less than 40 tonnes, strengthening will be required. In general as the deterioration progresses from phase (a) to (d) the complexity, cost and associated disruption arising from the maintenance increases substantially. Deterioration in phase (a) is called primary deterioration while in phase (b), (c) and (d) it is called secondary deterioration. An example of a system linking condition state and maintenance method for a bridge element suffering from reinforcement corrosion is given in Table 17.1. The criteria are based on the visual observations and tests carried out during a principal inspection.

Table 17.1 The link between condition state, deterioration stage and maintenance method for corrosion of reinforced concrete

Condition state	Stage of deterioration	Maintenance type
1	No visible or latent defects Neither chlorides nor carbonation have reached the reinforcement	Preventative e.g. Silanc, Waterproofing membranes
2	No visible or latent defects Chlorides or carbonation have reached the reinforcement and corrosion has started	Stop ongoing corrosion, e.g. cathodic protection, desalination, realkalization
3	Visible defects and/or latent defects e.g. cracking, delamination or spalling of concrete, macrocell corrosion	Repair damaged concrete by concrete replacement and and stop ongoing corrosion
4	Loss of steel-concrete bond or loss of cross-section of reinforcement is sufficient to produce a failure in the in the assessment of load- carrying capacity	Repair the damaged concrete, stop ongoing corrosion and strengthen the weakened element or Replace the element

The condition assessment for different elements of a bridge recorded in the Bridge Management System (BMS) at each inspection provides three useful pieces of information:

- it is the input for algorithms calculating the rate of deterioration and predicting the future condition.
- it indicates the type of maintenance that is appropriate and
- it confirms that the inspection has been carried out.

This information helps the bridge manager to decide when to carry out maintenance, which method to use and to confirm that inspections have been carried out on schedule. Inspections are the only routine way by which the bridge manager can assure himself that his bridges are in a safe condition and it is all too easy for the inspections of a few members of the stock to be overlooked. Using the BMS it is straightforward to list any bridges where an inspection is overdue so that the problem can be corrected.

In general the conditions of the individual elements and components of a bridge are not combined to give an overall condition for the bridge because it would be possible for poor condition of one element to be masked by generally good condition of other elements resulting in necessary maintenance being overlooked. An overall bridge condition can however be useful for assessing the general condition of the whole stock of bridges.

The frequency of general and principal bridge inspections is currently prescribed but as more knowledge is gathered about the rate of deterioration it maybe possible to vary the frequency of inspections based on the bridge age and type of construction. This would enable the resources for inspections to be targeted at bridges with rapidly deteriorating elements or components. The improved information obtained would enable the most appropriate time for maintenance to be determined with greater precision.

The final type of inspection is called a special inspection and is carried out on an as required basis. Special inspections are usually triggered by an adverse principal inspection report and are used to determine the cause and extent of deterioration. This information is needed prior to maintenance work to decide the extent and type of repair. Special inspections usually involve extensive non-destructive testing (NDT) and material sampling.

17.3.2 Assessment
The assessment of the load-carrying capacity (Highways Agency, 1997) of a bridge is usually made when deterioration is considered to be enough to have reduced the strength or when the loading standard is changed. An assessment failure is a vital piece of information because it implies that essential maintenance is needed. Essential maintenance means that one of the following options must be selected without delay:

- strengthen or replace the bridge so that it passes the assessment
- restrict the traffic with lane or weight restrictions so that only safe loads are carried
- monitor the bridge frequently; this option should only be considered if the failure would not be catastrophic and early indications of failure would be readily visible.

Clearly an assessment failure has a major influence on both the time when maintenance is carried out and the type of maintenance.

17.3.3 Testing
Sampling and NDT (Bungey, 1982) are used in special inspections to establish the cause of deterioration, to locate latent defects and to estimate their rate of development. Latent defects can occur inside concrete or under protective coatings such as waterproofing membranes on concrete or paint films on steel. Latent defects can also occur on parts of bridges that are not usually accessible for inspection such as foundations, half joints and deck ends. The most common latent defect is corrosion of reinforcing steel. Corrosion of reinforcing steel ultimately disrupts the concrete, but sometimes it remains hidden, for example in post-tensioning ducts or where pitting corrosion results from macro cell action. These tests are useful for confirming the phase of deterioration for a bridge element.

This information is needed primarily to determine appropriate maintenance methods and to find the extent of the repairs that are needed. The information on the rate of the deterioration process can also act as an input to the algorithm predicting the best time for maintenance. Examples of commonly used tests are given in Table 17.2.

Table 17.2 Examples of sampling and non-destructive tests used for special inspections

Test	Purpose
Paint film thickness	to check for inadequate paint coverage on structural steelwork
Ultrasonic thickness gauge	to check the thickness of structural steel sections
Pit depth gauge	to measure the depth of pits in structural steelwork or reinforcing steel
Chloride sampling	to measure the chloride content of concrete at different depths from the surface
Carbonation depth	to measure the thickness of the carbonated layer of concrete near the surface
Cover depth	to measure depth of the reinforcing steel
Delamination soundings	to locate areas of steel–concrete delamination
Half cell potential	to locate areas where the reinforcing steel is corroding
Corrosion probes	to assess the corrosivity of concrete and monitor the progressive ingress of chlorides and carbonation
Corrosion rate	to estimate the rate of loss of steel cross-section
Amount of corrosion	to estimate the total reduction in cross-section of reinforcement

17.3.4 Deciding maintenance requirements

The information from inspections, assessment and tests should be sufficient to determine the maintenance requirements (Kreugler *et al.*, 1986) in most cases. For example an assessment failure indicates the necessity for essential maintenance as previously described. The inspection data gives the condition and some estimate of the rate of deterioration, which can be used to indicate the type of maintenance (preventative or repair) and its urgency. Test results establish the cause and extent of deterioration and hence suggest where maintenance should be carried out and the appropriate maintenance technique.

The bridge management objective that controls maintenance requirements is that serviceability must be fully maintained throughout the design life of the structure at a minimum life-time cost. At one extreme it could be decided not to do any maintenance work. In this case no maintenance costs would be incurred, but the bridge may become unserviceable at some point during its design life. When the bridge fails its assessment traffic restrictions will be necessary in the absence of maintenance.

The costs associated with the management of traffic and the delays to users can be very large on busy roads and are often much greater than maintenance costs.

At the other extreme it may be decided to carry out the maintenance work as soon as deterioration is detected. This is not usually efficient because if the rate of deterioration is low there are few adverse consequences of delaying maintenance and the money saved can then be used to maintain another bridge which is deteriorating quickly and has a higher priority for maintenance.

Except in the situation when assessment failure means that essential maintenance is required immediately, the decision about maintenance requirements should be based on a consideration of the consequences of (a) not maintaining, (b) doing preventative maintenance and (c) doing repairs now and at various times in the future. For example if on the basis of the current condition and the rate of deterioration it is decided that the condition at the end of the design life will not lead to an assessment failure then maintenance can be deferred indefinitely. This decision is only valid for a limited period and does not commit the manager to not doing any maintenance on the bridge until the end of its design life. The circumstances are re-appraised from time to time when new evidence from inspections, assessments and tests can be considered and if the rate of deterioration is shown to have increased the previous decision can be altered.

The purpose of preventative maintenance or repairs is to increase the age of the bridge when essential maintenance eventually becomes necessary. In other words preventative maintenance and repairs reduce the rate of deterioration. To see the effect of preventative maintenance or repairs on the life-time maintenance costs look at example given in Figure 17.1. This shows a time line indicating the time of preventative maintenance or repair and the time of essential maintenance when preventative maintenance/repairs are and are not carried out. The figure also shows the costs of preventative maintenance or repairs, essential maintenance without previous preventative maintenance or repairs, and essential maintenance following preventative maintenance or repair. All these costs are based on the time when preventative maintenance or repair is carried out. Applying the discounted cash flow (see Section 17.5.1) technique the net present value of future essential maintenance events can be calculated and the difference between them represents the money saved by carrying out preventative maintenance or repair. This cost must be compared with the cost of preventative maintenance or repair in order to assess the usefulness of these procedures. If the cost of preventative maintenance or repairs is less than the difference in the essential maintenance costs then the life-time cost will be reduced providing a justification for doing preventative maintenance or repairs. This procedure could be repeated by varying the age of the bridge when preventative maintenance or repairs are carried out to find the age when the benefit is maximized.

Although this procedure for deciding when to carry out preventative maintenance or repairs appears quite straightforward and to be suitable for conversion to a computer algorithm that would produce an optimal maintenance programme for the bridge it must be remembered that detailed information about the rate of deterioration in the presence and absence of preventative maintenance or repairs is required.

Event	Cost	Age
Preventative maintenance/Repairs [PR]	A	X
Essential maintenance when earlier preventative maintenance/repairs have not been carried out [E]	B	Y
Essential maintenance when earlier preventative maintenance/repairs have been carried out [E(PR)]	C	Z

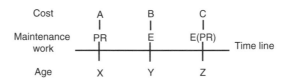

Maintenance work	Cost discounted to age X at 6% per year
PR	A
E	$B \times 1.06^{(X-Y)}$
E (PR)	$C \times 1.06^{(X-Z)}$

Benefit of using preventative maintenance or repairs = $B \times 1.06^{(X-Y)} - C \times 1.06^{(X-Z)}$

Cost of preventative maintenance or repairs = A

Cost : Benefit ratio = $A / (B \times 1.06^{(X-Y)} - C \times 1.06^{(X-Z)})$

Figure 17.1 Calculating the cost–benefit ratio of preventative maintenance or repairs.

This information can only be obtained by studying the deterioration of a cohort of bridges of similar type and age, treated with each of a range of preventative and repair treatments, including the do nothing option. At present this information is not available but with the computer storage of condition and maintenance data it should become available in the future. In the meantime the following guidelines should ensure a rational decision is made on the basis of limited information:

- if secondary deterioration has not commenced then most preventative maintenance methods are likely to be economically beneficial
- when secondary deterioration has started repairs are likely to be economically beneficial when essential maintenance would be needed within about 10 years, the increased life resulting from the repair work is thought to exceed 20 years, and the ratio of cost of repairs to the cost of essential maintenance is less than about 0.3.

17.3.5 Selecting an appropriate maintenance method

First, it is necessary to decide the type of maintenance (preventative, repair or essential strengthening) (Al-Subhi *et al.*, 1989) required on the basis of assessment and condition information. Secondly an appropriate maintenance method (OECD, 1981) needs to be selected and this depends on the cause and rate of deterioration, the current condition and the element of the bridge affected.

Essential maintenance generally involves strengthening or replacement of bridge elements. Strengthening techniques include welding, plate bonding and external post-tensioning which increase the stiffness of bridge decks. Replacement of elements has been used for deck slabs and beams, piers and columns. The primary purpose of essential maintenance is to increase the load carrying capacity and the reason for the inadequate capacity is secondary. If the reason is simply increased loading the maintenance can be limited to increasing the capacity, but if the reason is deterioration then maintenance must also include repairs and preventative maintenance.

The selection of the maintenance method for repairs and prevention depends primarily on the cause of deterioration. For steel construction the main cause of deterioration is corrosion and regular maintenance painting should be carried out to prevent the steel from corroding. If corrosion does occur then the only repair option is to grit blast back to shiny metal before repainting. An assessment of load carrying capacity should be carried out if corrosion has resulted in a significant reduction of steel section.

The selection of repair and prevention methods for concrete construction is more complex because there are numerous causes of concrete deterioration. The deterioration of reinforced concrete can be conveniently sub-divided into deterioration of the concrete and deterioration of the steel reinforcement. The main causes of concrete deterioration are sulphates, freeze–thaw cycles and alkali–silica reaction (ASR). Deterioration can also be related to poor mix design and construction processes such as compaction and curing. These types of deterioration can only be prevented by actions taken at the time of construction; there are no effective preventative actions that can be taken after construction. For example where the environment is known to contain significant quantities of sulphate or sulphide it is sensible to consider the use of sulphate resisting Portland cement. In regions experiencing large numbers of freeze–thaw cycles frost damage to concrete can be prevented by adding air entraining agent to the concrete mix. Frost damage is worse in concrete that is saturated with salty water so techniques such as waterproofing membranes and silane treatments may be helpful. Alkali–silica reaction between aggregates and the alkali in cement can be prevented by avoiding the most reactive types of aggregate and by keeping the alkali content of the cement below the designated limit. To set up damaging stresses in concrete the ASR requires water so procedures to reduce the water content of concrete such as waterproofing membranes and silane treatments may help. If these forms of concrete deterioration take place the only viable repair method is concrete replacement which may be extensive especially for ASR where entire sections can be affected. Sulphate and freeze–thaw damage normally occur only in the cover zone of the concrete. It is important to note that deterioration of the concrete will increase the risk of corrosion to the reinforcement because steel depassivators, like chlorides and carbon dioxide, will be able to move more easily through the concrete to the reinforcement.

Deterioration of the reinforcing steel is caused by corrosion and can be prevented by actions taken at the time of construction and for a period after construction. Preventative techniques that can be applied at construction include the use of epoxy coated mild steel, stainless steel or carbon or glass fibre reinforcement, inhibitors, cathodic protection, anti-carbonation coatings, silane treatments and waterproofing membranes. All of these techniques, except the last three, directly protect the reinforcement against corrosion and to date, have been used only occasionally largely on grounds of cost. Waterproofing membranes, silane treatments, and anti carbonation coatings are applied to the concrete and are designed to slow down the ingress of carbon dioxide and chlorides into the concrete thereby increasing the age of the structure when the reinforcement begins to corrode. These techniques can be used after construction because they are applied to the concrete surface and they should be effective, providing corrosion of the reinforcement has not already begun. It is important not to overlook the importance of well compacted and cured, low water:cement ratio concrete in preventing reinforcement corrosion.

When corrosion of the reinforcement occurs it results in a loss of steel section and/or cracking, spalling and delamination of the concrete due to the stresses produced as a result of the low density of rust compared with density of the steel. Reinforcement corrosion repair methods have two main functions, to stop the corrosion and to repair the damaged concrete. There are a number of techniques available:

• concrete replacement
• cathodic protection
• desalination
• realkalization.

Concrete replacement has to be used to repair the damage caused by corrosion regardless of which technique is used to stop corrosion. Concrete replacement can also be used to stop corrosion although this involves the removal of all the carbonated and chloride contaminated concrete even though it is physically sound. This often means that concrete repairs to stop corrosion are not economically viable. Cathodic protection can be applied at any time to stop corrosion caused by carbonation or chlorides. It functions by making the reinforcing steel cathodic with respect to an external anode system. Cathodic protection requires a permanent electrical installation. Desalination can be used to stop corrosion caused by chlorides and it works by migrating chloride ions towards an external anode and away from the reinforcing steel in an electric field; this process takes about 6 weeks. Realkalization stops corrosion caused by carbonation and it works by migrating sodium ions from an external anolyte into the concrete where in combination with the hydroxyl ions generated on the reinforcing steel due to the electric field, the alkalinity is raised to a level where the steel re-passivates. Realkalization takes about 4 weeks. Desalination, realkalization and concrete repair are normally used in conjunction with a preventative treatment such as silane or an anti-carbonation coating to increase the life of the repair. Cathodic protection does not require additional preventative measures because it is a permanent installation, but the anodes do require periodic replacement.

17.3.6 Bridge monitoring strategies

Previous sections of this chapter have described the types of information that can be obtained from inspections, assessments and non destructive testing, and how this can be used to decide the maintenance requirements and appropriate maintenance methods. In this section a monitoring strategy is described for concrete bridges suffering from salt induced reinforcement corrosion. This example is chosen because nearly all highway bridges are affected by reinforcement corrosion at some time during their life, reinforcement corrosion often results in latent defects and is mechanistically complex.

A major challenge for bridge managers is to make the best use of the resources available to achieve the objectives. If resources were unlimited there would be little difficulty in using inspections, assessments and tests to locate defects and apply remedial measures. In practice resources have been tightly constrained and the information on which to base decisions has been limited. Decisions based on limited information are usually either conservative or result in an increased risk of failure. Conservative decisions lead to expenditure on unnecessary maintenance work. An increased risk of failure can arise when latent defects are not detected. The philosophy of the maintenance strategy described below is the targetting of resources to generate more relevant information. It combines aspects of structural and materials performance and is summarized in the flow diagram shown in Figure 17.2.

The strategy is based on information from inspections, assessments and tests. Inspections would be used to locate all visible defects, to make an assessment of condition and to establish which parts of the structure are most vulnerable to deterioration. Assessments would be used to locate critical areas of a structure which are most likely to be involved in any failure. The redundancy of strength related to failure at these critical areas would also be estimated. Testing, which is a major consumer of resources, could then be targeted on those areas which are both structurally critical and most vulnerable to deterioration. The results of testing in these critical areas will provide a more reliable measure of condition which can be used to calculate a more accurate value for the load carrying capacity and hence to decide the most appropriate time and type of maintenance. The sequence of testing for reinforcement corrosion is designed such that simple tests are carried out on all the targeted areas, the more complex tests are only carried out when the simple tests indicate that corrosion is imminent and the tests for measuring the cross-section of reinforcement and steel concrete bond are only carried out in areas where previous tests have indicated the presence of localized and general corrosion. The simple tests involve the use of transducers embedded in the concrete which measure the corrosivity of the concrete and track the ingress of depassivators such as chlorides and carbon dioxide thereby providing an early warning of imminent corrosion. This enables the manager to set up remedial measures before significant deterioration has occurred,

Redundancy (R) = K - 0.91
where K is the reduction factor

Vulnerability (V) to deterioration = 3 (low) or 2 (moderate) or 1 (high)

The product, RV, is a measure of the priority for non destructive testing and sampling

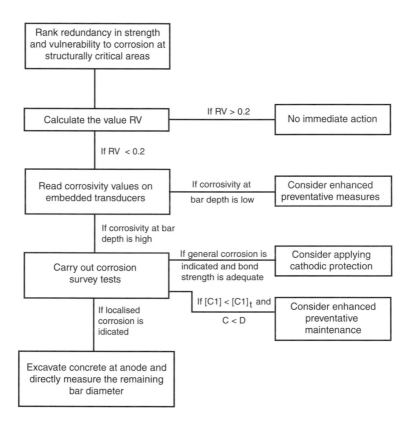

[C1]   Chloride content near the reinforcing bar
[C1]$_t$ Threshold chloride content
  C   Carbonation depth
  D   Depth of cover

Figure 17.2 Suggested flow diagram for monitoring deterioration of reinforced concrete.

preventing costly repair work. On those structures where corrosion is imminent a corrosion survey is carried out to find the cause, the precise location, the rate and the type of corrosion (localized or general). This information is used to decide the type of maintenance and maintenance method required. The concrete would be excavated from areas of localized corrosion so that the reinforcement can be examined and a direct measure of its cross-sectional area can be made. A delamination sounding survey would be carried out in areas of general corrosion. The information from the examination of reinforcement and the delamination survey would be used to calculate whether loss of steel section or reductions in steel-concrete bond have significantly affected the load-carrying capacity. This maintenance strategy will provide the information needed to determine the correct type of maintenance, best time for maintenance and the most appropriate maintenance method.

The targeting of inspection, assessment and testing resources enables the most vulnerable areas of the structure to be tested thoroughly at the most appropriate time.

17.4 Network-level bridge management

The development of relational database systems (Gusella *et al.*, 1996) during the past decade has enabled bridges to be managed as a stock. This has number of potential advantages:

- design features and materials associated with defects can be identified
- the cost-effectiveness of different maintenance methods can be established
- the influence of bridge maintenance on traffic flow can be examined
- the rate of deterioration of a particular bridge can make use of information on the deterioration of bridges of similar type and age elsewhere in the stock
- maintenance work on different bridges in the stock can be prioritized rationally
- maintenance programmes and budgets for the stock can be planned
- the performance of a maintenance programme can be evaluated in terms of change in the overall condition of the bridge stock.

Computer-based databases and their associated analysis algorithms are called Bridge Management Systems (BMS) (Federal Highways Administration, 1987). They consist of a number of modules such as:

- inventory
- inspection, assessment and test records
- maintenance records
- economics of maintenance methods
- economics of traffic disruption and management
- rate of deterioration
- optimized and prioritized maintenance programmes.

Originally BMS consisted only of an inventory whose primary function was the secure storage and easy retrieval of data on individual bridges. The primary function of these first-generation BMS was to facilitate the day-to-day management of bridges by generating database reports. An inventory is an essential feature of all BMs and is the basic store of data for all bridges in the stock.

A BMS inventory typically consists of about 200 data fields. Examples of data fields in a simple inventory are listed in Table 17.3. These data can be used to generate an almost unlimited number of reports to answer questions that may arise in the day-to-day management of bridges. A typical example of a report is given in Table 17.4.

Subsequent development responded to the need to store information about inspections, assessments, tests and maintenance work. These are second-generation BMSs. They enable the manager to check for bridges where planned maintenance work and inspections had been inadvertently overlooked, a situation that can easily occur when

Table 17.3 Typical inventory data fields

Inventory category	Examples
General	Bridge number, bridge name, structure type, grid reference, road number, obstacle crossed, year built, maintenance agent, main material and number of spans
Route	High vehicle route, heavy vehicle route, salted, weight restriction, traffic flow, traffic HGV, minimum width, minimum clearance
Span	Span number, span length and width, skew, deck type, main and secondary materials
Supports	Support reference, type, substructure type, foundation type
Public utilities	Type of utility, date notified
Components (joints, bearings, parapets, waterproofing membranes)	Component reference, span cross-reference, support cross-reference, component type, date of installation
Prestressing	Prestressing type, location

Table 17.4 Example of an inventory report from a simple BMS

Query: List the post-tensioned bridges built before 1960 that have no waterproofing membrane.

Report Fields: Bridge No., Bridge Name, Grid Ref, Year Built, Road, Post-tensioning type

Table 17.5 Example of a report from a second-generation BMS

Query: List all overbridges built before 1990 where a pier in the central reservation has a condition state less than 3 and has not been treated with silane

Report Fields: Bridge No., Bridge Name, Grid Ref, Road, Year built, Condition state

managing a stock of a thousand or more structures. This feature is essential for providing assurance that all the bridges in the stock are safe and serviceable. The information stored in these additional modules allows more complex reports to be generated (see Table 17.5) and they also provide vital input data for the algorithms used in third-generation BMS.

Third-generation BMS have the general objective of maintaining serviceability of the bridges in the stock at a minimum life time cost. This objective requires

(a) modules providing information on the economics (cost and lives) of maintenance methods and their impact on the flow of traffic and (b) algorithms for finding the rate of deterioration and optimized and prioritized maintenance programmes for the bridges in the stock. The optimized programme minimizes life-time costs whereas the prioritized programme minimizes life-time costs subject to the constraint that the budget available for maintenance work is less than the cost of the optimized maintenance programme. Most of the remainder of this chapter will concentrate on the data and algorithms required in order to predict future deterioration and generate optimized and prioritized maintenance programmes.

17.4.1 Rate of deterioration

In order to produce a maintenance programme for a bridge it is essential to know how quickly it will deteriorate. If deterioration is slow maintenance can either be deferred until the bridge is older or a simpler and cheaper type of maintenance can be adopted. Conversely if deterioration is rapid it may be necessary to bring the maintenance work forward. For a particular bridge with good inspection, maintenance and test records it should be possible to make an estimate of the rate of deterioration up to the present time and maybe slightly into the future using extrapolation. By using information from the inspection, maintenance and test records of other bridges in the stock, of a similar type of construction and material, when they were the same age as the bridge under consideration, it should be possible to make a better estimate of the rate of deterioration and predictions about the condition in the future.

Rates of deterioration (Hogg and Middleton, 1998) can, in principle, be determined either from inspection or test data. The results from physical tests offer the most direct approach, although it is far from straight forward. It is possible to devise models describing deterioration processes. These models are complex, reflecting the nature of the physical processes, and a specific model is required for each deterioration mechanism. Furthermore a deterioration process may involve a number of different mechanisms operating simultaneously or at different stages of the process. Consider for example, the deterioration caused by corrosion of reinforcing steel in concrete by de-icing salt. This process occurs in two distinct stages. In the first stage chloride ions from the de-icing salt slowly penetrate the concrete cover until they reach the reinforcement. In the second stage corrosion begins, when the chloride concentration of the concrete adjacent to the reinforcement exceeds the threshold value, and then propagates at a rate dependent on factors such as the concrete resistivity, potential difference between anodic and cathodic sites, cathode/anode area ratio, the rate of diffusion of oxygen and the electrochemical corrosion mechanism. In the first stage there are two mechanisms of chloride ingress (diffusion and absorption) and in the second stage corrosion can result in a non uniform reduction in reinforcement cross-section (localized corrosion) and/or a loss of bond between the steel and concrete due to cracking, spalling or delamination of the concrete (general corrosion). From this discussion it can be seen that at least four models would be needed in order to describe the reinforcement corrosion process. A large body of data would also be required and as this could only be obtained by expensive non-destructive testing, the inevitable conclusion is that the physical model approach is not feasible for finding the rate of deterioration of thousands of bridges. Nevertheless considerable work has been undertaken to establish physical models, but so far the results have been very inaccurate. The root cause is a lack of knowledge and real data about the physical processes involved, but the modelling work has demonstrated the variability of deterioration rates amongst nominally similar bridge elements.

The other approach (Saito and Sinha, 1990) is to use the condition state assessments made during bridge inspections. This data already exists and was cheap to collect. Measuring condition using the condition state assessments made by bridge inspectors is normally based on a set of discrete states which represent different stages in the deterioration process. There is a close association between the condition state and the appropriate type of maintenance which simplifies the interpretation of the condition measure. Different materials have different deterioration processes hence it is necessary to set up a condition state scale for all the common deterioration processes and materials of construction. A condition state scale consists of a number of states each of which is given a numeric value and a definition describing the stage of the deterioration process. The number of states is normally between 3 and 10. If the assessment is based entirely on visual observations, the number of states normally lies at the lower end of this range, whereas more states can be used if non-destructive tests and material sampling are used. When too few states are used the deterioration process is not adequately described, but if too many states are used it becomes difficult to differentiate between them, resulting in different inspectors making different condition assessments on the same element. Deterioration sometimes results in the formation of latent defects and in these cases it is recommended that the condition state should be based on visual observations, ndt and sampling. An example of a typical condition state scale for corrosion of reinforced concrete is shown in Table 17.1.

The general procedure is as follows:

(i)   Sub-divide the bridge stock into sets of bridge elements with characteristics that indicate that they should deteriorate by similar mechanisms. Factors that are most likely to influence the deterioration process are the construction material, geographic location and when and how the element was last maintained. As more information is obtained about deterioration processes the sub-division can be refined with the proviso that the number of bridges in each set is statistically significant.

(ii)   For each set of bridge elements the condition state vs age data for each element is aggregated to generate a function relating average condition state of the set with respect to age. This function cannot be used to predict the future condition of a particular element because its current (condition state, age) ordered pair is unlikely to lie on the average line.

(iii)   Carry out a Markov chain (Jiang *et al.*, 1989) for the deterioration process starting from a known situation such as the condition state at age zero has the value of 1. (For a five-point deterioration scale condition state 1 represents an undeteriorated state whereas condition state 5 represents an unserviceable state.) The Markov chain calculates the probability of a typical element being in a particular condition state at a particular age.

(iv)   An optimization procedure is carried out to minimize the difference between the condition states found for a given age by the procedures described in (ii) and (iii) above. The set of condition state probabilities from the Markov Chain associated with this minimal difference represent the condition state transition probabilities for the deterioration process based on the historic condition state vs age data used in (ii). These transition probabilities are the most accurate representation of the deterioration process based on the available data. A condition state transition probability is the probability that an element currently in state x will move to state y by the next inspection. Transition probabilities are used to represent the rate of deterioration because the condition state scale consists of discrete points rather than a continuum. Two reasonable assumptions are made to simplify the mathematics: (a) the condition state does not decrease with increasing age and (b) between consecutive inspections the condition state either remains constant or increases by one unit. These assumptions produce a relatively sparse transition state matrix which reduces the number of computations needed to predict future condition state values.

(v)   The predicted condition state in future years is obtained by matrix multiplication between the current condition state vector and the transition probability matrix. The condition state vector relates to the current condition of the particular bridge for which the prediction is being made. It is a five-element row vector representing the probability of the element being in each condition state. The transition matrix represents the historic evidence on how elements of a similar type and age have deteriorated. Thus predictions are based on the most recent data on the condition state of the particular element and the historic information about how similar elements have deteriorated. This is a rational basis for making predictions.

This description of how deterioration rates are obtained from inspection data is unavoidably complex but it is intended to give a flavour of the process for those with limited mathematics.

A mathematical description follows which some readers may prefer. The same step numbers are used. Step (i) is the same.

(ii)   the average condition state vs age function usually fits quite well to a polynomial equation such as:

$$C(t) = a + bt + ct^2 + dt^3$$

where $C(t)$ is the average condition state of the stock at time t years and a, b, c, d are polynomial coefficients.

(iii) The beginning of the Markov chain is represented by the tree diagram shown in Figure 17.3. The numbers at the nodes represent the condition state and the numbers associated with the branches of the tree such as p_{xy} represent the transition probability of going from state x to state y between consecutive inspections. Note the simplifying assumptions:

$$y \geq x \quad \text{and} \quad y = x \quad \text{or} \quad y = x + 1$$

Figure 17.3 can be used to determine the probability of being in a given state at a given time and to determine the average condition state at a given time. For example:

(a) probability of being in state 2 at the second inspection is given by $p_{11} p_{12} + p_{12} p_{22}$ and

(b) the average condition state at the second inspection is given by:

$$C_m(t, w) = p_{11}^2 + 2(p_{11}p_{12} + p_{12}p_{22}) + 3p_{12}p_{23}$$

where $C_m(t, w)$ is the average condition state at time t determined by the Markov chain and w is the set of probabilities P_{xx}.

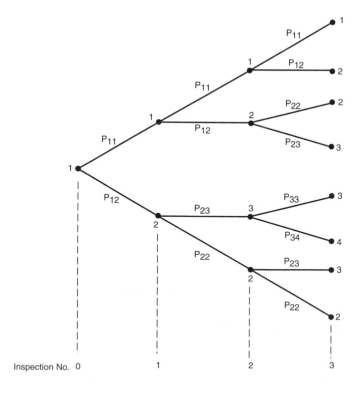

Inspection No.   0   1   2   3

Nodes   1, 2, 3, 4   condition states

Branches P_{11}, P_{12}, P_{22}, P_{23}, P_{33}, P_{34} transition probabilities

Figure 17.3 Markov chain diagram.

Thus in order to find $C_m(t, w)$ it is necessary to know the values for p_{11}, p_{22}, p_{33} and p_{44}.

For a five-point condition state scale $p_{55} = 1$, by definition. It follows from the simplifying assumptions that $p_{x,x+1} = 1 - p_{xx}$

The transition probabilities are usually combined in a matrix, P, called the transition state matrix.

Thus:

$$P = \begin{bmatrix} p_{11} & p_{12} & 0 & 0 & 0 \\ 0 & p_{22} & p_{23} & 0 & 0 \\ 0 & 0 & p_{33} & p_{34} & 0 \\ 0 & 0 & 0 & p_{44} & p_{45} \\ 0 & 0 & 0 & 0 & p_{55} \end{bmatrix}$$

Note for a five-point condition state scale the transition matrix is 5×5.

(iv) The optimization to find the transition state probabilities is based on the minimization of objective functions such as:

$$\min \sum_{t=0}^{t=4} \left| C(t) - C_m(t, w) \right|$$

The value of the set w corresponding to the minimization gives the transition probabilities. Thus:

$$w_{min} = \{ p_{11}, p_{22}, p_{33}, p_{44}, p_{55} \}$$

The minimization is obtained by cycling through the values of $p_{11}, p_{22}, p_{33}, p_{44}$ from 0 to 1 using a step size of 0.01 until the minimum is reached.

Consider a condition state row vector $V = [p_1\, p_2\, p_3\, p_4\, p_5\,]$, where $p_1, p_2, \ldots,$ p_5 are the probabilities of being in condition states $1, \ldots, 5$ at a particular time. For a five-point condition state scale the condition state vector has five elements.

The condition state vector after n inspections is given by:

$$V(n) = V(0)P^n$$

where $V(0)$ is the condition state vector at age zero, e.g. $[1, 0, 0, 0, 0]$ indicating that a new bridge element is normally in condition state 1 and not in any other condition state.

Multiplying the condition state vector $V(n)$ by a five-element column vector representing the condition states gives the predicted condition state after n inspections $C(n)$.

If $V(n) = [p_1\, p_2\, p_3\, p_4\, p_5]$ then $C(n) = V(n)S$, where $S = [1, 2, 3, 4, 5]^T$.

This completes the description of how the condition state of an element in future years can be predicted. It is important to remember that predictions become less reliable the further you go into the future. It is, however, only the predictions for the next few years that will have a bearing on immediate maintenance activities and these predictions should be reliable. The predictions should be recalculated periodically, including the most recent inspection assessments of condition state in the input data for the algorithm. The results of the predictions are normally represented as a condition state trajectory where a condition state is matched with each future inspection. An example of a condition state trajectory is:

Condition state	1	1	1	2	2	2	3	3	4	4	5
inspection number	0	1	2	3	4	5	6	7	8	9	10

The technique described above for predicting the condition of an element in future years is restricted, by the simplifying assumptions, to elements that have not received maintenance. Although the restriction includes most bridge elements there are a significant number that have been maintained and will be excluded. The technique described could be used for maintained elements if the assumptions were relaxed. In particular the assumption stating the condition state cannot decrease would need to be relaxed. The amount of condition data from bridges following maintenance is currently very limited and an alternative approach is needed at present. One approach is to associate two parameters with each maintenance method. These parameters are:

- The improvement in condition resulting from maintenance work:

$$\text{improvement } (I) = \text{condition state before maintenance} \\ - \text{condition state after maintenance}$$

- The life (L) which is the number of years following maintenance that the condition state remains constant, before deterioration re-commences.

For example, consider a bridge element of age 40 years with a condition state value of 3 that undergoes a maintenance treatment with $(I, L) = (1, 20)$, then it is predicted that the condition state will improve to state 2 $(3 - 1)$ following the maintenance and then maintain this value for 20 years until the element is 60 years of age. A comparison of condition state trajectories with and without maintenance provides a measure of the benefits of maintenance which can be compared with the costs.

Without maintenance condition state	3	4	4	4	5	5
Age (years)	40	45	50	55	60	65
With maintenance condition state	3	2	2	2	2	3

The experience of most maintenance methods is limited because they have only been used widely during the last decade, therefore while knowledge of their improvement is good, information about their life is less reliable at present. It is important to revise the values of improvement and life for maintenance methods regularly to take account of new data recorded in the inspection and maintenance modules of the BMS. Now that a procedure has been established to predict the future condition of bridge elements we are in a position to investigate the optimization of life time maintenance costs.

17.4.2 The optimal maintenance programme

The optimal maintenance programme (Jiang and Sinha, 1989) states which elements of which bridges should be maintained by specified methods each year such that the life-time maintenance cost of each bridge is minimized. An example of an optimal maintenance programme is shown in Table 17.6. The general approach to optimization and the various options that are available are best understood by considering the tree diagram in Figure 17.4. Each node of the tree is associated with a particular condition state, calculated using the techniques described in the previous section. Each vertically aligned set of nodes corresponds to a particular inspection of the bridge element. The branches emanating from a node, represent a number of maintenance options of which one is a 'no maintenance' option. The condition state of the node determines the type of maintenance that is appropriate and hence limits the number of possible maintenance methods to about two or three. The number of possible maintenance methods equals the number of branches emanating from a node. The number of branches from each node in Figure 17.4 is fixed for convenience only; in practice the number of branches can vary with the condition state. Each branch has an associated cost, discounted at a rate selected by the operator. Thus an important part of the optimization module is a library of maintenance costs. These costs are sub-divided into three types:

- engineering
- management of traffic
- traffic delays.

The operator can use any combination of these three types of cost. The optimization is based on the sum total of the cost types selected, although each cost can be itemized separately. These options for the operator are necessary because sometimes the influence of maintenance work on traffic delays will not be required. Engineering costs depend primarily on the maintenance method whereas the traffic-related costs depend mainly on an element being maintained and traffic density.

The optimization process consists, basically, of calculating the total cost of every path through the tree and identifying the path with the lowest total cost. The year and maintenance method associated with the branches of this path of minimum cost give the optimal maintenance programme. In practice this would result in an excessive number of computations taking a large amount of computer time. For example, if each node has three branches and optimization is required over the next 20 inspections then $3^{20} \approx 3 \times 10^9$ pathways would have to be calculated. This number can be reduced by a factor of about a million by using a technique called dynamic programming (Winston, 1991) together with the provision of some information about the last maintenance treatment.

Table 17.6 Example of part of an optimized or prioritized maintenance programme showing the data fields reported

Bridge No.	Element No.	Age	Year	Maintenance type	Maintenance method	Eng cost	TM cost	TD cost	Discounted total cost	CS
57	4	0	2000	Do nothing						1
57	4	3	2003	Preventative	Silane	20K	10K	25K	46K	1
57	4	4	2004	Do nothing						1
57	4	20	2020	Preventative	Silane	20K	10K	40K	22K	1
57	4	40	2040	Repair	Cathodic protection	150K	30K	100K	27K	3
57	4	41	2041	Do nothing						2

TM, traffic management; TD, traffic delay; CS, condition state.

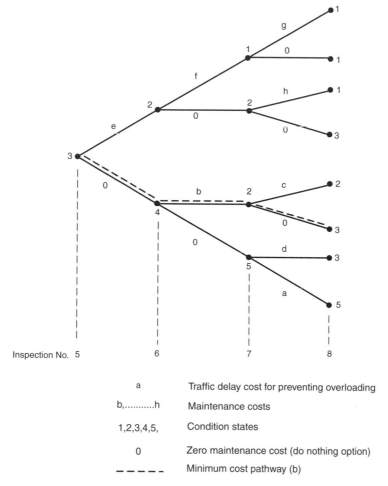

a   Traffic delay cost for preventing overloading

b,...........h   Maintenance costs

1,2,3,4,5,   Condition states

0   Zero maintenance cost (do nothing option)

– – – –   Minimum cost pathway (b)

Figure 17.4 Optimization tree diagram for two maintenance options per node.

The operator of the optimization module has the option of applying constraints to the optimization. When no constraints are used preventative maintenance and repairs are only carried out if they reduce the life time cost. Essential maintenance has to be done, but it could involve traffic restrictions instead of engineering work therefore it is important to include traffic management and delay costs if it is decided to do optimization without constraints. Optimization without constraints is the purest form of optimization and will deliver a programme that minimizes life-time costs.

Sometimes managers wish to use a constraint that the condition state must not exceed some threshold value, thereby ensuring that no bridge element will deteriorate beyond a certain level. The effect of this constraint on the tree diagram is that when at any time the condition state at the next inspection is predicted to exceed the threshold value the 'no maintenance' branch emanating from the node is eliminated. This ensures that some maintenance is carried out keeping the condition state below the threshold.

The important aspect of this optimization procedure is that the rate of deterioration forms an integral part of the procedure and in particular the effect of maintenance treatments on the rate of deterioration is fully taken into account.

The optimized maintenance programme represents an idealized situation and in some cases it is necessary to defer or bring forward maintenance work for a variety of reasons, such as to maintain the flow of traffic through the network. A frequent reason for having to modify the optimal programme is shortage of money; the budget is rarely large enough to carry out all the maintenance work specified in the optimal programme. In these circumstances it is necessary to prioritize the maintenance programme (Darby *et al.*, 1996). A general principle for prioritization is that the difference in the life-time costs of the optimal and prioritized maintenance programmes is minimized. A prioritized maintenance programme is by definition sub-optimal. Deferring optimal maintenance work is likely to result in more complex, extensive and costly maintenance in the future, even when discounting is taken into account. Furthermore if traffic disruption occurs, delaying maintenance will lead to more disruption due to the growth in traffic with time. The next section describes a prioritization procedure.

17.4.3 Prioritization

The procedure uses cost-benefit analysis to minimize the difference in life-time cost between the prioritized and optimized maintenance programmes. All maintenance work has a cost and a benefit. The prioritization algorithm investigates the consequences of delaying maintenance work specified in the optimal programme. Thus, in a particular year if a maintenance job, specified in the optimal programme, is not carried out there would be a benefit and cost. The benefit would be the money saved that year by not doing the work. The cost would be the increase in life-time cost that would occur if the work is not done that year compared with the life-time cost of the optimal programme.

For this exercise costs are discounted back to the year when the prioritization is carried out. In practice the optimization algorithm is run twice, in both cases starting at the year when the prioritization is required, but in one case maintenance work is allowed that year whereas in the other case it is not permitted. This procedure is carried out for every element that was scheduled for maintenance on the basis of the original optimized maintenance programme and the elements ranked in order of cost:benefit ratio. The elements with the lowest ratio have the highest priority for maintenance because the costs of deferring maintenance are relatively low compared with the benefits. The prioritized maintenance programme for the year is then obtained by maintaining bridges in order of increasing cost:benefit ratio until the budget is consumed. The bridges that are not maintained will have their deterioration calculated on the basis that no maintenance was carried out; they will be considered for maintenance again on the next occasion that the optimized programme specifies them. This will not necessarily be in the next year because the deferral of optimal maintenance the previous year could have resulted in a condition state transition which would necessitate more complex and expensive maintenance action. In these circumstances it is quite likely that the next optimal maintenance action could be some years in the future.

In the unlikely event that essential maintenance work specified in the optimal programme has too low a priority for the work to be done appropriate traffic restrictions must be imposed. The costs of such restrictions should be included in the cost–benefit analysis, therefore traffic delay costs should be included when the prioritization algorithm is used.

17.4.4 The effectiveness of a maintenance strategy

The question 'is the money spent on bridge maintenance providing good value?' is often asked. The answer depends on how closely the objective of the maintenance strategy is satisfied. The objective of the optimized maintenance strategy is to maintain serviceability of the bridges in the stock during their design life at a minimum life-time cost. The optimized maintenance programme produced by the algorithms described in Sections 17.4.1 and 17.4.2 will fully satisfy this objective because the mathematics are so defined. This is an ideal situation and when budgets are limited and maintenance work needs to be prioritized a different sort of objective is appropriate. The prioritized maintenance programme minimizes the difference in life-time costs between the prioritized and optimized programmes. This gives assurance that good value is being obtained from the money actually spent on maintenance, but it does not indicate the consequences of the reduced level of maintenance. In order to investigate the effect that the maintenance expenditure has on the condition of the bridge stock it is necessary to set objectives such as:

- the average condition of the stock in 20 years time will be in the range 2 to 3, for example, or
- the proportion of bridges requiring essential maintenance during the next 20 years will not exceed 0.5%.

One of the advantages of the algorithms described in Sections 17.4.1–17.4.3 is that they can be used in a hypothetical or 'what if' mode to investigate how closely the prioritized maintenance programmes for a range of budget values satisfy objectives of the type mentioned above. The prioritization algorithm ensures that any money spent on bridge maintenance achieves good value while an investigation of the consequences of a specific prioritized maintenance programme shows how close the programme gets to satisfying the condition target for the stock. The three graphs in Figure 17.5 show different ways of monitoring the effectiveness of prioritized maintenance programmes against practical objectives. In each case budget B2 satisfies the objective most closely. Budget B1 is clearly two small to meet the objective in each case whereas, although budget B3 satisfies the objective, it is larger than necessary.

This section concludes the discussion of the main algorithms employed in network level bridge management. The final section of the chapter briefly describes some other techniques used in the management of infrastructure.

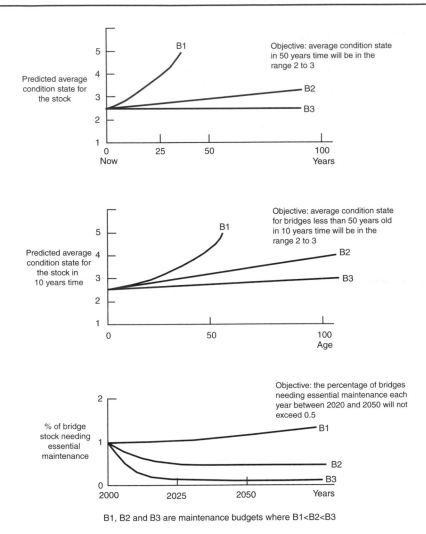

Figure 17.5 Varying the size of the maintenance budget to satisfy objectives.

17.5 Other techniques used in the management of bridges

The previous sections have described in some detail deterioration modelling, pre-diction of condition and the optimization and prioritization of maintenance. There are a number of other techniques that are often used:

- whole-life costing
- probabilistic modelling
- risk analysis.

These techniques are described briefly, in the following subsections.

17.5.1 Whole-life costing

Until the last decade most infrastructure investment decisions were based on lowest initial cost. Structures like bridges which are expected to have a long life inevitably require some maintenance during their life in order to achieve continuing serviceability. It was, therefore, considered to be more rational to base expenditure decisions on estimates of life-time cost instead of initial costs. The most important feature of whole-life costing (Vassie and Rubakantha, 1996) is the estimation of future costs. The main purpose of whole-life costing is to appraise the economics of a number of possible options to aid decision making. The type of option is unrestricted but the whole-life costing of bridges has usually been adopted to compare construction options for improving durability or to compare alternative maintenance methods. Since the objective is to appraise the economics of a range of options rather than to determine the actual expenditure needed over the life of a bridge the process of estimating future costs is simplified by neglecting the effects of inflation. The process of whole-life costing consists of a number of key steps:

- list the types of event leading to expenditure over the life of the structures
- estimate the cost of each of these events at today's prices
- establish whether the cost of each event will vary significantly with time over an above the effects of inflation
- estimate the age of the bridge when these events are likely to occur.

The types of event leading to expenditure over the life of a bridge could include:

- initial cost of design, land and construction
- preventative maintenance work
- repairs
- replacement of components such as expansion joints, parapets, bearings and water-proofing membranes
- strengthening
- replacement of elements such as decks or piers.

Some of these events may result in disruption to users of the bridge. The value of this disruption can be calculated using the computer program QUADRO (Department of Transport, 1982). These costs can be much higher than the maintenance costs for a bridge on a busy road. Their value is sensitive to the flow rate of vehicles and since vehicle flows have been increasing progressively over the last few decades, this provides an example of a cost which is increasing with time.

There is bound to be some uncertainty associated with some of these steps because decisions about what may happen in the future are involved. Whole-life costing models can be deterministic or probabilistic. Where considerable uncertainty about the values of the input variables to the model exist it is better to use a probabilistic model, where instead of inputting a single value for a variable such as the mean, a probability distribution is used as the input for the variable (see Section 17.5.2 for further details).

Whole-life costing is an economic analysis involving expenditure at different times. Economists consider that expenditure which can be deferred to a later date saves money. The idea is that the money could be invested during the period of deferral and therefore have an increased value when the expenditure is eventually made. This process is called discounted cash flow, but for maintenance work account must also be taken of any increased quantity of maintenance required at the later date due to continuing deterioration. The factor which defines the time value of money and hence the relative costs of work carried out at different times is called the discount rate (Spackman, 1991). The discount rate depends not only on investment interest rates, but also on the risk of a bridge not achieving its design life. This risk is evidently higher for bridges with a long required life than for a car tyre, for example, which has a relatively short life and value. Occasionally bridges have not reached their design life before demolition for reasons such as redundancy, road realignment, rapid deterioration or increased loading. Discount rates are also affected by commercial factors such as the rate of return required on the initial investment; a high rate of return is associated with a high discount rate. In the UK discount rates for infrastructure such as bridges have ranged between 6% and 8% per annum during the last decade. In other countries discount rates ranging from 2% to 20% have been used.

A discount rate of 6% per annum implies that costs incurred after an age of 40 years have a value of less than 10% of the value at age zero and are therefore often neglected because of the small contribution they make to the whole-life cost. This simplification is not always justified particularly where the base cost is high or is increasing with time. Nevertheless it is generally true to say that high discount rates do not favour options designed to improve durability and minimize future maintenance needs. This has resulted in much controversy and some people consider that infrastructure with a long design life should have a correspondingly low discount rate.

The base year for discounting is the year when the investment decisions are made. For durability measures introduced at the time of construction, the base year would be the year of construction whereas for maintenance options the base year would be the year when the maintenance work was carried out. The formula for calculating the net present value (NPV) of maintenance work carried out in the future is:

$$\text{NPV} = C_0(1 + 0.01i)^{-n}$$

where C_0 is the cost of the work in the base year, i is the discount rate expressed as a percentage and n is the year in which work is carried out, relative to base year.

The whole-life cost (WLC) consists of the summation of all the NPVs over the design life of the bridge:

$$\text{WLC} = \sum_{n=0}^{n=120} (\text{NPV})_n$$

where $(\text{NPV})_n$ is the net present value of any expenditure made in year n.

Whole-life costing is an economic analysis tool and does not specifically take account of factors such as energy consumption, the use and production of waste materials or pollutants, and the consumption of natural resources. These factors are important considerations for sustainable construction. It is likely that in the future a more comprehensive life-cycle analysis will be undertaken where sustainability issues are considered on an equal footing with economics.

17.5.2 Probabilistic methods

It is known that among a set of nominally similar bridges the rate of deterioration and the need for essential maintenance in a particular year, for example, vary markedly from bridge to bridge. This variability results from our limited knowledge of the complexities of the deterioration processes occurring in bridges. The uncertainty about the precise value of a variable means that it is often better to represent a variable by a probability distribution (Sundararajan, 1995) rather than by a single value such as the mean or a combination of mean and variance. The use of a mean value is clearly unsatisfactory since it provides no indication of the dispersion of values for individual bridges. The use of the mean and variance, although it provides a measure of dispersion, also assumes a Gaussian shaped distribution. A probability distribution that accurately describes the dispersion of individual results is the best option, but necessitates a considerable body of data. Simple distributions based on limited data can also be useful where great accuracy is not needed.

Probabilistic models have been used to predict whole-life costs, the number of bridges requiring essential maintenance each year and the cross-section of steel reinforcement remaining at different ages, when corrosion is occurring.

For example, triangular probability distributions have been used to estimate the number of bridges in a stock that will need essential maintenance each year. Only three pieces of information are required to set up a triangular distribution:

- the lowest age at which any bridge in the stock needs essential maintenance
- the highest age at which any bridge in the stock needs essential maintenance
- the mode of the distribution i.e. the age at which the need for essential maintenance is most likely.

Such approximate distributions will clearly only provide approximate results, but for broad distributions, where the result is not particularly sensitive to the shape of the distribution, the results can be adequate.

Probability distributions enable simulation techniques to be used of which Monte Carlo Simulation is the best known. Simulation techniques use a large number (typically about 1000) of 'what if' questions and express the answers as a probability distribution. A single 'what if' question could be, for example, 'what is the net present value of a particular maintenance action if it is carried out at age t years?' A distribution of ages when the maintenance action is needed can be randomly sampled to give an age when maintenance is needed. The usual procedure is to generate random numbers in the range 0 to 1, the same as the probability range, and then to read off the age corresponding to the random number from the cumulative version of the probability distribution. This age is then used to calculate one value of NPV (see Section 17.5.1). This procedure is repeated until the distribution of NPVs converges

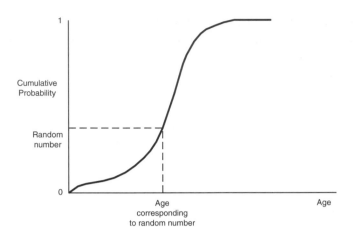

Figure 17.6 Random sampling of a cumulative probability distribution for the age of structure when a particular event occurs.

to within user specified limits. This sampling procedure, shown in Figure 17.6, ensures that the proportion of samples of a particular age corresponds to the value of the probability that maintenance will be needed at that age. Thus if the probability of maintenance at a particular age is low few cases corresponding to this age will be selected and vice versa.

17.5.3 Risk analysis

The risk (Shetty *et al.*, 1996) of an event occurring is generally defined as some combination, often the product, of the probability that the event will occur during a given time period and the consequences of it occurring. The probability of occurrence, although often difficult to calculate, has a clear meaning. The consequences of occurrence are, on the other hand, open ended and less well defined. For example the consequences of a bridge parapet failing due to vehicle impact could include injury to people in the impacting vehicle, injury to people underneath the bridge, damage to the bridge, damage to the features under the bridge, such as a road or railway, damage to the impacting vehicle and vehicles under the bridge, closure or partial closure of the bridge and the service underneath while repairs are effected. Consequences can be expressed as a monetary equivalent or on an arbitrary scale. It is not normally possible to make any more than a fairly rough estimate of the probability and consequences of an event occurring since they depend on may variables of uncertain value and interdependence.

The main advantage of carrying out a risk analysis is to ensure that the factors influencing the probability and the potential consequences are properly considered. This should identify the factors to which the probability of the event occurring is most sensitive and the circumstances in which the potential consequences are most severe. This information can then be used to introduce modifications to reduce risk when this is considered to be too high.

There are many ways, of varying thoroughness, of carrying out a risk analysis. The most simple type of risk analysis consists of a somewhat cursory, qualitative assessment of some of the factors affecting probability and some of the consequences. This type is only occasionally adequate. At the other extreme a full quantitative model, calculating the probability and consequences of the event occurring, is required. The events that engineers want to analyse are usually complex and the factors and their inter-relationships that affect the probability are poorly understood. Even when a reasonable probability model can be developed there is usually an insufficient availability of data to apply it. A full risk analysis is, therefore, often not feasible. A semi-quantatitive approach has been used with some success for a number of events. This approach uses a points scoring system for the probability and consequences and consists of a number of main steps:

(i)   set up a group of engineers with experience relating to the particular event
(ii)   get the group to list the factors affecting the probability and the potential consequences
(iii) get the group to rank the factors affecting probability and the consequences in order of importance; then allocate a weighting to each factor and consequence, by consensus
(iv)   set up criteria for each factor and consequence based on quantitative measures to provide an indicator of their importance in a particular situation; each criterion is associated with a points score.

The above steps set up the risk analysis system. Step (iii) provides an assessment of the importance each factor and consequence in general while step (iv) is an assessment of the importance of each factor and consequence in relation to a particular case. Consider the example in Table 17.7 that considers only two factors and two consequences for reasons of simplicity and presentation. In practice many more factors and consequences would be considered.

In the simplified example in Table 17.7 for a case where the membrane is 10 years old, 10% of tests indicated defective condition, the road is heavily salted and the cover depth is 40 mm then:

total probability score would be	4
normalized probability score would be	0.5
total consequences score would be	6
normalized consequence score would be	0.6
risk score would be	30

This example has taken no account of inter-relationships, although this is possible. For example if the road is not salted early corrosion would not occur. The risk system could take account of this by introducing the following restriction: 'If a zero point score is awarded for the consequence "salt entering the bridge deck" then a point score of zero must be awarded for the consequence "early corrosion of deck reinforcement".' A detailed example of a semi-quantitative risk analysis is provided in reference (Institution of Civil Engineers, 1998).

Table 17.7

Event		Failure of a bridge deck waterproofing membrane		
			Criteria	
Probability factors	Weighting	0 points	1 point	2 points
Age of membrane	1	<5 years	5–20 years	>20 years
Condition of membrane	3	<5% defective	5–15% defective	>15% defective
			Criteria	
Consequences	Weighting	0 points	1 point	2 points
Salt entering concrete deck	1	road not salted	No. of saltings per year <10	No. of saltings per year >10
Early corrosion of deck reinforcement	4	cover depth >50 mm	cover depth 30–50 mm	cover depth <30 mm

To apply the risk analysis the following steps are carried out for the particular case under consideration:

(a)   for each probability factor and consequence a points score is awarded depending on the criterion satisfied

(b)   the scores from (a) are multiplied by the weightings and summed to give a total probability score and a total consequence score

(c)   the total scores are normalized by dividing by the maximum possible probability and consequence scores giving a range of 0 to 1 for both totals

(d)   the risk is then obtained by multiplication of the normalized probability and consequence scores; further multiplication by 100 gives a risk scale of 0 to 100.

Bibliography

Al-Subhi K *et al*. Optimizing system level bridge maintenance rehabilitation and replacement decisions. FHWA/ NC 89-002. Washington, DC, 1989.

Bungey JH. *Testing of concrete in structures*. Surrey University Press, Guildford, 1982.

Darby JJ *et al*. Bridge management systems: the need to retain flexibility and engineering judgement. *Bridge Management* 3, 212–218, E&F Spon, London, 1996.

Das P. Bridge management objectives and methodologies. *Bridge Management 3*. 1–7. E & FN Spon, London, 1996.

Das P. Development of a comprehensive structures management methodology for the Highways Agency. *The Management of Highway Structures Conference*. Institution of Civil Engineers, London, 1998.

Department of Transport. *Quadro 2 User Manual*, 1982.

Department of Transport. *Bridge Inspection Guide*. HMSO, London, 1983.

Federal Highways Administration. *Bridge Management Systems*. FHWA DP-71-01. Washington, DC, 1987.

Frangopol DM and Estes AL. Lifetime bridge maintenance strategies based on system reliability. *Structural Engineering International IABSE* **7**, 193–198, 1997.

Gusella V *et al*. Information system for the management of bridges owned by the province of Perugia, Italy. *Bridge Management 3*, 592–602, E&F Spon, London, 1996.

Harding JE *et al*. eds. *Bridge Management: Inspection, Assessment, Maintenance and Repair*. London, 1990.

Harding JE *et al*. eds. *Bridge Management 2*. Thomas Telford, London, 1993.

Harding JE *et al*. eds. *Bridge Management 3*. E & FN Spon, London, 1996.

Highways Agency. *Inspection of Highway Structures*. BA63/94. *Design Manual for Roads and Bridges Volume 3*. HMSO, London, 1994.

Highways Agency. *The Assessment of Highway Bridges and Structures*. BD 21/97 and BA 16/97. *Design Manual for, Roads and Bridges Volume 3/Section 4*. HMSO, London, 1997.

Hogg V and Middleton CP. Whole life performance profiles for highway structures. *The Management of Highway Structures Conference*. Institution of Civil Engineers, London, 1998.

Institution of Civil Engineers. *Supplementary Load Testing of Bridges*. Thomas Telford, London, 1998: 59–63.

Institution of Civil Engineers. *The Management of Highway Structures Conference Proceedings*. Thomas Telford, London, 1998.

Jiang Y and Sinha KC. A dynamic optimization model for bridge management systems. *Transportation Research Board Annual Meeting*. Paper No. 88-0409, Washington, DC, 1989.

Jiang Y *et al*. Bridge performance prediction model using the Markov Chain. *Transportation Research Record* No. 1180, TRB, Washington, DC, 1989.

Kreugler J *et al*. Cost effective bridge management strategies. FHWA/RD-86/109. Washington, DC, 1986.

Organization for Economic Cooperation and Development. *Bridge Maintenance*. OECD, Paris, 1981.

Saito M and Sinha K. Timing for bridge replacement, rehabilitation and maintenance. *Transportation Research Board Annual Meeting*. Paper No. 890336, Washington, DC, 1990.

Shetty NK *et al*. A risk based framework for assessment and prioritisation of bridges. *Bridge Management 3*, 571–579, 1996.

Spackman M. Discount rates and rates of return in the public sector: Economic Issues. Government Economic Service Working Paper, No. 113, HM Treasury, 1991.

Sundararajan CR. *Probalistic Structural Mechanics Handbook*. Chapman & Hall, London, 1995.

US Department of Transportation, Federal Highways Administration. *Recording and Coding Guide for the Structure Inventory and Appraisal of the Nation's Bridges*. FHWA-ED-89-044. Washington, DC, 1988.

Vassie PR and Rubakantha SA. Model for evaluating the whole life cost of concrete bridges. *Corrosion of Reinforcement in Concrete Construction*. Society for Chemical Industry, 1996: 156–165.

Winston WL. *Operations Research*. PWS–Kent, Boston, MA, 1991.

18 Inspection, monitoring and assessment

C. ABDUNUR

In spite of their relatively simple structural systems and well-defined supports, bridges suffer heavily from the effects of age, climate and especially traffic that often exceeds the codified limits. Their strength and durability depend on the type and quality of the constituent materials, structural design, construction and maintenance. This latter factor is the only one that can be influenced for existing structures. Maintenance, however, can be effective only in the light of adequate inspection, monitoring and assessment procedures, facilitating, if need be, an appropriate preventive action (Calgaro and Lacroix, 1997). This chapter will consider: the main causes of damage, certain material and structural investigation methods, and assessment procedures. Only concrete and steel bridge decks will be discussed.

18.1 Main causes of degradation

The deterioration of bridges may have different causes and degrees of gravity, with or without visible signs. It can materialize in unusual cracks, excessive strain or changes in geometry. Other serious types of damage, such as embedded steel corrosion and concrete alkali–silica reaction, become apparent only when reaching an advanced stage. Before presenting investigation methods, the probable causes of the most frequently observed types of degradation will be classified. The classification is not exhaustive.

18.1.1 Deterioration of constituent materials
Concrete
While generally known for its durability, concrete suffers at times the consequences of gas and water penetration into its capillary system. Physical, physicochemical and chemical mechanisms may deteriorate its properties when external agents react with concrete hydrates forming expansive or soluble compounds. Concrete can also be subjected to destructive internal swelling due to chemically incompatible ingredients.

Physical or mechanical agents

Frost–thaw cycles act on the free water content in unprotected areas of concrete bridges. Scaling, swelling and often crack patterns affect the surface and deeper layers with a variable intensity depending on the porosity and degree of saturation. Road de-icing salts cause additional penetration of chloride ions, hence steel corrosion.

When surfaces of concrete decks and masonry piers are exposed to water flow, they are liable to abrasion and erosion; impacts aggravate the consequences.

Physicochemical agents

If no appropriate precautions are taken, concrete intrinsic shrinkage may favour various forms of cracks, in different directions, at consecutive formative stages of the material. The first cracks appear within hours from casting, due to packing down and sedimentation. After shuttering removal, other shallow widely spaced hair cracks may follow owing to self-desiccation shrinkage; they grow wider and deeper with high hydration heat. At a much longer term, drying shrinkage cracks may appear due to subsequent free water loss.

Chemical agents

The following illustrated reaction products are observed under the electron microscope (EM).

Carbonation: atmospheric carbon dioxide dissolves in water to form carbonic acid, which reacts with most cement hydrates (Figure 18.1). This reaction, called carbonation, may very slowly move into the unsealed concrete pores, cancel alkalinity and set up a steel corrosion process. In general, it would take many years and even decades for carbonation to reach a normally covered reinforcement.

Sulphate reaction: sulphates in concrete may be of external geological or biological origin, e.g. industrial pollution. They also have an internal origin, such as the use of gypseous or pyrite aggregates (Figure 18.2) and of quick hardening heat treatment, producing what is known as delayed ettringite formation (Divet *et al.*, 1998). The sulphate destructive action on cement constituents occurs in two steps: sulphates react

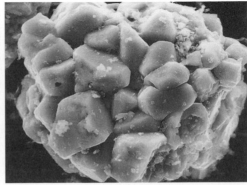

Figure 18.1 Carbonation: $CaCO_3$ crystals (EM) (Photo: LCPC).

Figure 18.2 Pre-existing pyrites (EM) (Photo: LCPC).

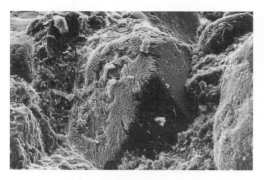

Figure 18.3 Expansive ettringite on aggregate surfaces (EM) (Photo: LCPC).

Figure 18.4 Destructive sulphate reaction in concrete (Photo: LCPC).

with calcium hydroxide to form secondary gypsum which, in turn, reacts with aluminates to form secondary ettringite (Figure 18.3), a very expansive crystalline substance inducing high internal pressures leading to concrete disintegration (Figure 18.4).

Marine environment: seawater attacks by its sulphates and chlorides. These go through a complex chain reaction with calcium hydroxide in the cement, to finally form the above-mentioned expansive ettringite (Figure 18.5). Chloride ions, remaining free after diffusion in the concrete, can attain and corrode the reinforcement.

Alkali–silica/aggregate reaction: in a concrete mix, the presence of incompatible ingredients can start a complex chemical reaction with destructive mechanical consequences (Hobbs, 1988; Swamy and Al-Asali, 1986). Among the known mechanisms, the most frequent is the alkali–silica reaction (ASR). It takes place between the interstitial alkaline solution, mainly from the cement, and a reactive type of silica particles from the aggregates. Favoured by lime and humidity, an alkali silicate gel is thus formed; it further combines with the calcium hydroxide of the mortar paste to produce a highly water absorbing hence swelling gel. The role of water was more thoroughly investigated in a recent study (Larive, 1998). The resulting concrete expansion and tendon or reinforcement over tension lead to crack patterns that may attain

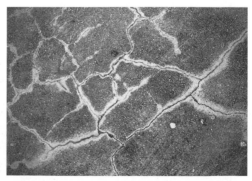

Figure 18.5 Various damaging products of marine attack (EM) (Photo: LCPC).

Figure 18.6 Concrete crack pattern due to alkali–silica reaction (Photo: LCPC).

Figure 18.7 Expansive ASR gel (EM)   *Figure 18.8 Expansive ASR crystals (EM)*
(Photo: LCPC).   *(Photo: LCPC).*

or exceed a 10-cm depth, with a 2–5 cm spacing (Figure 18.6). Observed under the scanning electronic microscope, the ASR products vary from a smooth gel (Figure 18.7) to different groups of crystals of crackling aspect (Figure 18.8). The gel is formed and often stays around the aggregates but may also migrate in the concrete mass and even leak out through surface cracks.

Steel

When in contact with atmospheric or chemical agents, steel can react and be dissolved. In civil engineering structures, this corrosion mechanism mainly results from an *electrochemical process* in solution. The corrosion intensity depends on the steel metallurgic and mechanical properties as well as on the reagent parameters, such as: composition, treatment, shape, stress, reagent concentration, pH, oxygen content, temperature (Engel and Klingele, 1981; Raharinaivo *et al.*, 1998).

Steel rusts when exposed to humid air ($\geqslant$50–70%) and to dust or other deposits especially sulphates. This atmospheric corrosion accelerates with certain polluting gases. In marine environments, chloride solutions create a thin permeable layer on the steel surface and allow a corrosion process to start.

In reinforced and prestressed concrete, embedded steel corrosion starts when the protective alkaline film is destroyed by carbonation or chloride ions (Figures 18.9 and 18.10). The process advances easily where the concrete cover lacks thickness and density; it also intensifies with available oxygen and humidity especially when the latter alternates with dryness. The resulting rust reduces the effective steel sections, their ductility and fatigue resistance. Through its expansion to about six times the reacting steel volume, rust also bursts the surrounding concrete. The inner steel layers may be less affected but the danger resides in their inaccessibility, hence difficult inspection and uncertain assessment.

Prestressing tendons are particularly sensitive to cover or grout injection quality; they are also vulnerable to other corrosion forms involving cracks, hydrogen embrittlement or fretting-fatigue mechanisms described below.

Figure 18.9 Reinforcement corrosion (Photo: LCPC).

Figure 18.10 Corrosion attaining tendon ducts (Photo: LCPC).

In many cases, as already mentioned, steel corrosion in concrete goes on without external signs or defects. However, where waterproofing is defective, water leakage may bring rust traces or efflorescence to the surface.

In suspension and cable stayed bridges, cables may deteriorate and fail either by section reduction or by cracking of constitutive wires. Each type of damage can result from both electrochemical and mechanical processes, respectively leading to corrosion and 'fretting'.

Wires corrode in two ways (Figure 18.11). *Pitting corrosion* occurs mainly at water retention points such as the lower parts of suspension cables, deviation saddles, anchorage zones and most other connections. *Stress cracking corrosion*, involves a reaction between surface oxides under tension and the steel, producing nascent hydrogen, hence metal embrittlement.

Small relative tangential displacements, between parallel touching wires of curved cables and tendons, can generate periodic, locally high frictional forces (Figure 18.12). This mechanism, called 'fretting', wears away or cracks the contact surfaces in the long term. The damage worsens with the continued presence of metal particles stemming from the process. Stress variation due to oscillatory friction can also cause fatigue cracks and lead to failure (Brevet and Siegert, 1996; Siegert, 1997).

(a)   (b)

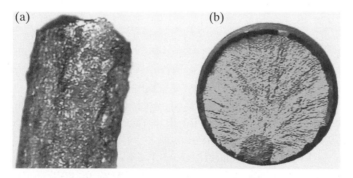

Figure 18.11 Two types of wire corrosion: (a) pitting, (b) stress cracking (Photo: LCPC).

Figure 18.12 'Fretting' on contact surfaces leading to 'fretting-fatigue' failure in tendon wires (Photo: LCPC).

Finally, under excessive and especially concentrated stress, high carbon steel with impurities suffers brittle failure while mild steel has a better-adapted ductile behaviour (Figure 18.13).

18.1.2 Degradation through applied forces

From a mechanical viewpoint, a structure is damaged when its static or dynamic equilibrium is disturbed. Failure occurs where the magnitude or repetition of the applied stresses or strains exceeds the allowable limits of the material.

Dead loads and live loads

Surprisingly, careless mistakes are sometimes committed in determining dead loads. Underestimating cubage or density and neglecting equipment weight, for example, reduce the safety factor but rarely are a direct source of damage.

On the other hand, during the multi-phase construction and subsequent life of a statically indeterminate prestressed concrete bridge, the combined time-dependent effects of concrete creep and steel relaxation have frequently caused serious structural cracks under dead load. These effects are now being accounted for in most design codes. However, given the possible differences between the assumed and actual viscoelastic behaviour, it is still advisable to weigh the support reactions right after construction, or as early as possible, for future reference.

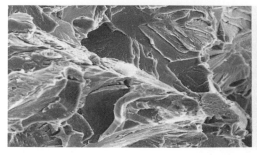

Figure 18.13 Failure of tendon wires: brittle (left), ductile (right) (EM) (Photo: LCPC).

Through their extreme values as well as their dynamic and fatigue effects, traffic loads are among the major causes of ageing. National loading codes have been periodically revised to cope, on a sounder scientific basis, with steadily increasing vehicle weights and traffic density. The evaluation of actual live load configurations is necessary for an existing bridge, regardless of the initial design considerations.

In road bridges, dynamic effects (other than fatigue) are incorporated by introducing impact coefficients depending on structural, surface, vehicle and velocity parameters. These coefficients proved higher near surface discontinuities and may explain, at least partially, the denser crack patterns often observed at the expansion joints of reinforced concrete bridge slabs. Additionally, vibration and longitudinal braking effects may displace and deteriorate inadequately designed bearing systems.

Most of railway bridges, constructed in the early 1900s using masonry or metal, were over-designed and thus able to cope with greater vehicle loads and speeds. However, heavier service increased their exposure to fatigue. Bridges of this type are equally subject to:

- rivet loosening due to impact effects on discontinued rails
- temperature effects of long rails on fixed supports
- resonance and lateral effects at high speed as well as horizontal forces due to speed changes.

Thermal effects
On the materials
By its level and fluctuation, temperature acts on material durability (Divet et al., 1998). Through its diffusion, it creates potential strains with local and global effects on the structure (Behr and Trouillet, 1998; Priestley and Blucke, 1979).

Concrete, cast in very cold weather, may deteriorate by the expansion of freezing water used in the mix. This should be set apart from the swelling and surface scaling effects of frost–thaw cycles on hardened concrete. Concrete, cast in hot weather ($>35°C$), hardens quickly and has reduced plasticity. Accelerated cement hydration also creates higher core-surface temperature gradients inducing tensile surface stresses on cooling. Extensive cracking may follow.

Another similar effect is caused by a completely different mechanism: high temperature and low atmospheric humidity favour free water loss, creating considerable hygroscopic gradients close to concrete surfaces. The resulting proportional potential strain, better known as drying shrinkage, induces very high tensile surface stresses exceeding the concrete tensile strength and often those due to live loads.

Where specific secondary reinforcement is insufficient, the superimposed effects of hydration and hygroscopic mechanisms cause extensive surface cracks, leading to concrete deterioration and steel exposure to corrosion.

In steel, while low temperature increases the yield point and strength, it also decreases ductility in certain types where brittle failure may occur with practically no plastic strain (Persy and Raharinaivo, 1987). High temperature may also catalyse certain chemical reactions causing material degradation.

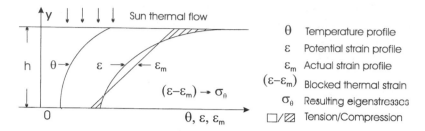

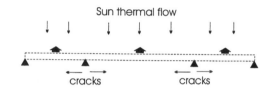

Figure 18.14 Temperature-induced strains and stresses over the height, h, of a concrete bridge deck (Photo: LCPC).

Figure 18.15 Temperature-induced tensile stresses in lower fibres, at intermediate supports (Photo: LCPC).

On the structure

In concrete bridges, heat from the sun diffuses throughout the deck, creating a potential non-linear strain field, ε, partially restricted by plane section conservation to an actual linear distribution ε_m. Over the section height, h, as illustrated in Figure 18.14, two longitudinal mechanical effects result:

• a non-linear self-balanced stress profile with maximum tension at mid-fibres
• a linear average strain distribution.

The former amplifies the local surface tension due to hydration heat and drying shrinkage; the latter affects continuous spans, inducing flexural stresses that, if not accounted for, may cause serious cracks in the lower fibres at the intermediate supports (Figure 18.15).

In composite bridges, a deterioration of the connectors was often observed due to the differential coefficient of thermal expansion between steel and concrete.

Miscellaneous destructive actions

The undermining of pier footings by water has been one of the principal causes of bridge failure. With the continual progress in deep foundation technology this hazard has now been practically eliminated in recent structures.

In seismic regions, a soil acceleration exceeding 0.3 g may seriously damage concrete bridges with heavy rigid elements, excessive longitudinal steel bars and/or insufficient transverse reinforcement. On the other hand, more satisfactory behaviour has been noticed in square, multi-span reinforced and prestressed concrete slab bridges as well as in prestressed beams on neoprene supports.

Ship and vehicle impact is less frequent but more serious on bridge decks than on piers, especially on light and slender structures (Calgaro, 1991).

Fire deteriorates concrete at 200°C, prestressing tendons at 175°C, and high strength steel bars at 350–450°C (Persy and Deloye, 1986).

18.1.3 Design errors

Design is mainly for material durability and structural adequacy. The degradation of bridges can result from mistakes made at different design stages, such as:

- inadequate theoretical models as far as the prediction of applied forces and assumptions on the local and mechanical behaviour especially fatigue
- insufficient anti-corrosion precautions
- wrong or inappropriate choice of the shape and arrangement of structural elements with regard to other durability aspects and access for maintenance.

Some specific design errors, linked with the chosen constituent materials and structural system, are considered below (Calgaro and Lacroix, 1997).

Reinforced and prestressed concrete bridge decks

Poorly designed waterproofing or drainage and a lack of compact concrete cover may seriously shorten the service life of a bridge.

In ordinary reinforced concrete bridge decks, structural design errors have relatively few consequences. Generally, damage may arise from neglected thermal stresses, unbalanced outward thrust of steel bars, secondary reinforcement inadequacy, lacking transversal ties of main steel layers , short anchor or splicing bar lengths and bearings too close to deck boundaries. Skew slab bridges may crack in their acute angle area if the required reinforcement density and direction are not observed. In continuous spans, wide laterally cantilevered slabs also crack at intermediate supports if longitudinal distribution reinforcement is inadequate.

In post-tensioned bridges insufficient anti-corrosion precautions constitute basic durability design errors, namely the *absence* of: tendon anchorage seals, watertight non-corrosive metal ducts, efficient deck waterproofing and drainage including the expansion joints, sufficient concrete cover preventing the degradation chain of reinforcement corrosion–concrete spalling–tendon exposure. Deck anchorages and multiple construction joints add to corrosion risks.

In structural design, the following errors have led to flexure cracks in box girders built by cantilever-balanced segments: discarding thermal gradient effects, underestimating prestress frictional losses and/or stress redistribution due to creep and shrinkage, stopping simultaneously several slab-anchored continuity tendons. Segment joints may also open when tendon couplings are improperly distributed or lack efficient de-bonding devices and complementary reinforcement.

Near the supports, shear cracks result from underdesigned active stirrups or an overestimated ability of longitudinal prestressing to reduce shear. Flexural and shearing inadequacies often combine to cause an intermediate form of cracks as illustrated in Figure 18.16.

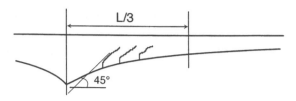

Figure 18.16 Shear cracks combined with flexural deficit in a box girder bridge (Photo: LCPC).

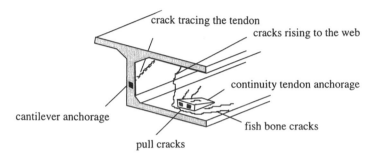

Figure 18.17 Principal types of cracks in anchorage zones.

Locally, underdesigned reinforcement around tendon anchorages causes diffusion or transfer cracks (Figure 18.17). In front of web-embedded anchorages, diffusion cracks retrace the tendons on the surface; although mechanically tolerated, they favour the infiltration of water that expands on freezing and worsens the case. At the bottom slab anchor blocks, diffusion cracks appear on both sides in fishbone arrangement and may extend to the webs with a ~45° inclination; transfer cracks develop transversally right behind the blocks.

The consequent discontinuities, mainly flexure cracks and opening joints, often develop a high fatigue risk for the tendons, depending on the number, intensity and variability of load cycles. Bond failure and irreversible deflections may follow, seriously affecting service life. Experience has shown, however, that fatigue damage hardly ever occurs where structural design succeeds in avoiding cracks in concrete sections.

Steel and composite bridge decks

In *steel bridges*, corrosion is by far the main cause of deterioration. This electrochemical process, described above, results from poor waterproofing or drainage and from other specific errors in the design. Mistakes include inappropriate arrangement and connections of structural elements, favouring water stagnation and condensation. For example, channel sections are often severely corroded when open upwards, but usually spared, even without maintenance, when inverted. Widely spaced rivets allow an initial and growing space for humidity and corrosion in between the connected plates. In a wet atmosphere, these plates may also be laminated by corrosion when metals of different electric potential are used at riveted points.

In steel road bridges, fatigue damage is often caused by live loads of intermediate magnitude and frequency. Orthotropic plate bridges are particularly vulnerable when the girder–stiffener welded junctions and the flexible overlying floor plate are not properly designed for fatigue. Examples are found in narrow, initially temporary orthotropic bridges with underdesigned floor plate thickness, thin surfacing and low quality welding.

In older railway bridges, which were not really designed for fatigue, the steady increase in traffic has led to serious deterioration, especially in short members, such as stringers and floor beams, directly receiving axle loads and often calculated with unrealistic boundary conditions.

Generally, fatigue damage shows in riveted and welded connections owing to underdesign, stress concentration of different origins and poor workmanship.

Errors in structural design include:

- non-concurrent truss bars developing substantial secondary moments at the joints and causing rivet or plate failure
- sudden change in section profile where, together with welding shrinkage, the cumulative stress concentration often causes cracks
- longitudinal rivet pull-out at the end angle connections of truss stringer webs, designed as simply supported on the floor beam webs, but constructed as continuous spans.

While design and execution of repairs is usually simpler in metal than in concrete, several mistakes may have serious consequences, for example:

- strengthening by welding without checking the state of the metal base
- failure to apply welding specifications on crack repair
- replacing truss bars without computing the dead load redistribution on the adjacent ones
- strengthening parts of an indeterminate structure without evaluating the new stiffnesses and their effects on moment and stress redistribution
- strengthening a riveted assembly by welds which, besides their risky execution, respond only to live loads and after rivet failure.

In composite bridges, durability and structural problems often arise from inadequate design of slab–girder connections with respect to delayed strains and local forces. Hair cracks first develop in the slab, around the connectors, due to restrained early-age concrete creep and shrinkage. The gradual growth of these cracks leads to local concrete plasticization, then to a relative slip between slab and girders, disturbing the required monolithic behaviour of the deck section.

These transverse discontinuities, worsened by negative moments at intermediate supports or diaphragms, allow water infiltration and accelerate material degradation, affecting service life.

18.1.4 Construction defects

Inadequate work drawings, non-observance of the rules of the art, lack of quality control and poor organization are among the main causes of construction defects

(Calgaro and Lacroix, 1997). The probable causes of some of the most serious and frequent defects are given below.

Reinforced and prestressed concrete bridge decks

Commonly, the quality and use of materials on the site are the most important factors.

In reinforced concrete, construction defects include irregular mixing and transportation from the plant, segregation due to excessive vibration or pouring from too high a point, unprepared formwork inner faces, poorly executed construction joints and abrupt thermal treatment. Inadequate curing accelerates and deepens drying shrinkage cracks, sometimes beyond the concrete cover, affecting reinforcement durability.

Precast concrete elements with inaccurate dimensions may not only cause matching problems but also defective force transmission and stress distribution.

As to reinforcement, a lack of quality control may lead to a bad choice of steel type and bending radii. Incompatible welding modifies the elastic modulus. Failure to maintain the reinforcement in the designed positions, before and during casting, often reduces concrete cover, causing steel corrosion and concrete spalling; it can also decrease the initially required structural capacity.

In prestressed concrete decks, specific construction defects are encountered. Those heavily affecting durability are a lack of waterproofing, incomplete grouting of tendon ducts in humid or aggressive environments and poor sealing of end, deck and transverse anchorages. In areas congested with ducts, especially flange soffits, concrete must be cast carefully or there is a risk of honeycombs, shrinkage cracks and spalling, with all the vulnerability they bring to the constituents.

On the structural level, misplaced tendon ducts are a major defect giving rise to lateral parasitic forces and outward thrusts that spall off concrete and even laminate the slab. A significant increase in frictional forces and wobble may also result and reduce the effective prestress, hence the structural capacity. In segmental post-tensioned bridges, especially of variable depth, the lack of interface glue and shear keys allows relative movements of the top and bottom slabs of successive units increasing the risk of water seepage and corrosion. Poor matching of these segments also generates longitudinal cracks starting from the contact points. In cantilever construction, very serious accidents may occur when the consecutive construction phases are not in accordance with the specifications.

Steel and composite bridge decks

The different parts of these bridges are usually prepared in factories where quality control is relatively easy to carry out. Assembly and protection are thus mainly concerned with possible construction defects on site. Metal bridges can be assembled in different ways: launching on a slipway, using pontoons or cranes, cantilevering, and so on. If not carefully tested for each project, these procedures may all be with some risk and jeopardize the reliability of the structure. Poorly executed welds can be another assembly defect; in cold weather, they are also liable to 'quenching'. Welding quality is usually tested by non-destructive methods, such as X- or γ-rays and ultrasound, if access is provided in the design. Defective protection against corrosion and difficult access for maintenance both lead to the extensive damage described earlier. Defects mostly concern joints, bar arrangements and other details favouring the condensation or stagnation of water and the accumulation of various humid deposits.

Composite bridge-decks may grow vulnerable due to uncontrolled concrete cracks resulting from certain construction defects:

- non-observance of the specified concrete mix and casting phases intended to minimize shrinkage
- excessive steel–concrete thermal gradient while concrete is setting
- too early formwork removal and insufficient curing.

18.1.5 Structural boundary equipment

Bearings and expansion joints grow old, wear out or deteriorate. In certain cases, they may reveal or induce abnormal structural behaviour by restricting or extending the designed boundary movement.

Bearings allow rotation with or without sliding. They are made of metal, concrete or laminated rubber. Their main defects are corrosion for metal, cracking and spalling for concrete and, for laminated rubber, sheet corrosion and relative slip with rubber cracks. Experience has shown that, in most cases, the corrosion of bearings is due more to their environment that to their material.

When incorrectly designed or executed, expansion joints and their supports are deteriorated by repetitive dynamic loads. A few millimetres' difference between the joint and road surfaces strongly amplifies the effect. Dynamic degradation includes failure of metallic elements and their welds, anchorage disorganization and failure or dislodging under vehicle braking or snowplough action. Frost–thaw cycles and de-icing salts also degrade concrete anchorages. In curved or skew bridges the use of toothed joints is often incompatible with their additional transversal movements.

18.2 Investigation methods

For the assessment of bridges, the classical scientific method still remains: gather facts and data, form a hypothesis, start investigating, check results or observations by other methods, establish a theory or draw conclusions.

The assessment of bridges thus begins with a thorough *desk study* of all drawings, records on the constituents, site reports, environmental conditions, and other available information. On site, these are coupled with a detailed *visual inspection* by trained eyes in search of apparent defects mentioned in Section 18.1. An 'intelligent' *guess* is then attempted as to the probable causes of the observed or suspected defects. Guided by this hypothesis, the *investigation* process starts. With reasonable selectivity and optimum cost, it should enable the evaluation of:

- the state and quality of the *constituent materials* with respect to established standards
- the actual *structural response* as compared to the theoretical behaviour of a sound bridge.

While each of these two objectives involves its own methodology, the frequent inter-action between material and structural defects often necessitates the association of both.

For the appraisal of the state of the materials, the methods involve laboratory analyses and experiments on samples as well as *in situ* tests on incorporated constituents.

For characterizing structural behaviour, a variety of techniques are used to evaluate: topographic and geometric variations, applied forces at key positions, the local mechanical state and the response to loading and to environmental fluctuations.

A substantial part of the task involves spotting eventual cracks and monitoring their size, geometry, direction, number and associated leakage of water and other substances.

18.2.1 Tests on the constituent materials
Investigation on samples
Extracted through partially destructive procedures, samples are usually limited in size and number and taken from the least structurally vital points of the bridge. Hence, while insufficiently representative, they can serve, after size effect correction, as calibration references completing documentary information and non-destructive test data.

Physical and mechanical tests
For concrete, physical tests on cored samples (Figure 18.18) include density, porosity, water content and sound velocity measurements. The data they supply, compared to established standards and reference points, help to delimit the extent of variation or degradation and, qualitatively, estimate their intensities. For instance, γ-densimetry is used in the laboratory on concrete samples to determine the percentage of water loss variation with depth and thus enable a correlation with drying shrinkage eigen-stresses. The density profile of a concrete slab can also help to estimate the depth of its degradation after exposure to fire. Porosity measurements usually reflect the quality of concrete cover protecting the reinforcement. Finally, before samples undergo destructive tests, it would be useful to measure their longitudinal and diametrical sound velocity (Figure 18.19), a technique discussed below.

For steel, samples are taken from already broken wires of suspension or stay cables and of post-tensioning tendons. The fracture surfaces can be examined to determine the cause of failure such as pitting or stress cracking corrosion and fretting-fatigue.

Concrete or steel samples are, respectively, subjected to compressive or tensile destructive tests to determine the corresponding ultimate strength of the material and, in the process, its stress–strain relationship. After ductile or brittle failure of steel samples, as already shown in Figure 18.13, microscopic examination of the fracture surfaces may partially reflect their potential mechanical properties. Although these mechanical tests are simple and conventional, concrete samples very rarely have the quality and dimensions of standard cylinders; results may hence be seriously misleading unless all defects and size effects are carefully accounted for.

Figure 18.18 Cored concrete samples (Photo: LCPC Lyon).

Figure 18.19 Sonic test on concrete sample (Photo: LCPC Lyon).

Petrography and metallography

Petrographic examination, under electronic and eventually optical polarizing micro-scopes, can give the mineralogical constitution of concrete ingredients and their degree of deterioration. As already illustrated in the figures of Section 18.1.1, these observations identify ASR and expansive ettringite destructive products.

Metallography provides complete data about the nature, structure, grain form and size (texture), impurities and formative stages of the metal, hence its properties (*De Ferri Metallographia*, 1979). It can be associated with chemical analyses.

Chemical and physicochemical tests

For these powerful methods and the data they supply, even small-sized samples can be sufficiently representative. However, this double advantage is partially tempered by high costs imposing as already stated, a judicious selection of available methods, guided by the hypothesis formed on the probable causes of observed or suspected flows.

Regarding *metals*, chemical analyses can identify the elementary constituents. They are completed by metallography. Electron microprobes are used to determine the nature and distribution of metallic or non-metallic inclusions.

Concerning the *concrete* material, the present methods include:

- elementary chemical analyses for oxides using a plasma torch
- X-ray diffractometry on concrete powder samples to identify its various minerals, superficial deposits and crystallized deterioration products
- differential thermal analysis and thermogravimetric analysis where the physico-chemical transformations and weight loss of the sample are respectively monitored while different components are successively modified or destroyed, each at a given temperature
- further scrutiny under the electron microscope to examine destructive reaction products.

In reinforced concrete and post-tensioned bridges, following field tests on undis-turbed material, samples for chloride detection are taken from grout in ducts and concrete at joints and anchorages. Accessible trapped water is also sampled for chlorides, other corrosive ions and pH level. The same is done for corrosion products observed on tendons or reinforcement.

Together, these methods supply most of the required information about concrete composition: the actual cement content and nature are compared with the initial construction data to detect an eventual deterioration. The initial dry mix, grading and chemically linked water can also be determined and help understand the degradation mechanism.

Investigation on in situ incorporated materials

In spite of the very useful data acquired through extracted samples, non-destructive tests (NDT) and even intrusive methods remain necessary on the same materials incorporated in the structure. Laboratory and field investigations are complementary: the former sets a reference, the latter facilitates a reasonable extrapolation to reality.

NDT on concrete

Tests, such as hammer strike and ultrasonic pulse velocity, are used to appraise the variation of material properties. Those involving impact-echo and radar were mainly developed to explore the inner geometry of the investigated volume. Radar can be an intermediate technique exploring concrete and certain inclusions associated with embedded steel.

The hammer test

A small metal projectile is flung by a spring, through a tube, against an anvil in contact with the concrete surface. On re-bouncing, the projectile contracts the spring through a distance proportional to the superficial concrete hardness, roughly reflecting its compressive strength. While unquestionably simple, this old method gives only qualitative results and relative values. It may be of interest for concrete areas with significant, clear-cut differences in quality.

Sonic tests

Theoretically, in a homogeneous medium, the propagation velocity of ultrasonic waves is a simple function of the elastic modulus, Poisson's ratio and density. In concrete and other heterogeneous materials, this relationship does not really hold but the distribution of measured relative velocities can reflect the quality and continuity of the assessed medium (Cote, 1996; Prost, 1974).

The ultrasonic pulse velocity (UPV) is among the best-known methods of this category. Using a specific piece of equipment, the velocity is obtained by measuring the wave propagation time between the emitting source E and the receiver R. These two devices, generally piezoelectric, are placed either on opposite faces or on the same surface of the structural element. The former arrangement, shown in Figure 18.20(a), can better assess inner layers but may be liable to errors. The latter is more often used, especially when only one face of a tabular member in easily accessible. As illustrated in Figure 18.20(b), the emitting source is fixed at the starting point while the receiver is moved along a straight line, at preferably equal intervals. The testing length may

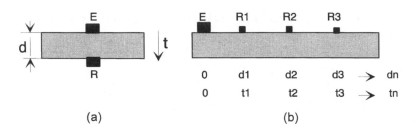

Figure 18.20 Measuring sound velocity (distance d/time t). (a) Through the thickness of a concrete flat element; (b) on the surface.

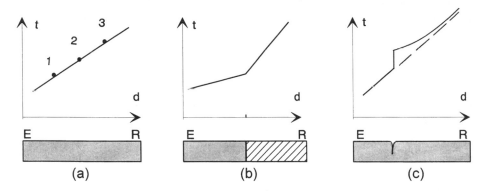

Figure 18.21 different time–space configurations using the ultrasonic pulse velocity. (a) Linear time–space relationship, representing a homogeneous continuum. (b) Curve discontinuity at a construction joint. (c) Crack effect.

be 1–2 m and the pace 10–20 cm. The propagation time is automatically measured using an electronic counter with microsecond accuracy. In a homogeneous continuum, the time–space diagram is invariably a straight line which, given the incremental approach, does not need to pass through the origin (Figure 18.21a). Measurements usually continue along several other parallel lines, forming a 'data grid', to map the assessed surface. UPV is thus used in concrete and masonry bridges to:

- spot discontinuities and local defects (Figures 18.21b, c)
- appraise the quality *variation* of the material over an area, by mapping the measured apparent velocity ranges (Figure 18.22).

It should be noted, however, that there is no close relationship between measured velocity and concrete compressive strength. In certain cases, this strength may be estimated, but only after calibration on extracted samples and fitting in correlation models.

Seismic tomography
UPV was successfully extended to enable, from surface observations, a more direct mapping of the velocity range contours, through a plane section of the investigated

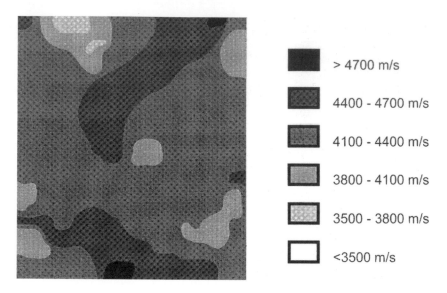

■	> 4700 m/s
▨	4400 - 4700 m/s
▨	4100 - 4400 m/s
▨	3800 - 4100 m/s
▨	3500 - 3800 m/s
□	<3500 m/s

Figure 18.22 Sound velocity ranges reflecting material quality variation.

object (Cote and Abraham, 1995). The method is based on the tomography principle, already practised by other means in the medical field. In civil engineering
applications, seismic waves are generated in the structural element and propagate
from a group of sources to a group of fixed receivers, placed on the surface with given
coordinates (Figure 18.23).

The wave propagation times, along their unknown respective source–receiver paths,
or *rays*, are recorded. From the acquired time–distance data, a computer-aided
solution of the inverse problem yields the velocity range contours of the assessed section. As already explained, the resulting map may be interpreted to appraise the relative variation of concrete or masonry properties. If requested, a three-dimensional
(3D) description of this variation can also be obtained either by repeating the pro-

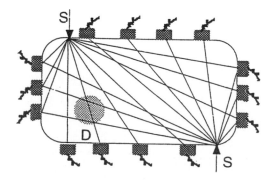

*Figure 18.23 The principle of seismic tomography. Only two source positions, S, are shown
out of several needed. The ray coverage quality depends on the number and direction of
rays intercepted by disc D, for all its positions in the investigated area. The D size is
closely related to the wavelength used.*

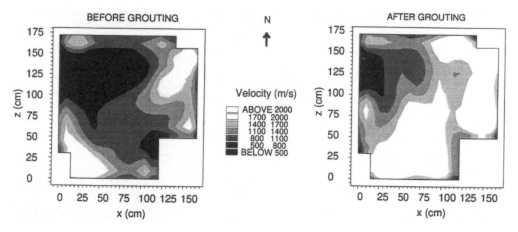

Figure 18.24 Grouting effectiveness in strengthening a masonry pier, shown by tomography.

cedure at successive parallel sections of the investigated volume or by a direct 3D imaging technique. Though relatively expensive, seismic tomography is fast and easy to apply. It may be best suited for checking the effectiveness of grouting evaluated through comparison of the velocity maps before and after the strengthening operation. An example is given in Figure 18.24. It is also worth reminding that such a comparative approach minimizes the inherent measurement error.

Impact-echo (IE)

Generated by mechanical impact, stress waves propagate within the concrete structure and are reflected back to the surface by internal discontinuities and external boundaries. Near the source, a transducer records the multiple reflection time signals which, in turn, are converted to frequency data. Analyses of the detected frequency peaks on the amplitude spectra, coupled with velocity calibration, locate and characterize existing concrete discontinuities. These can be voids, plate delaminations or lack of grout in post-tensioning ducts. Cornell University was the first to develop the IE method and has so far obtained the best results by adapting the impact force to the properties of the structure.

Ground penetrating radar (GPR)

Material discontinuities may also be detected by radar, a technique analogous to the sonic IE in its general approach (Cariou *et al.*, 1997). Through a concrete structure, electromagnetic waves can be emitted and their echoes received from interfaces, using a specific antenna and keeping the signals in the time domain. Reflection time signals, at successive antenna surface positions, form a 'radar profile' where echoes owing to material discontinuities can be spotted. If the wave velocities in the assessed medium are given or assumed, the measured time can be converted to depth, determining the echo position. In the same material, both GPR and IE methods have the same order of wave length and solving power.

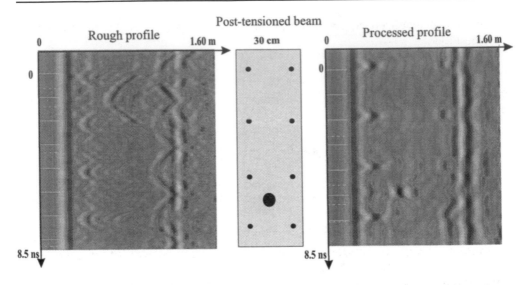

Figure 18.25 Radar profiles, before and after processing, obtained on a concrete beam with a single prestressing tendon between two layers of reinforcement bars.

GPR is mainly used to determine thicknesses and locate plate delaminations or relatively large voids, in beams containing tendon unlined or plastic ducts. Metal ducts, opaque to electromagnetic waves, can hence be indirectly located. The inclination of fractures may be estimated by establishing successive parallel radar profiles at a convenient spacing.

In the presence of reinforcement, wave diffraction by steel grids considerably complicates interpretation, unless the bars form a mesh larger than the wavelength and have a well-known geometry, with a sufficiently simple arrangement facilitating the correct orientation of the antenna set (Figure 18.25).

On masonry structures, applications are quite possible, but with great care when crossing relatively conductive materials such as clay.

Site tests on reinforced concrete materials

As already stated, reinforcement rusts when in contact with sufficiently high contents of aggressive reagents, coming from the environment, such as chloride ions and carbon dioxide (concrete carbonation). To check for reinforcement corrosion, and eventually predict its extension rate, specific NDT are conducted both on the steel and its protecting concrete cover. These tests are also supported by intrusive investigation and sampling methods.

Locating reinforcement

Specific magnetic devices or cover meters have been commercially developed to enable non-destructive location and identification of embedded steel awaiting investigation. Guided by construction drawings if available, the cover meter is run over the concrete surface to trace the actual reinforcement bar positions. To supply correct values, readings of concrete cover and bar diameters are first calibrated by direct scale measurements through a small-bore hole at a reference point.

Other more advanced metal detectors, developed for prestressed concrete, will be described below.

Detecting reinforcement corrosion

Half-cell potential. To locate corroding reinforcement, the most common procedure consists of measuring the steel half-cell potential E_c. This is the potential drop between a point on the reinforcement, reached through a small bore hole, and a mobile reference cell run over the concrete surface (ASTM, 1991; Brevet, 1983). A millivolt-meter is connected to each of these two poles (Figure 18.26). The reference cell is often either a copper–copper sulphate or a silver–silver chloride electrode. The optimum spacing of measurement points may range between 30 mm and 1 m, depending on the accuracy/cost ratio.

The measurement positions and the corresponding E_c values are mapped and the 'equipotential' contours plotted (Figure 18.26). Data can be acquired at other reinforcement levels to obtain section mapping.

To estimate the reinforcement 'corrosion risk', ASTM/C 876.91 recommendations divide the obtained E_c values into three 'classes'. For a reference cell of saturated copper–copper sulphate, the correlation is as follows:

- $E_c \sim -200$ mV$_{CSE}$: corrosion is unlikely (passivation)
- $-350 < E_c < -200$ mV$_{CSE}$: corrosion possibility
- $E_c < -350$ mV$_{CSE}$: high corrosion probability.

Hence, the corrosion risk increases with more negative potential drops.

It should be noted however, that the half-cell potential data do not necessarily convey a certainty but only a probability on the state of the reinforcement and that, for reliable measurements, certain conditions must be met: electrically continuous bars, sufficiently humid concrete cover and uncoated concrete surface. It obviously does not apply to post-tensioned structures with metal ducts. On the other hand, progress has been done to extend this method to subaquatic structures.

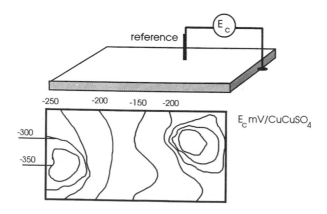

Figure 18.26 Principle and application example of the half-cell potential measurement E_c, reflecting steel corrosion probability in concrete.

Tests on concrete cover

Analysis of the concrete cover leads to the evaluation of the of steel corrosion risk or the identification of the reagents that may have already started the mechanism. Tests thus concern surface permeability, electrical resistance, carbonation and chloride contents.

Surface permeability, a property linked with durability, is measured on site by placing a bell-shaped container firmly against the concrete surface and pumping the air out to produce internal vacuum (Figure 18.27). The time, needed to restore the initial atmospheric pressure, reflects the permeability. This NDT procedure helps to appraise the resistance of concrete superficial layers to the infiltration of external aggressive reagents. Permeability values can either be compared at different points of the structure or monitored versus time at a given position.

The electric resistance of the concrete cover is inversely proportional to the moisture and salt contents, hence to the corrosion risk. It is measured by a NDT method for locating highly suspected areas. Four metal electrodes are placed on the concrete surface and a current I is induced between the two remote ones (Figure 18.28). The potential drop V, measured between the two neighbouring electrodes, is an inverse function of the corrosion risk.

The carbonation extent is determined by sampling cylindrical concrete cores, splitting each through its diametrical plane and applying on the two freshly exposed surfaces a coloured indicator (phenolphthalein). If a pink colour appears, no carbonation is supposed to exist and the pH exceeds 9.

The chloride content profile is traced by drilling at different depths through the concrete cover and analysing the extracted powder in the laboratory.

Figure 18.27 Top view of transparent vacuum bell for measuring the permeability of concrete surfaces (Photo: LERP).

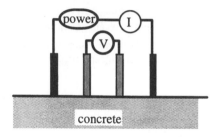

Figure 18.28 Measuring concrete electric resistance V/I, where V is the potential drop between the intermediate electrodes and I the current induced by the power source.

Predicting corrosion extension

Polarization resistance method. Several electrochemical methods are applied on site for measuring *corrosion rate* with polarization resistance techniques (Feliu *et al.*, 1989). A polarization test consists of passing a direct current *I* between the steel and a metal counterelectrode often placed on the adjacent concrete surface, above the tested reinforcement (Figure 18.29). Polarization spreads through the concrete, roughly as a truncated cone, from the counter-electrode down to the reinforcement plane. If *S* is the polarization-affected lateral area of the bar and if $i = I/S$ is the current density, the resulting steel potential change *E* can be measured against a reference electrode placed close to the reinforcement. The DC polarization resistance R_p is the slope of the *E* versus *i* curve. *I* is in A, *E* in V, *S* in cm^2 and R_p is in ohm cm^2.

For polarized reinforcement with neither cathodic protection nor stray current effect, the current density reflecting the corrosion rate i_{corr} is inversely proportional to R_p, hence:

$$R_p = B/i_{corr}$$

where constant $B = 0.026$ V if R_p is in ohm cm^2 and i_{corr} in A cm^{-2}. As the polarization resistance is usually measured only at few points of the bridge, these must be carefully chosen in the light of other investigations if possible. The lateral areas of the polarized bar segments should be well determined. Acquired data have to be corrected for the instantaneous effects of variable climatic factors, temperature and

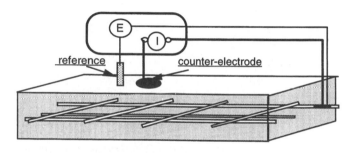

Figure 18.29 Polarization test scheme.

humidity. Several measurements at consecutive time intervals are therefore needed to obtain a reliable mean value of the corrosion rate.

Whatever the type of the equipment used, a corrosion rate i_{corr} greater than 10^{-6} A cm^{-2} is considered to be significantly high.

Built-in sensors

To monitor steel corrosion, sensors may be permanently placed in the concrete, at critical locations where corrosion is likely to occur. They are mainly of two types.

A *macro-cell sensor* consists of two embedded plain carbon steel bars, one near the concrete surface, the other at a deeper level (Figure 18.30a). As the shallow bar usually corrodes first, its half-cell potential will subsequently differ from that of the deeper one placed in normally sounder concrete. Hence, measuring the potential difference between the two bars detects corrosion initiation and sometimes even determines the corrosion current for estimating the rusting rate.

A *polarization resistance sensor* comprises three closely positioned elements. One is a steel bar placed near the reinforcement. The second is a steel piece used as a counter-electrode for measuring the polarization resistance of the first steel bar. The third element is a reference electrode (Figure 18.30b). Measuring the polarization resistance thus determines the steel corrosion rate, at the sensor location.

Site tests on prestressed concrete constituents

Post-tensioned bridges can be severely damaged without showing significant external signs. Surface visual inspection should hence be more detailed and accurate than for other types of concrete construction. From a material durability point of view, corrosion and its consequences on service life are the major concern. But while reinforcement corrosion is relatively simple to detect by the above-mentioned methods, deeply embedded tendons and their ducts often require direct visual examination through intrusive or partially destructive procedures. To make this internal inspection as selective and efficient as possible, a detailed desk study and the use of NDT methods should be carried out first.

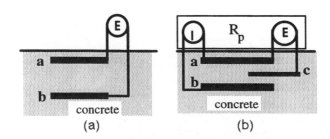

Figure 18.30 Sensors embedded in concrete. (a) Macro-cell sensor for detecting steel corrosion initiation: a and b are steel bars. (b) Polarization resistance sensor for determining steel corrosion rate: a and b are steel bars, c is a reference electrode.

A desk study of available drawings and other records facilitates the location of the usually most vulnerable parts to chloride attack: anchorages in their different positions and configurations, joints between segments, construction joints if poorly executed, high or low points along the duct profile where the former may be too close to the deck top and the latter a potential water collecting trough.

To better locate and identify tendon ducts and suspected defects, the specific NDT methods in use mainly include radiography, radioscopy, eventually the magnetic flux method and other lighter but less efficient techniques such as steel detectors, impact-echo and ground penetrating radar. The last two have already been considered in Section 18.2.1.

On these criteria, a very limited number of points are selected for visual examination and other associated tests, thus completing the field investigation.

In the following paragraphs, more details will be given on the equipment and procedures related to these methods.

Radiography and radioscopy

If a beam of γ- or X-rays strikes one side of the assessed concrete member and passes through its thickness, the emerging radiation energy will show up, with differential attenuation, on a photographic plate placed against the opposite side (Figure 18.31). Steel, denser than concrete, thus leaves a lighter trace while lacking grout or other voids give a darker image. This is the basic principle of conventional *radiography*.

Already applied in steel construction for checking the quality of welds, the technique has now been successfully extended to post-tensioned structures (Guinez *et al.* and COFREND Group, 1999).

Gamma radiography uses a radioactive source, iridium (^{192}Ir) or cobalt (^{60}Co). For a concrete thickness up to 0.5 m, it detects:

- incorrect positions of tendons, ducts, and reinforcement bars
- broken or slack wires or strands
- voids in ducts due to poor grouting (favouring tendon corrosion)
- defective ducts

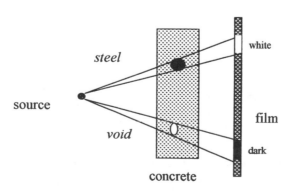

Figure 18.31 The radiography principle: the denser the element, the lighter the image.

- concrete bonding quality
- concrete discontinuities such as honeycombs, certain cracks, defective construction joints or density variation.

On the other hand, radiography cannot directly detect steel corrosion but only its advanced consequences: seriously reduced sections or fractured tendon elements.

High energy X-rays, generated by a linear accelerator, enable the investigation of thicker concrete members and with greater safety: no contamination or standby radiation hazards, shorter exposure time, penetration extended to a 1.2 m thickness and high image quality.

The exposure time varies directly with the square of the source–film distance, inversely with the source activity and exponentially with the investigated concrete thickness. Given these considerations, the source–film distance ranges in practice from 0.7 to 1.2 m. Through a 0.3 m concrete thickness, for example, the exposure time is 46 min for iridium, 2 min for cobalt and 20 s for a linear accelerator.

Radiographs (Figure 18.32) must be interpreted by qualified staff.

However, this method has several imperatives and limitations:

- The use of radiation sources is subject to *strict safety rules*, not only for the operating staff but also for all people present in the area. During the operation, entrance is strictly forbidden within a 20–100-m radius of the radiation source, depending on the type of the equipment and conditions of use.

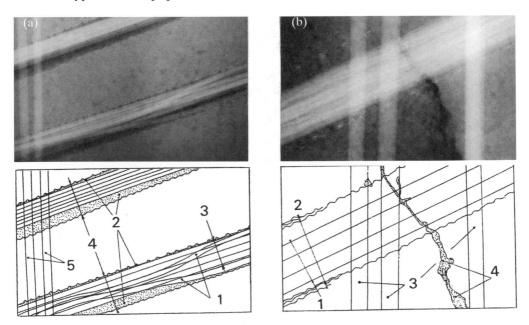

Figure 18.32 Two radiographs and their interpretation. (a) 1 – broken or slack wires; 2 – complete absence of grout; 3 – tendon parallel wires; 4 – tendon ducts; 5 – reinforcement bars. (b) 1 – sleeving ducts; 2 – tendon strands; 3 – reinforcement bars; 4 – construction joints (Photo: LRPC Blois).

- Occasionally, difficult access to both sides of the structural member may hinder the optimum source-and-film positioning and orientation.
- Certain configurations sometimes complicate the interpretation of radiographs, e.g. congested or horizontally parallel tendons.
- Radiography is relatively expensive, like many other specialized activities.

Further extension of the radiography principle has been achieved allowing:

- a continuous survey throughout the tendon lengths, with variable three-directional incident angles for easier interpretation
- a real-time image recording on videotape.

This advanced version is *radioscopy*, using X-rays generated by a linear accelerator placed on a mobile platform. It enables the inspection of a greater concrete thickness and with a shorter exposure time. Radioscopy however does not have the radiographic location accuracy. A conventional radiography is hence also integrated into the inspection system for more precise radiographs at points of particular interest prior to eventual drilling and regrouting.

Radiographic methods have greatly facilitated the location of poor grouting quality in many bridges thus liable to tendon corrosion.

Despite the stringent safety precautions they do require and the reluctance they may arouse, radiography and radioscopy remain at present among the most powerful non-destructive investigation methods for post-tensioned bridges.

The magnetic stray field method

Recent significant developments in procedure, metrology and signal analysis now enable a more reliable use of the magnetic stray field method for the detection of tendon flaws on site (Sawade *et al.*, 1997).

The investigated post-tensioned concrete member is wholly subjected to a magnetic field H_0. Leakage, due to tendon breakage or section reduction, locally disturbs the flux distribution and creates a corresponding magnetic stray field H_s, detected, without contact, by a mobile Hall-effect probe, advancing on a parallel longitudinal rail (Figure 18.33).

However, stray field signals are affected not only by tendon defects but also by the stirrups (and their eventual tying wires) close to the probe. To suppress the parasitic signals emanating from these mild steel elements, the new approach recommends

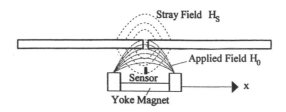

Figure 18.33 The principle of the magnetic stray field method.

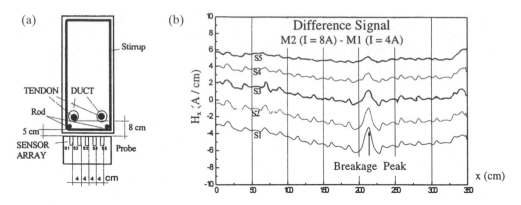

Figure 18.34 (a) Section of instrumented post-tensioned member. (b) Typical stray field signal of a single broken tendon.

several active field measurements $H(x)$, taken at *successively increasing magnitudes of the applied field* H_0. As H_0 rapidly decreases with distance, readings at its low values yield signals from the mild steel reinforcement near the surface. On the other hand, at high H_0 values causing field saturation, the magnetization increase is greater in the deeper steel layers than in those close to the surface. The signal increase at higher values of H_0 is hence mainly due to the deeply placed tendons.

Qualitatively, the detection of a tendon defect is based on the comparison of the shape of the signals at different H_0 values. Through quantitative analysis, the respective differences of these signals, obtained at different H_0 values, first suppress the parasitic effects of stirrups near the surface then amplify the impact of existing tendon flaws on the processed stray field profile, as illustrated in Figure 18.34.

Figure 18.34a shows the section of an instrumented post-tensioned member, investigated by the magnetic stray field method, in 350-cm segments. Several measurements were taken per segment, under increasing exciting field. Figure 18.34b presents the *difference signal* profiles for only two measurements, performed at respective exciting currents of 4 and 8 amps.

On these difference profiles, where the portion of the stirrups has been clearly suppressed, the tendon near the magnetometer S1 provokes a sharp breakage peak. The lateral decrease of the peak towards S5 suggests that the second tendon is intact. These results were confirmed later, after carefully demolishing the concrete.

Steel detectors

Further to the conventional covermeter described earlier, a more advanced surface operated version consists of a data acquisition scanner and an image monitor. Tendons and reinforcement can be detected, with a selective display at the requested depth, up to 18 cm approximately. A cursor indicates bar size and position.

Intrusive methods

Internal visual inspection takes place at very few points, carefully selected in the light of the desk study and NDT results, mainly situated where voids in ducts or other

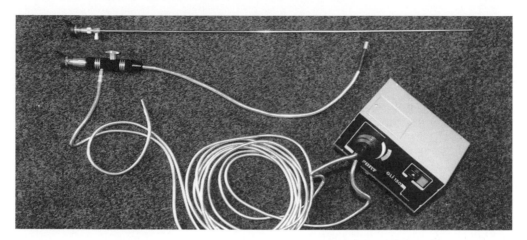

Figure 18.35 Flexible videoscope and rigid endoscope for internal inspection (Photo: LCPC).

defects have been detected or highly suspected. Cautious drilling, 25 mm in diameter, is usually acceptable for access to ducts; it may later be slightly overcored if required. Through the borehole, a flexible optical fibre videoscope is then inserted to examine tendon, duct and grout, in search for steel corrosion, pitting, wire fractures, or defective grouting (Figure 18.35). Inspection extends as far as the void dimensions allow the probe to reach (Figure 18.36). Although equally used for ducts, rigid endoscopes may be better recommended for external round-the-corner points, as in support bearing systems.

Void volume and leakage risk may be estimated to facilitate eventual re-grouting. Void volumes are deduced by applying gas pressure from a given container and using the equations of thermodynamics. To check for leakage, the gas pressure drop, if any, is monitored versus time.

Tension or slackness of tendon elements can be very locally and qualitatively confirmed by inserting a screwdriver between two adjacent wires and watching if it can be turned by hand.

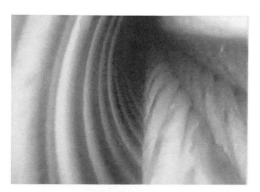 

Figure 18.36 Inspection of post-tensioning strands in two ungrouted corrugated metal ducts: one perfectly sound (left), the other slightly corroded (right) (Photos: LCPC).

Internal samples are taken through access ports, but only after completing all other *in situ* tests on undisturbed material. Sampling concerns the following, if present:

- grout, for chlorides
- trapped water, for chlorides, corrosive ions and pH level
- steel corrosion products
- fractured wires, for diagnosis and vulnerability estimation to stress corrosion.

External concrete samples can also be taken for chloride content at joints and anchorages.

After inspection, all boreholes must be thoroughly resealed according to the rules of the art.

Site tests on steel construction materials

In conventional steel bridges, investigation is mainly carried out on extracted samples. Field tests on the material are relatively rare but do exist. On the other hand, in suspension and cable-stayed bridges, laboratory tests are hardly possible to appraise the material condition of main structural members and field investigation is imperative.

Checking steel elements and welded assemblies

Material hardness tests are still practised, especially on old structures. When properly conducted on well-prepared corrosion-free surfaces, they detect metal quality variations. With the same precautions, the thickness of structural elements can be measured by ultrasonic methods.

In welded assemblies, several techniques are used for detecting cracks or other discontinuities: dye penetrant inspection, magnetic particle test, radiography and back-reflective ultrasonic methods. The first is the simplest for preliminary diagnosis. Also known as the back-percolation test, it consists of applying a special liquid over the cleaned assessed surface, letting it penetrate into eventual discontinuities and removing all remaining excess. An absorptive substance is then spread to draw back the infiltrated liquid. The visible absorption traces thus reveal and locate eventual defects. The liquid must have a very low surface tension and a marked colour. The magnetic particle test is roughly analogous to dye penetrant inspection; magnetic particles are spread and their re-arrangement pattern in a magnetic field reflects material discontinuities. The selective use of radiography and reflective ultrasonics for detecting defects is similar to that already presented for concrete.

Checking for corrosion damage in cables

In suspension or cable-stayed bridges, cables comprise concentric wire layers, either coated or threaded into ducts. While external wires can be easily inspected, inner ones may rust and break invisibly causing cable failure. For a more extensive evaluation of corrosion and its effects, two methods are usually applied: electromagnetism and acoustic emission.

Electro-magnetic survey for single-strand cables. For estimating the corrosion extent and effect on section loss, two half cylinders are placed around the cable and assem-

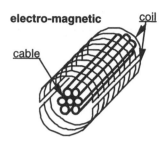

Figure 18.37 The electromagnetic survey principle: a two-element coil is placed around the cable and fed by AC current. The measured self-induction, L, depends on the cable corrosion.

bled to form a continuous coil (Figure 18.37) fed by a 10 kHz AC current. The resulting variable magnetic field creates eddy currents in the cable, with their corresponding self-induction opposed to that of the coil (Gourmelon and Robert, 1985). Corrosion develops a higher contact electric resistance between the wires and consequently allows a lower eddy current intensity, thus increasing the self-induction and hence the measured coil impedance. This quantity grows and stabilizes at a maximum when all wire layers have rusted.

In practice, the activated coil is moved along the whole cable and consecutive impedance variations are measured. For a reasonable evaluation of corrosion, prior laboratory calibration tests are required on typical cables of different degradation depths and intensities. For each configuration, a relationship can then be established between the oxidation ratio and the percentage impedance variation, taking a perfectly sound cable as a base value.

It should also be noted that the electromagnetic method cannot distinguish between pitting and stress cracking corrosion.

Attempts have been made to use electromagnetism for detecting local defects, especially wire breaks, in single-strand cables. One procedure consists of locally magnetizing the cable by a 100 Hz AC current through a surrounding solenoid. The receivers are two identical crossed coils, connected in opposite sense, thus detecting flux variation due to local defects while hardly responding to uniform oxidation. A peak on the recorded diagram may indicate a discontinuity, other geometric irregularities or a very localized oxidation. A clear signal is hence necessary but not sufficient to suspect a wire break, while a 'flat' diagram confirms the total absence of local defects in the cable.

Acoustic emission survey. In a corroded tensioned cable of a bridge, wires can break spontaneously. For a clearer assessment, it is often necessary to monitor the number and successive locations of these events over a period depending on the situation.

When a wire breaks, the released energy creates a transient elastic wave that propagates through the cable with identifiable amplitude and a practically constant velocity (~4700 m/s). Several sensors, S_i (i = 1, 2, etc.), distributed along the cable,

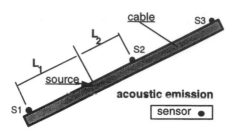

Figure 18.38 Principle of the acoustic emission technique.

can determine the respective arrival times t (S_i) of the signal reaching each of them from the same breaking point called *source*. The reference time 'zero' is the instant when the wave reaches the closest sensor to the source (Robert *et al.*, 1990).

In the schematic Figure 18.38, the given positions S_i and arrival times $t(S_i)$ yield the following: The distance S_2/S_3 and the arrival time difference $[t(S_3) - t(S_2)]$ determine v, the propagation velocity. The first two arrivals are at S_1 and S_2. The source is respectively situated at distances L_1 and L_2 from S_1 and S_2. L_1–L_2 is given by $v[t(S_1) - t(S_2)]$ and $L_1 + L_2$ is the distance S_1/S_2. L_1 and L_2, thus obtained, locate the source or breaking point.

Sensors are usually accelerometers, spaced at ~15 m in cables. They can also be fixed on the concrete surface of post-tensioned bridges, with a ~5-m spacing to monitor the tendons.

A specific data acquisition, processing and remote transmission system locates and reports cable wire breaks in real time.

Figure 18.39 gives examples of results obtained on two suspension bridges, A1 and A2. In A1, a 66 mm-diameter suspension cable broke suddenly. The remaining cables were monitored by acoustic emission during a 16-day strengthening period. Acoustic emissions were detected. After the cables were replaced, a detailed examination confirmed recent wire breaks. In bridge A2, only a part of one cable was monitored for 6 months. One acoustic emission per day was detected. After the cables were removed and replaced, visual inspection confirmed the acoustic emission results.

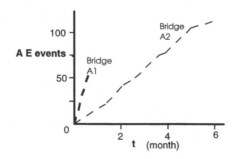

Figure 18.39 Examples of acoustic emission results, obtained on two suspension bridges, probably corresponding to wire breaks after further examinations.

18.2.2 Structural evaluation tests

For a given type of construction, structural assessment methods are usually chosen for their aptitude to evaluate the required parameters and for their suitability with regard to the mechanical system, constituent materials and the observed or assumed type of damage or deficiency.

The development of these methods, towards better performance and new concepts, did not always follow the chronological order of the structural designer's usual checking procedure. Stress, for example, the first parameter that conventional calculations usually yield, was one of the last made accessible to direct measurement on real structures.

In this section, field investigation methods will concern the general deformation of the bridge and the movement of its bearing points, the response of a sound or locally damaged bridge to loading, the forces resisting the loads, the local behaviour of certain areas or their discontinuities and finally the acting stresses.

A selective use of these techniques facilitates structural assessment and, eventually, optimum strengthening.

General deformation under dead load

Topographic monitoring of the structure's geometry helps to evaluate its general condition. It may disclose support settlement or movement, a modified cyclic response to temperature reflecting gradual damage, irreversible deformation of a concrete deck owing to under-estimated creep and other random effects.

Displacements are monitored by periodic precision levelling at several targets on the bridge, preferably referred to a network of benchmarks.

Rotations are measured with inclinometers and distances with invar tapes. Parallel temperature monitoring is necessary for thermal correction of all measurements.

In addition, a careful visual examination of the parapets or barrier walls can often detect permanent abnormal deformation of the deck.

Span deformation curve under test loading

Load-induced deflections, rotations and curvature are measured along the bridge spans and compared with their theoretical values, in order to analyse the actual general behaviour and detect eventual discontinuities. Customarily, *deflectometry* now stands for deflection metrology while *inclinometry* includes both rotation and curvature measurements.

Deflectometry

Traditionally, deflections under static test loading are among the first data obtained for the acceptance of a bridge. They set a reference for future assessment.

Before, during and after the loading tests, a detailed inspection must be carried out on the whole bridge by qualified staff. Access to all parts should be provided.

Loads are first applied to the supports, next to the spans one by one, and subsequently to different span combinations in accordance with the specifications.

In case of damage affecting the structural capacity, static test loads are cautiously applied and guided by prior calculations and/or direct stress measurements. If the load–deflection relationship deviates from its linear elastic path, the test is immediately stopped and the acquired data are analysed for a decision on possible new service load limit.

Deflections are usually measured at midspans and the supports. Other points may be included, depending on the instrument accuracy, often of the order of the millimetre.

There are various existing deflection-measuring techniques and devices, mainly:

- topographic levelling methods demanding highly qualified staff
- mechanical deflectometry (Figure 18.40), necessitating a fixed anchorage point below the bridge, which is not always possible
- electrical displacement gauges needing a rigid independent support
- laser deflectometry (Figure 18.41), the more recent, of delicate implementation, supplying continuous data, used for high bridges across rivers or railways.

The choice depends on the bridge dimensions, deflection range, access conditions and loading procedure. For data confirmation, two techniques are often used in parallel.

Under dynamic loading, at mid-span or pier heads, the induced vertical and horizontal deflection components are estimated either directly by seismographs or indirectly by accelerometers through double integration of the acquired data. However, while this method has some success in monitoring offshore structural elements, the deduced dynamic properties of a concrete bridge are not yet sufficiently sensitive to early damage. Data analysis seems to be more influenced by surface layer degradation than by real structural defects.

Figure 18.40 Mechanical deflectometer (Photo: LRPC Nancy).

Figure 18.41 Laser deflectometer: a stationary laser beam emission source (left) fixes the position of the target, a beam receiver cell mounted on a trolley (right). The trolley, thus brought to a standstill, has a sliding vertical track linked to a point on the bridge. After an initial reading, the relative sliding vertical distance is the deflection of that point on the bridge (Photos: LRPC Nancy).

Inclinometry

Under loading, through rotation or curvature measurements, inclinometry proved quite suitable for completing the bridge deformation curve and, in particular, for detecting and monitoring discontinuities such as flexure cracks or opening segment joints.

Rotation measurements along the span offer a complementary description of the structure's response to loading. They improve the deck deformation data and may extend them to the supports and other neighbouring members. Rotations are mainly measured by inclinometers, now commercially developed with a 10^{-6} to 10^{-8} rad sensitivity, easily fixed on the deck, without external links or reference points.

Over a vertical crack, two coupled inclinometers are placed on either side, as shown in Figure 18.42 (left). Differential readings, $\Delta\theta$, can then follow up the angular opening variation.

Figure 18.43 illustrates the influence of an increasing bending moment M on the behaviour of a detected rotation discontinuity $\Delta\theta$, monitored by differential readings of two coupled inclinometers, A and B, placed on either side of an existing flexure crack. The linear first part of the curve $M = f(\Delta\theta)$ represents the angular opening variation of the crack without any further growth, i.e. within its initial tip height limit z_0 before the present loading. The non-linear second part marks the crack growth, $z_0 + \Delta z$, under a greater moment M.

Curvature variation measurements are a further step completing the span deformation data under a test load and reflecting the response to flexure.

In sound sections, curvature variation is conventionally obtained from the strain profile measured by extensometry, using strain gauges or displacement sensors. For

Figure 18.42 Coupled inclinometers (left) and a curvature meter (right) (Photos: LREP).

most bridges with multiple cracks and inaccessible residual sections, a more global metrological approach is imperative as provided by inclinometry, where the developed specific instruments are now accurate, robust and mobile. They supply principal angular deformations and sieve out local disturbances, thus facilitating the analysis of damage effects on the general flexural behaviour.

Theoretically, the following expressions for curvature $\theta'(x)$ apply to a sound beam section x, of height h and flexural stiffness EI, under a bending moment $M(x)$ and a strain difference $\Delta\varepsilon$ between extreme fibres:

$$\theta'(x) = d\theta/dx = \Delta\varepsilon/h = M(x)/EI$$

where the $d\theta/dx$ term represents inclinometry and $\Delta\varepsilon/h$ extensometry.

For damaged sections, the extensometry expression is difficult to apply, owing to stress concentration and complex strain redistribution. The inclinometry expression remains reasonably valid.

The load-induced curvature diagram $\theta'(x)$ is determined by differentiating closely spaced rotation readings φ_i throughout the span, using either coupled inclinometers or a mobile curvaturemeter. The latter shown in Figure 18.42(a) is basically an instrument supplying automatic differential rotation readings, between two points, with a fixed spacing. Both types are usually placed on the uncracked, upper or lower, extreme fibre.

The *spacing* of inclinometric measurements is governed by several factors and requirements.

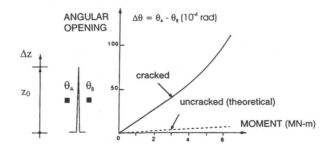

Figure 18.43 Influence of the bending moment on the angular opening of a bridge girder crack, measured by differential.

To trace the curvature variation diagram $\theta'(x)$ along a bridge, the required rotation measurements spacing, s_B, depends on the beam and eventual crack heights as well as on the accuracy of the instruments and expected results. From experience with inclinometry and bridge metrology in general, spacing varies from one-half to one-tenth the deck height, respectively, in sound and cracked configurations.

In *apparently sound* beams, the optimum spacing s_B, between two consecutive rotation readings φ_1 and φ_2, can be reasoned out as follows:

$$\text{load-induced local curvature } \theta' = (\varphi_2 - \varphi_1)/s_B$$

Assuming a negligible error in setting the base length s_B and an intrinsic instrumental uncertainty $\delta\varphi$ at each rotation measurement point, the maximum possible error in the differential rotation reading is $2\delta\varphi$ and the corresponding curvature uncertainty will be:

$$\delta\theta' = 2\delta\varphi/s_B$$

Along the same instrumented extreme fibre, situated at a normal distance, c, from the neutral axis, the load-induced stress σ is:

$$\sigma = Mc/I = EI\theta' \, c/I = cE\theta', \text{ hence } \delta\theta' = \delta\sigma/cE$$

Thus, to an uncertainty $\delta\theta'$ in the measured curvature, corresponds an uncertainty $\delta\sigma$ in the extreme fibre stress.

If the load-induced stress is required with an accuracy of $\delta\sigma = \delta\sigma_r$, then, from the last equations, the *optimum spacing s_B* of rotation measurements on sound beams, is given by:

$$s_B = c(2E\delta\varphi/\delta\sigma_r)$$

Example: an apparently sound concrete T-section of height h, with $c = 0.4h$, $E = 50\,000$ MPa, $\delta\sigma_r = 0.25$ MPa and $\delta\varphi = 10^{-6}$ rad, requires an inclinometer spacing of approximately $0.16h$.

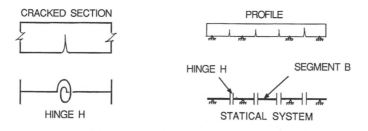

Figure 18.44 Modelling assumptions of cracked sections of a bridge.

In *damaged* sections, with sharp curvature variation, the maximum admissible inclinometer spacing should not exceed the tenth of the beam height, with one obligatory differential reading right over the crack. Furthermore, the influence length of a single crack may attain the order of the beam height h, on each side, hence a total of $l_c = 2h$. With multiple cracks over a length, q, the total range of sharp curvature variation is about $L_c = q + 2h$.

The same reasoning applies to the spacing of mobile curvature meter readings.

The moment–curvature method

The above mentioned theoretical suitability and developed inclinometry led to a field investigation method for detecting transverse discontinuities and determining the actual flexural response of sound or damaged bridges (Abdunur, 1997).

The bridge is equipped, throughout its spans, with inclinometry instruments supplying curvature variation data. Under convenient test load configurations, the measured curvature variation diagrams θ' are plotted for the whole length of the bridge.

If these diagrams are regular and reasonably follow the theoretical ones, then the flexural adequacy of the structure is probably maintained.

If, on the contrary, a sharp curvature redistribution appears at certain points, then transverse cracks or other discontinuities should be suspected and the structural system may be represented as illustrated in Figure 18.44.

The cracked sections and their disturbed vicinities are assimilated to a series of elastic or plastic hinges H, alternating with sound beam segments B and jointly setting up a new system in equilibrium.

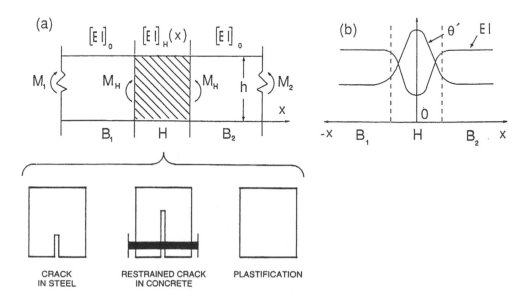

Figure 18.45 (a) Moments M and stiffnesses EI in cracked sections H and sound segments B. (b) Redistribution of curvature θ' and stiffness EI.

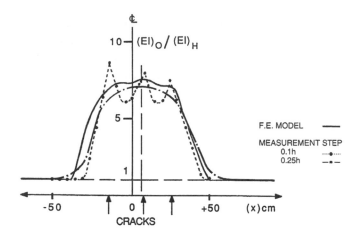

Figure 18.46 Theoretical and experimental redistribution curves of the inverse relative flexural stiffness over a beam segment with three cracks.

The main difficulty is the realistic determination of the relative residual flexural stiffness $[EI]_H$ of the hinge, where several variables and assumptions are involved. The proposed evaluation of $[EI]_H$, at a point x along the bridge, is hence experimental, based on the relationship between the applied bending moment $M(x)$, the resulting measured curvature $\theta'(x)$ and the flexural stiffness $EI(x)$. It proceeds as follows and as shown in Figure 18.45:

Under a test moment, as already stated, the resulting curvature diagram $\theta'(x)$ is known throughout the spans and in particular:

- $\theta'H(x)$, over the hinge H comprising the crack and its short influence zones on either side,
- $\theta'B(x)$, along the sound beam segments $B1$ and $B2$, especially their parts near the crack.

All sound beam segments B are assumed to conserve their given initial stiffnesses $[EI]_0$.

The bridge length is divided into modules (Figure 18.45a), each covering the hinge zone H of stiffness $[EI]_H(x)$ and the near parts of both adjacent sound segments $B1$ and $B2$ of given initial stiffness $[EI]_0$.

For each module $B1/H/B2$, the beam equation $M = EI . \theta'$ is applied at successive sections, more closely spaced in the hinge H zone where the moment M_H remains almost constant but the curvature θ' varies considerably and the local stiffness EI inversely to it (Figure 18.45b). The moment M_H is applied by the sound beam segments $B1$ and B_2. Hence, at any section x within the H length:

$$[EI]_H(x)\theta'_H(x) = M_H = [EI]_0\theta'_B$$

or

$$[EI]_H(x) = [EI]_0 \; \theta'_B/\theta'_H(x)$$

$[EI]_H (x)$, thus determined, is plotted versus x, as shown in Figure 18.45(b), giving the required equivalent residual rigidity of the damaged section or hinge. This quantity varies with x for an elastic hinge and with both position x and moment M_H for a plastic hinge. The hinge area H, in Figure 18.45(a), can have many possible configurations, among which the three illustrated, without modifying the above reasoning.

Most often, the *relative* residual rigidity $(EI)_H/(EI)_0$ is preferred for estimating the fractional remaining capacity of the bridge. Its *inverse*, $(EI)_0/(EI)_H$, is usually chosen for graphical representation, as in the real example shown in Figure 18.46.

The residual rigidities of the damaged sections H and those of the sound segments B now define the new structural system of the bridge and enable:

- the prediction of the real flexural response of the damaged structure to any given loading
- the evaluation of the residual load-bearing capacity and, eventually, the optimum needs for strengthening.

The whole procedure can be repeated at a later stage to verify the effectiveness of eventual repairs or simply to monitor a time-dependent mechanical change.

The moment–curvature method applies to steel or prestressed concrete bridges.

Towards new concepts in bridge acceptance procedures: in the light of the metrological possibilities discussed earlier, it may be imagined that, in the not too distant future, inclinometry could fully accompany deflectometry in the acceptance procedures for newly constructed or strengthened bridges.

Measurement of forces acting on a bridge

Support reactions and suspension or stay cable forces provide the external equilibrium of the bridge. Embedded tendons maintain the concrete stress profile within

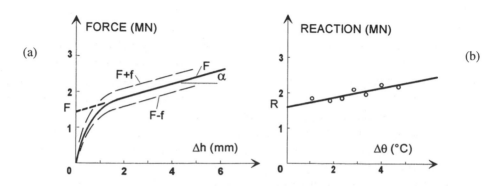

Figure 18.47 (a) Force–displacement curve giving a preliminary value of the support reaction. (b) Final value after correction for temperature gradient.

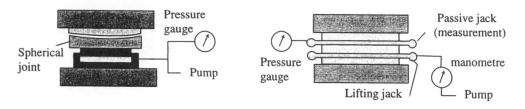

Figure 18.48 Flat piston jack (left) and double flat jack (right) for measuring support reactions. To put these devices in position, a 15-cm minimum clearance should be available between the deck and the bearing shelf. The bearing strength of the involved concrete surfaces must also be checked. At present, it is also possible to install bearing systems directly equipped with permanent force-measuring devices.

the allowable limits. The force, acting through each of these structural elements, has its own specific metrology.

Support reactions

In statically indeterminate bridges, support reactions are periodically measured to evaluate the developing forces during construction and, mainly, their *redistribution* throughout service life.

The measurement procedure consists of inserting a set of jacks around the existing bearing and using them to lift the deck. The force–displacement curve is plotted for the whole lifting-lowering cycle. The average curve thus eliminates frictional effects, f (Figure 18.47a). The first part of the graph marks the release of the original bearing. The second part, usually a straight line, represents the deck bending deflection; the slope gives the flexural stiffness. The reaction is deduced by extrapolating the straight segment to zero (Chabert and Ambrosino, 1983). However, this value closely follows the cyclic variations of thermal gradients. In fact, the present metrology was the first to reveal the intensity of these effects on continuous bridges, sometimes attaining the equivalent of full live load configurations. Consecutive parallel measurements of both the reaction and thermal gradient are hence obligatory for at least 24 hours to establish a relationship, as shown in Figure 18.47(b), and finally obtain the corrected reaction value at zero thermal gradient.

As the *total* reaction can only be measured, high precision instruments are imperative to reasonably detect differential values, of greater monitoring interest. A 1% error already limits reaction measurements to the abutments, where loads are much reduced and variations more detectable. Figure 18.48 shows two versions of the equipment now in use: the flat piston jack and the double flat jack.

In continuous spans, temperature-corrected reactions respond to dead load developments, differential settlement, stress profile modification in the deck (e.g. post-tensioning or differential creep) and specific or random consequences of various defects. Support reaction metrology can hence be used to detect abnormal dead load distribution, explain certain observed defects and, more basically, verify or modify structural design assumptions.

Forces in external cables

The vibration method. During the construction of suspension or stay bridges and of concrete structures with external post-tensioning, the individual cable forces are adjusted to achieve the required equality of stresses for the whole system under dead load. With time, this uniformity may be disturbed for various reasons. Periodic checking is hence necessary for appropriate re-adjustment. Theoretically, tensile cable forces could be measured with the type of jacks that initially applied them, but the operation and the equipment have both their drawbacks.

The vibration method offers a worthwhile alternative; it is simple, quick and has a wide application field (Robert *et al.*, 1991). The cable is modelled as a vibrating string of length *l* and mass μ per unit length, where the frequency *f* is related to the applied tension *T* by a well-known equation:

$$f_n = \frac{n}{2l}\sqrt{\frac{T}{\mu}}$$

where $n = 1, 2, 3, \dots$ is the rank of the harmonic.

The measured fundamental frequency and higher harmonics hence directly lead to the required force. The model is valid only when the cable flexural rigidity can be neglected, an assumption often acceptable, given the cable length and high tension. In practice, a 0.5% error is estimated if a linear relationship is confirmed between the order *n* of the successive harmonics and the corresponding measured frequencies. This linearity test is obligatory for each cable configuration to check the adequacy of the vibrating string model. If another 0.5% error is assumed for measurement instruments, the uncertainty on frequency would add to 1%.

The equipment comprises an accelerometer, with its associated electronics, and a data analyser immediately giving the successive frequencies. The accelerometer is mounted on the cable. Transverse oscillations are induced either by an abrupt tension or a supple tip hammer stroke, both at mid-length to favour the fundamental frequency. To reduce the error on the effective cable length *l*, the fixed points may also be equipped with accelerometers to check that no oscillations are detected on these vibration nods. Measurements take a few minutes.

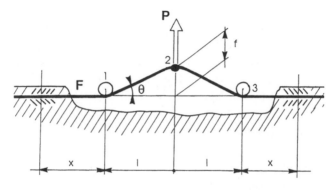

Figure 18.49 Principle of measuring the force in a prestressing wire.

The overall accuracy is ~5% of the tensile force, given the uncertainties on the measured frequency, length and mass. However, if the cable force *variation* is monitored, only the frequency error remains, which is of the order of 1% as already estimated.

Forces in embedded tendons

The cross-bow method. The actual prestressing force, determined through the measured concrete stresses in a section, can also be evaluated by direct action on the tendons, using the Crossbow method. This is based on the simple fact that the effort necessary to deflect a tight rope is proportional to its axial tensile stress. Figure 18.49 outlines the procedure, after carefully clearing the adjacent concrete, duct and grout. The resulting disturbed tendon length $2(l + x)$ is 60 cm approximately. A controlled perpendicular force P, coupled with a displacement sensor, then successively deflects prestressing wires through a distance f, limited to 4 mm. Theoretically, the tensile prestressing force F in each wire may be deduced by the formula:

$$P = 2(F + k)(f/l) + K(f/l)^3$$

where k and K are given constants.

In practice, the parasitic effects of friction, flexural stiffness, overstretching and random bond failure necessitate prior calibration tests on simulation models in the laboratory, using the same type of wires and the exact disturbed length $2(1 + x)$. A family of $P = g(f)$ reference curves are traced for different F-values. The curves established *in situ* could then be interpreted by direct comparison, leading to the actual prestressing force F in each wire. These data should undergo statistical treatment before they become reliable experimental results.

The incremental drilling method. Another way of measuring the tendon force is through minute incremental drilling, in the wire, of a hole 1.5 mm both in diameter and maximum depth. The strain release, measured by strain gauges on either side of the hole, leads to the total stress including the residual part due to the wire drawing process. This part can be estimated either by identical parallel tests on unstressed samples or by referring to the manufacturer's records or other sources. It thus gives access to the applied stress, hence the force.

Local behaviour of sections and discontinuities

Crack geometric survey under dead load. If correctly interpreted, cracks do not only indicate the positions and directions of excessive stresses, but can also be the external signs of the actual structural behaviour. In concrete structures for instance, visual inspection by experienced staff should already distinguish between active structural cracks, owing to intense forces or weak construction joints, from superficial cracks due to insufficient curing.

At successive inspections, eventual developing cracks are traced both on the structural drawings and the bridge itself, each group with different colours and consecutive dates. Crack openings are marked down to the tenth of a millimetre (Figure 18.50).

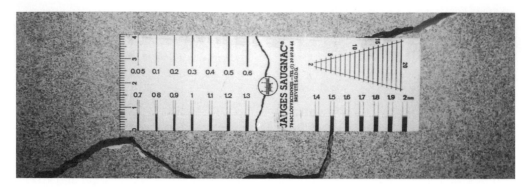

Figure 18.50 A specific scale for estimating crack widths (Photo: LRPC Bordeaux)

To check whether or not a crack goes through the structural element, coloured water or coring are used, the latter being more reliable but complicated in the presence of curved discontinuities.

Recording detailed data on the structural drawings facilitates interpretation and diagnosis, but does not dispense the analysts from carrying out their own visual inspection on site.

Local response to external loads

The above investigation methods, describing the general structural behaviour, should be locally completed under external loading by a detailed analysis at certain positions of particular interest. The local response to loading is generally characterized by multidirectional strains on continuous material surfaces and by relative displacements across cracks, opening joints or other discontinuities.

Strain. Under loading, strain is measured at several points of the structure not only to check deformability assumptions but also to calculate the corresponding stress variations. In linearly elastic media, a well-known strain–stress relationship is used, involving Young's modulus E and Poisson's ratio v. In the frequent cases of simple uniaxial loading, one strain direction is usually sufficient. Two orthogonal strain measurements give the respective stress variations along these same axes. Three directional strain measurements represent the general case of plane stress variation, determining in addition the principal stresses and their directions, as well as the tangential stresses.

While the deduced stress variations are usually reliable for steel, they may be less for concrete due to changing elastic properties and surface quality. The absence of a time parameter in the elasticity equation, as well as the frequent drift in strain gauges do not allow the interpretation of long-term creep effects on stress. By definition, relaxation cannot be monitored at all. Strain measurements are hence limited to load-induced instantaneous variations.

For most cases, the required strain measurement sensitivity is approximately 50×10^{-6} for steel and 2.5×10^{-6} for concrete, corresponding to 10 and 0.1 MPa stress variation accuracy, respectively.

Relative movement across discontinuities. Measuring the load-induced variation of a crack or joint opening is often useful for appraising over-tension in tendons or

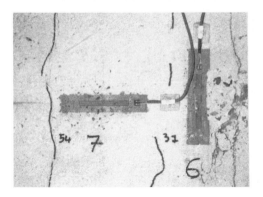 

Figure 18.51 Orthogonal strain gauges and marked surface cracks (left) and displacement sensors across a joint (right) (Photo: LCPC).

reinforcement bars crossing the discontinuity (Figure 18.51). However, overtension can be reasonably estimated only by associating constitutive laws on steel–concrete bonding. For reinforced concrete, these laws are relatively well known. For post-tensioned concrete, assumptions depend on grouting and duct quality. Hence, for a reliable estimation of overtension, prior non-destructive checking is necessary. The sensitivity required for measuring relative displacements is 10^{-3} mm for post-tensioned concrete and 10^{-2} for other types.

Instruments. The main types of strain or displacement measuring devices are:

- Electric resistance strain gauges, 1×10^{-6} overall sensitivity, used on steel and concrete surfaces, sometimes embedded in concrete. Drift is the main drawback.
- Vibrating wire strain gauges, 1×10^{-6} sensitivity, no significant drift, based on the oscillation frequency variation of a tight wire as a function of length, initially used embedded in concrete, now also on the surface, robust, need specific equipment for automatic measurement.
- Mechanical displacement gauges, 10^{-3}-mm sensitivity mainly on a 100-mm base, used for steel strain and concrete crack opening variation.
- Electric displacement transducers, mainly LVDT (linear voltage differential transformer), 10^{-3}-mm sensitivity, no particular drawbacks.

Coupled strain–displacement measurements. In cases where cracks or segmental joints do open under service loads, tendons are first checked both for corrosion by NDT techniques and for fatigue by existing numerical methods and other references. The discontinuity can sometimes be fully instrumented on both the concrete and the steel surfaces to estimate the local flexural capacity (Chatelain *et al.*, 1990; Godart, 1987). At several levels of the concrete section height, displacement sensors are placed across the discontinuity and aligned with long base strain gauges glued on one side at a sufficient distance from the edge to avoid local parasitic effects (Figure 18.52). Under a gradually applied decompression moment M, when the crack is still closed, both the sensors and the strain gauges usually vary linearly and proportionally (Figure 18.53). When the crack opens, at a moment M_1, the sensors show

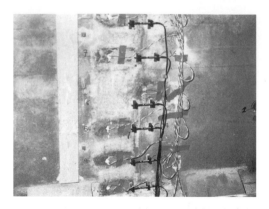

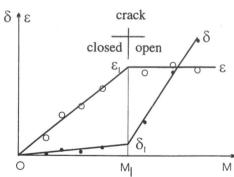

Figure 18.52 Strain gauges/displacement sensors coupled along the joint height (Photo: LCPC).

Figure 18.53 Joint openings δ and edge strains ε under a bending moment M.

a much higher extension rate while strain gauges no longer respond. Moment M_1 is hence an estimation of the residual flexural capacity. Small strain gauges are sometimes glued on the tendons crossing the discontinuity to give the steel over-tension.

Some guidelines for local field measurements
The success or failure of past field investigations led to certain guidelines.

1. *Define the framework of the formulated problem.* Through a detailed discussion between the design office and the experiment task group, both should:
   - recognize the difficulties of field metrology and work out a realistic programme
   - know the experiment stakes and provide optimum instrumentation.
2. *Determine the order of magnitude of the measured quantities.* Prior calculation of magnitudes, with high and low assumptions, may facilitate the choice of the best suited technology, reconcile accuracy with economy and fix the operating limits during tests.
3. *Use all necessary means:* While optimum cost remains a part of a carefully prepared investigation, blind economy reflex often proved too expensive in the end. Hence:
   - provide an adequate number of measurement points
   - use parallel available methods for confirmation
   - acquire complementary data for correction, especially regarding thermal cyclic variations.

Stress measurement and monitoring in concrete bridges
In the assessment of concrete or masonry bridges, the actual stress profile is a major parameter for estimating material and structural capacities. Under dead load, stress in concrete is a combination of an internal component, inherent to the material, and an external one due to thermal effects and applied forces, including prestressing.

If the total stress in a structure can be directly measured, it may immediately be compared with the allowable working values to estimate the material's safety margin during service life, eventual strengthening or partial reconstruction.

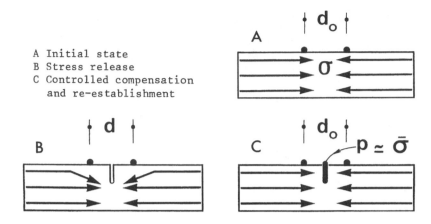

A Initial state
B Stress release
C Controlled compensation
  and re-establishment

Figure 18.54 Stages of direct stress evaluation.

Figure 18.55 Cutting and pressurizing a slot (Photo: LCPC).

Figure 18.56 A set of flat jacks (Photo: LCPC).

If the applied (or external) stress component can be isolated, it may lead to the real mechanical reserves in complex structural configurations such as ageing or damaged bridges with uncertain elastic properties and different degrees of indeterminacy.

For such cases, stress cannot be evaluated through conventional strain measurements. In the absence of cracks or clear discontinuities and with no prior knowledge of the actual absolute stresses, test loading may become arbitrary and destructive, undermining its own purpose.

In a fundamentally different approach for treating these cases, the partial stress release method was developed to *directly* measure the total stress and deduce its components (Abdunur, 1993, 1995). Indirect methods have also been developed (Ryall, 1996).

The release method

It is a local and partial release of stress, followed by a controlled pressure compensation, as in Figure 18.54. In practice, a reference displacement field is first set up

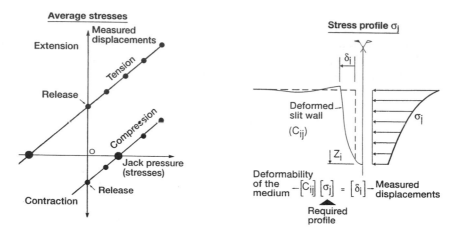

Figure 18.57 Access to average stresses by partial release and pressure compensation.　*Figure 18.58 Access to the total stress profile through a proposed model.*

on the surface. A slot 4-mm wide is then cut in a plane normal to the required stress direction (Figure 18.55). Finally, a special very thin flat jack (Figure 18.56) is introduced into the slot and used to restore the initial displacement field. The amount of the cancelling hydrostatic pressure gives the average total compressive stress normal to the slot. In the same way, with the same accuracy, tensile stresses are obtained by extrapolation as discussed later. The operating depth range is 80 mm.

Theoretical and technological studies have enabled the reconciliation of miniaturization and accuracy. The maximum error margin is 0.3 MPa. The measurement is 'direct' in the sense that the same physical quantity is involved (pressure for stress) and that none of the elastic properties of the material are needed. These are in fact determined in the process. Unlike conventional methods, no test loads are used. Post-measurement remedial techniques can restore the initial mechanical and aesthetic state of the investigated medium. The release method can thus be classified as globally and locally non-destructive.

Average compressive and tensile stresses

Compression is measured by the cancelling pressure as illustrated by the lower straight line of Figure 18.57. It is the jack pressure restoring the displacements to zero. In rare cases of damaged media, the compression line may be slightly curved and a special procedure leads to the right stress.

Tension can be estimated through measurements of supplementary forced extensions due to applied pressure increments, then by extrapolating down to zero displacement. This is illustrated by the upper line on Figure 18.57. For concrete, the relationship is usually linear, confirming the elastic-brittle tensile behaviour.

The slope of each line reflects the medium's deformability or the inverse actual *in situ* elastic modulus.

Separation of stress components

As already mentioned, after the evaluation of the average total stress estimating the material's safety margin, it is at least as important to correctly extract the applied (or external) stress reflecting the structural capacity. This requires:

• The precise determination of the total stress profile as a function of depth.
• A detailed knowledge of the internal stresses associated with the material.

The applied stress can be obtained by subtraction.

To obtain the *stress profile*, the operations, already described and shown in Figure 18.54, are repeated at increasing slot depths, z_i. Figure 18.58 summarizes the model-aided analysis of the acquired data. The required total stress profile, defined by the column matrix $[\sigma_j]$, can be determined by solving the linear system:

$$[C_{ij}] \, [\sigma_j] = [\delta_i]$$

δ_i is the surface displacement measured for a slot attaining depth z_i.

C_{ij} is the compliance matrix for a reference medium, adjusted to each case using the actual medium's response to the flat jack pressures.

The actual *in situ* modulus is thus determined in the process.

Internal stresses in concrete are mainly due to drying shrinkage. Using the release method itself, shrinkage was found to induce high superficial tension depending on the loading and exposure histories (Abdunur *et al.*, 1989). It falls rapidly with depth and changes to a moderate compression throughout the core. Like all other internal effects, shrinkage stresses are balanced over the section. In the presence of applied external forces, the superposition principle proved to be valid. Hence, given (1) their particular distribution, (2) superposition validity and (3) auto-equilibrium, the shrinkage eigenstresses can be identified and isolated from the total stress profile. A striking similarity in shape is usually noticed between the obtained shrinkage stress profile and the percentage water loss distribution by γ-densimetry (Abdunur, 1993).

Example

For a post-tensioned slab under dead load, Figure 18.59 gives a typical total stress profile across the thickness, traced from the stress release data after processing by the numerical model. The profile combines prestress and shrinkage effects. As shown, these two components are separated by using the three above-mentioned properties of shrinkage eigenstresses.

Stress redistribution and prestressing force

Further to the above example, the stress profile across the thickness is confirmed at three fibre levels of ageing post-tensioned T-beam sections (Figure 18.60): one in the upper flange and two in the web, at the theoretical neutral axis and nearest to the lower flange containing the tendons. The three average compression values, represented at their respective fibre levels (Figure 18.61), show a highly curved stress profile over the section height, fundamentally different to the conventionally plane distribution assumed or admitted by many codes of practice. This is probably the mechanical effect of differential non-linear creep and shrinkage, loading the bulky elements and relieving the thin ones (Abdunur, 1996).

Hence, in prestressed concrete multi-beam bridges, with thin exposed webs, there are two non-linear stress redistribution mechanisms occurring on two different scales: one across the section wall thicknesses and one over its height. It follows that the existing prestressing force cannot be estimated through one stress measurement point at the neutral axis. Several are needed at other fibre levels for appropriate integration over the section.

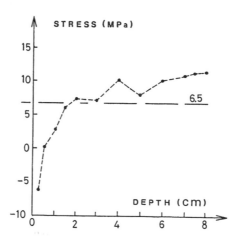

Figure 18.59 Total stress profile combining shrinkage stresses and a 6.5 MPa prestress.

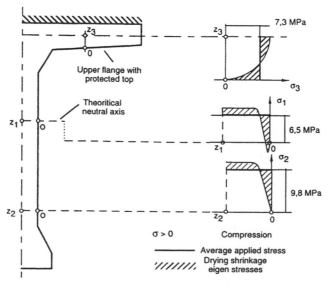

Figure 18.60 Stress distribution across beam wall thicknesses.

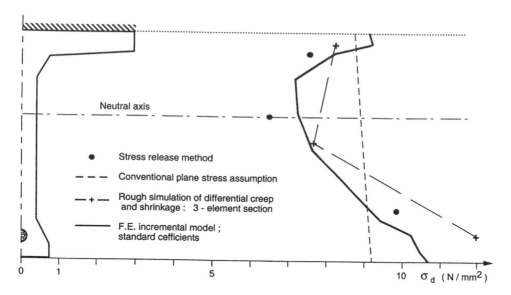

Figure 18.61 Measured and calculated dead load stresses in ageing post-tensioned concrete beams.

Recapitulation of acting stresses

The dead load stresses, their double redistribution due to concrete differential creep and shrinkage, the thermal gradient effects and live load stresses are summarized in Figure 18.62.

Stress monitoring

After the instantaneous evaluation of stress, a further and equally important step would be to monitor its cyclic and irreversible variations (Abdunur, 1992). In terms of the

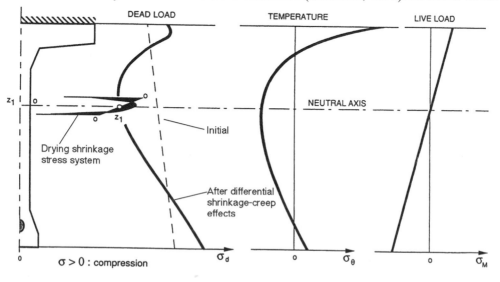

Figure 18.62 Main components of stress in ageing post-tensioned concrete bridges.

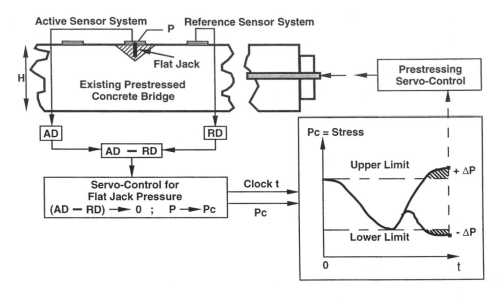

Figure 18.63 Processes of stress monitoring and eventual re-adjustment.

release method, stress monitoring may be defined as the time-dependent pressure of the flat jack, constantly ensuring the same local normal strain evolution that would occur if the medium were still free of any inclusion. The procedure is demonstrated in Figure 18.63. Once the initial stress is determined, the sensors over the flat jack will form the 'active system', measuring the time-dependent response of the surface displacement field in the artificially reconstituted continuum. Along the longitudinal axis of the active system, a second group of identical sensors, forming the 'reference system', is placed slightly beyond the range of the jack to follow up all displacement field variations in a close but undisturbed section. A specific regulator then automatically adjusts the flat jack pressure so that the displacement *variations* measured in the active system always equal those detected in the reference system. The time-dependent pressure p_c, thus obtained, gives the normal stress evolution in the equipped section. Monitoring can be remote-controlled and last for the lifetime of the structure.

In certain configurations, such as prestressed concrete bridges, direct stress monitoring may lead to corrective measures. If the flat jack pressure p_c clearly indicates a lasting stress outside the allowable limits, as indicated in the two lower shaded areas of Figure 18.63, then a selective action on external tendons may restore it to its optimum level.

Stress is thus gradually acquiring its metrological independence from strain, jointly improving the means of evaluating structures and exploring constitutive laws.

Commentary

In this section, the investigation methods regarding the materials and those concerning the structure were presented separately. The distinction is not really consistent since the final evaluation of the state of a bridge is necessarily global. It highlights however two parallel lines of approach that cannot always coincide: material properties, mainly studied in the laboratory, and structural analysis, mostly practised in the design office. This separation also reflects the present modelling limits in coupling material constitutive laws with structural analysis, to predict and describe their interaction, thus the reality.

On the other hand, the described methods must be selected with experience-based judgement. Many of them have to be used on distressed bridges because defects have been detected too late. Their cost is very high compared to that of a 'normal' management policy. Hence, it is never too early for a first detailed inspection, for steady periodic ones to follow and for appropriate action as soon as flaws are detected.

Finally, many bridge management specialists consider that the present investigation methods are still insufficient to easily tackle most cases. Further development is needed.

18.3 Assessment procedures

With increasing loading intensity, the structure's response grows gradually irreversible. The succession order of appearance of different defects depends on many factors. The 'loss of value' has two consecutive stages, corresponding to the *serviceability limit state* and the *ultimate limit state concepts*, SLS and ULS for short (Figure 18.64). Signs of damage usually indicate that certain serviceability limit states have been exceeded (Calgaro and Lacroix, 1997).

Structural assessment consists of appraising the physical and mechanical states. It is necessary when modifying a bridge to cope with increasing loads and when repairing it to eliminate detected flaws. One of the assessment components is structural reliability covering serviceability and structural safety, the former is related to the various SLS verifications, the latter to the ULS.

18.3.1 Stages of structural assessment

Generally, the evaluation of a bridge has two successive phases: Preliminary diagnosis and re-calculation.

Preliminary diagnosis is a brief assessment based on existing documents and first inspection results of the examined bridge. An experienced engineer then forms his idea about the probable causes and consequences of the observed or suspected damage. Exceptionally, cautious loading tests may be needed. This diagnosis is usually sufficient to work out a field investigation programme, regarding the dimensional, material and structural aspects, using several methods described in Section 18.2.

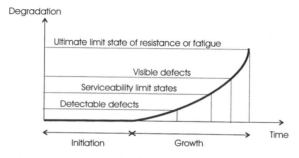

Figure 18.64 Degradation of a bridge with age.

The investigation results lead to the main re-calculation phase, essential for evaluating the structural reliability of the bridge. Re-calculation has a double objective:

- best possible estimation of the probable state of stress, accounting for observed defects
- evaluation of safety margins with respect to certain irreversible limit states or to failure.

Re-calculators should:

- thoroughly examine the bridge design construction and maintenance records
- be conscious of the usually scant available information on the past loading history, differential settlement and stress redistribution in the deck due to creep, sometimes re-adjusted by imposed strains through support level changes
- use all information on calculation assumptions and verification rules in force at the construction date as well as the present improved knowledge on material constitutive laws, e.g. creep and shrinkage, thermal gradients and fatigue.

At the end of this stage, different solutions are considered from technical and economic viewpoints to facilitate a decision on the future of the bridge: repair, strengthening, reconstruction or maintaining the present state with particular restrictions and monitoring.

18.3.2 Preparing for re-calculation

The preparatory steps include a theoretical approach through structural analysis and an experimental one supplying the actual values of certain involved parameters.

Structural analysis

The actual structural response is analysed, using conventional strength of materials or numerical methods, usually assuming a linear elastic behaviour. But non-linear models do help explain local and even general problems such as differential concrete creep and shrinkage. Other specific models, dealing for example with damage analysis and fracture mechanics, are required for cases of fatigue cracking in steel or composite bridges.

To be reasonably representative, the calculation must account for the actual accurate geometry of the structure, the consecutive construction phases and the eventual degradation of the constituent materials.

Actual geometry of the structure

Sonic tests determine the thickness of slabs or other elements and, indirectly, the weight of equipment. For old bridges with non-existent records, a complete geometric survey is required, using advanced numerical interpretation of photographic plates. Within the concrete, the tendon and reinforcement positions affect the section resistance and applied internal forces. If a difference between theoretical and actual positions is suspected, especially for the tendons, a check by radiography or radioscopy will be necessary.

Material degradation

Re-calculation should account for possible weakening of the constituent materials. Concrete may locally degrade by various physical or chemical attacks while steel sections may suffer a substantial reduction through corrosion.

In concrete bridges, cracks affecting section thicknesses also modify the distribution of forces in statically indeterminate configurations. In reinforced concrete, it is very difficult to evaluate the state of the steel stress. On the other hand, in seriously damaged steel decks, section loss can be estimated by local surface tests for corrosion.

Prestress force evaluation

One of the main uncertainties in evaluating tendon forces is the location of eventual wire failure points. These can be partially detected, using radiography, when the tendons are in relatively small bundles placed in thin-walled concrete sections.

Another uncertainty is how well and how far from the breaking point will the two new wire tips develop their re-anchorage, hence prestressing capacities.

In the absence of perfectly non-destructive measurements, two intrusive techniques can be used, the crossbow and incremental drilling. On the other hand, the stress release method can determine the prestressing force actually transmitted to the concrete.

Hence, for re-calculation, the average prestressing force is evaluated on the basis of the verified tendon profiles, assuming realistic friction coefficients usually associated with the adopted post-tensioning system. In case of doubt about the integrity of tendons, a percentage prestress loss is appraised in the light of field NDT results.

18.3.3 Structural re-calculation procedure

Only road bridges will be discussed here. Given the specific problems, whether structural or operational, railway bridges must usually be re-calculated by a specialized authority.

When a structure shows signs of damage, certain SLS are then supposed to have crossed the reversibility threshold. In a steel bridge, local defects due to fatigue reflect a steady development towards an ultimate limit state putting the structure out of service.

At this stage, the objective is to model the considered bridge, to set a reference state describing its probable behaviour and to provide a framework defining repair or strengthening criteria. This reference state also facilitates eventual further investigation.

Calculation codes and standards

Re-calculation is generally based on codified or standardized texts related to traffic loads and construction verification rules. For a bridge awaiting repair or strengthening, these texts offer a reasoning framework to estimate the present reliability level and define the intended new one.

On the whole, consecutive code amendments are constantly moving towards higher reliability. Thus, maintaining a structure within to the rules in force at the strengthening date prove more expensive than a mere defect correction in conformity with the codes that existed during its construction.

Present codes more accurately reflect reality, contain higher safety margins and offer a more reliable analysis than the old ones.

Checking a new project with respect to the ultimate limit state of resistance does not really correspond to a physical collapse process, but only secures a margin after exceeding the serviceability limit states.

Case of concrete bridges

Computations should refer to serviceability limit states (SLS), whether for re-calculating or for checking a repair or strengthening project.

In prestressed concrete bridges for example, unacceptable cracks occur long before general failure. The SLS computation bases are essentially the decompression limit states. At these discontinuities, the ultimate limit state does not usually govern except for fatigue of well-bonded tendons crossing active cracks.

Verifications are carried out for longitudinal and transversal bending, shear and torsion, prestress diffusion and eventual tendon outward thrusts. When all these act simultaneously, only a study of the physical equilibrium of the different section parts can give a satisfactory result. However, the risk of abrupt failure exists in regions of prevailing shear and corrosion-stricken tendons. In this respect, special attention should be given to tendons at post-tensioned beam extremities when there are doubts about the protecting grout quality.

In reinforced concrete bridges, the present codes afford a considerable place to ultimate limit states, hence the tangible effect after any reduction in the safety partial coefficients. In fact, the usual higher values of these coefficients are for crack limitation; if they are reduced, then specific precautions should be taken in this respect, independently from the failure risk. Here again, we should therefore refer to serviceability limit states.

Cases of steel and composite bridges

There are many steel bridges that now take loads of a much higher intensity than those for which they were designed. This previously mentioned paradox stems from the cautious old codes assuming a wider variety of load models and integrating random quality of metallurgic products, hence imposing very low allowable stresses. But while cables, girders or other main structural elements reasonably cope with the present traffic, secondary members, designed for much lighter local loads, should theoretically prove less effective. The observed resistance of many old stringers and floor beams to new axle loads is probably due to conservative old codes neglecting the favourable indeterminacy effects and, occasionally, thanks to secondary states of equilibrium achieved though friction with the slab.

Hence, many observations show the difficulty of evaluating the load-bearing capacity for an old bridge from the original calculations.

The probable structural state is thus determined as in the case of concrete bridges, but the repair or strengthening project refers rather to *ultimate limit state of resistance*. Under service loads, while elastic stress distribution in beam cross-sections is complicated due to local and tangential components, the evaluation of the ultimate bending moment is relatively reliable since most behaviour particularities disappear with a totally plasticized steel section. That is why the ULS are less conventional for steel and composite bridge sections than for their concrete counterparts and any reduction in the partial coefficients marks its full effect.

Concerning fatigue, the calculation of an existing structure's safety margin depends on assumptions strongly influencing the accuracy of Wöhler's curves, the choice of the detail category and other data. Several studies attempting to determine the reliability index with regard to fatigue did not yet reach conclusive results.

18.3.4 Re-calculation of a distressed bridge
State evaluation under dead load
The calculation of the structure under permanent loading enables the establishment and validation of the model that shall be used both for interpreting the observed defects and working out a repair or strengthening project.

Evaluating the structure under dead load consists of appraising the state of stress due to a 'quasi-permanent' combination of:

- average or probable dead weight, preferably checked by measurement
- average eventual prestressing force, also experimentally checked as described in Section 18.2
- a representative value of thermal gradients, depending on the return period chosen according to certain objectives.

Concerning thermal action, it is reminded that linear temperature gradients are only simplifying assumptions, used in some evaluation methods for new structures, but cannot describe the actual physical mechanism. Recent codes (e.g. Eurocode ENV1991-2.5) supply non-linear temperature gradients, between the extreme fibres of the main types of bridge decks, for a more accurate calculation of the structure's response.

Calculating a bridge under a quasi-permanent load combination should reflect the actual behaviour and check the experimental results, for example:

- calculated and measured support reactions must reasonably agree
- concrete regions where calculation shows a high tension or a drop in compressive stress should correspond to the observed cracks (Figure 18.65).

If this is not the case, investigation must start all over again. Any repair or strengthening carried out in a doubtful context may fail and cause additional defects.

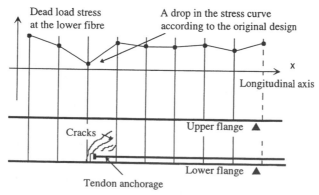

Figure 18.65 Profile of a post-tensioned bridge section. Concrete cracks appear where calculations show a sudden drop in compressive stress.

Calculation should also take into consideration:

- the consecutive construction phases, the exact schedule and any identified parasitic forces
- investigation results, e.g. geometric survey and strength tests
- the most representative constitutive laws of the involved materials, especially concerning steel relaxation, concrete creep and shrinkage.

Stress determination under a quasi-permanent load combination should enable the explanation or interpretation of eventual defects, especially those already visible under dead load.

In post-tensioned bridges, introducing a prestress margin may be justified if there are serious doubts about the integrity of tendons, or if a re-calculation based on a probable prestress value cannot adequately explain the observed structural behaviour. On the other hand, experience has shown that, even when an old bridge satisfies the latest codified recommendations, the often-slight section reinforcement generally fails to take up the developing tensile forces and to maintain the structure's safety with regard to brittle failure.

Response to service loads

After calculating the structure under quasi-permanent combination, the analysis of its behaviour under live loads has two objectives:

- improve diagnosis and achieve a positive and quantitative interpretation of the observed defects
- estimate the load-bearing capacity before working out a repair project.

As a reminder, codes defining service loads do not exactly represent the actual ones. They are models, accompanied with numerical values, intended to envelop the real load *effects*, in certain conditions.

For a realistic appraisal of the actual behaviour, both 'frequent' and 'characteristic' load combinations should be successively considered. These correspond to two distinct loading *levels*. The former is statistically taken for the traffic flow over a week or so, the latter over a much longer period. The structure's degree of damage is then estimated according to whether the observed defects are better interpreted on one level or the other.

The interpretation of calculation results should be thoroughly analysed. There are two possible cases:

- The calculation *enables* the interpretation of the observed defects. The elaboration of the repair project can then start, either to bring the bridge back to its initial service aptitude or, if the operation is too costly, to perform at best selective repair and limit the authorized live loads.
- The calculation model *does not enable* the explanation of the observed defects. Resorting to a more advanced model becomes then imperative, e.g. finite element analysis of the damaged parts. In many cases, these computations should be completed by field investigation tests described in Section 18.2. As certain defects may be due to construction or utilization errors, all possible information should be collected on these two items, including testimonies.

18.4 Conclusion

The elements and guidelines, briefly presented throughout the three sections of this chapter, are necessary but not sufficient to analyse the degradation, target the investigation and accomplish the assessment of existing bridges. These pieces of information supply some technical knowledge but cannot directly ensure judgement, only acquired through experience, given the frequently subtle interaction of parameters. For the safety of the public and the conservation of the cultural heritage and invested capital, it is essential to attain and/or maintain this level of competence, allowing better assessment and, eventually, optimum strengthening of bridges that internal and external factors have rendered structurally deficient or obsolete.

While present inspection procedures enable the assessment of bridges at a given time or period of their service life, further possibilities now open for a thorough monitoring and, possibly, active control of structural parameters such as stress and curvature redistribution. Incorporated instruments are being developed to measure and transmit the actual evolving parameters, analyse their time-dependent variation, disregard minor fluctuations but respond to irreversible deviations. The response may be a regulating action on applied forces such as the re-adjustment of external tendons or support levels. This approach towards integrated centralized actively reactive systems, operating in real-time, may pave the way to a generation of smart and sounder structures.

Bibliography

Abdunur C. Stress monitoring and re-adjustment in concrete structures. *1st European Conference on Smart Structures and Materials*, Glasgow, 1992.

Abdunur C. Direct access to stresses in concrete and masonry. *2nd International Conference on Bridge Management*, Guildford, 1993.

Abdunur C. Direct assessment and monitoring of stresses and mechanical properties of masonry arch bridges. *First International Arch Bridge Conference*, Bolton, 1995.

Abdunur C. Stress redistribution and structural reserves in prestressed concrete bridges. *3rd International Conference on Bridge Management*, Guildford, 1996.

Abdunur C. Monitoring the influence of transversal cracks in bridges. *Intelligent Civil Engineering Materials and Structures.* American Society of Civil Engineers, 1997.

Abdunur C, Acker P and Miao B. Surface shrinkage of concrete: evaluation and modelling. *IABSE Symposium*, Lisbon, 1989.

ASTM. Standard test method for half-cell potentials of uncoated reinforcing steel in concrete. ASTM Designation C876-91. *Annual Book of ASTM Standards*, American Society for Testing and Material, Philadelphia, PA, pp. 434–439, 1991.

Behr M and Trouillet P. Actions et sollicitations thermiques. *Bulletin de liaison des LPC* no. 155, pp. 57–72, 1988.

Brevet P. Application et interprétation des mesures de potentiel d'électrode des aciers enrobés de béton. *Bulletin de liaison des LPC* no. 125, pp. 125–128, 1983.

Brevet P and Siegert D. Fretting fatigue of seven wire strands, axially loaded, in free bending fatigue tests. *International Organisation of Studies on Endurance of Wire Ropes*, Bulletin no. 71, pp. 23–48, 1996.

Calgaro JA. Chocs de bateaux contre les piles de ponts. *Annales des Ponts et Chaussées*, no. 59–60, 1991.

Calgaro JA and Lacroix R. *Maintenance et réparation des ponts.* Chapters 1, 2, 3 and 12. Presses Ponts et Chaussées, Paris, 1997.

Cariou J, Chevassu G, Cote P, Derobert X and Le Moal JY. Application du radar géologique du en génie civil. *Bulletin de liaison des LPC* no. 211, pp. 117–131, 1997.

Chabert A and Ambrosino R. Pesées des réactions d'appui. *Association Française des Ponts et Charpentes*, National Conference, theme no. 3, pp. 31–46, Paris, 1983.

Chatelain J, Godart B and Duchêne JL. Detection, diagnosis and monitoring of cracked prestressed concrete bridges. *2nd NATO Workshop*, Baltimore, MD, 1990.

Cote P. NDT applied to concrete or masonry bridges. *Recent Advances in Bridge Engineering*. CIMME, Barcelona, 1996.

Cote P and Abraham O. Seismic tomography in civil engineering, *International Symposium on Nondestructive Testing in Civil Engineering*, Berlin, 1995.

De Ferri Metallographia, vol. V. Vorlag Stahleisen, Düsseldorf, 1979.

Divet L. *et al.* Delayed ettringite formation: the effect of temperature and basicity on the interaction between sulphates and the C-S-H phase. *Cement and Concrete Research* 28, 357–363, 1998.

Engel L and Klingele H. *An Atlas of Metal Damage*. Hanser Verlag, Munich. Translation Murray S, Wolfe Science Books, London, 1981.

Feliu S, Gonzalez JA, Andrade C and Feliu V. On-site determination of the polarisation resistance in reinforced concrete slabs. *Corrosion Science* 29, 105–113, 1989.

Godart B. Approche par l'auscultation et le calcul du fonctionnement de ponts en béton précontraint fissurés. *First Advanced Research US–European Workshop on Rehabilitation of Bridges*, CEBTP, Saint-Rémy-lès-Chevreuse, France, 1987.

Gourmelon JP and Robert JL. Détection de la corrosion dans les câbles de ponts suspendus par méthode électromagnétique. *8th European Congress on Corrosion*. Nice, 1985.

Guinez R *et al.* COFREND group, Principes généraux de l'éxamen radiographique à l'aide des rayons X et gamma des matériaux béton, béton armé et béton précontraint. Documentary manual/French Standards NF A 09-203, 1999.

Hobbs DW. *Alkali–Silica Reaction in Concrete*. Thomas Telford, London, 1988.

Larive C. Combined contribution of experiments and modelling to the understanding of alkali–silica reaction and its mechanical effects. Thesis 1997, Research Report no. 28, Laboratoire Central des Ponts et Chaussées (LCPC), 1998.

Persy JP and Deloye FX. Investigations sur un ouvrage en béton incendié. *Bulletin de liaison des LPC* no. 145, pp. 108–114, 1986.

Persy JP and Raharinaivo A. Etude de la rupture par temps froid d'éléments en acier provenant d'un pont suspendu. *Bulletin de liaison des LPC* no. 152, pp. 49–54, 1987.

Priestley M and Blucke I. Ambient thermal response of bridges. *Road Research*, Road Board, 1979.

Prost J. L'auscultation dynamique des structures. Son application à la patholohgie des ouvrages d'art. *Bulletin de liaison des LPC* no. 72, pp. 145–154, 1974.

Raharinaivo A. *et al. La Corrosion et la Protection des Aciers dans le Béton*. Collection LCPC, Presses Ponts et Chaussées, Paris, 1998.

Robert JL *et al.* Surveillance acoustique des câbles. *Bulletin de liaison des LPC* no. 169, pp. 71–78, 1990.

Robert JL *et al.* Mesure de la tension des câbles par méthode vibratoire. *Bulletin de liaison des LPC* no. 173, pp. 109–114, 1991.

Ryall MJ. A stressmeter for assessing the *in-situ* stresses in concrete bridge structures. *3rd International Conference on Bridge Management*, Guildford, 1996.

Sawade G, Krause H-J, Gampe U. 'Non-destructive examination of prestressed tendons by the magnetic stray field method. *Otto Graf Journal*, FMPA Baden-Württenberg (Otto Graf Institute), Vol. 8, pp. 140–150, Stuttgart, 1997.

Siegert D. Mecanismes de fatigue de contact dans les cables de haubanage du genie civil. Thesis, University of Nantes, Ecole centrale de Nantes, 1997.

Swamy RN and Al-Asali MM. *Alkalis in Concrete*, Dodson VH (Ed). STP930, American Society of Testing Materials, Philadelphia, PA, 1986.

19 Repair, strengthening and replacement

J. DARBY
WITH CONTRIBUTIONS FROM G. COLE, S. COLLINS
AND P. BROWN

19.1 Introduction

The resources devoted to repair, strengthening and replacement of bridges are now very significant, and continue to increase. In this climate it is easy to forget that this flourishing sector of industry is relatively new. As recently as the 1960s, maintenance was largely confined to painting of steelwork and pointing of masonry. At that time reinforced concrete was generally considered to be a maintenance free material. The growth in bridge repair, strengthening and replacement is due to the following factors:

- Large numbers of bridges built to expand the infrastructure are now reaching an age at which progressive deterioration becomes evident
- Salt was introduced for de-icing during the 1960s, with dramatic effects on the durability of concrete reinforcement
- There has been an increase in the weight of vehicles permitted unrestricted access on the highways, and an increase in the number of movements of abnormal loads requiring specific approval.

Increasing activity has brought about greater knowledge of deterioration mechanisms, and a wider choice of materials and methods designed to satisfy particular needs. However, selecting the right solution extends beyond consideration of purely technical matters. The need to minimize traffic disruption also has a major influence on contract planning and material selection.

Cost has always been the most important factor in determining the most appropriate remedial measures, but there is now a welcome move towards the consideration of life-cycle costing instead of considering only the lowest initial cost. Management systems incorporating these aspects are covered in Chapter 17. Attention to low-cost works covered in Chapter 16, such as waterproofing, drainage and impregnation, is likely to be particularly cost effective. However, there will inevitably remain a large number of structures requiring more extensive works of the kind discussed in this chapter.

Repair, strengthening and replacement covers such a wide area of specialist works that it is impossible to cover the subject exhaustively in this publication. Emphasis will therefore be given to factors influencing selection of the most appropriate maintenance action. Repair, strengthening and replacement should not be considered as unconnected topics. Instead, they should be seen as alternative interventions. The choice between them is based upon an evaluation of current defects, predicted deterioration and consequences, and the cost of remedial measures at each stage. The options could be simplified as: repair now, repair later, strengthen later, or ultimately replace. Delay in intervention will generally result in increased cost, but it would be incorrect to assume that early action is therefore always the most appropriate. Management decisions are also influenced by funding and operational factors.

This chapter will therefore concentrate on an outline of potential solutions, and the implications of their selection. The information will be provided in tabular form as far as possible. Whilst this should assist the initial selection, the technical and financial circumstances applicable to particular structures are essential factors in final evaluation and comparison of alternatives.

19.2 Repair and strengthening of concrete structures

19.2.1 Repair of concrete structures

The effective repair of concrete structures requires an understanding of the cause of deterioration, and an appreciation of the influence of the chosen repair technique on future durability. This should be based upon carefully planned testing of the existing structure (Concrete Bridge Development Group, in press).

The most common reason for concrete repair is corrosion of reinforcement, in which case the parameters of particular interest are as follows:

- reinforcement cover
- carbonation depth
- chloride contamination profile.
- concrete mix details
- age of concrete
- environmental factors influencing contamination and condition.

The testing programme and interpretation of results should lead to an understanding of both an extent of the existing defect, and the cause of deterioration. By consideration of these factors, the most appropriate repair interventions may be selected, as shown in Tables 19.1 and 19.2 in this chapter (see also the technical report by the Concrete Society, 1984).

Concrete repair may also be required for a number reasons other than reinforcement corrosion, such as that resulting from fire damage or frost damage. In such cases the same general advice applies, the extent of repair giving the initial indication of the most economic method of application, as shown in Table 19.3.

Table 19.1 Concrete repair interventions arising from chloride-induced corrosion

Extent of defect	Potential repair interventions					
	Mortar repair	Placed concrete	Sprayed concrete	Cathodic protection	Desalination	Impregnation
Spalling or contamination replacement over small areas	*					*
Spalling or contamination replacement over large areas <25 mm deep	*					*
Concrete replacement over large areas >25 mm deep		*	*	*	*	*

Table 19.2 Concrete repair interventions resulting from carbonation-induced corrosion

Extent of defect	Potential repair interventions					
	Mortar repair	Placed concrete	Sprayed concrete	Cathodic protection	Re-alkalization	Anti-carbonation coating
Spalling over small areas	*					*
Spalling or carbonation over large areas <25 mm deep	*					*
Carbonation over large areas >25 mm deep		*	*	*	*	*

Repair of concrete may take the form of crack sealing or structural crack repair. Such repair should only be undertaken if the cause of the cracking is understood,

Table 19.3 Concrete repair – application of repair materials

Extent of repair	Potential repair intervention		
	Mortar repair	Placed concrete	Sprayed concrete
Concrete replacement in small areas	*		
Concrete replacement over large areas <25 mm deep	*		
Concrete replacement over large areas >25 mm deep		*	*

the reasons for repair are well defined, and subsequent movements will not render the repair ineffective (Concrete Society, 1984).

Materials for the repair of concrete structures
The geometry of the repair will give initial guidance on the method of applying the repair material. There remains a wide choice of materials, and the need to meet any relevant client specifications. Table 19.4 summarizes the principal characteristic that may influence selection.

Methods of inhibiting corrosion
Concrete replacement is expensive because it is labour intensive, and may not be economic if the chloride contamination extends to a great depth. Provision of temporary support may also be impracticable or uneconomic. Alternative solutions are thus required.

Table 19.5 summarizes the methods of preventing reinforcement corrosion that do not require concrete removal.

19.2.2 Strengthening of concrete structures
Strengthening of concrete structures has been a difficult operation because changing the proportion of internal reinforcement and internal prestressing is impractical once the concrete member has been cast without its effective demolition. In these circumstances the choice may be limited to the following:

- increasing the depth of beams or slabs, including reinforcement in the extra concrete, and providing shear connection, perhaps by drilling and bonding
- increasing the width of beams, or more usually providing intermediate beams
- providing external prestress by external steel cables and anchorages or macalloy bars.
- drilling and grouting additional reinforcement, perhaps in the form of stressed bars
- providing buttresses or extra thickness to walls.

Table 19.4 Properties of materials influencing selection for the repair of concrete structures

Material	Properties	Main applications and comments
Concrete	Flow characteristics and strength designed to meet need (such as materials specified in BD 27/86)	Minimum practicable depth to flow >25 mm, and repair keyed beneath the reinforcement. Structural effects of major replacement require consideration. Access considerations control detailing, perhaps requiring increased local thickness.
Sprayed concrete	Sprayed concrete with aggregates above 10 mm is generally called 'Shotcrete', and with aggregates below 10 mm generally called 'Gunite'. 20–30 N/mm^2 typical strength for wet process, and 40–50 N/mm^2 for dry process Good density and bond, with low permeability with dry process. HA require characteristic strength	Minimum practicable depth 25 mm. Good access required. Continuous process requiring good workmanship, particularly with the dry process. Thick sections built up in layers, with the wet process when previous layer has stiffened, and in dry process when previous layer has gained strength. Smooth finish cannot be achieved, and the final coat is normally left unscreeded. Excess spray may be carefully sliced off, and second coat with fine sand aggregate produces more pleasant finish.
Hand-applied cementitious mortars and polymer modified cementitious mortars.	Polymer modified mortars are specified in BD 27/86 for highway works, and shall have a W/C ratio <0.4 and cement content >400kg/m^3	Well-placed cement sand mortars have proved durable over 25 mm thickness. Polymer modified mortars have improved properties and appropriate for thicknesses down to 12 mm. Of very high cost and rarely used. Have different expansion characteristics to cementitious materials, and rely on chemical bond.
Polyester resin and epoxy resin mortars.	Develop high mechanical strength within 24 to 48 hours. Provide impermeable coating to encapsulate reinforcement.	Rely entirely upon impermeability for protection, thus requiring good materials and workmanship. With lightweight fillers produce low-density thixotropic mortar suitable for overhead applications.
Epoxy resin for crack injection.	High modulus high strength resin. Available in low viscosity thixotropic systems.	Use for structural cracks where no further movement expected. Thixotropic system required for cracks which cannot be sealed on all sides.
SBR or acrylic latex emulsion for crack sealing	Low viscosity material disperses water into surrounding concrete to leave rubbery mass.	Appropriate for narrow cracks, or repeated application to wider cracks, to form water-resistant seal under low head where future movement limited.

Table 19.5 Comparison of alternative methods of inhibiting corrosion

Method	Basis of method	Advantages and disadvantages
Cathodic protection	Chloride-induced corrosion is loss of metal at anodic areas from which corrosion currents flow. A DC protection current is impressed between new anodes and the reinforcement, thus opposing and stopping the natural corrosion current.	The principle is well tried and proven to be effective. Prevents corrosion due to chlorides and carbonation. Systems are expensive to install, and require ongoing maintenance and ultimate replacement. Modern systems and materials extend life spans to over 30 years.
De-salination	The surface of the structure is covered with external anodes and an electrolyte reservoir. The negatively charged chloride ions migrate to the external electrolyte away from the negatively charged reinforcement.	Desalination provides a durable long-term solution only if future contamination is prevented. Concrete to be desalinated must lie between reinforcement cathodes and the anode. Desalination of concrete is completed in 8–13 weeks. De-salination is more expensive than cathodic protection.
Re-alkalization	The physical arrangement is similar to de-salination described above. An alkali electrolyte is used which causes alkaline ions to migrate to the reinforcement. This re-establishes the passive layer around the reinforcement and increases the pH value.	Re-alkalization provides a durable long-term solution only if further carbonation is prevented. The process is normally completed within 7 days.
Anti-carbonation surface treatments	Carbonation results from ingress of carbon dioxide into concrete, reacting with calcium hydroxide and reducing alkalinity. Surface treatments designed to exclude the carbon dioxide molecule yet allow passage of water vapour can inhibit carbon dioxide ingress.	Low-cost method of extending the life of structures which would otherwise suffer from corrosion due to carbonation. Cannot prevent corrosion if concrete already carbonated at reinforcement depth.
Migrating corrosion inhibitors	Coating applied to surface of concrete, the inhibitor penetrating by liquid and gaseous diffusion to form thin layer on reinforcement surface.	New technology and long-term effectiveness unproven. More expensive than silane, but likely to be cheaper than concrete removal or re-alkalization.
Impregnation surface treatments.	Materials such as silane when impregnated into concrete can prevent ingress of water in the liquid phase, but permit water vapour to exit the structure. Thus chlorides no longer enter carried by the liquid phase, and concrete slowly dries inhibiting corrosion.	Very low-cost solution. Only effective in inhibiting corrosion if water can also be prevented from entering the concrete from alternative faces. Limited no of instrumented trials demonstrating effectiveness.

Developments of new techniques and materials in recent years have opened up new possibilities. In particular, modern epoxy resin adhesives enable additional reinforcement to be added externally. This is equally durable and of significantly lower cost. The new methods may be categorized as follows:

- steel plate bonding
- unstressed composite plate bonding
- stressed composite plate bonding
- composite column wrapping.

This chapter will concentrate on these more recent developments, because they will have widespread application. Other methods will certainly have application in certain circumstances, perhaps in combination with external reinforcement, but that will depend upon the particular details of the structure.

Steel plate bonding

The technique of steel plate bonding for bridge strengthening was given official encouragement in 1994 with the publication by the Highways Agency of BA 30/94, 'Strengthening of Concrete Highway Structures using Externally Bonded Plates'. However, this was preceded by over 15 years during which major bridges in the UK were plated and the results monitored. Quinton and Swanley Interchange bridges on the M5 were plated in 1975 and 1977, respectively, followed by bridges on the M1 and A10 from 1982 to 1986. At that time there were many more applications overseas, particularly in Japan.

Steel plate bonding provides the section with additional tensile capacity at maximum eccentricity, where it is most effective. Structures that would otherwise have required demolition have been preserved. However, there are also significant limitations and disadvantages, as may be seen from Table 19.6.

Practical aspects of steel plate bonding

The effectiveness of plate bonding is entirely dependent upon the composite action between plates and parent concrete, and hence upon the adhesive bonding. High quality workmanship and supervision are essential. Both the steel and concrete surfaces must be grit blasted, dry, and free of dust. The steel will quickly deteriorate after grit blasting, and bonding should take place within four hours. Alternatively, the plates may be blasted, cleaned and primed away from the site. After transport within clean protection, they may be bonded after de-greasing and drying.

Unstressed fibre reinforced composite (FRC) plate bonding

Use of plate bonding for repair and strengthening is likely to increase as use of steel is superseded by fibre reinforced composites (FRC). All of the benefits outlined above are applicable, and disadvantages/limitations reduced. The plates are manufactured by the pultrusion process, and comprise carbon or glass fibres within a polymer matrix such as vinylester or epoxy. The number of applications worldwide in 1998 can probably be measured in hundreds. In the UK the largest research project into FRC bond-

Table 19.6 Factors influencing the selection of steel plate bonding for strengthening

Advantages/benefits	Disadvantages/limitations
Strengthening with minimal increase in dimensions	Sound substrate required, free of risk of expansive cracking due to reinforcement corrosion.
Tensile capacity at maximum eccentricity where most effective.	Steel plate liable to corrosion. Exposed surfaces require maintenance, and
External member relatively easy to inspect and check of effectiveness.	risk of adhesion failure if adhesive does not protect against rusting at interface.
	Section requires excess compressive capacity to avoid brittle failure.
	Extensive bolting required for temporary support and to resist end peeling.
	Risk of buckling in compression, resisted if stress is low by bolts.
	BA 30/94 restricts method to structures already able to support dead load and unfactored nominal live load.

ing to date has been the ROBUST project led by Mouchel Consulting. (Holloway and Leeming, 1999).

Figure 19.1 shows the application of adhesives and the plates to 18 m span test beams. Although the cost of composite plates is greater than steel plates at the present time, competition has shown that FRC plate bonding will compete successfully with steel plate bonding when all costs are taken into account.

The potential advantages of fibre reinforced composite (FRC) plate Bonding may be summarized as follows:

- Strength: FRC plates may be purpose designed with varying proportions of components.
- Weight: FRC plates are 20% of the density of steel, and 10% of the weight for the same ultimate strength.
- Transport: FRC plates may be easily transported due to low weight, and those of small thickness may be coiled for transport.
- Versatile design: the small thickness enables plates to cross over each other, one plate being kept within the adhesive thickness of the other.
- Surface preparation: steel plates must be quickly installed after preparation. In contrast, FRC plates can be manufactured with a protective ply that is peeled just before fixing to expose a prepared surface.
- Mechanical fixing: the need for fixing to counteract peel is reduced because of the reduced thickness compared with steel, as well as reduced weight to support in the short term.
- Durability: FRC plates do not corrode, and therefore do not suffer the same risks as steel of corrosion at the glue line or exposed surfaces. Maintenance costs are reduced.

Figure 19.1 Plate bonding during the ROBUST project. Application of the plates and adhesives to 18 m span experimental beams (courtesy of Mouchel).

- Fire resistance: composite systems have improved fire resistance compared to steel because of lower conductivity, but should be protected in vulnerable situations.
- Reduced construction costs and period: the low weight and easy handling and fixing lead to reduced construction periods and consequent costs.

Figure 19.2 End anchorages and prestressed composites providing strengthening to Hythe Bridge in Oxfordshire (courtesy of Mouchel).

Table 19.7 Comparison between stressed and unstressed FRC plate bonding

Advantages of stressed FRC plates	Disadvantages of stressed FRC plates
Resistance of materials below neutral axis is mobilized, such that all dead load effects may be neutralized before live loads are applied. The quantity of FRC plates required is greatly reduced compared with unstressed plates. May be used for applications where anchorage bond lengths are limited.	Additional cost of anchorage manufacture and fixing above cost of unstressed plates. Specification, supervision and workmanship must recognize the specialist nature of pre stressing and the precautions required for reliable bonding.

Stressed composite plate bonding

A very recent development has been the prestressing of advanced composite plates. This is likely to further extend the range of strengthening problems for which plate bonding is the most efficient and economic solution. A proprietary system has been developed in the UK (Darby *et al.*, 1999, 2000) in which anchorages are fixed to the structure and preformed tendons of pre-determined length are stressed until their end tabs engage with the anchorages. The first application of this method has been to a cast-iron bridge, Hythe Bridge in Oxfordshire. This is illustrated in Figure 19.2. Stressed composite plate bonding is equally applicable to concrete structures. Hythe Bridge was raised from a capacity of 8 tonnes to 40 tonne vehicles by the application of four plates each stressed to 16 tonnes. The end anchorages shown in Figure 19.2 were enclosed within a galvanized casing and fully grouted.

Stressing is particularly effective with FRC tendons because the system mobilizes the resistance of materials below the neutral axis, whilst retaining a high safety margin in the tendons below their very high ultimate strength. The advantages and disadvantages of stressed FRC plates compared with unstressed plates are compared in Table 19.7

Summary of FRC plate bonding

Plate bonding is a versatile solution to strengthening of bridge structures. The applications are summarized in Table 19.8.

Composite column strengthening

Systems have been developed for the external strengthening of columns using glass, carbon or aramid reinforcement within polymers. (Jolly *et al.*, 1998) Such systems enhance seismic behaviour by extending ductility, and research in N. America and Japan has been directed in particular towards meeting this need. However, ultimate strength and failure strain are also improved, and systems which fully enclose a column by wrapping will provide a protection against further ingress of water and chlorides.

Table 19.8 Structural deficiencies for which FRC plate bonding offers a potential solution

Structural deficiency	FRC plate bonding solution
Corrosion of reinforcement	Replacement of corroded reinforcement by external FRP, providing concrete remains sound.
Strengthening of structures in flexure.	External FRP reinforcement added providing there is sufficient section capacity in compression.
Uncertain durability of steel prestressing tendons.	External FRP may be added as a 'Safety net', and/or stress to replace prestress which may have been lost.
Inadequate stiffness or serviceability due to cracking under load.	Stressed external FRP to remove tensile stresses.
Strengthening of structures in shear.	Web may be reinforced externally, but this aspect less researched than other applications.

19.3   Repair and strengthening of metal structures

19.3.1 Metal bridges in the United Kingdom

The first metal bridge structures were constructed in the late eighteenth century. Large numbers of cast iron and, later on, wrought iron bridges were built during the canal and railway expansion of the nineteenth century. Many of these early examples remain in use to this day, often carrying loads greatly in excess of what they were designed for. Modern metal structures are constructed almost exclusively in steel, but thousands of bridges built of cast iron, wrought iron and early steels are still in service today on our highway and railway networks. Wrought iron and early steel structures were usually of riveted construction. Welded construction became established in IJK bridges in the period following World War II, while riveted construction, although becoming less common, did not finally pass out of use until the 1960s.

Metal bridge structures are generally very durable if properly maintained. The deterioration processes to which they are subject are well understood and can be readily rectified if caught early enough. The primary causes of metal bridge deterioration include the following:

- corrosion
- fatigue or other cracking
- deflections or distortion caused by loading or impact damage.

Of these, corrosion caused by breakdown of the protective system, often aggravated by chlorides from road salting, is by far the most common cause of deterioration.

Corrosion and fatigue damage are both often exacerbated by poor design detailing which creates corrosion traps and stress concentrations in members. Good inspection procedures will detect all forms of metal deterioration early enough for action to be taken before serious damage – affecting the load carrying capacity of the structure – has resulted. One advantage of metal bridges structures over concrete bridge structures is that deterioration (with the obvious exceptions of closed box members or enclosed details such as load bearing pins) is readily detectable on the surface of the members and visible to the trained eye with a minimum of equipment.

Ultrasonic gauges can be used to measure the thickness of plates and members during inspection. This equipment may also be used by specialists to detect the presence of cracking in elements that may not be accessible by other means, such as load bearing pins. Other methods of detecting cracking in metal structures include X-ray photography, and magnetic particle testing. These techniques should only be undertaken by specialist testing companies. Cracking which is not visible to the naked eye can be detected using dye penetrant testing, a simple technique which can be used by a bridge inspector.

If the strength and mechanical properties of the metal structure are not known they can be determined by cutting a small sample or coupon from a low-stressed or unimportant area such as a stiffener or flange near a bearing. The coupon can be laboratory tested for tensile strength and composition. If repairs or strengthening involving welding are contemplated the composition of the metal and its weldability must be established. This can be determined by a laboratory from a coupon or from shavings obtained by drilling.

19.3.2 Repair of metal structures
General
Repair of metal structures suffering from deterioration caused by one or more of the deterioration processes listed above should only be attempted when the cause of the problem has been clearly established and understood. In most cases the cause of deterioration will be immediately apparent, however simply repairing the defect may result in its re-occurrence if the original underlying cause is not identified. It may be that an inappropriate inspection and maintenance cycle has resulted in complete breakdown of the paint system which could have been avoided by earlier attention and overcoating. It is possible that localized corrosion has resulted from inadequate or faulty deck drainage, or the incorporation of a corrosion trap in the original design which traps dirt and debris leading to early onset of paint breakdown and severe corrosion.

In these cases, the design of remedial works and repairs should include the rectification, if practically and economically possible, of the original underlying cause of the damage to the structure to prevent its early re-occurrence. Other examples would include fatigue cracking caused by a poor connection detail resulting in out-of plane bending of a girder web plate. Simply repairing the damage without amending the detail would lead to the same type of cracking appearing again.

Economics of repair

An important consideration when designing repairs to a metal structure is the economics of repair against replacement – either of the affected member or the entire structure. Metal bridge repairs typically involve the use of very small quantities of material and large amounts of labour. It is therefore usually a false economy to concentrate on minimizing the material quantities in the design of repair works: of far greater importance is the ease of fitting on site. Repairs are thus often very expensive in comparison to the construction of a replacement structure with inherently lower future maintenance costs and a longer useful life. A 'life-cycle' cost comparison between the repair and replacement options will be necessary to determine the best option.

A further consideration is that details on metal bridges are very often repeated many times. If cracking or serious corrosion is found in one particular detail on a metal bridge, other locations where similar or identical details occur should be carefully investigated: similar problems may be detected in earlier stages at other positions. In this case a repair or modification which can be applied at all repetitions of the affected detail may be required, even those not currently exhibiting the defect. The costs of mobilization, providing access and possibly traffic management may make the wholesale modification of a suspect detail over an entire bridge during a single repair contract economically preferable to carrying out the work in smaller packages over several years.

Repair techniques

The most appropriate repair technique for an individual structure will depend on a large number of parameters, including:

- the material it is constructed from
- form of construction (cast, welded, bolted or riveted)
- degree of redundancy
- cause of defect (corrosion or fatigue for example)
- severity of damage
- costs of closure/ traffic management/ possessions
- extent or number of repairs.

Repairs by plating

Sections that have been weakened by corrosion can be repaired by applying new metal plates to replace the section loss. The new plates are bolted or welded over the corroded or under-strength areas, allowing sufficient overlap to ensure the effective transfer of stresses to and from the intact member. Designers should appreciate that new plating of this kind will not contribute to the dead load carrying capacity of the structure, and will itself add additional dead load to the structure being strengthened.

Plating can also be used to repair members that have suffered displacement and distortion of web and flange plates, perhaps by impact from over-height vehicles or vessels. New web or flange plating can be spliced over the distorted section. In severe cases the bent and distorted plates will have to be cut away prior to splicing in new sections of plate. If new plates are spliced over distorted areas of old plate care must be taken to fill and seal the areas between old and new plate to avoid the creation of a corrosion trap.

The practical difficulties of bolting new plates onto a corroded older structure should not be ignored: the surfaces of older sections are usually pitted and achieving good contact between the plates allowing satisfactory transfer of stress and avoiding the creation of corrosion traps can be difficult. Both surfaces to be joined should be painted with a primer with a sufficiently high coefficient of friction to allow HSFG bolts to work efficiently. The pitted plate should be filled with high-strength metal filled epoxy putty which will effectively fill and seal the gap between the plates and will allow full transfer of stresses between the old and new plates. Seating bolts in old pitted and uneven plates can also cause difficulties, which can be overcome by using metal filled putty to build up a flat seating surface. The design of the repair must of course allow for the weakening effect of the new bolt holes on the original section.

Plating by bolting is not suitable for cast iron structures: drilling and bolting through parts of cast iron members subject to tensile stresses is inadvisable. Much of this material is variable in composition; it may contain flaws or inclusions which, in conjunction with new bolt holes, could seriously reduce its ability to resist tensile stress.

The repair of riveted and bolted structures by plating requires particularly careful planning and design to maintain structural integrity and to effect trouble free site operations. Corroded rivets can be drilled out and replaced by bolts, however the member or entire structure must be relieved of load or the operation must be performed in a strictly controlled fashion, one rivet at a time, to avoid weakening the structure. Strengthening plates or sections for riveted structures must be detailed with clearance holes to clear rivet heads. Plating the underside of riveted bottom flanges over cover plates or splice plates may be impracticable without relieving the member of load, drilling out the rivets, removing the entire cover plate and replacing it with a heavier section.

Attaching new plates to corroded structures by welding is generally more problematic than by bolting. In the first place the designer must be satisfied that the material to be strengthened is weldable. Cast iron and wrought iron and many early steels are not of weldable quality. Secondly, the designer must ensure that the welded connection detail will not cause stress concentrations in a location subject to cyclic loading and lead to future fatigue problems. The effect of the welding process on the stresses in the member being strengthened must be carefully evaluated and the welding procedures carefully planned to avoid creating distortion and additional stresses.

In repairing members subject to tensile or compressive loading by plating, particularly those composed of built-up sections, care must be taken not to apply additional plating eccentrically as this could produce unforeseen eccentric load effects resulting in an effective reduction in section capacity, rather than the strengthening intended.

Repair of cracking

Cracking in metal bridges results from several causes depending on the type of structure and the material. Fatigue cracking is likely to be most serious in modern welded steel structures and is less common in riveted construction. Cast iron being a less ductile material than steel or wrought iron is more prone to cracking caused by impact and overstress; low temperatures can be a contributory factor.

Brittle cracking in cast-iron is difficult to repair effectively because of the problems of bolting and welding to the parent material. Proprietary systems for the repair of cracks in cast iron structures such as the 'Metalock' system can achieve good results. The system works by machining both edges of the crack to produce an interlocking profile and driving in a stitching element to connect the two edges. The technique depends on specialist equipment and licensed operators and may be difficult where access is a problem. The costs and likely success in terms of eventual load capacity of the member should be evaluated with care. The technique may be particularly appropriate for older decorative fascia beams when preserving the appearance of the structure is important, and loading can be relieved by additional structural members provided behind the fascia. An example of repair of cracking in cast iron by the 'Metalock' system is shown in Figure 19.3.

Fatigue cracking caused by stress concentrations or out of plane bending in web or other plates is usually more severe in welded construction, but is also found in riveted members. Cracks originating in welds will often propagate into the parent metal of the section if not repaired. Minor fatigue cracking in plates caused by stress

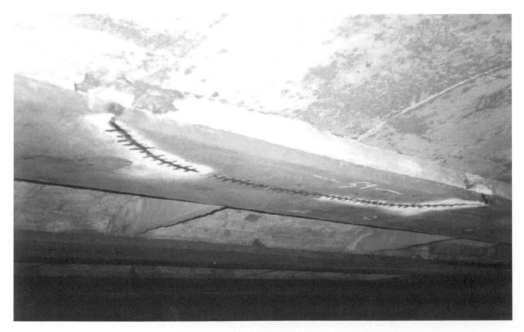

Figure 19.3 Repair of the cracks in the bottom flange of a cast-iron beam by the 'Metalock' system (courtesy of Mouchel).

concentrations and out of plane displacements can be effectively repaired by hole drilling at the end of the crack – thus reducing the stress concentration to a point at which no further propagation will occur. The hole can be filled with sealant and painted over. This technique may not be effective in stopping crack propagation if the out of plane displacements are large. A more effective long-term remedy may be to provide an alternative load path stiffening the connection and preventing the displacement of the affected plate. Alternatively it may be possible to increase the flexibility of the connection, provided this results in a reduction in the relative displacement of the affected plate. Major cracking which has seriously weakened the section may require a repair of the affecting area by cutting out and replacing the damaged plate, or by plating over the cracked section.

Cracking in welds caused by fatigue or faulty welding is usually repaired by grinding out the affected length of weld and re-welding. Some success in repair of small (less than 3 mm deep) weld cracks which have not propagated far has also been achieved by peening the weld with a mechanical air hammer making the weld more resistant to fatigue. Peening plastically deforms the weld and induces a local residual compressive stress zone in the weld and parent metal, preventing the formation and propagation of cracks

All of these repair techniques may be ineffective in the long term if the underlying cause of the fatigue damage is not addressed in the repair. It is essential that a thorough evaluation is made of the load effects and displacements which have caused the stress cycling at the affected area. It may be possible for alternative load paths to be provided to prevent displacements and stresses being induced. Alternatively, making a connection more flexible may actually reduce its proneness to fatigue damage, provided the relative displacements in the affected part are reduced. An example of measures taken in response to fatigue cracking caused by out of plane displacements is shown in Figure 19.4. Holes were drilled at the termination of existing cracks, rivets replaced, and new cover plates fitted to stiffen the joint and provide alternative load paths.

Member replacement

The most effective and economical repair strategy for severely damaged or corroded members may be complete replacement. Should replacement of an entire member be contemplated, the integrity and stability of the whole structure during removal and replacement will require careful consideration. Analysis of the original structure will indicate the levels of load to be expected in the affected member under dead load alone. Provided that the original member dimensions, properties and fixing details have been accurately duplicated, and the replacement scheme induces displacements across the whole structure which are identical to the originals, the loads in the replacement member and the other members will be the same as in the original structure. A means of checking the level of loading in the new member is often provided as part of the replacement scheme, by means of strain gauges, in order to ensure that the original distribution of loading between all members of the structure has been reproduced. In some cases it may be necessary to provide means of adjust-

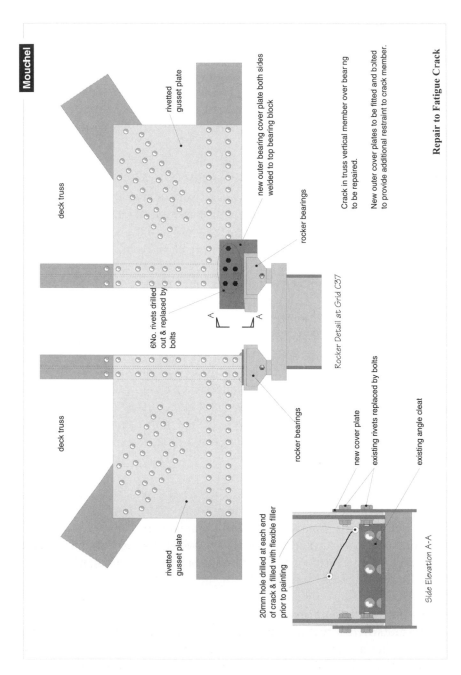

Figure 19.4 Repair to fatigue crack (diagram courtesy of Mouchel). This figure shows a repair to the connection at the end of the deck approach truss of the Puente Duarte bridge in the Dominican Republic. The cracking was caused by out of plane displacements of the vertical member web in a poorly detailed joint.

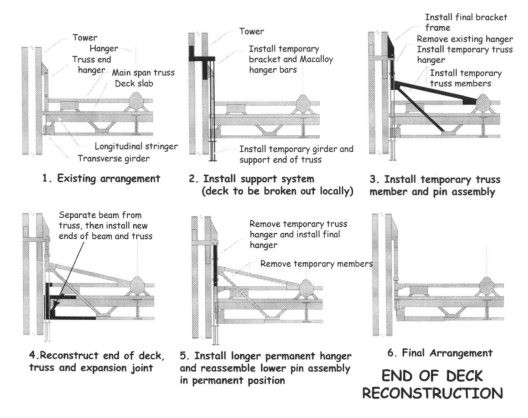

1. Existing arrangement

2. Install support system (deck to be broken out locally)

3. Install temporary truss member and pin assembly

4. Reconstruct end of deck, truss and expansion joint

5. Install longer permanent hanger and reassemble lower pin assembly in permanent position

6. Final Arrangement

END OF DECK RECONSTRUCTION

Figure 19.5 Example of member replacement (courtesy of Mouchel). Reconstruction of the end of the main deck truss of the Puente Duarte Bridge in the Dominican Republic. The reconstruction involved the replacement of parts of the truss longitudinal members and the end transverse girder. The end of the truss had to be supported at all times during reconstruction enabling the bridge to remain open to traffic.

ing loading in the replacement member, by jacking for example, to ensure the correct final distribution of loading between members.

An example of multiple member replacement is shown in Figure 19.5, illustrating the complex sequencing and temporary works required to replace severely corroded members forming part of the main deck truss of a suspension bridge whilst the bridge remains open to traffic.

19.3.3 Strengthening of metal structures

Strengthening of metal structures employs many of the same techniques as repair. Plating of under-strength members is often used to increase their live load capacity. As previously stated, the application of additional thicknesses of new plate to a member cannot increase its dead load capacity, and will actually contribute to the dead load to be carried by the original member. Strengthening against bending by plating in this way is therefore relatively inefficient. Designers may find that the additional capacity they seek cannot easily be achieved by plating, in members in which dead loads predominate, as the new plates increase dead load stresses so much that the gains in live load capacity are almost cancelled out.

Strengthening by plating

Strengthening plates can be fixed to the original member by bolting or welding. The detailing of fixings to the member is very important to avoid the creation of traps for water and debris which would lead to future corrosion problems. In the case of welded fixings there is the potential to create fatigue problems if the joint is incorrectly designed. Carrying plating into joints is also a problem. Bolted or riveted joints have usually been designed to allow only sufficient fasteners to be positioned to resist the originally anticipated loads. Strengthening joints can be a very difficult exercise often requiring partial or complete reconstruction of the joint to allow new and larger gusset plates to be inserted with more space for additional bolts.

Pre-stressing or load relieving

The problems of strengthening structures in which dead loading exceeds live loading can be addressed by stressing additional or replacement members into place, and thus altering the load distribution in the existing structure. Several different methods of achieving the desired result have been developed, for example jacking additional sections into box or built-up column sections in order to reduce the compressive stresses in the existing member and thus provide additional capacity. In contrast, large truss bridges have been strengthened by the addition of new pre-stressed tension members in place of existing unstressed members in order to alter the load distribution in the truss and reduce the levels of dead load stress in critical members.

The design of schemes involving the strengthening of compression members by jacking load into new sections which then become a part of a new composite member must pay particular attention to the local and global buckling of the new and existing members. The new member, unless particularly stocky, will need to be restrained by the existing member at points along its length to prevent buckling. The restraints must be designed to act effectively as lateral restraints, whilst allowing the new member to slide vertically as load is jacked into it. Use of this jacking technique, in conjunction with load measurement systems, allows a high degree of control in the proportion of vertical load carried by the new and existing members. Careful detailing of the baseplate and capping plate areas will be required to ensure even transfer and distribution of load. Flat jacks are often used as the means of loading the new members. This technique is illustrated in Figure 19.6, showing insertion of additional strengthening members in the man towers of a suspension bridge, preceding re-cabling and general strengthening works.

The principle of relieving load by pre-stressing new members into a structural system can also be used to strengthen individual members. Large steel plate girders have been strengthened in flexure by the addition of stressed bars positioned below their lower flanges. The bars were stressed from fabricated anchorages bolted to the underside of the beam's lower flanges and extended over the overstressed length of the central part of each span. Strengthening in this way relies not on the addition of large

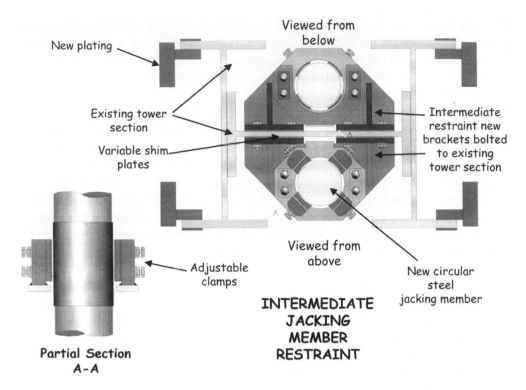

Figure 19.6 Strengthening of compression member (photographs courtesy of Mouchel).
The strengthening of the towers of the Puente Duarte bridge in the Dominican Republic
is illustrated. The figure shows the new circular members that will relieve the existing
section of load and the intermediate restraint brackets within the existing towers.

amounts of supplementary material, but on the alteration of the member's existing stress state. The prestress, by reducing the levels of tensile stress in the bottom flange of the beam effectively releases some of the 'locked-in' dead load stresses which can then be used to resist increased live loading.

A new variant of this technique has recently been used to strengthen a cast-iron highway bridge using pre-stressed advanced composite carbon fibre reinforced polymer plates. In this application fabricated steel anchorages were clamped to the lower flange of the cast iron beams. The composite plates were then stressed and attached to the anchorages which transferred the prestress force into the cast iron beam. The composite plates were also fully bonded to the underside of the flanges by epoxy adhesive and thus acted compositely with the beam.

A further means of altering the existing state of stress in a structure is by 're-profiling'. This reduces levels of tensile or compressive stress in critical members, and thus achieves an increase in structural capacity the structure, by inducing controlled amounts of load - generated by displacement of the structure – in critical members. For example, this principle has been used to increase the live load capacity of deck trusses in long span suspension bridges. By adjusting the lengths of hangers in order to produce a greater upward curvature of the deck under dead load alone, compressive forces are induced in the bottom chord of the truss, and tensile forces are induced in the top chord. These forces are in opposition to the forces resulting from dead and live loading, and therefore the net effect is to reduce the forces in the truss under dead loading alone, increasing the live load capacity of the truss. The forces induced in the truss can be readily calculated to enable the required amount of adjustment of deck levels to be determined. The operation must of course be carried out in a carefully controlled manner to avoid local overstress of truss members.

In a similar manner, if it is possible to adjust the level of supports of a continuous bridge girder or truss by jacking, moments can be induced over intermediate supports which can counteract the hogging effects of dead and live load. The usefulness of this technique is often limited by the practical difficulties of achieving the required amount of vertical movement at the abutments or intermediate supports within the constraints of highway or railway vertical alignments.

19.4 Repair and strengthening of masonry structures

19.4.1 Introduction

The masonry arch (see Figure 19.7) is the most common form of structure in the UK bridge stock amounting to some 40% of the total (Page, 1993). Most of these struc-

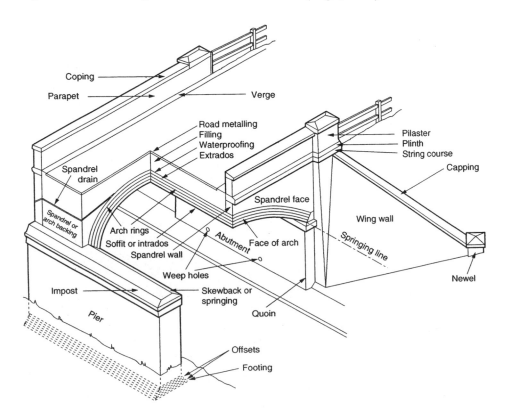

Figure 19.7 Typical construction of a masonry arch bridge (source Sowden, 1990).

tures have been subjected to prolonged exposure to environmental effects. They have also been subjected to loading well beyond the expectations of their original designers. Therefore, it is not surprising that these bridges suffer from a variety of defects. Effective repair and strengthening of masonry arch bridges is essential if their future is to be safeguarded. This part of the chapter has been written following a literature review and includes details of several recent innovative strengthening techniques. An extensive list of current references has also been provided.

19.4.2 The inspection and investigation process

Advice on how to carry out the inspection of masonry arches is well documented (HMSO, 1983; BA16, 1997). The inspection should seek to identify defects and their causes and helpful checklists of faults have been produced (HMSO, 1983; Welch, 1995). The presence of water plays a significant part in the deterioration process of masonry arch bridges and the management of water is vital to the successful maintenance of arch bridges (Pearson and Cunninghame, 1998). Therefore, it is particularly important to identify potential causes of water penetration, such as seepage from broken or blocked drainage systems. Seepage may be evident on an arch soffit, through cracks at the edges of the arch barrel (Newbegin and Shelley, 1995) or through spandrel walls (Figure 19.8).

The condition of the road, footpath and verge surfaces should be examined for faults that allow water into the fill. Service trenches cut adjacent to the spandrel and

Figure 19.8 Seepage through a spandrel wall (photo courtesy of Surrey County Council).

Table 19.9 Defects and remedial options (source: Page, 1993)

Defect	Possible solution
Accumulation of debris and vegetation	Routine maintenance
Deteriorated pointing and brickwork	Repoint Brickwork repairs Coating of masonry
Arch ring thickness assessed to be inadequate to carry required traffic loads	Saddle Sprayed concrete to soffit Prefabricated liner to soffit Retrofitting of reinforcement Grouted anchors Asphalt overlay
Internal deterioration of mortar, e.g. separation between rings of a multi-ring brick arch	Grout arch ring Radial anchors
Weak fill	Replace fill with concrete Grout fill if it is suitable
Foundation movement	Mini-pile Grout piers and abutments Underpin
Scour of foundations	Underpin Construct invert slab
Outward movement of spandrel walls	Fit tie bars Replace fill with concrete Take down and rebuild Grout fill if suitable
Arch ring – splitting below spandrel walls	Stitch (short tie bars spanning the crack)
Water leakage through arch ring	Waterproof road surface Waterproof arch ring extrados Improve drainage
Progressive deterioration	Waterproof and strengthen

wingwalls may allow water into a particularly vulnerable part of the structure (Page, 1996). Excavations relating to service company activity may also have damaged the arch ring itself. The effectiveness of historical road widening methods should also be noted.

A visual inspection should be supplemented by carrying out a 'hammer tap' test. This is an acoustic method of detecting separations between the brick rings of a masonry arch by striking the masonry with a hammer. Ring separation is indicated by a 'hollow sound'. The method is subjective and can vary within areas of any one arch, between arches of the same bridge and between different bridges. It can only pick up separation nearest to the intrados of the arch.

Further data can be obtained by coring through the arch ring to confirm the thickness of the arch barrel, which may vary across the width of the bridge. Particular care should be taken in this regard when assessing a bridge. The condition of the brick core will give a reasonable indication of the presence of ring separation but the coring method can disturb the parent material such that it becomes difficult to assess the condition of the brickwork itself. Information on fill material will need to be obtained from trial pits excavated in the carriageway. This activity causes traffic delay and care is required with the reinstatement to avoid the damaging effects of dynamic wheel impacts.

A useful tool for gathering information on masonry arch bridges is the impulse or ground-probing radar. The arch can be surveyed from the deck level alone but a more complete picture can be obtained by surveying the intrados. Information can be obtained on the thickness of different material boundaries and the thickness and extent of any saddling and haunching (Millar and Ballard, 1996). Ideally, the survey should be calibrated by trial pitting in non-sensitive areas and by carrying out a number of cores.

It is essential that the cause of deterioration is understood before the most effective repair or strengthening method can be determined. The effect of a repair on the behaviour of the existing structure must be considered. If the inherent articulation of the stonework or brickwork is lost as a result of the repair, it may have a long-term detrimental effect on the fabric of the structure. A series of defects and a range of possible repair and strengthening solutions (Page, 1993) have been set out (Table 19.9).

19.4.3 Identification of defects
Deteriorated pointing and brickwork
Masonry arches can suffer from a loss of pointing, mortar decay, spalling and delamination of brickwork. In the more severe cases splitting and disintegration may occur leading to complete loss of bricks. If left unchecked then there will be a loss of strength in the arch leading to barrel deformation and cracking. The main cause of this deterioration is often weathering and frost action assisted by water penetration (Newbegin and Shelley, 1995). The standard of bricks plays a significant part in the rate of deterioration. A good quality brick can often provide a second line of defence following the breakdown of a waterproofing and drainage system. Test cores can be used to determine the extent of frost action and damage, which may be limited to the outer courses of brickwork.

Sulfur compounds in the air or in rainwater can play an important part in the deterioration of brickwork. Mortars may be attacked by sulfates derived from the bricks themselves. Attack is gradual and occurs when the brickwork remains wet for long periods. Salts crystallize within the pores of the masonry causing stresses to develop, which can cause local fragmentation of the stone or brickwork.

Arch ring thickness assessed to be inadequate

The behaviour of masonry arches is well documented (Heyman, 1982; Page, 1993) and is described in Chapter 8. The strength of these structures is heavily influenced by the quality of the materials used, the quality of the workmanship involved in the construction and the geometric proportions of the design (HMSO, 1983).

It is the wide range of configurations and materials that have been used in the construction of these bridges, which makes it difficult to accurately assess their structural condition. Even though these bridges have inherent reserves of strength, the real problem is the progressive wear and tear of the structure. Consideration of assessment methods are detailed elsewhere (Hughes and Blackler, 1997) and are beyond the scope of this chapter.

Arch ring separation

Ring separation is a common problem with multi-ring brick arches and is associated with the loss of bond between successive rings caused by weathering and/or stress cycling of the mortar (Figure 19.9). Unless a detailed examination has been carried out, as described above, then engineers have only been able to make allowance for this defect through the use of a subjective condition factor. Tapping with a hammer is a simple but limited way of attempting to detect its presence.

Work at Bolton Institute (Melbourne, 1990; Melbourne and Gilbert, 1993) has attempted to determine the significance of defects on the load-carrying capacity of masonry arches by testing both models and full-scale two-ring brick arches. The

Figure 19.9 Example of arch ring separation (courtesy Surrey County Council).

conclusions were that ring separation caused a reduction of between 56% and 33% in the ultimate load-carrying capacity of the model and full-scale arch bridges, respectively, thereby confirming the importance of this defect.

Weak fill

The behaviour of fill material, particularly for deep arches and spandrel walls, is critical to the performance of the bridge. The major problem likely to effect fill is that the road surface waterproofing or the drainage breaks down and the fill becomes saturated. Fines may be washed out of the fill leading to voids. Water percolating through the arch ring is likely to lead to deterioration of the mortar. Saturated fill will substantially increase the lateral pressures on spandrel walls and even higher pressures if the fill freezes in winter, perhaps leading to outward displacement of the wall (Pearson and Cuninghame, 1998). Deformation of the fill will lead to an uneven pavement with consequent increase in dynamic wheel loads and possible damage to services, such as gas pipes, etc.

Foundation movement

Arch rings generate pressure on their abutments which may lead to outward movement. The fill behind abutments will resist this movement which may in itself cause inward movement. The effect on the arch ring will depend on the direction of movement and whether it is accompanied by rotation of the abutments. Transverse cracking in the arch ring is likely to occur. Recent cracks are a cause for concern as they indicate that fresh movement is taking place.

Movement of foundations is likely to cause the loss of bedding mortar between components of an arch and in severe cases to displacement or loss of brick and stone blocks (HMSO, 1983). If one edge of the bridge settles then longitudinal cracks will occur in the arch ring. This may be serious if the ring divides into effectively independent segments. If one abutment tilts relative to the other then diagonal cracks are likely to occur, starting near the side of the arch at a springing and spreading towards the centre of the barrel at the crown (Page, 1993).

Scour of foundations

An assessment of scour should be included in an overall review of an existing bridge. Scour is probably the most common cause of collapse of masonry arch bridges (Page, 1993). The foundations are generally shallow and therefore susceptible to scour. It is difficult to detect because it is likely to be at its worst when the river is in flood and access is impossible. However, proven methods are now available (Riddel, 1993; Meadowcroft and Whitbread, 1993). In particular, river bed profiles may be determined by poling or by a leadline.

Outward movement of spandrel wall

Spandrel walls suffer from the normal problems associated with exposed masonry such as weathering and loss of pointing. They are also frequently affected by dead and live load lateral forces generated through the fill or as a result of vehicle impact on the parapet or by freezing of the fill. The effect may be outward rotation, sliding on the arch ring, or bulging. Cracking of the arch ring beneath the inside edge of the spandrel wall is more likely to be caused by flexing of the ring as described below.

Spandrel walls are visually assessed. This is because no formal analysis method exists largely because of the very complex nature of their behaviour. It can be very difficult to ascertain the construction details. Lateral arching within the spandrel may increase its flexural strength. The strength of masonry has been shown (Thompson, 1995) to have a significant influence on the stability of the spandrel wall.

Arch ring – splitting below the spandrel walls

Spandrel walls stiffen the arch ring at its edges. The mechanism for cracking is thought to be differential movement between the relatively flexible brick arch and the stiff spandrel walls under live loading exacerbated by earth pressure on the spandrel walls. This type of failure may be assisted by rainwater getting into the structure at the parapet/surface joint and causing damage to the arch ring mortar where the spandrel wall meets the ring (Pearson and Cuninghame, 1998).

Water leakage through arch ring

The presence of water can cause loss of strength in arch rings. Seepage of water washes out mortar and, when associated with freeze–thaw cycles, is responsible for spalling and splitting. When the water carries salts it may evaporate from the surface leaving behind inorganic crystals which can push the surface off as flakes.

Progressive deterioration

There is considerable concern that some masonry arch bridges may be deteriorating rapidly as a result of high frequency loading caused by increasingly heavy traffic and axle loads (Pretlove and Ellick, 1990; Lemmon and Wolfenden, 1993). It has been reported (Powell, 1997) that bridges which have shown no signs of distress for many years can suddenly deteriorate. British Rail Research tests have also shown fatigue to be a problem, particularly when the masonry becomes wet. This can be avoided by limiting the serviceability limit state for brickwork compressive stress to 50% of the ultimate compressive stress (Lemmon and Wolfenden, 1995). There is little codified advice on the determination of the serviceability limit state but some recent work has been published (Choo and Hogg, 1995).

19.4.4 Repair techniques

It is vital that repair work is carried out which is sympathetic to the appearance, structural behaviour and existing materials of the bridge. There have been many examples in the past where unsuitable materials have been used for remedial work which were incompatible with the stonework or masonry (Ball, 1997). Timely intervention is also required. Typically, a combination of comparatively minor individual items such as inadequate pointing, open joints, outward movement of spandrels together with water damage from a burst water main over the barrels of the arches can leave a structure in urgent need of repair.

Routine maintenance

Routine maintenance involves modest expense compared with the possible consequences of neglect. The following items should be carried out during routine maintenance:

- keep road surface in sound condition to avoid dynamic loading
- maintain road profile to assist in shedding water
- maintain waterproofing system (where one exists)
- maintain drainage system
- remove vegetation from structure
- make good small areas of deteriorated mortar.

Repointing and mortar

Repointing is widely regarded as essential and may improve arch load capacity by restoring the structurally effective arch ring thickness to full depth. If properly done when it is needed, it may prevent the bridge from deteriorating to the point where it needs more expensive repair work. If incorrectly done it can accelerate deterioration of the structure. The mortar should not, for instance, be harder than the brick or stone. If it is too soft, the arch will continue to behave with a reduced effective thickness. The cost of repointing is modest compared with other techniques. However, if pointing is neglected too long, the cost of restoration will increase dramatically. The appearance of the bridge can be enhanced and the work need not disrupt traffic whilst being done.

It has been shown that, in general, the most effective method of repairing old brickwork and stone structures is to use lime mortars and renders (McDonald and Allan, 1997). Unfortunately, many current repair schemes are having to reverse inappropriate work that was carried out earlier this century, such as replacing very hard impervious cement pointing (Blackett-Ord, 1996). Older masonry structures were invariably built using lime mortar, which is flexible and porous, and any moisture that entered the structure was allowed to evaporate through the joints. The introduction of cement had the effect of sealing any moisture into the structure, keeping it saturated and susceptible to frost damage. In turn, water pressure built up behind the facework, forcing it outwards. Lime built structures can tolerate seasonal and minor structural movement without damage to masonry or joints. Any movement is taken up by minute adjustment within the flexible mortar beds over many courses of brick or stonework. Further advice can be found in the literature (Welch, 1995).

Lime mortar takes longer to harden than cement, and until it is fully set it is susceptible to frost damage. Its setting rate is reduced considerably as autumn advances. Lime gains strength slowly through re-absorbing carbon dioxide from the air. It will tend to dry out well before its full strength is reached which can cause two problems:

- shrinkage: this can be controlled by the use of sand and the addition of fibres for render
- rapid drying: shrinkage can be exacerbated if drying is not controlled.

A combination of curing under damp hessian and adequate wetting of the substrate can solve these difficulties.

Brickwork repairs

Where the surface of the stone or brickwork has been eroded the material can be reinstated using mortar repairs (Newbegin and Shelley, 1995) and the following points, taken from this reference, should be taken into account:

- employ experienced specialist firm
- highly skilled workforce is required
- prepare detailed specification
- grit blast masonry and wash down dust and contaminants
- soak with approved inhibitor to restrict fungal growth
- use a silicone enriched mortar repair reinforced with a polypropylene mesh
- reinforce repairs greater than 75 mm deep with stainless steel mesh
- secure mesh with stainless steel pins
- use dry mix with minimal shrinkage
- finish to profile by hand to match texture of adjacent masonry
- surface layers to be colour matched to adjacent masonry.

Brickwork delamination can be repaired by stitching the intrados back to sound brickwork using stainless steel reinforcing bars prior to filling the interspace with a free flowing, non-shrink, cementitious grout. This technique has been successfully used to restore a railway viaduct (Mathews and Paterson, 1993).

Coating of masonry

Generally, exterior stonework or brickwork is not coated or impregnated by materials such as silane. However, in certain circumstances, such a process can assist in the water management of a structure. Coatings must allow the masonry to breathe otherwise a build-up of water pressure may occur, leading to a failure of the coating or spalling of the masonry. The material should:

- be effective at reducing water uptake into stone
- be effective at allowing salts into solution to pass through it
- not alter the original colour of the stone.

It is recommended that specialist advice should be taken before any masonry coating systems are specified for use as a water-repellent (Pearson and Cuninghame, 1998).

Arch grouting

This technique is used to fill voids in the arch ring which are caused, for example, by ring separation. This will ensure that the full depth of section is available for load carrying. It does not affect the appearance of the bridge unless grout extrudes from cracks and is not removed. The grout, which is usually cementitious, needs to be carefully designed to avoid premature setting before it has completely filled the voids and to ensure that its properties are compatible with the existing arch material (Page, 1993). High pressure grouting may damage weak structures. In order to minimize this possibility, make sure that repointing of bricks is carried out and radial ties are installed (as appropriate) before carrying out the grouting operation.

Radial anchors

Grouted radial anchors have been used for many years to repair masonry structures. There use as a strengthening technique following a load assessment is considered further below. Radial anchors can be used to repair ring separation, arch face cracking and movement cracks between the arch barrel and the spandrel walls (see Figure19.10). They can also be used to stabilize spandrel walls and to reinforce a wide variety of masonry situations.

Diamond drilling techniques are used to core through masonry barrel arches, or alternatively through the spandrels and fill, to allow the introduction of a variety of anchor systems. These systems usually employ cementitious saturation anchors, or alternatively, cementitious anchors which utilize a polyester sock to prevent grout loss (Wardle, 1995). Anchors of different specifications can be mixed on the same project and the choice will depend on site conditions.

This type of anchor can also be used to repair longitudinal cracks. Typically, alternate voussoir stones are cored laterally (30 mm diameter) and the cores retained. A 30 mm hole is then drilled normal to the spandrel and at mid-depth of the arch ring, to a length of some 760 mm beyond the crack to be tied (up to a maximum of 12 m). The anchor is then installed and the grout injected down the bar fills the sock, expanding it to key into all recesses. The cracks are then pressure pointed and the ends of the stone cores reinserted to plug the holes at the arch face (Welch, 1995).

14.4.5 Parapets

At present there is no specific advice regarding the design of masonry parapets to contain errant vehicles to current loading requirements. As a result, some practitioners have constructed post-tensioned brick columns at each corner of the bridge and reinforced concrete torsion beams have been connected between each pair (Halsall, 1994). A reinforced brick parapet, which was capable of being repaired, was then built on each torsion beam. This is a complex solution and recent research (Gilbert *et al.*, 1995; Middleton, 1995) suggests that unreinforced masonry may be more than adequate to act as a parapet system. A method of assessing an existing parapet has been determined (CSS, 1995). It is recommended that this method is used for both assessment and design until such time as BS 6779 Part 4, which is a specification for parapets of reinforced and unreinforced masonry construction, is published.

Parapets constructed in the course of repair or reconstruction should have an adequate coping. The use of concrete copings to the top of the parapet will prevent the downward penetration of water and additionally will direct water clear of the wall below. The use of half-round bricks and creasing tiles perform a similar function and are aesthetically more pleasing. Capping such as brick on edge do not direct water clear of the wall but are less liable to vandalism.

19.4.6 Strengthening techniques

It should be appreciated that the various sections of this part of the chapter are not mutually exclusive. Many of the repair techniques listed will also be used in a strengthening contract. The combination of applications will depend on the extent of defects

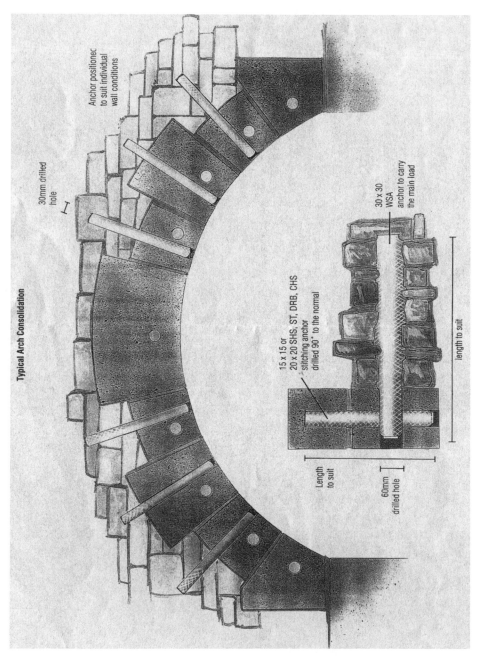

Figure 19.10 Radial anchors to repair arch face rings (courtesy of Cavity Lock Systems Ltd).

on each particular bridge. It is, of course, of primary importance that safety and arch stability are fully considered when evaluating a strengthening solution. A fuller treatment of techniques can be found elsewhere (Ashurst, 1992; Page, 1993, 1996). Careful programming is required to ensure that the various stages of the repair/strengthening project are carried out in the correct sequence.

A whole-life costing or life-cycle analysis philosophy is recommended. The objective should be to complete a durable solution with minimum traffic disruption at the time of construction.

Relieving slabs and saddling

The introduction of a secondary structure, such as saddling or a relieving slab (see Figure 19.11), may provide an excellent structural solution, particularly where historical widening can be integrated and the saddle can be waterproofed. Saddling involves removal of fill and casting an *in situ* concrete arch, which is often reinforced, on top of the existing arch. As such, the system will require relatively long periods to complete. The reasons for signs of distress and deterioration should be determined. For instance, movements of the abutments may cause cracking in the arch barrel. The addition of a saddle would lift the line of thrust, which may increase abutment movement and make the problem worse.

Codified rules (BD21, BA16) for the assessment of highway bridges indicate that arch barrels may be strengthened by means of a reinforced concrete saddle, however there is no guidance for the assessment, or indeed the design, of the strength of such arch bridges. Work carried out at the University of Nottingham (Choo *et al.*, 1995) showed that for the particular circumstances investigated a 33% increase in arch barrel thickness could double the load carrying capacity of the arch. The effectiveness of the technique was shown to be highly dependent on the bond condition between the brick and concrete. A parametric study was carried out using a two-

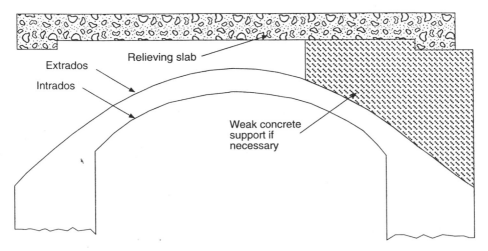

Figure 19.11 Arch relieving slab (source BA 16/93).

dimensional finite element program, which had previously been validated against tests carried out by the Transport Research Laboratory and others. The results were compared with experimental data from two tests on segmental circular arches of 2.5 m span. The project indicated that the results obtained from a modified MEXE analysis were extremely conservative. The authors cautioned that this method needed to be verified against further experimental data. In practice, good bond is provided between the concrete, the arch and the spandrel walls either by the use of ties and dowels, or by the rough state of the extrados brickwork.

Consideration should be given to the transverse behaviour of the saddle (Page, 1993). Unlike in the longitudinal direction, there is little or no induced compressive stress, in fact it is more likely to be in tension. The transverse restraint at the springings may be enough to cause cracking of the saddle. It is necessary, therefore, to give careful consideration to the sequence of casting.

At its simplest, overslabbing merely consists of providing a load-spreading slab to reduce local load intensity. It is of benefit to the barrel in that it allows the option of high-level waterproofing and it reduces lateral pressure on the spandrel walls (Welch, 1995). However, care is needed to prevent load being concentrated at the crown. When it is necessary to remove the intervening fill between slab and saddle then it should be replaced with weak concrete (7.5 N/mm^2), sufficiently weak to allow some movement and allow arch action to take place (Welch, 1995).

The extent of excavation required may preclude the use of these techniques because of traffic management, service company apparatus or safety grounds. However, the provision of a sound concrete surface does allow the arch to be effectively waterproofed. When this is carried out in conjunction with the provision of a maintainable drainage system then a lot of the water related defects described above could be avoided.

Sprayed concrete

Sprayed concrete, or guniting, is an ideal material to use on circular sections, such as the intrados of an arch, for increasing arch ring thickness. Pre-mixed concrete is sprayed at high velocity and it adheres on impact, filling crevices and compacting material already sprayed. A layer up to 300 mm thick may be applied. The concrete can be reinforced and with care, satisfactory finishes can be achieved (Minnock, 1997). Unfortunately, the addition of a skin of concrete will add unwanted weight and stiffness with reduced headroom. It does not enhance the appearance of the bridge and is very unlikely to be permitted on a listed structure.

Unfortunately, the lining may separate from the original arch by shrinkage of the concrete or by further deterioration of the arch material at the interface, which would mean that it would not increase the load capacity as much as if it were fully attached (Page, 1993). The method fails to address the most common cause of arch defects, water ingress from above. Rusting of the reinforcement would be a serious concern and every effort should be made to exclude water from the structure by other means.

Prefabricated liners

Arch ring thickness can be increased by attaching a corrugated metal or glass rein-forced cement lining to the soffit as permanent formwork, and filling the space between it and the arch ring with concrete or grout (Page, 1993). Care needs to be taken that the void is fully filled. It is quick to apply and involves no disruption to services, but it reduces the size of the arch opening and does not enhance the appear-ance of the bridge.

Embedded reinforcement

Embedded reinforcement is also known as retro-reinforcement. Originally developed by the University of Bradford, Brunel-Atkins Ltd. and Bersche-Rolt Ltd (Garrity, 1995a), it is a method of reinforcing existing masonry structures with little disrup-tion to the original exposed finishes. It involves the installation of small diameter stain-less steel reinforcing bars into grooves, up to 75 mm deep, that have been previously cut into the surface of the existing masonry. Each reinforcing bar is encapsulated in a proprietary grout. A series of unreinforced and reinforced 2 m span clay brick model arch bridges were tested in the laboratory at the University of Bradford (Garrity, 1995b). Retro-reinforcement was simulated in the tests using thin strips of steel glued to the surface of the brickwork with an epoxy adhesive. The test results offered the following conclusions:

- adding spandrel walls and parapets to an unreinforced arch ring produced an increase in the strength and stiffness
- reinforcement installed on the intrados strengthens and stiffens the arch by delay-ing the formation of hinges
- the small amounts of reinforcement used did not alter the fundamental behaviour of the model arch bridges.

Safety considerations associated with the design and installation of the temporary means of access and support required when carrying out any form of repair to a masonry arch bridge also apply with retro-reinforcement work. The reduction in struc-tural integrity caused by the cutting of grooves was a major concern and it may be necessary to install the reinforcement in comparatively narrow bandwidths and to provide temporary support to the adjacent unreinforced masonry. The need to use high strength grouts in order to develop the bond strengths required could result in the repaired zones being much stiffer than the surrounding unreinforced regions (Garrity, 1995b).

SSP Consulting Civil & Structural Engineers and Protec Industrial Ltd. have devel-oped a commercial retro-reinforcement system, known as Masonry Arch Repair and Strengthening (MARS). This system consists of cutting rebates both longitudinally and transversely, then fabricating an *in situ* cradle (see Figures 19.12 and 19.13) of 6 mm diameter stainless steel high yield rebar (Minnock, 1997). The bars are encap-sulated with a low modulus structural adhesive pumped under pressure to obtain high bond strength to the parent substrate and centralized reinforcement cradle. The

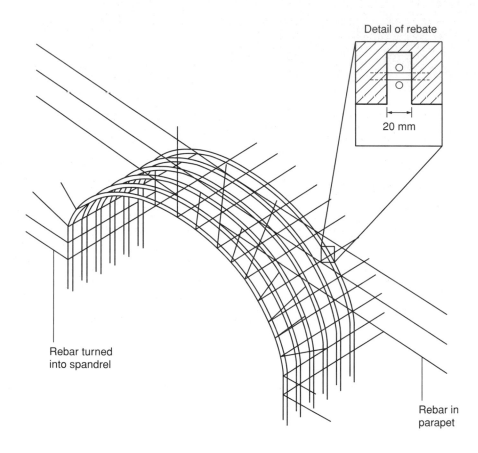

Figure 19.12 The MARS system of arch strengthening (source Sumon, 1999).

adhesive has been formulated and developed as a hybrid material based on polyurethane and epoxy resins with added lightweight fillers. The material is designed to bond to brick or stone in wet or dry conditions and is finished to provide a mastic/mortar joint appearance. There is an additional benefit that the spandrel walls and wing walls can be tied-in structurally to the arch. More recent developments of this system have included changes to the formulation of the adhesive used and it now incorporates radial ties into the arch ring.

A similar system has been developed by Helifix Ltd using their Helibeam masonry reinforcement system which is installed both circumferentially and transversely in slots cut in the brickwork. The Helibeam reinforcement is more flexible than the conventional steel section used in this type of system. As such, the system relies more heavily on the composite interaction between the reinforcement and the masonry and additional testing has been carried out at the Transport Research Laboratory to verify the material parameters involved. An advantage of a more flexible form of strength-

Figure 19.13 The MARS system under construction (courtesy of SSP/Protec).

ening is that, theoretically at least, the fundamental behaviour of the arch is not significantly altered. In a similar manner to the MARS system, Helifix also use radial ties through the arch ring.

Grouted anchors

Possibly the most recent system in the area of retro-fitted reinforcement to masonry arches is the Archtec system (see Figure 19.14). Developed jointly by Cintec Cavity Lock Systems, Gifford and Partners and Rockfield Software the system was evaluated on the test rig at the Transport Research Laboratory (see below). The system was initially evaluated and is now designed using an advanced discrete element technique (Owen *et al.*, 1998), which had not previously been applied to bridges in an industrial environment. Efficacy of the system lies in accurate structural analysis using this method. On occasions it has shown that there is unexpected hidden strength and strengthening is not required. More often, the analysis shows how reinforcement can be fitted to provide an optimum solution. The reinforcing components are standard Cintec polyester sock anchors. Very accurate rigs are used to diamond drill through the masonry so that anchors can be installed virtually along the thrust lines in the arch. The drilling can be undertaken from above or below the arch depending on access, services and traffic management considerations. The system has now (1999) been installed at over 20 sites in the United Kingdom.

In some cases retro-reinforcement will only be appropriate if used in conjunction with other remedial work such as pre-grouting of the masonry. It will not be possible

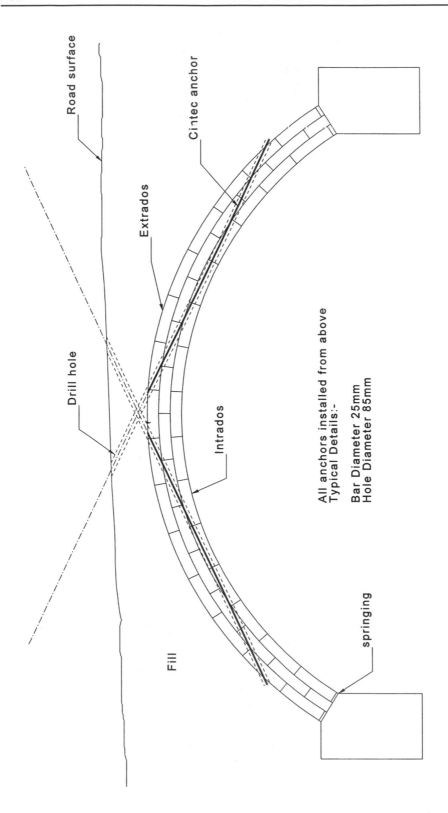

Figure 19.14 Elevation of typical Archtec arrangement for arch strengthening (courtesy Cintec Cavity Lock Systems).

to waterproof the arch by the use of these techniques alone. Responsible and experienced contractors should be used and good supervision should be provided.

Asphalt overlay

An alternative approach to strengthening arches by increasing the thickness of surfacing with a bituminous overlay has been proposed (Fairfield and Ponniah, 1996). A series of numerical analyses were carried out at the University of Edinburgh on a 2 m span brick arch using a variety of techniques. A 0.1 m overlay was shown to increase arch load capacity by 61%, for that particular geometry only. The extra cover causes the capacity increase by allowing increased stress dispersal and this was verified in accordance with the codified method, Boussinesq's analysis and elastic finite element analysis. This research work showed that this technique was cost-effective within certain constraints. If the arch shows signs of deterioration, as would be indicated by a reduced MEXE condition factor, then other methods of repair should be used. If the structure was otherwise safe then the authors of the paper suggest that capacity increases could be achieved by the use of the overlay method.

Grouting fill

Grouting of the contained ground above and behind an arch can be a useful measure. In suitably receptive grounds (not high in clay or silt) and in the absence of complications such as services, drainage systems, etc. the method is very effective and very economical. It increases the assessment fill factor to 0.9 and improves the arch ring condition factor by filling cracks and voids in the extrados. However, grout quantities can be hard to predict and considerable variation is to be expected (Welch, 1995).

Underpinning

This technique involves excavating material from beneath the foundations and replacing with mass concrete (Page, 1993). Alternatively, a series of mini-piles could be used. A sequence of work is followed to ensure stability of the existing structure is not compromised.

Invert slabs

This is a slab of concrete placed between the abutment walls or piers with its top surface at or below riverbed level (Page, 1993). It helps to prevent scour. If incorrectly installed, however, there is a risk of scour beneath the slab, particularly at its downstream end. This can be minimized by installing trench sheeting along the length of the invert slab. Where an existing invert, which acts as a strut between two supports, is being replaced then care should be taken to ensure that the stability of the bridge is not compromised.

Tie bars and pattress plates

The introduction of tie bars utilizing pattress plates or similar devices (Figure 19.15) to distribute loading in bridges has been common practice for many years. They consist of a bar passing through the full width of the bridge, with pattress plates at each

Figure 19.15 Arch strengthening with tie bars (courtesy of Surrey County Council).

end, generally secured by a nut and washer, to provide the restraint to the spandrel wall. Traditional methods of installation involve the excavation of trenches in the fill material between spandrels and a percussive method of forming holes through the masonry for the installation of the bars. However, the excavation of trenches across bridges is often impractical. Tie bars can rust if they are not adequately protected. If the arch ring requires strengthening at the same time a more common solution is to use a concrete saddle which will also relieve the spandrel wall of outward forces (Page, 1993).

Spandrel wall reinforcement

The replacement of a small amount of fill adjacent to a spandrel wall with concrete on a wide bridge could be used to help stabilize outward movement of spandrel walls. When the whole of the fill is replaced, the technique is akin to saddling and is likely to be used to deal with arch and wall problems at the same time. Care must be taken to avoid damaging the spandrels. Trials have been reported of the use of non-metallic reinforced fill which aims to reduce loading on the spandrel walls from the fill (Page, 1996). This technique can also be used to provide a drainage layer between the fill and the spandrel wall (Pearson and Cuninghame, 1998).

If there is sufficient, or a road closure is acceptable, then one solution would be to excavate behind the wall and rebuild it conventionally. To back the wall with mass concrete is a possibility, but to do so creates a deep, stiff beam edge to the arch, inconsistent in structural action with that of the arch. A more harmonious structural action results from incorporation of a reinforced earth system to support the fill. This prevents excessive pressure developing against the spandrel wall and the space between reinforced earth and back of wall is filled with single-size drainage material (Welch, 1995).

Water management

As noted above, it should be an objective of a repair and strengthening scheme to waterproof the arch and provide positive means of drainage. Points to consider are as follows:

- waterproofing should be tucked at the surface and above the arch ring
- the road surface should be maintained intact
- replace grass verges with paved surfaces
- service ducts should be encased in concrete with waterproofing above
- provide spare ducts
- seal cracks along kerbline, next to parapets and around street furniture
- ensure adequate reinstatement of service trenches and excavations
- realign and reprofile carriageways to shed water
- waterproof surfaces in contact with fill.

It is important to tackle the causes as well as the symptoms of drainage problems (Powell, 1990). Positive drainage outlets should be provided at the springing points of multi-span arches with an outlet through the piers. Holes can be cored through the brickwork (75–100 mm diameter) and a uPVC liner grouted in. The liner should project 25 mm from the stone or masonry face and be flush with the internal concrete. Perforated weep pipes can be provided within granular fill to assist in distributing water to drainage outlets. Weep holes should be provided in abutments and cut-off drains should be installed across the carriageway at the end of the bridge where topography dictates (Pearson and Cuninghame, 1998).

Part reconstruction

When arch ring damage is extensive, the only real recourse is to rebuild to a major extent. Construction is traditional in that it is necessary to build off centring, although several variations of constructional form have been adopted. It is best to 'knit' the reconstructed section into the remaining structure by removing alternate stones or bricks. However, where an arch has been historically widened it is best to leave this section intact and create a butt joint. The two sections would be united with a saddle which would also allow for the incorporation of a waterproofing membrane (Welch, 1995).

Evaluation of techniques

Six full-size model arch bridges have been tested to failure at the Transport Research Laboratory to investigate the effectiveness of a number of strengthening methods. A test frame has been constructed to enable a series of 5 m span, 2 m wide, three-ring-brick arches to be built and load tested to failure. Spandrel walls and road surfacing have been left out of the models in order to reduce the number of parameters being studied. The fill is retained by a steel box which has been designed not to restrain movement of the arch ring. The six types of arch tested, together with their reported failure load (Sumon, 1998, 1999), are as follows:

- an unstrengthened arch (242 kN)
- an arch strengthened with sprayed concrete (900 kN)
- an arch strengthened with a concrete saddle (701 kN)
- an arch with built in ring-separation (200 kN)
- a ring-separated arch strengthened using stainless steel mesh reinforcement (276 kN)
- a ring-separated arch strengthened using the Archtec System (410 kN).

The objectives of the test programme were as follows (Sumon, 1998):

- to examine experimentally the increase in capacity of the arch ring provided by a variety of repair strengthening methods
- to use the experimental results to develop and calibrate analytical models that could be used as part of the design process for repair and strengthening schemes
- if possible, to develop new methods of repairing and strengthening which are economic and overcome the problems of existing methods.

The tabulated results should be read in conjunction with the full reference. It should be noted that the results from proprietary systems are not directly comparable because they were tested at different stages in their development. Further tests have been carried out as part of a LINK project to investigate other strengthening methods but they have not yet (1999) been formally reported.

19.4.7 Reconstruction

The Highways Agency is encouraging the construction of new arch bridges and is in the process of preparing a Design Standard and Advice Note. The Brick Development Association has published a complementary guide (Cox and Halsall, 1996) to give practical advice on brick arch bridge design which was based on the experience of rebuilding Kimbolton Butts Bridge, Cambridgeshire. The behaviour of this bridge was extensively monitored and reported (Mann and Gunn, 1995; Prentice and Ponniah, 1995).

Until the Design Standard is published, it is generally necessary to adopt a geometric layout and then use one of the many assessment programmes (Page, 1993) to refine the design using an iterative approach. Longitudinal loading must be considered very carefully. On short single span bridges this load may be considered to be reacted by the carriageway construction, but on multi-spans or long spans this may not necessarily be the case. It may be possible to design the spandrels as shear walls to carry longitudinal forces into the piers and abutments (Halsall, 1994).

In some cases it may be preferable to entirely reconstruct an existing arch in masonry rather than consider a strengthening method. One example of such an application (Figure 19.16), where it was necessary to widen an existing bridge which was in a conservation area, has been reported (Sains, 1996). The choice of brick bonding is important. Ring separation can be a problem in header type arch rings. This can be eliminated by the use of English Bond. This type of bonding provides a shear key between the rings (Halsall, 1994).

If the mortar is too strong in the reconstruction it will encourage single cracks without distribution and if too weak it may fail in compression. It is not sufficient to specify only by a recipe mix or strength; the only satisfactory specification method is to specify a nominal mix, and maximum and minimum strength, the mix design being established on site, a flexible mix being preferable. It is inherent in structural brickwork construction that strict control will be kept on the materials and that the batching will be closely supervised. Brick strengths are seldom a problem but brick sizes can be. It is important for the detailer to remember that they are dealing with a natural material and that close dimensional specification may not be possible to achieve. There are distinct differences between 'metric' and 'imperial' brick sizes.

It is important to check that construction of centring does not permit undue deflection as the vaulting proceeds. Due allowance must also be taken of the moisture movement in the forms, and that they can be readily eased and removed (Halsall, 1994). The ideal way to proceed with the construction is to remove the centring, backfill and and complete the bridge before the pointing. There is a risk, however, that during the placing of the fill distortion of the arch may take place. The

Figure 19.16 English Bond arch rings on Mytchett Place canal bridge (courtesy of Surrey County Council).

centring should, therefore, be gradually eased rather than be instantaneously removed. If 10 mm of sand is used in place of mortar adjacent to the centring then the production of quality pointing is considerably eased.

A number of points should be taken into account when considering altering existing bridges:

• the refurbished bridge should be as good or better than the existing, aesthetically, environmentally and structurally (Wallsgrove, 1996)
• structures should not be cleaned more than is functionally necessary, i.e. grime should only be removed if it is damaging the fabric, or if architectural detail is no longer visible
• use the latest research to avoid unnecessary intervention
• avoid short-term solutions which cause ongoing liabilities
• work to be to satisfaction of English Heritage (for listed or scheduled structures).

It is very important to consult with the planning authorities when formulating works proposals to bridges that are listed buildings, scheduled ancient monuments, or in conservation areas. A detailed before and after record survey of the bridge may be required. Certain restrictions may be imposed which could, for example, prevent the removal of the existing fill - an alternative solution would then need to be devised. If at all possible, an aesthetically pleasing solution should be found.

19.4.8 Conclusion
Masonry arches have served the nation well over many years. The causes of defects should be determined before selecting appropriate repair and strengthening techniques. There are many new and innovative methods that allow effective intervention without causing widespread disruption to the users of the transport system. An appropriate use of materials will allow the aesthetic appeal of the masonry arch to be retained and enhanced. Careful and sensitive maintenance, including the effective management of water, will allow this satisfactory state of affairs to continue for many more years to come.

19.5 Replacement of structures
19.5.1 Introduction
The general trend towards increased loading requirements, and the consequent strength assessment programme, has dramatically increased the number of bridges known to be below capacity. Many of these structures are beyond economic strengthening; and others are reaching the end of their useful life because of durability problems. There is often no alternative to replacing the bridge.

The design of replacement structures is essentially the same as for new works, except that there are considerably more restraints on the type of structure which may be used. The demands of the existing network whether road, rail or waterway must be considered and these will normally dictate the type of construction. The disruption costs associated with bridge replacements can easily exceed the actual cost of the work. Minimizing the construction time is of prime importance in reducing the disruption. This situation has led to the development of new techniques and innovative methods to replace bridge decks.

The health and safety implications of working in close proximity to live traffic also require particular care both for the network user and the contractor. In many cases these requirements also dictate the design chosen.

19.5.2 General requirements and restraints on closure

The same structural types are open to the designer of replacement structures as for new works. However in most cases the type of structure chosen will depend on different factors to new construction. What the bridge carries, the time available for replacement, site restrictions and health and safety amongst others will dictate the type of structure chosen.

Constraints imposed by the existing road, rail and waterway networks

The structure being replaced will be on one or more existing networks, whether highway, railway or waterway and these will inevitably be affected by the construction of a replacement bridge.

The degree of disruption on the highway network is dependent on the traffic flow and suitability of alternative routes, and in some cases there may be no alternative. Diversions, road or lane closures cause delays to the user which have a cost in time or extra mileage; a notional value can be put on these delays to aid in comparison of different solutions. Traffic delay costs are evaluated using a computer program such as QUADRO (Highways Agency, 1999). This takes into account the way vehicles divert onto alternative routes when subject to delays on the main route by determining the cost of the additional time spent queuing. Delay and access costs are very variable, according to the route and local circumstances, but the importance of this aspect of planning may be judged from the typical costs shown in Table 19.10.

The cost of disrupting a railway line is more easily quantifiable in terms of lost income, loss of use of the line and provision of alternative services. Rail closures of more than 24 hours long will only be available at times when traffic is limited such as over the Christmas Period. Such closures need to be planned and booked with the railway authorities many months or even years in advance. Very short duration closures, typically up to 6 hours long, are available more easily at nights and weekends. The railway will normally provide their own staff to manage the possession, the cost of which needs to be added to the overall possession cost.

Waterways such as canals and navigable rivers usually have a closed season, typically from November to March. During this time it may be possible to close the canal or river and replace the structure, outside these times navigation may be restricted to certain hours by arrangement. At other times, protection of the canal traffic may be required, such as that shown in Figure 19.17.

The key to successful timing of structure replacements is consultation with the network operators at an early stage to determine what requirements they might have. This information will dictate the type of structure required.

Table 19.10 Typical delay and access costs at 1990 prices

Description	Traffic management	Traffic flow	Daily delay cost
Dual 3 lane motorway	Contra-flow carriageway closed	80 000 veh./day	£63 000
Dual 2 lane motorway	Contra-flow carriageway closed	60 000 veh./day	£106 000
Dual 2 lane road	Contra-flow carriageway closed	40 000 veh./day	£26 000
Single 7.3 m	Shuttle working under traffic lights	10 000 veh./day	£510
Railway	Overnight possession		£1500
Railway	48 hour closure		Varies dependent on line
Canal	Restrictions to waterway		£1000

Figure 19.17 Protection provided for canal traffic during deck replacement (courtesy of Oxfordshire County Council).

Public utilities and other services

Highway bridges carry a large number of underground and overhead services belonging to the utilities such as the water, electric, gas and telephone authorities. These can be found located in service ducts on more modern bridges, but on the older structures they are just as likely to be built into the deck or suspended from the side or soffit of the bridge.

At an early stage inquiries should be made with the utility companies to determine what services cross the bridge and whereabouts they are located. This information should be supplemented with hand-dug trial holes on or adjacent to the bridge to confirm the records and locate any other services not otherwise recorded. At this stage the condition of the pipes or ducts needs to be investigated to enable decisions to be made about how to accommodate the service or whether a diversion is required.

During the first half of the twentieth century cast iron was used extensively for pipe work and this can be very brittle and easily damaged, modern steel or plastic ducts are more robust and easier to handle and support.

The cost of diverting typical water or gas mains can be considerable, on one recent contract in Oxfordshire it cost around £20 000 to temporarily divert a water main over a length of 100 m. It is usually more economic wherever possible to support the service and work round it although there are then additional safety hazards associated with working adjacent to live services. Special support for services may be required as shown in Figure 19.18.

Utilities that cannot be diverted may need to be built in to the new structure thus limiting the possible design solutions.

Figure 19.18 Truss support provided to maintain services during deck reconstruction (courtesy of Oxfordshire County Council).

Early consultation is essential and in the UK this is normally carried out as part of the New Road and Street Works Act (NRSWA) procedures. Much of this work to accommodate services during bridge replacements may be undertaken in advance of the main contract possibly causing additional disruption.

Safety considerations

Working adjacent to live traffic is extremely hazardous and setting up a safe system of work is particularly important.

For highway work in the UK the Department of Transport has strict standards are laid out governing lane widths, road speeds and safety protection zones (Department of Transport, 1992). Minimum lane width is 3.0 m for all traffic reducing to 2.5 m for cars and light vehicles only. Safety zones are dependent on the road speed varying from 0.5 m up to 40 mph to 1.2 m over 40 mph. Working within the safety zone is not permitted. These restrictions will often dictate the extent of any partial reconstruction and need to be considered at an early stage. Temporary speed limits or road closures can take up to three months to arrange, so need to be considered at an early stage.

Railway Authorities have strict safety policies that dictate the method of working to ensure safety of the operatives and the line. The provision of railway safety staff is expensive and should be considered during estimating the costs.

The normal requirement of providing a workboat, life jackets, and so on, for working over waterways will apply. Navigable waterways will also require appropriate signing and also may require changes to the navigation, notice of such changes will be required some months in advance. Waterway authorities have strict guidelines for control of pollution, which should be considered at all stages, especially during the demolition phase.

It may be more economic to consider a temporary working platform below the structure being replaced to allow work to proceed when the road, railway or waterway below is still in use. Over rivers adequate clearance must be left below the platform to allow for debris to pass in times of flood. Platforms used over railways will require designing to allow for the suction caused by the passage of high speed trains.

19.5.3 Demolition of existing decks

The demolition of the existing bridge decks needs particular care and should be considered at the same time as the design of the replacement deck. When considering phased construction, where part of the bridge will remain under traffic, the structural integrity of the old construction after part has been demolished has to be maintained and should be checked at the design stage.

Working alongside live traffic and around live services such as gas mains and electric cables can be extremely dangerous and demolition work needs to be planned for safety. In the UK guidance has been published by the Health and Safety Executive (1984–1988), and British Standards Institution (1982). The utility companies have their own rules about the use of plant in close proximity to their apparatus. Similarly river authorities and waterway companies will require protection to be installed to stop water pollution. Disposal of the bridge once demolished will need to be considered in the context of recycling and reuse of materials.

The initial decision has to be made whether the bridge is to be totally demolished *in situ* or cut into transportable sections and removed to a more convenient location for breaking up. The time available and physical constraints may limit the choice. The number of methods of demolition available to the engineer has increased with the improvements in water jetting and diamond cutting techniques. Increases in the capacity of cranes and jacking have also made the removal of large sections a practical proposition.

Structural considerations

Before demolition of the deck or part of the deck commences, the bridge engineer should have a good understanding of how the existing structure carries its load and the effects of demolition of part of the structure. Most bridges known to be weak have some hidden reserves of strength, but this should not be relied upon after removal of part of the deck.

Masonry arches rely on the fill behind the abutments to resist the horizontal thrust, consequently the removal of fill either on or behind the arch should be undertaken with care. The sequence of demolition of multi-span arches is important for similar reasons, the adjacent arch often resisting the horizontal thrust. It may be necessary to prop multi-arches at springing level or alternatively plan for demolition of the complete structure.

Prestressed concrete can be extremely dangerous to demolish because of the large inbuilt prestressing forces. With prestressed beams wherever possible it is safest to remove the beams to an area where they can be broken up safely. With more complex post tensioned structures where this is not possible careful planning is essential, with reference to the original stressing sequences and details, if available (Lindsell, 1994). A demolition contractor with experience of this type of work should be used. Jack arch structures, much like multi arch bridges, rely on the adjacent jack arch to resist the horizontal thrust. Removal of half the deck may leave the remaining section unstable. An alternative means of restraining the horizontal forces is with tie bars. On many structures tie bars are continuous across the full width of the deck and cutting in half can reduce their effectiveness, leaving a potentially unstable deck. Slab type bridge decks often act as props between abutments thus ensuring the stability of the abutments. The extent of this propping action should be assessed and suitable props introduced before demolition commences.

Methods of demolition

The methods used to demolish a structure prior to replacement will depend on the time available, the proximity of services, traffic and other buildings, and the amount of the structure being demolished. With the traditional method of breaking up with a hydraulic breaker control of noise, dust, fumes and vibration is difficult; there is also the likelihood of overbreak onto the parts of the structure remaining. Special care and smaller breakers will be required in the vicinity of services.

The use of water jetting and hydro-demolition is suitable for use on small areas but is unlikely to be economic for use on complete decks. There is also the problem of dealing with the large amounts of water and slurry produced by this method, in particular over rivers.

Explosives are quick once the initial setting up has been completed and have been the best option for rapid removal of bridges over railways and motorways where possession times are short. Placing and identifying optimum locations for controlled explosives is a very specialist operation if success is to be guaranteed at the first attempt. Protection will be required for the railway or road onto which the bridge is dropped.

A number of techniques are available when a structure needs to be cut into a number of manageable portions. Diamond saw cutting either with a disc or a wire leave clean edges and are relatively quick but can be expensive. The method adopted to split the deck into discrete beams on the replacement of Botley Flyover in Oxfordshire UK (Figure 19.19) was to core a number of holes between the beams into which expanding jacks were inserted. On expanding the jacks the deck split along the line of holes. The beams were then easily craned out and removed.

On a small number of occasions, the removal of a complete bridge deck has been attempted using multi-wheeled transporter jacks, the removal of Ingst Bridge over the M4 in Avonmouth UK was one example (*New Civil Engineer*, 1992).

Figure 19.19. Demolition of Botley Flyover, Oxfordshire, by splitting pairs of beams (courtesy of Oxfordshire County Council).

19.5.4 Selection of superstructures

The selection of the construction type for replacement superstructures is determined by:

- how much of the road can be closed
- length of time available
- site area adjacent to the bridge being reconstructed
- overall dimensions of the bridge, in particular the span
- cost of disruption compared to the cost of the work
- fixed clearances under the bridge limiting the available construction depth.

Table 19.11 illustrates some of the factors to be taken into account when selecting the type of superstructure to use.

Table 19.11 Summary of deck types with advantages and disadvantages

Structural type	Disadvantages	Advantages
(1) Steel beam with *in situ* concrete deck	Speed of construction Construction of deck, curing of concrete, waterproofing	Easily craned into place. Suitable for roll-in. Light dead weight allows original substructure to be used
(2) Steel beam with precast concrete deck	Difficult to obtain full composite action between deck and beam	As (1). Bolt together sections allow rapid construction
(3) Precast concrete beams or arch units	Normally require an *in-situ* concrete deck with time for curing and waterproofing	Minimum disruption to road/rail/water under bridge can be installed in a single possession. Used with precast deck can offer a fast construction method
(4) *In situ* concrete	Slow construction period. Extensive falsework required	
(5) Precast concrete box	Large units can be heavy	Suitable when substructure requires replacement on smaller structures. Easily adaptable for jacking. Rapid Installation
(6) Corrugated steel	Time-consuming backfilling procedure. Over-wide excavation often required	Lightweight – easily craned
(7) Masonry arch	Slow construction method	

Aesthetics of the replacement bridge

The appearance of bridges is an emotive and subjective subject with many different opinions being expressed, nevertheless in recent years a number of books have been published with guidelines on appearance (Highways Agency, 1996). In designing a replacement the engineer needs to be aware of the importance of the structural form and how it fits in its setting.

Bridge decks can be replaced by copies of the original, but this will not always be possible or desirable. Higher loads have to be carried requiring deeper or stronger sections. Materials, such as cast iron cannot be economically copied in modern materials or arches are too labour intensive to reproduce quickly and economically. It may be the original bridge was inelegant or badly proportioned and the opportunity should be taken to improve on the original.

When trying to match a replacement deck and parapet with existing wingwalls in brick remember that metric bricks are a recent introduction. Imperial bricks or even a local non-standard size will have been used. Try to match where possible. Special bricks are still available for copings and string courses and can be made to order.

When replacing multi-span simply supported decks, the decks should be made continuous whenever possible with the advantages of reduced construction depth and less joints to leak improving the water management. Similarly the making the deck integral with the substructure will reduce future maintenance.

Wherever possible bridges with interesting original features should be strengthened and not replaced; retaining as many of the features as possible. However, in many cases it is not possible to retain, for example, an original wrought iron lattice truss or cast iron beam, because it does not have the structural capacity. On a number of bridges in Oxfordshire the original truss has been cleaned up and reused, with the loads being taken by new structural members (Figures 19.20 and 19.21). To the lay person the replacement bridge looks identical to the original.

The wrought iron edge girders were no longer structural in the reconstructed bridge, but retained the character of the structure.

New materials are available for the bridge engineer to use; consider advanced composite claddings to improve the appearance of the fascias. Soffit claddings can also hide and protect a steel beam deck.

Replacement of road bridges within an extended closure

If a bridge has to be replaced, undertaking this in one operation within an extended closure is often the cheapest option based on construction costs alone. Complete closure of a major road for an extended period would accrue considerable delay and disruption costs and would not normally be acceptable or economic. But where traffic is light and alternative routes short this method allows the contractor a site free from traffic together with easier construction methods. Health and Safety requirements may dictate full closure for bridges on narrow roads when there is not the width available to keep the a lane open to traffic whilst working on the other lane. Whenever road closures are used speed of construction is important to minimize delays and disruption.

Figure 19.20 Somerton Bridge Oxfordshire. Deck reconstruction during a road closure (courtesy of Oxfordshire County Council).

Figure 19.21 Somerton Bridge Oxfordshire. Replacement of wrought iron girders (courtesy of Oxfordshire County Council).

Complete closure is rarely acceptable and where access is still needed for cyclists and pedestrians, this can be provided by a small scaffold footbridge alongside the works.

For some types of bridge, such as through girders, replacement of the whole structure may be the only option available, either within an extended closure or short-term closure.

The design of a replacement bridge to be constructed during an extended closure is similar to new construction. The whole of the site is available to the contractor. The deck can be constructed as one without the need to ensure stability of a part constructed deck under traffic.

One major difference to new construction is working around services and utilities. Support gantries will be required to support services that have not been diverted, telephone ducts and the like can be supported by relatively simple scaffold structures. However gas and water mains will need more substantial support gantries for spans greater than about 5 m. Water mains also often have substantial thrust blocks immediately behind the abutments further limiting access. On older concrete bridges service ducts are often built into the deck and in many cases will need to remain at the same line and level, limiting the type of deck to *in situ* concrete in the area of the ducts.

Replacement of road bridges with short-term closures

When the high costs of delay and disruption make extended closures uneconomic, short-term closures take advantage of the lower traffic flows during the night or at weekends to reduce disruption. This type of contract needs thorough planning to max-

Figure 19.22 Installation of precast deck units covering half of the road width (courtesy of Oxfordshire County Council).

imize the use of often short closures, typically 8 hours for an overnight closure up to 48 hours for a weekend. Good public relations with plenty of advance notice for closures are a prerequisite for minimizing disruption and maximizing the amount of useful work. Such contracts are probably best undertaken in the summer months taking advantage of the warmer and shorter nights.

The choice of deck is limited, because of the need to ensure the carriageway is reinstated for use by the next morning. The choice has to be made between lifting or sliding in whole deck sections or lifting in smaller sections and stitching the sections together *in situ*. Figure 19.22 shows installation of precast units covering half of the road width. Note the need to prop abutments apart.

Time must be allowed for demolishing and removing the existing deck. Materials should be chosen for speed of construction; bolted joints where possible; rapid-setting concrete; etc. One type of transverse joint used by Oxfordshire County Council is illustrated in Figure 19.23.

Figure 19.23 Fast Connection of precast units with high strength friction grip bolts (courtesy of Oxfordshire County Council).

Minimizing the weight of replacement deck sections allows for easier craneage and or replacing larger sections of deck. New materials currently under development that give significant weight advantages decks are aluminium and advanced composites decking systems.

Replacement of road bridges in stages keeping the road open

To minimize delay and disruption the road should remain fully open. A temporary bridge can be constructed either alongside, if land is available, or over the structure being replaced. Adequate working space must be provided under a temporary bridge crossing over and this will also preclude the use of cranes, again limiting the structure type for the replacement deck to smaller units. A temporary bridge will cause some disruption due to the substandard alignment. There will also be disruption during the installation of the temporary bridge.

Replacement in two or more parts allows the road to remain open, with the traffic reduced to a single lane under traffic lights or as a single carriageway using a contraflow on dual carriageways. The traffic delay costs are reduced but the difficulties of construction adjacent to live traffic and in parts add to the overall cost of the contract. Continuous temporary barriers must be used to separate the work site from the live traffic lanes to protect the workforce and road user.

The existing structure will often dictate the location of the join; in particular on structures constructed with discrete beams and longitudinal jack arches. Similarly the new construction should be detailed so that reinforcement can be joined at the joint using couplers or deck joints occur at the correct location. Each part must be capable of taking full loading and should be designed such that the join between sections can be made under traffic. This may require propping under the newly constructed section.

The choice of deck construction is only limited by the space available for handling new sections of deck, considering the physical restraints of working adjacent to live traffic and around services. Steel beam and concrete slab, prestressed precast concrete beam infill decks and *in situ* reinforced concrete have all been used recently in Oxfordshire.

Railway bridges

For railway bridges the need to work within very short rail possessions has led to more innovative solutions being developed. Generally, except on heavily trafficked roads such as motorways it is not economical to consider these methods of construction.

- Roll or slide in decks can be constructed alongside the bridge to be replaced and slid in during a possession. Extensive temporary works are normally required adding to the cost of this method.
- Installation of complete deck sections is practical but requires large cranes, remembering that crane costs increase dramatically with size. Special attention is required to the location of the crane relative to the structure and offloading position to minimize the reach and hence crane capacity.

19.5.5 The selection of substructures

In many situations the substructure may be adequate to accommodate the replacement deck. In assessing its suitability for reuse the following considerations should be made.

- Is the existing substructure in good condition? Abutments and piers may show stability problems such as settlement or slippage. Structural problems such as cracking, poor brickwork, deteriorating concrete may also be apparent. Damage to reinforced concrete due to de-icing salts can be a problem under leaking deck joints.
- If the self weight of the replacement deck is greater than the original deck the imposed load on the substructure will clearly increase. Similarly increases in vehicle loading will increase the loads imposed. In this situation the substructure needs to be assessed for its suitability to take the increased loading.

The Design Code requirements for collision forces on, in particular piers, but also on abutments have increased dramatically in recent years. The pier should be assessed under collision loads.

Slab type bridge decks often act as props between abutments thus ensuring the stability of the abutments. The extent of this propping action should be assessed and replacement decks designed to accommodate it.

It is unlikely that the original bridge deck was designed to accommodate the current code requirements for longitudinal forces imposed by braking and skidding. The design of the bearings will need to allow these forces to be taken into the substructure. Alternatively, these forces can be taken off the structure into the fill behind the abutment by continuing the deck beyond the substructure and anchoring the deck.

In the majority of situations it is economic to retain the original abutments and piers, repairing as required to accommodate the new deck. Suitable methods for restoring the stability of abutments include the use of ground anchors and mini-piles.

Where the substructure needs to be replaced but deep excavations are not desirable due to the proximity of traffic the use of a contiguous bored pile or sheet pile wall should be considered.

19.5.6 Types of contract

The type of contract chosen needs to give an incentive for the contractor to complete the work in the minimum time possible while not reducing safety.

For UK Highways Agency Contracts this is usually achieved using a 'Lane Rental' type contract (Highways Agency, 1998(. This uses the incentive of a bonus for early completion coupled with a charge for late completion. The contractor, subject to a maximum time set by the designer, sets the programme duration. The amount of bonus/charge is normally calculated using QUADRO together with appropriate supervision costs, this is then directly related to the cost of traffic disruption.

Design and build contracts are commonly used in railway works capitalizing on the contractors' knowledge at an early stage in the design process. This teamwork or partnership approach also allows for better programming of rail closures using time-scales agreed by the contractor and client. It also allows more innovative methods to be used, sharing the risk between the contractor and client.

19.5.7 Public relations and planning

The key difference in new works and replacement works is in the level of consultation and planning required. It is important that for key structures on the network consultation with the network owners, public utilities and the users takes place at an early stage, sometimes years ahead of the work. This type of work cannot be considered in isolation from the rest of a network and should be part of an overall network management plan. it may be possible to make significant savings by combining contracts; sensible planning of other work on diversion routes can also reduce delays. To close the road needs a legal order and this will take about 3 months to process because of the large amount of consultation that must occur between the various interested bodies.

Operational needs of the railways and waterways will often dictate that firm date are set months in advance, with severe cost penalties for late changes.

Nearer the time of the works the media should be used to publicize the closure and give alternative means of travelling. Clear information boards should be displayed at and before the site giving clear information on diversions.

Consider the following means of getting the information to the people that will be affected by the disruption:

- radio announcements in traffic bulletins but make sure they are accurate
- press releases to the local papers, advertisements of closures
- consultation with the Freight Haulage Association, bus companies, etc.
- public meetings
- liaison with local authorities at county, district and parish level
- leaflets as a mail drop or available at libraries, garages, shops, etc.
- display boards on the roadside and well in advance of the work.

Good publicity can assist in the smooth running of a project. Road users appreciate an understanding of the reasons for the disruption, and opportunity to avoid the area where possible. The detailed planning required to forecast accurately can also result in minimized delays.

Bibliography

Ashurst D. An assessment of repair and strengthening techniques for brick and stone masonry arch bridges. Contractor Report 284. Transport Research Laboratory, Crowthorne, 1992.

BA16. The Assessment of Highway Bridges and Structures. Highways Agency Advice Note, May 1997.

Ball D. Waking a tired old lady. *Construction Repair*, 11, No. 2, 1997.

BD21, The Assessment of Highway Bridges and Structures. Highways Agency Standard. February 1997.

Blackett-Ord C. Repairs to Lambley Viaduct, Northumberland. *Construction Repair*, 10, No. 1, 1996.

British Standards Institution. BS6187 Code of Practice for Demolition, 1982.

Choo B and Hogg V. Determination of the serviceability limit state for masonry arch bridges. *Proceedings of the First International Conference on Arch Bridges*, Bolton, 1995.

Choo B, Peaston C and Gong N. Relative strength of repaired arch bridges. *Proceedings of the First International Conference on Arch Bridges*, Bolton, 1995.

Clear the decks. *New Civil* Engineer, 5 Mar, 1992

Concrete Bridge Development Group. Guide to testing and monitoring the durability of concrete structures. Concrete Bridge Development Group Guide No. 2, in press.

Concrete Society. Repair of concrete damaged by reinforcement corrosion. Concrete Society Technical Report No. 26, October 1984.

Cox D and Halsall R. *Brickwork Arch Bridges*. Brick Development Association, 1996.

CSS. The assessment and design of unreinforced masonry vehicle parapets, Vol. 1. County Surveyors' Society Bridges Group Report No. ENG/1-95, 1995.

Darby J., Luke S. and Collins S. Stressed and unstressed advanced composite plates for the repair and strengthening of structures. *Fourth Bridge Management Conference*, Surrey, April 2000

Darby, Brown and Haynes. Pre-stressed composite plates for the repair and strengthening of structures. *Structural Faults and Repair,* July, 1999

Department of Transport Highways and Traffic. BD 27/86. Materials for the Repair of Concrete Highway Structures. Department of Transport Highways and Traffic Departmental Standard.

Department of Transport. BA 30/94 Strengthening of Concrete Highway Structures using externally bonded Plates. Department of Transport Advice Note.

Department of Transport. Safety at street works. A code of practice, 1992.

Fairfield C and Ponniah D. A method of increasing arch bridge capacity economically. *ICE Proceedings, Structures and Buildings*, February 1996.

Garrity S. Retro-reinforcement – a proposed repair system for masonry arch bridges. *Proceedings of the First International Conference on Arch Bridges*, Bolton, 1995a.

Garrity S. Testing of small scale masonry arch bridges with surface reinforcement. *Proceedings of the Sixth International Conference on Structural Faults and Repair*, ECS Publications, Edinburgh, 1995a.

Gilbert M, Hobbs B and Molyneaux T. The response of masonry parapets to accidental vehicle impact. *Proceedings of the First International Conference on Arch Bridges*, Bolton, 1995.

Halsall R. Aesthetics of masonry arches. *Journal of the Institution of Highways and Transportation*, July 1994.

Health and Safety Executive. Health and safety in demolition work GS29, Parts 1 to 4, 1984–88.

Heyman J. *The Masonry Arch*. Horwood, Chichester, 1982.

Highways Agency. *Design Manual for Roads and Bridges,* Vol. 14, *Economic Assessment of Road Maintenance*. Highways Agency, 1997.

Highways Agency. The Appearance of Bridges and Other Highway Structures, 1996.

Highways Agency. Trunk Roads Maintenance Manual, Part 5, 1998.

HMSO. *Bridge Inspection Guide*. HMSO, London, 1983

Holloway and Leeming. *Strengthening of Reinforced Concrete Structures using Externally Bonded FRP Composites in Structural Engineering*. Woodhead Publishing, 1999.

Hughes T and Blackler M. A review of the UK masonry arch assessment methods. *ICE Proceedings, Structures and Buildings*, August 1997.

Jolly CK and Lillistone D. Stress-strain behaviour of confined concrete. *Concrete Communications Conference '98*. British Cement Association.

Lemmon C and Wolfenden PA. Developments in the analysis, testing and damage assessment of railway bridges. *ICE Proceedings, Transport*, November 1993.

Lemmon C and Wolfenden PA. Developments in the analysis, testing and damage assessment of railway bridges (discussion). *ICE Proceedings, Transport*, November 1995.

Lindsell P. Demolition and removal of existing prestressed concrete structures. *Bridge Modification Conference,* ICE, London, 1994.

Mann P and Gunn M. Computer modelling of the construction and load testing of a masonry arch bridge. *Proceedings of the First International Conference on Arch Bridges*, Bolton, 1995.

Mathews R and Paterson I. LMR to LRT: restoration of Cornbrook viaduct. *Proceedings of the Second International Conference on Bridge Management*, Guildford, 1993.

McDonald L and Allen J. Demystifying lime. *Construction Repair*, 11, No. 2, 1997.

Meadowcroft I and Whitbread J. Assessment and monitoring of bridges for scour. *Proceedings of the Second International Conference on Bridge Management*, Guildford, 1993.

Melbourne C and Gilbert M. A study of the effects of ring separation on the load-carrying capacity of masonry arch bridges. *Proceedings of the Second International Conference on Bridge Management*, Guildford, 1993.

Melbourne C. The assessment of masonry arch bridges. *Proceedings of the First International Conference on Bridge Management*, Guildford, 1990.

Middleton W. Research project into the upgrading of unreinforced masonry parapets. *Proceedings of the First International Conference on Arch Bridges*, Bolton, 1995.

Millar R and Ballard G. Non-destructive assessment of masonry bridges – the price:value ratio. *Construction Repair*, 10, No. 1, 1996.

Minnock K. Masonry arch repair and strengthening. *Construction Repair*, 11 No. 4, 1997.

Newbegin D. and Shelley J. Newton Cap Viaduct: conversion from railway to highway. *Construction Repair*, 9 No. 5, 1995.

Owen D. *et al*. Finite/discrete element models for assessment and repair of masonry structures. *Proceedings of the Second International Arch Bridge Conference*, Venice, 1998.

Page J. A guide to repair and strengthening of masonry arch highway bridges. Contractor Report 204. Transport Research Laboratory, Crowthorne, 1996.

Page J. *Masonry Arch Bridges – State of the Art Review*. HMSO, London, 1993.

Pearson S and Cuninghame J. Water management for durable bridges. Application Guide 33. Transport Research Laboratory, Crowthorne, 1998

Powell J. *Defects Due to Water – The Maintenance of Brick and Stone Masonry Arches*. (Ed. AM Sowden). E & F Spon, London, 1990.

Powell J. Report on progress and key issues arising out of the strengthening programme. *The Fifth Annual Surveyor Bridges Conference*, Nottingham, March 1997.

Prentice D and Ponniah D. Installation of data acquisition equipment in Kimbolton Butts Bridge. *Proceedings of the First International Conference on Arch Bridges*, Bolton, 1995.

Pretlove A and Ellick J. Serviceability assessment using vibration tests. *Proceedings of the First International Conference on Bridge Management*, Guildford, 1990.

Riddel J. Problems associated with assessing bridge structures for scour failure. *Proceedings of the Second International Conference on Bridge Management*, Guildford, 1993.

Sains A. Assessment and reconstruction of arch structures over the Basingstoke Canal. *Proceedings of the Third International Conference on Bridge Management*, Guildford, 1996.

Sowden AM (Ed.). *The Maintenance of Brick and Stone Masonry Structures*. E & F Spon, London, 1990.

Sprayed Concrete Association. Design and Specification.

Sumon S. New reinforcing systems for masonry arch rail bridges. *Proceedings of the Second International Conference on Railway Engineering*, London, 1999.

Sumon S. Repair and strengthening of three-ring-brick masonry arch bridges. *Proceedings of the 5th International Masonry Conference*, London, 1998.

Thompson D. Assessment of spandrel walls. *Proceedings of the First International Conference on Arch Bridges*, Bolton, 1995.

Wallsgrove J. Aesthetic aspects of widening and rehabilitating historic bridges. *Proceedings of the Third International Conference on Bridge Management*, Guildford, 1996.

Wardle J. Enhancement of Sonning Bridge and Cloud Bridge. *Construction Repair*, 9 No.3, 1995.

Welch P. Renovation of masonry arches. *Proceedings of the First International Conference on Arch Bridges*, Bolton, 1995.

Index